Advances in Corrosion Control of Magnesium and Its Alloys

Magnesium (Mg) and its alloys have received widespread acceptance in automobile industries and biomedical applications with substantial recent advancements made in their development. However, a significant limitation remains in their poor aqueous and galvanic corrosion resistance. This book covers both the fundamentals and recent advancements in two major corrosion protection strategies of magnesium and its alloys, namely, metal-matrix composites and protective coatings.

Key features:

- Covers all aspects of metal-matrix composites and protective coatings for magnesium alloys to improve their corrosion resistance, wear resistance, mechanical properties and biocompatibility
- Provides the most recent research advancements in the corrosion mitigation strategies of magnesium and its alloys

Complete with case studies specific to practical applications, this book serves as a ready reference for graduate students, researchers, engineers and industry professionals in the fields of materials science, corrosion science, magnesium technology, metal-matrix composites and protective coatings.

EDITOR

Viswanathan S. Saji is Research Scientist II/Associate Professor at the Interdisciplinary Research Center for Advanced Materials, King Fahd University of Petroleum and Minerals, Saudi Arabia. Before that, he was an Endeavour Research Fellow at the University of Adelaide, a Research Professor at Korea University, Senior Research Scientist at Ulsan National Institute of Science and Technology, Research Professor at Chosun University, Postdoctoral Researcher at Sunchon National University and Yonsei University, and Research Associate at the Indian Institute of Science and Indian Institute of Technology Bombay. Dr. Saji received his PhD degree from the University of Kerala. He has authored 90 journal publications and contributed to 11 books and 20 book chapters. His research interest lies in electrochemistry, corrosion science, and smart materials.

Advances in Corrosion Control
of Magnesium and Its Alloys
Metal Matrix Composites and
Protective Coatings

Edited by
Viswanathan S. Saji

CRC Press
Taylor & Francis Group
Boca Raton   London   New York

CRC Press is an imprint of the
Taylor & Francis Group, an **informa** business

First edition published 2024
by CRC Press
6000 Broken Sound Parkway NW, Suite 300, Boca Raton, FL 33487–2742

and by CRC Press
4 Park Square, Milton Park, Abingdon, Oxon, OX14 4RN

CRC Press is an imprint of Taylor & Francis Group, LLC

ISBN: 978-1-03233-468-4 (hbk)
ISBN: 978-1-03233-492-9 (pbk)
ISBN: 978-1-00331-985-6 (ebk)

DOI: 10.1201/9781003319856

Contents

PART I   Fundamentals

PART II   Magnesium Metal-Matrix Composites

Preface

Magnesium, magnesium alloys and magnesium composites have intense topical research attention due to their exceptional properties like high specific strength, good thermal and electrical conductivity, vibration and shock absorption, castability, weldability, biodegradability and biocompatibility, making them suitable for several industrial applications, including automotive, aerospace, electronics, biomedical, construction and defence. They are promising alternatives to aluminium alloys in automobile and aerospace applications and biodegradable polymers in implant applications. The feeble aqueous and galvanic corrosion resistance is the primary hurdle for the extensive application. Various intrinsic (alloy and composite modifications, design considerations etc.) and extrinsic (surface modifications, protective coatings etc.) methods have been considered for enhancing their corrosion resistance. This book addresses the two most important corrosion mitigation strategies: intrinsic – metal-matrix composites and extrinsic – protective coatings.

Chapters 1, 2 and 3 of Part I of the book provide concise accounts of the fundamentals of magnesium corrosion and the two mitigation strategies: magnesium metal-matrix composites and protective coatings. Parts II and III of the book are exclusively dedicated to the most recent updates on metal-matrix composites and protective coatings, respectively. Chapter 4 delivers a good account of solid-state and liquid-state processing methods of magnesium metal-matrix composites, whereas Chapter 5 explains additive-manufactured magnesium alloys and composites. Chapters 6, 7 and 8 describe magnesium-matrix composites with incorporated carbon nanostructures, rare earth elements and calcium phosphate/hydroxyapatite, respectively. Chapter 9 provides an interesting case study on tungsten carbide and graphite-reinforced magnesium-matrix composites. The last chapter of Part II (Chapter 10) delivers a good account of advancements in biodegradable magnesium-matrix composites.

Part III entirely focuses on the different types of protective coatings. Here, the chapters are arranged in the following sequence: sol-gel nanocomposite coatings (Chapter 11), hydrothermal and layered double hydroxide coatings (Chapter 12), chemical conversion coatings (Chapter 13), electrochemical conversion coatings (Chapter 14), electrodeposited nanocomposite coatings (Chapter 15), electroless nanocomposite coatings (Chapter 16), electrophoretic coatings (Chapter 17), CVD and PVD coatings (Chapter 18), thermal and cold spray coatings (Chapter 19) and polymer nanocomposite coatings (Chapter 20). Chapters 21 and 22 describe superhydrophobic and self-healing coatings, respectively. Chapter 23 provides a good description of various biomedical-grade coatings. The book's last chapter (Chapter 24) accounts for multi-layer coating strategies.

This book results from our genuine effort to collectively present the research advancements on the two important corrosion-control strategies. We hope the present book will be a handy reference tool for academic and industrial researchers working on corrosion-control strategies for magnesium and its alloys and composites.

Viswanathan S. Saji
King Fahd University of Petroleum and Minerals

Acknowledgements

We thank all the authors for their valued contributions to this book. We would also like to express our gratitude to all those granting us the copyright permissions for reproducing illustrations. We acknowledge the support received from King Fahd University of Petroleum and Minerals (KFUPM) during the writing phase of the book. We sincerely thank the CRC team for evolving this book into its final shape.

Contributors

Akeem Yusuf Adesina
Interdisciplinary Research Center for
    Advanced Materials, King Fahd University
    of Petroleum and Minerals
Dhahran 31261, Saudi Arabia

K.S. Akshay
Nanomaterials Research Laboratory, School
    of Materials Science and Engineering,
    National Institute of Technology Calicut
Kerala 673601, India

S. Aravindan
Department of Mechanical Engineering, Indian
    Institute of Technology
Delhi 110016, India

Andrej Atrens
The University of Queensland, School of
    Mechanical Engineering, Centre for
    Advanced Materials Processing and
    Manufacturing (AMPAM)
    http://researchers.uq.edu.au/researcher/141
St. Lucia, Qld., 4072 Australia

Sharath Babu
Department of Mechanical Engineering,
    National Institute of Technology Calicut
Kerala 673601, India

Sudip Banerjee
Department of Mechanical Engineering,
    National Institute of Technology
Sikkim, Ravangla

Carlos Henrique Michelin Beraldo
Federal University of Santa Catarina, Rua Eng.
    Agronômico Andrei Cristian Ferreira, s/n,
    Trindade
Florianópolis, SC, 88040–900, Brazil

Xiao-Bo Chen
School of Engineering, RMIT University,
    Melbourne
Victoria 3000, Australia

Yonghua Chen
College of Materials Science and Engineering,
    Chongqing University
Chongqing 400044, China
National Engineering Research Center for
    Magnesium Alloys,
    Chongqing University
Chongqing 400044, China

Thiago Ferreira da Conceição
Federal University of Santa Catarina,
    Rua Eng. Agronômico Andrei Cristian
    Ferreira, s/n
Trindade
Florianópolis, SC, 88040–900,
    Brazil
College of Materials Science and Engineering,
    Chongqing University
Chongqing 400044, China

Jiahao Deng
College of Materials Science and Engineering,
    Chongqing University,
Chongqing 400044, China

Liju Elias
Department of Chemistry, National Institute
    of Technology Karnataka, Surathkal
    Srinivasnagar
Mangalore 575025, India

Dahai Gao
College of New Materials and Chemical
    Engineering, Beijing Key Lab of
    Special Elastomeric Composite Materials,
    Beijing Institute of Petrochemical
    Technology
Beijing 102617, China

Shaokang Guan
School of Materials Science and Engineering &
    Henan Key Laboratory of Advanced
    Magnesium Alloy, Zhengzhou
    University
Zhengzhou, 450001, PR China

Enyu Guo
Key Laboratory of Solidification Control and
    Digital Preparation Technology (Liaoning
    Province), School of Materials Science
    and Engineering, Dalian University of
    Technology
Dalian 116024, PR China

Manoj Gupta
Department of Mechanical Engineering,
    National University of Singapore, 9
    Engineering Drive 1
Singapore, Republic of Singapore

A.C. Hegde
Department of Chemistry, National Institute
    of Technology Karnataka, Surathkal
    Srinivasnagar
Mangalore 575025, India

Song-Jeng Huang
National Taiwan University of Science and
    Technology, No. 43, Keelung Rd., Sec. 4,
    Da'an Dist.
Taipei City 106335, Taiwan (R.O.C.)

Mohamed Abdrabou Hussein
Interdisciplinary Research Center for
    Advanced Materials, King Fahd University
    of Petroleum & Minerals (KFUPM)
Dhahran 31261, Saudi Arabia

M.A. Joseph
Department of Mechanical Engineering,
    National Institute of Technology Calicut
Kerala 673601, India

Sathiyalingam Kannaiyan
National Taiwan University of Science and
    Technology, No. 43, Keelung Rd., Sec. 4,
    Da'an Dist.
Taipei City 106335, Taiwan (R.O.C.)

A.S. Khanna
Chairman SSPC India, Mumbai 400079
Maharashtra, India

Shrinivas Kulkarni
School of Engineering, RMIT University,
    Melbourne
Victoria 3000, Australia

A. Madhan Kumar
Interdisciplinary Research Center for
    Advanced Materials, King Fahd University
    of Petroleum & Minerals
Dhahran, Kingdom of Saudi Arabia

T.S. Sampath Kumar
Medical Materials Laboratory,
    Department of Metallurgical and Materials
    Engineering, Indian Institute of Technology
    Madras
Chennai, 600036, India

Asma Lamin
School of Engineering, RMIT University,
    Melbourne
Victoria 3000, Australia

Jiaxin Li
College of Materials Science and Engineering,
    Chongqing University,
Chongqing 400044, China

Di Mei
School of Materials Science and
    Engineering & Henan Key Laboratory of
    Advanced Magnesium Alloy, Zhengzhou
    University
Zhengzhou, 450001, PR China

T.S.N. Sankara Narayanan
Department of Analytical Chemistry,
    University of Madras, Guindy
    Campus
Chennai-600 025, India

Nasirudeen Ogunlakin
Interdisciplinary Research Center for
    Advanced Materials, King Fahd
    University of Petroleum and
    Minerals
Dhahran 31261, Saudi Arabia

Yibo Ouyang
Key Laboratory of Solidification
    Control and Digital Preparation
    Technology (Liaoning Province),
    School of Materials Science and
    Engineering, Dalian University of
    Technology
Dalian 116024, PR China

Fusheng Pan
College of Materials Science and Engineering,
    Chongqing University
Chongqing 400044, China
National Engineering Research Center for
    Magnesium Alloys, Chongqing University
Chongqing 400044, China

K. Ponappa
Department of Mechanical Engineering
IIITDM
Jabalpur, India

V.P. Muhammad Rabeeh
Nanomaterials Research Laboratory, School
    of Materials Science and Engineering,
    National Institute of Technology Calicut
Kerala 673601, India

Shebeer A. Rahim
Department of Mechanical Engineering,
    National Institute of Technology Calicut
Kerala 673601, India

P.V. Rao
Department of Mechanical Engineering, Indian
    Institute of Technology
Delhi 110016, India

K. Ravichandran
Department of Analytical Chemistry,
    University of Madras, Guindy Campus
Chennai-600 025, India

Prasanta Sahoo
Department of Mechanical Engineering,
    Jadavpur University
Kolkata 700032, India

Viswanathan S. Saji
Interdisciplinary Research Center for
    Advanced Materials, King Fahd University
    of Petroleum and Minerals
Dhahran 31261, Saudi Arabia

Sankaranarayanan Seetharaman
Advanced Remanufacturing & Technology
    Centre (ARTC), Agency for Science,
    Technology and Research (A*STAR),
    3 Cleantech Loop, #01/01 CleanTech Two
Singapore 637143, Republic of Singapore

Department of Mechanical Engineering, National
    University of Singapore, 9 Engineering Drive 1
Singapore, Republic of Singapore

Vijay Sisarwal
School of Engineering, RMIT University,
    Melbourne
Victoria 3000, Australia

Xuhui Tang
School of Materials Science and Engineering &
    Henan Key Laboratory of Advanced
    Magnesium Alloy, Zhengzhou University
Zhengzhou, 450001, PR China

G. Vedabouriswaran
Department of Mechanical Engineering, Netaji
    Subhas University of Technology
Delhi, India

Augusto Versteg
Federal University of Santa Catarina, Rua Eng.
    Agronômico Andrei Cristian Ferreira, s/n,
    Trindade
Florianópolis, SC, 88040–900, Brazil

Kai-Rui Wang
School of Engineering, RMIT University,
    Melbourne
Victoria 3000, Australia.

Liguo Wang
School of Materials Science and Engineering &
    Henan Key Laboratory of Advanced
    Magnesium Alloy, Zhengzhou University
Zhengzhou, 450001, PR China

Tongmin Wang
Key Laboratory of Solidification Control and
    Digital Preparation Technology (Liaoning
    Province), School of Materials Science
    and Engineering, Dalian University of
    Technology
Dalian 116024, PR China

Kai Wei
College of New Materials and Chemical
    Engineering, Beijing Key Lab of Special
    Elastomeric Composite Materials, Beijing
    Institute of Petrochemical Technology
Beijing 102617, China

Liang Wu
College of Materials Science and Engineering,
    Chongqing University
Chongqing 400044, China
National Engineering Research Center for
    Magnesium Alloys, Chongqing
    University
Chongqing 400044, China

Wenhui Yao
College of Materials Science and Engineering,
    Chongqing University
Chongqing 400044, China
National Engineering Research Center for
    Magnesium Alloys, Chongqing
    University
Chongqing 400044, China

Yujie Yuan
College of New Materials and Chemical
    Engineering, Beijing Key Lab of Special
    Elastomeric Composite Materials, Beijing
    Institute of Petrochemical Technology
Beijing 102617, China

You Zhang
College of New Materials and Chemical
    Engineering, Beijing Key Lab of Special
    Elastomeric Composite Materials, Beijing
    Institute of Petrochemical Technology
Beijing 102617, China

Yuhan Zhang
College of Materials Science and Engineering,
    Chongqing University
Chongqing 400044, China

Zhe Zhang
College of New Materials and Chemical
    Engineering, Beijing Key Lab of Special
    Elastomeric Composite Materials, Beijing
    Institute of Petrochemical Technology
Beijing 102617, China

Shijie Zhu
School of Materials Science and Engineering &
    Henan Key Laboratory of Advanced
    Magnesium Alloy, Zhengzhou University
Zhengzhou, 450001, PR China

Part I

Fundamentals

1 Magnesium Alloys – Corrosion Fundamental

Andrej Atrens

CONTENTS

1.1 INTRODUCTION

Mg is the most active of the engineering metals. The films that form on the surface of Mg during corrosion are only partially protective in the technologically important chloride solutions (Song and Atrens 1999, 2003; Atrens et al. 2013, 2015, 2022). Mg corrosion in such chloride solutions has been widely studied to gain insight into the corrosion of Mg in auto service. The knowledge regarding the corrosion behaviour of Mg in chloride solutions has provided the foundation for the understanding of Mg corrosion in the more complex solutions used to understand Mg biocorrosion for the use of Mg in biodegradable implants (Atrens, Liu and Abidin 2011; Atrens et al. 2018; Mardina, Venezuela, Dargusch et al. 2022; Mardina, Venezuela, Maher et al. 2022). The low protectivity of the typical corrosion product film means that the intrinsic Mg corrosion rate is typically substantial, 0.3 mm/y in chloride solutions as measured by weight loss for high-purity Mg (Atrens et al. 2020). There have been significantly different approaches to the understanding of the Mg corrosion mechanism. It is suggested that the interested reader might consider the discussions by (Atrens et al. 2015; Atrens, Chen and Shi 2022) that (i) summarize compelling evidence for the Mg corrosion mechanism based on the unipositive Mg^+ ion as explained next and also (ii) explain the evidence for the existence of the unipositive Mg^+ ion (and how this ion has been shown to react with water in milliseconds in gas phase studies).

The highly active nature of Mg has two obvious consequences on the Mg corrosion reaction, determining both the cathodic reaction and the details of the anodic reaction. The open circuit potential of Mg during aqueous corrosion is so negative that the cathodic partial reaction is hydrogen evolution, in contrast to the oxygen reduction reaction, which is the cathodic reaction of most other technological metals, for example, the corrosion of steels in natural waters. The second consequence is that there is sufficient energy for there to be a reaction intermediate, Mg^+ between metallic Mg and Mg^{2+}, and furthermore, that the reaction intermediate is sufficiently energetic to chemically split water (Atrens, Gentle and Atrens 2015). In addition, quantum mechanics forbids the simultaneous exchange of two electrons (Bockris and Reddy 1977), and first principles studies have indicated

DOI: 10.1201/9781003319856-2

that the sequential loss of two electrons is energetically easier than the simultaneous transfer of two electrons (Ma et al. 2017, 2019).

The Mg anodic partial reaction:

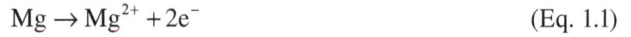

$$Mg \rightarrow Mg^{2+} + 2e^- \tag{Eq. 1.1}$$

occurs in two steps by the sequential loss of the two electrons. The loss of the first electron produces the uni-positive Mg^+ ion by the following reaction:

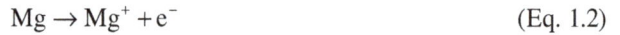

$$Mg \rightarrow Mg^+ + e^- \tag{Eq. 1.2}$$

A fraction k of Mg^+ ions lose the second electron to produce the Mg^{2+} ion that is stable in water by the following reaction:

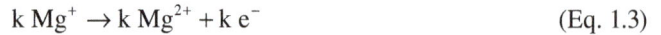

$$k \, Mg^+ \rightarrow k \, Mg^{2+} + k \, e^- \tag{Eq. 1.3}$$

The complement chemically splits water, producing hydrogen by the following chemical reaction:

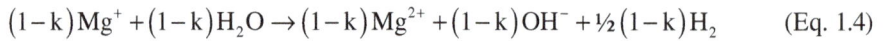

$$(1-k)Mg^+ + (1-k)H_2O \rightarrow (1-k)Mg^{2+} + (1-k)OH^- + \tfrac{1}{2}(1-k)H_2 \tag{Eq. 1.4}$$

The two steps of the anodic partial reaction are balanced, at the free corrosion potential, by the cathodic partial reaction, which can be written as follows:

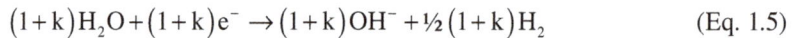

$$(1+k)H_2O + (1+k)e^- \rightarrow (1+k)OH^- + \tfrac{1}{2}(1+k)H_2 \tag{Eq. 1.5}$$

Reactions (Eq. 1.2) to (Eq. 1.5) sum to give the well-known overall Mg corrosion reaction as follows:

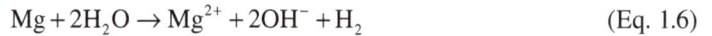

$$Mg + 2H_2O \rightarrow Mg^{2+} + 2OH^- + H_2 \tag{Eq. 1.6}$$

It is also worth mentioning that the chemical splitting of water (Eq. 1.4) is expected to occur at the Mg surface because gas phase studies have indicated that Mg^+ reacts with water in milliseconds, and the energetics (Ma et al. 2017, 2019) is such that Mg^+ is expected to be the surface species on the Mg surface because the following reaction is energetically favourable and is expected to occur essentially instantaneously because it involves the transfer of one electron.

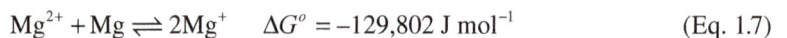

$$Mg^{2+} + Mg \rightleftharpoons 2Mg^+ \quad \Delta G^o = -129{,}802 \text{ J mol}^{-1} \tag{Eq. 1.7}$$

This Mg corrosion mechanism is designated herein as the unipositive Mg^+ ion corrosion mechanism and is illustrated in Figure 1.1. Evidence was reviewed in (Atrens et al. 2015; Atrens, Chen and Shi 2022). Compelling recent evidence is provided in section 1.6. The alternative mechanism is also discussed herein (the enhanced catalytic activity mechanism). This enhanced catalytic activity mechanism is contradicted by the experimental evidence as explained in what follows.

1.2 QUANTIFICATION OF MG CORROSION

The overall Mg corrosion reaction is summarised by Eq. (1.6). This reaction indicates that the Mg corrosion rate can be conveniently measured by the measurement of the evolved hydrogen during free corrosion, as suggested by (Song, StJohn and Atrens 2001). The technique is illustrated in Figure 1.2. The hydrogen from the corroding Mg specimen is funnelled into an upturned solution-filled burette. The evolved hydrogen volume can be easily measured periodically by the displacement of the solution in the burette.

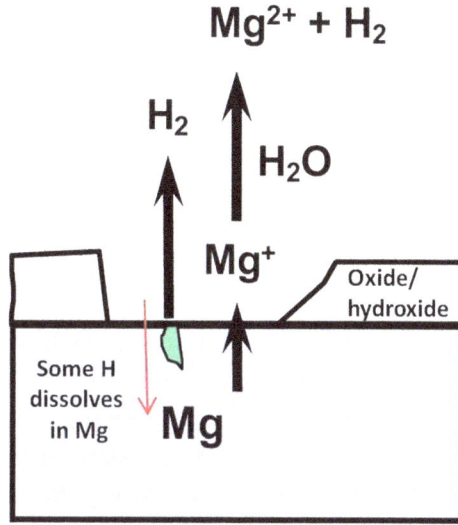

FIGURE 1.1 Schematic of the uni-positive Mg⁺ ion corrosion mechanism (Song and Atrens 2003). Corrosion preferentially occurs at breaks in the partially protective surface film. The cathodic partial reaction is the hydrogen evolution reaction, which occurs preferentially on second-phase particles and on Fe-rich impurity particles. The anodic partial reaction occurs by the sequential loss of two electrons to produce the Mg^{++}, which is stable in solution. The anodic reaction intermediate Mg^+ is sufficiently energetic to chemically split water to produce anodic hydrogen. *Reproduced with permission from (Song and Atrens 2003) © 2003 Wiley-VCH Verlag GmbH & Co KG, Weinheim.*

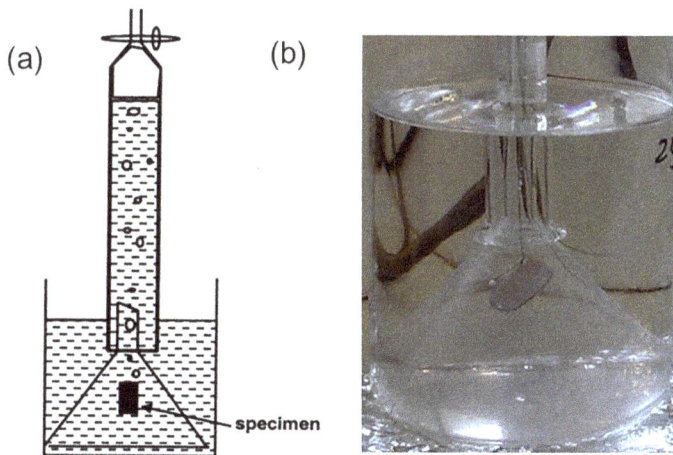

FIGURE 1.2 (a) The hydrogen evolution method for the measurement of the Mg corrosion rate (Song, StJohn and Atrens 2001). The hydrogen from the corroding Mg specimen is funnelled into an upturned solution-filled burette. The evolved hydrogen volume can be easily measured periodically. (b) The fishing line specimen which can be used to measure the evolved hydrogen and the corrosion rate via weight loss (Shi and Atrens 2011). *Reproduced (a) with permission from (Song, StJohn and Atrens 2001) © 2016, The Minerals, Metals & Materials Society and (b) from (Shi and Atrens 2011) © 2010 Elsevier Ltd.*

The Mg corrosion rate, as a penetration rate, P_H (mm/y), can be evaluated from (Shi and Atrens 2011; Zainal Abidin et al. 2011):

$$P_H = 2.279V_H \qquad (Eq.\ 1.8)$$

where V_H (mL/cm²/day) is the hydrogen evolution rate (with the hydrogen volume converted to that at 0°C using the ideal gas law) or given by (Qiao et al. 2012; Cao, Shi, Hofstetter et al. 2013; Cao, Shi, Song et al. 2013; Shi et al. 2013):

$$P_H = 2.088V_H \qquad (Eq.\ 1.9)$$

where V_H (mL/cm²/day) is the hydrogen evolution rate (with the hydrogen volume measured at 25°C).

The Mg corrosion rate so measured from the hydrogen evolution rate is in good agreement with the measured weight loss rate, as shown in Figure 1.3, provided that the Mg corrosion rate is substantial, typically greater than ~3 mm/y. For low corrosion rates, the corrosion rate measured from the evolved hydrogen is typically too small, maybe by orders of magnitude at small corrosion rates, because the amount of hydrogen dissolved in the solution introduces a substantial error. It should be noted that the hydrogen evolution method also does not provide good measures of the Mg corrosion

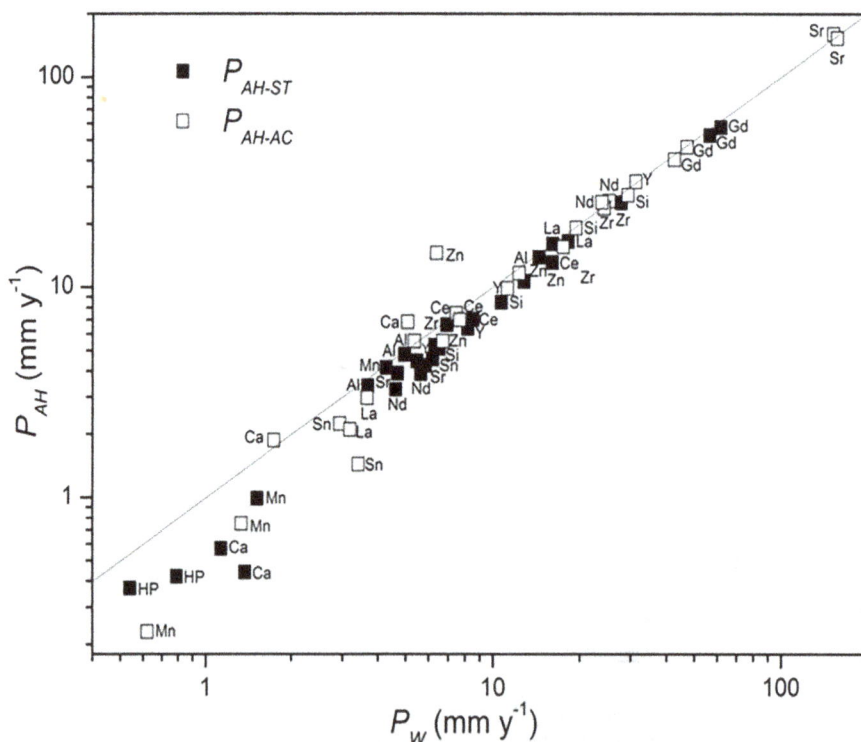

FIGURE 1.3 Cross plot of independent measurement of the corrosion rate measured using hydrogen evolution and by weight loss (Atrens et al. 2015). The Mg corrosion rate so measured from the hydrogen evolution rate is in good agreement with the measured weight loss rate for substantial corrosion rates. For low corrosion rates, the hydrogen evolution method gives values that are too low. Data from (Cao et al. 2013; Shi et al. 2013). *Reproduced with permission from (Atrens et al. 2015) © 2015 Wiley-VCH Verlag GmbH & Co. KGaA, Weinheim.*

in the synthetic body fluids used to simulate biodegradation of medical implants. Again, the issue seems to be the dissolution of hydrogen in the solution.

The corrosion rate measured from weight loss, P_W (mm/y) can be evaluated from (Shi and Atrens 2011):

$$P_W = 2.10W_L \qquad \text{(Eq. 1.10)}$$

where W_L is the normalized weight loss rate (mg cm^{-1} d^{-1}) of the specimen after solution immersion and after removal of the corrosion products compared with the normalized weight of the specimen before solution exposure:

$$W_L = (W_b - W_a) / At \qquad \text{(Eq. 1.11)}$$

where W_b (mg) is the normalized specimen weight before solution exposure, W_a (mg) is the normalised specimen weight after solution exposure after removal of all corrosion products without removal of any M metal, A (cm^2) is the area of the specimen, and t (d) is the solution exposure duration.

The Mg corrosion rate can also be measured from the corrosion current density, i_{corr} (mA cm^{-2}), typically evaluated from polarization curves using Tafel extrapolation (typically in the range 70 mV to 200 mV from the corrosion potential, E_{corr}). This corrosion rate, P_i, can be evaluated using (Shi and Atrens 2011):

$$P_i = 22.85 i_{corr} \qquad \text{(Eq. 1.12)}$$

Figure 1.4a shows a Mg specimen encapsulated in metallurgical resin, with one face available for solution exposure, with an encapsulated electrical connection to the specimen. Such a specimen is easy to prepare and is convenient to use for electrochemistry experiments. However, experience

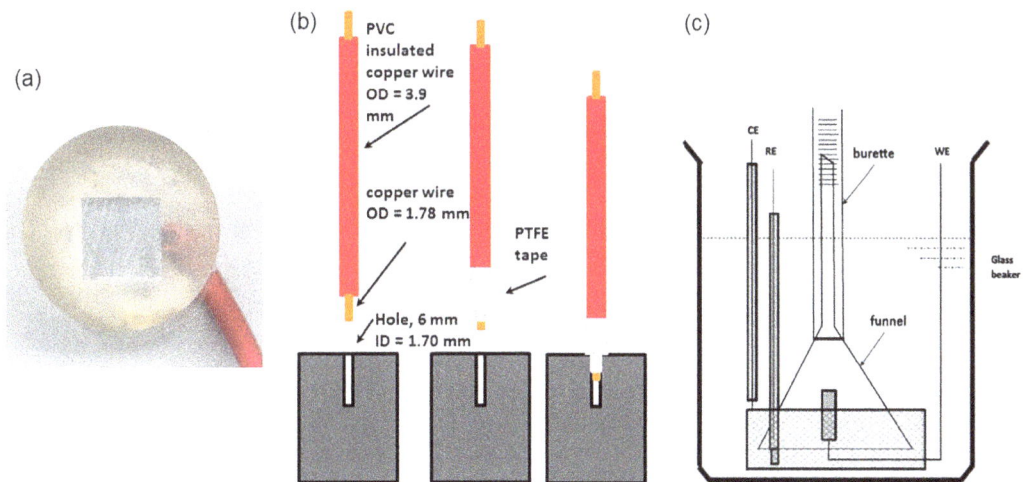

FIGURE 1.4 (a) Encapsulation of a Mg specimen for electrochemical measurements (Shi and Atrens 2011). (b) the assembly of a plug-in specimen. (c) The plug-in specimen can be used for weight loss measurements, hydrogen evolution measurements, and electrochemical measurements (Atrens et al. 2015). *Reproduced (a) with permission from (Shi and Atrens 2011) © 2010 Elsevier Ltd and (b &c) from (Atrens et al. 2015) © 2015 Wiley-VCH Verlag GmbH & Co. KGaA.*

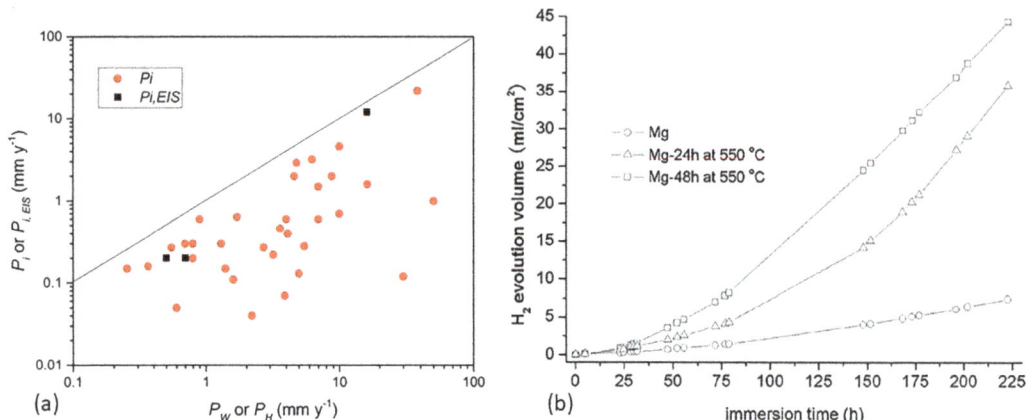

FIGURE 1.5 (a) The corrosion rate measured using electrochemistry, P_i or $P_{i,EIS}$, typically does not provide a good measurement of the steady-state corrosion rate as measured by weight loss or hydrogen evolution (Atrens et al. 2020). (b) The hydrogen evolution during corrosion of pure Mg after heat treatment at 550°C for 48 hours (Liu et al. 2009). The corrosion rate at any time is given by the slope of the curve. The initial slope (and initial corrosion rate) was comparatively low, much lower than the steady-state corrosion rate. The corrosion rate of the Mg after heat treatment was significantly higher than for the cast Mg without heat treatment due to the precipitation of Fe-rich precipitates as predicted by the Mg-Fe equilibrium phase diagram (Liu et al. 2009). *Reproduced (a) with permission from (Atrens et al. 2020) © 2020 Published by Elsevier B.V. on behalf of Chongqing University, and (b) from (Liu et al. 2009) © 2008 Elsevier Ltd.*

indicates that weight loss determinations are not always accurate. Figures 1.4b and c illustrate the plug-in specimen (Shi and Atrens 2011). This has been successfully used to measure weight loss, hydrogen evolution and to conduct electrochemical measurements.

Figure 1.5(a) (Atrens et al. 2020) provides an overview of recent measurements of the corrosion rate, P_i, compared with the corrosion rate measured in the same paper using weight loss or hydrogen evolution. This shows that the corrosion rate, P_i, typically does not provide a good measurement of the steady state corrosion rate as measured by weight loss or hydrogen evolution. This is because electrochemical measurements typically measure the corrosion rate soon after specimen immersion in solution, before establishment of steady-state conditions. This can be understood by the data in Figure 1.5(b) for the hydrogen evolution during corrosion of pure Mg after heat treatment at 550°C for 48 hours (Liu et al. 2009). The corrosion rate at any time is given by the slope of the curve. The initial slope (and initial corrosion rate) was comparatively low, much lower than the steady-state corrosion rate.

1.3 Mɢ CORROSION RATES

Figure 1.6 (Atrens et al. 2020) provides a recent evaluation of Mg corrosion rates, both for pure Mg and for Mg alloys in chloride solutions and synthetic body fluids. The corrosion rates have been plotted against total alloying content for convenience and to spread out the data. There is clearly no general relationship between corrosion rate and total alloying content, nor is there any expectation that there should be a relationship.

The green diamonds indicate that the corrosion rates of Mg-Al alloys in atmospheric corrosion are significantly lower than for Mg alloys immersed in solution. These lower corrosion rates are attributed to more protective corrosion products due to the periodic drying during atmospheric exposure.

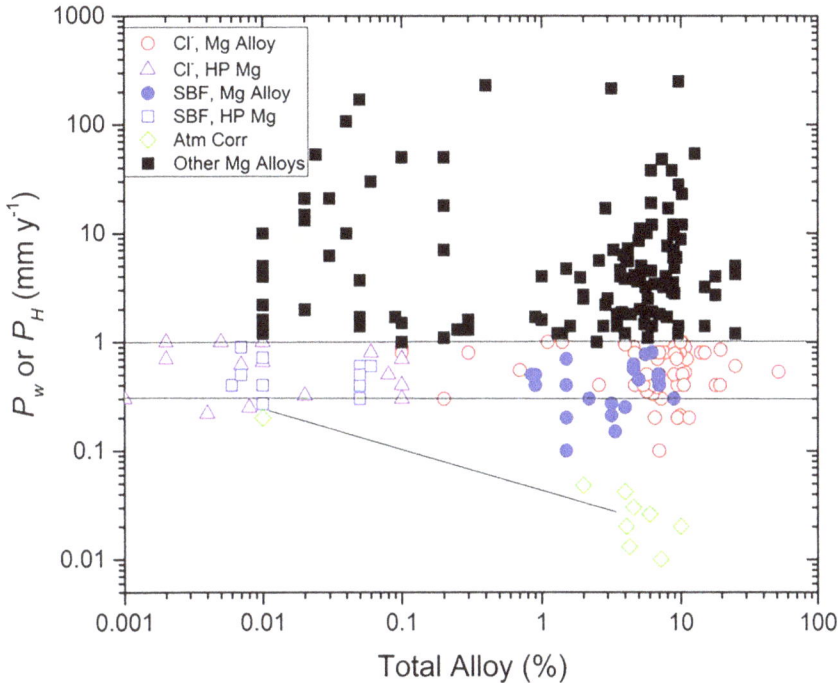

FIGURE 1.6 A recent evaluation of Mg corrosion rates (Atrens et al. 2020), both for pure Mg and for Mg alloys in chloride solutions and synthetic body fluids. The corrosion rates have been plotted against total alloying content for convenience and to spread out the data. There is clearly no general relationship between corrosion rate and total alloying content. *Reproduced with permission from (Atrens et al. 2020) © 2020 Published by Elsevier B.V. on behalf of Chongqing University.*

A horizontal line was drawn in Figure 1.6 for a corrosion rate of 0.3 mm/y. This corrosion rate represents the lowest reliable corrosion rate measured using weight loss for pure Mg and high-purity Mg and is designated as the intrinsic Mg corrosion rate in chloride solutions (as measured by weight loss). There are only a few Mg alloys that have provided a reliable measurement of a corrosion rate in a chloride solution that is less than the intrinsic Mg corrosion rate, despite the large amount of research aimed at producing a Mg alloy with a low corrosion rate. This includes much research that claims alloying has produced a Mg alloy with a lower corrosion rate. Typically, these alloys have a corrosion rate higher than the intrinsic Mg corrosion rate even though the Mg alloy has a lower corrosion rate than that of the alloy without the alloying addition.

There are many data for pure Mg and Mg alloys between the two horizontal lines drawn on Figure 1.6 at 0.3 mm/y and 1 mm/y. This includes Mg alloys. It also includes the Mg-Li alloy (Xu et al. 2015; Frankel 2015) claimed to have corrosion rate much better than previous alloys, despite the fact that lower corrosion rates had been previously published.

Figure 1.6 shows that there are many Mg alloys with corrosion rates higher than the intrinsic Mg corrosion rate of 0.3 mm/y in a chloride solution as measured by weight loss. The typical second phase in Mg alloys causes microgalvanic acceleration of the corrosion, because the second phase typically has a potential more positive than that of the Mg matrix, and the second phase provides a more efficient surface for the cathodic hydrogen evolution reaction. The corrosion rate of a Mg alloy is typically greater than the intrinsic Mg corrosion rate of 0.3 mm/y in chloride solutions as measured by weight loss because of the microgalvanic corrosion acceleration and because the alloying typically does not cause the formation of a surface film that is much more protective than the film formed on high-purity Mg on solution exposure.

1.4 SECOND PHASES

Song and Atrens (1999) explained the influence of microstructure, particularly the presence of second phases, on the corrosion behaviour of Mg alloys, relying heavily on the prior research of (Song et al. 1998, 1999). The typical second phase in Mg alloys has an electrochemical potential more positive than that of the alpha-Mg matrix. This means that the second phase has the possibility to cause microgalvanic corrosion acceleration. This concept is illustrated in Figure 1.7(a). Any other technical metal has an electrochemical potential more positive than that of Mg metal, so the other metal causes galvanic corrosion of Mg if in contact with Mg and both metals are in contact with the same electrolyte. Exactly the same principle applies for a Mg alloy containing second phases, as illustrated in Figure 1.7(b).

Figure 1.7(c) illustrates the electrochemistry for Mg-Al alloys which contain sufficient Al to produce the beta-phase. The polarisation curves are presented for the matrix (the alpha-Mg) and the second phase (the beta-phase). For each phase, immersed in the solution separately, there is a corrosion potential, ~ −1630 mV for the alpha-Mg and ~ −1260 mV for the beta-phase. There is a potential difference of ~ 470 mV between the corrosion potentials of the alpha-Mg and the beta-phase. This potential difference accelerates the corrosion of the alpha-Mg matrix. Note also that the cathodic reaction, the hydrogen evolution reaction, is much quicker on the beta-phase, as is clear from the higher current density for the cathodic reaction on the beta-phase compared with that on the alpha-phase. Thus, in general, the second phase accelerates the corrosion rate of the Mg alloy because (i) there is a potential difference between the second phase and the alpha-Mg matrix and (ii) the second phase facilitates the cathodic partial reaction on its surface.

The typical microstructure of diecast AZ91D is shown in Figure 1.8(a) (Song et al. 1999). The microstructure consists of a primary alpha-Mg matrix and an eutectic microconstituent consisting of eutectic alpha-Mg and the beta-phase ($Mg_{17}Al_{12}$). Figure 1.8(b) presents the appearance after a

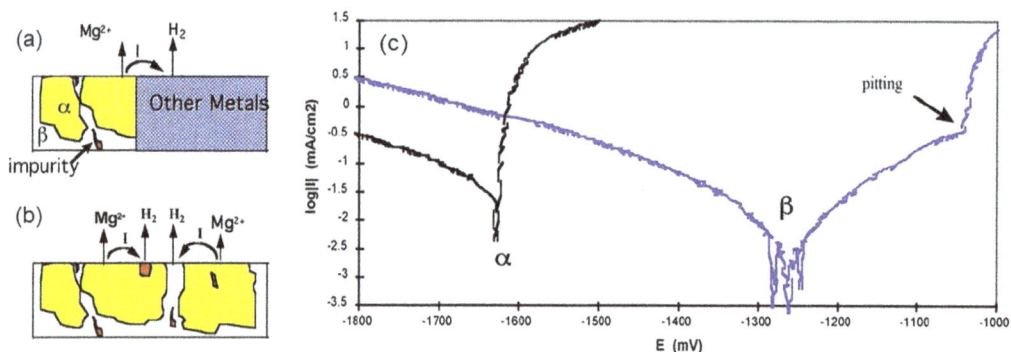

FIGURE 1.7 (a) Coupling Mg metal in a solution to any other technological metal causes galvanic acceleration of the Mg alloy. (b) Similarly, there is micro-galvanic corrosion acceleration of the alpha-Mg matrix in a Mg alloy by the second phases because the second phases have a more positive potential and typically provide a more catalytic surface for the cathodic hydrogen evolution reaction. (c) The polarization curves for the matrix (the alpha-Mg) and the second phase (the beta-phase) (Song and Atrens 1999). For each phase, immersed in the solution separately, there is a corrosion potential, ~−1630 mV for the alpha-Mg and ~−1260 mV for the beta-phase. There is a potential difference of ~470 mV between the corrosion potentials of the alpha-Mg and the beta-phase. This potential difference accelerates the corrosion of the alpha-Mg matrix. Note also that the cathodic reaction, the hydrogen evolution reaction, is much quicker on the beta-phase, as is clear from the higher current density for the cathodic reaction on the beta-phase compared with that on the alpha-phase. Thus, in general, the second phase accelerates the corrosion rate of the Mg alloy because (i) there is a potential difference between the second phase and the alpha-Mg matrix and (ii) the second phase facilitates the cathodic partial reaction on its surface. *Reproduced with permission from (Song and Atrens 1999) © 1999 WILEY-VCH Verlag GmbH, Weinheim, Fed. Rep. of Germany.*

short time immersion in a chloride solution, and Figures. 1.8(c) and 1.8(d) provide cartoons of cross-sections illustrating the details of the corrosion. Micro-galvanic acceleration of the large primary alpha-Mg grain in the centre left of Figure 1.8(a) caused this grain to corrode away completely, as illustrated in the left part of the section in Figures. 1.8(c) and 1.8(d). Similarly, the eutectic alpha has also corroded away, undermining some of the eutectic beta.

Figure 1.9 illustrates the case where the beta-phase is interconnected (Song et al. 1999). The cross-section in Figure 1.9(a) indicates the top surface of the alloy in contact with the solution. Figure 1.9(b) indicates that the alpha-phase in contact with the solution corrodes until the corrosion

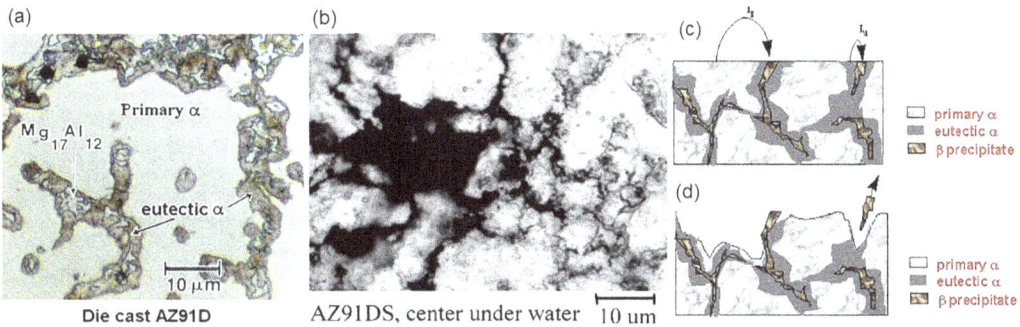

FIGURE 1.8 (a) The typical microstructure of die cast AZ91D consisted of a primary alpha-Mg matrix and an eutectic microconstituent consisting of eutectic alpha-Mg and the beta-phase ($Mg_{17}Al_{12}$). (b) The appearance of the diecast AZ91 after a short time immersion in a chloride solution (Song and Atrens 1999). The cartoons (c) and (d) provide cross-sections illustrating the details of the corrosion. Micro-galvanic acceleration of the large primary alpha-Mg grain in the centre left of (a) caused this grain to corrode away completely, as illustrated in the left part of the section in (c) and (d). Similarly, the eutectic alpha has also corroded away, undermining some of the eutectic beta. *Reproduced with permission from (Song and Atrens 1999) © 1999 WILEY-VCH Verlag GmbH, Weinheim, Fed. Rep. of Germany.*

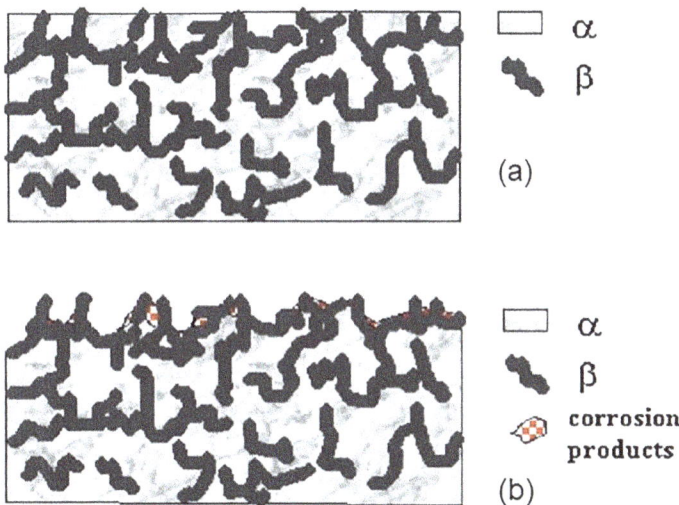

FIGURE 1.9 Corrosion when the beta-phase is interconnected (Song and Atrens 1999). The cross-section in (a) indicates the top surface of the alloy in contact with the solution. (b) indicates that the alpha-phase in contact with the solution corrodes until the corrosion front reaches the beta-phase, whereupon there is essentially only the beta-phase in contact with the solution, the corrosion rate slows down significantly and becomes essentially that of the beta-phase. *Reproduced with permission from (Song and Atrens 1999) © 1999 WILEY-VCH Verlag GmbH, Weinheim, Fed. Rep. of Germany.*

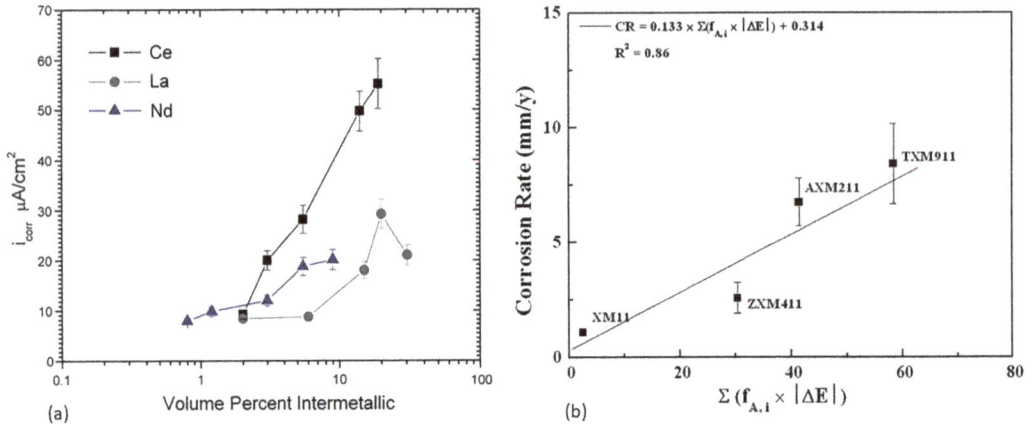

FIGURE 1.10 (a) Increase of corrosion rate (measured in terms of the corrosion current density, i_{corr}, from polarization curves) in a chloride solution with increasing volume percent intermetallic phase based on Mg-Ce, Mg-La and Mg-Nd alloys (Birbilis et al. 2009). (b) The corrosion rate for a MgCaMn alloy (XM11), and that alloy containing 4%Zn (ZXM411), 2%Al (AXM211), and 9%Sn (TXM911) plotted against $\Sigma\left(f_i\left|\Delta E_i\right|\right)$, (Bahmani, Arthari and Shin 2019), where f_i is the fraction of the ith second phase in each microstructure and ΔE_i is the absolute value of the volta-potential difference between the second phase and the matrix. There was good agreement, indicating that this formulation provided a good explanation for the magnitude of the microgalvanic corrosion acceleration. *Reproduced (a) with permission from (Birbilis et al. 2009) © 2008 Elsevier Ltd, and (b) from (Bahmani, Arthari and Shin 2019) © 2019 Published by Elsevier B.V. on behalf of Chongqing University.*

front reaches the beta-phase, whereupon there is essentially only the beta-phase in contact with the solution; the corrosion rate slows down significantly and becomes essentially that of the beta phase.

This established the principles governing the corrosion of Mg alloys containing second phases. The corrosion of the Mg alloy is accelerated for isolated second phases because of the potential difference between the second phase and the alpha-Mg matrix and because the second phase is typically a more efficient cathode for the cathodic hydrogen evolution reaction. Furthermore, the microgalvanic corrosion acceleration also increases with increasing volume fraction of the second phase until there is sufficient second phase to form a network, whereupon the corrosion is as illustrated in Figure 1.9. These principles have been verified by subsequent research. Figure 1.10(a) (Birbilis et al. 2009) illustrates the increase of corrosion rate (measured in terms of the corrosion current density, i_{corr}, from polarization curves) in a chloride solution with increasing volume percent intermetallic phase based on Mg-Ce, Mg-La and Mg-Nd alloys. The increases of corrosion rate can be correlated with the electrochemical properties of the individual intermetallic phases, in particular (i) the difference between their free corrosion potentials and that of the alpha-Mg matrix, and (ii) the facilitation of the hydrogen evolution reaction on the second phase.

Figure 1.10(b) (Bahmani, Arthari and Shin 2019) presents the corrosion rate for a MgCaMn alloy (XM11), and that alloy containing 4%Zn (ZXM411), 2%Al (AXM211), and 9%Sn (TXM911) plotted against $\Sigma\left(f_i\left|\Delta E_i\right|\right)$, where f_i is the fraction of the ith second phase in each microstructure and ΔE_i is the absolute value of the volta-potential difference between the second phase and the matrix. There was good agreement, indicating that this formulation provided a good explanation for the magnitude of the microgalvanic corrosion acceleration.

1.5 IMPURITY EFFECT

It has been known since the pioneering work of (Hanawalt, Nelson and Peloubet 1942) in 1942 that the corrosion of Mg is sensitive to the impurity elements Fe, Ni, Cu, Co. Figure 1.11 (Liu et al. 2009)

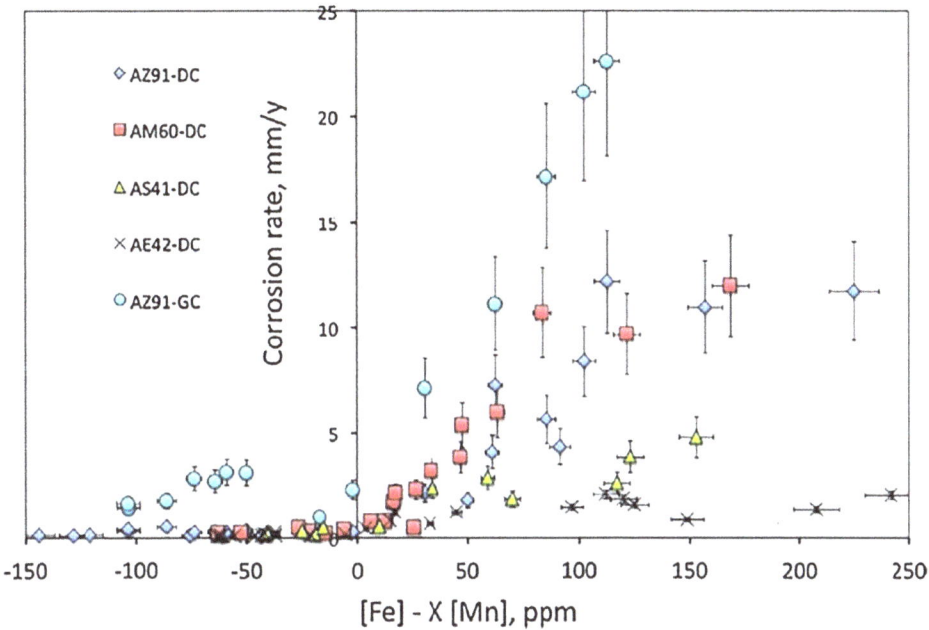

FIGURE 1.11   The corrosion rate is substantial above the compositional dependent tolerance limit (indicated as the zero point of the X-axis) and small for compositions below the tolerance limit. *Reproduced with permission from (Liu et al. 2009) © 2008 Elsevier Ltd.*

shows that the corrosion rate is substantial above the compositional dependent tolerance limit (indicated as the zero point of the X-axis) and small for compositions below the tolerance limit.

The Fe tolerance limit can be understood from the Mg-Fe equilibrium phase diagram as presented in Figure 1.12 (Liu et al. 2009). The Mg-Fe equilibrium phase diagram is an eutectic. The eutectic concentration corresponds to an Fe concentration of 0.018wt.% or 180 wt ppm, which corresponds to the Fe tolerance limit for a pure Mg casting. The formation of the Fe-rich BCC phase occurs first during (equilibrium) solidification of pure Mg containing a greater Fe concentration, so that the alloy, on cooling to room temperature, contains the HCP alpha-Mg matrix and the Fe-rich BCC second phase, which causes the high corrosion rates due to microgalvanic corrosion acceleration. Normal solidification produces only an alpha-Mg matrix for pure Mg containing a lower Fe concentration, and low corrosion results because of the intrinsic Mg corrosion without any second phases present. However, the Mg-Fe phase diagram indicates that such a solidified pure Mg is in a meta-stable state. Heating can cause approach to equilibrium, the precipitation of the Fe-rich BCC phase, and a higher corrosion rate. This is illustrated by the data in Figure 1.5(b). The corrosion rate of the pure Mg was low and constant. Heating at 550°C causes a significantly higher corrosion rate due to the precipitation of the Fe-rich second phase.

The Mg-Fe equilibrium diagram of Figure 1.12 indicates that the Fe tolerance limit is ~ 2 wt ppm for Mg that has been heat treated, e.g. a heat-treated casting or an extrusion that was heated during the extrusion process. This indicates that the Fe tolerance effect is caused by a special case of microgalvanic corrosion, by the Fe-rich second phase.

Figure 1.13 (Taltavull et al. 2014) shows by the weight loss measurements that there is no microgalvanic acceleration for the corrosion rate of pure Mg in chloride solutions with increasing chloride concentration. This is in contrast to two-phase Mg alloys, where the corrosion rate increases with increasing chloride concentration.

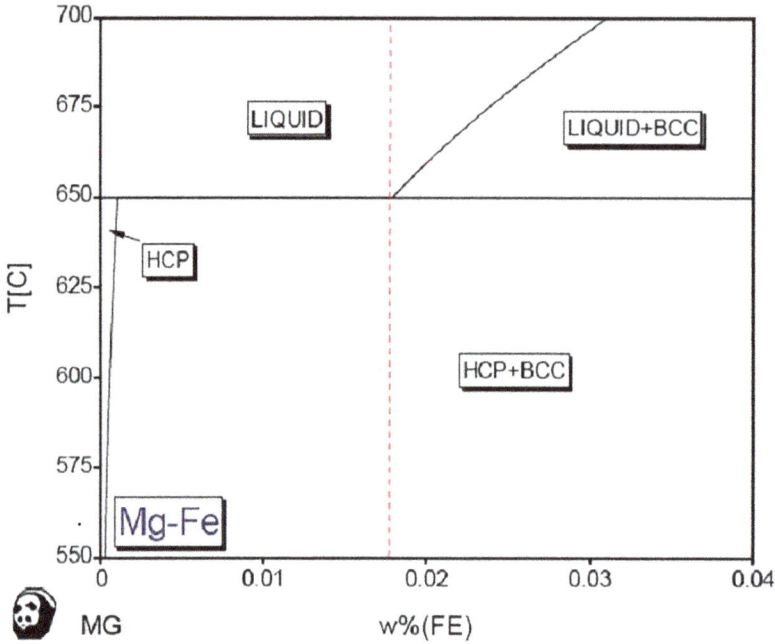

FIGURE 1.12   The Mg-Fe equilibrium phase diagram (Liu et al. 2009) is an eutectic. The eutectic concentration corresponds to an Fe concentration of 0.018wt.% or 180 wt ppm, which corresponds to the Fe tolerance limit for a pure Mg casting. The Mg-Fe equilibrium diagram also indicates that the Fe tolerance limit is ~ 2 wtppm for Mg that has been heat treated, e.g. a heat-treated casting or an extrusion that was heated during the extrusion process. *Reproduced with permission from (Liu et al. 2009) © 2008 Elsevier Ltd.*

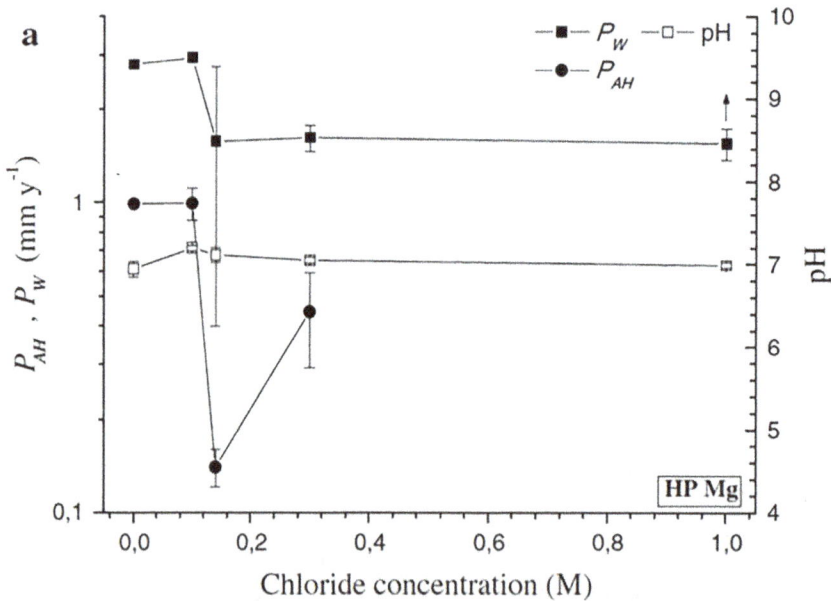

FIGURE 1.13   There is no microgalvanic acceleration for the corrosion rate as measured by weight loss of pure Mg in chloride solutions with increasing chloride concentration. This is in contrast to two-phase Mg alloys, where the corrosion rate increases with increasing chloride concentration. *Reproduced with permission from (Taltavull et al. 2014) © 2013, Springer Science Business Media New York.*

1.6   ANODIC HYDROGEN EVOLUTION

The negative difference effect (NDE) is a particular aspect of the Mg corrosion mechanism. The rate of hydrogen evolution on a Mg surface decreases with positive polarization when the polarization is started at a negative potential (this is the cathodic partial reaction of hydrogen evolution). The hydrogen evolution rate decreases to a low rate close to the corrosion potential. The hydrogen evolution rate again increases for anodic polarization to more positive potentials. This anodic hydrogen evolution for anodic polarization is the negative difference effect. Song 2005 showed that the anodic hydrogen was evolved at sites different to those sites at which hydrogen was evolved during the cathodic hydrogen partial reaction. This observation has been confirmed by Fajardo et al. 2016. This provides strong evidence that the anodic hydrogen is evolved by a mechanism different to that for the evolution of the cathodic hydrogen.

The anodic hydrogen evolution is easily understood in terms of the unipositive Mg^+ ion corrosion mechanism as introduced in section 1.1. Mg corrosion mechanism. Anodic polarization decreases the speed of the cathodic hydrogen evolution reaction (Eq. 1.5) rapidly to zero so that there is little to no hydrogen produced by the cathodic hydrogen evolution reaction. At the same time, anodic polarization increases the speed of the anodic partial reaction, so that more Mg^+ is produced, which produces a greater amount of hydrogen by the splitting of water by the chemical reaction given by Eq. (1.4).

Figure 1.14 presents the apparent valence of Mg, V, and the value of k (as defined in Eq. (1.3)) during anodic polarization from the work of (Li et al. 2021). V is evaluated from:

$$V = N_{e,a} / N_w \qquad \text{(Eq. 1.13)}$$

where N_w (mmol cm^2 h^1) is the Mg atom flux corresponding to the weight loss rate and $N_{e,a}$ (mmol cm^2 h^1) is given by:

$$N_{e,a} = \frac{I_a}{F} = \frac{I_{applied} + I_c}{F} \qquad \text{(Eq. 1.14)}$$

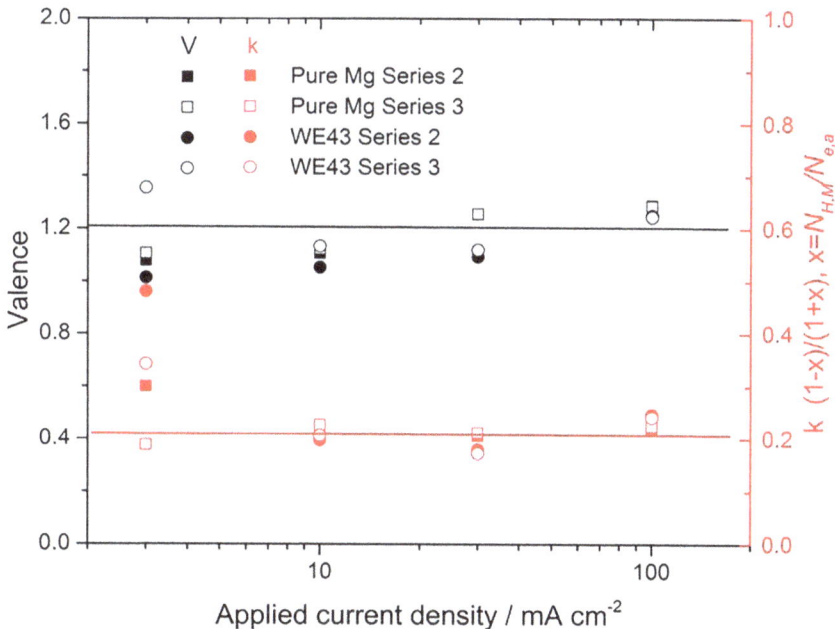

FIGURE 1.14   The apparent valence of Mg, V, and the value of k (as defined in Eq. (1.3)) during anodic polarization from the work of (Li et al. 2021). *Reproduced with permission from (Li et al. 2021) © 2021 Chongqing University. Publishing services provided by Elsevier B.V. on behalf of KeAi Communications Co. Ltd.*

where $N_{e,a}$ is the electron flux corresponding to the anodic partial reaction, I_a, corresponding to the applied anodic current density, $I_{applied}$, I_c is the cathodic current density, and F is the Faraday.

The value of k is evaluated from

$$k = (1-X)/(1+X), X = N_H / N_{e,a} \qquad \text{(Eq. 1.15)}$$

where N_H (mmol cm^2 h^1) is the corresponding flux of hydrogen atoms.

The value of $V = 1.2$ is consistent with expectations of the unipositive Mg$^+$ ion corrosion mechanism that the value of V is between 1.0 and 2.0. The value of $V = 1.2$ implies that k = 0.2, in excellent agreement with the data in Figure 1.14.

The reaction sequence given by Eqs. (1.2) to (1.6) of the unipositive Mg$^+$ ion corrosion mechanism imply that:

$$N_H / N_w = (1-k)/2 \qquad \text{(Eq. 1.16)}$$

which indicates $N_H/N_w = 0.4$ for k = 0.2. Figure 1.15 provides a plot of N_H against N_w from the data of Li et al 2021. The symbols indicate the data. The line drawn through the data is that for $N_H/N_w = 0.4$, indicating that this provides a good description of the data.

Thus the independent measurements of N_w, $N_{e,a}$, and N_H are in excellent agreement with the expectations of the unipositive Mg$^+$ ion corrosion mechanism, providing compelling evidence for the Mg corrosion mechanism.

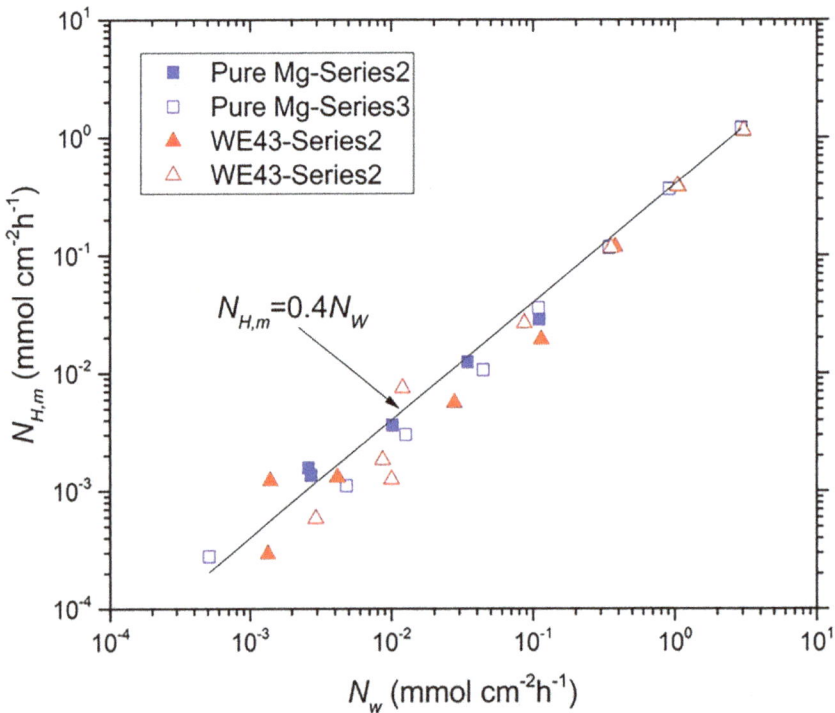

FIGURE 1.15 A plot of N_H against N_w from the data of (Li et al. 2021). The symbols indicate the data. The line drawn through the data is that for $N_H/N_w = 0.4$, indicating that this provides a good description of the data. *Reproduced with permission from (Li et al. 2021) © 2021 Chongqing University. Publishing services provided by Elsevier B.V. on behalf of KeAi Communications Co. Ltd.*

1.7 ENHANCED CATALYTIC ACTIVITY MECHANISM

The experimental evidence in Figures. 1.14 and 1.15 compellingly contradicts the enhanced catalytic activity mechanism. This mechanism assumes that the overall Mg partial anodic reaction occurs in one step, as indicated in Eq. (1.1), despite the fact that quantum mechanics forbids the simultaneous transfer of two electrons (Bockris and Reddy 1977) and that first principles studies indicate that it is easier to lose the electrons sequentially (Ma et al. 2017, 2019). The enhanced catalytic activity mechanism assumes that the apparent Mg valence $V = 2$, whereas the experimental data in Figure 1.14 indicates that $V = 1.2$. Furthermore, the enhanced catalytic activity mechanism assumes that there should be no relationship between the quantities in Figure 1.15.

If the cathodic partial reaction, such as given by Eq. (1.5), is governed by Tafel kinetics, the current density, $i_{cHER,T}$, at an applied potential, E, is given by

$$i_{cHER,T} = i_{OH} \exp\left\{-\left(E - E_H\right)/b_H\right\} \qquad \text{(Eq. 1.17)}$$

where i_{OH} is the exchange current density of the hydrogen evolution reaction (a constant), E_H is the equilibrium potential of the hydrogen evolution reaction, and b_H is the Tafel constant for the hydrogen evolution reaction. For anodic polarization, the exponential term decreases rapidly so that the cathodic hydrogen evolution rate quickly becomes exceedingly low and negligible.

The main assumptions (Frankel, Samaniego and Birbilis 2013) of the enhanced catalytic activity mechanism are that (i) the cathodic partial reaction is that given by Eq. (1.5), (ii) the anodic hydrogen evolution reaction is produced by this cathodic hydrogen evolution reaction, and (iii) the cathodic hydrogen evolution reaction rate, $i_{cHER,c}$, at an applied potential, E, is given by

$$i_{cHER,c} = i_{OH,c} \exp\left\{-\left(E - E_H\right)/b_H\right\} \qquad \text{(Eq. 1.18)}$$

where the terms have the same meaning as those in Eq. (1.16) except that $i_{OH,c}$ is the constant i_{OH} for cathodic polarizations BUT increases rapidly for anodic polarisations, increases so rapidly that it exceeds the decrease of the exponential terms, so that the current density increases with anodic polarization. This means that the enhanced catalytic activity mechanism predicts that the cathodic hydrogen evolution rate increases with anodic polarization to explain the measured anodic polarization rate. Birbilis et al. 2014 proposed that the enhanced catalytic activity mechanism could be validated by the measurement of the catalytically enhanced rate of the cathodic hydrogen evolution reaction.

Li et al. 2021 followed this suggestion, measured the "catalytically enhanced rate of the cathodic reaction," and found that the experimental data contradicted this mechanism: the "catalytically enhanced rate of the cathodic reaction" was much less than the anodic hydrogen evolution rate and thus could not explain the anodic hydrogen evolution.

Figure 1.16 provides the summary of the data of Li et al. 2021. The full symbols indicate the rate of anodic hydrogen evolution (HE) for specimens anodically polarized at the given applied anodic current density. Both pure Mg and the Mg alloy WE43 showed increased anodic hydrogen evolution rate with increasing applied anodic current density. Typical data is presented in Figure 1.17(a) for the evolved hydrogen volume on pure Mg for the various applied current densities. The hydrogen evolution rate is given by the slope of each line, and these anodic hydrogen evolution rates are plotted in Figure 1.16. The potential was recorded during each of the experiments of Figure 1.17(a), and cathodic polarization curves were measured immediately after each of the experiments of Figure 1.17(a). Typical cathodic polarization curves are shown in Figure 1.17(b). These are designated by the anodic current density applied in the preceding measurement of anodic hydrogen evolution. The cathodic polarization curves measured the cathodic hydrogen evolution rate. These cathodic polarization curves were extrapolated to the potential measured during the prior experiments measuring the anodic hydrogen evolution rate. This extrapolation of the cathodic polarization curves measured the "catalytically enhanced cathodic hydrogen evolution rate," corresponding to

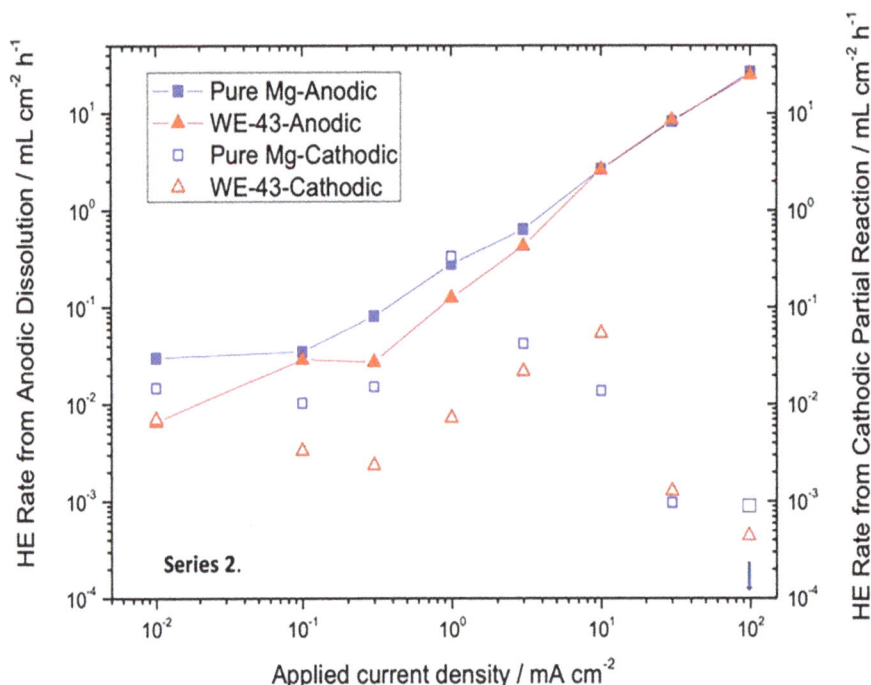

FIGURE 1.16 A summary of the data of Li et al. 2021. The full symbols indicate the rate of anodic hydrogen evolution (HE) rate for specimens anodically polarized at the given applied anodic current density. Both pure Mg and the Mg alloy WE43 showed increased anodic hydrogen evolution rate with increasing applied anodic current density. The measured "catalytically enhanced cathodic hydrogen evolution rate" are also plotted for comparison. These were measured from the polarization curves as shown in Fig. 1.17b. The main proposal of the enhanced catalytic activity mechanism is that the measured "catalytically enhanced cathodic hydrogen evolution rate" would explain the anodic hydrogen evolution rate. The data shows that the measured "catalytically enhanced cathodic hydrogen evolution rate" (open symbols) is typically less than the anodic hydrogen evolution rate (closed symbols) and is much less than the anodic hydrogen evolution rate for applied current densities greater than 1 mA cm^{-2}. This indicates that this experimental data contradicts the main proposal of the enhanced catalytic activity mechanism. *Reproduced with permission from (Li et al. 2021) © 2021 Chongqing University. Publishing services provided by Elsevier B.V. on behalf of KeAi Communications Co. Ltd.*

the measured anodic hydrogen evolution rate. Figure 1.16 provides a plot of the measured "catalytically enhanced cathodic hydrogen evolution rate" (open symbols) plotted against the anodic applied current density applied during the prior experiments to measure the anodic hydrogen evolution rate. The main proposal of the enhanced catalytic activity mechanism is that the measured "catalytically enhanced cathodic hydrogen evolution rate" would explain the anodic hydrogen evolution rate. Figure 1.16 shows that the measured "catalytically enhanced cathodic hydrogen evolution rate" (open symbols) is typically less than the anodic hydrogen evolution rate (closed symbols) and is much less than the anodic hydrogen evolution rate for applied current densities greater than 1 mA cm^{-2}. This indicates that this experimental data contradicts the main proposal of the enhanced catalytic activity mechanism.

1.8   CONCLUSIONS AND OUTLOOK

The following are the important aspects of Mg corrosion.

- The highly active nature of Mg determines both the cathodic reaction (the cathodic hydrogen evolution reaction) and the details of the anodic reaction. There is sufficient energy

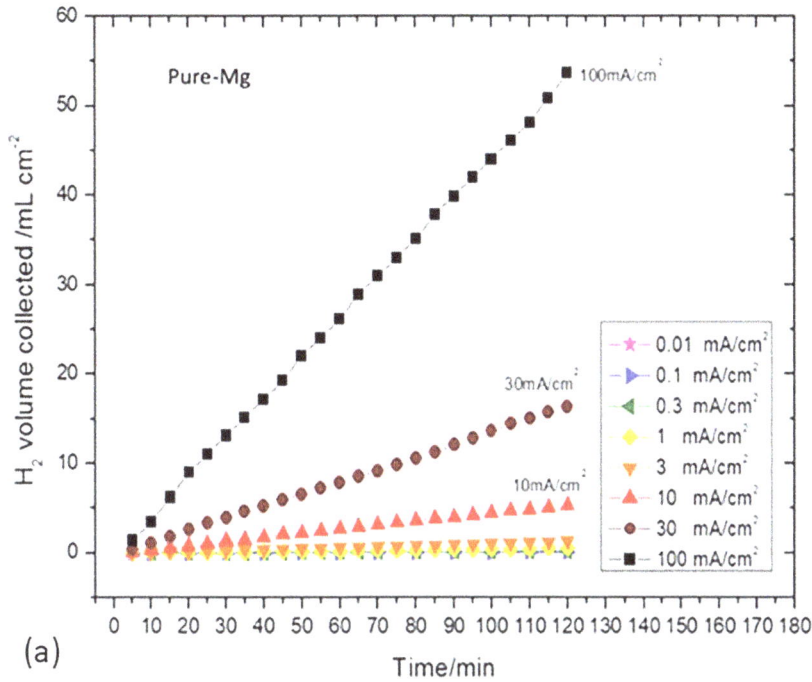

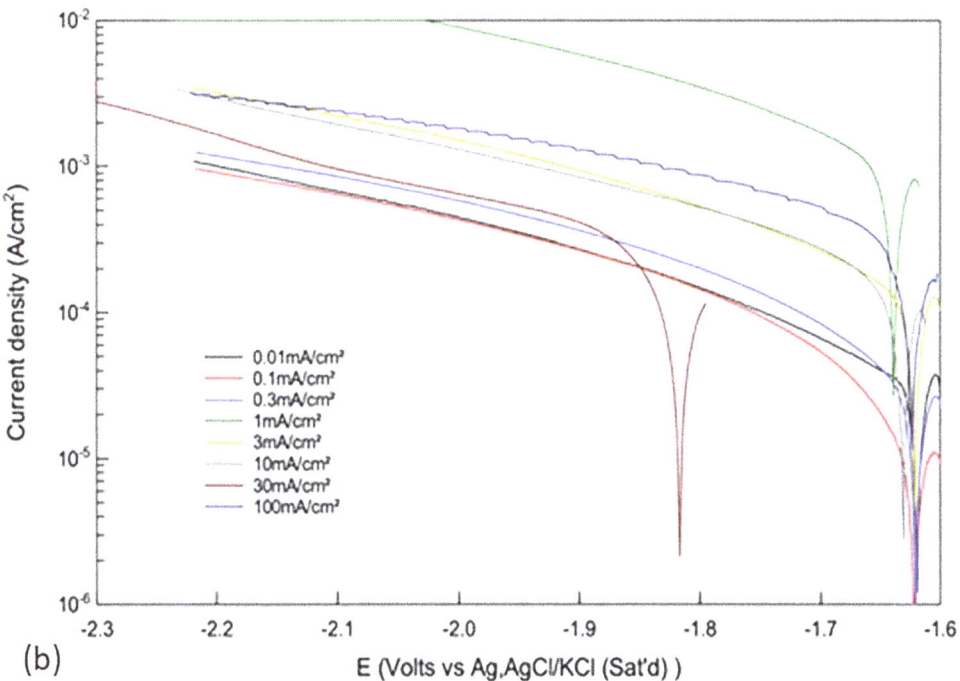

FIGURE 1.17 (a) The evolved hydrogen volume on pure Mg for the various applied current densities (Li et al. 2021). The hydrogen evolution rate is given by the slope of each line, and these anodic hydrogen evolution rates are plotted in Fig. 1.16. Fig. 1.17b shows typical cathodic polarization curves (Li et al. 2021) designated by the anodic current density applied in the preceding measurement of anodic hydrogen evolution. The cathodic polarization curves measured the cathodic hydrogen evolution rate. These cathodic polarization curves were extrapolated to the potential measured during the prior experiments measuring the anodic hydrogen evolution rate. This extrapolation of the cathodic polarization curves measured the "catalytically enhanced cathodic hydrogen evolution rate," corresponding to the measured anodic hydrogen evolution rate. The measured "catalytically enhanced cathodic hydrogen evolution rate" are also plotted in Fig. 1.16 (as the open symbols). *Reproduced with permission from (Li et al. 2021) © 2021 Chongqing University. Publishing services provided by Elsevier B.V. on behalf of KeAi Communications Co. Ltd.*

available during the anodic partial reaction for there to be a reaction intermediate Mg^+ between metallic Mg and Mg^{2+} and for Mg^+ to chemically split water (in milliseconds).

- The films formed on the surface of Mg alloys during aqueous corrosion are only partially protective in the technologically important chloride solutions. The films on Mg alloys are typically not much more corrosion resistant than those films formed on high-purity Mg.
- The typical second phase in Mg alloys causes microgalvanic acceleration of the corrosion of the Mg alloy, because the second phase typically has a potential more positive than that of the Mg matrix, and the second phase provides a more efficient surface for the cathodic hydrogen evolution reaction.
- The corrosion rate of a Mg alloy is typically greater than the intrinsic Mg corrosion rate of 0.3 mm/y in chloride solutions as measured by weight loss.
- The Mg corrosion rate should be measured using weight loss or hydrogen evolution.
- Electrochemical methods typically give Mg corrosion rates that do not agree with the steady-state Mg corrosion rate.
- The unipositive Mg^+ ion corrosion mechanism explains Mg corrosion. This mechanism is supported by compelling experimental evidence.
- The enhanced catalytic activity mechanism has received much attention but is contradicted by the experiment.

ACKNOWLEDGEMENTS

The many co-workers are thanked who have helped in developing our understanding of Mg corrosion including Sean Johnston, Zhiming Shi, Catalina Taltavull, Nor Ishida Zainal Abidin, Aleks D Atrens, Darren Martin, Akif Soltan, Matt Dargusch, Syeda Mehreen, Guang-Ling Song, Ming Liu, Xingrui Chen.

REFERENCES

Atrens, A, X Chen, Z Shi, Mg corrosion – recent progress, *Corrosion and Materials Degradation* 3 (2022) 566–597. https://doi.org/10.3390/cmd3040031

Atrens, A, I Gentle, A Atrens, Possible dissolution pathways participating in the Mg corrosion reaction, *Corrosion Science* 92 (2015) 173–181

Atrens, A, S Johnston, Z Shi, MS Dargusch, Viewpoint – understanding Mg corrosion in the body for biodegradable medical implants, *Scripta Materialia* 154 (2018) 92

Atrens, A, M Liu, NI Zainal Abidin, Corrosion mechanism applicable to biodegradable magnesium implants, *Materials Science and Engineering B* 176 (2011) 1609–1636

Atrens, A, Z Shi, S Mehreen, S Johnston, Gl Song, X Chen, F Pan, Review of Mg alloy corrosion rates, *Journal of Magnesium and Alloys* 8 (2022) 989–998

Atrens, A, GL Song, F Cao, Z Shi, PK Bowen, Advances in Mg corrosion and research suggestions, *Journal of Magnesium and Alloys* 1 (2013) 177–200

Atrens, A, G-L Song, M Liu, Z Shi, F Cao, MS Dargusch, Review of recent developments in the field of magnesium corrosion, *Advanced Engineering Materials* 17 (2015) 400–453

Bahmani, A, S Arthari, KW Shin, Corrion behaviour of Mg-Mn-Ca alloy: Influences of Al, Sn and Zn, *Journal of Magnesium and Alloys* 7 (2019) 38–46

Birbilis, N, MA Easton, AD Sudholtz, SM Zhu, MA Gibson, On the corrosion of binary magnesium-rate earth alloys, *Corrosion Science* 51 (2009) 683–689

Birbilis, N, A King, S Thomas, G Frankel, J Scully, Evidence for enhanced catalytic activity of magnesium arising from anodic dissolution, *Electrochimica Acta* 132 (2014) 277–283

Bockris, J, A Reddy, *Modern electrochemistry*, 6th printing ed, Plenum Press, New York, 1977

Cao, F, Z Shi, J Hofstetter, PJ Uggowitzer, G Song, M Liu, A Atrens, Corrosion of ultra-high-purity Mg in 3.5% NaCl solution saturated with $Mg(OH)_2$, *Corrosion Science* 75 (2013) 78–99

Cao, F, Z Shi, GL Song, M Liu, A Atrens, Corrosion behaviour in salt spray and in 3.5% NaCl solution saturated with $Mg(OH)_2$ of as-cast and solution heat-treated binary Mg-X alloys: X = Mn, Sn, Ca, Zn, Al, Zr, Si, Sr, *Corrosion Science* 76 (2013) 60–97

Fajardo, S, CF Glover, G Williams, GS Frankel, The source of anodic hydrogen evolution on ultra-high purity magnesium, *Electrochemical Acta* 212 (2016) 510–521

Frankel, GS, Ready for the road, *Nature Materials* 14 (2015) 1189

Frankel, GS, A Samaniego, N Birbilis, Evolution of hydrogen at dissolving magnesium surfaces, *Corrosion Science* 70 (2013) 104–111

Hanawalt, JD, CE Nelson, JA Peloubet, *Transactions of the American institute of mining and metallurgical engineers*, vol. 147, The American Institute of Mining And Metallurgical Engineers, Pennsylvania, 1942, p. 273

Li, Y, Z Shi, X Chen, A Atrens, Anodic hydrogen evolution on Mg, *Journal of Magnesium and Alloys* 9 (2021) 2049–2062

Liu, M, PJ Uggowitzer, AV Nagasekhar, P Schmutz, M Easton, G Song, A Atrens, Calculated phase diagrams and the corrosion of die-cast Mg-Al alloys, *Corrosion Science* 51 (2009) 602–619

Ma, H, X-Q Chen, R Li, S Wang, J Dong, W Ke, First-principles modeling of anisotropic anodic dissolution of metals and alloys in corrosive environments, *Acta Materialia* 130 (2017), 137–146

Ma, H, M Liu, W Chen, C Wang, X Chen, J Dong, W Kei, First principles study on the effects of twin boundaries an anodic dissolution of Mg, *Physical Review Materials* 3 (2019) 053806, 1–10

Mardina, Z, J Venezuela, MS Dargusch, Z Shi, A Atrens, The influence of the protein bovine serum albumin (BSA) on the corrosion of Mg, Zn, and Fe in Zahrina's simulated interstitial fluid, *Corrosion Science* 199 (2022) 110160

Mardina, Z, J Venezuela, C Maher, Z Shi, MS Dargusch, A Atrens, Design, mechanical and degradation requirements of biodegradable metal mesh for pelvic floor reconstruction, *Biomaterials Science* 10 (2022) 3371–3392. DOI: 10.1039/d2bm00179a

Qiao, Z, Z Shi, N Hort, N Zainal Abidin, A Atrens, Corrosion behaviour of a nominally high purity Mg ingot produced by permanent mould direct chill casting, *Corrosion Science* 61 (2012) 185–207

Shi, Z, A Atrens, An innovative specimen configuration for the study of Mg corrosion, *Corrosion Science* 53 (2011) 226–246

Shi, Z, F Cao, GL Song, M Liu, A Atrens, Corrosion behaviour in salt spray and in 3.5% NaCl solution saturated with Mg(OH)₂ of as-cast and solution heat-treated binary Mg-RE alloys: RE = Ce, La, Nd, Y, Gd, *Corrosion Science* 76 (2013) 98–118

Song, GL, Recent progress in corrosion and protection of magnesium alloys, *Advanced Engineering Materials* 7 (2005) 563–586

Song, GL, A Atrens, Corrosion mechanisms of magnesium alloys, *Advanced Engineering Materials* 1 (1999) 11–33

Song, GL, A Atrens, Understanding magnesium corrosion mechanism: A framework for improved alloy performance, *Advanced Engineering Materials* 5 (2003) 837–858

Song, GL, A Atrens, M Dargusch, Influence of microstructure on the corrosion of diecast AZ91D, *Corrosion Science* 41 (1999) 249–273

Song, GL, A Atrens, D St. John, An hydrogen evolution method for the estimation of the corrosion rate of magnesium alloys, *Magnesium Technology* (2001) 255–262

Song, GL, A Atrens, X Wu, B Zhang, Corrosion behaviour of AZ21, AZ501 and AZ91 in sodium chloride, *Corrosion Science* 40 (1998) 1769–1791

Taltavull, C, Z Shi, B Torres, J Rams, A Atrens, Influence of the chloride ion concentration on the corrosion of high-purity Mg, ZE41 and AZ91 in buffered Hank's solution, *Journal of Materials Science: Materials in Medicine* 25 (2014) 329–345

Xu, W, N Birbilis, G Sha, Y Wang, JE Daniels, Y Xiao, M Ferry, A high-specific-strength and corrosion-resistant magnesium alloy, *Nature Materials* 14 (2015) 1229–1235

Zainal Abidin, NI, AD Atrens, D Martin, A Atrens, The corrosion of high purity Mg, Mg2Zn0.2Mn, ZE41 and AZ91 in Hank's solution at 37°C, *Corrosion Science* 53 (2011) 3542–3556

2 Magnesium Metal-Matrix Composites – Types and Fabrication Approaches

K. Ponappa, S. Aravindan and P.V. Rao

CONTENTS

2.1 INTRODUCTION

Improvement in fuel efficiency and reduction in gas emissions are becoming the top priority goals of the automotive and aerospace sectors. This is gaining more importance due to the unpredictable long-term availability of non-renewable energy sources. Weight reduction can be achieved either by design optimization or by direct substitution of heavier steel components with lightweight metals. Light alloys (aluminum, titanium, and magnesium alloys) and their composites are finding applications in aircraft and automotive industries. Owing to fabrication limitations, these composites are presently used in such industries in smaller volumes. Large volume of lightweight materials is going to be employed for improved fuel economy and environmental friendliness. Not only weight reduction but also the important factors such as safety, environmental friendliness, durability, corrosion resistance, recyclability, and ease of processing and joining are to be considered for such light weight applications.

Compared to other lightweight structural materials, magnesium is lighter in weight, i.e. magnesium has 30% lower weight than aluminum, 75% lighter than zinc, and 70% lighter than steel (Avedesian and Baker 1999). Magnesium possesses the best strength-to-weight ratio of any of the commonly available metals. Owing to its specific modulus (stiffness), magnesium is used in applications like transmission cases and transfer cases, where stiffness is an important consideration.

DOI: 10.1201/9781003319856-3

The damping property makes magnesium likely to be a part in instrument panels, beams, and brackets. Compared to aluminum and iron, magnesium can be easily cast into more complex shapes even with thin walls (Eliezer, Aghion, and Froes 1998), due to its increased fluidity. High die-casting rates, good electromagnetic interference (EMI) shielding property, good weldability under a controlled environment, dimensional accuracy, and excellent machinability are the noteworthy advantages of magnesium (Caton 1991).

Owing to its poor mechanical and physical properties, pure magnesium is not at all useful in structural applications. In order to enhance the properties, addition of alloying elements to magnesium is essential (Mordike and Ebert 2001). In order to cater to the need of high demand for a material possessing high specific strength and stiffness, magnesium alloy based metal-matrix composites (MMCs) have been developed (Chua, Lu, and Lai 1999). Addition of hard reinforcement particles in magnesium matrix decreases the inherent limitations such as loss of strength and poor creep resistance at elevated temperatures. Compared to conventional un reinforced parts, these materials provide improved strength, corrosion, and wear resistance (Xi et al. 2005). In spite of these better properties of magnesium than other monolithic materials, fire hazard and surface degradation are the problems associated with the processing such as melting, machining, and heat treatment. Protective gases like argon, SF_6, etc., are used in casting of magnesium rather than the usage of flux. Thus the quality of castings is improved (Blawert, Hort, and Kainer 2004).

Improvement in properties of the existing composite materials is constantly taking place by the way of utilizing the newer processing methodologies. There are several methods available for fabricating magnesium alloy–based composites, such as stir casting (Wang et al. 2008), powder metallurgy (Nieh and Wadsworth 1995), squeeze casting (Jayalakshmi, Satish, and Seshan 2002), mechanical alloying (Lu, Thong, and Gupta 2003), disintegrated melt deposition (Hassan and Gupta 2003), and rheo casting.

2.2 EFFECT OF ALLOYING ELEMENTS

Addition of zinc and aluminum impart solid solution strengthening. Zinc helps in refinement of precipitates in aluminum-containing magnesium alloys whereas aluminum greater than 6 wt.% promotes precipitation hardening. Zinc with zirconium, rare earths (RE), or thorium produces precipitation hardening alloys. Addition of silver enhances the precipitation hardening. Zirconium addition is for grain refinement (Polmear 2006; Czerwinski 2008). However, zirconium is not used in aluminum alloys because of the formation of Zr_3Al_2 brittle compounds. Addition of manganese improves the corrosion resistance. Fluidity of casting alloys can be increased by the addition of silicon, but it impairs the corrosion resistance in the presence of iron. Thorium and yttrium improve the creep resistance. The RE elements are potential solid solution strengtheners. Mischmetal or didymium is generally added as rare earth element addition in magnesium alloys. Mischmetal has 50% cerium, and the rest is mainly lanthanum and neodymium. Didymium has around 85% neodymium and 15% praseodymium. In order to minimize corrosion, the impurity elements nickel, iron, and copper are to be kept at low levels.

Magnesium-based alloys can be categorized into two groups, namely magnesium-aluminum and magnesium-zirconium alloys. The most common example for Mg-Al alloy is AZ91 (9% Al, 1% Zn, 0.2% Mn). Mg-Al group alloys have some disadvantages also. The mechanical properties of these alloys tend to deteriorate above 200°C, and they exhibit variation in mechanical properties with section thickness and outcropping microporosity (Czerwinski 2008; Zhang et al. 2006).

Zirconium specifically provides a grain refining effect to magnesium matrix that is not available to Mg-Al alloys. Due to this grain refinement, magnesium-zirconium alloy exhibit improved mechanical properties with reduced porosity and uniform properties irrespective of section thickness.

However, compared to other structural metals, magnesium alloys have a relatively low absolute strength, especially at elevated temperatures. Their applications are usually limited to temperatures up to 200°C. Efforts are there to obtain high-temperature magnesium materials, which, in turn, lead to the development of several new alloy systems. However, this progress has not prompted wide applications of these magnesium alloys in the automotive industry.

2.3 PROCESSING TECHNIQUES FOR MAGNESIUM ALLOYS AND COMPOSITES

Magnesium alloys and composites are materials that are commonly used in various industries. Both materials have their own unique properties and characteristics, and their properties determine the types of applications they are best suited for. This section explains the difference between magnesium alloys and composites and what makes them unique.

Magnesium alloys are made by combining magnesium with other elements such as aluminum, manganese, zinc, and other metals. The addition of these elements to magnesium enhances its mechanical and physical properties, such as strength, toughness, and resistance to corrosion. Magnesium alloys are known for their low weight, high specific strength, and good ductility. These properties make them ideal for use in a wide range of applications, including aerospace, automotive, and consumer electronics. Composites, on the other hand, are materials made by combining two or more materials to create a new material with properties that are different from those of the individual components. The two main components of a composite are the matrix and the reinforcement. The matrix is the material that surrounds and holds the reinforcement in place, while the reinforcement provides the strength and stiffness to the composite. Composites can be made from a wide range of materials, including metals, ceramics, and polymers. They are known for their high strength, stiffness, and light weight, which makes them ideal for use in many applications, such as aerospace, sporting goods, and construction.

One of the main differences between magnesium alloys and composites is the way in which they respond to loading. Magnesium alloys deform plastically when subjected to loading, meaning that they will permanently deform when subjected to a load that exceeds their yield strength. Composites, on the other hand, will deform both plastically and elastically, meaning that they will return to their original shape after the load is removed. This makes composites well suited for applications where cyclic loading is a concern, such as in sporting goods or in aircraft structures. Another difference between magnesium alloys and composites is the way in which they are manufactured. Magnesium alloys are typically cast or forged, while composites are manufactured using a process known as layering, in which layers of matrix and reinforcement materials are built up to form the final product. This layering process allows composites to be tailored to meet specific strength and stiffness requirements, making them well suited for applications where custom properties are required.

Magnesium alloys and composites can be processed in several ways, including solid-state processing, liquid-state processing, and additive manufacturing. Solid-state processing involves the reshaping of a solid material through techniques such as forging, rolling, and extrusion. Liquid-state processing involves the shaping of a material while it is in a liquid state, such as casting. Additive manufacturing involves the building up of a material layer by layer using a 3D printing process. These different processing methods offer different advantages and disadvantages, and the choice of which method to use depends on the desired properties of the final product and the specific requirements of the application.

2.3.1 POWDER METALLURGY

The metal alloy powder and the ceramic short fibers/whisker particles are blended either in dry form or in liquid form. Compaction, canning, degassing, and high temperature consolidation by

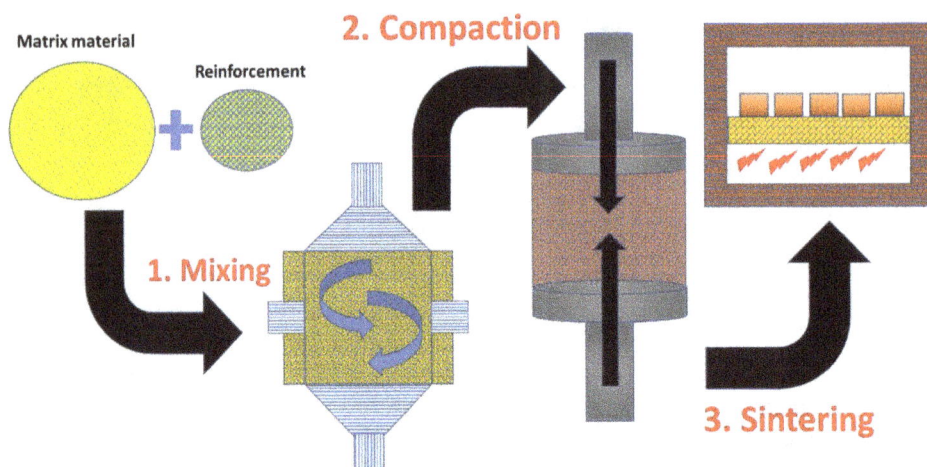

FIGURE 2.1 Powder metallurgy process steps.

hot isostatic pressing (HIP) or extrusion follow the blending process. Mixing of the prepared powder, compacting the mixed powder in a specifically shaped die, and subsequent sintering in a furnace are the steps associated in the PM process. Secondary processing operations may be performed for improving the properties or to achieve the required dimensional precision. PM-processed composites contain oxide particles which act as dispersion-strengthening agents and have strong influence on the matrix properties particularly during heat treatment. PM process is used to fabricate a wide range of composites with restricted size. Due to its high reactive nature, magnesium has been blended under SF_6/CO_2 protective atmosphere, and this makes the process a tedious one. Moreover, a secondary process is needed for powder metallurgical components. Damage to the reinforcing particles can be induced by a secondary process like extrusion, which affects the composites' physical and mechanical properties. **Figure 2.1** illustrates the steps followed in a PM process.

The PM route requires less temperature, and it is highly cost-effective. This process can easily fabricate a large variety of common and complex geometrical parts, allowing for mass production at a lower cost. The mechanical properties of the fabricated material are improved by the PM process because reinforcement particles are mixed properly within the matrix. A relationship has been established between the amount of reinforcement, the size of micro/nanoparticle reinforcement, the type of reinforcement, the surface type of reinforcement, and the properties of PM-produced composites. In general, fabrication of such composites with nanoscale reinforcements is difficult, but the properties of the final nanocomposite components would be beneficial.

Microwave sintering and spark plasma sintering of magnesium-based composites are found to be potential methods for improved product performance through faster sintering (energy savings) and grain refinement.

2.3.2 SPRAY DEPOSITION

Molten metal is sprayed as droplets together with the reinforcing phase, and the same is collected on a substrate. Molten metal is sprayed over the pre-formed reinforcement, which is kept on a substrate. The process parameters in spray processing are the initial temperature, size distribution and velocity of the molten metal droplets, temperature, and feeding rate of the reinforcement, and the position, nature, and temperature of the substrate. In this process, the molten metal is atomized using gases to achieve fine droplets having diameter up to 300 μm.

2.3.3 INFILTRATION PROCESS

A pre-formed porous phase of the reinforcing material is held within a mold, and the same is infiltrated with molten metal. The molten metal fills the interstices and pores to produce the composite. When a ceramic is infiltrated by liquid metal, wetting of the ceramic does not happen. The process is completed by way of applying a force to infiltrate the liquid metal into the ceramic metal. The applied force overcomes the capillary and drag forces. The morphology, composition, volume/weight fraction of reinforcement, the initial temperature of the infiltrating metal, and the nature and magnitude of the externally applied mechanical or hydraulic force are to be optimized to produce a sound composite.

2.3.4 SQUEEZE CASTING

Squeeze casting combines the two processes, namely forging and casting. Sometimes, this process is also called a liquid forging process. Molten metal solidifies under pressure in this process. With or without the use of pressure, reinforcement is mixed into the molten metal matrix. The process involves the following steps: pre-heating the chosen metal and reinforcement, pouring the molten metal, applying pressure to the molten metal, and mixing and removing the prepared composite material from the punch and die setup. In order to produce quality composites, the process parameters such as mold pressure, duration of pressure application, mould temperature, time delay in pressurizing the melt, pouring temperature, die temperature, and filling velocity are to be controlled effectively. It is much easier to fabricate defect-free cast composites using this technique if all the process parameters are optimized. **Figure 2.2** explains the squeeze casting process.

The molten metal is poured into the lower half of the pre-heated die. As the metal starts solidifying, the upper half closes the die, and it delivers the required pressure during the solidification process. Since the process uses liquid metal, the amount of pressure applied is significantly lower than the forging pressure, and parts of great detail can be produced. The porosity is low, and the mechanical properties are improved. Both ferrous and non-ferrous materials can be produced using this method. Squeeze casting is also used for fabricating magnesium-based composites with limited pre-form size. Metal yield is very low, and the reinforcement particle will experience severe damages. It cannot be easily adapted in conventional magnesium foundries.

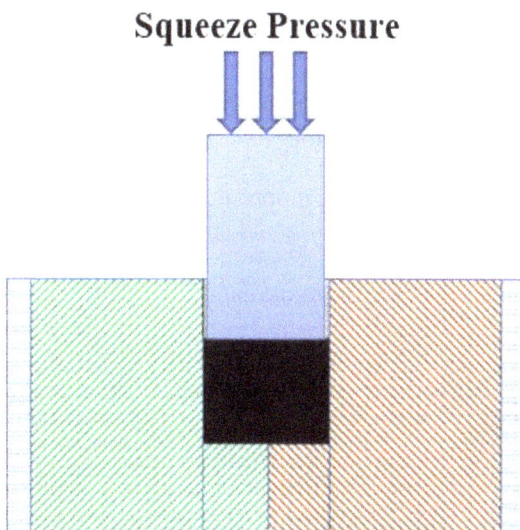

Squeeze Pressure

FIGURE 2.2 Schematic diagram showing pressure employed on the melt for a squeezing process.

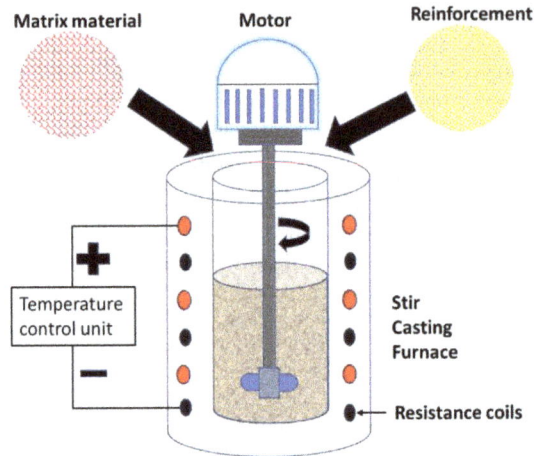

FIGURE 2.3   Conventional resistance stir casting.

2.3.5   STIR CASTING

The conventional stir casting process has been used to produce discontinuous particle reinforced MMCs. Proper wetting of reinforcement particles by liquid metal as well as getting a homogeneous dispersion of ceramic particles are the salient points in the process. Stir casting is a liquid-state MMC development technique in which the required reinforcement is added or mixed into the molten metal matrix through mechanical stirring. The most common form of reinforcement material is powder. Mechanical stirring of the melt and reinforcement is necessary. Finally, the resulting mixture can be moulded or cast to create the desired shape. **Figure 2.3** depicts the entire stir casting process.

The reinforcement is incorporated into the molten metal and allowed to solidify. The problem of poor wetting between the particulate reinforcement and the molten metal is overcome by the mechanical force through stirring. In the vortex method, the liquid metal is strongly stirred, and the particles are allowed to enter the vortex. Through this method, incorporation of up to 30% ceramic particles in the size range 5 to 100 µm in a variety of molten magnesium alloys is possible. In a process known as compo-casting, the particles and the liquid metal are mixed while the matrix metal is kept at a temperature between the solidus and liquidus. The increase in viscosity of the slurry leads to preventing the buoyant migration of particles, which, in turn, leads to uniform distribution of reinforcement particles.

Near net shape components can be made by this stir casting process. In preparing MMCs by the stir casting method, the manufacturers certainly come across some problems, as mentioned here.

- Attaining identical distribution of the reinforcement
- Wettability of reinforcement particle by liquid metal
- Porosity between reinforcement and matrix
- Undesirable chemical reactions between the molten metal and reinforcement

The method of introducing particles into the matrix melt plays a role in getting uniform dispersion of the reinforcement. The following are the adopted methods to have the reinforcement particles in the matrix melt (Zhang et al. 2006; Rohatgi, Asthana, and Das 1986; Krishnan, Surappa, and Rohatgi 1981).

1 Injection gun is used to introduce the particles entrained in an inert carrier gas into the melt
2 Pre-placement of reinforcement particles in the mould itself

3 Pushing the particles in melt using reciprocating arrangement
4 Similar to spray casting
5 Dispersion of fine particles in the melt utilizing centrifugal action
6 Pre-infiltrating a packed bed of particles to form pellets of a master alloy
7 Using high-intensity ultrasound for improving dispersion
8 Ultra-high vacuum with elevated temperature for a prolonged duration

In the vortex method, after the matrix material is melted, it is stirred vigorously to form a vortex at the surface of the melt, and the reinforcement material is then introduced at the side of the vortex. The pressure difference between the inner and the outer surface of the melt sucks the particles and transfer them into the liquid. Insufficient wetting of reinforcement by the liquid metal and non-homogeneous dispersion of the ceramic particles are the major impediments. Several structural defects such as porosity, particle clusters, oxide inclusions, and interfacial reactions are to be avoided by proper casting methodology.

2.3.6 TWO-STEP STIR CASTING

The reinforcement particles tend to float in spite of their larger specific density because of the high surface tension and poor wetting between the particles and the molten metal. When the gas layers are broken and the particles are wetted, the particles will tend to sink to the bottom (Krishnan, Surappa, and Rohatgi 1981; Hashim, Looney, and Hashmi 1999). In order to provide the particle distribution and to increase the wettability, two-step mixing is necessary, i.e. to heat the slurry to a temperature above the liquidus and then stirring. In an attempt to enhance the applicability of magnesium for a wide spectrum of performance and critical applications, the addition of reinforcement to the alloy is essential. Stir and squeeze casting process parameters influencing the mechanical properties of MMCs are:

i **Squeeze pressure:** It is the prominent factor, as it improves the wettability and interfacial bonding between the matrix and reinforcement. Solidification under the application of pressure reduces the porosity by way of suppressing the gas bubbles. The cooling rate also is enhanced through the contact of die.

ii **Reinforcement size:** Particle size affects the properties of the developed composites. Smaller-sized reinforcements leads to superior mechanical properties.

iii **Stirring speed:** Depending on the viscosity of molten magnesium melt, the distribution of reinforcement particles in the matrix can also be controlled. The viscosity should be in such a way that it should not restrict the movement of reinforcement particles during stirring. It should not also allow the particles to float/suspend the particles in the melt. The stirring speed also depends on the various profiles of the stirrer blade, such as three bladed, four-bladed propeller, helical ribbon, and axial and radial turbine blades, for proper mixing.

iv **Stirring time:** A homogeneous distribution of the particles is always required to maximize the mechanical properties. Longer stir time provides uniform distribution. This again depends on the blade profile (shape).

v **Melt temperature:** Generally, from the phase diagram of an alloy system and from the spiral fluidity test data, the melting temperature for the matrix system can be chosen. The high melt temperature improves the wetting ability of the melt, at the expense of its viscosity. Low melt temperature may lead to improper filling of the complete mould cavity, and it may lead to the agglomeration of particles.

vi **Stirrer blade material:** Graphite stirrer blades or stainless-steel stirrer blades coated with zirconia are generally used as a stirrer blade materials. The material choice of the stirrer should be such that the stirrer should not react with the magnesium melt. A niobium stirrer is preferred nowadays in the magnesium composite casting process.

2.3.7 ULTRASONIC-ASSISTED CASTING (UAC)

Ultrasonic energy with a frequency of 20 KHz to 18 KHz is used in the ultrasonic-assisted casting (UAC) process. When an ultrasonic wave of this frequency passes through a liquid, it causes successive dilation and compression. Mechanical vibrations of 18 KHz frequency are generated in this wave. Microbubbles form in the liquid during this process. Because the bubbles are unable to absorb enough energy, they explode (cavitation). Cavitation's explosion action separates clumps of particles and distributes them evenly throughout the liquid. Because of the low wettability and high tendency of nanoparticles to cluster, the UAC process is effective in dissolving agglomeration formation that occurs in nanocomposites. Bubbles explode in less than 6 to 10 seconds during each cavitation cycle, with the 'hot spot' reaching temperatures of 5000°C, pressures of 1000 atm, and heating/cooling rates of > 1010 K/S during the microseconds transient. **Figure 2.4** shows the schematic of the ultrasonic-assisted casting process.

2.3.8 FRICTION STIR PROCESSING

Friction stir processing (FSP) is a technique for altering the properties of a material by causing localized, extreme plastic deformation. This deformation is accomplished by forcing a non-consumable tool into the workpiece and stirring it as it is pushed laterally through it. FSP is a new technique for altering the microstructure of sheet metal to improve its properties. Even at high temperatures, a single FSP pass can achieve significant grain refinement and homogenization, resulting in improved formability. FSP is a solid-state process in which the material in the processed area undergoes intense plastic deformation, resulting in dynamically recrystallized grain structure. The research conducted on FSP is primarily focused on aluminum alloys. In a variant of A356, a significant increase in mechanical properties, strength, ductility, and fatigue life have been reported in the nugget zone. FSP may be a viable method for improving local mechanical properties and repairing casting defects in magnesium casting. Multiple-pass FSP had no discernible effect on the grain size of the previous pass in recent work on Mg-Y-Zn cast alloy.

Secondary phase particles can be integrated and dispersed in a material during FSP to make surface MMCs (surface composite fabrication with FSP). A diagram of the FSP is shown in **Figure 2.5**. Another advantage of FSP is that unlike adding secondary phase particles, it is possible to reduce grain size on the bottom. Small grooves or holes are created on the surface of the sheets, and the secondary phase particles are filled. This process can be scaled to cover a larger area by overlapping

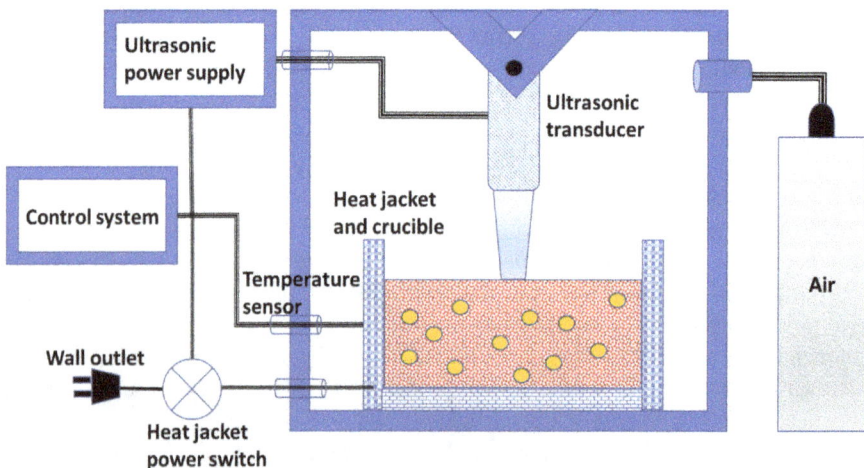

FIGURE 2.4   Schematic diagram of ultrasonic-assisted casting (UAC).

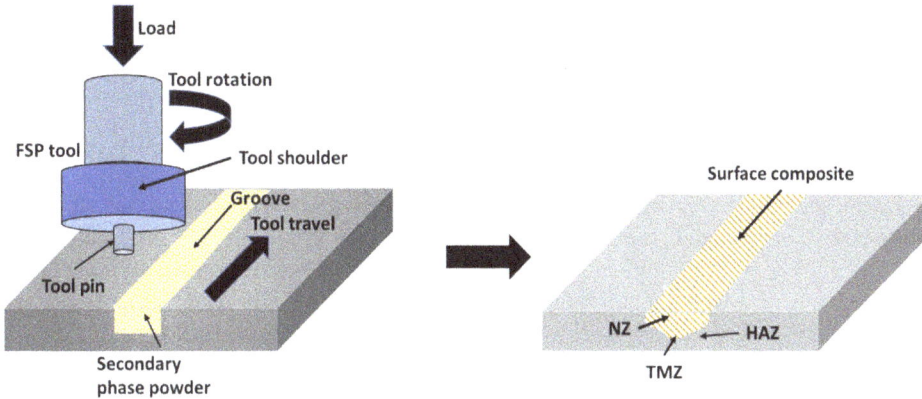

FIGURE 2.5   Schematic representation of surface MMCs fabrication by FSP.

passes. The concept of overlapping passes is especially favourable towards attaining superplasticity through microstructural modification.

Several methods for introducing secondary phase particles into the matrix material during FSP have been developed. Making a groove to the required depth and subsequently filling the groove with reinforcement particles and closing the grooves with a probeless tool can make the process viable to make surface composites of magnesium. Drilling a hole and filling it with the required micro/nano reinforcement particles for subsequent FSP is also recommended. Similar to the layering approach, laminate of reinforcement is sandwiched between metal foils to generate magnesium-based composites.

2.3.9   POWDER INJECTION MOULDING

Injection moulding is a manufacturing process used to create complex parts by injecting a thermoplastic or thermosetting material into a mould. Injection moulding of MMCs is a process in which the metal-matrix composite material is heated and then injected into a mould under high pressure. The parts produced by this process have high strength and good thermal properties. This process also allows to produce complex shapes and sizes that cannot be easily produced by other manufacturing methods. Powder injection moulding (PIM) is a technology that combines powder metallurgy and plastic injection moulding. The technique evolved from the injection moulding approach used to produce plastics. However, the technology for PIM is more sophisticated than that of plastic injection moulding due to the necessity to remove the binder as well as densify and reinforce the part. Ceramic or metal powders are used in the PIM process. Depending on whether metal or ceramic powder is used, the method is known as metal injection moulding (MIM) or ceramic injection moulding (CIM). Despite its growing popularity, PIM technology remains cloaked in secrecy and protected by patents. The method is frequently utilized in the bulk manufacture of near-net form parts. PIM is now used in a wide range of applications, including small, complicated, high-volume, and high-value products. It is used in gun and armament parts, office machinery, computer peripherals, medical and dental tools, orthodontic devices, and food and beverage components, among other things (Aghion, Bronfin, and Eliezer 2001; Strong 1991; Kass 1991; Farzin-nia 1991).

The four typical PIM process phases (German 1990) are mixing, injection moulding, debinding, and sintering. Initially, an appropriate binder formulation is combined with the metal or ceramic powder. The mixture is pulverized into a size appropriate for injection moulding (known as feedstock). Heat and pressure are used to mould the item into the desired shape during moulding. This is performed by heating the feedstock to a temperature above which it can flow but below which the binder will degrade. The binder is subsequently removed from the moulded item during the debinding procedure. Following debinding, the part is subjected to sintering, which densifies the part to provide the necessary mechanical qualities.

2.3.10 Wire-Arc Additive Manufacturing

Conventional and additive manufacturing techniques represent two distinct approaches to material processing. Conventional manufacturing techniques, such as casting, forging, and machining, involve the removal or reshaping of material to form a desired product. Additive manufacturing, on the other hand, involves the building up of a material layer by layer to form a final product. Conventional manufacturing techniques are generally well established and have been used for many years in various industries. They offer advantages such as high production rates and the ability to produce large components with complex shapes.

Additive manufacturing, also known as 3D printing, is a relatively new approach to material processing that offers a number of unique advantages. It allows for the production of complex shapes and structures that are difficult to produce using conventional manufacturing methods. It also offers the ability to produce small quantities of components quickly and inexpensively, making it well suited for prototyping and low-volume production. The processing methods available with additive manufacturing include fused deposition modelling (FDM), stereo lithography (SLA), and selective laser sintering (SLS). Conventional and additive manufacturing represent two distinct approaches to material processing, each with its own advantages and disadvantages. The choice of which approach to use depends on the specific requirements of the application, including the size and complexity of the component, the desired properties of the final product, and the volume of components required.

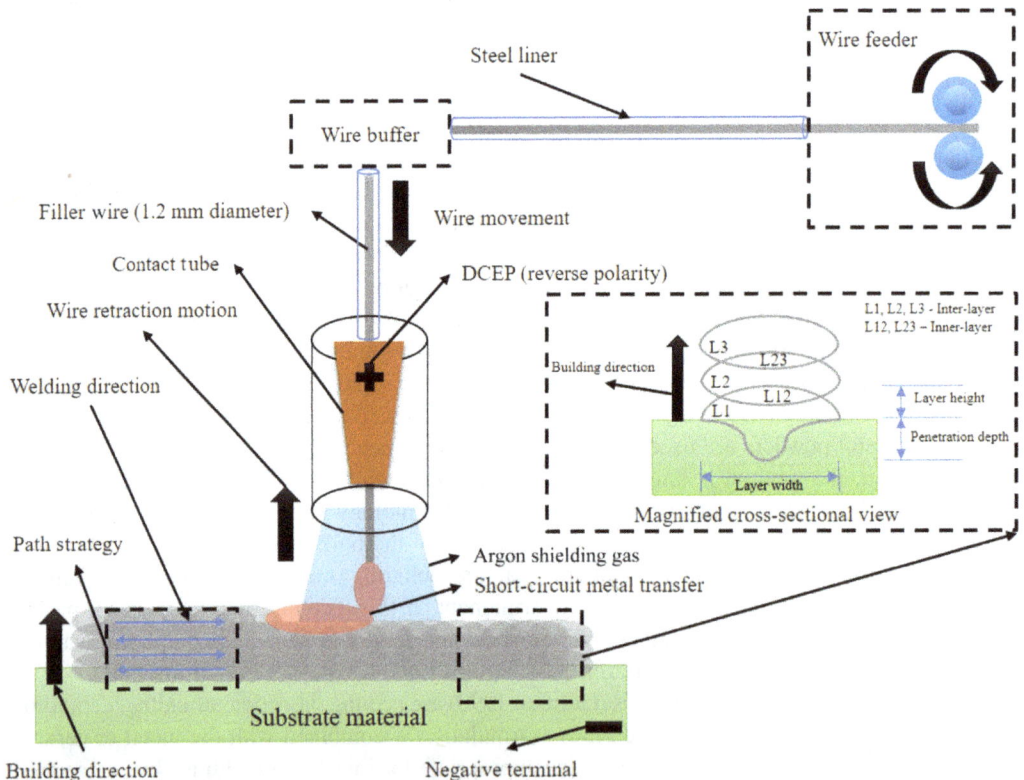

FIGURE 2.6   Schematic representation of a WAAM process.

Recently, the methodology of wire-arc additive manufacturing has become a popular method to manufacture large industrial components. Wire-arc additive manufacturing (WAAM) processes are nowadays used to develop magnesium-based alloys. A typical layer-by-layer assembly schematic is shown in **Figure 2.6**. Gas tungsten arc welding, gas metal arc welding, especially in cold metal transfer mode, and plasma arc welding are used for this layer-by-layer assembly (Mugada, Aravindan, and Ravikumar 2020). Specifically, usage of nano-treated filler wires would be helpful in development of magnesium-based composites.

2.4   CONCLUSIONS AND OUTLOOK

Owing to energy savings and lightweight applications, magnesium metal-matrix composites are the best materials in many industrial applications. This chapter provided brief information on types and fabrication approaches used for magnesium metal-matrix composites. Microwave sintering, powder injection moulding, ultrasonic-assisted stir casting, and wire-arc additive manufacturing technologies are useful fabrication methodologies for industrial component development. Fast sintering such as microwave and spark plasma sintering are preferred for magnesium-based composite development owing to the inherent advantages of these methodologies. Two-step stir cast with ultrasonic assisted process in a controlled environment is highly useful to manufacture magnesium based composites. Newer methodologies such as wire-arc additive manufacturing followed by subsequent heat treatment processes, friction stir processing in a controlled environment, and powder injection moulding are highly suited for the development of larger components of magnesium-based composites.

REFERENCES

Aghion, E., B. Bronfin, and D. Eliezer. "The role of the magnesium industry in protecting the environment." *Journal of Materials Processing Technology* 117 (2001) 381–385.

Avedesian, Michael M., and Hugh Baker (Eds.). *ASM Specialty Handbook: Magnesium and Magnesium Alloys.* Almere Materials Park, OH: ASM International, 1999.

Blawert, C., N. Hort, and K. U. Kainer. "Automotive applications of magnesium and its alloys." *Transactions of the Indian Institute of Metals* 57 (2004) 397–408.

Caton, P. D. "Magnesium-an old material with new applications." *Materials & Design* 12 (1991) 309–316.

Chua, B. W., L. Lu, and M. O. Lai. "Influence of SiC particles on mechanical properties of Mg based composite." *Composite Structures* 47 (1999) 595–601.

Czerwinski, Franciszek. *Magnesium Injection Molding.* New York: Springer, 2008.

Eliezer, D., E. Aghion, and F. H. Froes. "Magnesium science, technology and applications." *Advanced Performance Materials* 5 (1998) 201–212.

Farzin-nia, Farrokh. "Unique ability of MIM to produce complex parts." Proceedings of the Sixth Gorham Advanced Materials Institute MIM Meeting, San Diego, CA, 1991.

German, Randall M. *Powder Injection Molding.* Princeton, NJ: Metal Powder Industries Federation, 1990.

Hashim, J., L. Looney, and M. S. J. Hashmi. "Metal matrix composites: Production by the stir casting method." *Journal of Materials Processing Technology* 92 (1999) 1–7.

Hassan, S. F., and M. Gupta. "Development of high strength magnesium copper based hybrid composites with enhanced tensile properties." *Materials Science and Technology* 19 (2003) 253–259.

Jayalakshmi, S., Satish V. Kailas, and S. Seshan. "Tensile behaviour of squeeze cast AM100 magnesium alloy and its Al2O3 fibre reinforced composites." *Composites Part A: Applied Science and Manufacturing* 33 (2002) 1135–1140.

Kass, A. "Utilization of MIM at Kodak." Proceedings of the Sixth Gorham Advanced Materials Institute MIM Meeting, San Diego, CA, 1991.

Krishnan, B. P., M. K. Surappa, and P. K. Rohatgi. "The UPAL process: A direct method of preparing cast aluminum alloy-graphite particle composites." *Journal of Materials Science* 16 (1981) 1209–1216.

Lu, L., K. K. Thong, and M. Gupta. "Mg-based composite reinforced by Mg2Si." *Composites Science and Technology* 63 (2003) 627–632.

Mordike, B. L., and T. Ebert. "Magnesium: Properties – applications – potential." *Materials Science and Engineering: A* 302 (2001) 37–45.

Mugada, K., S. Aravindan, and D. Ravikumar. "Wire arc additive manufacturing of metals: State of the art and challenges." In *Additive Manufacturing Applications for Metals and Composites*, edited by K. R. Balasubramanian and V. Senthilkumar. Hershey: IGI Global, 2020.

Nieh, T. G., and J. Wadsworth. "Superplasticity in a powder metallurgy magnesium composite." *Scripta metallurgica et materialia* 32 (1995) 1133–1137.

Polmear, I. J. *Light Alloys: From Traditional Alloys to Nanocrystals.* Oxford: Elsevier/Butterworth-Heinemann, 2006.

Rohatgi, P. K., R. Asthana, and S. Das. "Solidification, structures, and properties of cast metal-ceramic particle composites." *International Metals Reviews* 31 (1986) 115–139.

Strong, S. "The gun that won the west fires: How colt came to use MIM." Proceedings of the Sixth Gorham Advanced Materials Institute MIM Meeting, San Diego, CA, 1991.

Wang, X. J., X. S. Hu, K. Wu, K. K. Deng, W. M. Gan, C. Y. Wang, and M. Y. Zheng. "Hot deformation behavior of SiCp/AZ91 magnesium matrix composite fabricated by stir casting." *Materials Science and Engineering: A* 492 (2008) 481–485.

Xi, Y. L., D. L. Chai, W. X. Zhang, and J. E. Zhou. "Ti-6Al-4V particle reinforced magnesium matrix composite by powder metallurgy." *Materials Letters* 59 (2005) 1831–1835.

Zhang, X., L. Liao, N. Ma, and H. Wang. "Mechanical properties and damping capacity of magnesium matrix composites." *Composites Part A: Applied Science and Manufacturing* 37 (2006) 2011–2016.

3 Surface Modifications and Protective Coatings – Types and Approaches

A.S. Khanna

CONTENTS

3.1 INTRODUCTION

Surface engineering is the mother of almost all technologies and can be the first or final step of several industrial and day-to-day applications. Its range can start from a simple example of a glowing face to a sophisticated component of an aeroplane. A fashion-conscious person is always bothered about how long her makeup will look fresh, for which she explores not only a good make up foundation and top cream but also for a good cleansing solution, which needs to be applied before makeup, to clean any dust or grease on the face. This is very similar to the surface preparation of a steel part before application of coating for corrosion protection. Many industrial components require coatings for improving either aesthetics or corrosion protection, but their application requires, first, a good surface preparation using a suitable chemical cleaning or mechanical cleaning using a blast machine (TNEMEC, 2015). The final step in any component fabrication is its outer look and aesthetics. This requires a proper cleaning followed by a suitable coating. This is true for our several daily use items, such as many consumer durables, car body, our mobiles and electronic watch etc.

The term 'surface engineering' actually can alternatively be called 'surface treatment' or 'surface modification' or 'creating an engineered surface using various surface modification methods.' Surface treatment can be simply a change due to washing or cleaning by a chemical which can either remove dust or grease or a chemical reaction which can change the surface composition to a

DOI: 10.1201/9781003319856-4

different chemical entity on the surface. In surface modification, it can be a combination of surface cleaning followed by application of a different material which protects it from many environmental pollutants. In surface engineering, it can be the role of some physical process which alters the properties of the surface. Surface treatment, using shot-blasting, shot-peening by high impact, or allowing accelerated particles to hit on the surface to create a new surface, can be examples of surface engineering to create clean and stress-free surfaces, respectively (Hutchings and Shipway, 2017). To simplify, it is better to classify surface engineering in three main categories, as discussed in what follows.

3.1.1   SURFACE MODIFICATION WITHOUT CHANGING THE MATERIAL CHEMICALLY

This type of surface modification can be achieved by thermal or mechanical means, altering metallurgy of the surface or surface texture.

- *Thermal processes*: Surface heat treatment, particularly those that undergo phase transformations like the martensitic reaction hardening of carbon steels, low-alloy steels, and cast irons – laser, flame, induction.
- *Mechanical processes*: Cold working – by shot-peening, shot blasting, explosive hardening, or other specialized machining processes induce compressive stresses, increasing hardness and fatigue resistance.
- *Changing surface texture using machining and blasting.*
- *Other processes*: Modification of surfaces by chemical/electro-etching, laser engraving, use of various chemical, solvent, and ultrasonic cleaning processes could also be included here.

3.1.2   SURFACE TREATMENT BY ALTERING SURFACE CHEMISTRY

These processes involve either [SVD], diffusion of elements from the coating material to the substrate, or result in a chemical reaction of the substrate with the coating material when exposed to a chemical liquid or vapour at high temperatures.

- *Thermochemical diffusion processes*
  - a Carburizing
  - b Carbonitriding
  - c Nitriding/nitro-carburizing
  - d Boronizing
- *Chemical vapour deposition*
  - a Aluminizing (calorizing, alonizing)
  - b Chromizing
  - c Siliconizing
- *Electrochemical processes* (anodizing)
- *Chemical conversion coatings* (phosphating, chemical blacking, chromating, etc.)
- *Ion implantation processes* (impingement of accelerated ions on the surface entering up to depth of nano-dimension)

3.1.3   SURFACE MODIFICATION BY ALTERING SURFACE CHEMISTRY BY APPLYING A FOREIGN MATERIAL (COATINGS) ON THE SURFACE

The coated material usually has a mechanical bond with the substrate or can form a chemical bond with the surface. Examples include corrosion protection coatings using paints, thermal spray, physical vapour deposition, chemical vapour deposition, electroplating, dip coating, etc. (Tanya

Galvanizers, 2016). The following properties of the surface can be changed as a result of these surface modification processes:

- *Aesthetic*: Clean surface, free of dust, grease, salts, and corrosion products
- *Mechanical properties*: Hardness, impact, friction and wear, erosion, and abrasion
- *Chemical*: Corrosion, oxidation
- *Conductivity*: Conductive surfaces, anti-static surfaces
- *Smart surfaces*: Hydrophobic, dust free, textured, anti-bacterial

The next question is the basis of selecting various surface treatment/modification techniques.

- Based upon change in function property
- Substrate characteristics
- Thickness of the modified surface
- Throughput of the process (slow, fast)
- Requirement of vacuum
- Geometry of the component
- Economics of process

One of the simplest ways to select a technique or process is based on the thickness of change on the surface you are looking for. Figure 3.1 is a schematic of the range of various techniques based

Classification of Surface Treatments

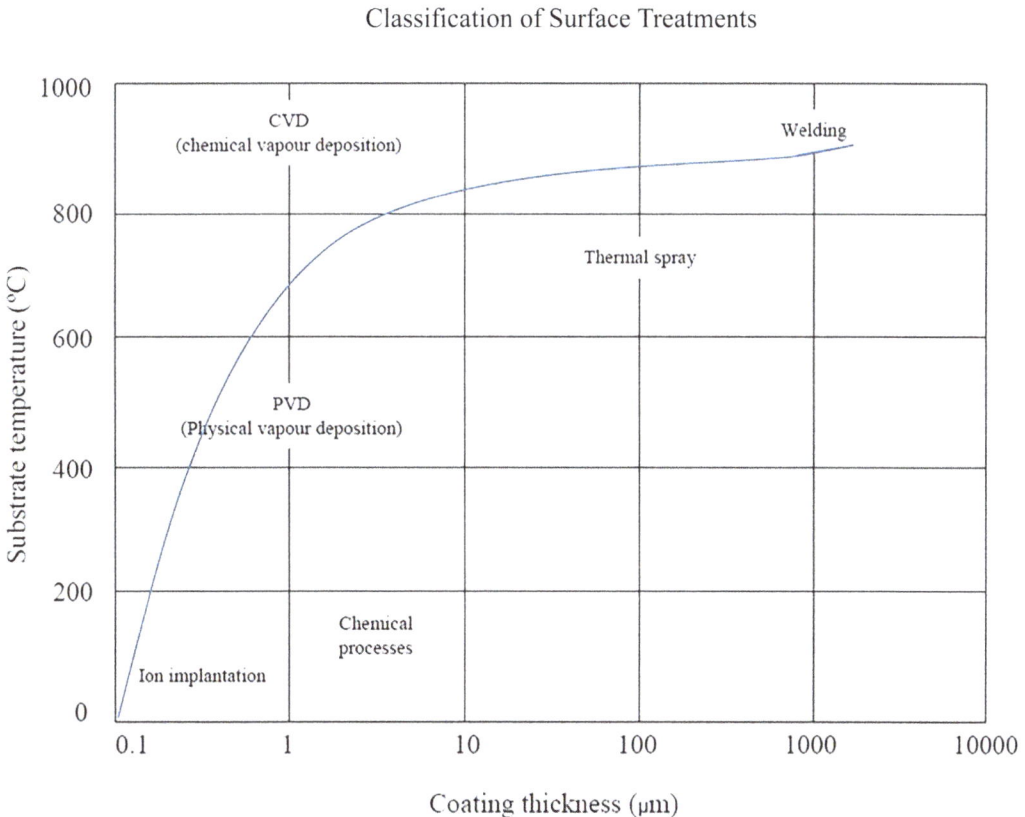

FIGURE 3.1   Schematic of various surface treatment techniques based upon the thickness of change on surface and the temperature of the process. *Modified after (Billard et al., 2018).*

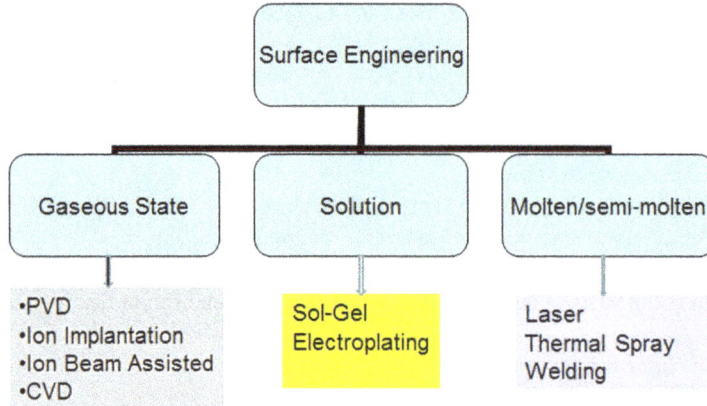

FIGURE 3.2 Classification of various surface modification techniques based upon the state of the coating material.

on thickness of the coating and the temperature up to which a process can reach during application (Billard et al., 2018).

Thus, based on **Figure 3.1**, it is clear that if one is interested in a surface modification by coating at a low thickness, lower than 10 µm, the physical vapour deposition (PVD) and chemical vapour deposition (CVD) are the only possibilities. If one is looking for larger thicknesses from a fraction of a mm to more than a mm, then thermal spray is perhaps the answer. Also, both PVD and CVD require vacuum for initiating the process, white thermal spray processes work in normal environments such as arc, flame, and atmospheric plasma. Further, PVD and CVD require costly equipment's, while thermal spray is relatively a cheaper technique.

There are several other techniques such as paint coating which requires a simple brush and roller for application, but for big areas, it requires spray guns. Electroplating is again simple but requires electrochemical baths, and the coating can form from very thin to thick coating. Dip coatings are very common, for example, galvanization – zinc coating on steel, which is formed by dipping steel into molten zinc metal at temperatures above 450°C.

There are still sophisticated processes such as laser surface modification where the desired material is added through a laser radiation which melts the substrate and allows the desired material to react with the substrate, simultaneously creating a layer of the material to be coated.

There is another classification of various surface treatment processes based upon whether the process is in a gaseous/vapour state, solution, or solid/melt. This is given in **Figure 3.2.**

Another very powerful classification of various surface modification techniques is based upon their mechanism of bonding: whether it is mechanical bonding or chemical bonding. Table 3.1 classifies various techniques in these two categories:

Now we would discuss various coating processes.

3.2 METALLIC COATINGS

3.2.1 Hot-Dip Coatings

The metallic coating is one of the biggest sectors, where surfaces are coated with various metals or alloys using a host of different processes. The simplest one is the hot-dip method, which is a very old process and is based upon dipping the substrate in a molten metal for a few seconds, where it partially reacts with the substrate and forms a chemically bonded coating. The most common metallic coating is the zinc coating, which is applied by dipping steel sheets in liquid zinc melt, maintained at a temperature above its melting point for a few seconds. The dipping is done after the steel

TABLE 3.1

Classification of Coatings Based upon Their Bonding to Substrate

Mechanical-bonded	Chemical-bonded (diffusion thru substrate)
Paint coatings	Sol-gel coatings
PVD	Hot-dip galvanization
Thermal spray	CVD
	Laser alloying
	Nitriding, carburizing

surface is thoroughly cleaned and activated in several activation processes. The coating formed is a true metallic coating with a multilayer structure with pure zinc on top and another three layers with lower zinc alloyed with iron from steel. Galvanized coatings have the biggest market used for several day-to-day and industrial applications.

On the other hand, the aluminizing is carried out by an entirely different method of thermo-chemical diffusion process at high temperatures, typically in the range of 800°C to 1000°C with a prolonged soaking time to aid the diffusion potential. It is a chemical diffusion treatment wherein the surface layer of the material is impregnated with aluminium in the form of pack having aluminium and chromium in a matrix of aluminium oxide. This process is called the pack aluminizing process. It is primarily used on steels but also on nickel- and cobalt-based alloys to obtain greater creep resistance, hardness, and corrosion resistance (Total Materia, 2019).

3.2.2 THERMAL SPRAY COATINGS

The other most frequent method of doing metallic coating is thermal spray. Thermal spray is a process in which either the powder of the said metal/alloy or metal in the form of wire is melted using a strong heat source, flame, arc, or plasma. Liquid drops so formed are forced at a very high speed on the surface to be coated, where they are splat cooled, resulting in a strong coating. Four different thermal spray processes differ in terms of their heat sources: heating the source by a flame called flame spray process, heating by an arc is known as arc spray process, and where heating is initiated by plasma, it is called plasma spray process. If the heating is carried out by detonation process in the gun, it is called detonation spray process. All four thermal spray processes give different morphology of the coating which differs in the density/porosity and hence in their effectiveness in preventing corrosion, oxidation, and wear and abrasion (**Figure 3.3**) (Kuroda et al., 2008).

Typical applications of thermal spray cover a huge range of components, either as part of the original manufacturing process or as a reclamation or re-engineering technique for a wide range of rotating and moving parts from machines of all kinds, including: road and rail vehicles, ships, aircraft, pumps, valves, printing presses, electric motors, paper making machines, chemical plants, food machinery, mining and quarrying machinery, earthmovers, machine tools, power generation and aerospace turbine repair, landing gear (chrome replacement), and virtually any equipment which is subject to wear, erosion, or corrosion. This is done using either arc spray, flame spray, or high-velocity oxygen fuel (HVOF) systems to spray steels, nickel alloys, carbides, stainless alloys, bronzes, copper. Hard coatings were made using the HVOF process (Khanna et al., 2009).

A few typical industrial applications which confirm increased durability, modified electrical properties, improved corrosion protection, higher hardness, and/or improved wear and abrasion properties:

* *Wind turbines*: To prevent atmospheric corrosion damage to wind turbines.
* *Oil & gas industry*: Pipes, shafts, which are often exposed to harsh environments.

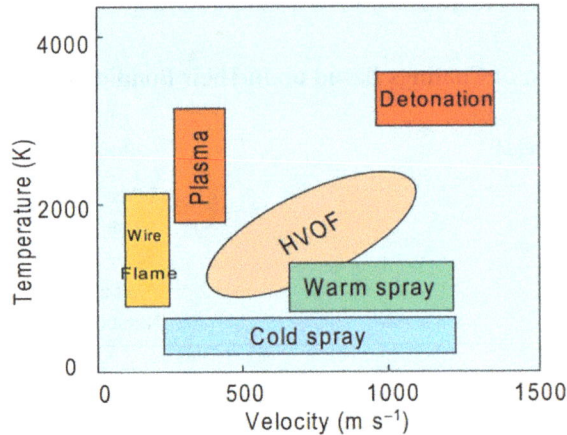

FIGURE 3.3 Thermal spray process – description in terms of the intensity of heat source and the velocity on substrate. *Reproduced with permission from (Kuroda et al., 2008) © 2008 National Institute for Materials Science.*

- *Bridges*: Metal (zinc or aluminium) spraying to increase its durability due to cathodic protection.
- *Petrochemical plants*: Proper metallic coating of pipes in the inner layer of these petrochemical pipes.
- *Infrastructural industry*: The structure of an infra project is build up by steel coated with metallic coatings to provide strength to the surface of a structure.
- *LPG cylinders*: Metal spraying is used to protect LPG, propane gas, or butane gas bottles against corrosion.
- *Architectural coatings:* Metal spraying is preferred to hot-dip galvanizing in some industries.
- *Film industry sets:* Artistic coatings for shooting the shots.

3.2.3 PHYSICAL VAPOUR DEPOSITION

Another method of applying thin metallic coatings is PVD. It includes a host of processes which have the basic principle to bring the desired material to be coated in vapour form and then allow it to fall on the substrate. Simple methods of vapour deposition to ion plating to sputter deposition and electron beam deposition are well known. The coating thickness ranges from fraction of a micron to a few microns. Most of the coatings made are usually to enhance wear and friction of the substrate. Electron beam deposition became one of the most important processes for thermal barrier coatings for turbine blade application.

In the electron beam evaporation method, the desired metal to be coated can be evaporated using high-energy electrons in the form of an intense beam. A hot filament is used to get thermionic emission of electrons, which can, after acceleration, provide sufficient energy for evaporating any material. In a typical case involving 1 A of emission accelerated through a 10 kV voltage drop, 10 kW is delivered upon impact (Wikipedia, PVD) (Green, 2022).

One of the very important applications of electron beam evaporation coatings (EBPVD) is turbine blades, which results in giving an elongated grain structure of coating instead of an equiaxed structure. The former helps in releasing stress in the coating and saves it from cracks and damages.

3.3 MODIFICATIONS BY PAINT COATINGS

Paint coating appears to be the most trivial method of surface modification. It is usually applied using a brush to give a barrier coating of paint, which is generally an insulating material, which helps in preventing corrosion by restricting the flow of electrons. However, paint coatings cannot be considered as a trivial technique when applied on industrial systems, where durability and its functionality become very important (Khanna, 2020). There are various varieties of paint coatings based upon its utility, purpose, and function. In a very simple classification, it can be divided into two main types: decorative and industrial. Decorative paints are usually applied to enhance the aesthetics of the structure/component, while industrial paints are mainly used for corrosion protection.

The second-best classification of paint coatings is based upon the type of resin (binder) it uses. Based upon the resin type, such as epoxy, urethane, vinyl, alkyd, or polyester, it is designated as epoxy coating, urethane coating, or vinyl-, alkyd-, or polyester-based coating, respectively. The main components of a paint coating are solvent, binder, pigment, and additives. A proper mixture of all these four makes a paint coating. The paint manufacturing is a simple mixing process where first the binder is added into the solvent and nicely dispersed. It is then followed by systematic addition of pigments and additives till a uniform mixture is obtained. The quality of paint depends upon the types of mixing methods used such as agitation, blending, attritions, ball, and bead mill, etc. (Made How, 2022).

The paint coatings are of a single component and of two components. Single-component paints are usually a mixture of solvent resin and some pigments and additives, while two-component paint systems have a resin part which is made using solvent, resin, pigments, and additives, and the second component is called catalyst or hardener, which is mixed with the resin part just before application. The role of catalyst/hardener is to harden the coating with time till it fully dries. This is achieved by chemical reaction between the resin and the hardener, leading to a high level of crosslinking, which provides superior anti-corrosion properties to the paint coating, especially low permeability and strong barrier protection. For example, for epoxy resins, amines or amides are used as hardeners. For two-component polyurethane coatings, cynates are used as hardeners.

3.3.1 GREEN COATINGS

Another important classification of paint coating is eco-friendliness of paint coatings. As discussed, the main components of paint coatings are solvent, binders, pigments, and additives. Based upon the amount of resin in the solvent and pigment and additive concentration, the paint has a volume solid percentage which decides the thickness of the coating after drying. As per the simplest mechanism of the paint drying process, the solvents, which are mostly volatile in nature, evaporate and leave the dried coating on the substrate. Since the most common solvents are benzene, xylene and toluene, which are toxic in nature, and hence, when they evaporate they pollute the environment and especially affect the health of applicators. Thus, a better paint system is that which either has very low quantity of solvent or has no solvent or uses alternative solvent such as water. This results in two additional types of classifications: solventless coatings and waterborne coatings. There are two additional advantages of solventless coatings: apart from eco-friendliness, they can give higher thickness from 500 to 2000 microns in one or more coat. Second, several high-performance and functional coatings used in industry are solventless with addition of several pigments such as fibres or glass flakes which enhance the durability of such coatings to a very long duration. **Figure 3.4** summarizes various paint coating classifications discussed here.

3.3.2 FUNCTIONAL/SMART COATINGS

Another classification of the coatings can be on its functional action and its smart behaviour. We have several coatings which come under the heading of smart coatings. These coatings, in addition

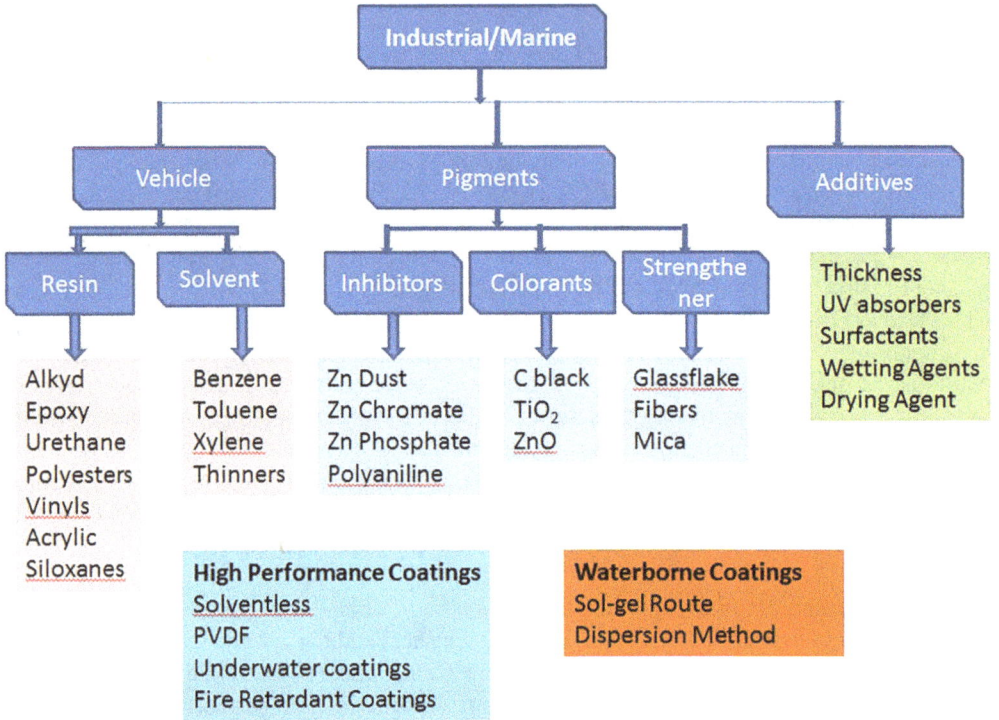

FIGURE 3.4 Classification of various paint coatings.

to doing the normal function of anti-corrosion, also do a specific function, such as self-cleaning (Verma et al., 2013), self-healing (Thanawala et al., 2014), or anti-graffiti (Adapala et al., 2015), creating a conducting surface and perhaps many more. The biggest role played to make a coating smart is by the addition of nanoparticles or to make a nano-coating by processes such as the sol-gel process.

3.3.3 Nano-Modified Coatings

The role of nanotechnology can be understood from **Figure 3.5**. As discussed, initial selection of coating is made based upon the chemistry of the resin, which decides how durable a coating is in a particular environment. Additional properties are decided by the choice of pigments and various additives. It is now established that as the particle size of the pigments/additives decreases, there is enhancement in the basic properties of the coating such as corrosion resistance, mechanical properties, and many specific properties such as fire resistance, waterproofing tendency, etc. It is now well established that size and shape of the nanoparticles is very important. The next important thing is the concentration of nanoparticles (Dhoke et al., 2009) (Gaur and Khanna, 2015).

Let us now first compare an epoxy coating where almost 10% micro-sized ZnO was added to take care of UV blocking resistance. Distribution of such particles in epoxy matrix is as shown in **Figure 3.6a**, which shows that even 10% of ZnO is unable to protect the epoxy matrix and it gets deteriorated when UV light falls on the matrix. Now consider the same case by taking now, 0.1% of nano ZnO particles with the achieved distribution as shown in **Figure 3.6b**. You would see that the matrix is almost covered by ZnO particles uniformly distributed in the resin matrix, so the chance of sunlight falling on the matrix is very low. Now take the third examples of 0.01% graphene particles (2D structure). They cover the whole surface in a still better way thus, showing that flat 2D graphene nanoparticles are even more effective, that too at a still lower concentration (**Figure 3.6c**).

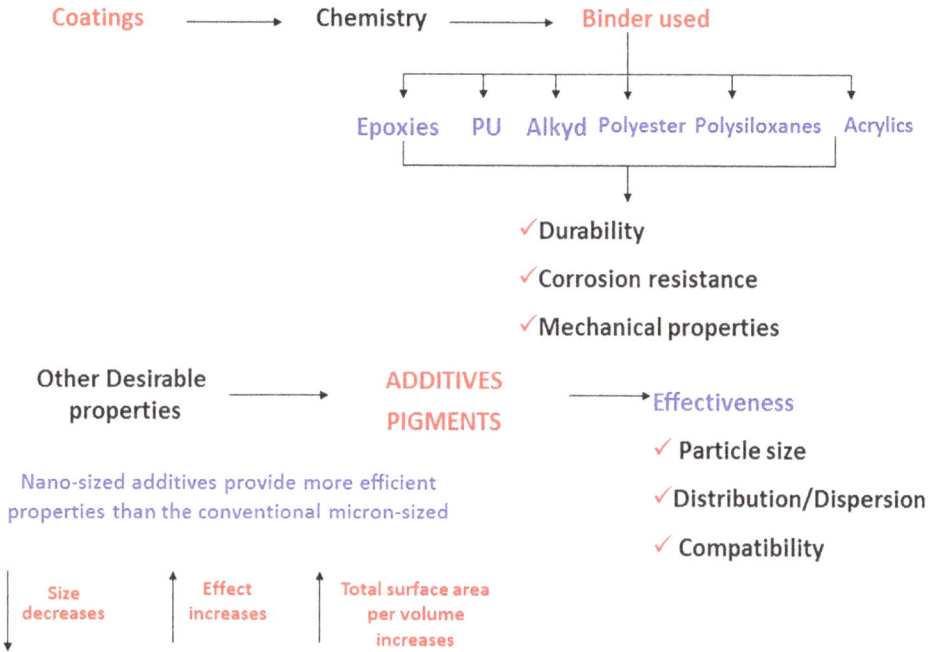

FIGURE 3.5   Concept of nanotechnology in paint coatings.

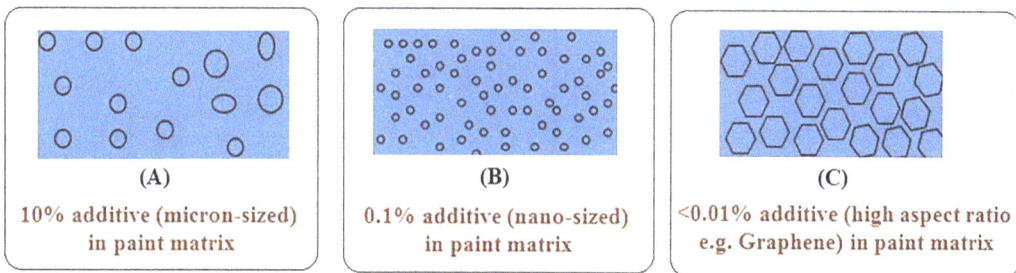

FIGURE 3.6   A simple way to show the effect of nanopigment additions, loading, and size effect.

Figure 3.7 shows some of the nanoparticles with different shapes and size, and **Figure 3.8** shows the effect of size and shape of nanoparticles in changing the effect of yellowness and % gloss due to UV light. It is very clear both size and shape and concentration affect the properties of the coating.

Another powerful method to develop high-performance coatings is to develop inorganic-organic thin coatings using the sol-gel method, which give excellent performance and functional applications such as water repellence, hydrophobicity, etc. (Pathak et al., 2006) (Wankhede et al., 2013).

3.4   CONCLUSIONS AND OUTLOOK

Surface engineering is the most powerful and essential branch of materials science, without which all material fabrications are incomplete. Its scope starts from cleaning to surface preparation and various kinds of surface treatments with and without changing the surface composition of the substrate. Various kinds of techniques are used for this purpose, starting from various cleaning

a) ZnO - 30 nm Spherical   b) ZnO - 90 nm Spherical   c) ZnO - Flakes

d) Rutile TiO_2 - 130 nm   e) Fe_2O_3 - 75 nm Sphere   f) Fe_2O_3 Rods - 180 nm

FIGURE 3.7   SEM images showing the size and shapes of different nanoparticles (Dolai and Khanna, 2021).

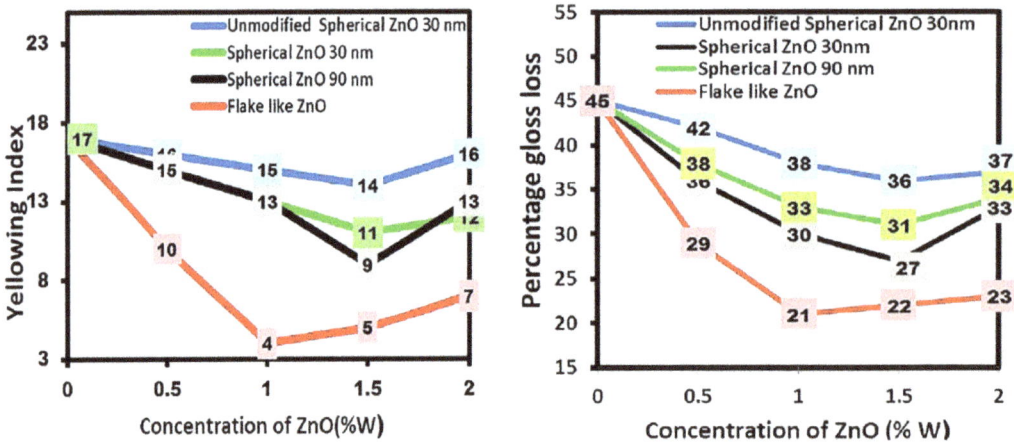

FIGURE 3.8   Effect of size, shape, and loading on the weathering properties of the paint coatings (Dolai and Khanna, 2021).

methods to surface treatments by heating, flame, lasers and all kinds of surface coating methods ranging from hot-dip to chemical conversion to PVD/CVD, thermal spray, and paint coatings.

REFERENCES

Adapala P, Gaur S, Puri RG, Khanna AS, Development and evaluation of nano-silica dispersed polyurethane based coatings for improved anti-graffiti and scratch resistance, *Open J. Appl. Sci.* 5, 2015, 808–818.

Billard A, Maury F, Aubry P, Balbaud Célérier F, Bernard B, Lomello F, Maskrot H, Meillot E, Michau A, Schuster F, Emerging processes for metallurgical coatings and thin films, *Comptes Rendus Physique.* 19, 2018, 755–768.

Dhoke SK, Khanna AS, Sinha TJM, Effect of nano-ZnO particles on the corrosion behavior of alkyd-based waterborne coatings, *Prog. Org. Coatings*. 64, 2009, 371–382.

Dolai SS, Khanna AS, Development of functional coatings using nanoparticles, CIA, Paint coating special issue, March 2021, pp. 56–60.

Gaur S, Khanna AS, Functional coatings by incorporating nanoparticles, *Nano Res. Appl*. 1, 2015, 1–8.

Green J, An overview of electron beam evaporation, *Stanford Advanced Materials*, 2022, www.sputtertargets.net/an-overview-of-e-beam-evaporation.html, Access date December 20, 2022.

Hutchings I, Shipway P, Surface engineering, in *Tribology – Friction and Wear of Engineering Materials*. 2nd Ed., 2017, pp. 237–281, Elsevier, ISBN: 978-0-08-100910-9.

Khanna AS, Coating: A vanguard to protect from corrosion invasion, *Industrial Product Finder*, September 2020, pp. 48–51.

Khanna AS, Kumari S, Kanungo S, Gasser A, Hard coatings based on thermal spray and laser cladding, *Int. J. Refract. Metal. Hard Mater*. 27, 2009, 485–449.

Kuroda S, Kawakita J, Watanabe M, Katanoda H, Warm spraying-a novel coating process based on high-velocity impact of solid particles, *Sci. Technol. Adv. Mater*. 9, 2008, 033002.

Made How, www.madehow.com/Volume-1/Paint.html, Access date December 20, 2022.

Pathak SS, Khanna AS, Sinha TJM, Sol gel derived organic-inorganic hybrid coating: A new era in corrosion protection of material, *Corros. Rev*. 24, 2006, 281–306.

Tanya Galvanizers, *Hot Dip Galvanizing Process*, 2016, www.galvanizers.co.in/blog/hot-dip-galvanizing-process/, Access date December 20, 2022.

Thanawala K, Mutneja N, Khanna AS, Singh Raman RK, Development of self-healing coatings based on linseed oil as autonomous repairing agent for corrosion resistance, *Materials*. 7, 2014, 7324–7338.

TNEMEC Application Guide, 2015, www.tnemec.com/documents/341/APP_GUIDE_General_Surface_Prep_Guide.pdf, Access date December 20, 2022.

Total Materia, *The Aluminizing Process*, 2019, www.totalmateria.com/page.aspx?ID=CheckArticle&site=ktn&NM=446, Access date December 20, 2022.

Verma G, Swain S, Khanna AS, Hydrophobic self-cleaning coating based on siloxane modified waterborne polyester, *Int. J. Sci. Eng. Technol*. 2, 2013, 192–200.

Wankhede RG, Morey S, Khanna AS, Birbilis N, Development of water-repellent organic – inorganic hybrid sol – gel coatings on aluminum using short chain perfluoro polymer emulsion, *Appl. Surf. Sci*. 283, 2013, 1051–1059.

Wikipedia, PVD, https://en.wikipedia.org/wiki/Electron-beam_physical_vapor_deposition, Access date December 20, 2022.

Part II

Magnesium Metal-Matrix Composites

4 Magnesium-Based Metal-Matrix Composites by Solid-State and Liquid-State Processing

Sankaranarayanan Seetharaman and Manoj Gupta

CONTENTS

4.1 INTRODUCTION

The automotive and aviation sectors are increasingly looking for lightweight materials like aluminium (Al) and magnesium (Mg) to minimise the fuel energy consumption and carbon emission (Zhao et al., 2021). In this regard, Mg being the lightest structural metal, it is an ideal candidate for weight critical applications (Gupta & Seetharaman, 2017; Powell, 2022). While Al alloys are widely studied and applied to various engineering structures, Mg offers an additional ~33% weight-saving potential with similar physical properties and mechanical strength (Bamberger & Dehm, 2008). Mg also possesses excellent damping capacity, machinability and castability, making it suitable for a range of applications (Gupta & Seetharaman, 2017). It also exhibits suitable characteristics such as non-toxic, biodegradable nature, comparable mechanical properties as human bone etc. to be suitable for making temporary body implants (Wang et al., 2020). Despite benefits as stated, the load-bearing attributes and thermal stability of Mg remains inferior due to the inherent limitations of HCP crystal structure. These properties can be improved by making metal-matrix composites (MMCs), viz. through the addition of reinforcements of desired volume fraction (Seetharaman et al., 2021).

DOI: 10.1201/9781003319856-6

TABLE 4.1
Properties of Common Reinforcement Materials (Kainer, 2006)

Metal	Crystal structure	Density (g/cm³)	Melting point (°C)	Modulus (GPa)	Thermal expansion coefficient	Thermal conductivity
Al_2O_3	Hex.	3.9	2050	410	8.3	25
B_4C	Rhom.	2.52	2450	450	5–6	29
SiC	Hex.	3.21	2300	480	4.7–5	59
TiB_2	Hex.	4.5	2900	370	7.4	27
TiC	Cub.	4.93	3140	320	7.4	29
TiN	Cub.	5.24	2950	600	9.4	29

In general, hard and strong ceramic particles are considered for reinforcements (Kainer, 2006; Seetharaman & Gupta, 2021). Table 4.1 lists the properties of commonly used ceramic reinforcements. Based on the morphology, reinforcements can be classified as either fibres or particle types. The fibre reinforcement can be further classified as either (i) long (continuous fibres with L > 100D), or (ii) short fibres (L < 100D). With respect to particles, they can be dispersed either randomly or in a well-defined pattern. While the addition of continuous fibres improves the strength and elastic modulus, the mechanical properties depend on the fibre orientation and loading direction. On the other hand, particle reinforcements offer benefits such as low cost and simple production process in addition to the isotropic properties (Kainer, 2006; Seetharaman & Gupta, 2021). Reinforcements primarily act as a force-bearing body and effectively contribute towards property enhancement. The general matrix-strengthening mechanisms include: (i) load transfer effects, (ii) thermal mismatch strengthening, (iii) grain size strengthening and (iv) Orowan strengthening (Seetharaman & Gupta, 2021).

Several published literatures highlight the improvements in mechanical properties of Mg due to the addition of ceramic reinforcements (Bharathi & Sampath Kumar, 2022; D. Kumar et al., 2021; S. S. Kumar & Mohanavel, 2022; Saranu et al., 2022; Seetharaman et al., 2020a), especially the nanosized reinforcements (Gupta & Wong, 2015; Seetharaman et al., 2020a). **Figure 4.1** shows the mechanical properties of Mg-MMCs reinforced with different particle reinforcements. A careful review of these studies indicates that the benefits of reinforcements predominantly depend on their effective dispersion, which is mainly affected by the processing methods. Further, the available literature also highlights the improvement in other engineering properties such as wear resistance (Saranu et al., 2022) and thermal stability (R. Kumar et al., 2019), even as comparable to Al alloys and composites.

While the inclusion of ceramic reinforcements in micro- or nano-length scale improves the mechanical properties of Mg, the corrosion response in general depends on the characteristics of particle matrix interface and the protective film (Seetharaman et al., 2020b). **Figure 4.2** shows the schematic of corrosion of Mg alloys and composites in aqueous solution. While hydrogen gas evolution and protective layer formation tend to decline the corrosion rate, the breakdown of weak protective layers allows faster degradation. In this regard, recent studies indicate that the addition of nano-length scale reinforcements improves the corrosion resistance of Mg (Seetharaman et al., 2020b), and the same can be further improved by applying suitable surface coatings (Nazeer & Madkour, 2018). Hence, extensive research works are ongoing to explore all these aspects to enhance the adoption of Mg composites (Khorashadizade et al., 2021; Nazeer & Madkour, 2018; Seetharaman et al., 2020b).

4.2 PROCESSING OF MAGNESIUM-BASED MMCS

A variety of methods can be employed to fabricate Mg-MMCs, and they can be broadly grouped under either liquid- or solid-state processing methods as shown in **Figure 4.3**.

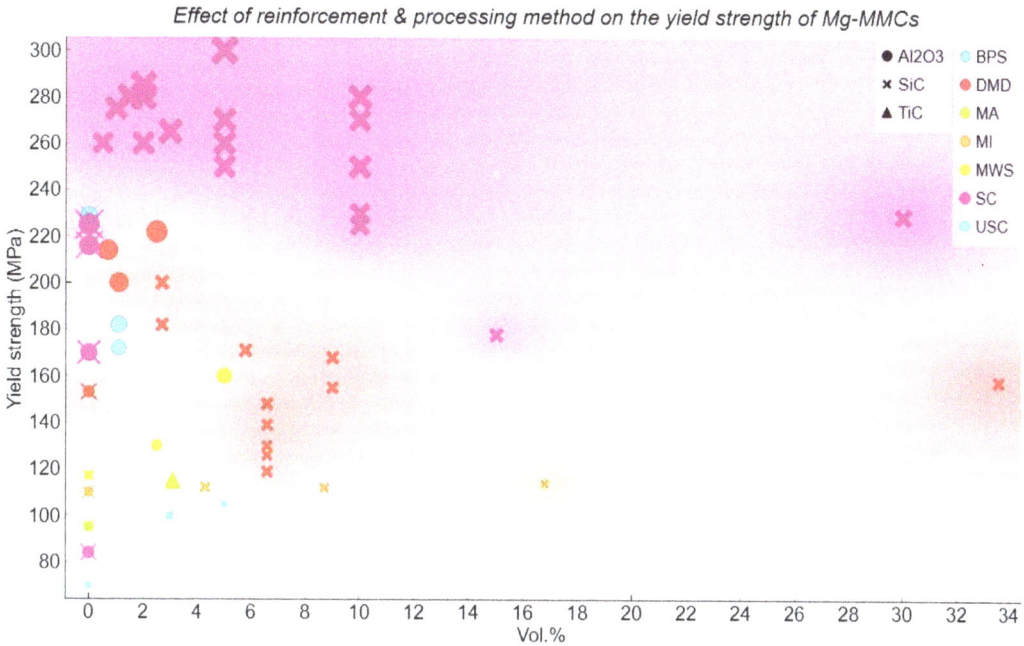

FIGURE 4.1   Mechanical properties of Mg-MMCs.

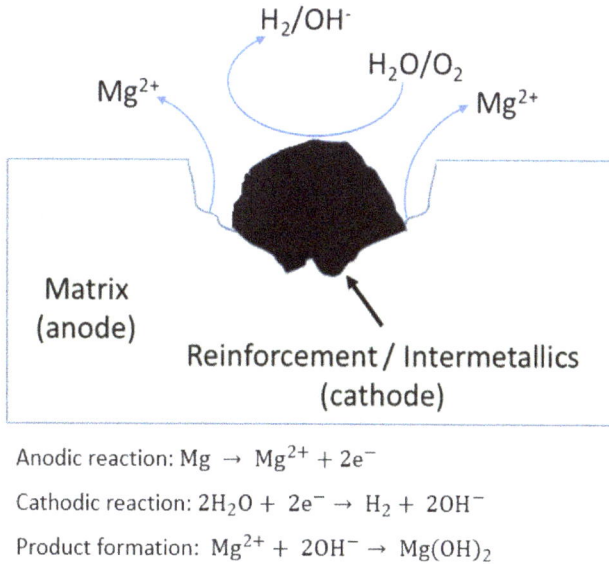

Anodic reaction: $Mg \rightarrow Mg^{2+} + 2e^-$

Cathodic reaction: $2H_2O + 2e^- \rightarrow H_2 + 2OH^-$

Product formation: $Mg^{2+} + 2OH^- \rightarrow Mg(OH)_2$

FIGURE 4.2   Schematic of corrosion in Mg alloys and composites. *Reproduced with permission from (Seetharaman et al., 2020b) © IOP Publishing Ltd 2020.*

4.2.1   LIQUID-STATE PROCESSING METHODS

Liquid-state processes involve the dispersion of reinforcement in a molten matrix and solidification of the composite slurry into required shapes. Based on the mode of reinforcement dispersion, the liquid-state processing methods can be further classified into: (1) melt infiltration, (2) stir casting or compo-casting and (3) melt deposition (Seetharaman & Gupta, 2021).

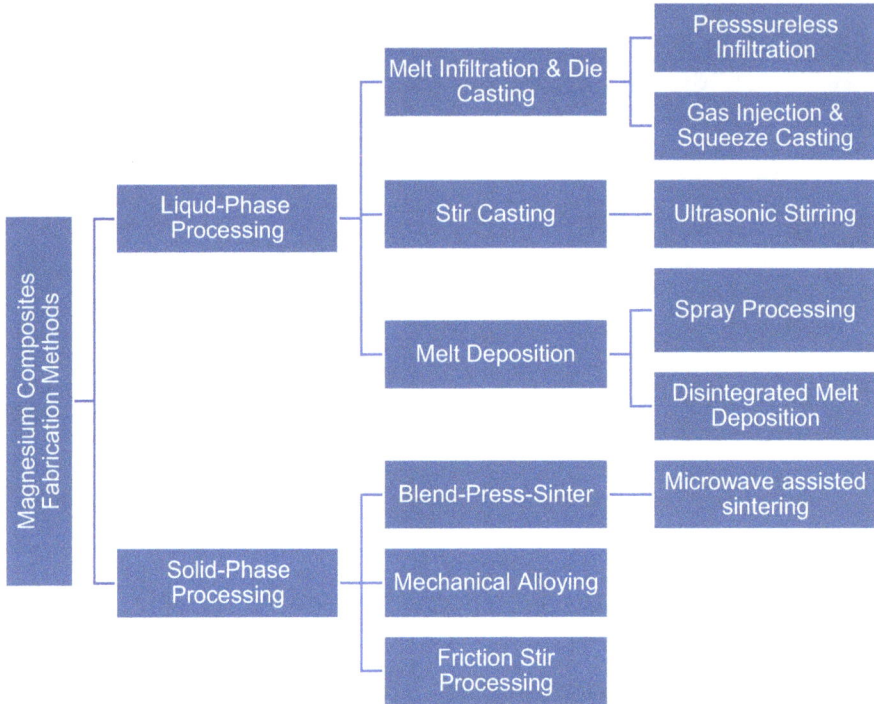

FIGURE 4.3   Classification of fabrication methods applicable to Mg composites.

4.2.1.1   Melt Infiltration

In melt infiltration, a liquid metal alloy is either forced or spontaneously infiltrated into porous preforms containing up to 70% reinforcements. When the melt can spontaneously infiltrate through the mould without any external force, it is defined as pressureless infiltration. This process is effective when the wettability between matrix and the reinforcing phase is good, as poor wettability delays the infiltration process and results in the formation of undesirable reaction products. In this regard, the wettability of reinforcements can be improved by activators or processing in a nitrogen atmosphere. In case of forced infiltration, the molten metal is force injected into the porous preform by means of an external pressure (e.g. pressure die infiltration) or mechanical force (e.g. squeeze casting, **Figure 4.4**), which helps to overcome the issues associated with poor wetting. As the melt solidifies under high pressure, the MMCs fabricated by forced infiltration are often free from defects like porosity and shrinkage. Therefore, high precision capability and low cost are some of the key benefits of forced infiltration (Seetharaman et al., 2022).

4.2.1.2   Stir Casting

In stir casting, either a mechanical or an ultrasonic stirrer is used to disperse the reinforcements in a Mg alloy matrix (**Figure 4.5**), and the molten composite slurry is then solidified by die casting. Here, the distribution of reinforcements in the Mg matrix depends on the melt temperature and stirring conditions, including the stirrer geometry. As reinforcements tend to agglomerate, careful attention must be paid to ensure their effective dispersion in the Mg matrix and, in some cases, wetting agents are also applied on to the reinforcement to improve the interfacial bonding and to avoid any unwanted reaction products (Seetharaman et al., 2022).

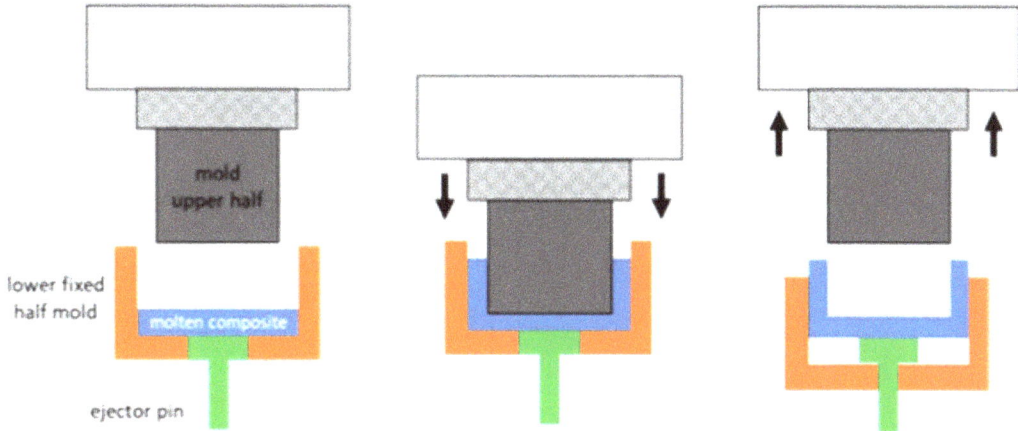

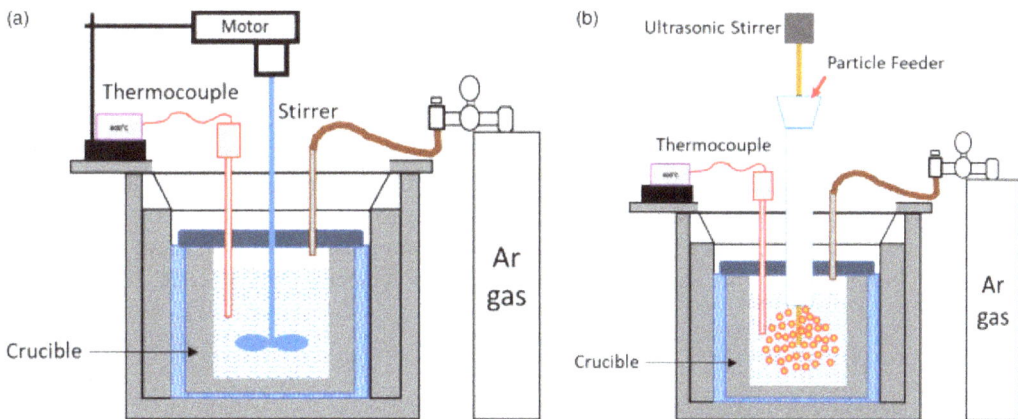

4.2.1.3 Melt Deposition

While several deposition methods including immersion and electroplating, chemical vapour deposition (CVD), physical vapour deposition (PVD) are applicable for fibre-reinforced composites, spray deposition methods including disintegrated melt deposition (DMD) are suitable to produce particle reinforced Mg composites.

4.2.1.3.1   Spray Deposition

Spray deposition is a promising method for producing particle-reinforced Mg composites. It involves the injection of particle reinforcements into a spray of molten metal before being deposited onto a substrate. The resulting composite is then subjected to suitable post-processing techniques of scalping, consolidation and secondary finishing for densification. **Figure 4.6** shows the schematic of spray deposition or forming process in which a spray gun is used to atomise the molten matrix metal into which the reinforcement particles are injected (Seetharaman et al., 2022).

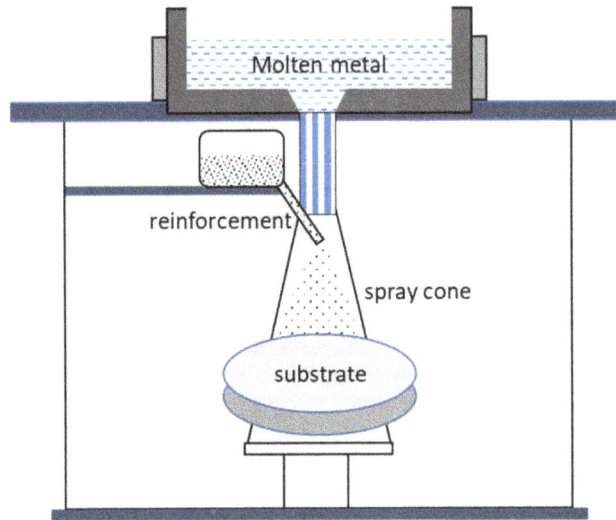

FIGURE 4.6 Spray deposition. *Reproduced with permission from (Seetharaman et al., 2022) © 2022 by the authors. Licensee MDPI, Basel, Switzerland.*

4.2.1.3.2 Disintegrated Melt Deposition

Disintegrated melt deposition (DMD) combines the advantages of cost-effective stir casting and spray processing methods, as it involves the vortex mixing of reinforcements and the deposition of molten slurry onto a metallic substrate after disintegration by jets of inert gases (**Figure 4.7**). Unlike spray deposition, this liquid-state method employs lower impinging velocity to achieve a bulk composite with fine grain structure and low segregation of reinforcements (Malaki et al., 2019).

4.2.2 Solid-State Processing Methods

Solid-state methods applicable for Mg composites generally involve the preparation of a composite blend, which is then consolidated by compaction and sintering. Here, the blend can be prepared by either simple blending or mechanical alloying techniques (Seetharaman & Gupta, 2021).

4.2.2.1 Simple Blending of Powder Particles

In this method, the composite powder blend is prepared by simple mixing of matrix and reinforcement powders. The prepared blend is then hot or cold compressed into a billet of required dimensions, which is then canned, degassed and sintered at a temperature closer to the solidus temperature of the matrix alloy (**Figure 4.8**). Here, it is important to note that the blend-press-sinter method is more effective for producing particle-reinforced Mg composites, as the reinforcement fibres can often get damaged under the high pressure during pressing. For fibre-reinforced Mg composites, the fibres are infiltrated by the dry matrix powder first, which is then followed by compaction and sintering or hot isostatic pressing (Malaki et al., 2019; Seetharaman et al., 2022).

4.2.2.2 Mechanical Alloying

In mechanical alloying, the powder particles are subjected to repeated cold welding, fracturing and re-welding, which results in local melting and consolidation due to the frictional heat developed at the particle interface (Suryanarayana, 2019; Suryanarayana et al., 2001). The mechanically alloyed powders can be then densified by either cold- or hot-pressing techniques. The schematic is shown in

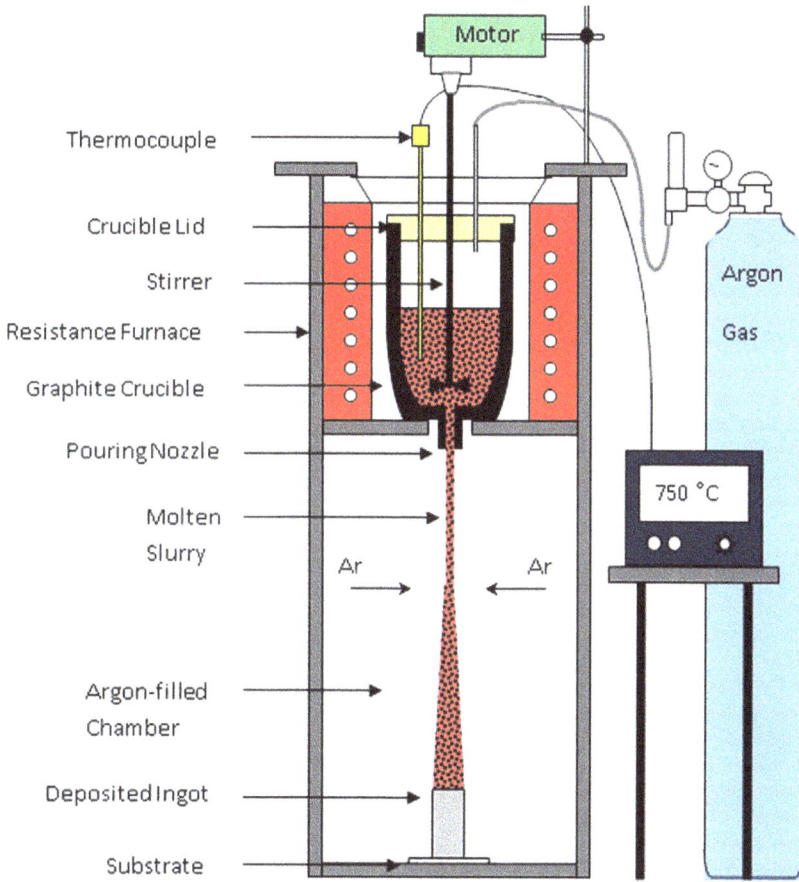

FIGURE 4.7   Disintegrated melt deposition. *Reproduced with permission from (Malaki et al., 2019) © 2019 by the authors. Licensee MDPI, Basel, Switzerland.*

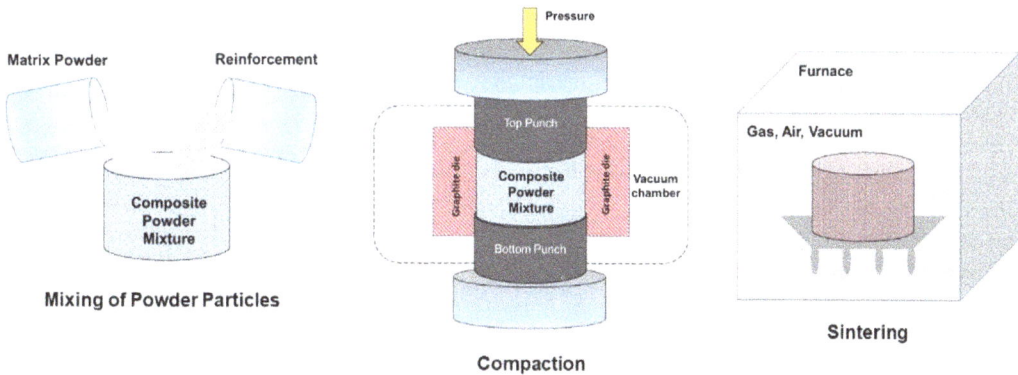

FIGURE 4.8   Schematic showing simple blend-press-sinter method. *Reproduced with permission from (Seetharaman et al., 2022) © 2022 by the authors. Licensee MDPI, Basel, Switzerland.*

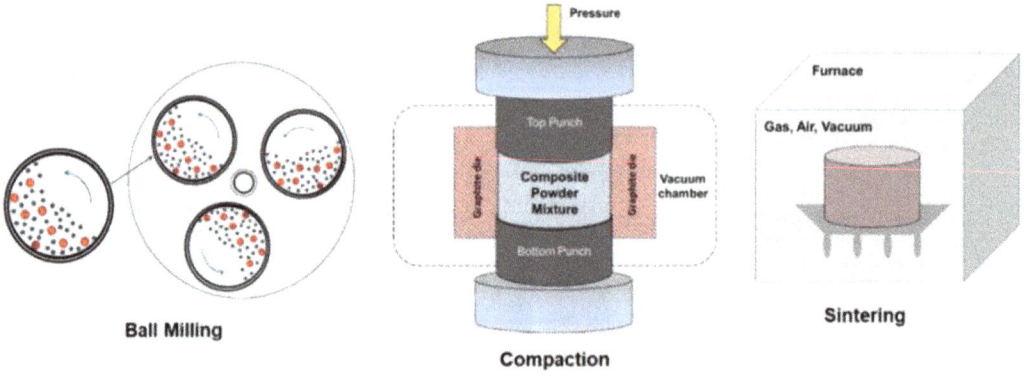

FIGURE 4.9 Schematic showing mechanical alloying–assisted powder consolidation. *Reproduced with permission from (Seetharaman & Gupta, 2021) © 2021 Elsevier Inc.*

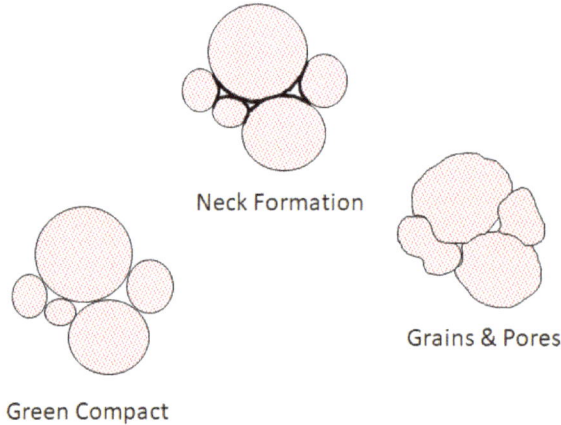

FIGURE 4.10 Stages in sintering: Stage I: Particle bonding in green compact, Stage II: Growth of contact points into 'neck' and Stage III: final microstructure showing grains and pores. *Reproduced with permission from (Seetharaman & Gupta, 2021) © 2021 Elsevier Inc.*

Figure 4.9. This method can be used to develop a range of equilibrium/non-equilibrium alloys and composites, as it ensures the homogenous distribution of reinforcing constituents and the generation of a high volume of dislocation densities.

4.2.2.3   Conventional Sintering of Powder Materials

The sintering process involves heating the green compact billets prepared by simple powder blending and mechanical alloying methods to a temperature closer to the solidus line of the matrix alloy (**Figure 4.10**). Here, the atomic diffusion facilitates the formation of inter-particle bonds between the powder particles. In most cases, the sintering of green powder compact also facilitates the microstructural recrystallisation for strengthening alongside densification and removal of residual lubricant.

4.2.2.4   Spark Plasma Sintering

This process utilizes a uniaxial force and a pulsed (on-off) direct electrical current (DC) to consolidate the powder raw materials (**Figure 4.11**), and it involves three major stages: (i) plasma heating, (ii) joule heating and (iii) plastic deformation (Mamedov, 2002). During plasma heating, a localised

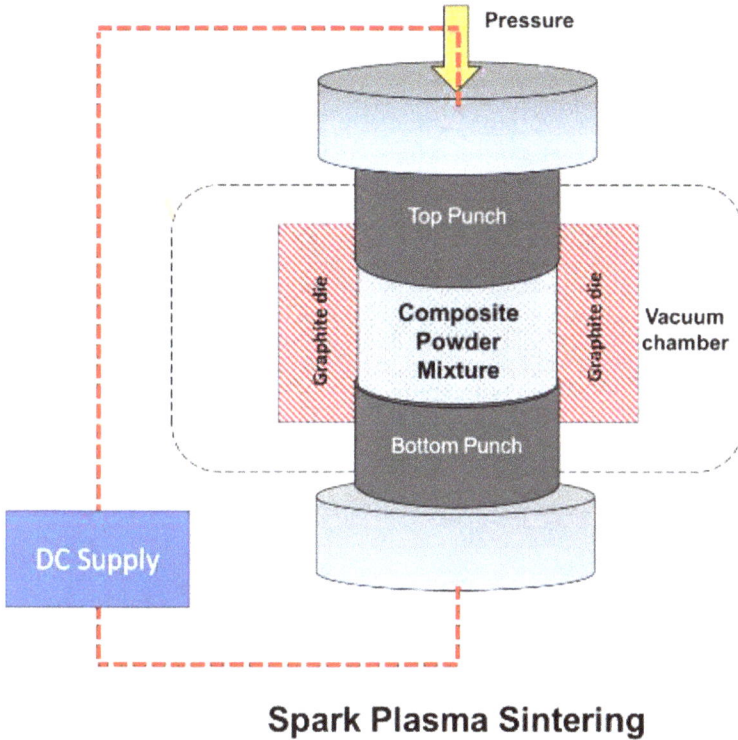

Spark Plasma Sintering

FIGURE 4.11   Schematic of spark plasma sintering.

and momentary heating of particle surfaces occurs due to the electrical discharge between powder particles. The flow of DC current between the particles then results in necking due to the Joule heating effect, which increases the diffusion of atoms at the particle interface. The heated material becomes soft in the final stage and deforms under the application of uniaxial force. Therefore, the spark plasma sintering (SPS) technique combines the benefits of atomic diffusion and plastic deformation to achieve densification of powder compact by up to 90%. It is also important to note that the SPS process is usually carried out at a low atmospheric pressure to ensure rapid consolidation.

4.2.2.5   Microwave-Assisted Rapid Sintering
Microwave sintering has emerged as an energy-efficient technique to consolidate metal powders. It involves the self-heating of the material core due to dielectric and magnetic losses resulting from the interaction between the electric and magnetic fields and the subsequent transfer of heat from the core to the surface of the material (Matli et al., 2016). As microwaves exhibit an inverse temperature distribution, the microwave heating is generally rapid, thereby reducing the processing time by > 80%. This is unlike conventional heating where the transfer of heat by conduction, convection and radiation occurs from the surface to the interior of the material and hence is relatively more time consuming (**Figure 4.12**). While microwave processing has been largely limited to ceramics in the past, there are several recent papers confirming the feasibility of processing metallic materials including micro, nano and amorphous MMCs.

4.2.2.6   Friction Stir Processing
Friction stir processing (FSP) is a solid-state welding method used to fabricate Mg-based composite materials through surface modification (Li et al., 2019). In this process, the material undergoes severe plastic deformation as shown in the schematic (**Figure 4.13**), which results in a homogeneous

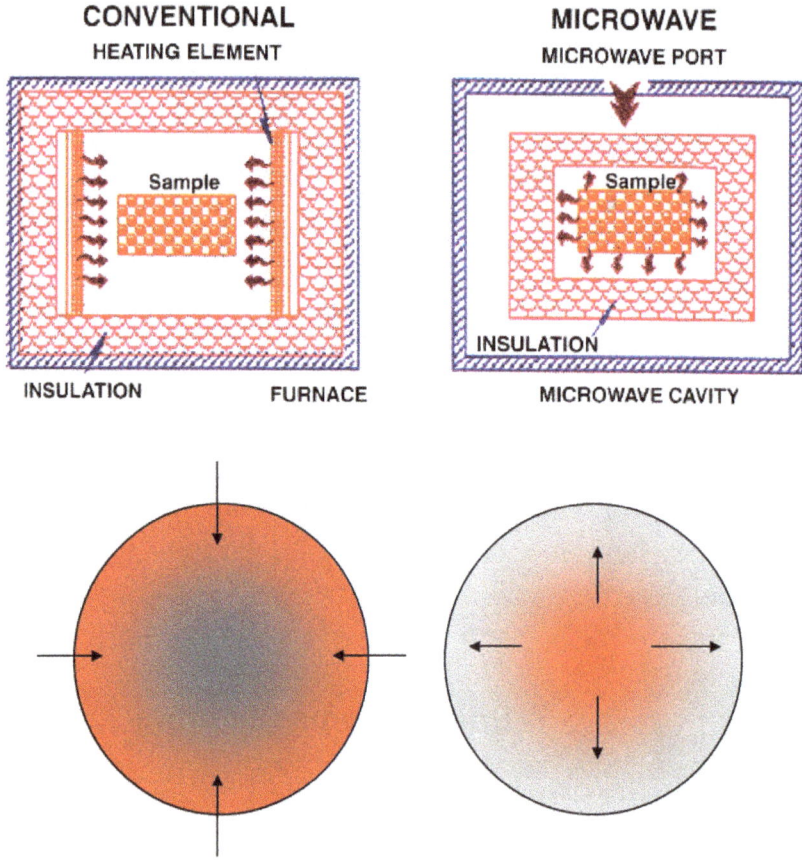

FIGURE 4.12 Schematic of microwave sintering and heat transfer principle. *Reproduced with permission from (Matli et al., 2016) © 2016 by the authors; licensee MDPI, Basel, Switzerland.*

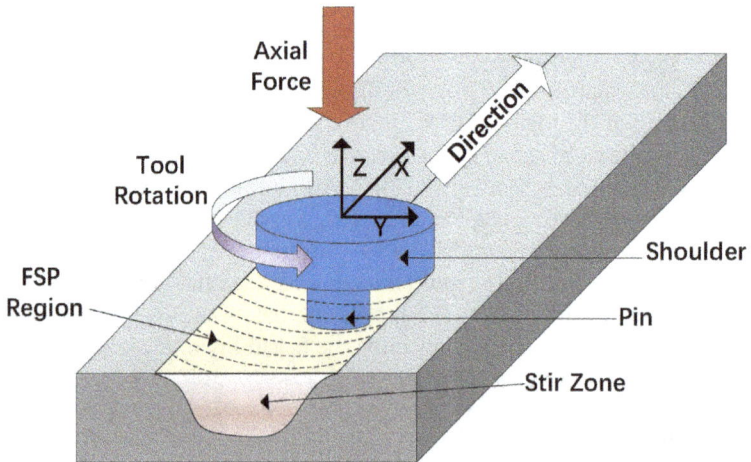

FIGURE 4.13 Schematic of friction stir processing. *Reproduced with permission from (Li et al., 2019) © 2019 by the authors. Licensee MDPI, Basel, Switzerland.*

fine-grained microstructure. Being a solid-state method, FSP does not involve the melting of materials, and it effectively avoids defects like porosities and hot cracks that are commonly observed during the solidification of molten composite slurry. Here, the effective dispersion of reinforcements depends on the frictional heating at the interface between a rotating tool and the matrix material.

The benefits and limitations of each of the discussed processing methods are outlined in Table 4.2.

TABLE 4.2

Benefits and Limitations of Different Processing Methods Applicable to Mg-MMCs

Fabrication technique	Benefits	Limitations
Liquid-state processing		
Pressureless Infiltration	• Simple process • Economical	• Difficult to produce intricate shapes • Undesirable interfacial reaction products
Gas Injection	• Reasonable uniform distribution of reinforcement • Able to produce complex parts with reasonable dimensional accuracy • Suitable for larger volume of reinforcement	• Damage to reinforcement due to injection pressure • Porosity and unstable solidification • Use of gases • Process efficiency uncertain
Squeeze Casting (Mechanical Force–Assisted Infiltration)	• Suitable for complex parts • Effective dispersion reinforcement • Capable of using larger reinforcement quantity than stir casting (up to 40%–50%) • Economical for large quantity production	• Damage to reinforcement • Clustering of reinforcement
Pressure Die Casting	• Suitable for complex parts and large quantity production • Effective dispersion of reinforcement • Better dimensional accuracy • Economical	• Undesirable reaction products • Damage to reinforcement
Stir Casting	• Suitable for mass production • Suitable for larger reinforcement quantity (up to 30%) • Economical	• Damage to reinforcement • Clustering of reinforcement • Undesirable brittle interfacial reaction products • Difficult to produce complex shapes • Unstable vortex • Gas entrapment
Ultrasonic Stirring	• Better dispersion of reinforcement	• Unstable vortex • Gas entrapment • Damage to reinforcement
Spray Processing	• Finer microstructure due to faster solidification rates	• Unsuitable for complex and intricate shapes • Expensive due to the use of gases • Porosity
Disintegrated Melt Deposition	• Combine the benefits of stir casting and spray processing • Flexible process in terms of reinforcement types and volume fractions • Effective distribution of reinforcement • Lesser chance for segregation • Finer grain structure due to faster cooling rates	• Not suitable for intricate shapes

(Continued)

TABLE 4.2 (Continued)

Benefits and Limitations of Different Processing Methods Applicable to Mg-MMCs.

Fabrication technique	Benefits	Limitations
Solid-State Processing		
Simple Blend-Press-Sinter	• Simple and economical processes • Lesser interfacial reaction	• Not suitable for complex shapes • Poor dispersion of reinforcement • Chances of contamination due to binders
Mechanical Alloying	• Suitable for different alloys including non-equilibrium alloys • High strength, strengthening due to high dislocation density • Effective dispersion of reinforcement	• Not suitable for complex parts • Not suitable for mass production • Increased reactivity of powder materials hence raising safety issues
Spark plasma sintering	• Simple and rapid process	• Damage to reinforcement • Unable to do complex shapes • Energy intensive
Microwave Sintering	• Energy efficient • Faster process	• Scaling up can be an issue due to unavailability of industry level equipment
Friction Stir Processing	• Solid-state methods without melting • Relatively easy to control the process • Effective dispersion of reinforcement by controlling the FSP parameters	• Process efficiency uncertain • Bulk composites of larger dimensions may not be possible beyond surface modification

4.3 CONCLUSIONS AND OUTLOOK

To summarise, there is an increasing interest in magnesium composites due to the increasing availability of different types of reinforcements and sophisticated fabrication methods. It is clear that the state-of-the-art liquid- and solid-state methods like ultrasonic stir casting, DMD, SPS and microwave sintering are reasonably effective in dispersing the hard and strong reinforcement phases in the Mg matrix. Some of the heavily exploited reinforcements in the Mg matrix include SiC, Al_2O_3 and TiC, which caused a tremendous improvement in mechanical strength although the ductility of those composites is often compromised except for the addition of nanoparticle reinforcements. While other engineering properties such as wear and thermal stability have also improved due to the addition of ceramic reinforcements, the corrosion resistance tend to be inferior, and the same can be improved by suitable surface coatings. While these published properties of Mg composites encourage their use in critical engineering applications, further extensive studies are underway covering their overall engineering aspects for effective technology transfer.

The outlook for Mg-based MMCs looks extremely promising based on the results obtained through active research conducted over last two decades involving several types of matrices and reinforcements. These results have unquestionably validated that Mg-based MMCs can lead to superior performance under many service conditions and in multiple engineering sectors. Perhaps the biggest challenge is the scaling up of primary processing techniques for industrial viability. With active and willing industry partners, the solutions for energy conservation, sustainability and reliability can easily be achieved through Mg-based MMCs.

REFERENCES

Bamberger, M., & Dehm, G. (2008). Trends in the development of new Mg alloys. *Annual Review of Materials Research*, *38*(1), 505–533, https://doi.org/10.1146/annurev.matsci.020408.133717.

Bharathi, P., & Sampath Kumar, T. (2022). Latest research and developments of ceramic reinforced magnesium matrix composites – a comprehensive review. *Proceedings of the Institution of Mechanical Engineers, Part E: Journal of Process Mechanical Engineering*, doi: 10.1177/09544089221126044.

Gupta, M., & Seetharaman, S. (2017). Magnesium based nanocomposites for cleaner transport. In *Nanotechnology for Energy Sustainability*. Raj, B., Van de Voorde, M., & Mahajan, Y. (Eds.), pp. 809–830, Wiley-VCH Verlag GmbH & Co. KGaA, doi: 10.1002/9783527696109.ch33.

Gupta, M., & Wong, W. L. E. (2015). Magnesium-based nanocomposites: Lightweight materials of the future. *Materials Characterization*, *105*, 30–46.

Kainer, K. U. (2006). Basics of metal matrix composites. In *Metal Matrix Composites: Custom-Made Materials for Automotive and Aerospace Engineering*. Kainer, K. U. (Ed.), pp. 1–54, Wiley-VCH Verlag GmbH & Co. KGaA, doi: 10.1002/3527608117.ch1.

Khorashadizade, F., Abazari, S., Rajabi, M., Bakhsheshi-Rad, H. R., Ismail, A. F., Sharif, S., Ramakrishna, S., & Berto, F. (2021). Overview of magnesium-ceramic composites: Mechanical, corrosion and biological properties. *Journal of Materials Research and Technology*, *15*, 6034–6066.

Kumar, D., Phanden, R. K., & Thakur, L. (2021). A review on environment friendly and lightweight magnesium-based metal matrix composites and alloys. *Materials Today: Proceedings*, *38*, 359–364.

Kumar, R., Suri, N. M., & Khatkar, S. K. (2019). Recent advancements in creep resistant magnesium alloys and composites: A review. *Pramana Research Journal*, *9*(6), 631–636.

Kumar, S. S., & Mohanavel, V. (2022). An overview assessment on magnesium metal matrix composites. *Materials Today: Proceedings*, *59*, 1357–1361.

Li, K., Liu, X., & Zhao, Y. (2019). Research status and prospect of friction stir processing technology. *Coatings*, *9*(2), 129.

Malaki, M., Xu, W., Kasar, A. K., Menezes, P. L., Dieringa, H., Varma, R. S., & Gupta, M. (2019). Advanced metal matrix nanocomposites. *Metals*, *9*(3), 330.

Mamedov, V. (2002). Spark plasma sintering as advanced PM sintering method. *Powder Metallurgy*, *45*(4), 322–328.

Matli, P. R., Shakoor, R. A., Amer Mohamed, A. M., & Gupta, M. (2016). Microwave rapid sintering of Al-metal matrix composites: A review on the effect of reinforcements, microstructure and mechanical properties. *Metals*, *6*(7), 143.

Nazeer, A. A., & Madkour, M. (2018). Potential use of smart coatings for corrosion protection of metals and alloys: A review. *Journal of Molecular Liquids*, *253*, 11–22.

Powell, A. (2022). A magnesium clean energy ecosystem vision. In *Magnesium Technology 2022*. Maier, P., Barela, S., Miller, V. M., & Neelameggham, N. R. (Eds.), pp. 121–126, The Minerals, Metals & Materials Series, Springer, Cham, doi: 10.1007/978-3-030-92533-8_20.

Saranu, R., Chanamala, R., & Putti, S. (2022). Wear behaviour of magnesium metal matrix hybrid composites – a detailed review. *International Journal of Mechanical Engineering*, *7*(1), 1488–1499.

Seetharaman, S., & Gupta, M. (2021). Fundamentals of metal matrix composites. In *Fundamentals of Metal-Matrix Composites*, Brabazon, D. (Ed.), pp. 11–29, Elsevier, https://doi.org/10.1016/B978-0-12-819724-0.00001-X.

Seetharaman, S., Subramanian, J., Singh, R. A., & Gupta, M. (2021). An insight into magnesium based metal matrix composites with hybrid reinforcement. In *Encyclopedia of Materials: Composites*. Brabazon, D. (Eds.), Vol. 1, pp. 52–77, doi: 10.1016/B978-0-12-819724-0.00098-7.

Seetharaman, S., Subramanian, J., Singh, R. A., Wong, W. L. E., Nai, M. L. S., & Gupta, M. (2022). Mechanical properties of sustainable metal matrix composites: A review on the role of green reinforcements and processing methods. *Technologies*, *10*(1), 32.

Seetharaman, S., Tekumalla, S., & Gupta, M. (2020a). Introduction to magnesium-based nanocomposites. In *Magnesium-Based Nanocomposites: Advances and Applications*. Seetharaman, S., Tekumalla, S., & Gupta, M. (Eds.), IOP Publishing, Bristol, doi: 10.1088/978-0-7503-3535-5ch1.

Seetharaman, S., Tekumalla, S., & Gupta, M. (2020b). Wet corrosion and biocompatibility of magnesium nanocomposites. In *Magnesium-Based Nanocomposites: Advances and Applications*. Seetharaman, S., Tekumalla, S., & Gupta, M. (Eds.), IOP Publishing, Bristol, doi: 10.1088/978-0-7503-3535-5ch10.

Suryanarayana, C. (2019). Mechanical alloying: A novel technique to synthesize advanced materials. *AAAS Research*, *2019*, 4219812, doi: 10.34133/2019/4219812.

Suryanarayana, C., Ivanov, E., & Boldyrev, V. V. (2001). The science and technology of mechanical alloying. *Materials Science and Engineering: A*, *304*, 151–158.

Wang, J., Xu, J., Hopkins, C., Chow, D. H., & Qin, L. (2020). Biodegradable magnesium-based implants in orthopedics – a general review and perspectives. *Advanced Science*, *7*(8), 1902443.

Zhao, F., Li, H., Zhao, Z., & Liu, X. (2021). Research progress of lightweight metallic materials and their processing technologies for modern transportation. *International Journal of Computational Materials Science and Surface Engineering*, *10*(2), 69–87, https://doi.org/10.1504/IJCMSSE.2021.118498.

5 Additive Manufacturing for Magnesium Alloys and Composites

Shrinivas Kulkarni, Vijay Sisarwal, Kai-Rui Wang, Asma Lamin and Xiao-Bo Chen

CONTENTS

5.1 INTRODUCTION

Additive manufacturing (AM) has been developing massively worldwide, from rapid prototyping over the last decade to reforming the ways of design and manufacturing. Different from subtractive manufacturing skills, AM exhibits an additive nature, with one thin layer of feeding materials placed on the right spots followed by melting and solidifying (commonly cooling) as guided by a digital model or digital design. The manufacturing process is progressive, with programmed manipulation to yield a final three-dimensional build. The advantages of such a net-shape technique are obvious, including simplified operations, no need for expensive tools and sophisticated skills, minimal consumption of feedstocks, the diverse and complex geometry of final builds, and customised designs and services. On the other hand, it is inevitably associated with several limitations such as low production rate, requirements of high-quality power (e.g. metal), heterogenous microstructure, and poor mechanical properties. As such, optimised procedures and appropriate post-treatments have been developed to additively manufacture a variety of metal components with desired surface features and mechanical properties.

As the lightest metallic materials, magnesium (Mg) alloys (density of 1.7 g/cm^3) and their composites have been designed and optimised as substitutes for aluminium (2.7 g/cm^3) and steel ($\approx$ 7.85 g/cm^3) components to satisfy growing demands of the reduction in carbon emissions imposed by global legislative bodies (Polmear, 1995), in particular in the transportation sector (i.e. vehicles, trains, and aircraft) – a principal attributed to the drastic climate change (Esmaily et al., 2017; Fyfe et al., 2016). A rule of thumb is that, for each percentage of automotive weight reduction, there is a corresponding percentage drop in fuel or electricity consumption and, thus, greenhouse gas emissions (Kelly et al., 2015). Traditional options to fabricate Mg-based parts are die-casting, extrusion, and a number of mechanical processing techniques (e.g. sawing, punching, drilling, milling, and turning). Although the automotive industry holds the largest market share of Mg alloys, biomedical

DOI: 10.1201/9781003319856-7

applications hold the premise as the most prolific end users. For such value-adding service, custom-ised design, the feasibility of complex geometries and surface quality of final products are vital. AM greatly outperforms its competitors in all these aspects.

Over the last few years, the AM of Mg alloys and composites (primarily in the scaffold format) has been advancing faster than traditional methods, which has enabled delicate control over porosi-ties, mechanical properties, and corrosion properties (Kuah et al., 2022). However, both scientific and technical challenges remain in AMed Mg alloys, such as safety issues related to handling the powder-shape of Mg even in an inert atmosphere (Figure 5.1) (Esmaily et al., 2020), the low evapora-tion temperature (1091 °C), and poor understanding of the composition-processing-microstructure-property relationship in AM-Mg and appropriate post-treatment prior to the expected structural and functional services.

Of the family of Mg alloys, the branch of Mg-rare-earth- (RE) based alloys has proved to have great potential with their high achievable density. A great deal of effort was cast upon WE43 (Mg-3.5Y-2.3Nd-0.5Zr in wt.% as a primary constituent), a high-strength, corrosion-resistant, and

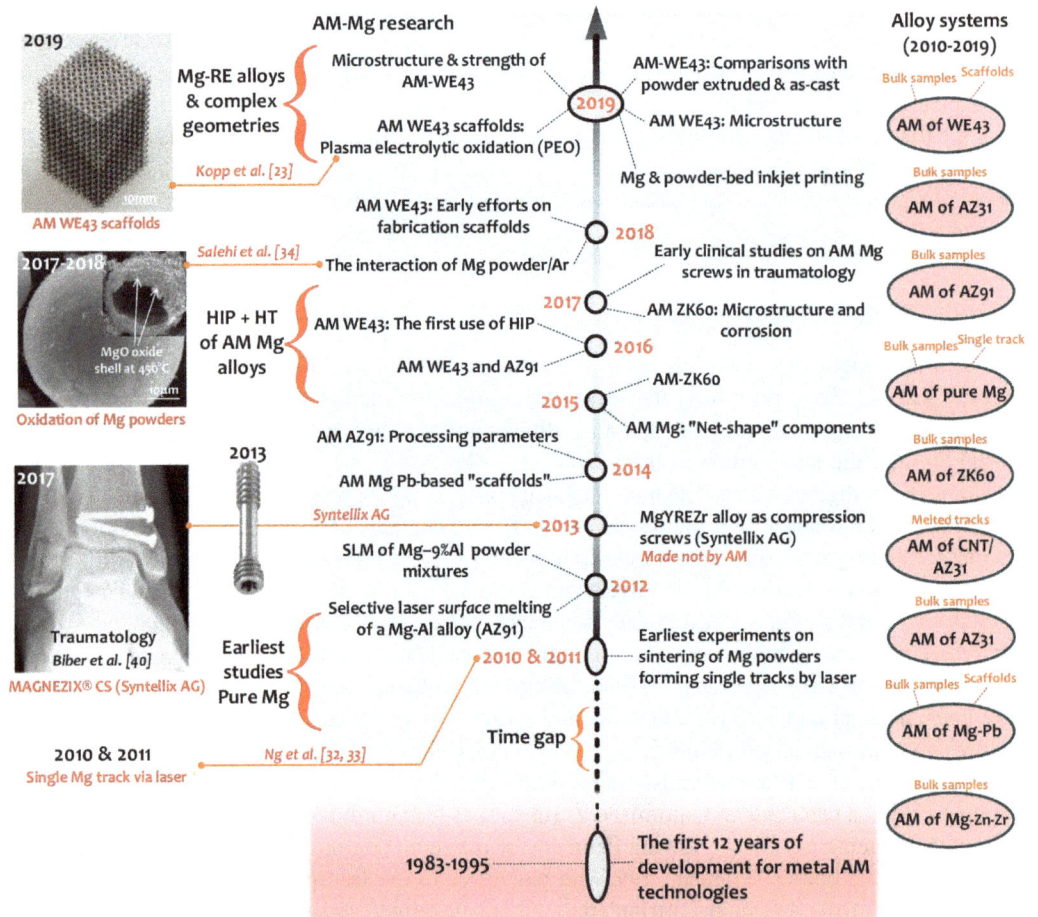

FIGURE 5.1 Timeline displaying a historical background of AM-Mg research and development, indicating "landmarks" since the first scientific study on utilisation of AM to sinter Mg powder from literature. Note the difference in the geometrical complexity that can be accomplished by AM (*i.e.* scaffold) as compared to other state-of-the-art manufacturing technologies (*i.e.* compression screw). *Reproduced with permission from (Esmaily et al., 2020) © 2020 Elsevier B.V.*

biocompatible material which many regulatory bodies have approved in some particular clinical trials and operations. The mechanical properties of AM-built Mg alloy WE43 parts are highly dependent on porosity and microstructure (Gangireddy et al., 2019). However, the corrosion performances were polarised in literature (Esmaily et al., 2020; Suchý et al., 2021). A dense structure of WE43 alloys fabricated by selective laser melting (SLM) demonstrated that microstructural factors such as silicon inclusions (Suchý et al., 2021) and/or distribution of secondary phases (Esmaily et al., 2020; Suchý et al., 2021) and porosity are key to high susceptibility to corrosion. It is common that Mg alloys accommodate oxides (Y, Nd, Zr based) as inclusions in the microstructure during the build process. Nevertheless, limited insights were provided regarding the roles of those oxide inclusions in regulating corrosion and mechanical properties.

AM of Mg-based composites remains in its infancy, but biodegradable Mg-based composite devices with compatible mechanical and physical properties to those of human bones are greatly expected to take a vital share of the global orthopaedic implants market, which will reach USD 88.79 billion by 2026 (Market, 2023). Currently, countries including Australia are experiencing an exceedingly higher demand for functional bone grafts due to the ageing population. However, no reports exist on clinical evaluation of Mg scaffolds. Efforts have been cast upon the development of Mg-based multi-material structures with satisfactory mechanical properties, excellent biocompatibility, controlled degradability, and drug delivery ability for bone tissue engineering.

In the following parts of this chapter, the basic concepts and state-of-the art progresses regarding AM Mg alloys and composites will be introduced, in particular for biomedical services, followed by the key challenges and opportunities for the future.

5.2   WHAT IS KNOWN

AM is the concept of using raw materials, including metals, polymers, ceramics, and hydrogels, to achieve a customised component with irregular shape/geometry through a layer-by-layer manner, guided by a pre-designed digital model and driven by varying mechanics. AM enables the manufacturing of highly complex geometries that are either difficult or technically impossible to build through conventional machining processes. It is of great importance in biomedical devices with complex internal and external geometries which are required to promote cell growth, proliferation, and bone regeneration. A large variety of technical processes in this regard have been explored, spanning laser beam power bed fusion (L-PBF) (Gieseke et al., 2016; Niu et al., 2018), wire-arc additive manufacturing (WAAM) (Guo et al., 2019), friction stir processing (Palanivel et al., 2015) (Figure 5.2), paste extrusion deposition (Farag and Yun, 2014), and cold spray and inkjet methods (Salehi, Maleksaeedi, Nai et al., 2019; Salehi, Maleksaeedi, Sapari et al., 2019). Electron beam power bed fusion, the mainstream AM technique for Ti and Al components, remains a technical challenge to AM Mg alloys because E-beam melting must be conducted in high vacuum, in which Mg evaporates vastly. Each individual technique exhibits unique merits and disadvantages in terms of safety, cost, surface finish, wear/corrosion resistance, and mechanical properties. Based on the physical nature of Mg, particularly the powder shape, a small number of these options are attempted in order to fabricate high-quality Mg parts.

5.2.1   Operational and Health Safety

The raw metal materials for AM are available in powder or wire forms. The storage and use of Mg, specifically Mg powder for AM processing, is concerning due to its highly reactive nature. In this state, the surface energy of the Mg powder increases and poses a higher risk of reacting with atmospheric oxygen to enable combustion, which is a key roadblock to the development of additive manufacturing processes for Mg and their alloys (and composite). In those cases, PBF is embedded with the highest standards of risk management. Specific facilities and accessories that are capable of handling Mg powders in the less active or fully inert atmosphere are required, along with

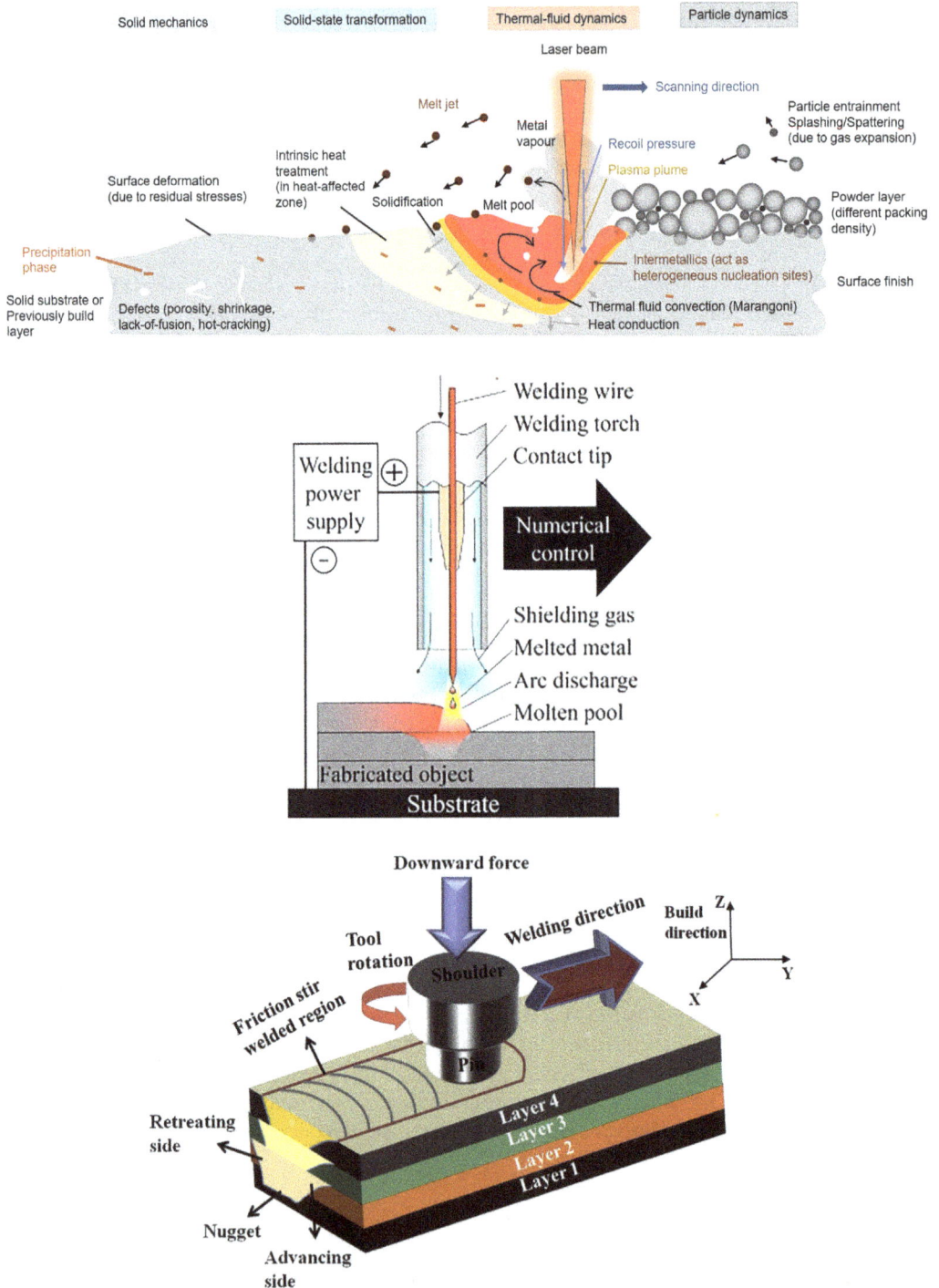

FIGURE 5.2 Schematic illustration of the additive manufacturing processes. (a) Laser beam power bed fusion (Kotadia et al., 2021); (b) Wire-arc additive manufacturing (Takagi et al., 2018); and (c) Friction stir welding (Palanivel et al., 2015). *Reproduced with permission from (a) (Kotadia et al., 2021) Crown Copyright © 2021 Published by Elsevier B.V.; (b) (Takagi et al., 2018) © 2018 Elsevier B.V. and (c) (Palanivel et al., 2015) © 2014 Elsevier Ltd.*

appropriate safety measures. A high potential fire and explosion risk exists in Mg powder manufacturing plants and related facilities. High purity and ultrafine powders in sizes ranging from about 1 to 100 nm have been explored for usage in industrial and research fields. Information on explosion characteristics of Mg powder is necessary to predict the likelihood and severity of explosions and subsequently design explosion prevention and mitigation measures. The evaluation of flammability and explosion risks of Mg powders vary dramatically from micro- to nano-dimensions. The extensive and systematic investigation is desirable to determine the explosion characteristics of Mg powders that are being used in AM process.

5.2.2 MECHANICAL PROPERTIES

Mg alloy WE43 consisting of Y and Nd as the main alloying ingredients is a well-established material for biomedical implants. Given its good printability and a large processing window to reduce porosity, the commercial availability of WE43 alloy has been massively expanded by additive manufacturing to create high-quality components to fit the unique and complex bone geometry of individual patients. However, the high cooling rate and subsequent heat treatment incur a microstructure dramatically different from that of their counterparts fabricated through conventional techniques (powder extrusion and machining). Therefore, it remains a challenge to understand the working mechanisms and regulate mechanical properties and corrosion performance of the final built product (Bar et al., 2019).

Given the high oxygen affinity of RE elements, it is impossible to exclude the presence of RE oxide in WE43 powder. Zener-pinning of yttrium oxide (Y_2O_3) particles in feedstock powder is anticipated to suppress grain growth. Therefore, the microstructure of AMed WE43 alloy varies greatly in literature. L-PBF can generate WE43 parts with refined and uniformly equiaxed grains (230 W energy input) (Zumdick et al., 2019) as well as coarse and strong-basal-textured grains (100 W energy input) (Bar et al., 2019), which evidently leads to different mechanical properties. In contrast, it is recognised that grain growth during solidification is governed by the type and concentration of solute atoms (StJohn et al., 2005). If WE43 powder is significantly oxidised during manufacturing, storage, and transport, the solute concentration of RE in the powder is much lower than its fresh state. The low concentration of solute atoms in powder cannot restrict grains from preferential growth and thus leads to large and basal-orientated grains.

Tensile properties of laser-assisted AM Mg alloys were summarised in a recent review against the cast and wrought (rolled and extruded) counterparts (Zeng et al., 2022). The reported yield strength of AM builds varies between 200 and 350 MPa, which satisfies the requirements of most structural applications, although ductility (less than 5%) remains a concern. High residue stress in the as-built parts suffering rapid solidification, embrittlement of grain boundaries (enrichment of intermetallic phases) and weak bonding between powder were postulated to be the primary causes. Record high ductility (12.2%) was reported in a laser additively manufactured Mg alloy WE43 (Zumdick et al., 2019). Appropriate post-treatment such as annealing at high temperatures and friction stir processing can tackle the poor ductility to some degree (Hyer et al., 2020) because of refined grain size, alleviated residue stress, and redistributed intermetallics in the microstructure (Deng et al., 2021).

Non–laser-based manufacturing approaches have encountered different challenges towards optimal mechanical properties of Mg builds. Sintering-based AM methods require multiple operations to yield final 3D parts and an analogous geometry of the final (green) parts, followed by sintering at high temperature to consolidate powders. Sinter-based AM provides unique opportunities for interconnected porous Mg scaffolds that are favourable for tissue engineering; however, porosity is the key to determining mechanical properties of the final products. A median pore diameter at ~15 μm and pore openness at ~95% were reported in binder jet printed Mg parts after sintering to a relative density of 71% (Salehi et al., 2021).

Friction stir–based AM is a solid-state technique founded on the well-established process of friction stir welding to additively manufacture large-scale near-net-shaped components. A defect-free and well-consolidated WE43 alloy built by a friction stir-based AM method exhibits linear relation to grain size in its microhardness profile, which is ascribed to the spatial changes of plastic deformation and the thermal cycles that occurred during the build process. Mg alloy WE43 built by friction stir–based methods display superior mechanical properties than those reported for as-cast WE43 (yield strength, YS = 137 MPa, ultimate tensile strength, UTS = 217 Mpa, and elongation, EL = 8.2%) (Xiang et al., 2018), which mainly originated from its remarkably finer grain size. Friction stir–based AMed WE43 has its YS and UTS close to but higher ductility than L-PBF AMed WE43 due to the greater density and more equiaxed grain microstructure (Zeng et al., 2022). In-depth understanding is necessary to correlate texture with the anisotropy of friction stir–based additively manufactured Mg alloys for optimised mechanical properties.

5.2.3 Corrosion Performance

The utilisation of Mg-based alloys in a variety of engineering applications has been greatly limited in part due to an inherently high corrosion rate in aqueous or atmospheric systems (Wu et al., 2015; Chen et al., 2014). Mg alloys serving in aqueous conditions form a mildly protective coating of $Mg(OH)_2$, which turns unstable and porous in chloride environments due to the formation of soluble $MgCl_2$ (Chen et al., 2011a). As a result, the protective hydroxide film on Mg becomes porous and leads to severe pitting corrosion. Corrosion kinetics is fundamentally governed by surface chemistry, stress, and topography.

Rapid but well-controllable corrosion benefits the physiological environment (Song et al., 2022; Wang et al., 2020). A limited number of commercial implant devices made of Mg-alloy components have been approved by major health and safety regulatory agencies in countries and regions such as Korea and the EU, where essential guidelines for approval are not yet fully established. It is apparent that sound controls over the high degradation rate using appropriately configured approaches are key prior to the wider commercial utilisation of safe and functional Mg-based degradable implants. Consequently, the critical criteria and the state-of-the-art strategies for controlling degradation of Mg alloys fabricated through conventional processes have been systemically reviewed (Song et al., 2019). The following section will focus on the corrosion performance and biodegradation control strategy of AMed Mg-based biomaterials.

(a) Impurity. Given the active nature of Mg powders, oxides are inevitably involved in the microstructure of final components built through L-PBF, which is greatly detrimental to the strength, ductility, and corrosion resistance. It is recognised that oxygen pickup of sintered zones is concomitant with the high affinity of Mg for any residual or impurity oxygen (Ng et al., 2011). In contrast, binder jetting is an excellent alternative AM approach to minimise oxide involvement in the microstructure and improve corrosion resistance. In the first step, Mg alloy powders are operated at room temperature to form the required shape and structure so that no violent oxidation incurs even in the air atmosphere. Sintering printed green parts in a protective atmosphere also suppress oxidation reactions. Therefore, such a two-step AM technique is particularly suitable for AM Mg alloys.

A study led by Esmaily tried to correlate microstructure, electrochemical behaviour, and corrosion response of L-PBF AMed WE43 with varying post-treatment methods (say as-SLM, SLM + hot isostatic pressing (HIP), and SLM + HIP + heat treatment) in comparison with cast WE43 as reference (Esmaily et al., 2020). Oxide particles are mainly sourced by powder. The electrochemical characterisation in a dilute 0.01 M NaCl revealed accelerated cathodic kinetics of the SLM prepared WE43, which was ascribed to the pronounced re-distribution of Zr and Y. The results suggested that the corrosion of L-PBF-prepared Mg alloys could be optimised through quality-control of powder.

(b) Microstructure. The microstructural heterogeneities due to directional solidification and formation of different secondary phases due to alloying elements influence not only the strength but also the corrosion mechanism of Mg and its alloys. The corrosion resistance of Mg alloys can be improved by processes that refine grains in the microstructure. Grain refinement leads to a low corrosion rate given the formation of corrosion products uniformly covering the entire surface (Yan

et al., 2021). The role of crystallographic plane, specifically the strong texture with growth direction <1010>, was evident in L-PBF Mg and caused severe corrosion. The high-volume fractions of the secondary phase supported galvanic corrosion of both Mg-Zn-Ca alloy (Lu et al., 2015) and Mg alloy AZ91 (Guan et al., 2010). Such impurity-dominated corrosion acceleration was not exclusive to bulky Mg parts but their scaffold form (Li et al., 2018). Mg alloy WE43 scaffolds were prepared through L-PBF, which aimed to provide an alternative solution to tailor the biodegradation profile through topological design and to develop multifunctional bone-substituting materials, enabling full regeneration of critical-size load-bearing bony defects (Figure 5.3). The lower corrosion rate (0.17 ml/cm^2) than that of most cast and extruded WE43 competitors (0.3–2 ml/cm^2) is ascribed to the adsorption of protein on Mg surfaces as a result of the addition of 5% fetal bovine serum in the culture media.

Time-resolved control on degradation is feasible when corrosion resistance is well oriented spatially throughout a volume using hybrid-AM. Ultrasonic peening was employed on Mg WE43 alloy every 20 layers during L-PBF to form interlayers. Interlayer ultrasonic peening formulates localised severe plastic deformation, leading to grain refinement and compressive residual stress (Hyer et al.,

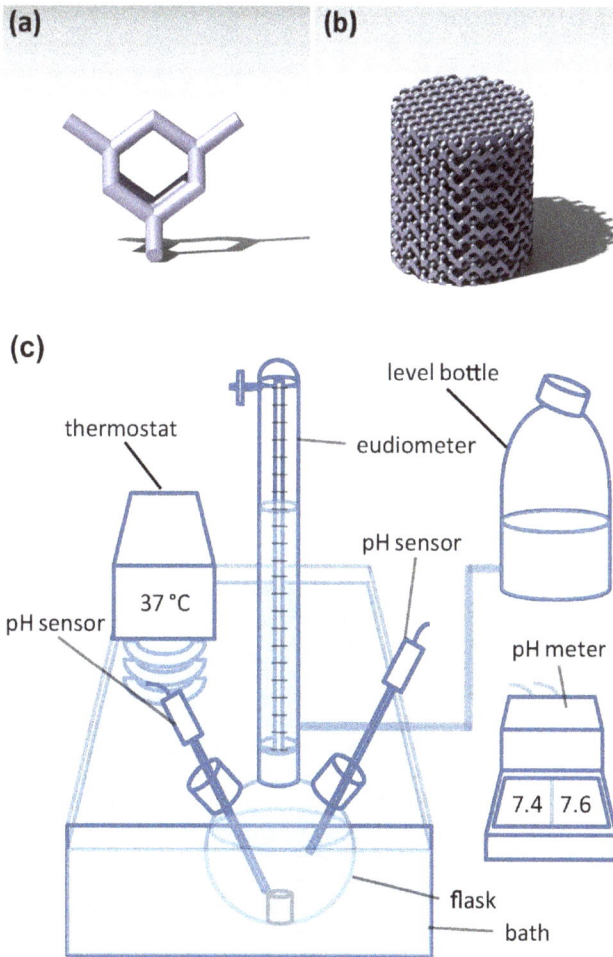

FIGURE 5.3 Design of WE43 scaffolds and degradation study setup: (a) diamond unit cell, (b) CAD model of the scaffold and (c) *in vitro* degradation eudiometer system (Li et al., 2018). *Reproduced with permission from (Li et al., 2018) © 2017 Acta Materialia Inc. Published by Elsevier Ltd.*

2020). The use of interlayer during AM enables functionalised interfacial properties within Mg and mitigates pitting and uniform corrosion within print cells (Sealy et al., 2021).

Corrosion challenges are unique and profound in Mg-based composite materials, regardless of the ways of manufacturing. Shape, spatial distribution, and electrochemical nature of the regular inclusions in Mg matrix, such as inorganic compounds and metallic ingredients, play a profound role in their corrosion responses. (Tiwari et al., 2007). The incorporation of chemically and electrochemically noble fibres can significantly increase the strength of Mg over their particulate equivalents whilst leading to accelerated corrosion. The high corrosion kinetics is ascribed to the galvanic effect (Hihara and Kondepudi, 1993; Hihara and Kondepudi, 1994) between fibre inclusions and Mg matrix, which is determined by solution oxygenation and the exposed region of the mono-filaments (Nunez-Lopez et al., 1996). In addition, the presence of matrix–ceramic interfaces may result in higher dislocation densities and inhomogeneities. It could transform the surface film dramatically and incur vulnerability to film breaking down. The heterogenous surface conditions are also a great challenge to the formation of a compact protective coating to tackle corrosion attacks. Sound pre-treatment approaches are therefore required to homogenise the surface chemistry (Chen et al., 2011b; Yang et al., 2012).

(c) Physical defects. The geometrical design, such as porosity and surface roughness, affects the corrosion rate. L-PBF-built parts are characteristic of porosity which is governed by different mechanisms (Figure 5.4a) (Oliveira et al., 2020; Zeng et al., 2022). Of those, processing parameters,

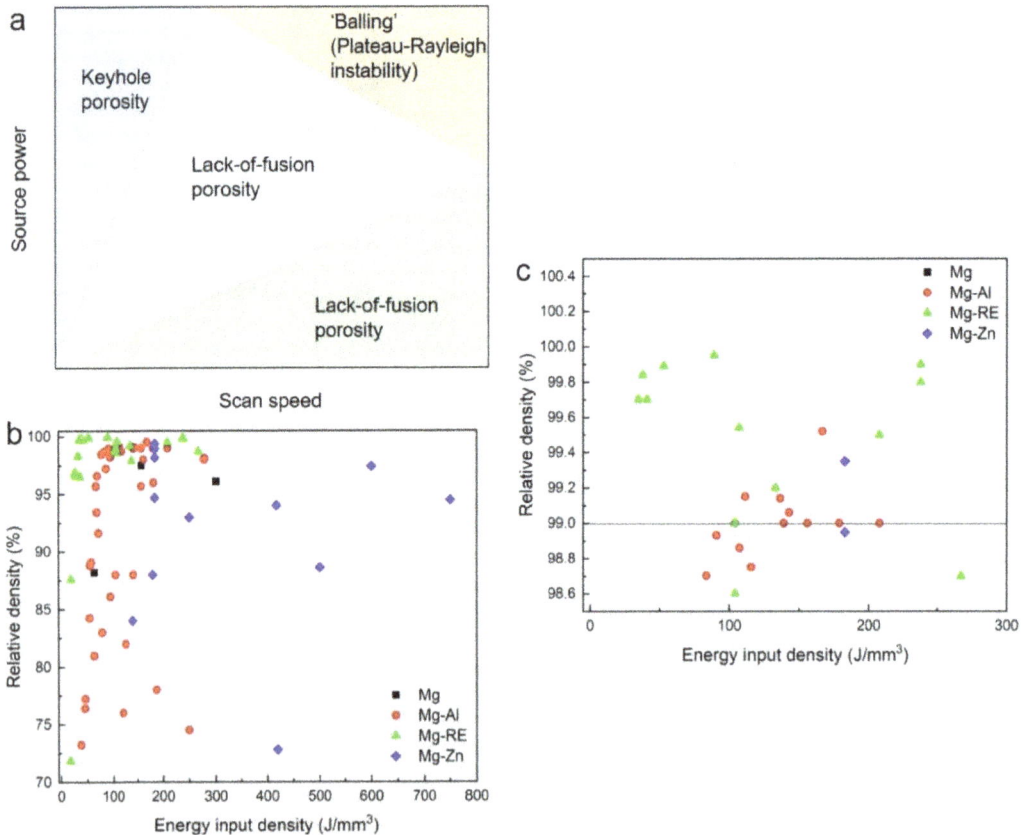

FIGURE 5.4 (a) Schematic illustration of processing window and relevant defect, (b) relative density as a function of energy input density of L-PBF-Mg alloys reported to date, and (c) the samples with high relative density (≥ 99%). *Reproduced with permission from (Zeng et al., 2022) © 2022 Chongqing University. Publishing services provided by Elsevier B.V. on behalf of KeAi Communications Co. Ltd.*

in particular, energy input density (E_v), play a vital role in the formation of porosity or density (Figure 5.4b and c). Although an optimal E_v value/range exists for each individual alloy system, high relative density remains a challenge through using blended elemental metal powders, given the significant local incompatibility during rapid solidification as a result of a mismatch in thermal properties of each individual element.

Physical defects such as pores are greatly detrimental to the corrosion performance of all metal materials including Mg alloys (Khan et al., 2008, 2011). Corrosion-induced surface defects play a larger part than pits in fatigue crack initiation and propagation. Stress is accumulated around defects, which are vulnerable to the formation of micro-galvanic couples (anode specific) and boost corrosion processes. The formation of corrosion products fills up the pore sites and leads to volume expansion. With time, micro-cracks are created and accommodate electrolytes for corrosion. Pre-existing pores or defects are ideal locations for pitting because of the high surface energy and imperfect coverage of the oxide film upon the Mg surface.

It is documented that the corrosion of Mg alloys is initiated from film-free regions (Song and Atrens, 1999; Liu et al., 2014). The presence of aggressive aqueous species breaks down the mildly protective surface film and leads to high corrosion rates. To understand the localised corrosion of Mg, commonly described as pitting, within the range of micro and nano-metres, scanning electrochemical microscopy (SECM) was employed to visualise the evolution of electrochemical active surfaces of Mg in NaCl electrolyte (Liu et al., 2014). Their results demonstrated that the surface film formed on Mg alloys at the beginning of immersion exhibits sound protectiveness in the initial corrosion. The concentrated chloride ions accelerate the dissolution and hydration of surface films, leading to a high number of active spots. In addition, the dissolution of the surface film was accelerated as a function of Cl^- concentration. Effects of common industrial pre-treatments such as grinding, polishing, and sandblasting are profound on the corrosion performance of Mg alloys. It was found that sandblasting severely deteriorates the corrosion performance of Mg alloy AZ31 sheets. Surface analysis indicated that the changes in corrosion behaviour of AZ31 by these pre-treatments are associated with the alteration of its impurity contamination, surface state, grain size, and intermetallic particles (Song and Xu, 2010).

5.3 CHALLENGES AND OPPORTUNITIES

Though substantial progress in the discipline of AM-Mg has been evident over the last decade, challenges remain towards high-quality Mg prints to satisfy the basic requirements for commercial engineering services. In the laser power melting domain, processing settings as critical as quality control cover pre-alloyed powder to build strong, ductile- and corrosion-resistant Mg parts in a feasible manner. Significant effort is worthwhile for exploring and optimising strategies and approaches to conduct Mg powder blending for L-PBF processing.

Health and safety are of great concern related to storage and handling of active Mg powders. Whilst safety is the top priority in the production of Mg-based materials, in the case of AM-Mg in relation to AM specifically, the following are of great importance as overviewed previously (Zeng et al., 2022): (1) minimising the viable volume of stock powder; (2) placing strict regulations over powder handling and storage; and (3) managing environments in all production and processing procedures. Given the high demand for lightweight structural components with complex topography and geometry, AMed Mg alloys are expected to display comparative strength, ductility, and fatigue responses to those of their AMed Al competitors. Mechanistic roles of rapid solidification in microstructure and strengthening mechanisms are key to designing and additively manufacturing AMed Mg alloys for commercial applications. Post-treatment is a common option to remove residual stress from and reduce porosity of AMed parts, but their contribution to those of Mg alloys remains unclear.

The biomedical market has been urgently waiting for the commercialisation of biocompatible Mg alloys (in particular at load-bearing sites) with a controlled degradation profile to serve as temporary implants. The unique biodegradation nature of Mg alloy implants implies that it would eliminate the need for a secondary removal surgery when the expected supporting tasks are accomplished, as they could safely "disappear" over time. Nonetheless, the inherent drawback of the notorious wild degradation is an issue that must be addressed appropriately for Mg implants and is consequently given unique attention. The corrosion control is more challenging in AMed than in cast and extruded Mg alloys, given their unique melting-cooling characteristics. From the perspectives of bulk Mg alloys, the design of alloying composition and optimisation of microstructure (phases, grains, etc.) are vital to suppressing corrosion tendency and minimising corrosion kinetics in the aggressive physiological service environments. In addition, the surface of Mg ligands of AMed parts is vulnerable to corrosion initiation and propagation. Based on the irregular (curved, porous, and asymmetric) shapes of biomedical devices, chemical conversion coating is a sound technique to build up a dense and protective coating upon the entire surface. New coating designs are desired to provide additional protection to AM-ed Mg parts against degradation.

5.4   CONCLUSIONS AND OUTLOOKS

There is no doubt that additive manufacturing has opened a new chapter in the history of structural materials design and fabrication in a net-shape and customised fashion. Such advanced technology can revolutionise the world in a broad range of aspects, including environment, safety, economy, and health. AMed Mg alloys and composite materials embrace an unseen era particularly in biomedical applications, though challenges remain with regards to scientific and technical domains. As an emerging industry, significant progress has been made and relevant knowledge paradigms have been established as a result of consistent and productive effort. There are a number of ways to push the boundaries of the existing understandings as articulated in the earlier section. The present chapter has sought to focus on aspects of AMed Mg alloys and composites through the lens of microstructure, processing settings, and corrosion management. It is also acknowledged that this chapter cannot be fully comprehensive owing to the emerging breadth of the field. However, the many exceptional works to date (and apologetically, the many exceptional works that are not included herein) have provided enough insight and concrete evidence that there is much promise in AMed Mg-based materials with respect to promising mechanical properties and high corrosion resistance. Furthermore, broader research into such a particular discipline has demonstrated progress at the frontiers of AMed Mg alloy and composite materials.

ACKNOWLEDGEMENTS

The authors are grateful for the financial support from the Australian Research Council (ARC) through the Linkage Scheme (LP150100343) and the Australian Commonwealth government and RMIT University through the RTP Fee Offset program (S.K.).

REFERENCES

Bar, F., Berger, L., Jauer, L., Kurtuldu, G., Schaublin, R., Schleifenbaum, J.H. & Loffler, J.F. 2019. Laser additive manufacturing of biodegradable magnesium alloy WE43: A detailed microstructure analysis. *Acta Biomaterialia*, 98, 36–49.
Chen, X.B., Birbilis, N. & Abbott, T.B. 2011a. Review of corrosion-resistant conversion Coatings for magnesium and its alloys. *Corrosion*, 67, 035005-1-035005-16.
Chen, X.B., Birbilis, N. & Abbott, T.B. 2011b. A simple route towards a hydroxyapatite – Mg(OH)$_2$ conversion coating for magnesium. *Corrosion Science*, 53, 2263–2268.
Chen, X.B., Nisbet, D.R., Li, R.W., Smith, P.N., Abbott, T.B., Easton, M.A., Zhang, D.H. & Birbilis, N. 2014. Controlling initial biodegradation of magnesium by a biocompatible strontium phosphate conversion coating. *Acta Biomaterialia*, 10, 1463–1474.
Deng, Q., Wu, Y., Su, N., Chang, Z., Chen, J., Peng, L. & Ding, W. 2021. Influence of friction stir processing and aging heat treatment on microstructure and mechanical properties of selective laser melted Mg-Gd-Zr alloy. *Additive Manufacturing*, 44, 102036.

Esmaily, M., Svensson, J.E., Fajardo, S., Birbilis, N., Frankel, G.S., Virtanen, S., Arrabal, R., Thomas, S. & Johansson, L.G. 2017. Fundamentals and advances in magnesium alloy corrosion. *Progress in Materials Science*, 89, 92–193.

Esmaily, M., Zeng, Z., Mortazavi, A.N., Gullino, A., Choudhary, S., Derra, T., Benn, F., D'elia, F., Müther, M., Thomas, S., Huang, A., Allanore, A., Kopp, A. & Birbilis, N. 2020. A detailed microstructural and corrosion analysis of magnesium alloy WE43 manufactured by selective laser melting. *Additive Manufacturing*, 35, 101321.

Farag, M.M. & Yun, H.-S. 2014. Effect of gelatin addition on fabrication of magnesium phosphate-based scaffolds prepared by additive manufacturing system. *Materials Letters*, 132, 111–115.

Fyfe, J.C., Meehl, G.A., England, M.H., Mann, M.E., Santer, B.D., Flato, G.M., Hawkins, E., Gillett, N.P., Xie, S.P., Kosaka, Y. & Swart, N.C. 2016. Making sense of the early-2000s warming slowdown. *Nature Climate Change*, 6, 224–228.

Gangireddy, S., Gwalani, B., Liu, K., Faierson, E.J. & Mishra, R.S. 2019. Microstructure and mechanical behavior of an additive manufactured (AM) WE43-Mg alloy. *Additive Manufacturing*, 26, 53–64.

Gieseke, M., Noelke, C., Kaierle, S., Wesling, V. & Haferkamp, H. 2016. Selective laser melting of magnesium and magnesium alloys. *In:* Hort, N., Mathaudhu, S.N., Neelameggham, N.R. & Alderman, M. (eds.) *Magnesium Technology 2013*, Cham, Springer International Publishing.

Guan, Y.C., Zhou, W., Zheng, H.Y. & Li, Z.L. 2010. Solidification microstructure of AZ91D Mg alloy after laser surface melting. *Applied Physics A*, 101, 339–344.

Guo, Y., Pan, H., Ren, L. & Quan, G. 2019. Microstructure and mechanical properties of wire arc additively manufactured AZ80M magnesium alloy. *Materials Letters*, 247, 4–6.

Hihara, L.H. & Kondepudi, P.K. 1993. The galvanic corrosion of SiC monofilament/ZE41 Mg metal-matrix composite in 0.5 M $NaNO_3$. *Corrosion Science*, 34, 1761–1767, 1769–1772.

Hihara, L.H. & Kondepudi, P.K. 1994. Galvanic corrosion between SiC monofilament and magnesium in NaCl, Na_2SO_4 and $NaNO_3$ solutions for application to metal-matrix composites. *Corrosion Science*, 36, 1585–1595.

Hyer, H., Zhou, L., Benson, G., Mcwilliams, B., Cho, K. & Sohn, Y. 2020. Additive manufacturing of dense WE43 Mg alloy by laser powder bed fusion. *Additive Manufacturing*, 33, 101123.

Kelly, J.C., Sullivan, J.L., Burnham, A. & Elgowainy, A. 2015. Impacts of vehicle weight reduction via material substitution on life-cycle greenhouse gas emissions. *Environmental Science & Technology*, 49, 12535–12542.

Khan, S.A., Bhuiyan, M.S., Miyashita, Y., Mutoh, Y. & Koike, T. 2011. Corrosion fatigue behavior of die-cast and shot-blasted AM60 magnesium alloy. *Materials Science and Engineering: A*, 528, 1961–1966.

Khan, S.A., Miyashita, Y., Mutoh, Y. & Koike, T. 2008. Fatigue behavior of anodized AM60 magnesium alloy under humid environment. *Materials Science and Engineering: A*, 498, 377–383.

Kotadia, H.R., Gibbons, G., Das, A. & Howes, P.D. 2021. A review of laser powder bed fusion additive manufacturing of aluminium alloys: Microstructure and properties. *Additive Manufacturing*, 46, 102155.

Kuah, K.X., Salehi, M., Ong, W.K., Seet, H.L., Nai, M.L.S., Wijesinghe, S. & Blackwood, D.J. 2022. Insights into the influence of oxide inclusions on corrosion performance of additive manufactured magnesium alloys. *NPJ Materials Degradation*, 6, 36.

Li, Y., Zhou, J., Pavanram, P., Leeflang, M.A., Fockaert, L.I., Pouran, B., Tümer, N., Schröder, K.U., Mol, J.M.C., Weinans, H., Jahr, H. & Zadpoor, A.A. 2018. Additively manufactured biodegradable porous magnesium. *Acta Biomaterialia*, 67, 378–392.

Liu, W., Cao, F., Xia, Y., Chang, L. & Zhang, J. 2014. Localized corrosion of magnesium alloys in NaCl solutions explored by scanning electrochemical microscopy in feedback mode. *Electrochimica Acta*, 132, 377–388.

Lu, Y., Bradshaw, A.R., Chiu, Y.L. & Jones, I.P. 2015. Effects of secondary phase and grain size on the corrosion of biodegradable Mg-Zn-Ca alloys. *Materials Science and Engineering: C*, 48, 480–486.

Market. 2023. Orthopedic implant market, forecast till 2031. *Straight Research*. https://straitsresearch.com/report/orthopedic-implants-market

Ng, C.C., Savalani, M.M., Lau, M.L. & Man, H.C. 2011. Microstructure and mechanical properties of selective laser melted magnesium. *Applied Surface Science*, 257, 7447–7454.

Niu, X., Shen, H. & Fu, J. 2018. Microstructure and mechanical properties of selective laser melted Mg-9 wt%Al powder mixture. *Materials Letters*, 221, 4–7.

Nunez-Lopez, C.A., Habazaki, H., Skeldon, P., Thompson, G.E., Karimzadeh, H., Lyon, P. & Wilks, T.E. 1996. An investigation of microgalvanic corrosion using a model magnesium-silicon carbide metal matrix composite. *Corrosion Science*, 38, 1721–1729.

Oliveira, J.P., Lalonde, A.D. & Ma, J. 2020. Processing parameters in laser powder bed fusion metal additive manufacturing. *Materials & Design*, 193, 108762.

Palanivel, S., Nelaturu, P., Glass, B. & Mishra, R.S. 2015. Friction stir additive manufacturing for high structural performance through microstructural control in an Mg based WE43 alloy. *Materials & Design*, 65, 934–952.

Polmear, I.J. 1995. *Light Alloys: Metallurgy of the Light Metals*, New Your, J. Wiley & Sons.

Salehi, M., Maleksaeedi, S., Nai, S.M.L., Meenashisundaram, G.K., Goh, M.H. & Gupta, M. 2019. A paradigm shift towards compositionally zero-sum binderless 3D printing of magnesium alloys via capillary-mediated bridging. *Acta Materialia*, 165, 294–306.

Salehi, M., Maleksaeedi, S., Sapari, M.A.B., Nai, M.L.S., Meenashisundaram, G.K. & Gupta, M. 2019. Additive manufacturing of magnesium – zinc – zirconium (ZK) alloys via capillary-mediated binderless three-dimensional printing. *Materials & Design*, 169, 107683.

Salehi, M., Seet, H.L., Gupta, M., Farnoush, H., Maleksaeedi, S. & Nai, M.L.S. 2021. Rapid densification of additive manufactured magnesium alloys via microwave sintering. *Additive Manufacturing*, 37, 101655.

Sealy, M.P., Karunakaran, R., Ortgies, S., Madireddy, G., Malshe, A.P. & Rajurkar, K.P. 2021. Reducing corrosion of additive manufactured magnesium alloys by interlayer ultrasonic peening. *CIRP Annals*, 70, 179–182.

Song, G.L. & Atrens, A. 1999. Corrosion mechanisms of magnesium alloys. *Advanced Engineering Materials*, 1, 11–33.

Song, G.-L. & Xu, Z. 2010. The surface, microstructure and corrosion of magnesium alloy AZ31 sheet. *Electrochimica Acta*, 55, 4148–4161.

Song, M.-S., Li, R.W., Qiu, Y., Man, S.M., Tuipulotu, D.E., Birbilis, N., Smith, P.N., Cole, I., Kaplan, D.L. & Chen, X.-B. 2022. Gallium – strontium phosphate conversion coatings for promoting infection prevention and biocompatibility of magnesium for orthopedic applications. *ACS Biomaterials Science & Engineering*, 8, 2709–2723.

Song, M.-S., Zeng, R.-C., Ding, Y.-F., Li, R.W., Easton, M., Cole, I., Birbilis, N. & Chen, X.-B. 2019. Recent advances in biodegradation controls over Mg alloys for bone fracture management: A review. *Journal of Materials Science & Technology*, 35, 535–544.

Stjohn, D.H., Qian, M., Easton, M.A., Cao, P. & Hildebrand, Z. 2005. Grain refinement of magnesium alloys. *Metallurgical and Materials Transactions A*, 36, 1669–1679.

Suchý, J., Klakurková, L., Man, O., Remešová, M., Horynová, M., Paloušek, D., Koutný, D., Krištofová, P., Vojtěch, D. & Čelko, L. 2021. Corrosion behaviour of WE43 magnesium alloy printed using selective laser melting in simulation body fluid solution. *Journal of Manufacturing Processes*, 69, 556–566.

Takagi, H., Sasahara, H., Abe, T., Sannomiya, H., Nishiyama, S., Ohta, S. & Nakamura, K. 2018. Material-property evaluation of magnesium alloys fabricated using wire-and-arc-based additive manufacturing. *Additive Manufacturing*, 24, 498–507.

Tiwari, S., Balasubramaniam, R. & Gupta, M. 2007. Corrosion behavior of SiC reinforced magnesium composites. *Corrosion Science*, 49, 711–725.

Wang, W., Song, M.-S., Yang, X.-N., Zhao, J., Cole, I.S., Chen, X.-B. & Fan, Y. 2020. Synergistic coating strategy combining photodynamic therapy and fluoride-free superhydrophobicity for eradicating bacterial adhesion and reinforcing corrosion protection. *ACS Applied Materials & Interfaces*, 12, 46862–46873.

Wu, R., Yan, Y., Wang, G., Murr, L.E., Han, W., Zhang, Z. & Zhang, M. 2015. Recent progress in magnesium-lithium alloys. *International Materials Reviews*, 60, 65–100.

Xiang, C., Gupta, N., Coelho, P. & Cho, K. 2018. Effect of microstructure on tensile and compressive behavior of WE43 alloy in as cast and heat treated conditions. *Materials Science and Engineering: A*, 710, 74–85.

Yan, C., Xin, Y., Chen, X.B., Xu, D., Chu, P.K., Liu, C., Guan, B., Huang, X. & Liu, Q. 2021. Evading strength-corrosion tradeoff in Mg alloys via dense ultrafine twins. *Nature Communications*, 12, 4616.

Yang, H.Y., Chen, X.B., Guo, X.W., Wu, G.H., Ding, W.J. & Birbilis, N. 2012. Coating pretreatment for Mg alloy AZ91D. *Applied Surface Science*, 258, 5472–5481.

Zeng, Z., Salehi, M., Kopp, A., Xu, S., Esmaily, M. & Birbilis, N. 2022. Recent progress and perspectives in additive manufacturing of magnesium alloys. *Journal of Magnesium and Alloys*, 10, 1511–1541.

Zumdick, N.A., Jauer, L., Kersting, L.C., Kutz, T.N., Schleifenbaum, J.H. & Zander, D. 2019. Additive manufactured WE43 magnesium: A comparative study of the microstructure and mechanical properties with those of powder extruded and as-cast WE43. *Materials Characterization*, 147, 384–397.

6 Carbon Nanotubes-Reinforced Magnesium Metal-Matrix Composites

Song-Jeng Huang and Sathiyalingam Kannaiyan

CONTENTS

6.1 INTRODUCTION

The automotive and aerospace industries have been doing more research to improve vehicles' performance in recent years, and this has been an increasing trend. There have been many challenges that researchers have encountered in order to find a solution to enhance the ratio between high strength and light weight. As composites get lighter and stronger, it is expected that they will perform better under a variety of testing conditions and in a wide range of engineering applications. Composites that are lighter and stronger will perform better under diverse testing conditions. A magnesium alloy is lighter than high-strength steels and aluminum alloys among various advanced materials (Prasad et al. 2022). The weight of magnesium alloys is approximately 33% less than that of aluminum alloys and 75% lighter than that of steel alloys. Due to their excellent castability, machinability, and damping capacity, magnesium-based alloys have gained much attention. However, the low strength, poor ductility, and toughness and weak aqueous corrosion resistance limit their widespread applications (Meher et al. 2022; Song-jeng, Sathiyalingam, and Murugan 2022).

DOI: 10.1201/9781003319856-8

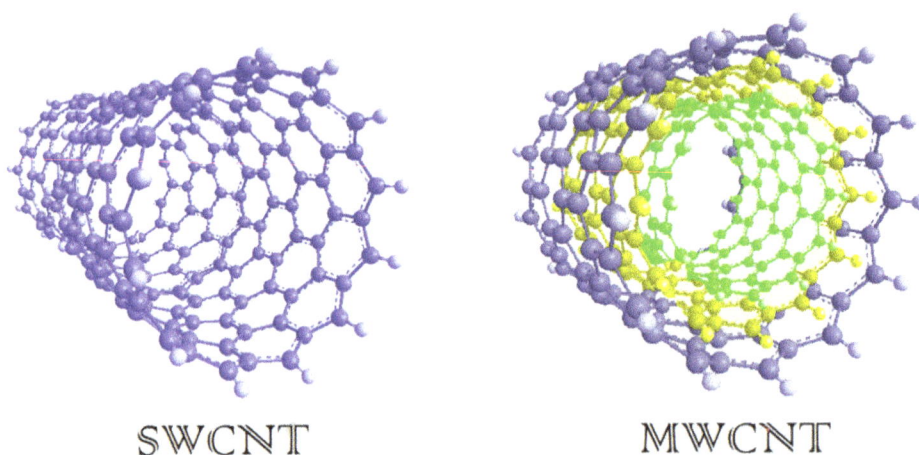

The addition of nanoparticles to magnesium-matrix composites significantly improves the mechanical properties of the composites and improves their toughness, promoting the development of magnesium-matrix composites as a whole (Yang et al. 2021). The advantages of nanoscale reinforcements are mainly attributed to their superior mechanical properties as well as to the unique characteristics of their interface with the matrix, which allows them to achieve outstanding performance when compared to conventional composite materials (Huang, Subramani, and Chiang 2021). A number of nanocarbon materials have been studied as potential reinforcement materials for metal-matrix composites (MMCs) due to their outstanding mechanical properties and include carbon nanofibers (CNFs), carbon nanotubes (CNTs), and graphene nanoplatelets (GNPs) (S. R. Bakshi, Lahiri, and Agarwal 2010; Ogawa and Masuda 2021).

As a result of their high specific strength, high specific stiffness, and low density, CNTs are a very viable candidate to be used in Mg MMCs. The versatility of CNTs, their sp^2 hybridization, and their ability to align their atoms into rope-like structures make them the most widely explored nanostructured materials (Radhamani, Lau, and Ramakrishna 2018)(Abbas and Huang 2020). **Figure 6.1** shows the two different morphologies of CNTs, single-walled (SWCNT) and multi-walled (MWCNT) carbon nanotubes.

6.2 FABRICATION METHODS OF MG-CNTS COMPOSITES

There are five different fabrication processes used to develop Mg-CNT composites. Based on the processing method used by previous researchers, fabrication methods can be classified into solid-state, liquid-state, and others as discussed in what follows (Shirvanimoghaddam et al. 2017; Upadhyay et al. 2022).

6.2.1 POWDER METALLURGY

The use of powder metallurgy (PM) is usually considered one of the easiest and most cost-effective routes for fabricating composites based on magnesium and CNT when it comes to solid-state processing methods. The mixture of Mg and CNT powders is obtained by ball milling, and milled composite powder compacted, sintered, and extruded. As part of further sintering, a number of other processes are also applied in order to achieve more refined characteristics, including hot

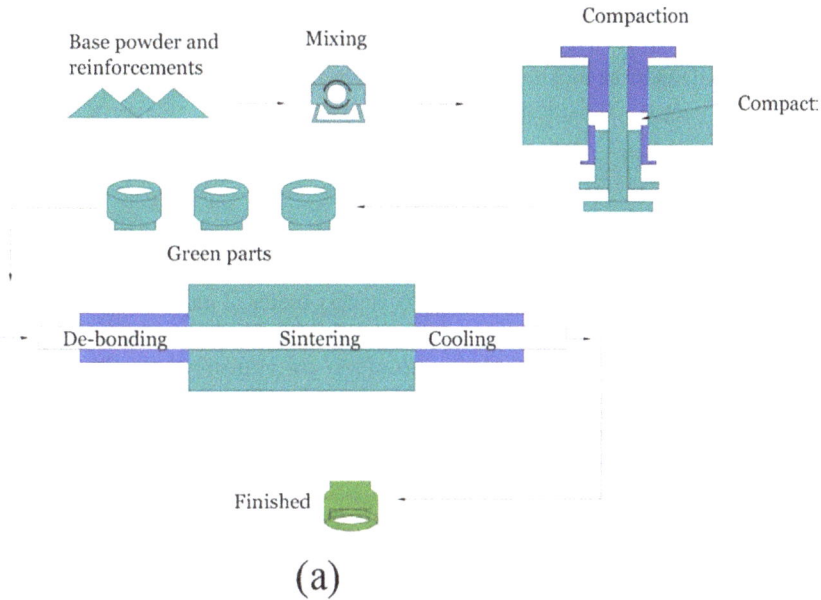

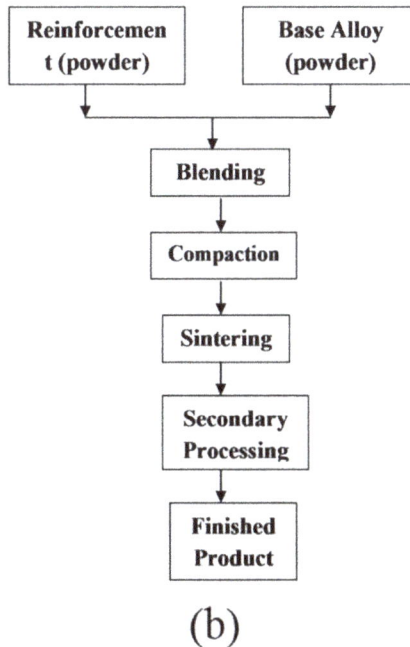

FIGURE 6.2 Schematic of powder metallurgy. *Reproduced with permission from (Thakur 2021) © 2021 The Author(s), under exclusive license to Springer Nature Singapore Pte Ltd.*

pressing, cold pressing, and spark plasma sintering, for example. The schematic representation of powder metallurgy process is shown in **Figure 6.2**. A high-energy ball milling (HEBM) setup was employed in the development of Mg/CNT nanocomposites that were shielded by an argon gas atmosphere. Stearic acid was added to the ball milling process as a controlling agent. CNTs were uniformly dispersed within the magnesium matrix due to the process, and subsequently, the uniform mixture was subjected to hot press sintering. Nanocomposite showed superior strength and malleability in mechanical testing. The significance of powder metallurgy processing is homogeneous dispersion, interfacial reaction, and bonding with matrix of the composite (Rashad et al. 2015; Ding et al. 2020; Upadhyay et al. 2022; Shirvanimoghaddam et al. 2017).

6.2.2   SEMI-POWDER METALLURGY

During ball milling, CNTs lose their geometry and structure, thus lessening their strength. The semi-powder metallurgy (SPM) technology was developed to solve these issues. Unlike in ball milling, where planetary clustering is used to mix matrix material and reinforcing agent, in SPM, the matrix material and reinforcing agent are mixed with a liquid solvent. The paste-like composite powders with CNTs were dried and heated in an argon gas atmosphere to thermally decompose the residual surfactant elements from the powder surface. In addition to improved microstructures due to functional groups associated with CNTs, there was an increase in nucleation sites and an enhancement in the strength of the mechanisms, which are responsible for load transfer. As a result of the studies, scientists found that SPM and extrusion were more effective methods of producing Mg-CNT composites (Shirvanimoghaddam et al. 2017; Fukuda et al. 2011; Rashad et al. 2014).

6.2.3   FRICTION STIR PROCESSING

FSP is a solid-state processing method that is primarily applied to the development of surface composites in which the matrix material does not have to be melted, making it a promising method for developing matrix-free surface composites. The process and schematic of friction stir processing is shown in **Figure 6.3**. In the FSP process, preliminary work consists of preparing a groove in the base metal as the first step. In the next step, the groove is filled with reinforcements using rotating tools that are non-consumable. Then, the tool-containing pin is stirred to ensure that the reinforcement is dispersed throughout the matrix. As a result of FSP, grain refinement occurs in the processed region, altering properties that are structure dependent (Azizieh, Kokabi, and Abachi 2011; Morisada et al. 2006; Sharma et al. 2019).

6.2.4   MELT STIRRING

A Mg-CNT MMC can be developed using a stir casting method, as shown in **Figure 6.4**. Here, temperature was gradually increased with a 100°C increment after a 15-minute stabilizing period. In order to avoid burning, carbon dioxide (CO_2) and sulfur fluoride (SF6) were used at 400°C, while argon was used at 700°C in order to prevent oxidation. The molten slurry was stirred at 300 rpm for 5 minutes after stabilizing at 760°C for 30 minutes. A steel ingot was placed in a lower chamber to receive the mixture. However, stirring can easily lead to CNTs agglomeration and oxidation of molten magnesium. In addition to these problems, mechanical agitation also destroys the unique crystallographic properties of CNTs, resulting in poor mechanical properties (Abbas et al. 2020).

6.2.5   SQUEEZE STIRRING

As part of the squeeze casting process (**Figure 6.5**), the mold needs to be pre-heated, and the matrix needs to melt, then the matrix is poured into the mold, and then the mold has to be compressed. A preform of reinforcement phase (CNT) is placed into the lower fixed mold half. This involves placing the preform in the die and applying specific amounts of the molten matrix. Once the die has been pressed, a high vacuum pressure of 70 to 100 MPa is applied under it. A smaller, highly dense CF-MMC part can be made using this technique. A squeeze casting process could eliminate porosity and shrinkage, improve mechanical properties, and shape articles almost perfectly. High pressure is usually used when squeeze casting is undertaken in order to achieve the maximum

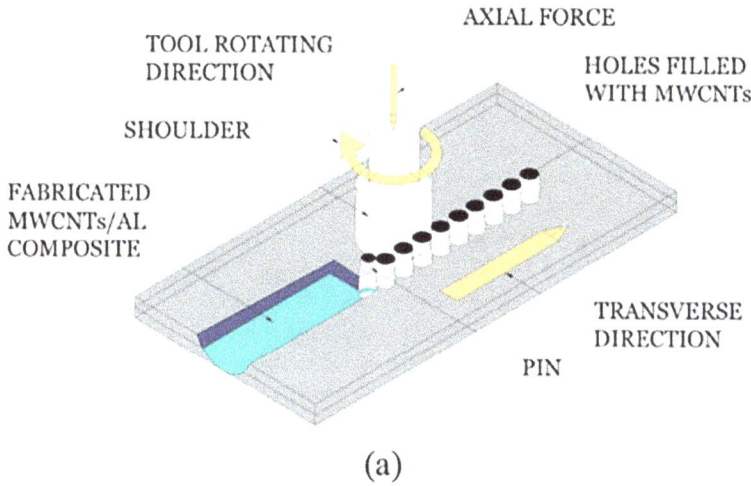

(a)

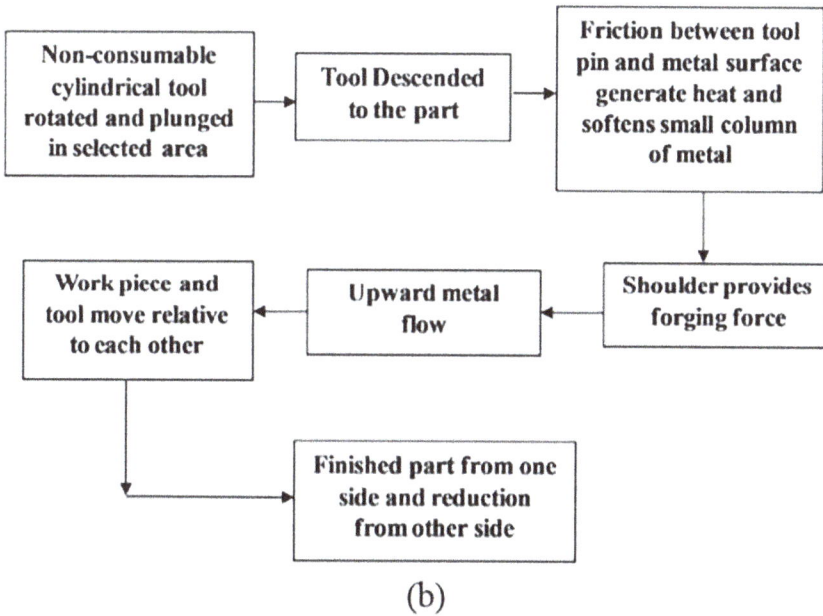

(b)

FIGURE 6.3 Schematic of friction stir process. *Reproduced with permission from (Thakur 2021) © 2021 The Author(s), under exclusive license to Springer Nature Singapore Pte Ltd.*

amount of infiltration into the preforms, which could damage them (Shirvanimoghaddam et al. 2017; Upadhyay et al. 2022).

6.2.6 Deposition of Disintegrated Melt

The schematic of deposition of disintegrated melt method is shown in Figure.6.6. A graphite crucible was used to melt the magnesium alloy and then to superheat it at 750°C. In order to add the CNT to the molten magnesium, the vibratory feeder was used and an axial flow impeller was used to stir the molten metal. It is important to position the impeller centrally during the mixing of CNT and magnesium. A period of 0 to 15 minutes was taken to superheat the composite melt after adding the CNT reinforcement. After mixing, the melt was poured from the bottom of the crucible. During

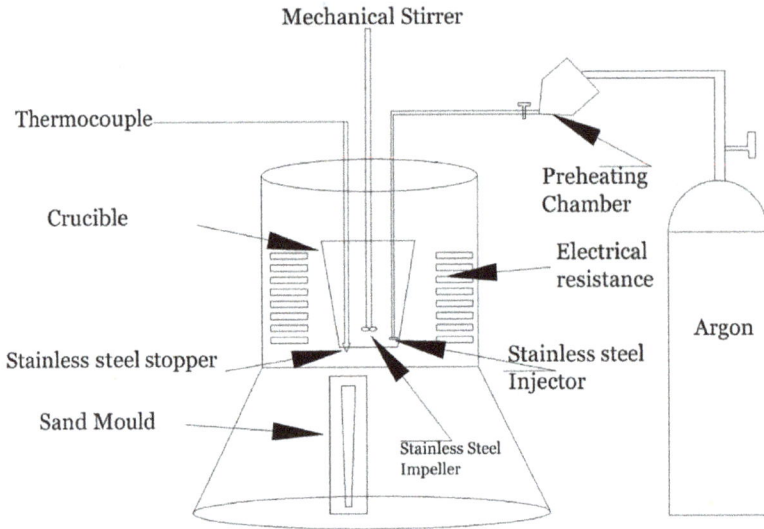

(a)

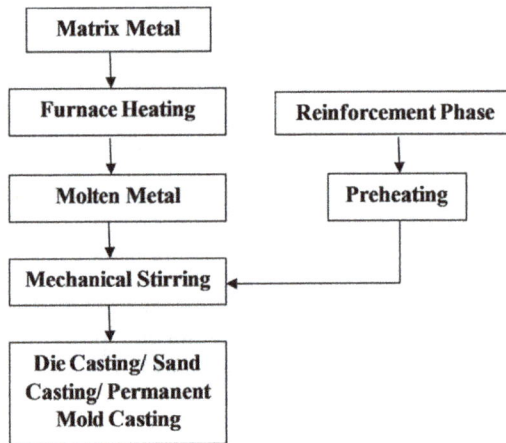

(b)

FIGURE 6.4 Schematic of stir casting. *Reproduced with permission from (Thakur 2021) © 2021 The Author(s), under exclusive license to Springer Nature Singapore Pte Ltd.*

the pouring process, stirring continued. As a result of jets of argon gas pointing normal to the melt stream, the composite melt stream was disintegrated. Metallic substrates were deposited with the disintegrated composite melt (Tham, Gupta, and Cheng 1999; Upadhyay et al. 2022).

6.2.7 Cold Gas Dynamic Spraying

Cold gas dynamic spraying (CGDS) is one of the most advanced methods of material processing that makes use of the self-consolidation of particles in order to achieve superior results. High-velocity impact induces self-consolidation, while solid particles remain solid while they are joined. Metallic or composite powder is often accelerated by using high-pressure gas, resulting in

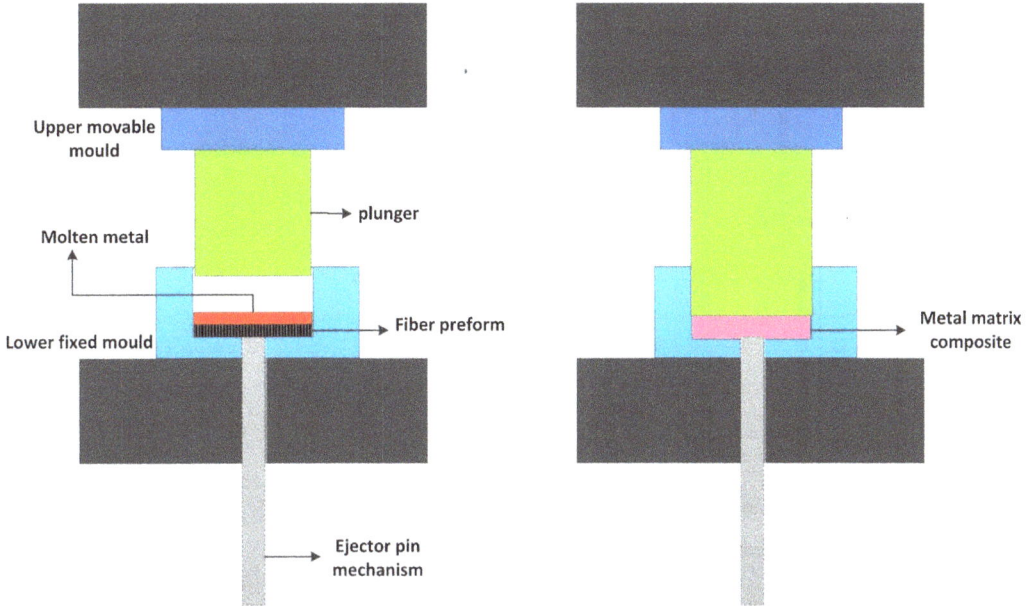

FIGURE 6.5 Schematic of squeeze casting. *Reproduced with permission from (Shirvanimoghaddam et al. 2017) © 2016 Elsevier Ltd.*

high-speed collisions, which deposit onto a substrate. A CGDS process utilizes the difference in pressure between the nozzle's inlet and outlet. As shown in **Figure 6.7**, the nozzle design significantly impacts the velocity of gas flow. This pressure difference is significant because it affects the velocity of gas flow. In a CGDS process, particles have adhered to the substrate and the layer the particles are deposited on. With CGDS, the dispersion on the surface of the substrate material is homogeneously distributed (Srinivasa R. Bakshi et al. 2008; Zhang et al. 2020).

6.3 STRENGTHENING MECHANISM

6.3.1 LOAD TRANSFER

Mg-composite is formed when the matrix material, reinforced by a CNT material, and a reinforcing agent are interfacial bonded together in a way that allows the matrix material to transfer stress from the reinforcing agent to the matrix material. It is therefore expected that the increased interfacial stress level in the metallic matrix will result in a reduction in stress level in the metallic matrix as the load is efficiently transferred from the elastically softer metal matrix to the strengthened CNT (Chen et al. 2017; P. Li et al. 2021; Upadhyay et al. 2022).

6.3.2 OROWAN MECHANISM

In composite materials, plastic deformation exceeds the limit of their yield strength. The Orowan mechanism is responsible for the plastic deformation of composites caused by the crystallographic movement of dislocations. In the presence of dispersed CNTs in the Mg matrix, loops of dislocation begin to form as the integrated agents slip, bend, and sidestep according to the Orowan mechanism. Because the residual loops emerged sequentially, the composite was strengthened as the reinforcing agents were dispersed within the metal matrix. Consequently, CNTs have been found to act as

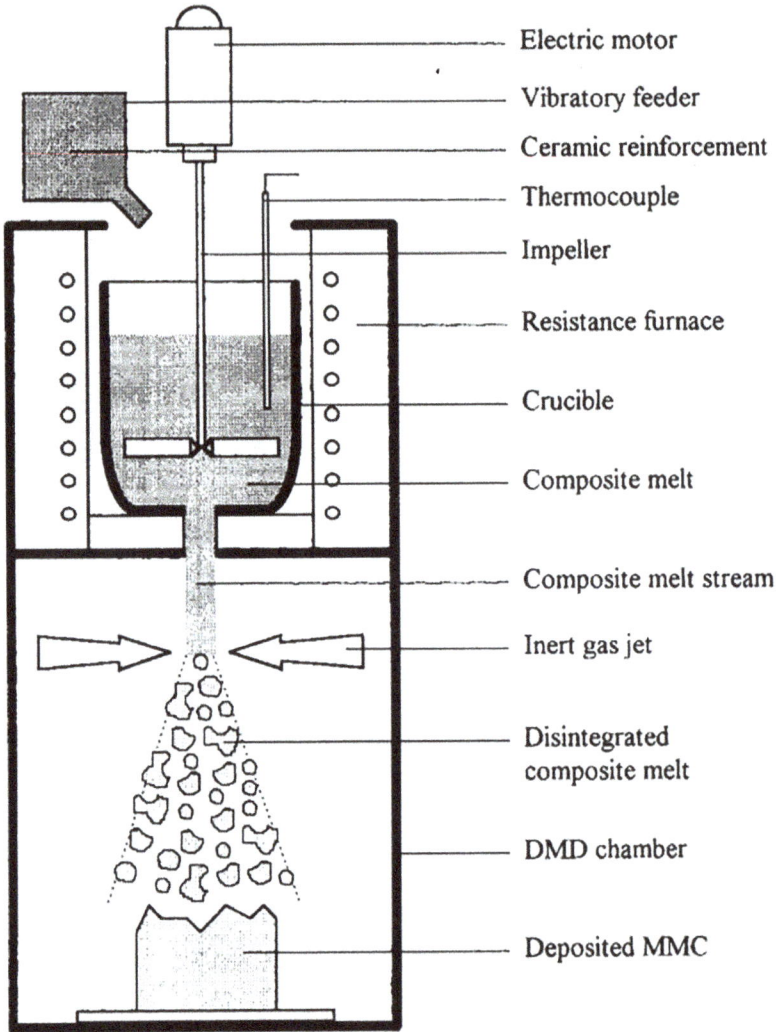

FIGURE 6.6 Deposition of disintegrated melt. *Reproduced with permission from (Tham, Gupta, and Cheng 1999) © 1999 IoM Communications Ltd, Rights managed by Taylor & Francis.*

impenetrable barriers, much like precipitates that are non-shearable, and this may lead to higher yield stress through an Orowan mechanism. In addition, there are no specific studies at the moment that are examining the interfacial processes at play during the interaction of a dislocation with a CNT. There is also uncertainty as to whether Orowan loops actually are required for dislocation motion or whether plastic deformation or fracture of the CNT will accommodate the strain caused by shearing of the metal around the CNT, leading to a quasi-cutting mechanism of dislocation movement (Qianqian Li et al. 2009; Upadhyay et al. 2022).

6.3.3 THERMAL INCONGRUITY

As a result of dense dislocations surrounding the reinforcing agent in the reinforcement matrix, thermal incongruity occurs. Apparently, this is because of the fact that there is a significant difference in coefficients of thermal expansion between the reinforcement materials and the matrix materials, and this may have caused this discrepancy. The difference in coefficient of thermal expansion

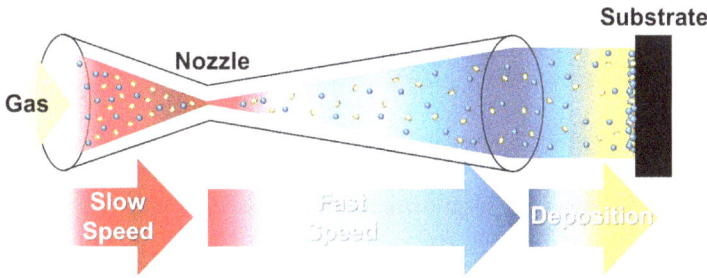

FIGURE 6.7 Schematic of cold gas dynamic spraying. *Reproduced with permission from (Yoo et al. 2023) © 2022 Elsevier Ltd.*

(CTE) between the reinforcing agent and matrix results in thermal and internal stress (Q. Li et al. 2009; Upadhyay et al. 2022).

6.3.4 GRAIN BOUNDARY STRENGTHENING

Magnesium/CNT-reinforced composites are strengthened by strengthening their grain boundaries. As a result of grain refinement, the composite's compressive strength is enhanced. Experimental investigations have also shown that grain boundary strengthening occurs when the grain size is refined during composite material fabrication. As CNTs have a high aspect ratio, they can be used to fabricate composites containing finite voids to lock dislocations even with very low weight percentages of CNT, and so composite mechanical properties are improved (Q. Li et al. 2009; Upadhyay et al. 2022).

6.4 ANTI-CORROSION PROPERTIES

The susceptibility of Mg and its alloy/composites to aqueous and galvanic corrosion is the major obstacle for their mega-scale use in the industry (Aydin et al. 2020; Thostenson and Chou 2004; Namilae and Chandra 2006).

The corrosion behavior is generally measured by immersion weight-loss and electrochemical methods in NaCl corrosive medium. In an aqueous solution, monolithic magnesium undergoes the following reaction (Q. Li et al. 2012):

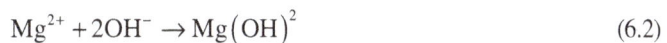

$$Mg \rightarrow Mg^{2+} + 2e^- \tag{6.1}$$

$$Mg^{2+} + 2OH^- \rightarrow Mg(OH)_2 \tag{6.2}$$

$Mg(OH)^2$ converts into $MgCl_2$ salt in the presence of Cl^- ions, which is the key factor of corrosion enhancement of magnesium. The following reaction shows the magnesium dissolution:

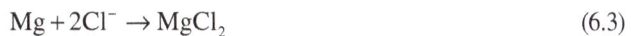

$$Mg + 2Cl^- \rightarrow MgCl_2 \tag{6.3}$$

Generally, the particulate reinforcement of ceramics and carbon fiber categories leads to fissure and pitting corrosion at the interface of matrix and reinforcement (Wu et al. 2017). The incorporation of CNT as a reinforcement promotes galvanic corrosion behavior due to its higher electrode potential than magnesium. In addition, the intermetallic β-phase such as $Mg_{17}Al_{12}$, Al-Mn, and Al-Mn-Fe act as cathode terminals and promote micro-galvanic corrosion. If we talk about some of the different fabrication routes, the following outcomes were observed so far. Magnesium reinforced with MWCNT by friction stir welding showed a reduction in anti-corrosion properties due to galvanic coupling. It had been observed in the powder metallurgy process, the corrosion property degraded with the increase in CNT weight percentage.

The AZ category of magnesium alloy shows almost similar characteristics when reinforced with CNT. For instance, in the case of AZ31, a 0.7 V positive shift was observed when 0.89 wt.% of CNT was added, which denotes the progression in galvanic corrosion (Hamdy, Alfosail, and Gasem 2013). Similarly, reinforced AZ61 and AZ91 show approximately the same corrosion behavior in 3.5% NaCl solution up to 0.2 wt.% of CNT reinforcement. Further addition of CNT decreases the corrosion resistance, as seen in Figure 6.8. The figure concludes that the addition of CNT (0.1 or 0.2 wt.%) has better corrosion resistance. The study conducted so far showed that the addition of CNT above 2 wt.%, in the AZ series of alloy worsens the corrosion resistance (Fukuda et al. 2011).

The adaptation of surface coating methods like plasma electrolytic oxidation (PEO) and polycaprolactone (PCL) coating can restrict the corrosion initiation (Upadhyay et al. 2022). The corrosion behavior of AZ31B-H24 magnesium alloy reinforced with MWCNTs composite coated with PEO/PCL materials was explored. As a result of applying a thick and thin PCL layer on top of the MWCNTs-PEO coating, it was observed that the corrosion performance was significantly improved (Daavari, Atapour, Mohedano, Sánchez et al. 2021). Mg/MWCNT composites were fabricated by the PM method and subsequently subjected to PEO. The study reported a significant decrease in corrosion rates, ~1.6 times those for the uncoated composite (Aydin et al. 2020). The incorporation of MWCNTs into the PEO coating was helpful to increase the roughness, create a thicker inner barrier layer, and reduce hydrophilicity (Daavari, Atapour, Mohedano, Arrabal et al. 2021).

6.5   WEAR AND FRICTION PROPERTIES

The weak tribological characteristics of Mg like low hardness and wear resistance are also a serious concern. However, the tribological properties can be significantly improved by proper selection of the fabrication procedure and the addition of optimum reinforcing agents. However, a greater extent of reinforcing agent could cause clustering, which could act as plowing agents, and that can enhance the abrasion (Tjong 2013; Jamshidijam et al. 2013). **Figure 6.9** evidences the positive effect of the CNT reinforcing agent on the tribological property of Mg MMC. The improvement in wear characteristics observed after being reinforced with CNT is due to several phenomena. Firstly, the addition of reinforcement up to CNT has shown a reduction in grain size, and an increase in hardness which implies an extraordinary reduction in wear-resistant characteristics. Secondly, the self-lubricating character of CNT reduces the coefficient of friction. Depending on the thickness of the oxidation layer that develops on the composites, the wear-out characteristics of the composite will be significantly influenced. During the wear cycle of the composites, the wear rate was lowered because the surfaces could not slide due to the thicker oxidation layer. In the presence of CNTs, the anti-wear properties and coefficient of friction of the composites were drastically improved (Abbas et al. 2020).

6.6   OTHER CARBON NANOMATERIALS AS REINFORCEMENTS

Various other carbon nanostructures including carbon fibers (C_f), graphene nanoplatelets (GNP), graphene nanosheets (GNS), etc. are also used as reinforcements for multifunctional MMNCs. A stir-casting method for fabricating graphene nanoplatelets-AZ31 composites with improved high-temperature formability was reported by (Rashad et al. 2016). Wang et al. fabricated composites using a pressureless infiltration process with 45 vol.% continuous carbon fibers (C_f) in a magnesium matrix. C_f and Mg matrix showed excellent interfacial bonding with no visible debonding or microcracks. Due to the interfacial reaction, the volume expansion in the interfacial layer could result in large compressive stress. It led to nanostructured defects at the edges, such as dislocations

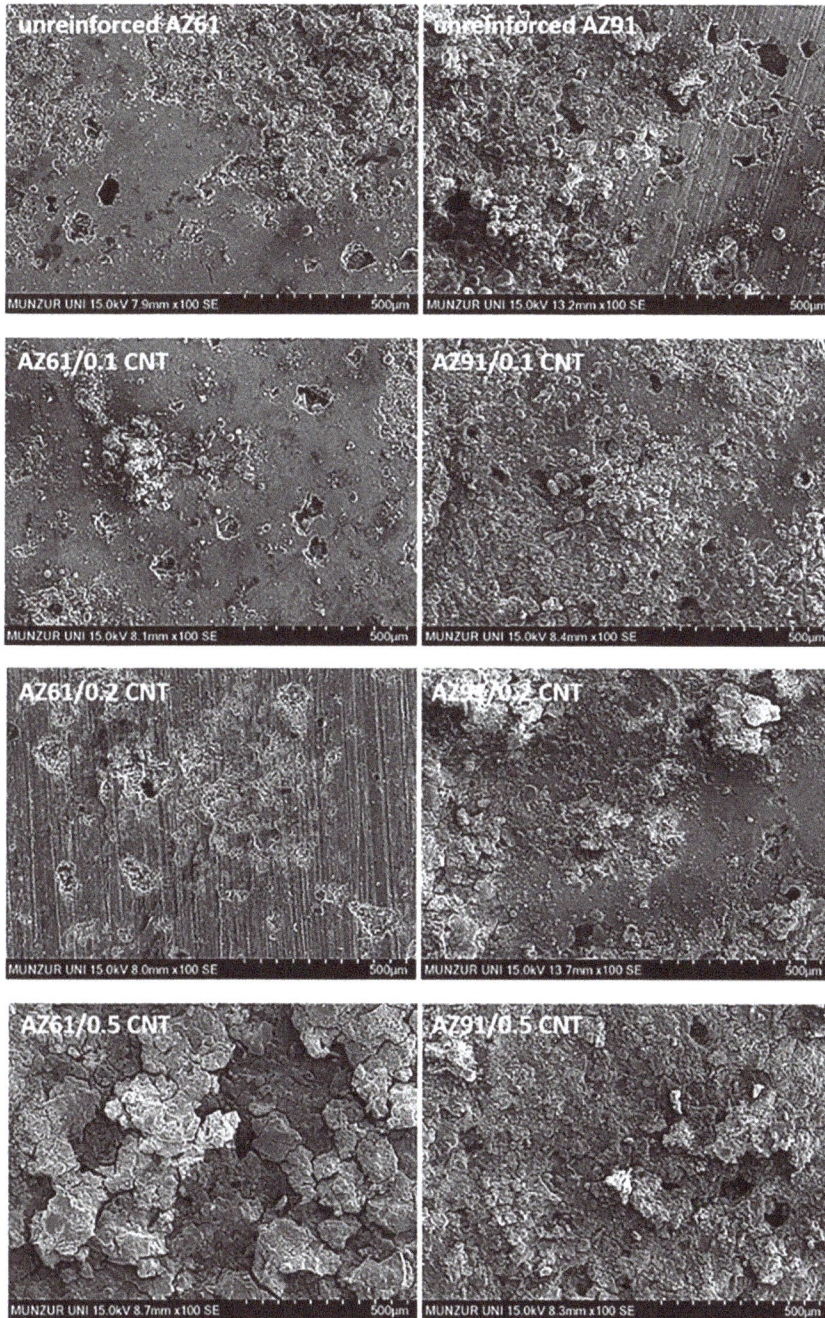

FIGURE 6.8 The surface SEM images of the unreinforced alloys and composites reinforced with different CNT ratio after the corrosion test in 3.5% NaCl (at two different magnifications). *Reproduced with permission from* (Say, Guler, and Dikici 2020) © *2020 Elsevier B.V.*

(Wang, Xiao, and Ma 2012). In recent years, GNPs/Mg composite has replaced CNT/Mg composite because GNPs are easily dispersed in many different types of solvents and matrix. The reinforcement distribution in the matrix alloy was dispersed homogeneously and evenly with the help of PM and melt stirring followed by hot extrusion. As a result of the addition of GNP to the Mg (such as

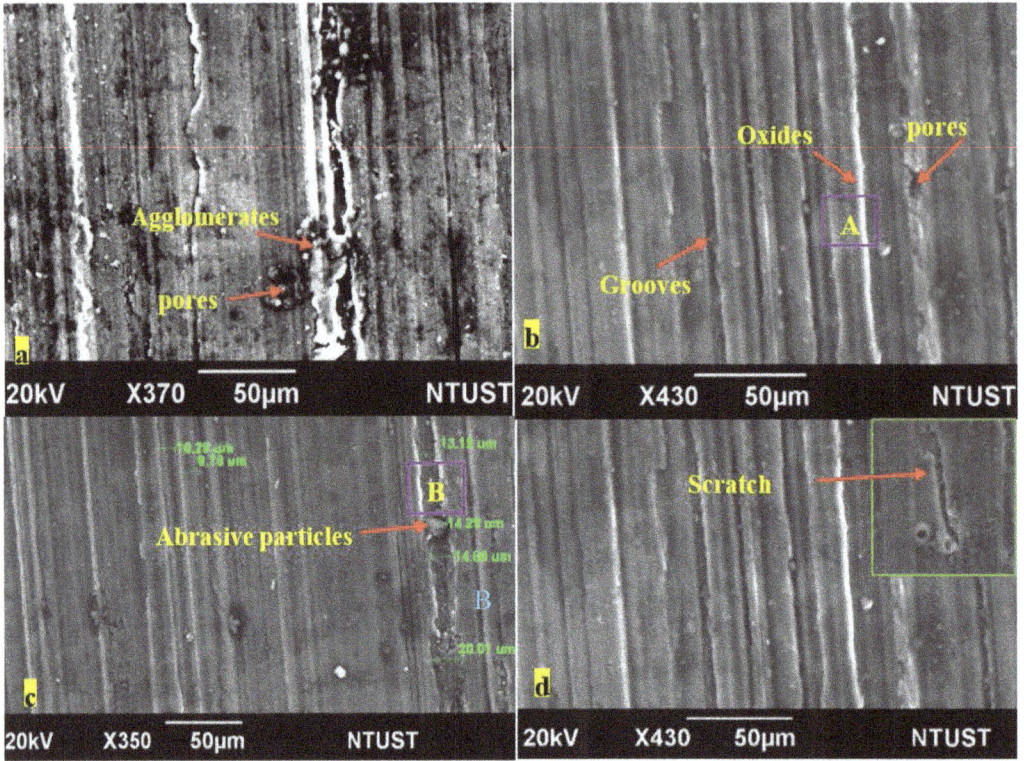

FIGURE 6.9 Pictorial observations of external worn-out structure of AZ31-based CNT-reinforced composites through SEM composed with CNTs wt.% of (a) 1.0, (b) 0.5, (c) 0.1, and (d) 0, in dry circumstances. *Reproduced with permission from (Abbas et al. 2020) © 2019 Elsevier Ltd.*

ZK 60 alloy) matrix, thermal conductivity, compressive strength, and ductility were significantly increased (Guler and Bagci 2020).

6.7 CONCLUSIONS AND OUTLOOK

- The properties of CNTs are remarkable in terms of their mechanical as well as thermal properties. Because of these characteristics, they are considered ideal materials for many applications. Nevertheless, their use is limited to a few industries due to disadvantages, including various impurities, uneven morphology, and a lack of precision characterization techniques.
- Due to the agglomeration of CNTs in composites reinforced with CNTs, their properties are significantly reduced. There is still a need for further research to identify the optimal content of CNT reinforcement in order to avoid this phenomenon.
- As a simple and inexpensive technique, melt processing such as stir casting is the most promising process for preparing CNT-reinforced nanocomposites. In order to prepare CNT-reinforced nanocomposites on a large-scale using melt processing, it is necessary to enhance the mixing, wettability, and dispersion of CNTs in the molten matrix.
- MMCs reinforced with CNTs have highly enhanced mechanical, thermal, and tribological properties. On the other hand, there hasn't been much discussion of how to evaluate the fatigue qualities and corrosion resistance of these materials.

- Fabrication routes can have a significant effect on the coefficient of friction and wear resistance of composites depending on how they are manufactured. With an increase in CNT content in the magnesium matrix, self-lubrication properties, mechanical characteristics, and wear resistance could be enhanced.
- An optimum CNT-incorporated composite could provide significantly enhanced corrosion resistance. However, microscale galvanic corrosion could occur with higher CNTs embedded in magnesium-matrix composites in severe aggressive solutions. The adaptation of various surface coating methods like plasma electrolytic oxidation (PEO), and polycaprolactone (PCL) coating can restrict the corrosion initiation.

REFERENCES

Abazari, Somayeh, Ali Shamsipur, Hamid Reza Bakhsheshi-Rad, Ahmad Fauzi Ismail, Safian Sharif, Mahmood Razzaghi, Seeram Ramakrishna, and Filippo Berto. 2020. "Carbon Nanotubes (CNTs)-Reinforced Magnesium-Based Matrix Composites: A Comprehensive Review." *Materials* 13 (19): 1–38. https://doi.org/10.3390/ma13194421.

Abbas, Aqeel, and Song-Jeng Huang. 2020. "Qualitative and Quantitative Investigation of As-Cast and Aged CNT/AZ31 Metal Matrix Composites." *The Journal of The Minerals, Metals & Materials Society (TMS)* 72 (6): 2272–2282. https://doi.org/10.1007/s11837-020-04114-7.

Abbas, Aqeel, Song-Jeng Huang, Beáta Ballóková, and Katarína Sülleiová. 2020. "Tribological Effects of Carbon Nanotubes on Magnesium Alloy AZ31 and Analyzing Aging Effects on CNTs/AZ31 Composites Fabricated by Stir Casting Process." *Tribology International* 142 (February). https://doi.org/10.1016/j.triboint.2019.105982.

Aydin, Fatih, Aysun Ayday, M. Emre Turan, and Hüseyin Zengin. 2020. "Role of Graphene Additive on Wear and Electrochemical Corrosion Behaviour of Plasma Electrolytic Oxidation (PEO) Coatings on Mg – MWCNT Nanocomposite." *Surface Engineering* 36 (8): 791–799. https://doi.org/10.1080/02670844.2019.1689640.

Azizieh, M., A. H. Kokabi, and P. Abachi. 2011. "Effect of Rotational Speed and Probe Profile on Microstructure and Hardness of AZ31/Al2O3 Nanocomposites Fabricated by Friction Stir Processing." *Materials and Design* 32 (4): 2034–2041. https://doi.org/10.1016/j.matdes.2010.11.055.

Bakshi, S. R., D. Lahiri, and A. Agarwal. 2010. "Carbon Nanotube Reinforced Metal Matrix Composites – A Review." *International Materials Reviews* 55 (1): 41–64. https://doi.org/10.1179/095066009X12572530170543.

Bakshi, S. R., V. Singh, K. Balani, D. Graham McCartney, S. Seal, and A. Agarwal. 2008. "Carbon Nanotube Reinforced Aluminum Composite Coating via Cold Spraying." *Surface and Coatings Technology* 202 (21): 5162–5169. https://doi.org/10.1016/j.surfcoat.2008.05.042.

Chen, B., J. Shen, X. Ye, L. Jia, S. Li, J. Umeda, M. Takahashi, and K. Kondoh. 2017. "Length Effect of Carbon Nanotubes on the Strengthening Mechanisms in Metal Matrix Composites." *Acta Materialia* 140: 317–325. https://doi.org/10.1016/j.actamat.2017.08.048.

Daavari, Morteza, Masoud Atapour, Marta Mohedano, Raul Arrabal, Endzhe Matykina, and Aboozar Taherizadeh. 2021. "Biotribology and Biocorrosion of MWCNTs-Reinforced PEO Coating on AZ31B Mg Alloy." *Surfaces and Interfaces* 22 (November 2020): 100850. https://doi.org/10.1016/j.surfin.2020.100850.

Daavari, Morteza, Masoud Atapour, Marta Mohedano, Hugo Mora Sánchez, Juan Rodríguez-Hernández, Endzhe Matykina, Raul Arrabal, and Aboozar Taherizadeh. 2021. "Quasi-in Vivo Corrosion Behavior of AZ31B Mg Alloy with Hybrid MWCNTs-PEO/PCL Based Coatings." *Journal of Magnesium and Alloys* (xxxx). https://doi.org/10.1016/j.jma.2021.09.010.

Ding, Yunpeng, Jilei Xu, Jinbiao Hu, Qiancheng Gao, Xiaoqin Guo, Rui Zhang, and Linan An. 2020. "High Performance Carbon Nanotube-Reinforced Magnesium Nanocomposite." *Materials Science and Engineering A* 771 (October 2019): 138575. https://doi.org/10.1016/j.msea.2019.138575.

Fukuda, Hiroyuki, Katsuyoshi Kondoh, Junko Umeda, and Bunshi Fugetsu. 2011. "Fabrication of Magnesium Based Composites Reinforced with Carbon Nanotubes Having Superior Mechanical Properties." *Materials Chemistry and Physics* 127 (3): 451–458. https://doi.org/10.1016/j.matchemphys.2011.02.036.

Guler, Omer, and Nihal Bagci. 2020. "A Short Review on Mechanical Properties of Graphene Reinforced Metal Matrix Composites." *Journal of Materials Research and Technology* 9 (3): 6808–6833. https://doi.org/10.1016/j.jmrt.2020.01.077.

Hamdy, Abdel Salam, Feras Alfosail, and Zuhair Gasem. 2013. "Electrochemical Behavior of a Discontinuously A6092/SiC/17.5p Metal Matrix Composite in Chloride Containing Solution." *Electrochimica Acta* 88: 129–134. https://doi.org/10.1016/j.electacta.2012.10.079.

Huang, Song-Jeng, Murugan Subramani, and Chao-Ching Chiang. 2021. "Effect of Hybrid Reinforcement on Microstructure and Mechanical Properties of AZ61 Magnesium Alloy Processed by Stir Casting Method." *Composites Communications* 25 (April): 100772. https://doi.org/10.1016/j.coco.2021.100772.

Jamshidijam, Mahdiyeh, Ali Akbari-Fakhrabadi, Seyyed Morteza Masoudpanah, Gholam Heydar Hasani, and Ramalinga V. Mangalaraja. 2013. "Wear Behavior of Multiwalled Carbon Nanotube/AZ31 Composite Obtained by Friction Stir Processing." *Tribology Transactions* 56 (5): 827–832. https://doi.org/10.1080/10402004.2013.804969.

Li, Q., M. C. Turhan, C. A. Rottmair, R. F. Singer, and S. Virtanen. 2012. "Influence of MWCNT Dispersion on Corrosion Behaviour of Their Mg Composites." *Materials and Corrosion* 63 (5): 384–387. https://doi.org/10.1002/maco.201006023.

Li, Qianqian, Andreas Viereckl, Christian A. Rottmair, and Robert F. Singer. 2009. "Improved Processing of Carbon Nanotube/Magnesium Alloy Composites." *Composites Science and Technology* 69 (7–8): 1193–1199. https://doi.org/10.1016/j.compscitech.2009.02.020.

Li, Pubo, Wanting Tan, Mangmang Gao, and Keren Shi. 2021. "Strengthening of the Magnesium Matrix Composites Hybrid Reinforced by Chemically Oxidized Carbon Nanotubes and in Situ Mg2Sip." *Journal of Alloys and Compounds* 858: 157673. https://doi.org/10.1016/j.jallcom.2020.157673.

Meher, Arabinda, Manas Mohan Mahapatra, Priyaranjan Samal, and Pandu R. Vundavilli. 2022. "A Review on Manufacturability of Magnesium Matrix Composites: Processing, Tribology, Joining, and Machining." *CIRP Journal of Manufacturing Science and Technology* 39: 134–158. https://doi.org/10.1016/j.cirpj.2022.07.012.

Morisada, Y., H. Fujii, T. Nagaoka, and M. Fukusumi. 2006. "MWCNTs/AZ31 Surface Composites Fabricated by Friction Stir Processing." *Materials Science and Engineering A* 419 (1–2): 344–348. https://doi.org/10.1016/j.msea.2006.01.016.

Namilae, S., and N. Chandra. 2006. "Role of Atomic Scale Interfaces in the Compressive Behavior of Carbon Nanotubes in Composites." *Composites Science and Technology* 66 (13): 2030–2038. https://doi.org/10.1016/j.compscitech.2006.01.009.

Ogawa, Fumio, and Chitoshi Masuda. 2021. "Fabrication and the Mechanical and Physical Properties of Nanocarbon-Reinforced Light Metal Matrix Composites: A Review and Future Directions." *Materials Science and Engineering A* 820 (June): 141542. https://doi.org/10.1016/j.msea.2021.141542.

Prasad, S. V. Satya, S. B. Prasad, Kartikey Verma, Raghvendra Kumar Mishra, Vikas Kumar, and Subhash Singh. 2022. "The Role and Significance of Magnesium in Modern Day Research-A Review." *Journal of Magnesium and Alloys* 10 (1): 1–61. https://doi.org/10.1016/j.jma.2021.05.012.

Radhamani, A. V., Hon Chung Lau, and S. Ramakrishna. 2018. "CNT-Reinforced Metal and Steel Nanocomposites: A Comprehensive Assessment of Progress and Future Directions." *Composites Part A: Applied Science and Manufacturing* 114 (August): 170–187. https://doi.org/10.1016/j.compositesa.2018.08.010.

Rashad, Muhammad, Fusheng Pan, Yanglu Liu, Xianhua Chen, Han Lin, Rongjian Pan, Muhammad Asif, and Jia She. 2016. "High Temperature Formability of Graphene Nanoplatelets-AZ31 Composites Fabricated by Stir-Casting Method." *Journal of Magnesium and Alloys* 4 (4): 270–277. https://doi.org/10.1016/j.jma.2016.11.003.

Rashad, Muhammad, Fusheng Pan, Aitao Tang, Muhammad Asif, and Muhammad Aamir. 2014. "Synergetic Effect of Graphene Nanoplatelets (GNPs) and Multi-Walled Carbon Nanotube (MW-CNTs) on Mechanical Properties of Pure Magnesium." *Journal of Alloys and Compounds* 603: 111–118. https://doi.org/10.1016/j.jallcom.2014.03.038.

Rashad, Muhammad, Fusheng Pan, Jianyue Zhang, and Muhammad Asif. 2015. "Use of High Energy Ball Milling to Study the Role of Graphene Nanoplatelets and Carbon Nanotubes Reinforced Magnesium Alloy." *Journal of Alloys and Compounds* 646: 223–232. https://doi.org/10.1016/j.jallcom.2015.06.051.

Say, Yakup, Omer Guler, and Burak Dikici. 2020. "Carbon Nanotube (CNT) Reinforced Magnesium Matrix Composites: The Effect of CNT Ratio on Their Mechanical Properties and Corrosion Resistance." *Materials Science and Engineering A* 798 (June): 139636. https://doi.org/10.1016/j.msea.2020.139636.

Sharma, Sanjay, Amit Handa, Sahib Sartaj Singh, and Deepak Verma. 2019. "Influence of Tool Rotation Speeds on Mechanical and Morphological Properties of Friction Stir Processed Nano Hybrid Composite of MWCNT-Graphene-AZ31 Magnesium." *Journal of Magnesium and Alloys* 7 (3): 487–500. https://doi.org/10.1016/j.jma.2019.07.001.

Shirvanimoghaddam, Kamyar, Salah U. Hamim, Mohammad Karbalaei Akbari, Seyed Mousa Fakhrhoseini, Hamid Khayyam, Amir Hossein Pakseresht, Ehsan Ghasali, et al. 2017. "Carbon Fiber Reinforced Metal Matrix Composites: Fabrication Processes and Properties." *Composites Part A: Applied Science and Manufacturing* 92: 70–96. https://doi.org/10.1016/j.compositesa.2016.10.032.

Song-Jeng, Huang, Kannaiyan Sathiyalingam, and Subramani Murugan. 2022. "Effect of Nano-Nb 2 O 5 on the Microstructure and Mechanical Properties of AZ31 Alloy Matrix Nanocomposites." *Advances in Nano Research, an International Journal* 4: 407–416. https://doi.org/10.12989/anr.2022.13.4.407.

Thakur, Vijay Kumar. 2021. "Materials Horizons: From Nature to Nanomaterials Series Editor, Springer." https://doi.org/https://doi.org/10.1007/978-981-33-4550-8.

Tham, L. M., M. Gupta, and L. Cheng. 1999. "Influence of Processing Parameters during Disintegrated Melt Deposition Processing on Near Net Shape Synthesis of Aluminium Based Metal Matrix Composites." *Materials Science and Technology* 15 (10): 1139–1146. https://doi.org/10.1179/026708399101505185.

Thostenson, Erik T., and Tsu Wei Chou. 2004. "Nanotube Buckling in Aligned Multi-Wall Carbon Nanotube Composites." *Carbon* 42 (14): 3015–3018. https://doi.org/10.1016/j.carbon.2004.06.012.

Tjong, Sie Chin. 2013. "Recent Progress in the Development and Properties of Novel Metal Matrix Nanocomposites Reinforced with Carbon Nanotubes and Graphene Nanosheets." *Materials Science and Engineering R: Reports* 74 (10): 281–350. https://doi.org/10.1016/j.mser.2013.08.001.

Upadhyay, Gaurav, Kuldeep K. Saxena, Shankar Sehgal, Kahtan A. Mohammed, Chander Prakash, Saurav Dixit, and Dharam Buddhi. 2022. "Development of Carbon Nanotube (CNT)-Reinforced Mg Alloys: Fabrication Routes and Mechanical Properties." *Metals* 12 (8). https://doi.org/10.3390/met12081392.

Wang, W. G., B. L. Xiao, and Z. Y. Ma. 2012. "Evolution of Interfacial Nanostructures and Stress States in Mg Matrix Composites Reinforced with Coated Continuous Carbon Fibers." *Composites Science and Technology* 72 (2): 152–158. https://doi.org/10.1016/j.compscitech.2011.10.008.

Wu, Liqun, Ruizhi Wu, Legan Hou, Jinghuai Zhang, Jianfeng Sun, and Milin Zhang. 2017. "Microstructure and Mechanical Properties of CNT-Reinforced AZ31 Matrix Composites Prepared Using Hot-Press Sintering." *Journal of Materials Engineering and Performance* 26 (11): 5495–5500. https://doi.org/10.1007/s11665-017-2971-5.

Yang, Yan, Xiaoming Xiong, Jing Chen, Xiaodong Peng, Daolun Chen, and Fusheng Pan. 2021. "Research Advances in Magnesium and Magnesium Alloys Worldwide in 2020." *Journal of Magnesium and Alloys* 9 (3): 705–747. https://doi.org/10.1016/j.jma.2021.04.001.

Yoo, Sung Chan, Dongju Lee, Seong Woo Ryu, Byungchul Kang, Ho Jin Ryu, and Soon Hyung Hong. 2023. "Recent Progress in Low-Dimensional Nanomaterials Filled Multifunctional Metal Matrix Nanocomposites." *Progress in Materials Science* 132 (November 2020): 101034. https://doi.org/10.1016/j.pmatsci.2022.101034.

Zhang, Yingpeng, Qun Wang, Gang Chen, and Chidambaram Seshadri Ramachandran. 2020. "Mechanical, Tribological and Corrosion Physiognomies of CNT-Al Metal Matrix Composite (MMC) Coatings Deposited by Cold Gas Dynamic Spray (CGDS) Process." *Surface and Coatings Technology* 403 (September): 126380. https://doi.org/10.1016/j.surfcoat.2020.126380.

7 Magnesium Metal-Matrix Composites Reinforced with Rare Earth Elements

G. Vedabouriswaran and S. Aravindan

CONTENTS

7.1 INTRODUCTION

Mg alloys and composites are highly influential in the fabrication of lightweight structural members to aircraft passenger seats, helicopter gearbox housing to medical applications such as knee bracelets, and laptop casing to military and surveillance gadgets. Mg alloys having low density than Al alloys becomes a natural selection/choice over Al alloys and hence plays a vital role in engineering and biomedical applications. Nevertheless, in biomedical implants, where biocompatibility and biodegradability are required, Mg alloys with rare earth (RE) elements prove to be the finest choice in orthopedic applications. While Mg alloys are developed to the demand of such a wide variety of applications, there are specific engineering applications wherein tailoring of mechanical properties such as increased hardness, improved wear and corrosion resistance are in demand. Such specific requirements necessitate the development of Mg composites by introducing hard ceramic particles whereby the developed composite exhibits phenomenal increase in hardness, wear and corrosion capabilities is possible. Hence the development of Mg composites has gained attraction since such Mg metal-matrix composites (MMCs) also cater to various industrial applications requiring a high strength-to-weight ratio.

Hence there arises an immense requirement in the development of Mg composites which are need-based requirements of various industrial applications. The choice of selection of hard ceramic particles, their size (micro/nano-sized) becomes the major factors to be considered in the development of MMCs. Nevertheless, Mg alloys and Mg composites altogether remain as two different major choices to be considered during selection of a prospective material that could satisfy need-based engineering applications.

The ASTM standards classify magnesium (Mg) alloys as (i) Mg-Mn (M), (ii) Mg-Al-Mn (AM), (iii) Mg-Al-Zn-Mn (AZ), (iv) Mg-Zr (K), (v) Mg-Zn-Zr (ZK), with RE (ZE), (vi) Mg-RE-Zr (EZ), (vii) Mg-Ag-RE-Zr (QE), (viii) Mg-Y-Zr (WE), (ix) Mg-Zn-Cu-Mn (ZC), (x) Mg-Al-Si-Mn (AS) and (xi) Mg-Al-Sr (AJ). The physical properties of Mg alloys are influenced by the alloying elements and their chemical composition. Table 7.1 gives an idea about the various alloying elements and their influences and effects achievable while alloying with Mg.

Addition of these elements results in achieving specific mechanical properties in Mg alloys. These elements form intermetallics and helps in improved hardness, corrosion, strengthening, grain

TABLE 7.1
Influence and Effects of Alloying Elements in Mg (Vedabouriswaran 2020)

Element	Influences and effects
Al	Improves the strength and hardness of the alloy and increases the melting range
Be	Decreases surface oxidation during melting/casting
Ca	Increases grain refinement and reduces oxidation. Improves the rolling characteristics of the alloy and creep resistance. Mg alloyed with Zn and Ca finds its usage in biodegradable medical devices.
Ce	Improves the corrosion resistance and plastic deformation ability. Increases the work hardening.
Cu	Improves the strength characteristics at elevated and room temperature
Fe	A harmful impurity of the alloy, which reduces the corrosion resistance
Li	Increases the ductility and formability, reduces the strength
Mn	Increases the saltwater corrosion capabilities
Ni	Increases the yield strength as well as ultimate strength but affects the ductility and corrosion resistance
Nd	Improves the material strength
Rare earth metals	Increases the high temperature creep resistance, corrosion resistance and strength. Improves the castability by reducing the freezing range and hence porosity reduces. Lessens weld cracking. Rare earth metals are added in the form of misch metal (mixture of Ce, La and Nd).
Si	Increases the fluidity of the melt
Sr	Used in conjunction with other elements to increase the creep resistance
Ag	Increases the mechanical properties
Th	Increases the creep strength at elevated temperature, reduces the formation of cracking during forging
Y	Enhances high temperature strength and creep
Zn	Most common alloying element next to Al. Increases the fluidity of the melt and improves corrosion resistance. In conjunction with Zr and rare earth metals, the strength and the precipitation hardening improves.
Zr	Grain refiner in conjunction with Zn and other rare earth metals

refining. Figure 7.1 gives information about various Mg alloys and the effect of various alloying elements.

7.2   Mg METAL-MATRIX COMPOSITES WITH RE ELEMENTS

Various combinations of Mg metal-matrix composites are possible by changing the alloying element, and thereby, developing a specific Mg composite can cater to the specific needs. Development of the magnesium metal-matrix composites by alloying with RE element can be majorly classified under (a) binary system, (b) ternary system and (c) higher alloying systems.

The binary system of Mg alloys comprises addition of RE elements, namely yttrium, cerium, gadolinium, lanthanum, erbium, neodymium, dysprosium individually, and hence Mg binary alloys were developed were as follows.

1 Mg-Y
   Alloying of Mg with yttrium (Y) has been carried out by researchers. Strengthening in Mg-Y alloy is due to solid solution strengthening and precipitation hardening The atomic radii of Mg is 145 pm and Y has an atomic radii of 212 pm. Due to the large difference in the atomic radii of the alloying element, such strengthening prevails.

2 Mg-Ce
   Addition of cerium (Ce) in Mg refines the grain structure and improves fire-retarding capacity.

Mg-Al-Mn
Mg-Al-Zn

Mg-Al

Mg-Al-Ca
Mg-Al-Ca-RE
Mg-Al-Si
Mg-Al-Sr
Mg-Al-RE

Mg-Al-Cu-Mn

Mg-Y-RE

Mg-Zn-RE
Mg-Zn-Zr
Mg-Zn-Cu
Mg-Zn-Th
Mg-Zn-Mn
Mg-Zn-Al-Ca
Mg-Zn-Ca

Mg + Al
↑ Strength
↑ Hardness
↑ Castability

Mg + Y, Mg + RE, Mg + Zr
↑ Creep resistance
↑ Corrosion resistance
↑ Grain refinement

12 **Mg** Magnesium 24.3050

Mg-RE-Zn
Mg-RE-Ag
Mg-Nd-Gd

Mg-RE

Mg + RE
↑ Creep resistance

Mg + As, Mg + Ge
↑ Strength

Mg-As Mg-Ge

Mg + Zn
↑ Strength

Mg-Zn

Mg + Ca
↑ Biocompatibility

Mg-Ca

Mg-Ca-Ag
Mg-Ca-RE

Mg + Th
↑ Creep resistance
↑ Castability
↑ Radioactive

Mg-Th

Mg-Th-Zr
Mg-Th-Zn
Mg-Th-Mn

Mg + Li
↑ Ductility
↓ Strength
↓ Density

Mg + Ag
↑ Hardness
↑ Creep resistance
↑ Biocompatability

Mg-Ag

Mg-Ag-Th
Mg-Ag-RE

Mg-Li

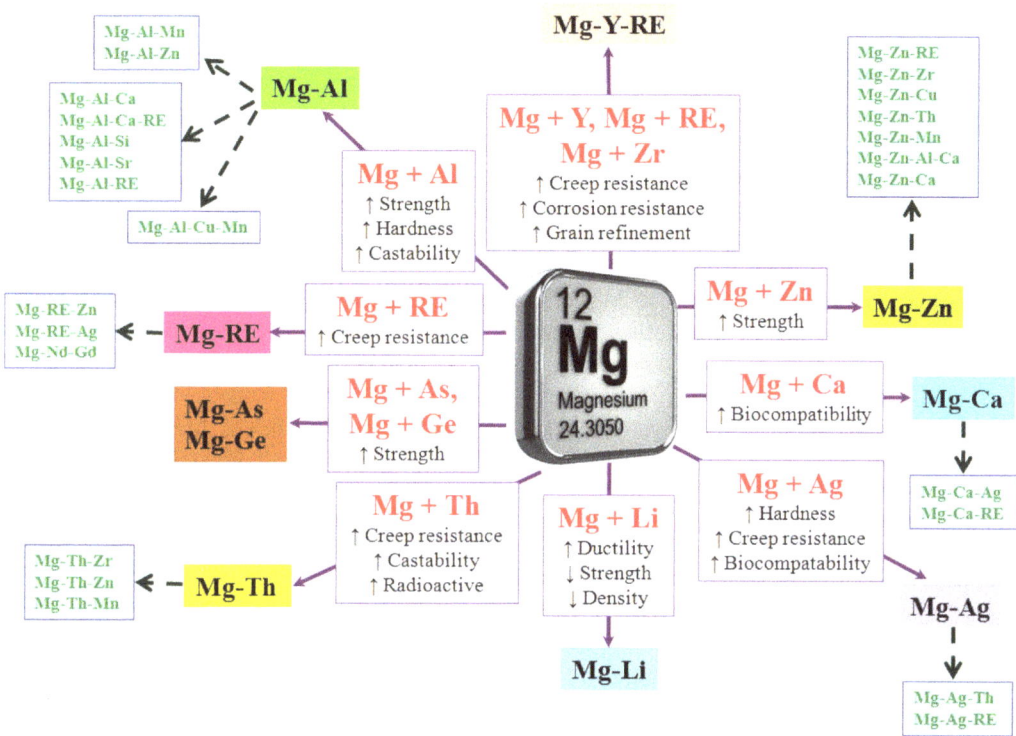

FIGURE 7.1   Various Mg alloys developed by alloying (Vedabouriswaran 2020).

3 Mg-Gd

The solubility of gadolinium (Gd) in Mg is around 23.4 wt.% (at eutectic temperature) and hence helps in the solid solution strengthening of Mg when alloyed. Researchers have reported that Mg-Gd alloys demonstrate better elongation to fracture upon comparison with other metal implant materials such as stainless steel (Hort et al. 2010).

4 Mg-La

Lanthanum (La) exhibits limited solid solubility with a high eutectic temperature of 612°C. Owing to this poor solubility of La in Mg, the Mg-La alloys do not undergo age hardening.

5 Mg-Er

The rare earth element that has a good solubility in Mg is erbium (Er).

6 Mg-Nd

Due to high solubility of neodymium (Nd) in Mg, it exhibits the highest solid solubility in Mg with lowest eutectic temperature (552°C) and exhibits best response to age hardening when alloyed with Mg. Formation of Mg_3Nd phase, a very hard phase, contributes for the better strength of Nd-containing Mg alloys than Ce- and La-containing Mg alloys upon comparison (Hort et al. 2010) (Chia et al. 2009).

7 Mg-Dy

Dysprosium (Dy) exhibits a high solid solubility in Mg. The melting point of the $Mg_{24}Dy_5$ intermetallic is 560°C. Addition of Dy improves the room-temperature mechanical properties of the Mg alloy by solid solution strengthening and precipitation hardening.

The ductility of Mg-Dy alloy decreases with the increase in Dy content, and hence, the alloy has very poor ductility at room temperature. Mg-10Dy alloys are developed for biomedical applications such as bone fixtures due to their low ductility and corrosion properties, whereas in the ternary system, several compositions of the alloying elements, specifically the RE elements, being the major alloying elements, are developed. Such developed alloys were referred to as Mg-RE based ternary systems, such as (i) Mg-Al-RE, (ii) Mg-Zn-RE, (iii) Mg-Zr-RE and (iv) Mg-Sn-RE.

In the case of higher alloy systems, more than three elements were alloyed. The alloying elements in such type of higher alloy systems could be RE elements, Al, Zr, Zn, Li, Mn, Sn, V, Ca and Cu.

Based on research on Mg-RE systems, among the binary system, it can be said that Mg-Y alloys demonstrated better strength characteristics, whereas Mg-Er alloys demonstrates better ductility. With respect to Mg ternary alloy systems, Mg-Zn-RE showcases higher strength and ductility than other ternary systems. While in the higher alloy systems, Mg-Zn-RE based high alloy system exhibited better strength and ductility (Tekumalla et al. 2015).

Various Mg ternary alloy systems available are, (i) Mg-RE-Zn, (ii) Mg-RE-Ag, (iii) Mg-Nd-Gd, (iv) Mg-Y-RE, (v) Mg-Ag-Th, (vi) Mg-Ag-RE, (vii) Mg-Ca-RE, (vii) Mg-Al-RE.

Bulk and surface composites of Mg alloys containing RE elements are developed by various researchers. In surface composites, the top surface of the Mg alloy is subjected to surface modification through friction stir processing (FSP); thereby, a variety of surface composites are developed. The surface processing of the alloy helps in tailoring the surface properties of the material while the bulk remains unaltered. Ceramic reinforcements are introduced into the stir zone and dispersed without any agglomeration of the particles. Such surface composites developed can be used in reducing sliding wear of the mating components, as the case depth of the developed composites is sufficiently of higher thickness (approx. 3 to 5 mm is achievable).

Friction stir processing methodology being a solid-state processing of materials, RZ5 alloy was chosen for the development of MMCs by introducing hard ceramic particles at the stir zone. The developed surface composites containing hard ceramic particles exhibited improved mechanical and corrosion characteristics. During FSP, different hard ceramic particles were introduced in the stir zone of RZ5 alloy. The hard particles introduced for the development of surface MMCs were (i) B_4C, (ii) multi-walled carbon nanotubes (MWCNT) and (iii) a mixture of zirconia and alumina particles (80:20 ratio) thereby three different types of surface MMCs were developed. The particles introduced at the stir zone were found to be dispersed uniformly. Such developed composites are named MMC-A (which contains B_4C particles), MMC-B (which contains MWCNT particles) and MMC-C (which contains a mixture of zirconia and alumina). The developed MMCs possessed refined grains at the stir zone.

Introduction of these hard ceramic particles contributed to the increased microhardness of the processed RZ5 alloy. The base material (RZ5) had a microhardness of 81 HV, while the MMCs, namely MMC-A, MMC-B and MMC-C demonstrated substantial increase in the microhardness, and the microhardnesses were 403 HV, 125 HV and 293 HV respectively. Phenomenal increase in the hardness of the composites is directly correlated to the presence of hard reinforcement particles. Hence selective increase in the microhardness of the RZ5 alloy is feasible by FSP, and FSP proved to be a viable method for development of surface MMCs. The introduced hard particles dispersed uniformly, and the friction stir processed zones of all MMCs are found to be defect free. Added advantage of FSP was refinement of microstructure. The coarse grain of RZ5 alloy refined due to dynamic recrystallization. Figure 7.2 (a) and (b) gives an idea of the grain sizes before and after FSP of RZ5 Mg alloy (Vedabouriswaran 2020).

Improved hardness of the material is influential in wear behavior of the MMC. Hence understanding the sliding wear behavior of MMC is essential. Dry sliding wear characterization studies on all the MMCs revealed different wear mechanisms at various loads. The wear pins of all the MMCs were subjected to a sliding wear test (pin-on-disk setup) at various normal loads that ranged between 10N and 75N. The coefficient of friction of the tribo pair was monitored for 1000 m of

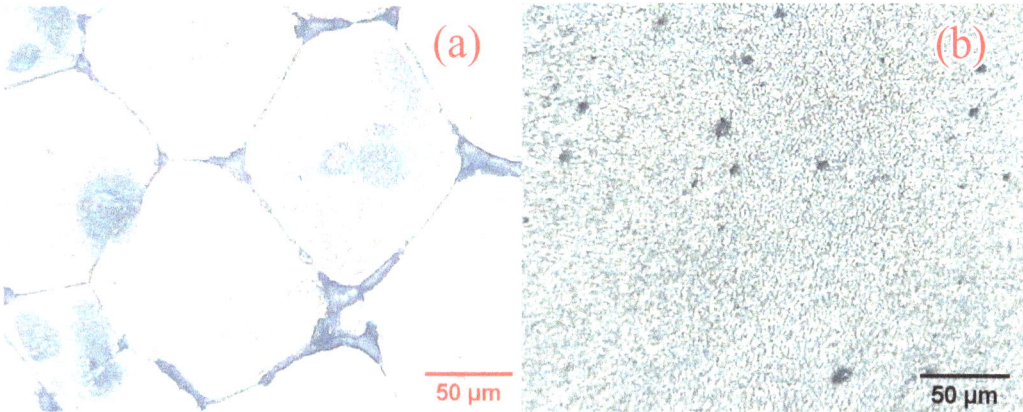

FIGURE 7.2 (a) Microstructures of RZ5 (a) before FSP and (b) after FSP (Vedabouriswaran 2020).

TABLE 7.2
Wear Rate and Wear Resistance of RZ5 and Other MMCs (Vedabouriswaran 2020)

	Applied normal load							
	10 N		25 N		50 N		75 N	
Name of the specimen	Wear rate mm³/m	Wear resistance Nm/mm³	Wear rate mm³/m	Wear resistance Nm/mm³	Wear rate mm³/m	Wear resistance Nm/mm³	Wear rate mm³/m	Wear resistance Nm/mm³
BM	0.0365	273.80	0.0529	472.27	0.0629	795.15	0.0659	1137.67
MMC-A	0.0193	518.51	0.0330	758.51	0.0404	1237.37	0.0473	1585.76
MMC-B	0.0240	417.37	0.0404	618.71	0.0422	1185.31	0.0539	1391.24
MMC-C	0.0187	535.27	0.0297	842.26	0.0437	1143.45	0.0500	1500

sliding distance. Wear pins having different particles such as B_4C, MWCNT and a mixture of zirconia and alumina exhibited different and distinctive coefficients of friction under various normal load than the RZ5 alloy (base material). Abrasion, adhesion and oxidative wear mechanisms were identified during the conduct-of-wear test. Table 7.2 gives an idea about the wear rate of the MMC upon comparison with the base material (RZ5) (Vedabouriswaran 2020).

Composites exhibiting good corrosion resistance are essential in such application. Here mitigation of corrosion is essential. Chemical, oil and refinery process industries and marine applications require material to be resistant to corrosion, and hence composites developed should be sufficiently resistant to corrosion. Surface RZ5 MMC hence developed was also subjected to corrosion studies.

Immersion corrosion testing of all MMCs was also conducted. The MMCs were immersed in a 3.5% NaCl solution for 24 hours. Pitting corrosion occurred during the immersion test. The MMCs exhibited an enhanced corrosion resistance than the RZ5 alloy, which is essentially due to finer grains and presence of hard particles. RZ5 (base material) exhibited severe pitting corrosion than the metal-matrix composites. Potentiodynamic polarization test was conducted on all the samples for the estimation of corrosion rate. Tafel extrapolation (Figure 7.3) gave the estimate of corrosion current density (i_{corr}) and potential (E_{corr}) for all the samples, and corrosion rates were estimated. MMCs exhibited improved corrosion resistance, and hence less corrosion rate prevailed compared to the RZ5 alloy (base material). (Vedabouriswaran 2020).

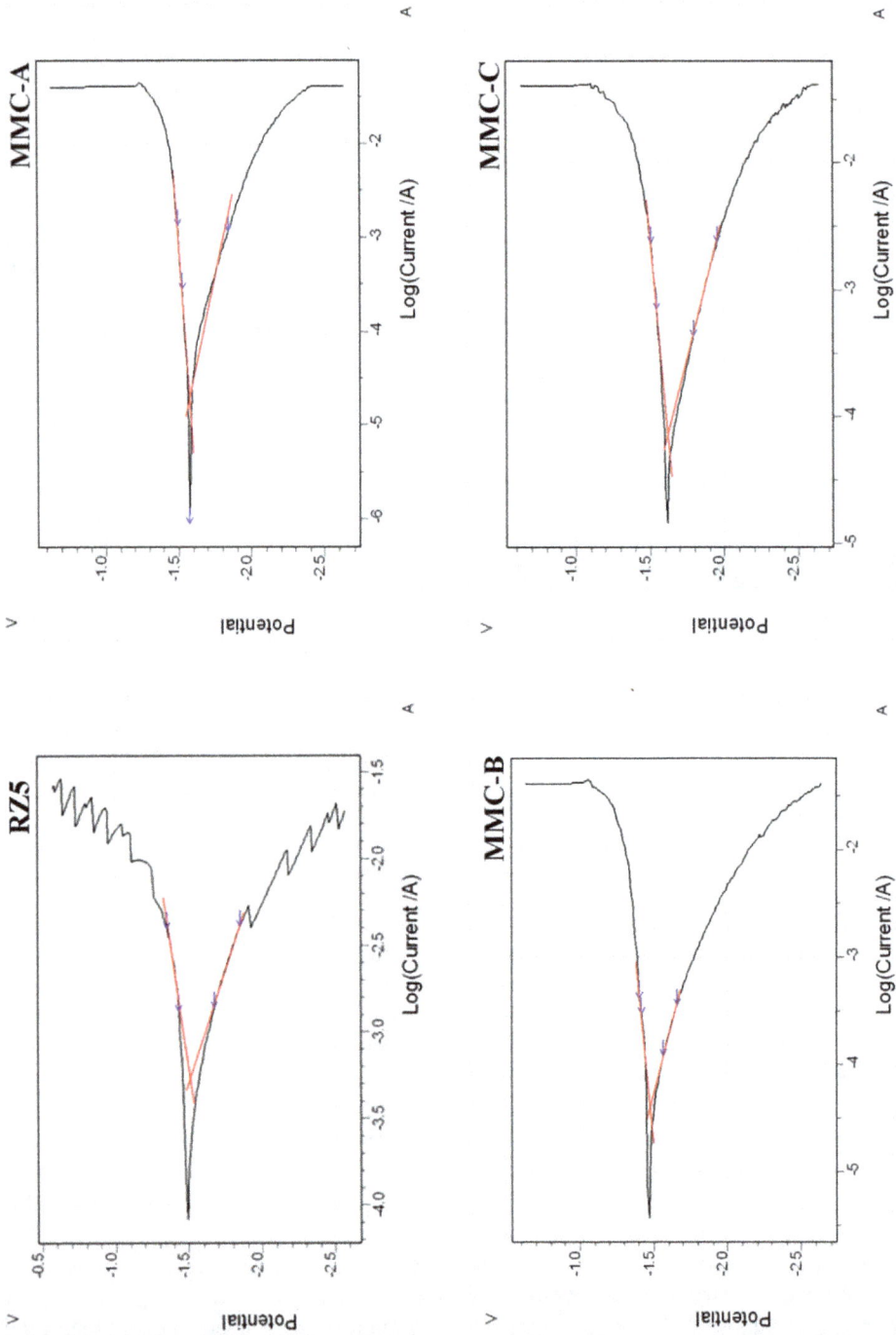

FIGURE 7.3 Potentiodynamic polarization curves of RZ5 and other composites (Vedabouriswaran 2020).

Carbon fibers reinforced in Mg-Al-RE metal-matrix composites were developed in AE 44 alloy by liquid-solid extrusion. AE44 alloy contains 4% Al and 4% RE (mischmetal), a typical Mg-Al-RE alloy, and poses good mechanical properties at room and elevated temperatures. The developed composites exhibited good dispersion of carbon fiber without any apparent porosity. (Li et al. 2016).

TiB_2-reinforced magnesium RZ5 alloy-based *in situ* MMCs were synthesized by stir casting. During the synthesis of composite, temperature, holding time and casting environment were the major variables. The hardness of the RZ5 alloy-based composites increases by 7.12%, 17.06% and 32.07% with the addition of 4, 6 and 8 wt.% of TiB_2 reinforcements, respectively (Meher et al. 2020).

Effect of SiC particle addition on the microstructure and texture evolution of WE 43 Mg upon FSP revealed refinement in the microstructure of the surface composite. The refinement in the microstructure is attributed to the particle-stimulated nucleation mechanism and pinning effect of SiC particles. SiC-reinforced WE43 composite exhibited a novel RE texture component and conventional B-fiber shear texture component. (Mosayebi et al. 2022).

Alloying of Mg alloy with combination of RE element and FSP demonstrated to be favourable in the improvement of the corrosion resistance of Mg-9Al alloys. The corrosion resistance of the fabricated Mg-9Al-xRE alloys improved upon addition of RE element, and the corrosion resistance improved further with increasing the amount of RE addition. The alternation in the corrosion behavior of the alloy is due to the refined and redistributed precipitates in α-Mg matrix. The refinement of cathodic precipitates enlarges the local area ratio of anode and cathode and hence retards severe galvanic corrosion tendency in the magnesium alloys. In addition to this, the presence of large amount of fine Al-RE precipitates favors accumulative growth of the corrosion product layer during the corrosion propagation and thus improves the stability and compactness of the protective layer (Li et al. 2016).

Influence of La and Ce mixed rare earth metal on abrasion and corrosion resistance of AM60 Mg alloy reveals that upon adding La and Ce mixed RE metal, the AM60 Mg alloy's electrochemical properties, friction and wear characteristics improved. The hydrogen evolution corrosion current density, polarization current density and capacitance decrease, while the corrosion potential and polarization resistance increase. The friction coefficient and specific wear rate decreased as addition of La and Ce resulted in refining the grains of α phase and $Mg_{17}Al_{12}$ phase and forming new Al_2Ce and $Al_{11}La_3$ phases, which inhibits the progress of cathodic depolarization, while the added Ce and La forms a composite anodic film on the surface of the alloy, which blocks the dissolution and hence improves the corrosion resistance. (Chenghao et al. 2015).

Metal-matrix composite coatings on light alloys are in demand in the transportation sector so as to reduce the weight of vehicles without reducing mechanical properties. To negate this, composite coatings in various mixtures of Al, Si, Ti and SiC can be a viable option, but the high reactivity between the melted Al and the reinforcement must be avoided and is feasible by laser cladding composite coatings on ZE41 Mg alloys. By laser cladding, addition of different alloying elements (silicon or titanium) to the composite matrix is feasible, and the formation of Al_4C_3 is prevented. Dilution of Mg from the substrate in the Al coating matrix occurs, and it produces an important effect in the matrix reinforcement reactions. Formation of Al_4C_3 is avoided and the mechanical properties improved (Riquelme et al. 2019).

Mg-Gd-Y-Nd-Zr alloys prepared by mold casting were investigated for the effects of Nd/Gd ratios on the microstructures and mechanical properties. Cast alloys consist mainly of α-Mg and β-Mg_5(GdYNd). The volume fractions of the secondary phase increased, and the grains were slightly refined with the increasing Nd/Gd ratio when the alloying addition is equal. Meanwhile, fibers of the second phase also increase in the extruded alloys when the Nd/Gd value increased. However, the Nd/Gd ratio could hardly influence the mechanical properties of the extruded alloys. The aging hardening response also decreases with the increasing Nd/Gd ratios. Peak-aged extruded alloy Mg-10Gd-3Y-4Nd-0.6Zr (wt.%) exhibited highest ultimate tensile stress and tensile yielding stress at room temperature and the elevated temperature of 200°C, which are 379 MPa and 332 MPa at room temperature and 342 MPa and 260 MPa at 200°C, respectively (Liu et al. 2016).

7.3 CONCLUSIONS AND OUTLOOK

With wide applications in industrial and medical fields, it is evident that Mg MMCs will be futuristic materials in all areas where weight reduction, corrosion and wear, biocompatibility and biodegradability of medical implants are the prime requirements. Mg with rare earth elements is a viable solution to cater to these needs.

Development of Mg composites by the introduction of ceramic particles could be a viable solution to be chosen. Proper selection of hard particles and manufacturing methodology adapted in the development of MMCs shall play a pivotal role in satisfying the diverse industrial, engineering, defense and biomedical applications. The demand for MMCs is phenomenal owing to the demand in replacing conventional material such as Al by Mg alloys and its MMCs, as these composites prove to be a futuristic material.

The challenge to reduce carbon emissions, development of lightweight structural materials and biocompatibility and biodegradability requirements in medical applications can be addressed by Mg alloy and Mg MMCs and hence could augment the future requirements.

REFERENCES

Chenghao, Liang, Wang Shusen, Huang Naibao, Zhang Zhihong, Zhang Shuchun, and Ren Jing. "Effects of lanthanum and cerium mixed rare earth metal on abrasion and corrosion resistance of AM60 magnesium alloy." Rare Metal Materials and Engineering 44 (2015): 0521–0526.

Chia, T.L., M.A. Easton, S.M. Zhu, M.A. Gibson, N. Birbilis, and J.F. Nie. "The effect of alloy composition on the microstructure and tensile properties of binary Mg-rare earth alloys." Intermetallics (2009): 481–490.

Hort, N., et al. "Magnesium alloys as implant materials – principles of property design for Mg-RE alloys." Acta Biomater (2010): 1714–1725.

Li, Shaolin, Lehua Qi, Ting Zhang, Jiming Zhou, and Hejun Li. "Interfacial microstructure and tensile properties of carbon fiber reinforced Mg-Al-RE." Journal of Alloys and Compounds 663 (2016): 686–692.

Liu, Xuan, Zhiqiang Zhang, Qichi Le, and Lei Bao. "Effects of Nd/Gd value on the microstructures and mechanical properties of Mg-Gd-Y-Nd-Zr alloys." Journal of Magnesium and Alloys 4 (2016): 214–219.

Meher, Arabinda, Manas Mohan Mahapatra, Priyaranjan Samal, and Pandu R. Vundavilli. "Study on effect of TiB2 reinforcement on the microstructural and mechanical properties of magnesium RZ5 alloy based metal matrix composites." Journal of Magnesium and Alloys 8, no. 3 (2020): 780–792.

Mosayebi, M., A. Zarei-Hanzaki, M. Tahaghoghi, H.R. Abedi, A. Moshiri, and A. Ghaderi. "Effect of second phase particles on the microstructure and texture of rare earth elements containing magnesium matrix surface-composite produced by friction stir processing." Journal of materials research and technology 18 (2022): 2428–2434.

Riquelme, Ainhoa, Pilar Rodrigo, María Dolores Escalera-Rodríguez, and Joaquín Rams. "Characterisation and mechanical properties of Al/SiC metal matrix composite coatings formed on ZE41 magnesium alloys by laser cladding." Results in Physics 13 (2019): 1–11.

Tekumalla, Sravya, Sankaranarayanan Seetharaman, Abdulhakim Almajid, and Manoj Gupta. "Mechanical properties of magnesium-rare earth alloy systems: A review." Metals (2015): 1–39.

Vedabouriswaran, G. "Friction stir processing of magnesium alloy." Ph.D. thesis, New Delhi: Indian Institute of Technology, 2020.

8 Calcium Phosphate/ Hydroxyapatite-Based Magnesium Metal- Matrix Composites

*Xuhui Tang, Di Mei, Shijie Zhu,
Liguo Wang and Shaokang Guan*

CONTENTS

8.1 INTRODUCTION

Inorganic calcium phosphates, represented by HA, have excellent bioactivity, biocompatibility, and osteoconductivity. Their chemical composition is similar to that of the hard tissues in the human body (Benedini et al. 2020; Antoniac et al. 2021). Thus, implantation of HA into the human body does not cause any irritation and rejection (Fidancevska et al. 2007; Kamali et al. 2018). HA promotes the formation of strong bones by combining with primitive bone tissues (Przekora et al. 2021). However, HA suffers from poor toughness and cannot meet application requirements when used alone.

In the human body, the element Mg is involved in all stages of bone metabolism. The depletion of Mg would result in cessation of bone growth, osteopenia and bone fragility, and reduced osteoblast activity (Shan et al. 2022). As one type of biomedical metallic material, Mg alloys show similar Young's modulus and density to that of natural bone, which causes less stress shielding effect when they are used as bone fixation implants (Witte et al. 2008; Zheng et al. 2014). Most importantly, the biodegradability of Mg alloys makes them unique from other biomedical metallic materials, such

DOI: 10.1201/9781003319856-10

as titanium and stainless steel, avoiding secondary surgery injuries after their service is completed (Ehtemam-Haghighi et al. 2016; Talha et al. 2013). Thus, the employment of Mg alloys as bone implants attracts much attention in the related field.

For better load bearing, the improvement of mechanical properties of Mg alloys is one of the research focuses in the scientific community. In addition, the uncontrollable rapid degradation of Mg and its alloys also critically influences their applicability as bone implant materials. In order to make up for these deficiencies and also improve the bioactivity of Mg alloys, Mg-CaP/HA composites are gradually gaining attention. In this chapter, the manufacturing methods of Mg-CaP/HA and the influence of reinforcement's shape on their mechanical properties and their degradation behaviors are briefly overviewed.

8.2 MANUFACTURING METHODS

The performance, applications, and cost of composite materials are closely related to their manufacturing methods and processes. The type of manufacturing technology is a key issue affecting their development and wide applications. The different available manufacturing methods of Mg-CaP/HA composite are roughly divided into solid-state methods, liquid-state methods, and other methods. Solid-state manufacturing technology is a type of method for mixing new composite materials with reinforcing materials when the base metal is in a solid state, including powder metallurgy, hot pressing, hot isostatic pressing, rolling technology, extrusion technology, friction stir processing technology, etc. In liquid-state technology, the matrix metal is mixed with a reinforcing material to form a composite when it is in a molten state. Examples include vacuum pressure infiltration, squeeze casting, stirring casting, liquid metal infiltration, and co-spray deposition, etc. In addition, other methods, such as physical vapour deposition, chemical vapour deposition, and thermal spraying technologies, have been also employed for preparing metal-matrix composite. In the following text, several frequently employed methods that are used for manufacturing Mg-CaP/HA composite are briefly introduced.

8.2.1 SOLID-STATE METHODS

8.2.1.1 Powder Metallurgy

Powder metallurgy is the earliest method used to manufacture metal-matrix composites. This method is mainly used to manufacture particle- or whisker-reinforced metal-matrix composites. The representative process of this manufacturing method is as follows: the matrix and reinforcement powders are first mixed and placed into a mold of the desired shape. The mixing process can be done in the dry state or in suspension; pressure is then applied to make the powder denser (cold pressing); the compact is then heated to a temperature below the melting point but sufficient to produce significant solid-state diffusion (sintering). The blended mixture can be directly hot pressed or hot isostatic pressed (HIP) to achieve higher density; the consolidated composite can be used for secondary deformation processing.

The content of particles (whiskers) in the composite produced by powder metallurgy technology has no limitations, and their size can also be altered in a wide range. However, the cost of this method is relatively high, and it is also difficult to produce the parts and billets in large sizes. In this manufactory process, uniform mixing and the prevention of oxidation of the matrix and reinforcement powder should be critically controlled for better properties of composites.

Due to maximum homogeneity and low cost, powder metallurgy has been used to fabricate Mg metal-matrix composites. By using powder metallurgy methodology, the uniform distribution of reinforcements in the metal matrix can be accomplished with the assistance of powder without melting with or without chemical reactions between the calcium phosphate/hydroxyapatite and the Mg matrix. The process of powder metallurgy is shown in **Figure 8.1**.

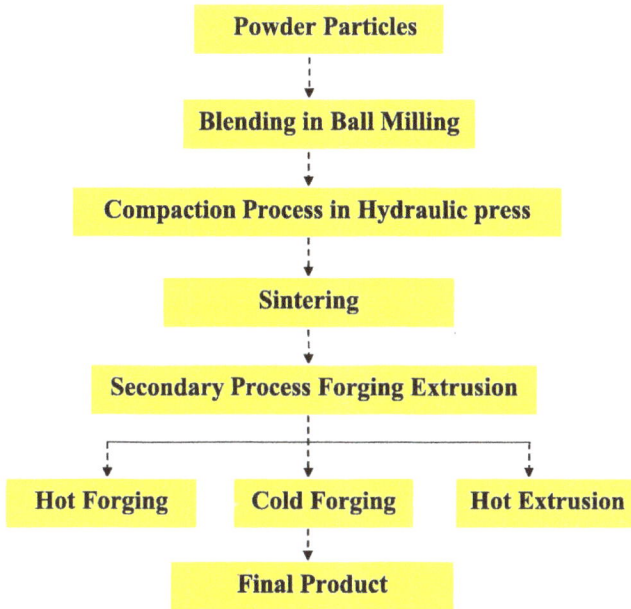

FIGURE 8.1   Powder metallurgy process flow. *Reproduced with permission from (Jayasathyakawin et al. 2020); © 2019 Elsevier Ltd.*

With the rapid development of related technologies, emerging sintering methods have been used in powder metallurgy. There is a method termed spark plasma sintering, which is utilized in the powder metallurgy process. Compared with the casting and traditional powder metallurgy processes, the spark plasma sintering (SPS) process has not only the capability of fast densification and controlling grain growth but also possesses sufficient metallurgical reaction (Prakash et al. 2018). It has emerged as one of the most efficient and convenient methods for a variety of metals, alloys, and composites (Zhang et al. 2017). As demonstrated by Cui et al., pure Mg, Zn, and HA powders were used as raw materials for fabricating Mg-matrix composites by SPS-assisted powder metallurgy technology (Cui et al. 2019). The mixed powder was put into a graphite mold, and after being compacted, it was heated and sintered using SPS equipment under a vacuum. The stress shielding effect can be effectively avoided. Mg-5.5Zn/HA composite can also provide robust mechanical properties. However, with more addition of HA, more pores were shown in the composite. During the corrosion test, the nano-HA falls off as the immersion test progresses, causing gaps and micropores (as shown in **Figure 8.2(d)** and (**f**)) (Prakash et al. 2018). The corrosion rate of Mg-5.5Zn/HA composites can be effectively reduced only by adding a reasonable content of nano-HA, making the HA distribution uniform and increasing the density of the composites. The results showed that after adding 10 wt.% nano-HA particles, spark plasma sintering of Mg-5.5Zn/10HA showed better corrosion resistance than that of Mg-Zn alloys. Therefore, the uniform mixing of the matrix powder and the particle (whisker) reinforcement and the prevention of oxidation of the matrix powder critically affect the entire process.

Microwave radiation is considered another novel sintering method that can reduce the processing time. Microwave heating can ensure uniform heating with low energy consumption and allows rapid heating without thermal gradients (Kutty et al. 2015). As compared with other traditional methods, a more homogeneous structure can be obtained by microwave radiation (Menezes and Kiminami 2008; Yadoji et al. 2003). It is also applicable to Mg matrix composites (Wan et al. 2016). Xiong et al. mixed Mg powder and HA powder by ball milling for 4 hours. The mixture was cold-pressed in a steel cylindrical mold at a pressure of 50 MPa. The green compacts were densified by

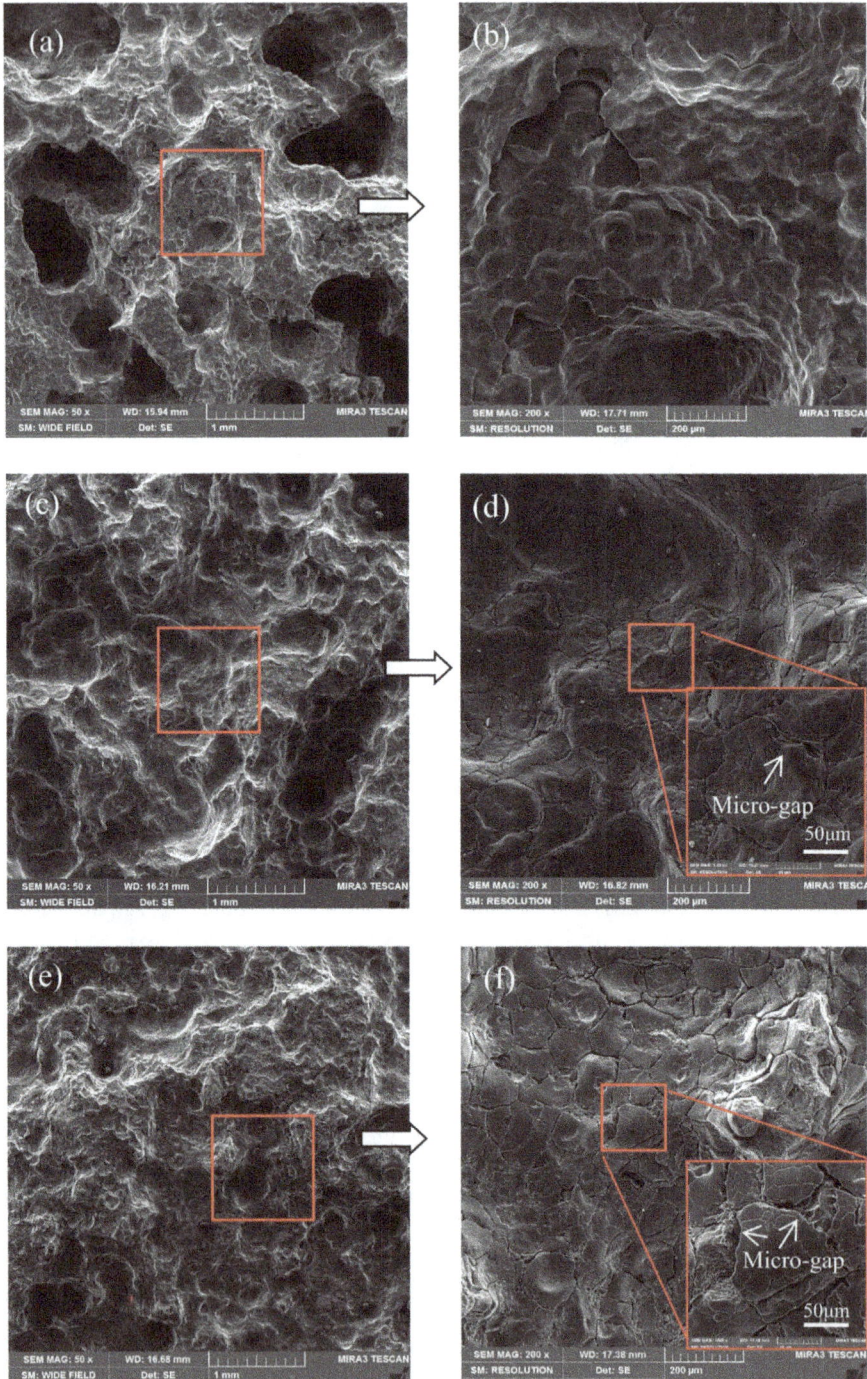

FIGURE 8.2 SEM images of corroded samples (removed corrosion products) after soaking in SBF for 85 h (magnified area enclosed by red square): (a), (b) Mg-5.5Zn alloy, (c), (d) Mg-5.5ZN/5HA composite, (e), (f) Mg-5.5Zn/10HA composite. *Reproduced with permission from (Cui et al. 2019). © 2019 Elsevier Inc.*

sintering at 500°C for 10 minutes in an argon atmosphere using a microwave oven and then cooled to room temperature in the furnace (Xiong et al. 2016). Microstructural observations and density measurements showed that the HA/Mg composites could be prepared by microwave-assisted sintering technique. It is worth noting that the sintering time in the work was only 10 minutes, which is much shorter than the traditional sintering process (usually about 1 hour (Ashuri et al. 2012)). More importantly, the as-prepared HA/Mg composites showed better cytocompatibility and bioactivity than pure Mg. The related literatures demonstrated that the HA/Mg composites prepared by the rapid microwave sintering technique are biocompatible, and their biological behavior can be controlled by adjusting the HA content in the composites (Xiong et al. 2016; Aboudzadeh et al. 2018). Shamami et al. mixed the Mg and HA powders through a mechanical mixer and pressed them into a steel mold using a press. The prepared briquette was placed in the microwave for sintering 12 minutes and then cooled to room temperature in a microwave oven (Shamami et al. 2021). The results indicated that the prepared composites showed higher compressive strength and microhardness compared to that of pure Mg. The employment of microwave radiation can effectively densify pure Mg and Mg/hyaluronic acid composites, but the higher heating rate and shorter processing time of rapid microwave sintering increase the possibility of the formation of defects, such as cracks and pores (Champion 2013).

8.2.1.2 Friction Stir Processing

Friction stir processing (FSP) is a solid-state machining technology developed from friction stir welding (FSW) (Mishra and Ma 2005). The stirring action of the FSP tool increases the local temperature inside the processed material and then softens the matrix material. The combination of increased temperature and mechanical actions allows better incorporation between reinforcements and matrix. Traditional powder metallurgy and casting might lead to severe particle agglomeration, resulting in the inconsistent performance of metal-matrix composites (Gu et al. 2010; Feng and Han 2011). During the FSP process, a rotating tool with a specially designed pin and shoulder is inserted into the plate and then moved laterally along a predetermined line. Since such processing is accompanied by severe plastic deformation and localized heating, FSP can weaken particle agglomeration to a certain extent. At the same time, the combined effects of mechanical stress and temperature rise could lead to the recrystallization and grain refinement of matrix (Ma et al. 2003; Ma 2008).

Ratna Sunil et al. employed friction stir machining for processing pure Mg sheets by using a tapered pin tool with a shoulder diameter of 15 mm, under the load of 5000 N, the traverse speed of the rotating tool along the traverse axis is 12 mm/min, and the rotation speed is 1200 rpm. To produce the FSP-Mg-HA composite, before FSP, a shallow groove 1 mm wide and 2 mm deep was machined on the surface of the Mg sheet using a milling cutter, and the groove was filled with HA powder (Ratna Sunil et al. 2014). Better corrosion resistance was observed for FSP-Mg-HA compared to Mg, where the corrosion rate reduction is due to the rapid passivation layer formation (Hamu et al. 2009) and the reduced intensity of galvanic couple between grain interior and grain boundary (Argade et al. 2012). The improved corrosion resistance of FSP-Mg-HA composite and its lower weight loss originated from the enhanced biomineralization on the composite surface, which reduced further degradation in aggressive environments.

The AZ31B-HA surface composites were formed by FSP using hydroxyapatite powder and AZ31B alloy plate, and the HA powder was packed inside the AZ31B plate in a layer-by-layer manner (as shown in **Figure 8.3**) (Ho et al. 2020). Ring-shaped selected area diffraction patterns were obtained in the case of all FSP Mg-HA composite samples in TEM analysis (as shown in **Figure 8.4**). Central dark field imaging revealed that the HA phase was uniformly distributed within the α-Mg grains at the nanoscale (19 ± 7 nm). Considering the starting average particle size (10 μm) of the HA powder, it is interesting to observe that the incorporation of the HA phase in the Mg matrix ranges from nanometers to micrometers. This may be due to the mechanical decomposition of the original HA particles (10 μm) during the FSP process to form nanoscale particles. The grain refinement of the Mg matrix achieved by FSP leads to the enhanced corrosion resistance.

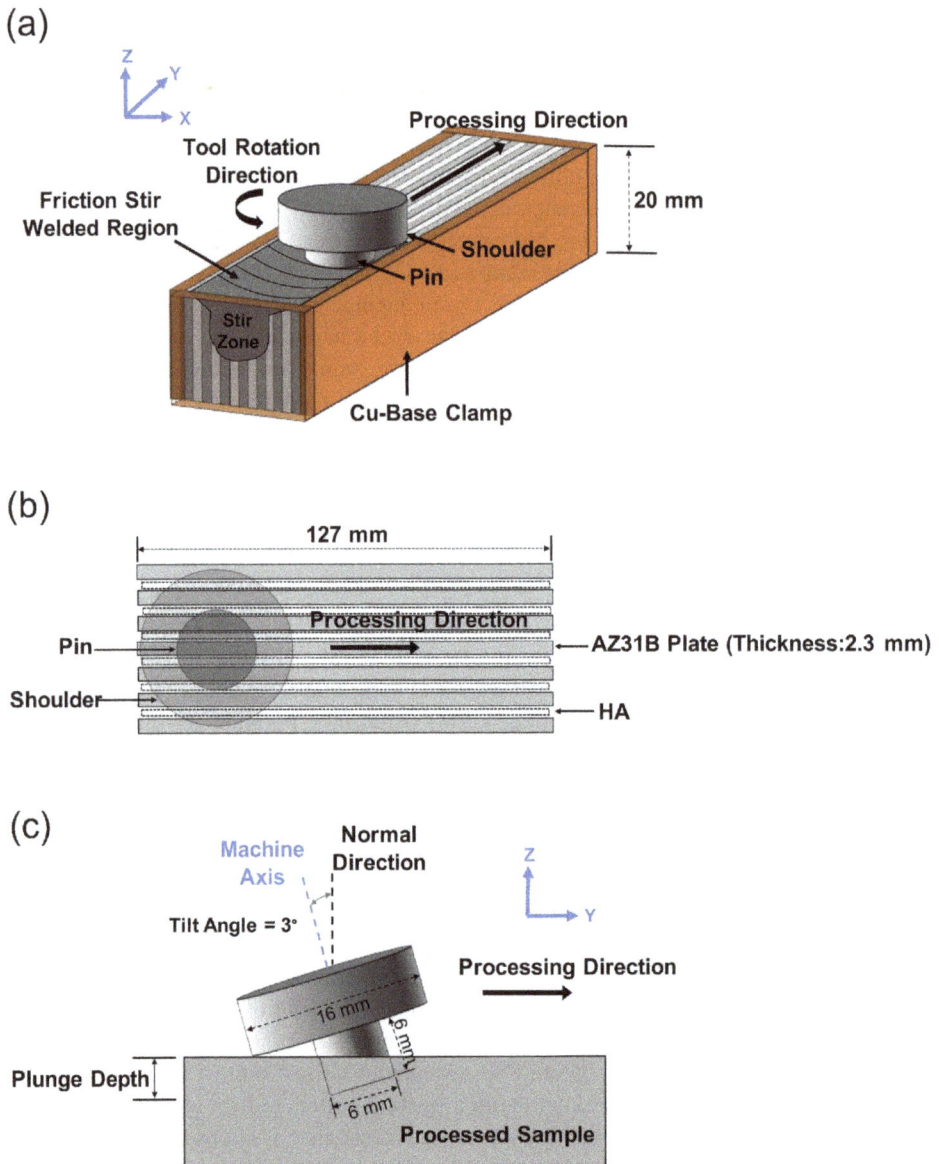

FIGURE 8.3 Schematic of FSP additive manufacturing of AZ31B Mg-HA composite (a) isometric view of the FSP, (b) top view, and (c) side view of the FSP procedures. *Reproduced with permission from (Ho et al. 2020) © 2020 Elsevier B.V.*

The main reason for the improvement of corrosion resistance is that during the thermomechanical processes, recrystallization and recovery leads to a refined microstructure reducing the dislocation density and residual stress (Jang et al. 2007). The grain refinement is dominant in the stabilization of the corrosion product layer of FSP Mg alloys (Argade et al. 2012).

In summary, FSP can be used to fabricate biodegradable Mg-HA composites. Since the process is a solid-state technology, the dispersed phase stabilization commonly encountered in melting and sintering methods can be avoided when developing new composites. The economic viability of this technique can be addressed by designing multi-pass experiments with a single setup to increase productivity using optimized tool geometry (especially shoulder diameter, pin profile, and length)

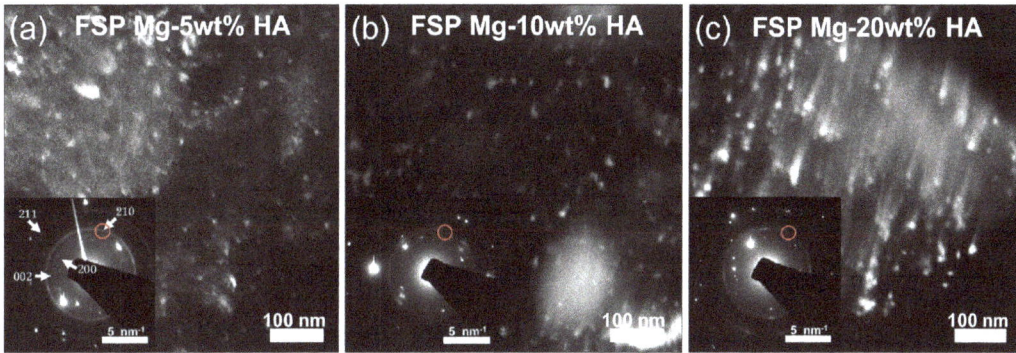

FIGURE 8.4   TEM high-magnification centered darkfield images of FSP (a) Mg-5 wt.% HA, (b) Mg-10 wt.% HA, and (c) Mg-20 wt.% HA composites. The insets in each subfigure provide corresponding SAD pattern as well as the location on the diffraction ring marked by red circle used for performing centered DF imaging. *Reproduced with permission from (Ho et al. 2020) © 2020 Elsevier B.V.*

(Ratna Sunil et al. 2014). However, limited by the technical characteristics of this process, it is more suitable for the manufacture of gradient composite materials and is powerless for the manufacture of devices with complex structures by using itself alone.

8.2.1.3   Others

In addition to the solid-state methods mentioned earlier, researchers have used a combination of various solid-state manufacturing techniques to prepare composite materials. Dubey et al. fabricated Mg-3Zn/HA composites by a powder processing route followed by spark plasma sintering (Dubey et al. 2021). Turan et al. used Mg and fullerene powders (size ~100 μm) and Al powder to prepare Mg-fullerene composites. A semi-powder metallurgy method combined with ultrasonic treatment, stirring, drying, and hot pressing was used to avoid the agglomeration of nanoparticles (Turan et al. 2018).

8.2.2   Liquid-State Methods

Liquid manufacturing technologies for metal-matrix composites include vacuum pressure infiltration technology, squeeze casting technology, liquid metal stirring casting technology, liquid metal infiltration technology, co-spray deposition technology, and thermal spraying technology. At present, the liquid metal infiltration, stirring casting, and pressure-less infiltration methods are mainly used to prepare Mg-CaP/HA. For successfully preparing metal-matrix composites by liquid manufacturing technologies, the following points need to be achieved: (1) uniformly distributed reinforcements, (2) sufficient wettability between the two components, (3) low porosity, and (4) minimum chemical interaction between the reinforcement and the matrix (Hashim et al. 1999).

Liquid metal stirring casting technology is one of the main methods that are suitable for industrial-scale production of particle-reinforced metal-matrix composites, with a simple process and low manufacturing cost. The basic principle is to uniformly disperse the particles in the metal melt by stirring in a certain way and then pouring them into ingots, castings, etc. There are still some drawbacks in the manufacture of particle-reinforced metal-matrix composites by liquid metal stirring casting technology. Small particles are frequently employed as reinforcements for improving the reinforcement effect. However, the particles with the size between 10 and 30 μm have poor wettability with the metal melt and are not easily evenly dispersed in the metal melt. It is easy to induce agglomeration. In addition, strong stirring during the manufacturing process would cause the oxidation of the metal melt and a large amount of air inhalation in the final products. Therefore, effective strategies must be taken to improve the wettability of the particles in metal melt and to prevent the oxidation and gettering of the metal. Khalili et al. fabricated an Mg/HA surface metal-matrix nanocomposite for bone implants using liquid processing techniques combined with stir-centrifugal casting and post-thermomechanical processing (Khalili et al. 2021).

Liquid metal infiltration technology is a method to obtain composite materials by infiltrating preformed porous reinforcement billet with metal melt. Among the currently available techniques for processing metal-matrix composites, liquid metal infiltration is more suitable to achieve a high volume fraction (> 50%) of reinforcements in metal-matrix composites (Mattern et al. 2004; Zeschky et al. 2005). This processing method combines the advantages of vacuum suction casting and pressure casting. One of the advantages of this technology is that it makes the distribution of reinforcement more uniform and eliminates the residual porosity and interface reaction between the reinforcement and the matrix (Mattern et al. 2004). Another advantage of the process is the direct fabrication of final or near-final shaped composite components, minimizing the general machining difficulties of composites (Kevorkijan 2004). Finally, when an appropriate barrier is used to prevent the penetration of metal melt, it is also possible to produce selectively reinforced parts, where the matrix is reinforced only where it is needed (Kevorkijan 2004; Peng et al. 2004). To prepare porous Mg-HA/tricalcium phosphate (TCP) scaffolds, Gu et al. immersed polyurethane (PU) foam in a ceramic slurry (a mixture of HA and b-TCP) and then dried it at room temperature. The ceramic slurry-coated PU foam was sintered in a furnace. The MgCa-HA/TCP composite scaffolds were fabricated by vacuum-driven molten MgCa alloy infiltration and held for 2 minutes while the melt solidified (Gu et al. 2011). The prepared composite scaffold is about 200 times stronger than the pristine porous HA/TCP scaffold but retain only half the mechanical properties of the bulk MgCa alloys. Its corrosion rate in Hanks' solution is slower than that of bulk MgCa alloy alone.

At present, there are relatively new studies using directional freezing and pressureless infiltration technology to prepare Mg-CaP/HA. Yang et al. prepared porous HA scaffolds by directional freezing and added a certain amount of SiO_2 to them by doping or soaking. The prepared porous HA scaffolds were then subjected to pressureless infiltration using Mg alloy melt (Yang et al. 2021). The Mg/HA composite obtained by this method exhibited a biomimetic layered structure and showed higher compressive strength and wear resistance. Furthermore, the composite exhibited slowed degradation and surface apatite mineralization in SBF, indicating its potential application in biomedical applications.

8.2.3 OTHERS

In addition to the mentioned methods, some other processing methods have also been used to manufacture Mg-based composites. Khanra et al. developed Mg-HA composites through melting and extrusion routes. Different amounts of chemically synthesized HAP powders were added to the Mg melts, and mechanical stir was used to disperse the HAP particles in the matrix. The as-cast billets were homogenized and then extruded (Khanra et al. 2010). Razavi and Huang prepared Mg-1.61Zn-0.18Mn-0.5Ca/1HA nanocomposites by using a novel technique combining high shear solidification and hot extrusion, followed by heat treatment (Razavi and Huang 2019).

8.3 TYPES OF REINFORCEMENTS

The mainly used types of reinforcements for CaP/HA-Mg composites include fibers, whiskers, and particles, which are selected according to the needs of the required performances of the composite material.

8.3.1 GRANULAR REINFORCEMENTS

Granular CaP/HA as a reinforcement material is a kind of non-metallic particle with high strength, good heat resistance, wear resistance, and high-temperature resistance. Particulate reinforcements tend to improve wear resistance, heat resistance, strength, and modulus in metal matrix in the fine particles (< 50μm). In Mg alloys, powder metallurgy (Xiong et al. 2016), liquid metal stirring (Kevorkijan 2004), pressure infiltration (Peng et al. 2004), *in situ* synthesis (Lei et al. 2012), etc. are mainly used to manufacture particle-reinforced metal-matrix composites. Due to the low cost of granular CaP/HA and the isotropicity of final products, granular CaP/HA is widely used as a reinforcement for Mg metal-matrix composites.

The previous studies demonstrated that the use of optimal shape and size of HA can provide remarkable improvements in the mechanical and biodegradable properties of Mg metal-matrix composites (Roohani-Esfahani et al. 2010; Wu et al. 2018). In addition, the bioactive behavior of Mg-HA composites can also be tuned by altering the appropriate shape and size of HA, which contribute to its enhanced osteoconductivity and bone-regeneration capabilities.

Jaiswal et al. synthesized Mg-3Zn/HA composites by adding 15 wt.% HA with two different shapes and sizes, circular (R-HA) and cylindrical (C-HA) (as shown in **Figure 8.5**), via SPS technology. The effects of the shape and size of HA on the compressive strength, elastic modulus, and hardness of Mg-HA composites were investigated. In addition to this, the biodegradation of the two composites was evaluated in a modified SBF by static immersion studies (Jaiswal et al. 2020). For the mechanical properties, the hardness, elastic modulus, compressive strength, and ultimate compressive strength were all improved after adding HA to the Mg-3Zn matrix. The addition of 15 wt.% C-HA in the Mg-3Zn matrix showed higher hardness and elastic modulus compared to that of 15 wt.% R-HA. Compared with R-HA, the compressive strength of C-HA is increased by about 21%, and the toughness of the C-HA composite is also higher than that of the R-15HA composite. The corrosion tests indicated that the addition of HA reduced the corrosion rate compared to Mg-3Zn alloy. Furthermore, C-HA has enhanced corrosion resistance compared to R-HA reinforcement.

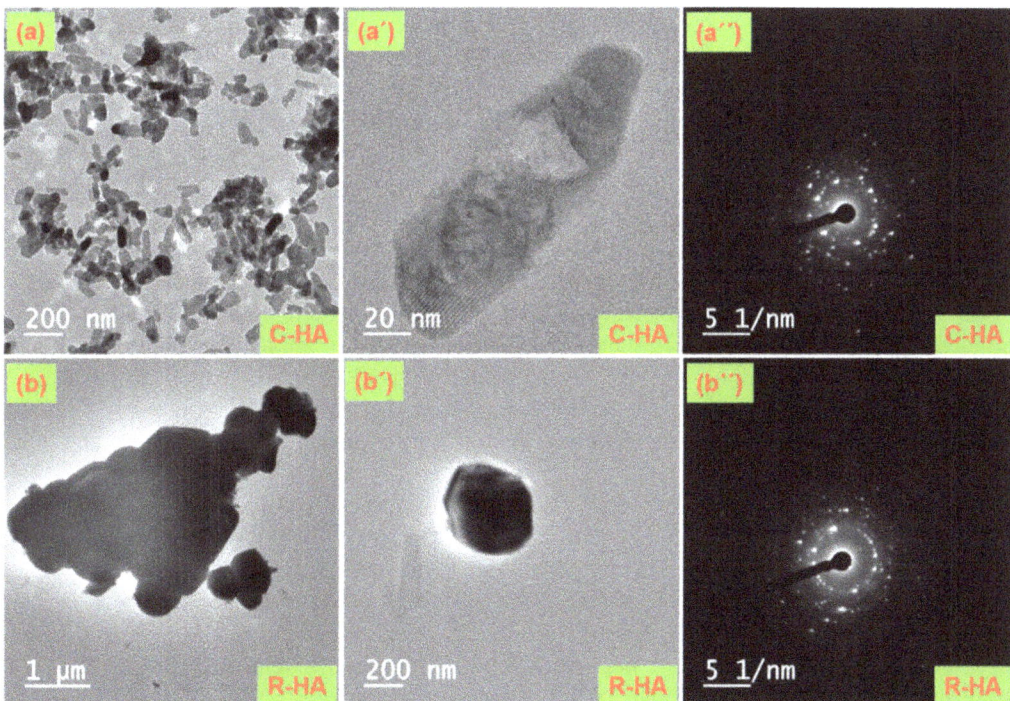

FIGURE 8.5 Transmission electron microscopy images of (a) agglomerated cylindrical HA (C-HA), (b) agglomerated round shape HA (R-HA), (a′ and b′) morphology of individual R and C-HA particle, and (a″ and b″) diffraction pattern of R and C-HA powder. *Reproduced with permission from (Jaiswal et al. 2020) © 2020 Elsevier Ltd and Techna Group S.r.l.*

8.3.2 FIBER REINFORCEMENTS

There are two types of fiber reinforcements, continuous long fibers and short fibers, which are used in metal-matrix composites. The long fibers generally have high strength and elastic modulus along the axial direction. In contrast, the properties of short fibers in metal-matrix composites are non-directional. The fibers are manufactured by the spray method with low production cost and high production efficiency. When they are used as reinforcements, it is necessary to first make fibers into preforms, felts, cloths, etc. and then use extrusion casting, pressure infiltration, and other methods to manufacture fiber-reinforced metal-matrix composite products.

Commonly used fiber reinforcements mainly include carbon fiber, alumina, boron nitride fiber, etc., among which carbon fiber (CF) has gradually shown its advantages in enhancing the properties of Mg alloys. Carbon fiber has the characteristics of low density, high modulus, high specific strength, high temperature resistance, and acid resistance. Carbon fiber also has good electrical conductivity, excellent biocompatibility, and mechanical strength. It is considered a promising reinforcing material for bone tissue engineering applications (Wu et al. 2013; Kobayashi and Kawai 2007).

Since Mg-doped HA has a lower sintering temperature and higher decomposition temperature, CF-reinforced Mg-doped HA composites can be processed at lower temperatures, which will reduce the oxidative damage of CF in the composites, thereby improving the mechanical properties of composites (Yedekçi et al. 2021). Although the mismatch of thermal expansion coefficients between CF and Mg-HA composites might affect the reinforcement effect of CF, this problem can be alleviated by coating Mg-HA on CF by electrochemical deposition. The existence of the interfacial transition layer on the Mg-HA-coated CF can reduce the adverse effect of the thermal expansion coefficient difference between the CF and the Mg-HA matrix, effectively improving the interfacial bonding and improving the mechanical properties of the Mg-HA-coated CF reinforced Mg-HA composites (Yi et al. 2021). Zhao et al. prepared rod-like Mg-HA on the CF surface by electrochemical deposition and then sintered the Mg-HA-coated CF reinforced Mg-HA composite by pressureless sintering at a low temperature of 700°C (as shown in **Figure 8.6**) (Zhao et al. 2022). In addition, the Mg-HA coating can isolate the oxygen generated from the decomposition of HA from CF. The oxidative damage of CF can be effectively prevented (Fang et al. 2021).

8.3.3 WHISKER REINFORCEMENTS

Whiskers are tiny single crystals grown under artificial conditions. The whiskers are used as reinforcements in composite materials by powder metallurgy, squeeze casting, and other methods. However, the manufacturing and sorting process of whiskers is more complicated, and the cost is much higher than that of particles and fibers. The diameter of whiskers is rather small (0.2–1 μm), and the atoms are highly ordered, so their intensity is close to the theoretical value for a complete crystal. It not only has excellent high temperature resistance and corrosion resistance but also has good mechanical strength, light weight, high elastic modulus, high hardness, and other characteristics.

Until now, few works focused on Ca-P/HA whisker reinforced Mg metal-matrix composites. While some researchers have tried to synthesize HA whiskers (HAW) by a hydrothermal method (Nouri-Felekori et al. 2019), they have not been used in Mg metal-matrix composites. In HA, several cations (Na^+, K^+, Mg^{2+}, Zn^{2+}, Sr^{2+} et.,) and anions (F^-, Cl^-, CO_3^{2-}, etc.) can replace the HA lattice ions, which will eventually affect the crystal structure (Nouri-Felekori et al. 2019). The biological advantages and cellular responses of some ion-substituted HAs have been extensively studied (Wang, Li, Ito, Watanabe, Sogo et al. 2016; Wang, Li, Ito, Watanabe, and Tsuji 2016; Wang et al. 2013). Among the ions used for this purpose, divalent cations such as Mg^{2+} (Farzadi et al. 2014; Stipniece et al. 2014), Zn^{2+} (Guerra-López et al. 2015; Tang et al. 2009), and Sr^{2+} (Landi et al. 2007; Boyd et al. 2015) have attracted great attention. These ions, as essential trace elements in the human body, can provide useful properties, especially to accelerate bone mineralization and formation (Kaygili and

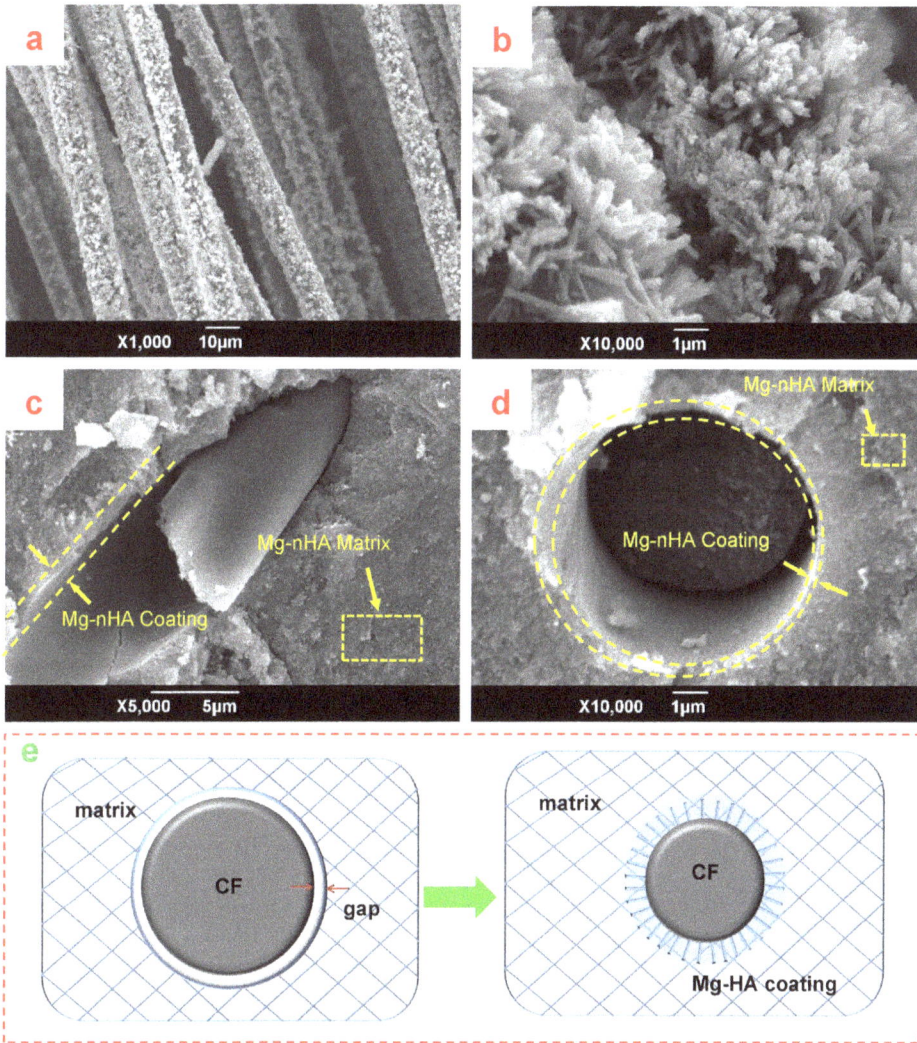

FIGURE 8.6 The SEM images of (a, b) Mg-HA coating on CF, (c, d) CF/Mg-HA/Mg-HA composites sintered at 700°C, (e) schematic diagram of CF and CF/Mg-HA/Mg-HA composites. *Reproduced with permission from (Zhao et al. 2022) © 2021 Elsevier Ltd and Techna Group S.r.l.*

Keser 2015). Considering the well-known enhancing ability of whiskers and the biological advantage of incorporating cations into the HA structure, we believe that the employment of modified HA whiskers as reinforcements for Mg metal-matrix composites is worthy of further investigation.

8.4 CORROSION BEHAVIORS

When the Mg alloys are exposed to a corrosive environment, the following equations (Eqs. 8.1–8.3) can be used to describe their basic corrosion reactions:

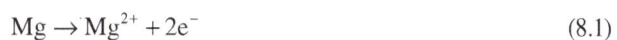

$$Mg \rightarrow Mg^{2+} + 2e^{-} \qquad (8.1)$$

$$2H_2O + 2e^- \rightarrow H_2 \uparrow + 2OH^- \tag{8.2}$$

$$Mg^{2+} + 2OH^- \rightarrow Mg(OH)_2 \tag{8.3}$$

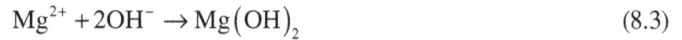

The uncontrollable corrosion behavior of Mg alloys critically affects their availability as biodegradable implant materials. Generally, alloying and melt purification, deformation process, and surface modification are regarded as the main approaches for improving the corrosion resistance of Mg alloys (Ali et al. 2019; Kasaeian-Naeini et al. 2022; Pan et al. 2015). However, with the development of Mg metal-matrix composites, it has also shown potential for enhancing the anti-corrosion properties (Sezer et al. 2018). Table 8.1 briefly summarizes the related investigations.

Dubey et al. co-precipitated Mg- and Zn-doped HA via the wet precipitation method. Mg/co-precipitated HA composites were prepared by spark plasma sintering. After adding 10 wt.% co-precipitated HA, the corrosion resistance of Mg was improved by about 178% after immersion in SBF for 7 days. The reduced corrosion rate of the composite originates from the homoepitaxial

TABLE 8.1

Summary of the Research Results about the Corrosion Behavior of Mg-CaP/HA Metal-Matrix Composites

Ref.	Composite	Processing method	Corrosion medium	Outcome
(Gu et al. 2010)	Mg/HA (10, 20 and 30 wt.%)	Powder metallurgy	SBF	The corrosion rate of Mg/HA composites increased with the HA content.
(Gu et al. 2011)	MgCa-HA/TCP	Liquid metal infiltration	Hanks	The average corrosion rate of MgCa-HA/TCP is slower than that of the bulk MgCa alloy alone.
(Campo et al. 2014)	Mg-HA (5, 10, and 15 wt.%)	Powder metallurgy and extrusion	PBS	Mg-5HA exhibits the best resistance to corrosion.
(Sunil et al. 2014)	Mg-HA	Friction stir processing	SBF	Grain refinement and corrosion resistance is improved.
(Khalajabadi et al. 2015)	Mg/HA/MgO nanocomposites	Blend-cold press-sinter powder metallurgy technique	SBF	Decreasing the corrosion rate with an increase in MgO wt.% up to 10% and HA wt.% up to 12.5%.
(Xiong et al. 2016)	HA/Mg	Microwave assisted sintering	SBF	Mechanical properties, corrosion resistance, and biological behavior can be properly controlled by adjusting HA content.
(Ahmadkhaniha et al. 2016)	Mg/HA	Friction stir processing	DPBS	Corrosion resistance of FSPed Mg is significantly increase.
(Ho et al. 2020)	AZ31B Mg-HA (5,10, and 20 wt.%)	Friction stir additively manufactured	SBF	Friction stir process Mg-5wt.%HA composite exhibits the highest corrosion resistance compared to untreated AZ31B.
(Prakash et al. 2018)	Mg-Zn-Mn-HA	Mechanical alloying and spark plasma sintering	SBF	The degradation rate is reduced from 1.98 mm/year to 0.97mm/year.
(Cui et al. 2019)	Mg-5.5Zn/HA	Spark plasma sintering	SBF	Addition of 10 wt.% HA can decrease the corrosion rate by 49%.
(Ho et al. 2020)	Mg-HA (5, 10 and 20 wt.%)	Friction stir processing	SBF	Mg-5 wt.% HA composite demonstrated the highest corrosion resistance.

nature and uniform distribution of the enhanced co-precipitated HA that facilitated the nucleation of apatite growth (Dubey et al. 2021). However, it does not mean that the high content of HA is helpful for the anti-corrosion property of Mg metal-matrix composites. Gu et al. fabricated Mg-xHA (10, 20, 30 wt.%) composites using the powder metallurgy process. The degradation rate of metal-matrix composites increased with the increase in HA content (Gu et al. 2010). Viswanathan et al. reported similar results that Mg-xHA (10, 20 wt.%) composites exhibited lower corrosion resistance with increasing HA content (Viswanathan et al. 2013). Cui et al. prepared Mg-5.5 wt.% Zn and nano-HA (0, 5, 10 wt.%) composites by SPS method. The results indicate that the addition of nano-HA with a reasonable content of 10 wt.% and an increase in the density of the composites effectively reduced the corrosion rate of the Mg-5.5Zn/HA composites (Cui et al. 2019). Ho et al. fabricated AZ31B/HA (5, 10, and 20 wt.%) composites using FSP. As the HA content increases, the micro/nano galvanic couple also increases, resulting in the deterioration of corrosion resistance of the composites. However, all the fabricated Mg/HA composites have better corrosion resistance than the untreated AZ31B alloy.

Razavi and Huang prepared two different shapes of HA nanoparticles (i.e. spherical and needle-like) and used a novel high shear solidification technique combined with severe plastic deformation to prepare nanocomposites (Mg2Zn-0.2Mn-0.5Ca/1HA). It was found that the composites with spherical HA nanoparticles sowed better corrosion resistance compared with acicular HA nanoparticles (as shown in **Figure 8.7**) (Razavi and Huang 2020).

The corrosion resistance of Mg-CaP/HA metal-matrix composites in pseudo-physiological environment can be improved within a certain range by adding reinforcements. The improvement of corrosion resistance partially originates from the accelerated formation of the protective apatite layer caused by the induction effect of CaP/HA on the surface of metal-matrix composites. Another possibility is derived from the grain refinement of Mg matrix caused by processing. When adding more than a certain amount of CaP/HA, the corrosion resistance of Mg-CaP/HA composite will decrease instead. This may be related to the change of the aggregation state with the increase of CaP/HA or the change of the microstructure and the appearance of texture.

8.5 CONCLUSIONS AND OUTLOOK

In this chapter, the development and recent advances of calcium phosphate/hydroxyapatite-based Mg metal-matrix composites (Mg-CaP/HA) for biomedical applications were introduced. For the biomedical applications of Mg alloys, the main breakthrough limitation is its insufficient strength and related rapid corrosion rate. By reasonably selecting the reinforcement and the corresponding processing method, Mg-CaP/HA composites could overcome these deficiencies to a certain extent. In the existing research, it can be found that adding appropriate content of CaP/HA can effectively improve the corrosion resistance of Mg metal-matrix composites. At the same time, the advanced processing method can improve the density and strength of the composites. Using finer and uniform CaP/HA particles or altering their shape is also an effective way to improve the corrosion resistance and strength of the composites. However, it should be noted that although the corrosion resistance of Mg-CaP/HA composites was improved to a certain extent after adding an appropriate amount of CaP/HA, it still could not reach the ideal degradation rate. In future work, the current problems can be solved by properly selecting Mg matrix and the size of CaP/HA reinforcements combined with suitable processing with optimum process parameters. The corrosion resistance of Mg metal-matrix composites can be further improved by introducing the additional surface coatings (Zhang et al. 2020). In addition, it is necessary to explore the relationship between interfacial microstructure, properties, and comprehensive properties of composites in terms of surface properties, morphology, the interaction between reinforcements and matrix, interfacial reaction, interfacial characterization, and so on for optimizing the interface and improving the comprehensive properties of composites.

Most importantly, although CaP/HA were regarded as biocompatible implant materials, endowing it with richer biological functions through the modification of the CaP/HA reinforcements will open up a broader space for their biomedical applications.

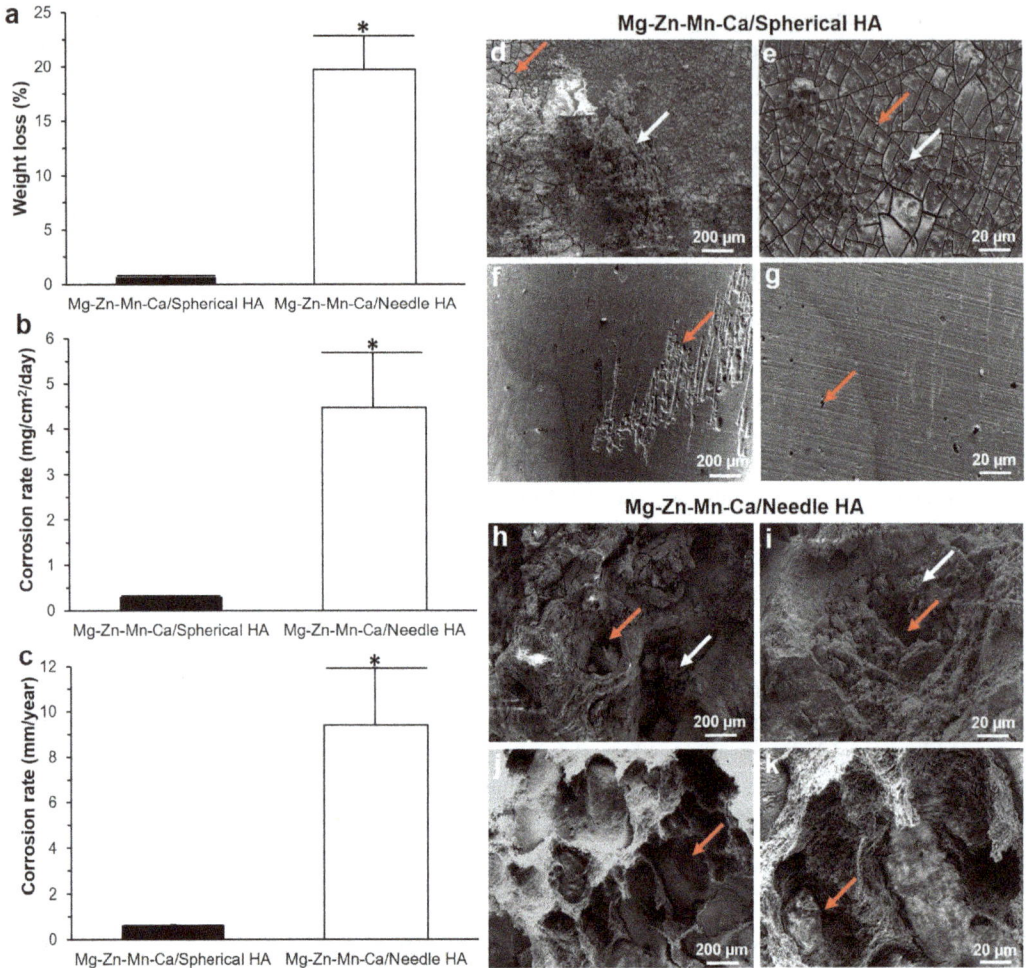

FIGURE 8.7   Results of immersion tests of Mg-Zn-Mn-Ca/Spherical HA and Mg-Zn-Mn-Ca/Needle HA (a–c) and SEM images from their corroded surfaces (d–k). *Reproduce with permission from (Razavi and Huang 2020) © 2020 Elsevier B.V.*

REFERENCES

Aboudzadeh N, Dehghanian C, and Shokrgozar MA. In vitro degradation and cytotoxicity of Mg-5Zn-0.3Ca/nHA biocomposites prepared by powder metallurgy. Transactions of Nonferrous Metals Society of China 2018, 28, 1745–1754.

Ahmadkhaniha D, Fedel M, Sohi MH, Hanzaki AZ, and Deflorian F. Corrosion behavior of magnesium and magnesium – hydroxyapatite composite fabricated by friction stir processing in Dulbecco's phosphate buffered saline. Corrosion Science 2016, 104, 319–329.

Ali M, Hussein MA, and Al-Aqeeli N. Magnesium-based composites and alloys for medical applications: A review of mechanical and corrosion properties. Journal of Alloys and Compounds 2019, 792, 1162–1190.

Antoniac IV, Antoniac A, Vasile E, Tecu C, Fosca M, Yankova VG, and Rau JV. In vitro characterization of novel nanostructured collagen-hydroxyapatite composite scaffolds doped with magnesium with improved biodegradation rate for hard tissue regeneration. Bioactive Materials 2021, 6, 3383–3395.

Argade GR, Panigrahi SK, and Mishra RS. Effects of grain size on the corrosion resistance of wrought magnesium alloys containing neodymium. Corrosion Science 2012, 58, 145–151.

Ashuri M, Moztarzadeh F, Nezafati N, Ansari Hamedani A, and Tahriri M. Development of a composite based on hydroxyapatite and magnesium and zinc-containing sol – gel-derived bioactive glass for bone substitute applications. Materials Science and Engineering: C 2012, 32, 2330–2339.

Benedini L, Laiuppa J, Santillán G, Baldini M, and Messina P. Antibacterial alginate/nano-hydroxyapatite composites for bone tissue engineering: Assessment of their bioactivity, biocompatibility, and antibacterial activity. Materials Science and Engineering: C 2020, 115, 111101.

Boyd AR, Rutledge L, Randolph LD, and Meenan BJ. Strontium-substituted hydroxyapatite coatings deposited via a co-deposition sputter technique. Materials Science and Engineering: C 2015, 46, 290–300.

Campo RD, Savoini B, Muñoz A, Monge MA, and Garcés G. Mechanical properties and corrosion behavior of Mg – HAP composites. Journal of the Mechanical Behavior of Biomedical Materials 2014, 39, 238–246.

Champion E. Sintering of calcium phosphate bioceramics. Acta Biomaterialia 2013, 9, 5855–5875.

Cui ZQ, Li WJ, Cheng LX, Gong DQ, Cheng WL, and Wang WX. Effect of nano-HA content on the mechanical properties, degradation and biocompatible behavior of Mg-Zn/HA composite prepared by spark plasma sintering. Materials Characterization 2019, 151, 620–631.

Dubey A, Jaiswal S, Garg A, Jain V, and Lahiri D. Synthesis and evaluation of magnesium/co-precipitated hydroxyapatite based composite for biomedical application. Journal of the Mechanical Behavior of Biomedical Materials 2021, 118, 104460.

Ehtemam-Haghighi S, Prashanth KG, Attar H, Chaubey AK, Cao GH, and Zhang LC. Evaluation of mechanical and wear properties of TixNb7Fe alloys designed for biomedical applications. Materials & Design 2016, 111, 592–599.

Fang C, Hu P, Dong S, Feng JX, Xun LC, and Zhang XH. Influence of hydrothermal carbon coating on the properties of CF/ZrB2/SiBCN prepared by slurry injection. Journal of the European Ceramic Society 2021, 41, 84–91.

Farzadi A, Bakhshi F, Solati-Hashjin M, Asadi-Eydivand M, and Osman N. Magnesium incorporated hydroxyapatite: Synthesis and structural properties characterization. Ceramics International 2014, 40, 6021–6029.

Feng A, and Han Y. Mechanical and in vitro degradation behavior of ultrafine calcium polyphosphate reinforced magnesium-alloy composites. Materials & Design 2011, 32, 2813–2820.

Fidancevska E, Ruseska G, Bossert J, Lin Y-M, and Boccaccini AR. Fabrication and characterization of porous bioceramic composites based on hydroxyapatite and titania. Materials Chemistry and Physics 2007, 103, 95–100.

Gu XN, Wang X, Li N, Li L, Zheng YF, and Miao X. Microstructure and characteristics of the metal-ceramic composite (MgCa-HA/TCP) fabricated by liquid metal infiltration. Journal of Biomedical Materials Research–Part B: Applied Biomaterials 2011, 99, 127–134.

Gu XN, Zhou WR, Zheng YF, Dong LM, Xi YL, and Chai DL. Microstructure, mechanical property, biocorrosion and cytotoxicity evaluations of Mg/HA composites. Materials Science and Engineering: C 2010, 30, 827–832.

Guerra-López JR, Echeverría GA, Güida JA, Viña R, and Punte G. Synthetic hydroxyapatites doped with Zn(II) studied by X-ray diffraction, infrared, Raman and thermal analysis. Journal of Physics and Chemistry of Solids 2015, 81, 57–65.

Hamu GB, Eliezer D, and Wagner L. The relation between severe plastic deformation microstructure and corrosion behavior of AZ31 magnesium alloy. Journal of Alloys and Compounds 2009, 468, 222–229.

Hashim J, Looney L, and Hashmi MSJ. Metal matrix composites: production by the stir casting method. Journal of Materials Processing Technology 1999, 92–93, 1–7.

Ho Y-H, Joshi SS, Wu T-C, Hung C-M, Ho N-J, and Dahotre NB. In-vitro bio-corrosion behavior of friction stir additively manufactured AZ31B magnesium alloy-hydroxyapatite composites. Materials Science and Engineering: C 2020, 109, 110632.

Jaiswal S, Dubey A, and Lahiri D. The influence of bioactive hydroxyapatite shape and size on the mechanical and biodegradation behaviour of magnesium based composite. Ceramics International 2020, 46, 27205–27218.

Jang Y-H, Kim SS, Yim CD, Lee CG, and Kim S-J. Corrosion behaviour of friction stir welded AZ31B Mg in 3·5%NaCl solution. Corrosion Engineering, Science and Technology 2007, 42, 119–122.

Jayasathyakawin S, Ravichandran M, Baskar N, Anand Chairman C, and Balasundaram R. Magnesium matrix composite for biomedical applications through powder metallurgy – Review. Materials Today: Proceedings 2020, 27, 736–741.

Kamali S, Shemshad S, Khavandi A, and Azari S. Construction novel hydroxyapatite-nitinol nanocomposite for hard tissue applications. Materials Chemistry and Physics 2018, 220, 331–341.

Kasaeian-Naeini M, Sedighi M, and Hashemi R. Severe plastic deformation (SPD) of biodegradable magnesium alloys and composites: A review of developments and prospects. Journal of Magnesium and Alloys 2022, 10, 938–955.

Kaygili O, and Keser S. Sol – gel synthesis and characterization of Sr/Mg, Mg/Zn and Sr/Zn co-doped hydroxyapatites. Materials Letters 2015, 141, 161–164.

Kevorkijan V. Mg AZ80/SiC composite bars fabricated by infiltration of porous ceramic preforms. Metallurgical and Materials Transactions A 2004, 35, 707–715.

Khalajabadi SZ, Kadir MRA, Izman S, and Ebrahimi-Kahrizsangi R. Fabrication, bio-corrosion behavior and mechanical properties of a Mg/HA/MgO nanocomposite for biomedical applications. Materials & Design 2015, 88, 1223–1233.

Khalili V, Moslemi S, Ruttert B, Frenzel J, Theisen W, and Eggeler G. Surface metal matrix nano-composite of magnesium/hydroxyapatite produced by stir-centrifugal casting. Surface and Coatings Technology 2021, 406, 126654.

Khanra AK, Jung HC, Hong KS, and Shin KS. Comparative property study on extruded Mg – HAP and ZM61–HAP composites. Materials Science and Engineering: A 2010, 527, 6283–6288.

Kobayashi S, and Kawai W. Development of carbon nanofiber reinforced hydroxyapatite with enhanced mechanical properties. Composites Part A: Applied Science and Manufacturing 2007, 38, 114–123.

Kutty MG, Bhaduri SB, Zhou H, and Yaghoubi A. In situ measurement of shrinkage and temperature profile in microwave- and conventionally-sintered hydroxyapatite bioceramic. Materials Letters 2015, 161, 375–378.

Landi E, Tampieri A, Celotti G, Sprio S, Sandri M, and Logroscino G. Sr-substituted hydroxyapatites for osteoporotic bone replacement. Acta Biomaterialia 2007, 3, 961–969.

Lei T, Tang W, Cai S-H, Feng F-F, and Li N-F. On the corrosion behaviour of newly developed biodegradable Mg-based metal matrix composites produced by in situ reaction. Corrosion Science 2012, 54, 270–277.

Ma ZY. Friction stir processing technology: A review. Metallurgical and Materials Transactions A 2008, 39, 642–658.

Ma ZY, Mishra RS, Mahoney MW, and Grimes R. High strain rate superplasticity in friction stir processed Al-Mg-Zr alloy. Materials Science and Engineering: A 2003, 351, 148–153.

Mattern A, Huchler B, Staudenecker D, Oberacker R, Nagel A, and Hoffmann MJ. Preparation of interpenetrating ceramic-metal composites. Journal of the European Ceramic Society 2004, 24, 3399–3408.

Menezes RR, and Kiminami RHGA. Microwave sintering of alumina-zirconia nanocomposites. Journal of Materials Processing Technology 2008, 203, 513–517.

Mishra RS, and Ma ZY. Friction stir welding and processing. Materials Science and Engineering: R: Reports 2005, 50, 1–78.

Nouri-Felekori M, Khakbiz M, and Nezafati N. Synthesis and characterization of Mg, Zn and Sr-incorporated hydroxyapatite whiskers by hydrothermal method. Materials Letters 2019, 243, 120–124.

Pan YK, He SY, Wang DG, Huang DL, Zheng TT, Wang SQ, Dong P, and Chen CZ. In vitro degradation and electrochemical corrosion evaluations of microarc oxidized pure Mg, Mg-Ca and Mg-Ca-Zn alloys for biomedical applications. Materials Science and Engineering: C 2015, 47, 85–96.

Peng LM, Cao JW, Noda K, and Han KS. Mechanical properties of ceramic-metal composites by pressure infiltration of metal into porous ceramics. Materials Science and Engineering: A 2004, 374, 1–9.

Prakash C, Singh S, Verma K, Sidhu SS, and Singh S. Synthesis and characterization of Mg-Zn-Mn-HA composite by spark plasma sintering process for orthopedic applications. Vacuum 2018, 155, 578–584.

Przekora A, Kazimierczak P, and Wojcik M. Ex vivo determination of chitosan/curdlan/hydroxyapatite biomaterial osseointegration with the use of human trabecular bone explant: New method for biocompatibility testing of bone implants reducing animal tests. Materials Science and Engineering: C 2021, 119, 111612.

Ratna Sunil B, Sampath Kumar TS, Chakkingal U, Nandakumar V, and Doble M. Friction stir processing of magnesium-nanohydroxyapatite composites with controlled in vitro degradation behavior. Materials Science and Engineering: C 2014, 39, 315–324.

Razavi M, and Huang Y. A magnesium-based nanobiocomposite processed by a novel technique combining high shear solidification and hot extrusion. Recent Patents on Nanotechnology 2019, 13, 38–48.

Razavi M, and Huang Y. Effect of hydroxyapatite (HA) nanoparticles shape on biodegradation of Mg/HA nanocomposites processed by high shear solidification/equal channel angular extrusion route. Materials Letters 2020, 267, 127541.

Roohani-Esfahani S-I, Nouri-Khorasani S, Lu Z, Appleyard R, and Zreiqat H. The influence hydroxyapatite nanoparticle shape and size on the properties of biphasic calcium phosphate scaffolds coated with hydroxyapatite-PCL composites. Biomaterials 2010, 31, 5498–5509.

Sezer N, Evis Z, Kayhan SM, Tahmasebifar A, and Koç M. Review of magnesium-based biomaterials and their applications. Journal of Magnesium and Alloys 2018, 6, 23–43.

Shamami DZ, Rabiee SM, and Shakeri M. Use of rapid microwave sintering technique for the processing of magnesium-hydroxyapatite composites. Ceramics International 2021, 47, 13023–13034.

Shan ZM, Xie XH, Wu XT, Zhuang SY, and Zhang C. Development of degradable magnesium-based metal implants and their function in promoting bone metabolism (A review). Journal of Orthopaedic Translation 2022, 36, 184–193.

Stipniece L, Salma-Ancane K, Borodajenko N, Sokolova M, Jakovlevs D, and Berzina-Cimdina L. Characterization of Mg-substituted hydroxyapatite synthesized by wet chemical method. Ceramics International 2014, 40, 3261–3267.

Sunil BR, Kumar TS, Chakkingal U, Nandakumar V, and Doble M. Friction stir processing of magnesium-nanohydroxyapatite composites with controlled in vitro degradation behavior. Materials Science and Engineering: C 2014, 39, 315–324.

Talha M, Behera CK, and Sinha OP. A review on nickel-free nitrogen containing austenitic stainless steels for biomedical applications. Materials Science and Engineering: C 2013, 33, 3563–3575.

Tang Y, Chappell HF, Dove MT, Reeder RJ, and Lee YJ. Zinc incorporation into hydroxylapatite. Biomaterials 2009, 30, 2864–2872.

Turan ME, Sun Y, and Akgul Y. Mechanical, tribological and corrosion properties of fullerene reinforced magnesium matrix composites fabricated by semi powder metallurgy. Journal of Alloys and Compounds 2018, 740, 1149–1158.

Viswanathan R, Nagumothu R, Kennedy S, Sreekanth D, Venkateswarlu K, Sandhya Rani M, and Muthupandi V. Plasma electrolytic oxidation and characterization of spark plasma sintered magnesium/hydroxyapatite composites. Materials Science Forum 2013, 765, 827–831.

Wan YZ, Cui T, Li W, Li C, Xiao J, Zhu Y, Ji DH, Xiong GY, and Luo HL. Mechanical and biological properties of bioglass/magnesium composites prepared via microwave sintering route. Materials & Design 2016, 99, 521–527.

Wang X, Li X, Ito A, Watanabe Y, Sogo Y, Hirose M, Ohno T, and Tsuji NM. Rod-shaped and substituted hydroxyapatite nanoparticles stimulating type 1 and 2 cytokine secretion. Colloids and Surfaces B: Biointerfaces 2016, 139, 10–16.

Wang X, Li X, Ito A, Watanabe Y, and Tsuji NM. Rod-shaped and fluorine-substituted hydroxyapatite free of molecular immunopotentiators stimulates anti-cancer immunity in vivo. Chemical Communications (Camb) 2016, 52, 7078–7081.

Wang X, Li X, Onuma K, Sogo Y, Ohno T, and Ito A. Zn- and Mg- containing tricalcium phosphates-based adjuvants for cancer immunotherapy. Scientific Reports 2013, 3, 2203.

Witte F, Hort N, Vogt C, Cohen S, Kainer KU, Willumeit R, and Feyerabend F. Degradable biomaterials based on magnesium corrosion. Current Opinion in Solid State and Materials Science 2008, 12, 63–72.

Wu JC, Ruan CS, Ma YF, Wang YL, and Luo YF. Vital role of hydroxyapatite particle shape in regulating the porosity and mechanical properties of the sintered scaffolds. Journal of Materials Science & Technology 2018, 34, 503–507.

Wu MY, Wang QY, Liu XQ, and Liu HQ. Biomimetic synthesis and characterization of carbon nanofiber/hydroxyapatite composite scaffolds. Carbon 2013, 51, 335–345.

Xiong GY, Nie YJ, Ji DH, Li J, Li CZ, Li W, Zhu Y, Luo HL, and Wan YZ. Characterization of biomedical hydroxyapatite/magnesium composites prepared by powder metallurgy assisted with microwave sintering. Current Applied Physics 2016, 16, 830–836.

Yadoji P, Peelamedu R, Agrawal D, and Roy R. Microwave sintering of Ni-Zn ferrites: Comparison with conventional sintering. Materials Science and Engineering: B 2003, 98, 269–278.

Yang L-K, Jin Q, Guo R-F, and Shen P. Bio-inspired lamellar hydroxyapatite/magnesium composites prepared by directional freezing and pressureless infiltration. Ceramics International 2021, 47, 11183–11192.

Yedekçi B, Tezcaner A, Alshemary AZ, Yılmaz B, Demir T, and Evis Z. Synthesis and sintering of B, Sr, Mg multi-doped hydroxyapatites: Structural, mechanical and biological characterization. Journal of the Mechanical Behavior of Biomedical Materials 2021, 115, 104230.

Yi L-F, Yamamoto T, Onda T, and Chen Z-C. Orientation control of carbon fibers and enhanced thermal/mechanical properties of hot-extruded carbon fibers/aluminum composites. Diamond and Related Materials 2021, 116, 108432.

Zeschky J, Lo J, Höfner T, and Greil P. Mg alloy infiltrated Si-O-C ceramic foams. Materials Science and Engineering: A 2005, 403, 215–221.

Zhang L, He ZY, Tan J, Calin M, Prashanth KG, Sarac B, Völker B, Jiang YH, Zhou R, and Eckert J. Designing a multifunctional Ti-2Cu-4Ca porous biomaterial with favorable mechanical properties and high bioactivity. Journal of Alloys and Compounds 2017, 727, 338–345.

Zhang YB, Yang HL, Lei SQ, Zhu SJ, Wang JF, Sun YF, and Guan SK. Preparation of biodegradable Mg/β-TCP biofunctional gradient materials by friction stir processing and pulse reverse current electrodeposition. Acta Metallurgica Sinica (English Letters) 2020, 33, 103–114.

Zhao XN, Yang Z, Wang WY, Jiang GZ, Wei SS, Liu A, Guan JX, and Wang PF. Preparation of carbon fiber/Mg-doped nano-hydroxyapatite composites under low temperature by pressureless sintering. Ceramics International 2022, 48, 674–683.

Zheng YF, Gu XN, and Witte F. Biodegradable metals. Materials Science and Engineering: R: Reports 2014, 77, 1–34.

9 Magnesium Matrix Hybrid Composites
A Case Study with WC and Graphite Reinforcements

Prasanta Sahoo and Sudip Banerjee

CONTENTS

9.1 INTRODUCTION

Nowadays, metal-matrix composites (MMCs) are getting immense attention from the research fraternity due to different attractive and noticeable specific properties. Typically, insertion of ceramic-based reinforcements in matrix metal is found to be a majestic solution to obtain preferable specific properties which cannot be obtained by the base material alone. Presently, MMCs are used in the electronics industry, aerospace industry, chemical industry, automobile industry, etc. (Polmar et al. 2017; Luo and Pekguleryuz 1994; Tun 2009; Kainer 2006). However, the ever-increasing demand for different specific properties has pressed researchers to develop new materials by fortifying reinforcements in different base matrices. In this context, Mg alloys are considered an excellent choice as matrix material because of low density, excellent machinability, outstanding strength-to-weight ratio, superior castability, etc. (Suneesh and Sivapragash 2018; Banerjee et al. 2021a, 2021b, Banerjee, Sarkar et al. 2022). Due to all these advantages, the automotive and aerospace industries are replacing different ferrous components (transmission case, piston, steering shaft, brake parts, etc.) with Mg-based materials. Some well-established applications of Mg-MMCs are tabulated in

Table 9.1. Literatures reveal that Mg matrices may be fortified with a wide variety of ceramic-based materials. Few notable reinforcing materials are alumina (Al_2O_3, whiskers and particulates), silicon carbide (SiC, whiskers and particulates), boron carbide (B_4C), tungsten carbide (WC), titanium carbide (TiC), yittria (Y_2O_3), carbon nanotube (CNT), and titanium diboride (TiB_2). It is also observed in literature that commonly employed reinforcing agents are ceramic in nature. Detail classifications of properties of commonly employed reinforcing agents are presented in Table 9.2. Properties of MMCs also depend on particle size and type of reinforcements (Dey and Pandey 2015). Ceramic-based reinforcements are normally in the form of whisker, fibres, or particulate. However, particulate form is more suitable, as whiskers are more agglomeration prone, while end properties of fibre reinforced composites depend on direction of load (Tun 2009; Kainer 2006; Lloyd 1994; Gupta et al. 2011; Chawla 2006; Clyne and Withers 1995). Gopal et al. (2017) reported a noticeable role of particle size as well as wt.% of reinforcements in controlling hardness and tribological characteristics of Mg-CRT-BN composites (CRT – cathode ray tube). Huang et al. (2011) observed significant change in tribological behaviour of Mg-MMCs with respect to change in particle size. Dey and Pandey (2015) reviewed different behaviours of Mg-MMCs and found that particle size makes a significant

---

TABLE 9.1

Well-Established Applications of Mg MMCs (Kainer and Mordike 2000)

Attributes	Reinforcement	Applications
Light weight	Low density ceramic, cenosphere, fly ash	Batteries, frame members
High strength	CNT, Al_2O_3, WC, SiC	Brake parts, connecting rods
High thermal conductivity	CBN, WC, high-conductivity carbon	Brake components, electronic packaging, cylinder liners, super/turbocharger parts
Wear resistance	Al_2O_3, WC, SiC, graphite	Brake components, cylinder liners, bearing surfaces, cam shafts, piston
Self-lubricating	TiB_2, graphite, hexagonal BN, MoS_2	Gear surface, bearing journals, CV joints, pistons, cylinder liner
Low cost	Fly ash	Water pumps, accessory brackets, valve covers, intake manifolds, alternator covers, oil pans

---

TABLE 9.2

General Classification of Ceramic Reinforcements with Their Properties (Lloyd 1994; Bau 2009)

	Ceramic particles								
Properties	Carbides				Oxides		Nitrides		Borides
	SiC	WC	TiC	B_4C	Al_2O_3	Y_2O_3	BN	AlN	TiB_2
Density (g/cm³)	3.3	15.63	4.9	2.5	3.9	5	2.2	3.2	4.5
Young's Modulus (GPa)	480	530–700	320	450	410	120	90	350	370
Conductivity (W/m°C)	59	28	29	29	25	27	25	10	27
Melting point (°C)	2300	3000	3140	2450	2050	2425	3000	2300	2900
CTE (µ/°C)	4.7–5	4.5	7.4	5–6	8.3	8.1	3.8	6	7.4
Crystal type	Hex	Hex	Cub	Rhomb	Hex	Cub	Hex	Hex	Hex

impact in controlling properties. Moreover, fewer particles is needed to achieve the required prop-erties when nanoparticles are incorporated in matrix instead of micro-particles (Casati and Vedani 2014). Accordingly, researchers are incorporating different nanoparticles, i.e. WC, TiC, B_4C, AlN, SiC, Al_2O_3, graphite, TiB_2, ZnO, etc. in Mg matrix to achieve various desired properties (Nguyen et al. 2015; Alam et al. 2011; Erman et al. 2012; Meenashisundaram and Gupta 2014; Selvam et al. 2014; Kaviti et al. 2018; Dalmis et al. 2016; Banerjee et al. 2019a, 2019b; Karuppusamy et al. 2019a, 2019b; Banerjee et al. 2020, 2021c; Al-Maamari et al. 2019; Narayanasamy et al. 2018).

However, ever-increasing demand for energy efficient materials has pressed the researchers to extend further as it is difficult for single reinforcement to fulfil required demands. Subsequently, researchers are aiming towards optimizing different properties economically and endorsed a new approach to develop new class material called hybrid composite. Typically, hybrid composites con-sist of two or more reinforcements in the form of primary reinforcement and secondary reinforce-ment. The presence of two or more constituents can provide the prosperity of both reinforcements in the base matrix. Usually, incorporation of ceramic-based harder particles as primary reinforce-ment and softer particles as secondary reinforcement is more common. Typically, harder parti-cles provide better strength and enhance mechanical properties, whereas softer particles help to achieve lower density and better tribological properties (Macke et al. 2012). Furthermore, hybrid composites can also provide better mechanical properties, microstructural properties, enhanced wettability, mass production provision, and advantage of reinforcing localized regions (Suneesh and Sivapragas 2018). Some commonly used reinforcement combinations are SiC-Graphene (Gr), SiC-Al_2O_3, SiC-CNT, WC-Gr, and CNT-Al_2O_3 (Thakur, Kwee et al. 2007; Wong et al. 2006; Thakur, Srivatsan et al. 2007; Radhika and Subramanian 2013; Feng et al. 2008; Zhang et al. 2006; Banerjee and Sahoo 2022; Banerjee, Sutradhar et al. 2022). Girish et al. (2015) investigated the effect of incorporation of SiC and graphite particles in AZ91 matrix to enhance wear resistance of AZ91 alloy. Similarly, Thakur, Srivatsan et al. (2007) examined the Mg-SiC-CNT hybrid composite and observed enhanced mechanical and thermal properties. Prakash et al. (2016) scrutinized mechani-cal and tribological behaviour of AZ91-SiC-graphite composites and reported that microhardness and wear resistance have enhanced by 300% and 80%, respectively. Babu et al. (2010) reported bet-ter mechanical properties of hybrid composites compared to base matrix while incorporating Al_2O_3 and graphite particles in Mg matrix. Banerjee, Sahoo et al. (2022; Banerjee, Sutradhar et al. 2022) have developed AZ31-WC-graphite, which yielded better mechanical, corrosion, and tribological properties compared to AZ31 alloy. Khatkar et al. (2018) have reported that hybridization using graphite as secondary reinforcement in Mg matrix provides positive impact on mechanical as well as tribological properties.

In the present study, emphasis is given to WC as primary reinforcement and graphite as second-ary reinforcement in Mg matrix. WC is selected as primary reinforcement because of noticeable properties like higher melting point, higher elastic modulus, high hardness, better thermal con-ductivity, and high oxidation resistance. Graphite is taken as secondary reinforcement due to its self-lubricating properties, low density, etc. Moreover, properties of MMCs depend on different parameters like type and size of reinforcement, fabrication technique, etc. Judicious selection of fabrication method is also essential to achieve homogeneous distribution and obtain the desired matrix-reinforcement interface. Typically, wettability and particle distribution are main concerns for development of MMCs. In this regard, different fabrication methods i.e. stir casting, ultrasonic vibration–associated stir casting, disintegrated melt deposition (DMD), and powder metallurgy are considered in literature. In stir casting, shape, speed and location of stirrer are main factors. But the main problem of the stir casting technique is agglomeration of reinforced particles. This problem can be encountered by providing ultrasonic vibration in molten slurry (Casati and Vedani 2014; Dieringa 2018; Malaki et al. 2019). However, considering the importance of Mg-based hybrid com-posites in the aviation and aerospace sectors, the present chapter details recent advances in this area with particular emphasis on Mg-WC-graphite hybrid nanocomposites. The main points discussed in this chapter are the fabrication methods of Mg-WC-graphite composites, their microstructural and mechanical properties, and tribological and corrosion characteristics. Finally, discussion on the potential of Mg-WC-graphite hybrid nanocomposites as future prospects of automotive and aerospace material is presented.

9.2 REINFORCEMENT SELECTION

Judicious selection of reinforcement is of immense importance for developing MMCs. Utmost care is needed while selecting reinforcement. The base matrix, required properties, and development techniques are prime governing factors in this regard. Typically, the criteria for selection of reinforcement depend on the following characteristics (Mingbo et al. 2008; Ding et al. 2008; Gunther et al. 2006; Morisada et al. 2006):

- Low density value of reinforcement
- Mismatch of thermal expansion coefficient between matrix and reinforcement to minimize thermal impact
- Bonding ability with matrix material
- Ease of availability
- Superior wettability
- Low corrosion
- Good mechanical properties

Usually, reinforcements are found in fibres, whisker, or particulate form. Superior final properties can be obtained for any developed composite if (a) reinforcement is compatible with base matrix, (b) minimum porosity is attained, (c) reinforcement is homogeneously distributed in matrix, (d) selected reinforcement possesses sufficient strength and stiffness, and (e) superior interfacial integrity is attained between matrix and reinforcement (Ibrahim et al. 1991). Accordingly, in the current study, emphasis is given to WC and graphite as reinforcement due to their different positive properties.

9.3 FABRICATION TECHNIQUES

Different fabrication techniques are employed by researchers to develop Mg-MMCs. These processes are also useful to develop Mg-based metal-matrix nanocomposites. Commonly employed processing techniques used to develop Mg-MMCs may be classified in the following manner:

- *Liquid phase–based processing methods:* Stir casting, disintegrated melt deposition (DMD), squeeze casting, ultrasonic vibration–associated stir casting, compo-casting, melt infiltration, etc. (Banerjee et al. 2021b; Poddar et al. 2007; Gui et al. 2004; Kumar and Dhindaw 2007; Ramalingaiah and Ray 2011; Contreras et al. 2004; Manor et al. 1993; Balart and Fan 2014; Asano and Yoneda 2002; Yoshida et al. 1999; Hu 1998; Ghomashchi and Vikhrov 2000; Girot et al. 1987)
- *Solid phase–based processing methods:* Powder metallurgy (PM), hot rolling, mechanical alloying, physical/chemical vapour deposition, diffusion bonding, etc. (Rathee et al. 2018; Lee et al. 1997; Goh et al. 2005; Jiang et al. 2005; Tunber and Vervoot 1995; Baldwin et al. 1996; Diplas et al. 1998)
- *Processing methods based on both the phases (solid/liquid):* Rheocasting and spray atomization (Ridder et al. 1981)
- *Other processing techniques:* Reactive hot pressing, friction stir processing, rapid solidification technique, and reactive spontaneous infiltration (Ahsan et al. 2016; Wang et al. 2004; Cao et al. 2007; Zuo et al. 2015; Kevorkijan et al. 2002; Ünal 1992)

Here, a few commonly used techniques are discussed briefly.

9.3.1 Liquid State–Based Methods

Liquid metallurgy–based processing techniques are found to be widely used for Mg-MMCs. In liquid metallurgy processes, reinforcement phases are infiltrated in molten base matrix to obtain a homogeneous mixture. Infiltration of reinforcements in molten matrix forms composite slurry, which may be obtained with the help of mechanical agitation produced by centrifugal dispersion–based mechanical stirring or injection gun (Schwartz 1984). Liquid metallurgical techniques mainly vary depending on particular infiltration processes. After typical infiltration of reinforcements, molten slurry is guided to produce required products. Depending on requirements, infiltration techniques are modified, and further development is made. These techniques are economical and suitable for mass production. Most dominant liquid metallurgical techniques (Banerjee et al. 2021a) are stir casting, DMD, ultrasonic vibration–associated stir casting, and squeeze casting.

9.3.1.1 Stir Casting

The stir casting technique is one of the most economical and commonly used fabrication techniques of MMCs. In this process reinforcement phases are added in molten matrix followed by continuous stirring to achieve homogeneous dispersion of reinforcements in the matrix. Usually, mechanical stirring is provided at high temperature. After sufficient stirring, molten slurry is allowed to solidify through different casting methods like sand casting, permanent mould casting, die casting, etc. Researchers have used this technique to fortify Al_2O_3, WC, and SiC particles in Mg matrix. Sometimes, secondary operations are also conducted to achieve homogeneous dispersion and reduce porosity-related issues (Saravanan and Surappa 2000; Amirkhanlou and Niroumand 2010). Recently, Aatthisugan et al. (2017) have developed Mg-based hybrid composites (AZ91D-B_4C-Gr) using the stir casting technique. Developed hybrid composites possessed superior wear resistance and mechanical properties. Schematic representation of the stir casting technique followed by Aatthisugan et al. (2017) is presented in **Figure 9.1**.

However, premature failure of castings may happen due to distribution-related issues of secondary reinforcements. This issue is quite normal for stir casting. But for good strengthening, uniform distribution of secondary reinforcement is also required. Another important observation in stir casting is the agglomeration-related problem of reinforcements because of sedimentation of surfacing during pouring (Chen and Yao 2014; Qiang 2010). Even oxidation-proneness of Mg may get enhanced due to gas entrapment during stir casting. Therefore, the need of further modification in stir casting is felt by researchers. Recently, two-level mixing processes, ultrasonic vibration–assisted stir casting, etc. are getting attention. In the two-level mixing technique, matrix material is held in semisolid condition, and reinforcement is mixed. Then that slurry is again heated, and proper stirring is done (Goh et al. 2010).

On the other hand, ultrasonic vibration–assisted stir casting is a modification of the stir casting process. In this process, initially particles are incorporated in molten matrix, and mechanical stirring is done while ultrasonic vibration is provided afterwards. Ultrasonic vibration provides shock waves, acoustic streaming, and very high temperature, which helps to de-agglomerate the particle

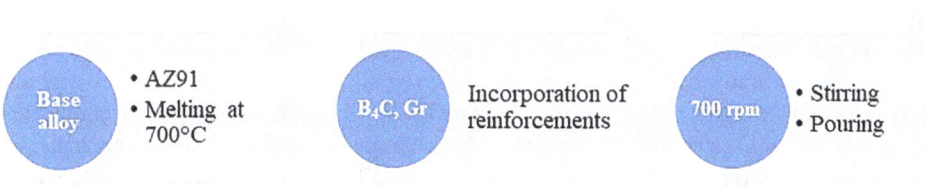

FIGURE 9.1 Schematic representation of stir casting technique (Aatthisugan et al. 2017).

FIGURE 9.2 Schematic representation of ultrasonic vibration–assisted stir casting technique (Banerjee and Sahoo 2022; Banerjee, Sutradhar et al. 2022).

clusters. As a result, reinforcement phase is homogeneously distributed in matrix (Banerjee et al. 2021a). Recently, several researchers have taken advantage of this technique to develop Mg-MMCs. Erman et al. (2012) have incorporated SiC nanoparticles in Mg-matrix with the help of ultrasonic treatment. Casati and Vedani (2014) have reviewed that researchers are successfully developing Mg-based nanocomposites using ultrasonic treatment. Banerjee et al. (Banerjee et al. 2019a, 2019b, 2019c, 2019d; Banerjee, Sahoo et al. 2021) have successfully incorporated WC nanoparticles in AZ31 matrix using the ultrasonic vibration–assisted stir casting technique. Banerjee et al. have also developed AZ31-WC-graphite hybrid nanocomposites using the same technique and observed superior mechanical, tribological, and corrosion properties. Schematic representation of the ultrasonic vibration–assisted stir casting technique followed by Banerjee et al. (2022; Banerjee, Sutradhar et al. 2022) is presented in **Figure 9.2**.

9.3.1.2 Squeeze Casting
In the squeeze casting technique, sufficient pressure is employed while injecting molten metal in a preform. Preforms are usually of short particles of fibres. Subsequently, molten metal will entrap in the pores of that preform and form the desired composite. Reinforcement preforms may be formed by vacuum or press forming. This process is normally employed for soft matrix (Al, Mg) having low melting point. Recently, researchers have successfully developed different composites (i.e. AZ91-SiC, ZK51A-SiC, AM100-Al_2O_3, etc.) using the squeeze casting technique (Zheng et al. 2001; Jayalakshmi et al. 2002; Lo et al. 2004). Svoboda et al. (2007) have employed squeeze casting to fabricate Mg-based hybrid composites using QE22 and AZ91 as base alloys with SiC/pure Si and carbon fibres as reinforcements. Squeeze casting is useful to incorporate larger volume of reinforcements (Yong and Clegg 2005).

9.3.1.3 DMD Process
The DMD process is a unique combination of conventional casting process and spray forming. In this process, low impinge jet velocity and superheat temperatures are employed. In this technique, solidification of atomized melt happens rapidly, and required distribution of particles is achieved. Literature reveals that Mg-based composites can be successfully developed through this technique. Hassan and Gupta (2005, 2006, 2007) have incorporated Al_2O_3, ZrO_2, and Y_2O_3 particles in Mg matrix using the DMD technique and observed excellent distribution of particles. Similarly, Nguyen and Gupta (2008) have also used this technique to fortify Al_2O_3 particles in Mg matrix. Recently, Wong and Gupta (2005) have incorporated hybrid reinforcements (carbon fibres and galvanised iron wire mesh) in pure Mg using the DMD process and reported excellent interfacial bonding as well as superior mechanical properties. Sankaranarayanan et al. (2014) have also fabricated Mg-based hybrid composites through the DMD process by incorporating B_4C and Ti particulates.

It is observed in literature that that DMD technique typically generates uniformly distributed composite having equiaxed grains and minimum porosity. As a result, better specific properties are also achieved for DMD-generated composites. This technique is also economical, as material waste is significantly low. But enhanced viscosity of the slurry is a matter of concern for the DMD technique, as distribution of larger amount of reinforcements through stirring is difficult. Hence, incorporation of a larger amount of reinforcements (micro particles > 30% & nanoparticles > 3%) is not recommendable for this technique (Suneesh and Sivapragash 2018).

FIGURE 9.3 Schematic representation of fabrication of Mg-0.2Y$_2$O$_3$-0.6Cu through PM technique (Tun et al. 2010).

9.3.2 SOLID STATE–BASED METHODS

Solid-phase processing methods are another alternative for fabricating Mg-MMCs. Main advantages of solid-state techniques are minimal interfacial reaction, lower diffusion rate, and low temperature. In these techniques, matrix and reinforcements are in either powder or particle form. Powders are mixed and integrated to end products through different processes. In solid phase–based methods, powder metallurgy and diffusion bonding are some widely accepted techniques.

9.3.2.1 Powder Metallurgy (PM) Technique

The powder metallurgy technique is one of the oldest and best-accepted solid-phase processing methods. In this technique, matrix and reinforcements are blended and mixed thoroughly. Then compaction is done in a mould at high pressure, and green compact is formed. Afterwards, sintering of green compact is done at the required temperature (below the melting point of the base matrix) so that minimal interfacial reaction will occur. Researchers have incorporated different hybrid reinforcements such as micro- and nano-SiC, SiC & CNT, SiC & Al$_2$O$_3$, Y$_2$O$_3$ & Ni, SiC & MWCNT, micro- and nano-Al$_2$O$_3$, Al & CNT, TiC & MoS$_2$, CNT & Al$_2$O$_3$, etc. (Thakur, Kwee et al. 2007; Ahsan et al. 2016; Habibi et al. 2011; Thakur, Balasubramanian et al. 2007; Wong and Gupta 2006; Tun and Gupta 2009; Wong et al. 2005; Selvakumar and Narayanasamy 2016). Schematic representation of fabrication of Mg-0.2Y$_2$O$_3$-0.6Cu through the PM technique is presented in **Figure 9.3**. Tun and Gupta (2009) have developed Mg-Y$_2$O$_3$-Ni hybrid composites through a microwave sintering–associated PM technique. It is observed that tensile properties have been enhanced significantly due to incorporation of hybrid reinforcements.

9.4 MG-WC-GRAPHITE HYBRID NANOCOMPOSITES

9.4.1 MICROSTRUCTURAL PROPERTIES

Microstructures of composites are mainly regulated by distribution of reinforcing particulates and fabrication procedures. As researchers have used a wide variety of fabrication techniques, microstructures of the developed composites also vary. Incorporation of reinforcements typically reduces grain size and refines grain structure. Researchers have fortified different nanoparticles (TiC, BN, SiC, Y$_2$O$_3$, Al$_2$O$_3$, graphite, CNT, WC, etc.) in Mg matrix to improve microstructural properties. Researchers have also studied microstructural changes of Mg-based hybrid MMCs (Selvakumar and Narayanasamy 2016; Tun et al. 2010). Tun et al. (2010) have examined the microstructure of Mg-Y$_2$O$_3$-Cu hybrid composites and observed noticeable reduction in grain size. Zhou et al. (2012) have incorporated varying wt.% of SiC and CNT particulates in AZ91 matrix. Optical micrographs reveal that grain structure is refined significantly up to 1 wt.% addition of reinforcements. Researchers have also found that the size of the reinforced particles also has significant impact on maintaining structural integrity of composites. Hassan and Gupta (2005) have found that incorporation of nano-sized particles has shown better impact on different properties compared to micron-sized reinforcement. Recently, Banerjee et al. (Banerjee and Sahoo 2022; Banerjee et al. 2021c; Banerjee, Sutradhar et al. 2022) have incorporated WC (1 & 2 wt.%) and Gr nanoparticles (1 & 2 wt.%) in AZ31 matrix and scrutinized the microstructural changes with the help of optical micrographs and SEM images. Optical micrographs and SEM images of AZ31-WC-graphite

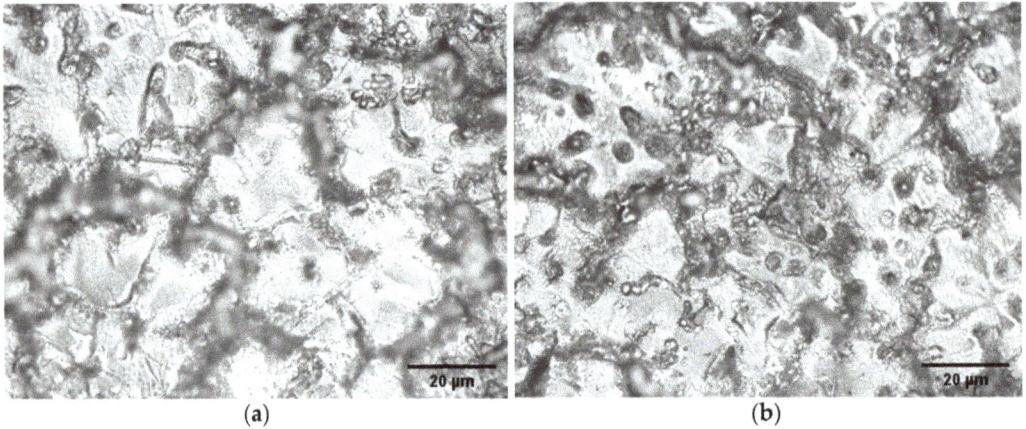

FIGURE 9.4   Optical micrograph of hybrid nanocomposites (a) Mg-1WC-1Gr, (b) Mg-2WC-1Gr. *Reproduced with permission from (Banerjee and Sahoo 2022) © 2022 by the authors. Licensee MDPI, Basel, Switzerland.*

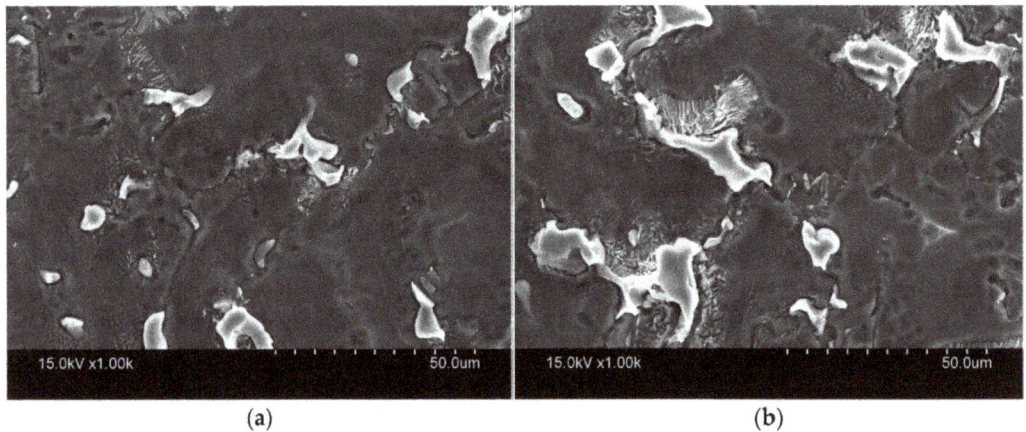

FIGURE 9.5   SEM micrograph of hybrid nanocomposites (a) Mg-1WC-1Gr, (b) Mg-2WC-1Gr. *Reproduced with permission from (Banerjee and Sahoo 2022) © 2022 by the authors. Licensee MDPI, Basel, Switzerland.*

hybrid composites are presented in **Figure 9.4** and **Figure 9.5**, respectively. Figure 9.4 discloses that optical micrographs mainly consist of α-phase (α-Mg) and β-phase (β-$Mg_{17}Al_{12}$). Existence of the β-phase mainly helps to reduce the grain growth, resulting in reduced grain size. Incorporation of reinforcements mainly generates interconnecting chains, which have blackish structures. These blackish structures increase with increase in wt.% of graphite reinforcements. It is also observed that the presence of WC particles forms spherical structures, and graphite forms flakelike structures. Equiaxed grains having interfacial bonding is also observed. Availability of equiaxed grains indicates that reinforced particulates have successfully stimulated nucleation at interfaces and persuaded recrystallization. **Figure 9.5** reveals that WC and graphite particulates are uniformly distributed in AZ31 matrix. No significant cluster formation is found in SEM micrographs.

9.4.2   MECHANICAL PROPERTIES

Mechanical properties of composites normally depend on wt.%/vol.% of reinforcement, their shape, size, and dispersion. Properties of reinforcing particles may have positive or negative impact on mechanical properties of the developed composite. Apart from properties of reinforcement, interfacial bond types also have considerable impact on mechanical properties of composites. Composites having weak interfacial bonding between matrix reinforcements show poor properties. Typically, composites possess heterogeneous elastic-plastic behaviour under mechanical loading, whereas reinforcement exhibits elastic deformation, while matrix exhibits plastic deformation (Suneesh and Sivapragash 2018). Zhang et al. (2014) have incorporated Al_2O_3 particles and fibre forms as reinforcement in AM60 matrix and scrutinized tensile behaviour. Elastic modulus and yield strength were enhanced by 40% and 80%, respectively. Hassan and Gupta (2003) have incorporated micro- and nanosized Cu particulates in Mg matrix and observed 0.2% enhancement in yield strength. Rashad et al. (2015) have developed Mg-Cu-Gr hybrid composites. It was reported that ultimate tensile strength and yield strength had enhanced significantly for the nanocomposites. Nguyen and Gupta (2010) have examined compressive behaviour of $AZ31–1.5Al_2O_3$-Ca hybrid composites and observed superior compressive characteristics compared to AZ31 alloy. Sankaranayanan et al. (2011) have developed $Mg-Ti-Al_2O_3$ hybrid composites and reported that ultimate compressive strength and compressive yield strength have enhanced by 45% and 20%, respectively, compared to the base matrix.

Recently, Banerjee et al. (2021c) have developed AZ31-WC-graphite hybrid nanocomposite and conducted microhardness and nanoindentation tests. The microhardness test revealed that incorporation of 1 wt.% of graphite nanoparticles with AZ31-WC nanocomposite resulted in enhancement in microhardness, while further incorporation of graphite had a negative effect. AZ31-2WC-1Gr shows highest microhardness, and around 78% enhancement in microhardness value is achieved compared to AZ31 matrix. The effect of hybrid reinforcement on microhardness is presented in **Figure 9.6a**.

Banerjee et al. further investigated nanohardness and elastic modulus of AZ31-WC-Gr hybrid composites with the help of a nanoindentation test. The loading-displacement curve of different hybrid nanocomposites is analysed using the Oliver-Pharr method to find the elastic modulus and nano-hardness of the samples. Details of nanohardness and elastic modulus of AZ31-WC-graphite composites are shown in Table 9.3. The effect of hybrid reinforcements on nanohardness is presented in **Figure 9.6b**. It was observed that nanohardness and elastic modulus are enhanced due to incorporation of 1 wt.% of graphite nanoparticles, but further incorporation (2 wt.%) of graphite results in detriment. AZ31-2WC-1Gr showed the best result among developed hybrid nanocomposites. This study provides an interesting result that both elastic modulus and nanohardness were enhanced simultaneously (Banerjee et al. 2021c).

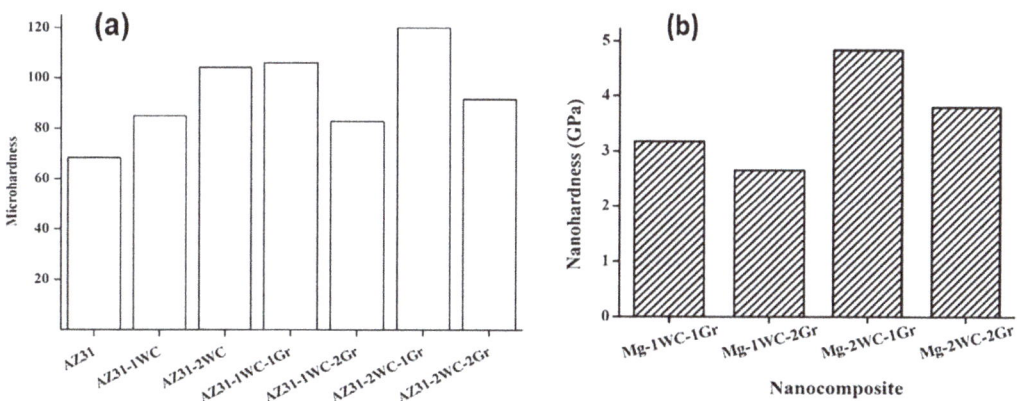

FIGURE 9.6   (a) Microhardness plot and (b) nanohardness plot of base alloy and Mg-WC-Gr hybrid nanocomposite. *Reproduced (b) with permission from (Banerjee et al. 2021c) © 2020, American Foundry Society.*

TABLE 9.3

Nanohardness and Elastic Modulus of AZ31-WC-Graphite Hybrid Composites.
Reproduced with permission from (Banerjee et al. 2021c) © 2020, American Foundry Society.

Material	Nanohardness (GPa)	Elastic modulus, E (GPa)
AZ31-1WC-1Graphite	3.17	70.25
AZ31-1WC-2Graphite	2.65	51.27
AZ31-2WC-1Graphite	4.83	122.84
AZ31-2WC-2Graphite	3.79	101.97

9.4.3 TRIBOLOGICAL PROPERTIES

It is already observed that AZ31-WC-Gr hybrid composites are showing noticeable enhancement in mechanical properties, making them a novel option as well-accepted tribological material. Typically, tribological characteristics rely on mechanical behaviour, microstructure, load, sliding distance, sliding speed, etc. Literature also reveals that fortification of hybrid reinforcements significantly enhances tribological characteristics of Mg/Mg alloys. Recently, Banerjee et al. (2021c) developed AZ31-WC-Gr hybrid nanocomposites by incorporating varying amounts of WC (1 & 2 wt.%) and graphite (1 & 2 wt.%) nanoparticles. Tribological behaviour of the developed hybrid nanocomposites are examined by carrying out tribological tests for different load (10–40N) and sliding speed (0.1–0.4 m/s) in a pin-on-disc tribotester, considering EN8 steel disc as counter-surface.

Effect of load and sliding speed on wear rate is depicted in **Figure 9.7**. It is clear from Figure 9.7 that presence of 1 wt.% of graphite reduced wear rate significantly, while further addition of graphite enhances wear rate. Typically, this is due to the microhardness of hybrid composites. It is observed in Figure 9.6a that incorporation of 1 wt.% of graphite enhances microhardness, while negative effect is there for 2 wt.% of graphite.

Banerjee et al. (2021c) correlated this result with Archard's wear law (Archard 1953). It is observed that ceramic-based WC particles help to increase load bearing capability of the composite, while graphite acts as a solid lubricant, which creates a tribo-layer between sample and counter-face. On the other hand, addition of 2 wt.% of graphite enhances brittle fracture tendency, resulting increase in wear rate. It is also observed that wear rate enhances significantly with increase in load, while moderate enhancement in wear rate is observed for hybrid composites. Basically, the protective tribolayer generated due to frictional heating helps to control wear rate for composites. But further enhancement in load causes delamination and thermal softening, which leads towards enhancement in wear rate. Similarly, wear rate enhances continuously with increase in sliding speed. Base alloy shows maximum wear rate, while hybrid composites show moderate increment. The effect of incorporation of hybrid reinforcements on coefficient of friction (COF) for varying load and sliding speed is presented in **Figure 9.8**. It was observed that COF decreases continuously with increase in applied load, while COF value decreases moderately with increase in sliding speed.

Banerjee, Sutradhar et al. (2022) have examined abrasive wear behaviour of AZ31-WC-graphite hybrid composites for different sliding distances (track diameter – 30, 40, & 50 mm) and different abrasive grit size (400, 600, & 800 grit). The effect of hybridization on wear rate with respect to sliding distances and abrasive grit size is presented in **Figure 9.9 (a)** and **Figure 9.9 (b)**, respectively. It is obvious from Figure 9.9a that fortification of 1 wt.% of Gr nanoparticles in AZ31-WC composites reduces wear rate, while fortification of 2 wt.% Gr increases wear rate. Moreover, wear rate increases linearly with increase in sliding distance. Additionally, it is distinct from Figure 9.9b that wear rate reduces significantly with increase in abrasive grit size for all the materials.

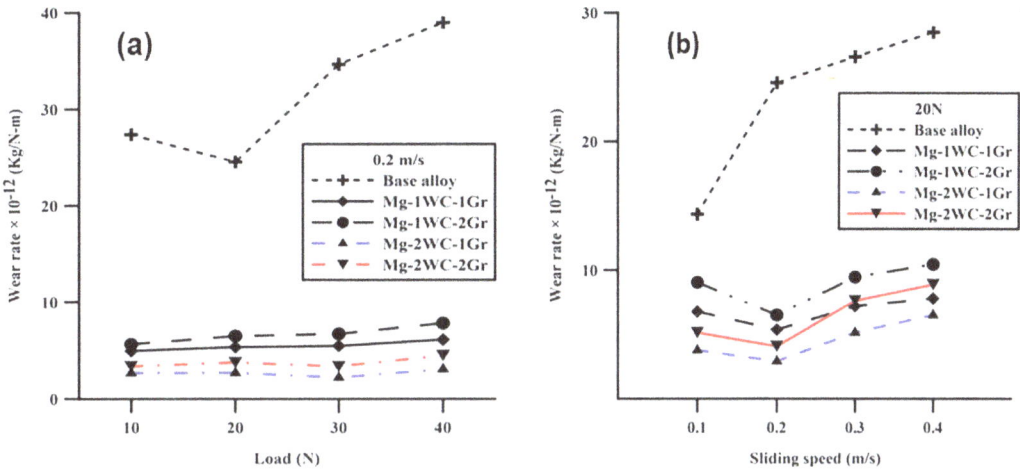

FIGURE 9.7 Wear rate of hybrid nanocomposites with respect to (a) load and (b) sliding speed *Reproduced with permission (Banerjee, Sutradhar et al. 2022) © Taylor & Francis.*

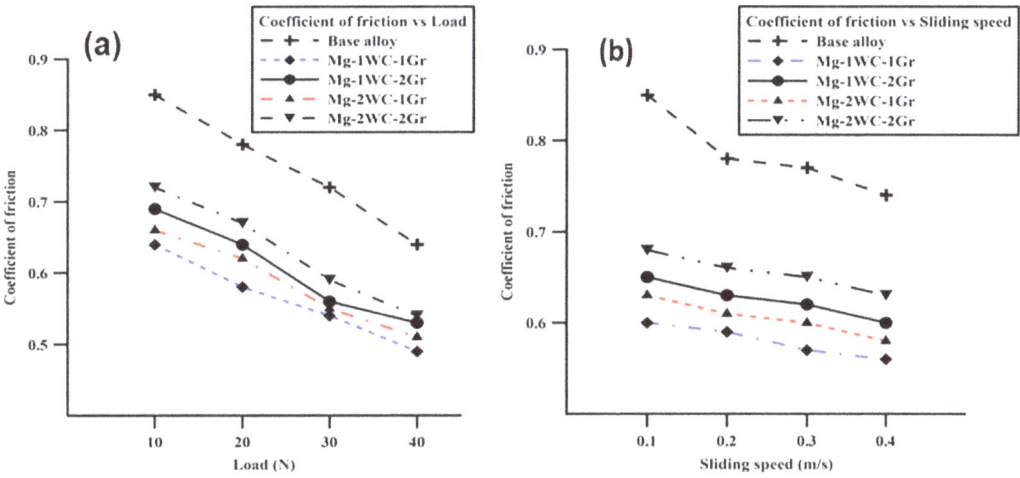

FIGURE 9.8 COF of hybrid nanocomposites with respect to (a) load and (b) sliding speed. *Reproduced with permission (Banerjee, Sutradhar et al. 2022) © Taylor & Francis.*

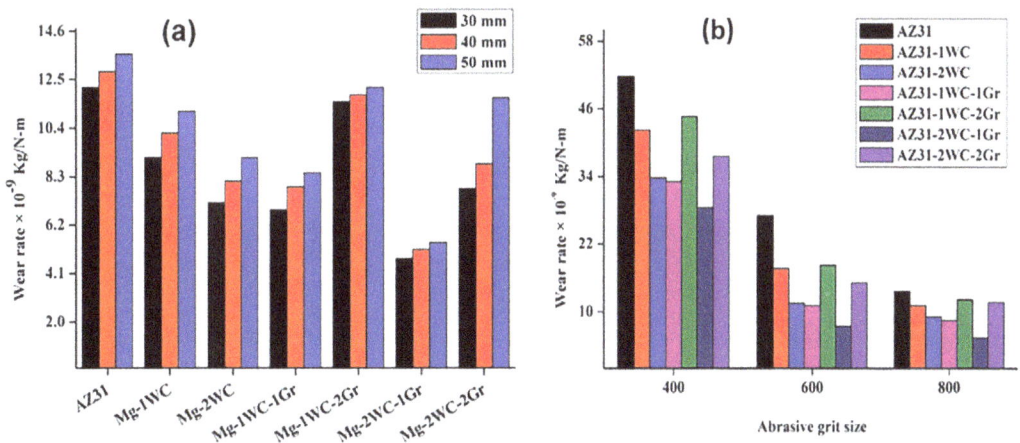

FIGURE 9.9 Effect of hybridization on wear rate with respect to (a) sliding distance and (b) abrasive grit size. *Reproduced with permission from (Banerjee and Sahoo 2022) © 2022 by the authors. Licensee MDPI, Basel, Switzerland.*

9.4.4 CORROSION CHARACTERISTICS

Researchers are very much concerned about corrosion characteristics of Mg-based hybrid composites because of the corrosion-proneness of Mg. Some researchers observed positive impact while some others observed negative impact of reinforcement on corrosion characteristics. Banerjee et al. (2021c) have investigated corrosion characteristics of Mg-WC-graphite hybrid composites in 3.5% NaCl solution. Results of corrosion tests were presented in the form of Nyquist (**Figure 9.10a**) and

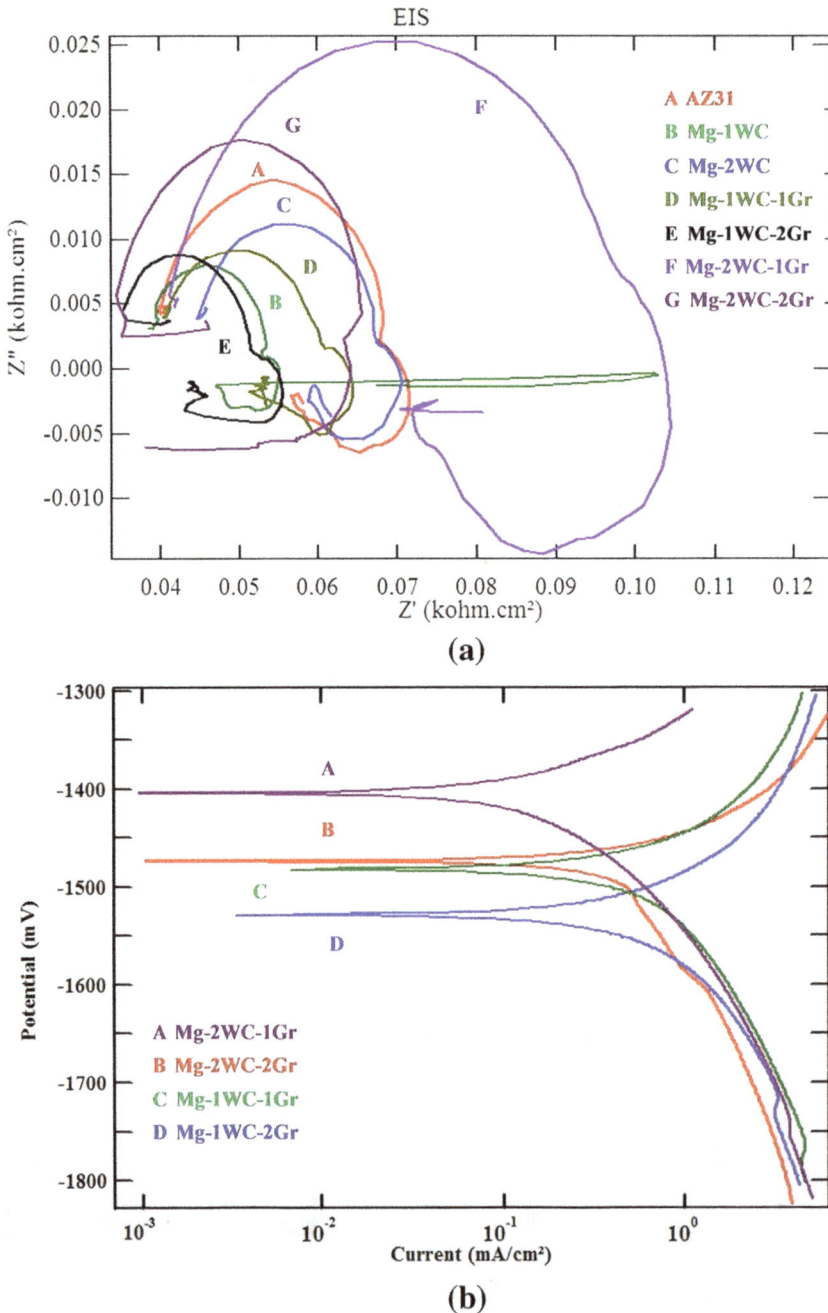

FIGURE 9.10   (a) Nyquist and (b) Tafel plots. *Reproduced with permission from (Banerjee et al. 2021c) © 2020, American Foundry Society.*

TABLE 9.4

Tafel Extrapolation Result. *Reproduced with permission from (Banerjee et al. 2021c) © 2020, American Foundry Society.*

Material	i_{corr} (mA/cm²)	E_{corr} (V)
Base matrix	0.07	−1.43
Mg-1WC-1Gr	0.09	−1.48
Mg-1WC-2Gr	0.15	−1.53
Mg-2WC-1Gr	0.06	−1.41
Mg-2WC-2Gr	0.08	−1.47

Tafel (**Figure 9.10b**) plots. Nyquist plots were comprised of two capacitive loops and one inductive loop for all the materials. It is also observed in the Nyquist plot that the diameter of capacitive loop of Mg – 2 wt.%WC – 1 wt.% graphite is maximum among the tested materials. This finding establishes that Mg-2WC-1graphite is the most corrosion-resistive material. The results of the Tafel extrapolation (Table 9.4) confirm this result, as Mg-2WC-1graphite possessed the highest corrosion potential (E_{corr}) and lowest corrosion current density (i_{corr}) values.

9.5 CONCLUSIONS AND OUTLOOK

This chapter provides a concise account of the synthesis, microstructure, mechanical properties, and tribological and corrosion characteristics of the Mg-WC-graphite hybrid composites. Primarily, dispersion capability and economic aspects of different fabrication techniques have been discussed with the help of existing literature. The effects of incorporating WC and graphite nanoparticles and the dispersion of these reinforcements in the Mg matrix have been detailed. Literature reveals that WC and graphite particles could be successfully fortified in the Mg matrix using the ultrasonic vibration–assisted stir casting technique. Ultrasonic vibration helps to avoid cluster formation of reinforcements by homogeneously dispersing reinforcements. The mechanical properties like microhardness, elastic modulus, and nanohardness have been enhanced noticeably by incorporating 2 wt.% of WC and 1 wt.% of graphite particles. Tribological characteristics like wear and friction behaviour of hybrid composites at dry sliding and abrasive conditions also yielded the best result for the composite. The enhancement in the corrosion characteristics of Mg-WC-Gr hybrid nanocomposites has also been discussed.

The future prospects of hybrid composites are wide, as they possess benefits of combined properties of different reinforced particles. Presently, ceramic-based particles are mostly used as reinforcement. Sometimes, the cost of ceramic reinforcements is found to be higher. To develop cost-effective hybrid composites, industrial or agricultural wastes can be considered as reinforcement. Recycling of those waste materials can also be beneficial in the context of sustainability. Waste products like rice husk, red mud, coconut husk, fly ash, etc. may be considered as feasible reinforcing agents for MMCs.

Magnesium-based hybrid composites are very useful for their different properties. However, widespread utilization of hybrid MMCs will be possible if all the aspects of these materials are disclosed in detail. Some literature is available on mechanical properties, but further studies related to other characteristics could be more detailed. Hence, further studies on corrosion and tribological aspects are needed. Evaluation of the effect of secondary processing (rolling, machining, welding, forging, extrusion, etc.) is also essential. Further studies are imperative considering recyclability, economic aspects, and reliability.

REFERENCES

Aatthisugan, I, A Razal Rose, D Selwyn Jebadurai, 2017, Mechanical and wear behaviour of AZ91D magnesium matrix hybrid composite reinforced with boron carbide and graphite. Journal of Magnesium and Alloys, 5(1), 20–25.

Ahsan, Q, ZW Tee, S Rahmah, SY Chang, M Warikh, 2016, Wear and friction behaviour of magnesium hybrid composites containing silicon carbide and multi-walled carbon nanotubes. Advances in Materials and Processing Technologies, 2(2), 303–317.

Alam, ME, S Han, QB Nguyen, AMS Hamouda, M Gupta, 2011, Development of new magnesium based alloys and their nanocomposites. Journal of Alloys and Compounds, 509(34), 8522–8529.

Al-Maamari, AEA, AA Iqbal, DM Nuruzzaman, 2019, Wear and mechanical characterization of Mg – Gr self-lubricating composite fabricated by mechanical alloying. Journal of Magnesium and Alloys, 7(2), 283–290.

Amirkhanlou, S, B Niroumand, 2010, Synthesis and characterization of 356-SiC$_p$ composites by stir casting and compocasting methods. Transactions of Nonferrous Metals Society of China, 20, s788–s793.

Archard, J, 1953, Contact and rubbing of flat surfaces. Journal of Applied Physics, 24(8), 981–988.

Asano, K, H Yoneda, 2002, Microstructure and mechanical properties of Al-Cu-Mg alloy matrix hybrid composites fabricated by squeeze casting. International Journal of Cast Metals Research, 14(4), 199–205.

Babu, JSS, KP Nair, G Unnikrishnan, CG Kang, HH Kim, 2010, Fabrication and properties of magnesium (AM50)-based hybrid composites with graphite nanofiber and alumina short fiber. Journal of Composite Materials, 44(8), 971–987.

Balart, MJ, Z Fan, 2014, Surface oxidation of molten AZ91D magnesium alloy in air. International Journal of Cast Metals Research, 27(3), 167–175.

Baldwin, KR, DJ Bray, GD Howard, RW Gardiner, 1996 Corrosion behaviour of some vapour deposited magnesium alloys. Materials Science and Technology, 12(11), 929–944.

Banerjee, S, S Poria, G Sutradhar, P Sahoo, 2019a, Dry sliding tribological behavior of AZ31-WC nanocomposites. Journal of Magnesium and Alloys, 7(2), 315–327.

Banerjee, S, S Poria, G Sutradhar, P Sahoo, 2019b, Corrosion behavior of AZ31-WC nano-composites. Journal of Magnesium and Alloys, 7(4), 681–695.

Banerjee, S, S Poria, G Sutradhar, P Sahoo, 2019c, Tribological behavior of Mg-WC nano-composites at elevated temperature. Materials Research Express, 6(8), 0865c6.

Banerjee, S, S Poria, G Sutradhar, P Sahoo, 2019d, Nanoindentation and scratch resistance characteristics of AZ31–WC nanocomposites. Journal of Molecular and Engineering Materials, 7(03n04), 1950007.

Banerjee, S, S Poria, G Sutradhar, P Sahoo, 2020, Abrasive wear behavior of WC nanoparticle reinforced magnesium metal matrix composites. Surface Topography: Metrology and Properties, 8(2), 025001.

Banerjee, S, S Poria, G Sutradhar, P Sahoo, 2021a, Mg-WC nanocomposites – recent advances and perspectives. Recent Advances in Layered Materials and Structures, 199–228.

Banerjee, S, S Poria, G Sutradhar, P Sahoo, 2021b, Understanding fabrication and properties of magnesium matrix nanocomposites. In Recent Advances in Layered Materials and Structures (pp. 229–252), Springer. Singapore.

Banerjee, S, S Poria, G Sutradhar, P Sahoo, 2021c, Nano-indentation and corrosion characteristics of ultrasonic vibration assisted stir-cast AZ31–WC – graphite nano-composites. International Journal of Metalcasting, 15(3), 1058–1072.

Banerjee, S, P Sahoo, 2022, Fabrication and investigation of abrasive wear behavior of AZ31-WC-graphite hybrid nanocomposites. Metals, 12(9), 1418.

Banerjee, S, P Sahoo, JP Davim, 2021, Tribological characterisation of magnesium matrix nanocomposites: A review. Advances in Mechanical Engineering, 13(4), 16878140211009025.

Banerjee, S, P Sahoo, JP Davim, 2022, Tribological performance optimization of Mg-WC nanocomposites in dry sliding: A statistical approach. Frontiers in Materials, 9.

Banerjee, S, G Sutradhar, P Sahoo, 2022, Effect of incorporation of graphite nanoparticles on wear characteristics of Mg-WC nano-composites in dry sliding condition. Transactions of the IMF, 1–11.

Banerjee, S, P Sarkar, P Sahoo, 2022, Improving corrosion resistance of magnesium nanocomposites by using electroless nickel coatings. Facta Universitatis. Series: Mechanical Engineering, 20(3), 647–663.

Bau, NQ, 2009, Development of Nano-Composites Based on Magnesium Alloys System AZ31B. Doctoral Dissertation, National University of Singapore. Singapore.

Cao, W, CF Zhang, TX Fan, D Zhang, 2007, In situ synthesis of TiB$_2$/Mg composites by flux-assisted synthesis reaction of the Al-Ti-B system in molten magnesium. Key Engineering Materials, 351, 166–170.

Casati, R, M Vedani, 2014, Metal matrix composites reinforced by nano-particles – a review. Metals, 4(1), 65–83.

Chawla, KK, 2006, Metal Matrix Composites, Wiley-VCH Verlag GmbH & Co. KGaA. Weinheim.

Chen, L, Y Yao, 2014, Processing, microstructures, and mechanical properties of magnesium matrix composites: a review. Acta Metallurgica Sinica (English Letters), 27(5), 762–774.

Clyne, TW, PJ Withers, 1995, An Introduction to Metal Matrix Composites, The Press Syndicate, Cambridge University Press. Cambridge.

Contreras, A, VH Lopez, E Bedolla, 2004, Mg/TiC composites manufactured by pressure less melt infiltration. Scripta Materialia, 51(3), 249–253.

Dalmis, R, H Cuvalci, A Canakci, O Guler, 2016, Investigation of graphite nano particle addition on the physical and mechanical properties of ZA27 composites. Advanced Composites Letters, 25(2), 37.

Dey, A, KM Pandey, 2015, Magnesium metal matrix composites-a review. Reviews on Advanced Material Science, 42, 58–67.

Dieringa, H, 2018, Processing of magnesium-based metal matrix nanocomposites by ultrasound-assisted particle dispersion: A review. Metals, 8(6), 431.

Ding, SX, WT Lee, CP Chang, LW Chang, PW Kao, 2008, Improvement of strength of magnesium alloy processed by equal channel angular extrusion. Scripta Materialia, 59(9), 1006–1009.

Diplas, S, P Tsakiropoulos, RMD Brydson, 1998, Development of Mg – V alloys by physical vapour deposition Part 1–Bulk and surface characterisation. Materials Science and Technology, 14(7), 689–698.

Erman, A, J Groza, X Li, H Choi, G Cao, 2012, Nanoparticle effects in cast Mg-1 wt% SiCnano-composites. Materials Science and Engineering: A, 558, 39–43.

Feng, YC, L Geng, PQ Zheng, ZZ Zheng, GS Wang, 2008, Fabrication and characteristic of Al-based hybrid composite reinforced with tungsten oxide particle and aluminium borate whisker by squeeze casting. Materials & Design, 29(10), 2023–2026.

Ghomashchi, MR, A Vikhrov, 2000, Squeeze casting: An overview. Journal of Materials Processing Technology, 101(1–3), 1–9.

Girish, BM, BM Satish, S Sarapure, DR Somashekar, Basawaraj, 2015, Wear behavior of magnesium alloy AZ91 hybrid composite materials. Tribology Transactions, 58(3), 481–489.

Girot, FA, LOUIS Albingre, JM Quenisset, ROGER Naslain, 1987, Rheocasting Al matrix composites. JOM, 39(11), 18–21.

Goh, CS, KS Soh, PH Oon, BW Chua, 2010, Effect of squeeze casting parameters on the mechanical properties of AZ91–Ca Mg alloys. Materials & Design, 31, S50–S53.

Goh, CS, J Wei, LC Lee, M Gupta, 2005, Development of novel carbon nanotube reinforced magnesium nanocomposites using the powder metallurgy technique. Nanotechnology, 17(1), 7.

Gopal, PM, KS Prakash, S Nagaraja, NK Aravinth, 2017, Effect of weight fraction and particle size of CRT glass on the tribological behaviour of Mg-CRT-BN hybrid composites. Tribology International, 116, 338–350.

Gui, MC, JM Han, PY Li, 2004, Microstructure and mechanical properties of Mg – Al9Zn/SiC$_p$ composite produced by vacuum stir casting process. Materials Science and Technology, 20(6), 765–771.

Günther, R, C Hartig, R Bormann, 2006, Grain refinement of AZ31 by (SiC) P: Theoretical calculation and experiment. Acta Materialia, 54(20), 5591–5597.

Gupta, M, SNM Ling, 2011, Magnesium, Magnesium Alloys, and Magnesium Composites, John Wiley & Sons Inc. Hoboken, NJ. Materials and Manufacturing Processes, 1339.

Habibi, MK, M Paramsothy, AMS Hamouda, M Gupta, 2011, Using integrated hybrid (Al+ CNT) reinforcement to simultaneously enhance strength and ductility of magnesium. Composites Science and Technology, 71(5), 734–741.

Hassan, SF, M Gupta, 2003, Development of high strength magnesium copper based hybrid composites with enhanced tensile properties. Materials Science and Technology, 19(2), 253–259.

Hassan, SF, M Gupta, 2005, Enhancing physical and mechanical properties of Mg using nanosized Al$_2$O$_3$ particulates as reinforcement. Metallurgical and Materials Transactions A, 36(8), 2253–2258.

Hassan, SF, M Gupta, 2006, Effect of particulate size of Al$_2$O$_3$ reinforcement on microstructure and mechanical behavior of solidification processed elemental Mg. Journal of Alloys and Compounds, 419(1–2), 84–90.

Hassan, SF, M Gupta, 2007, Development of nano-Y$_2$O$_3$ containing magnesium nanocomposites using solidification processing. Journal of Alloys and Compounds, 429(1–2), 176–183.

Hu, H, 1998, Squeeze casting of magnesium alloys and their composites. Journal of Materials Science, 33(6), 1579–1589.

Huang, SJ, YR, Jeng, VI Semenov, YZ Dai, 2011, Particle size effects of silicon carbide on wear behavior of SiCp-reinforced magnesium matrix composites. Tribology Letters, 42(1), 79–87.

Ibrahim, IA, FA Mohamed, EJ Lavernia, 1991, Particulate reinforced metal matrix composites – a review. Journal of Materials Science, 26(5), 1137–1156.

Jayalakshmi, S, SV Kailas, S Seshan, 2002, Tensile behaviour of squeeze cast AM100 magnesium alloy and its Al2O3 fibre reinforced composites. Composites Part A: Applied Science and Manufacturing, 33(8), 1135–1140.

Jiang, QC, HY Wang, BX Ma, Y Wang, F Zhao, 2005, Fabrication of B_4C particulate reinforced magnesium matrix composite by powder metallurgy. Journal of Alloys and Compounds, 386(1–2), 177–181.

Kainer, KU (Ed.), 2006, Metal Matrix Composites: Custom-Made Materials for Automotive and Aerospace Engineering, Wiley-VCH Verlag GmbH & Co. KGaA. Weinheim.

Kainer, KU, BL Mordike (Eds.), 2000, Magnesium Alloys and Their Applications (pp. 1–46), Wiley-Vch. Weinheim.

Karuppusamy, P, K Lingadurai, V Sivananth, 2019a, To study the role of WC reinforcement and deep cryogenic treatment on AZ91 MMNC wear behavior using multilevel factorial design. Journal of Tribology, 141(4).

Karuppusamy, P, K Lingadurai, V Sivananth, 2019b, Influence of cryogenic treatment on as-cast AZ91+ 1.5 wt.% WC Mg-MMNC wear performance. In Advances in Materials and Metallurgy (pp. 185–197), Springer. Singapore.

Kaviti, RVP, D Jeyasimman, G Parande, M Gupta, R Narayanasamy, 2018, Investigation on dry sliding wear behavior of Mg/BN nanocomposites. Journal of Magnesium and Alloys, 6(3), 263–276.

Kevorkijan, V, T Kosmač, K Kristoffer, 2002, Spontaneous reactive infiltration of porous ceramic preforms with Al – Mg and Mg in the presence of both magnesium and nitrogen – new experimental evidence. Materials and Manufacturing Processes, 17(3), 307–322.

Khatkar, SK, NM Suri, S Kant, 2018, A review on mechanical and tribological properties of graphite reinforced self-lubricating hybrid metal matrix composites. Reviews on Advanced Materials Science, 56(1), 1–20.

Kumar, SM, BK Dhindaw, 2007, Magnesium alloy – SiC_p reinforced infiltrated cast composites. Materials and Manufacturing Processes, 22(4), 429–432.

Lee, DM, BK Suh, BG Kim, JS Lee, CH Lee, 1997, Fabrication, microstructures, and tensile properties of magnesium alloy AZ91/SiCp composites produced by powder metallurgy. Materials Science and Technology, 13(7), 590–595.

Lloyd, DJ, 1994, Particle reinforced aluminium and magnesium matrix composites. International Materials Review, 39(1), 1–23. DOI: 10.1179/imr.1994.39.1.1.

Lo, J, G Shen, R Santos, 2004, Preform cracking in squeeze cast magnesium based composites – effects of tooling temperature. International Journal of Cast Metals Research, 17(4), 213–219.

Luo, A, MO Pekguleryuz, 1994, Cast magnesium alloys for elevated temperature applications. Journal of Materials Science, 29(20), 5259–5271.

Macke, A, BF Schultz, P Rohatgi, 2012, Metal matrix composites. Advanced Materials and Processes, 170(3), 19–23.

Malaki, M, W Xu, AK Kasar, PL Menezes, H Dieringa, RS Varma, M Gupta, 2019, Advanced metal matrix nanocomposites. Metals, 9(3), 330.

Manor, E, H Ni, CG Levi, R Mehrabian, 1993, Microstructure evolution of $SiC/Al_2O_3/Al$-Alloy composites produced by melt oxidation. Journal of the American Ceramic Society, 76(7), 1777–1787.

Meenashisundaram, GK, M Gupta, 2014, Low volume fraction nano-titanium particulates for improving the mechanical response of pure magnesium. Journal of Alloys and Compounds, 593, 176–183.

Mingbo, Y, P Fusheng, C Renju, S Jia, 2008, Comparison about effects of Sb, Sn and Sr on as-cast microstructure and mechanical properties of AZ61–0.7 Si magnesium alloy. Materials Science and Engineering: A, 489(1–2), 413–418.

Morisada, Y, H Fujii, T Nagaoka, M Fukusumi, 2006, MWCNTs/AZ31 surface composites fabricated by friction stir processing. Materials Science and Engineering: A, 419(1–2), 344–348.

Narayanasamy, P, N Selvakumar, P Balasundar, 2018, Effect of weight percentage of TiC on their tribological properties of magnesium composites. Materials Today: Proceedings, Elsevier Publications, 5(2), 6570–6578.

Nguyen, QB, M Gupta, 2008, Increasing significantly the failure strain and work of fracture of solidification processed AZ31B using nano-Al_2O_3 particulates. Journal of Alloys and Compounds, 459(1–2), 244–250.

Nguyen, QB, M Gupta, 2010, Improving compressive strength and oxidation resistance of AZ31B magnesium alloy by addition of nano-Al_2O_3 particulates and Ca. Journal of Composite Materials, 44(7), 883–896.

Nguyen, QB, YHM Sim, M Gupta, CYH Lim, 2015, Tribology characteristics of magnesium alloy AZ31B and its composites. Tribology International, 82, 464–471.

Poddar, P, VC Srivastava, PK De, KL Sahoo, 2007, Processing and mechanical properties of SiC reinforced cast magnesium matrix composites by stir casting process. Materials Science and Engineering: A, 460, 357–364.

Polmear, I., D St John, JF Nie, M Qian, 2017, Light Alloys: Metallurgy of the Light Metals, Butterworth-Heinemann. Oxford.

Prakash, KS, P Balasundar, S Nagaraja, PM Gopal, V Kavimani, 2016, Mechanical and wear behaviour of Mg – SiC – Gr hybrid composites. Journal of Magnesium and Alloys, 4(3), 197–206.

Qiang, Z, 2010, Development of Hybrid Mg-based composites. Electronic Theses and Dissertations, 204 University of Windsor, Canada.

Radhika, N, R Subramanian, 2013, Effect of reinforcement on wear behaviour of aluminium hybrid composites. Tribology-Materials, Surfaces & Interfaces, 7(1), 36–41.

Ramalingaiah, BS, S Ray, 2011, Microstructure and mechanical properties of cast composites of steel wool infiltrated by magnesium and AZ91 alloy. Materials and Manufacturing Processes, 26(9), 1173–1178.

Rashad, M, FS Pan, M Asif, A Ullah, 2015, Improved mechanical properties of magnesium – graphene composites with copper – graphene hybrids. Materials Science and Technology, 31(12), 1452–1461.

Rathee, S, S Maheshwari, AN Siddiquee, M Srivastava, 2018, A review of recent progress in solid state fabrication of composites and functionally graded systems via friction stir processing. Critical Reviews in Solid State and Materials Sciences, 43(4), 334–366.

Ridder, SD, S Kou, R Mehrabian, 1981, Effect of fluid flow on macrosegregation in axi-symmetric ingots. Metallurgical Transactions B, 12(3), 435–447.

Sankaranarayanan, S, S Jayalakshmi, M Gupta, 2011, Effect of ball milling the hybrid reinforcements on the microstructure and mechanical properties of Mg–(Ti+ n-Al$_2$O$_3$) composites. Journal of Alloys and Compounds, 509(26), 7229–7237.

Sankaranarayanan, S, RK Sabat, S Jayalakshmi, S Suwas, M Gupta, 2014, Microstructural evolution and mechanical properties of Mg composites containing nano-B$_4$C hybridized micro-Ti particulates. Materials Chemistry and Physics, 143(3), 1178–1190.

Saravanan, RA, MK Surappa, 2000, Fabrication and characterisation of pure magnesium-30 vol.% SiC$_P$ particle composite. Materials Science and Engineering: A, 276(1–2), 108–116.

Schwartz, MM, 1984, Composite Material Handbook, McGraw-Hill Book Company. New York.

Selvakumar, N, P Narayanasamy, 2016, Optimization and effect of weight fraction of MoS$_2$ on the tribological behavior of Mg-TiC-MoS$_2$ hybrid composites. Tribology Transactions, 59(4), 733–747.

Selvam, B, P Marimuthu, R Narayanasamy, V Anandakrishnan, KS Tun, M Gupta, M Kamaraj, 2014, Dry sliding wear behaviour of zinc oxide reinforced magnesium matrix nano-composites. Materials & Design, 58, 475–481.

Suneesh, E, M Sivapragash, 2018, Comprehensive studies on processing and characterization of hybrid magnesium composites. Materials and Manufacturing Processes, 33(12), 1324–1345.

Svoboda, M, M Pahutová, K Kuchařová, V Sklenička, KU Kainer, 2007, Microstructure and creep behaviour of magnesium hybrid composites. Materials Science and Engineering: A, 462(1–2), 220–224.

Thakur, SK, K Balasubramanian, M Gupta, 2007, Microwave synthesis and characterization of magnesium based composites containing nanosized SiC and hybrid (Si C+ Al$_2$O$_3$) Reinforcements. Journal of Engineering Materials and Technology, 194–199.

Thakur, SK, GT Kwee, M Gupta, 2007, Development and characterization of magnesium composites containing nano-sized silicon carbide and carbon nanotubes as hybrid reinforcements. Journal of Materials Science, 42(24), 10040–10046.

Thakur, SK, TS Srivatsan, M Gupta, 2007, Synthesis and mechanical behavior of carbon nanotube – magnesium composites hybridized with nanoparticles of alumina. Materials Science and Engineering: A, 466(1–2), 32–37.

Tun, KS, 2009, Development and characterization of new magnesium based nanocomposites. Doctoral dissertation, National University of Singapore, Singapore.

Tun, KS, M Gupta, 2009, Development of magnesium/(yttria+ nickel) hybrid nanocomposites using hybrid microwave sintering: Microstructure and tensile properties. Journal of Alloys and Compounds, 487(1–2), 76–82.

Tun, KS, M Gupta, TS Srivatsan, 2010, Investigating influence of hybrid (yttria+ copper) nanoparticulate reinforcements on microstructural development and tensile response of magnesium. Materials Science and Technology, 26(1), 87–94.

Tunberg, T, P Vervoort, 1995, Extrusion of spray deposited QE22 magnesium alloy and composite. Powder Metallurgy, 38(2), 98–102.

Ünal, A, 1992, Rapid solidification of magnesium by gas atomization. Material and Manufacturing Process, 7(3), 441–461.

Wang, HY, QC Jiang, YG Zhao, F Zhao, 2004, In situ synthesis of TiB$_2$/Mg composite by self-propagating high-temperature synthesis reaction of the Al – Ti – B system in molten magnesium. Journal of Alloys and Compounds, 379(1–2), L4-L7.

Wong, WLE, M Gupta, 2005, Using hybrid reinforcement methodology to enhance overall mechanical performance of pure magnesium. Journal of Materials Science, 40(11), 2875–2882.

Wong, WLE, M Gupta, 2006, Effect of hybrid length scales (micro+ nano) of SiC reinforcement on the properties of magnesium. Solid state phenomena, 111, 91–94. Trans Tech Publications Ltd.

Wong, WLE, M Gupta, CYH Lim, 2006, Enhancing the mechanical properties of pure aluminum using hybrid reinforcement methodology. Materials Science and Engineering: A, 423(1–2), 148–152.

Wong, WLE, S Karthik, M Gupta, 2005, Development of hybrid Mg/Al$_2$O$_3$ composites with improved properties using microwave assisted rapid sintering route. Journal of Materials Science, 40(13), 3395–3402.

Yong, MS, AJ Clegg, 2005, Process optimisation for a squeeze cast magnesium alloy metal matrix composite. Journal of Materials Processing Technology, 168(2), 262–269.

Yoshida, M, S Takeuchi, J Pan, G Sasaki, N Fuyama, T Fuj, H Fukunaga, 1999, Preparation and characterization of aluminum borate whisker reinforced magnesium alloy composites by semi-solid process. Advanced Composite Materials, 8(3), 259–268.

Zhang, XN, L Geng, GS Wang, 2006, Fabrication of Al-based hybrid composites reinforced with SiC whiskers and SiC nanoparticles by squeeze casting. Journal of Materials Processing Technology, 176(1–3), 146–151.

Zhang, XN, Q Zhang, H Hu, 2014, Tensile behaviour and microstructure of magnesium AM60-based hybrid composite containing Al$_2$O$_3$ fibres and particles. Materials Science and Engineering: A, 607, 269–276.

Zheng, M, K, Wu, C Yao, 2001, Effect of interfacial reaction on mechanical behavior of SiC$_w$/AZ91 magnesium matrix composites. Materials Science and Engineering: A, 318(1–2), 50–56.

Zhou, X, D Su, C Wu, L Liu, 2012, Tensile mechanical properties and strengthening mechanism of hybrid carbon nanotube and silicon carbide nanoparticle-reinforced magnesium alloy composites. Journal of Nanomaterials, 2012.

Zuo, M, DG Zhao, ZQ Wang, HR Geng, 2015, Investigation on WC – Al composite coatings of AZ91 alloy by mechanical alloying. Materials Science and Technology, 31(9), 1051–1057.

10 Advancements in Magnesium Metal-Matrix Composites for Degradable Implant Applications

Shebeer A. Rahim, V.P. Muhammad Rabeeh,
Sharath Babu, M.A. Joseph and T. Hanas

CONTENTS

10.1 INTRODUCTION

Medical implants are made of biocompatible materials and are expected to substitute or/assist diseased tissue or organs. These implants can be divided into two classes based on their period of usage; permanent and temporary. Permanent implants are those which are expected to be functioning lifelong in the human body (e.g. hip joint implant, knee cap etc.). The temporary implants are used for a limited duration to support the tissue and are to be removed from the body after fulfilling

DOI: 10.1201/9781003319856-12

their function (e.g. orthopaedic screws, pins, plates etc.). It is always preferred to develop temporary implants using biodegradable materials, as they can be tuned to disappear from the site after the tissue is healed. Though there are many polymeric- and ceramic-based biodegradable materials, limited numbers of metals and alloys can be used for this purpose. However, the load-bearing temporary implant applications demands the use of metallic materials to support the tissue (Davis et al. 2022; Shuai et al. 2019). Currently, bioinert materials such as titanium (Ti) alloys, cobalt-chromium (Co-Cr) alloys, and stainless steel (SS) are used as conventional implant materials for temporary implant application. These conventional implants are non-biodegradable and necessitate a second surgery to remove them from the deceased site after the tissue is recovered (Wang et al. 2020). In addition, there are several other complications with such metallic implants, such as inadequate osseointegration and potential toxic ion leaching after an extended duration (Eliaz 2019). It is also noted that these implant materials have a higher elastic modulus (EM) than human bone and thus generate a mismatch in EM between bone and the implant. Such a large mismatch results in the stress shielding effect, which leads to loss of bone mass, decreased bone growth and acute discomfort in patients (Al-Tamimi et al. 2017; Rahim et al. 2022).

The issues associated with the aforesaid bioinert implant materials can be effectively resolved by using a compatible biodegradable metallic implant material. A "biodegradable" implant has the potential to degrade safely and efficiently without any residue in the host. In recent years, biodegradable metals (BMs) have garnered considerable interest, especially for orthopaedic applications and becoming next-generation biomedical materials with multifunctional aspects. Mg, zinc (Zn), and iron (Fe)-based materials have been intensively researched for use in biodegradable metallic implants (Li, Guo et al. 2020). In contrast to Mg, which degrades more rapidly in the physiological system, Fe degrades very slowly. Zn-based materials have only recently been introduced due to their moderate degradation rate. Both the quick degradation of Mg and the extremely slow degradation of Fe are undesirable for temporary implants. However, a good number of attempts that focus on Mg for developing orthopaedic implants have been reported in recent years.

Mg is one of the essential elements and is involved in different enzymatic reactions in the human body. Furthermore, the majority of Mg is found in bone tissue, making it an important component of the skeletal system. Moreover, Mg has density and EM close to the human bone. Though there are many such benefits, the rapid degradation is a major challenge that needs to be addressed. The high degradation rate of Mg in the physiological environment reduces the mechanical integrity and causes premature failure of the implant and the release of a large volume of hydrogen gas that causes subcutaneous bubbles. In light of these limitations, Mg alloys and Mg-based composites are developed for implant applications. Figure 10.1 depicts the key benefits of Mg and the ideal performance

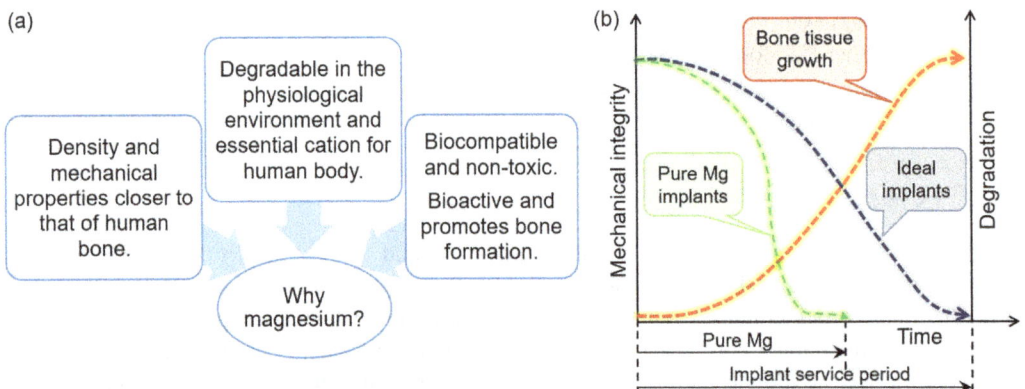

FIGURE 10.1   (a) Key benefits of Mg; (b) ideal degradable implant behaviour.

of biodegradable implant materials. Only a selected group of Mg alloys with non-toxic alloying elements (Ca, Zn, Sr, Si, Sn, Mn, Zr) can be employed in biomedical applications (Chen et al. 2019; Zhang et al. 2013). In addition to these Mg-based alloys, Mg metal-matrix composites (Mg-MMCs) with a wide choice of dispersed phases also show adaptable mechanical, biodegradable and bioactive properties.

10.2 FABRICATION OF BIODEGRADABLE MAGNESIUM-MMCS

The popular methods to fabricate biodegradable Mg-MMCs can be classified as liquid-state processing, solid-state processing and additive manufacturing (Arora et al. 2022; Bajakke, Malik, and Deshpande 2019). The fabrication techniques determine the microstructure as well as bulk properties influencing the mechanical and biodegradation behaviour.

10.2.1 LIQUID-STATE PROCESSING

Dispersing the secondary phase in the molten metal is one of the predominant techniques used for developing MMCs. The key aspects of liquid-state processing are cost-effectiveness, simplicity for manufacturing complex shapes and large-scale manufacturing. In order to prevent particle agglomeration and gas entrapment during the addition of reinforcement, it is essential to optimize the liquid-state process parameters. The popular liquid-state processing methods for biodegradable Mg-MMCs are stir casting and disintegrated melt deposition.

10.2.1.1 Stir Casting

The dispersion achieved by either a mechanical stirrer or ultrasonic probe immersed in the molten metal is popularly called stir casting. It is generally considered one of the cost-effective methods for manufacturing MMCs. The technique uses mechanical stirring to incorporate reinforcing phases such as ceramic particles, short fibres and whiskers into the molten metal. Schematic representation of stir casting is shown in Figure 10.2. There is a significant risk of particle agglomeration during fabrication using mechanical stirring. To obtain better degradation behaviour and mechanical properties, it is necessary to maintain a uniform dispersion of reinforcement particles without agglomeration. In general, particle agglomeration occurs due to density variation and poor wettability of the reinforcements. It is possible to improve the homogenous dispersion of reinforcement particles by controlling the stir-casting process parameters such as stirring time and speed, stirrer blade angle, pouring temperature, solidification rate and reinforcement characteristics such as composition, preheating temperature size, and wettability. Techniques such as the use of surfactants, surface treatments and altering the viscosity of the melt etc., can be attempted to get a better dispersion (James et al. 2021; Kumar et al. 2017). The use of high-temperature ultrasonic probes is also reported to effectively disperse nanosized particles in the molten metal. The transient cavitation effects due to ultrasonic stirring produce implosive effects to break up the agglomerates and disperse them uniformly throughout the molten matrix (Idrisi and Mourad 2019). Pasha et al. (2022) obtained homogeneous dispersion of Si_3N_4 nanoparticles in pure Mg by stir casting followed by ultrasonication carried out at 20 kHz frequency for 5 minutes in an inert atmosphere with a power of 2 kW. The composite exhibited excellent improvement in degradation resistance and bioactivity as well as mechanical properties due to the homogeneous dispersion of Si_3N_4 reinforcement and grain refinement.

Khalili et al. (2021) utilized stir-centrifugal casting to manufacture Mg-MMC with nano-HA (nHA) as the reinforcement. During solidification, the nHA particles were added to the Mg melt using the pressure of inert gas and centrifugal force. The reinforcement was preheated in an electric oven for 2 hours at 400°C to minimize the dissimilarity between the molten Mg matrix and nHA particle during wetting and casting.

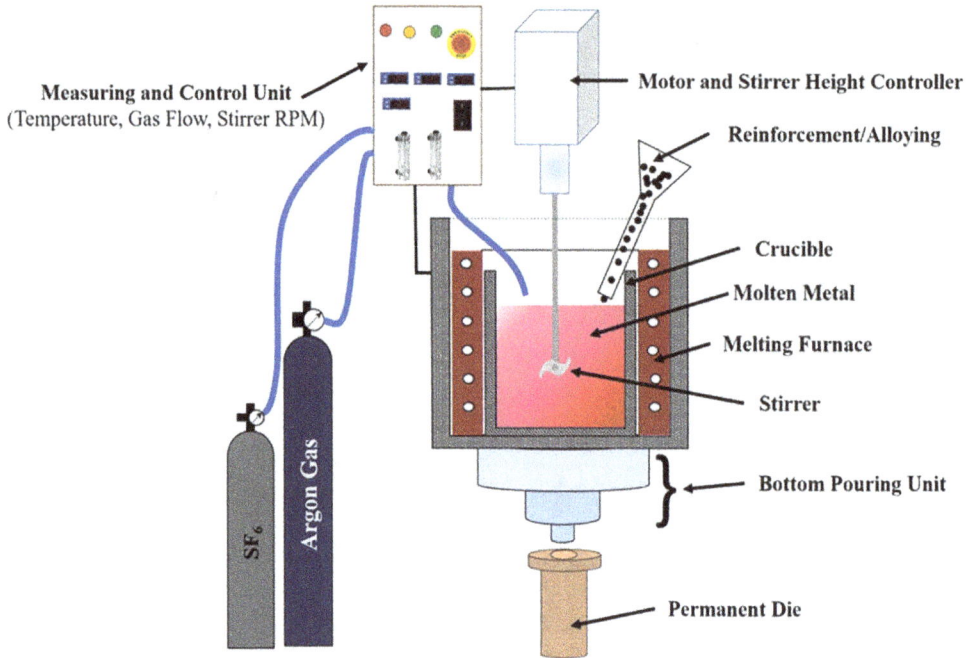

FIGURE 10.2 Schematic representation of stir casting. *Reproduced with permission from (Verma et al. 2022), © 2022, Elsevier.*

10.2.1.2 Disintegrated Melt Deposition

Disintegrated melt deposition (DMD) technique is another liquid-state process to fabricate Mg-MMCs. The schematic representation of DMD is shown in Figure 10.3. The process involves stirring the melt at high speed (450 rpm) for roughly 5 minutes after heating the Mg chips and rein-forcement particles to 750°C in an inert environment in a graphite crucible. The stirring was done to make it easier for the reinforcement particles to be mixed and distributed evenly throughout the metallic matrix. After stirring, the composite melt was poured into the graphite crucible through a centrally located hole of 10 mm in diameter. Two linear argon gas jets were then used to disintegrate the stream that was created at a distance of 200 mm from the pouring point. Following this, the disintegrated composite melt slurry was deposited on a metallic circular substrate that was situ-ated 242 mm away from the gas disintegration point (Gupta, Lai, and Saravanaranganathan 2000). Parande et al. (2020) synthesized Mg-Zn-Si/HA nanocomposites by DMD led to grain refinement, obtaining greater yield strength while maintaining ductility. The produced composite also displayed superior degradation resistance and biocompatibility, indicating the benefit of the DMD technique for degradable orthopaedic applications.

10.2.2 SOLID-STATE PROCESSING

In solid-state processing, the matrix and reinforcement are processed in the solid form. The com-mon techniques mostly use the matrix in either solid or powder form, and secondary phases are gen-erally in powder form. Solid-state techniques such as powder metallurgy and friction stir processing are used to disperse the powder in the matrix.

10.2.2.1 Powder Metallurgy

Powder metallurgy (PM) is a standard method for making Mg-MMCs. The key benefits of this route are homogeneous particle distribution and lower processing temperature. Particle agglomeration

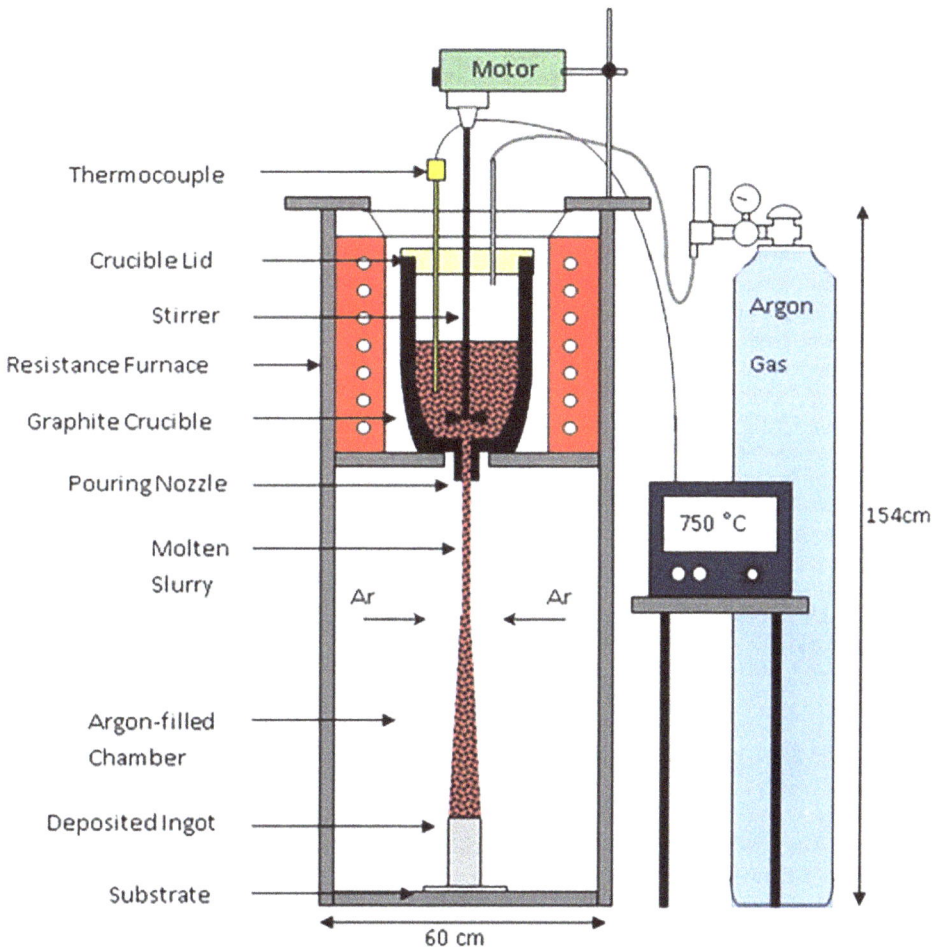

FIGURE 10.3 DMD process. *Reproduced with permission from (Gupta and Wong 2015), © 2015, Elsevier.*

caused by the poor wettability of the secondary phase during liquid-state processing can be evaded by using the PM route. The process includes three major steps: powder mixing, green compaction, and sintering. Predetermined quantities of pure Mg and dispersed phase are taken in powder form. Techniques such as ball milling, attritor milling and mechanical alloying are used to blend for a specific time so as to ensure proper dispersion of the phases. Razzaghi et al. (2020) obtained uniform dispersion of nano-NiTi particles in the Mg-Zn-Ag alloy due to the repeated fracturing and cold welding that occurred during mechanical alloying. Once the powder is dispersed, compaction at high pressure (50–500 MPa) is carried out using the appropriate dies (Kumar and Pandey 2020). The compaction parameters will be well controlled to get sufficient green strength for the components for further handling during the fabrication process. The compacted powder mix is then sintered at elevated temperatures (400–500 °C) in an inert atmosphere for a duration of ~10 minutes to achieve the required density. The parameters affecting the performance of the composite include the percentage of reinforcement, compaction pressure, sintering temperature, heating rate, soaking time etc. In some cases, the two-step sintering technique is used to facilitate grain refinement and improve material properties. Additionally, advanced sintering methods like hot press sintering and spark plasma sintering (SPS) allow the simultaneous application of pressure and temperature to form sintered products (Somasundaram et al. 2022). However, the high cost of PM setup and raw

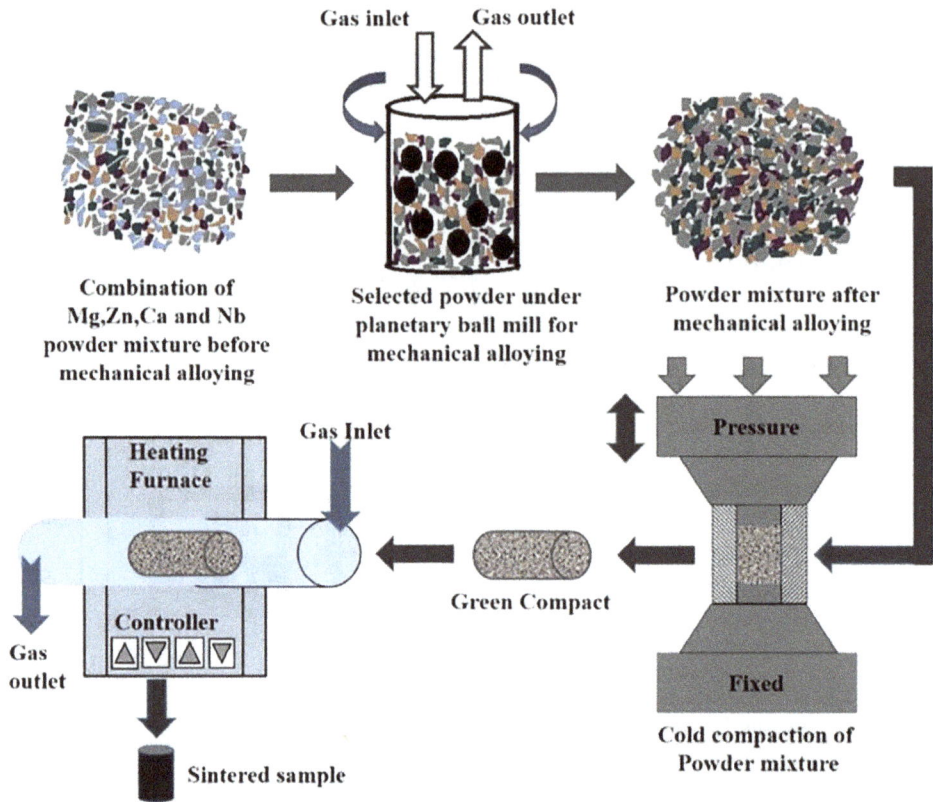

FIGURE 10.4 Solid-state fabrication by powder metallurgy. *Reproduced with permission from (Kumar and Pandey 2020), © 2020 Published by Elsevier B.V. on behalf of Chongqing University.*

materials, the difficulty of obtaining complex geometries and the low density of the PM products are major limitations.

10.2.2.2   Friction Stir Processing

The friction stir processing (FSP) technique, inspired by friction stir welding (FSW), has become a promising method for fabricating Mg-MMCs and changing the bulk metallic microstructure in the solid state. FSP can be used to overcome the difficulty in processing bioceramics and Mg-matrix that arise from the mismatch in their physical properties. The stirring action during FSP causes a localized temperature rise that softens the matrix material and facilitates the incorporation of reinforcements (Ho et al. 2020). Additionally, the combined action of plastic deformation and elevated temperature during FSP causes precipitate dissolution, grain refinement and closure of porosity, resulting in a fine, homogeneous, and pore-free structure (Ma 2008). During FSP, a cylindrical rotating tool with a pin-and-shoulder design is inserted into the narrow groove of the matrix filled with reinforcement (Hanas et al. 2018), as shown in Figure 10.5. The tool is then rotated and moved along the desired length in the traverse direction while an appropriate load is applied. As a result, the material in the stir zone experiences significant plastic deformation and starts to recrystallize dynamically, which leads to the formation of microstructure with ultrafine grain size (Butola et al. 2022; Liu et al. 2021). The FSP has limitations, such as expensive equipment costs, low productivity and the inability to fabricate complex geometries.

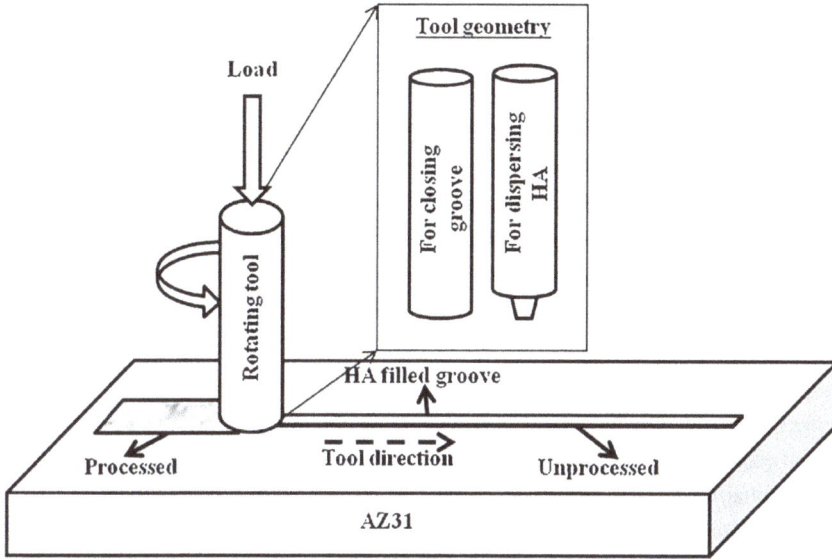

FIGURE 10.5   FSP process. *Reproduced with permission from (Hanas et al. 2018), © 2018, Elsevier.*

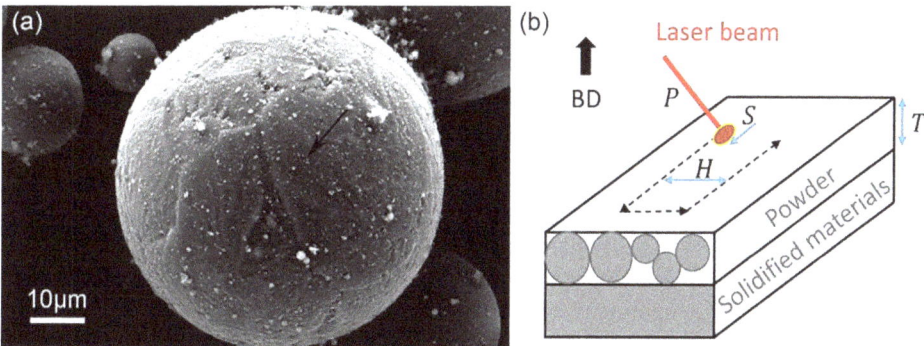

FIGURE 10.6   (a) Nano-SiC-coated AZ91D powder; (b) schematic representation of L-PBF process. *Reproduced from (Niu et al. 2021), under Creative Commons Attribution License (CC BY 4.0).*

10.2.3   ADDITIVE MANUFACTURING

Additive manufacturing (AM) opens up great opportunities in the fabrication of patient-specific implants through a controlled and precise layer-by-layer deposition. The scaffolds prepared through AM are extensively investigated, as they possess unique extracellular matrix structures that can promote cell adhesion and proliferation (Jahr et al. 2021; Kumar et al. 2019). The greatest benefit is its ability to directly create complicated solid structures, which is crucial for biological applications (Su et al. 2020). Usually, the computer-aided manufacturing model of a patient-specific tissue is constructed by transferring raw anatomical data from medical imaging equipment to a 3D printing system. Among AM methods, laser powder-bed fusion (L-PBF) is a feasible technique to rapidly fabricate orthopaedic implants of complex geometries. Niu et al. (2021) blended AZ91D powder with nano-SiC particles of size 50 nm using a ball mill to obtain nanoparticle-coated AZ91D powder for L-PBF. The laser beam was then used to fuse the nano-SiC-coated AZ91D powder layer by layer in an argon atmosphere, as shown in Figure 10.6. The L-PBF improved the mechanical properties due to the formation of refined columnar grains and homogeneous distribution of reinforcement.

10.3   MAJOR CHALLENGES IN FABRICATING MAGNESIUM-MMCS

Porosity, agglomeration and uniform dispersion of reinforcement in the composite are among the major challenges (Bagheri et al. 2022; Pasha et al. 2022) faced while developing metal-matrix composites. Non-uniform dispersion of the secondary phase and porosity accelerate the degradation rate and consequently deteriorate the performance of Mg-MMCs in the physiological environment (Dubey et al. 2021). Air bubbles that enter the melt or adsorbed gases on the surface of dispersed particles are primarily responsible for the porosity of MMCs. In addition, the difference in wettability between the molten matrix and the dispersed particles causes agglomeration, which further contributes to an increase in porosity. Porosity is also caused by the shrinkage of the composite during solidification (Banijamali et al. 2022). Thermomechanical processing such as rolling, extrusion and equal channel angular pressing (ECAP) are performed on the fabricated Mg-MMCs to overcome these challenges. Banijamali et al. (2022) used hot rolling to obtain uniform distribution of B_4C reinforcement in the WE43 matrix with minimum porosity. Further, the dynamic recrystallization that occurred during the hot rolling of the composite resulted in grain refinement and hence improvement in mechanical properties. Sun et al. (2022) fabricated Mg-MMC with tricalcium phosphate as reinforcement using ultrasonic-assisted stir casting followed by extrusion. Further, multi-pass ECAP was processed on the composite to obtain grain refinement by dynamic recrystallization and homogenous dispersion of reinforcement. The enhanced characteristics of Mg-MMCs can be attributed to the homogenous dispersion of reinforcement and the effective transfer of load between the matrix and reinforcement.

10.4   BIODEGRADABLE MAGNESIUM-BASED MMCS TYPES

Biodegradable Mg-MMCs reported so far differ from each other mainly based on the secondary phases. The primary phase is usually pure Mg or an Mg-based alloy. In addition to the processing techniques, the dispersed phase also plays a decisive role in the development of Mg-MMCs for biodegradable implants. Though many such phases are available as reinforcements for structural applications, biodegradable implant applications necessitate non-toxic characteristics from the components. The fundamental requirement of such a material is to control the rapid degradation while maintaining adequate mechanical integrity during the tissue healing period (Dutta, Gupta, and Roy 2020). Additionally, the selected reinforcement should be biocompatible and osteoconductive to enhance bone healing. Recent studies have shown increased interest in the following bioresorbable ceramics as secondary phases.

10.4.1   MAGNESIUM/HA COMPOSITE

Hydroxyapatite (HA) or hydroxylapatite [$Ca_5(PO_4)_3OH$] is a naturally occurring calcium phosphate (CaP) mineral essential for bone growth. It has a hexagonal structure and is the most stable form of CaP at normal temperatures, with a Ca/P ratio of 1.67 (Kim et al. 2002; Mostafa and Brown 2007). The biocompatibility and chemical similarity of HA to that of natural bone have made it an exceptional candidate for orthopaedic applications. Many works are already reported on dispersing HA in Mg matrices for improving the bioactivity and reducing the degradation rate. The low solubility in the physiological environment combined with the excellent biocompatibility of nanocrystalline HA can simultaneously enhance the degradation resistance and mechanical stability of the implant (Fulmer et al. 2002; Zhou and Lee 2011). The size and shape of the HA can also influence biodegradation behaviour. Jaiswal, Dubey, and Lahiri (2020) reported that the cylindrical shape HA incorporated into Mg-3Zn alloy by SPS resulted in efficient mixing and uniform distribution due to

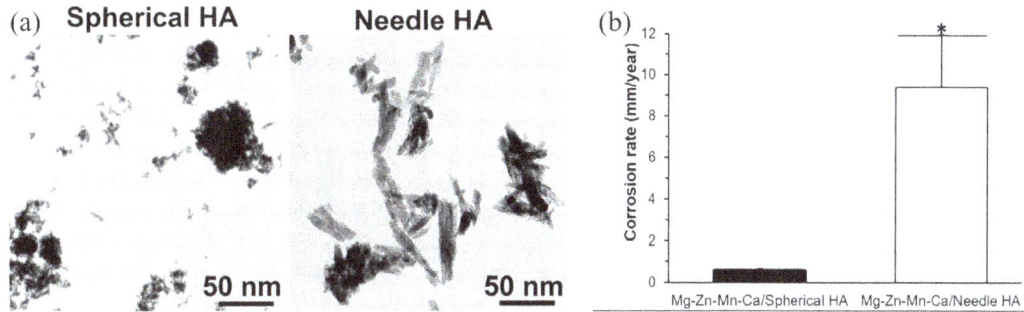

FIGURE 10.7 (a) TEM images of spherical and needle-like HA nanoparticles in Mg matrix; (b) degradation rate. *Reproduced with permission from (Razavi and Huang 2020), © 2020, Elsevier.*

the large surface area-to-volume ratio. As a result, the degradation resistance was increased by 63% and ultimate tensile strength (UTS) by 40%. The addition of HA also improved the cell viability by more than 2 times. Razavi and Huang (2020) found that spherical nano-HA had significantly higher degradation resistance than irregular needle-shaped nano-HA, as shown in Figure 10.7. The poor degradation resistance for irregular-shaped HA reinforcement is due to the low wettability caused by the intricate geometric features. Further, the spherical nano-HA developed bioactive deposits, offering a protective layer on the Mg substrate. Lopes et al. (2020) also obtained a significant improvement in degradation resistance on HA-incorporated Mg due to the formation of a bioactive layer that was rich in Ca, P and O. Although SPD methods can be used to disperse HA reinforcement uniformly, incorporation of more than 10 wt.% of HA can result in uncontrolled degradation due to significant particle agglomeration (Guo et al. 2021). Though dispersing HA has a lot of advantages, it comes with certain challenges too, the primary being the effective dispersion of HA in the Mg matrix (Huang, Li, and Zhou 2018). While many authors reported solid-state techniques such as PM route or FSP, there are a few reports on stir casting too. The agglomeration of HA particles in the liquid melt and porosity are among the major concerns to be addressed while following the stir casting route.

10.4.2 MAGNESIUM/FA COMPOSITE

The fluorine ions replace the hydroxyl group of HA to form a hard CaP phase with increased thermal stability-Fluorapatite (FA), $[Ca_5(PO_4)_3F]$ (Bulina et al. 2020). It is a highly insoluble and crystalline form of CaP mineral that is an essential constituent of tooth enamel (Rey et al. 2008). The low solubility, superior cell adhesion compared to HA and enhanced phosphate action of FA make it an important material for biodegradable implant applications. Razavi et al. (2010a) obtained a reduced degradation rate and accelerated apatite formation, contributing to enhanced osteoconductivity by dispersing nano-FA (Figure 10.8 (a)) in AZ91 alloy blend-press sintering (BPS) method. After 72 hours of immersion in SBF, the SEM micrographs of the composite showed white cauliflower deposits, while AZ91 alloy showed intergranular corrosion without such deposits (Figure 10.8 (b), (c)). Vidal et al. (2022) fabricated Mg/HA/FA composited by upward FSP. After immersion in SBF, the composite samples showed an apatite layer rich in fluoride. The osteoblastic cell proliferation was enhanced due to the presence of FA. Preparation of the Mg-nFA composite by the BPS technique not only reduced the degradation rate but also significantly improved the microhardness and elastic modulus (Razavi, Fathi, and Meratian 2010b). The FA particles in the composite acted as nucleation sites for the formation of phosphates to control the degradation while increasing the bioactivity. However, a high FA content resulted in poor mechanical properties due to agglomeration. Hence, optimization of FA content is vital to combine good mechanical properties and controlled degradation behaviour.

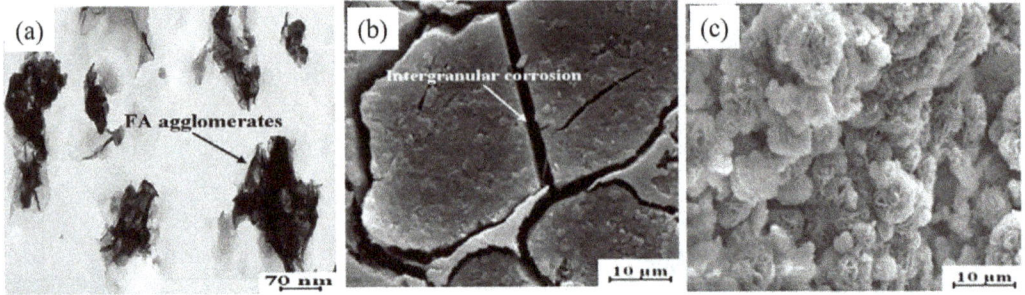

FIGURE 10.8   (a) TEM of FA in AZ91 matrix; (b) after immersion bare; (c) after immersion AZ91/FA nanocomposite. *Reproduced with permission from (Razavi, Fathi, and Meratian 2010a),* © 2010, Elsevier.

10.4.3   MAGNESIUM/β-TCP COMPOSITE

Tricalcium phosphate (TCP), $[Ca_3(PO_4)_2]$ is the highly soluble form of CaP with a Ca/P ratio of 1.50. It has three polymorphs: high-temperature α-TCP, α′-TCP and low-temperature β-TCP phase (Carrodeguas and De Aza 2011). At normal temperatures, β-TCP is the most stable phase and is popular for bone regeneration due to its osteointegration, biocompatibility, high solubility and superior wettability (Jeong et al. 2019). According to Ruiz-Aguilar et al. (2018), the method of synthesis strongly influences the size and morphology of β-TCP (Figure 10.9 (a), (b)). Parande et al. (2018) prepared Mg/β-TCP with porosity less than 1% by PM technique and obtained better mechanical properties than pure Mg, as shown in Figure 10.9 (c). Additionally, the incorporation of β-TCP controlled the degradation rate by accelerating the nucleation and growth of bone-like apatite on the surface during immersion in SBF. However, there are reports that Ca from β-TCP diffuses into the Mg matrix to form galvanic couples and accelerate local corrosion (Narita et al. 2019). Such a type of accelerated degradation is more common during high-volume fractions of the secondary phase. Watanabe et al. (2019) attempted microstructural modification by extrusion of components prepared by the PM route and obtained equiaxed fine grain size (1.3 μm) composite with 10 wt.% β-TCP. But the composite exhibited a lesser strength than the alloy without β-TCP, indicating the poor strengthening effect for higher weight fractions due to particle agglomeration.

10.4.4   MAGNESIUM/CCP COMPOSITE

Calcium polyphosphate (CPP), $[Ca(PO_3)_2]$, also known as calcium metaphosphate, is formed by long chains of phosphate units connected by Ca atoms (Jackson et al. 2005). CPP often exists in a glassy state and is a suitable material for biodegradable orthopaedic applications due to its high biocompatibility, bioresorption and osteoconductivity (Jackson et al. 2005; Hatt et al. 2019). Feng and Han (2011) uniformly distributed 2.5, 5.0, 7.5 and 10.0 wt.% of CPP in the ZK60A matrix through the PM route to improve the degradation resistance. During the immersion test in physiological saline, the weight loss and pH variation of the samples decreased significantly with an increase in CPP concentration. Even though the inclusion of CPP improved the yield strength and elastic moduli, the CPP content above 5 wt.% caused particle agglomeration and reduced the mechanical properties (Figure 10.10). The PM route also produced uniform dispersion of CPP with excellent interfacial bonding with the ZK60 matrix and improved degradation resistance due to accelerated HA formation during immersion in SBF (Feng and Han 2010).

10.4.5   MAGNESIUM/BG COMPOSITE

Bioactive glasses (BG) are composed of silicate-based amorphous materials that are compatible with the physiological environment. They are known for their bioactivity and can adhere to bone

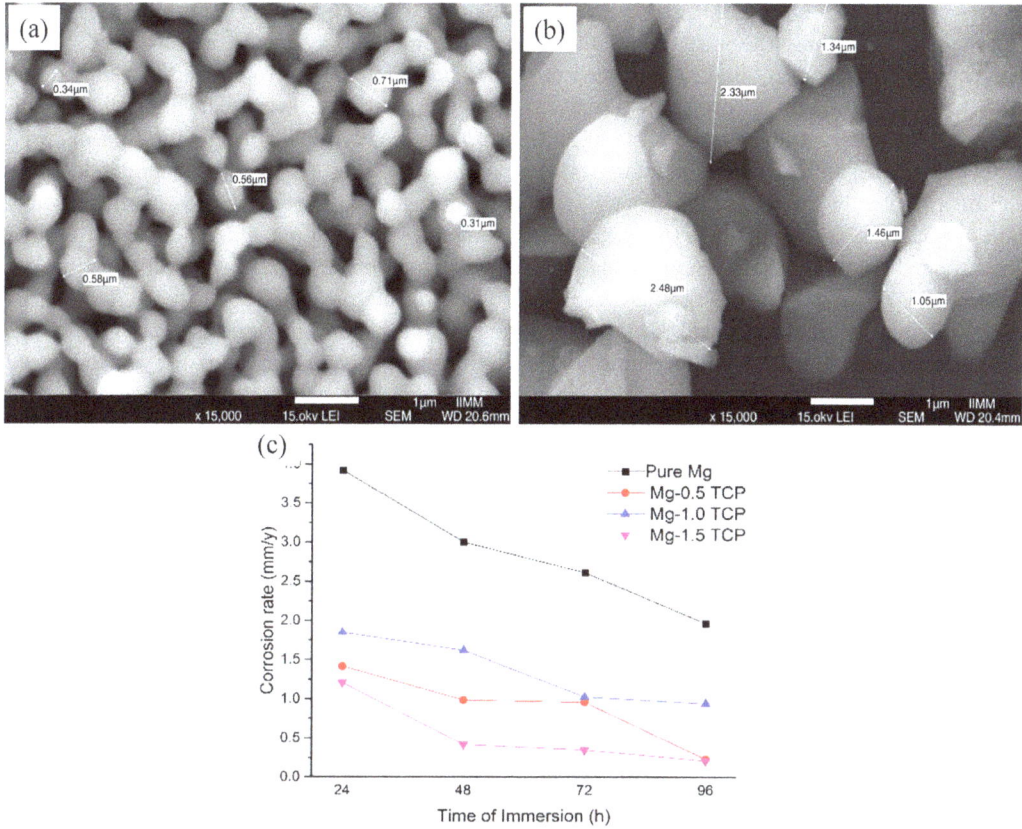

FIGURE 10.9   β-TCP powder by (a) mechanosynthesis; (b) sigma reagent powders. *Reproduced with permission from (Ruiz-Aguilar et al. 2018), © 2018, Elsevier.* (c) Corrosion rate for different composites. *Reproduced from (Parande et al. 2018), under Creative Commons Attribution License (CC BY 4.0).*

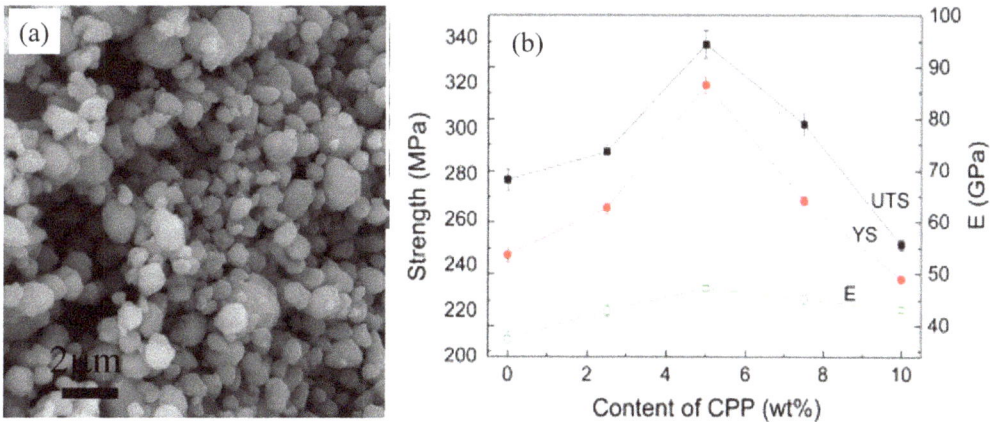

FIGURE 10.10   (a) CCP; (b) ultimate strength, yield strength and elastic modulus of the composite decreases after 5 wt.%. *Reproduced with permission from (Feng and Han 2011), © 2011, Elsevier.*

tissues as well as promote the regeneration of the bone while completely degrading over time. Consequently, they have the ability to rebuild damaged or injured bone to its normal state and function (bone regeneration). Bioglass has been proven to develop a very strong chemical interaction with bone by facilitating the growth of a surface of hydroxycarbonate apatite. The mineral hydroxycarbonate apatite resembles bone mineral and promotes osseous repair. It was the first material observed to form an interfacial interaction with surrounding tissues, outstanding osteoconductivity and controllable biodegradability (Hench et al. 1971). Later, it was referred to as bioglass. Many improvements have been made to the original BG to enhance degradation and bone integration. BG has attracted great interest as a reinforcing material for biocomposites due to its ability to stimulate bone regeneration and prevent rapid degradation of the Mg matrix. BG-Mg-MMCs developed by extrusion, powder metallurgy, spark plasma sintering, hot press sintering, microwave sintering etc. have already been reported. These composites have better mechanical characteristics and enhanced degradation resistance with lower H_2 evolution and exhibited high cell proliferation in the *in vitro* cell culture studies (Li, Han et al. 2020; Khodaei et al. 2019). However, increasing the amount of BG in a matrix of Mg decreases its mechanical characteristics due to segregation and interfacial cracking (Dutta et al. 2019). Duan et al. (2020) implanted BG-magnesium phosphate cement (BGC), and BGC reinforced with Mg-Zn-Ca alloy rod (BGC-Mg). In addition to higher HA production in SBF and superior cytocompatibility, the *in vivo* study on the New Zealand white rabbit displayed excellent osteoconductivity and bone regeneration, as shown in Figure 10.11.

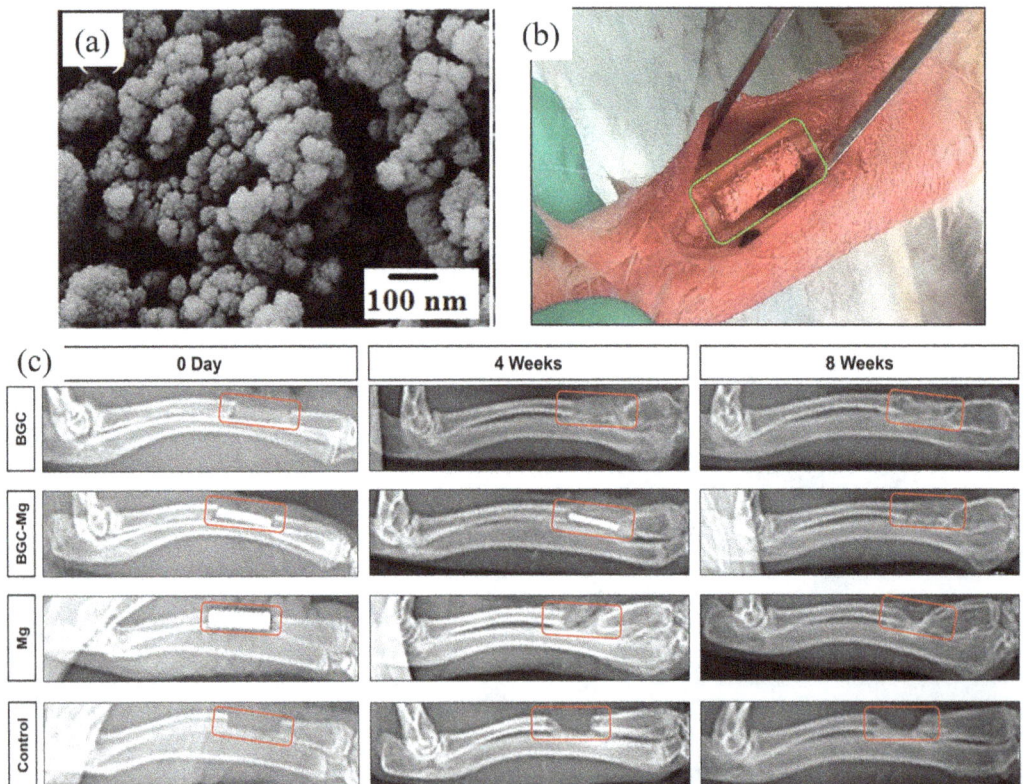

FIGURE 10.11   (a) Bioglass. *Reproduced with permission from (Khodaei et al. 2019), © 2019, Elsevier.* (b) Implantation in rabbits; (c) X-ray images of radius regeneration; bone defects in control group were kept empty. *Reproduced from (Duan et al. 2020), under Creative Commons Attribution License (CC BY 4.0).*

10.4.6 MAGNESIUM/TiO₂ COMPOSITE

Titanium dioxide (TiO_2) is one of the stable bioceramic materials which exist in three forms (anatase, rutile and brookite). A unique composition named Aeroxide TiO_2 P25, having a composition of anatase 75 wt.% and rutile 25 wt.%, is employed in biological applications. Previously, TiO_2 was considered the ideal coating material for Mg-based alloys to improve degradation resistance and bioactivity. Recently, TiO_2 has been employed in biodegradable Mg matrixes as a reinforcement material in order to improve their corrosion protection and mechanical characteristics (Khorashadizade et al. 2021; Elumalai et al. 2021). Ong et al. (2017) suggested that Mg-MMC having 2.5 vol% TiO_2 is a promising material for orthopaedic implants. The low volume percentage of TiO_2 has no significant influence on the cytotoxicity, and hence further research is needed before clinical trials. Nevertheless, Bolokang et al. (2015) and Khoshzaban et al. (2014) observed a reduction in yield strength and ductility due to nano-TiO_2 agglomeration (Figure 10.12). However, subsequent extrusion processing of the composites minimized particle agglomeration, resulting in enhanced tensile properties.

10.4.7 MAGNESIUM/ZrO₂ COMPOSITE

Zirconia or zirconium dioxide (ZrO_2) is a biocompatible ceramic, which was predominantly used as an important bioinert material in the field of dental implantation due to its biocompatible nature. Due to their exceptional mechanical characteristics and strong chemical and biological tolerance, zirconia-based materials have attracted interest as biomaterials for reconstructing hard tissues. Recently, ZrO_2 has been used as an effective material for improving the characteristics of degradable implant materials. ZrO_2 is currently studied either as a coating on the surface or as a dispersed phase in Mg-based metal matrices (Barabás et al. 2021; Qiao et al. 2022). Jalilvand and Mazaheri (2020) reinforced spherical nano ZrO_2 particles of average size 40 nm (Figure 10.13 (a)) in the AZ31 matrix by FSP to obtain improved mechanical properties through significant grain refinement. Multiple FSP passes to limit the possibility of ZrO_2 particle agglomeration and produce a homogeneous dispersion with grain refinement, as shown in Figure 10.13 (b), (c). Qiao et al. (2022) also obtained uniform dispersion with grain refinement after four FSP passes on AZ31/

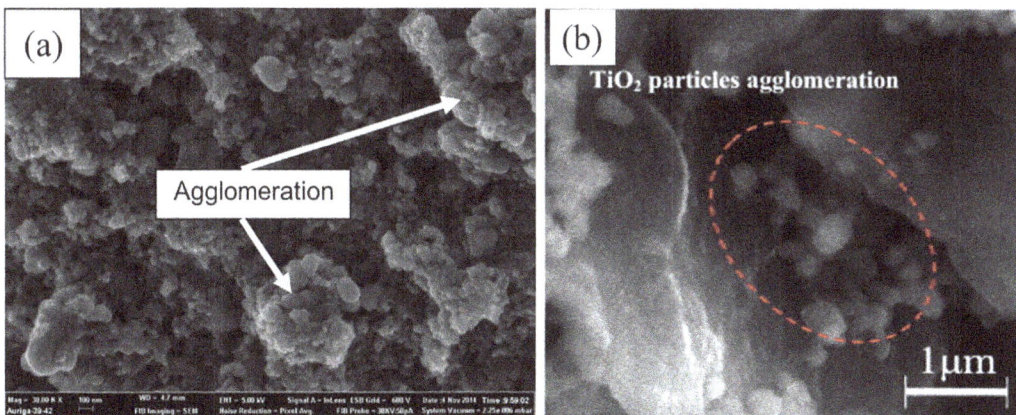

FIGURE 10.12 (a) SEM images of unmilled and nitrided Mg/TiO₂. *Reproduced with permission from (Bolokang et al. 2015), © 2015, Elsevier.* (b) Particle agglomeration of TiO₂. *Reproduced with permission from (Khoshzaban et al. 2014), © 2014, Elsevier.*

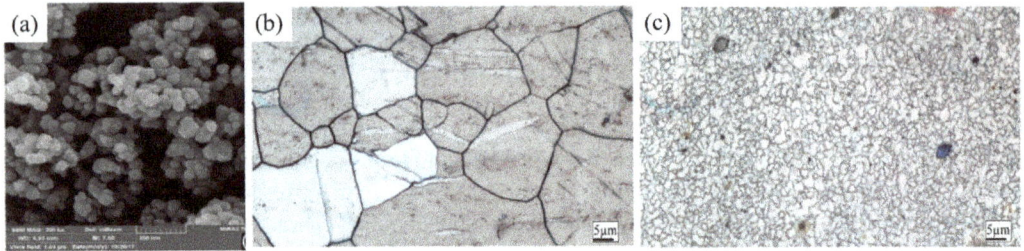

FIGURE 10.13   (a) nano ZrO$_2$; (b) AZ31; and (c) composite. *Reproduced with permission from (Jalilvand and Mazaheri 2020), © 2020, Elsevier.*

ZrO$_2$ composite. However, the MC3T3-E1 cell activity was slightly decreased due to the incorporation of ZrO$_2$. Further, in vitro cell-culture studies are necessary to understand the cytocompatibility of Mg/ZrO$_2$ composites.

10.4.8   MAGNESIUM/GNP COMPOSITE

Due to the excellent biocompatibility as well as unique physicochemical characteristics, carbon materials like graphene and nano-diamond have recently arisen as a candidate for biomedical applications. Among all carbon compounds, graphene has garnered the most attention in the field of biomaterials due to its diverse beneficial features (Dutta, Gupta, and Roy 2020). Graphene nano-platelets (GNPs) are made up of multiple graphene layers linked together through intense van der Waals force and covalent bonding to nearby atoms/layers. Inhibiting corrosion and acting as a barrier to the surroundings on an atomic scale, a single graphene sheet of graphene can function as a barrier. Furthermore, GNPs have antibacterial characteristics and outstanding biocompatibility. Therefore, GNPs can be utilized as reinforcement in the Mg matrix to enhance the mechanical properties, degradation resistance and biological characteristics (Munir, Wen, and Li 2020). Kumar et al. (2018) reinforced GNPs of thickness 2 to 10 nm on Mg-Al alloy by PM route followed by hot extrusion to reduce porosity (Figure 10.14). The GNP reinforcement reduced grain size, resulting in an improvement in mechanical properties. Saberi et al. (2020) observed strong interfacial interaction across Mg and GNPs due to sufficient attraction and high specific surface area. Under compressive loads, the high stresses are transferred from matrix to reinforcement, hence improving the loading stress of the Mg-GNP composite at lower concentrations (less than 0.5 wt.%). The GNP lower concentration of GNP reinforcement also enhanced the MG63 cell compatibility by more than 2 times.

10.4.9   OTHER REINFORCEMENTS

In addition to the reinforcements listed earlier, many more materials compatible with biological systems were utilized to strengthen the Mg matrix in order to achieve controlled degradation, acceptable biocompatibility, and mechanical characteristics. Bredigite (Ca$_7$MgSi$_4$O$_{16}$) is a bioceramic substance found in natural bone with similar mechanical properties to cortical bone. The high biocompatibility, osteoblast adhesion and proliferation-promoting capabilities made it to be a suitable reinforcement material. Dezfuli et al. (2017) reported homogenous dispersion of bredigite in pure Mg by PM route reduced the degradation rate in DMEM by 24 times solution and enhanced bioactivity as well as biocompatibility. Nano zinc oxide (ZnO) is another reinforcement employed to improve mechanical, degradation and biological performance (Esmaeilzadeh et al.

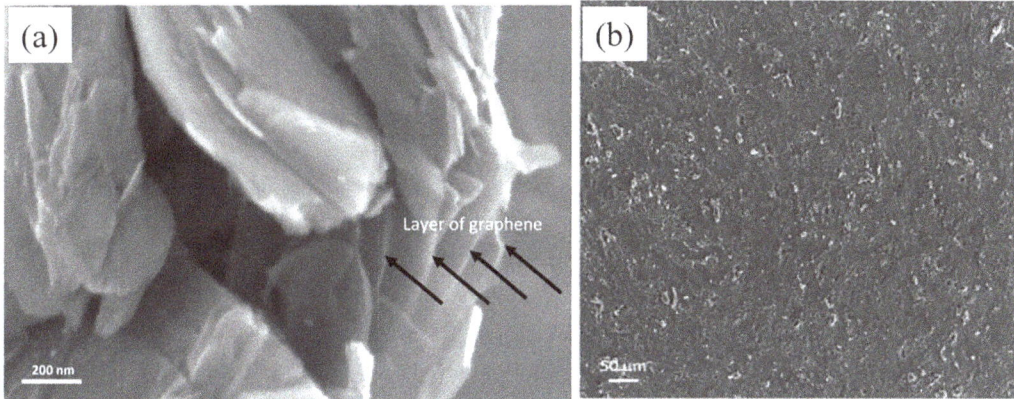

FIGURE 10.14　(a) GNPs; (b) Mg-GNP composite. *Reproduced with permission from (Kumar et al. 2018), © 2018, Taylor & Francis.*

2022). Similarly, MgO and coated MgO are also ideal reinforcements for fabricating biodegradable Mg-MMCs due to their excellent biocompatibility and the ability to release Mg ions favourable for metabolic activities (Zhang et al. 2021; Shuai et al. 2020). As a result, the integration of MgO into the Mg matrix enhances the mechanical and physiological properties. The increased characteristics of MgO/Mg composites can also be attributed to their strong interfacial interaction (Shuai et al. 2020).

10.5　CONCLUSIONS AND OUTLOOK

In this chapter, biodegradable magnesium metal-matrix composites emphasizing degradation, biocompatibility and mechanical properties were briefly discussed. The improvement in biodegradation and mechanical behaviour with the use of bioactive reinforcements is a promising approach toward degradable metallic implant development. The literature highlights that the size and distribution of the reinforcement, as well as the fabrication method, are crucial for targeting an ideal implant material. The conventional casting method appears unsuccessful in ensuring desirable implant behaviour. However, this limitation has led to the exploration of advanced fabrication techniques such as squeeze casting, disintegrated melt deposition, hot press sintering, spark plasma sintering and additive manufacturing techniques. Moreover, the biodegradability and mechanical properties obtained by these innovative approaches were encouraging. The key aspects of advancements in biodegradable Mg-MMCs are summarized in Table 10.1.

The *in vitro* corrosion studies of degradable Mg-MMCs performed using electrochemical tests and immersion tests proposed that homogeneous distribution of reinforcement is desirable. Hence, research should focus on nondestructive scanning methods to investigate the distribution of reinforcements and their influence on degradation behaviour. Also, such analysis can effectively predict the degradation mechanism. Investigations along these lines would offer significant findings in the field of biomaterials. Additionally, to screen out unsuitable materials for animal trials, the investigations reported so far have been narrowed to *in vitro* studies. However, it is important to develop a correlation between *in vitro* and *in vivo* studies. Because *in vivo* investigations are required to study the complex degradation behaviour and the immunogenic response of degradable Mg composites. Modelling and simulation of degradable implant behaviour based on the available in vivo results can provide a better understanding of future implant development.

TABLE 10.1
Key Findings of Mg-MMCs from the Literature

Composite & fabrication	Mechanical properties	Degradation rate		Biocompatibility	Reference
		Electro-chemical test	Immersion test		
Mg-3Zn, 15 wt.% HA (cylindrical), PM by SPS.	UCS increased by ~ 40%, E increased by ~ 24%.	CR decreased by 63%	DR decreased by 270%, 14 days in SBF.		(Jaiswal, Dubey, and Lahiri 2020).
ZK61, 10 wt.% HA, PM with hot extrusion.	YS increased by ~ 50%, HV increased by ~ 35%	CR decreased more than 4 times	HER reduced ~ 4 times, 72 h in SBF.	Improved MC3T3-E1 cell viability ~ 25%.	(Guo et al. 2021).
AZ91, 20 wt.% FA, PM by BPS. AZ31-HA/FA upward FSP.	YS increased by ~ 16%, HV increased by ~ 16%	I_{corr} decreased 2 orders of magnitude	DR decreased by ~ 60%, 72 h in SBF.	Metabolic activity of MG63 cells by ~ 30%.	(Razavi, Fathi, and Meratian 2010b), (Vidal et al. 2022).
Pure Mg, 1.5 Vol.% β-TCP, PM with hot extrusion.	UCS increased by ~ 53%, YS increased by ~ 34%.		DR decreased by ~ 80%, 96 h in Hanks' solution.		(Parande et al. 2018).
Pure Mg, 10 wt.% β-TCP, PM with SPS.		I_{corr} decreased only by ~ 40%	WL reduced ~ 20%, Mg ion release reduced ~ 40%, 11 days in SBF.	BMSC cell adhesion density increased by ~ 30%.	(Narita et al. 2019).
ZK60A, 20 wt.% CPP, PM with hot extrusion.	UCS increased by < 10%, YS decreased by ~ 12%.	I_{corr} decreased only by ~ 25%	WL reduced ~ 35%, 10 days in SBF.		(Feng and Han 2010).
Pure Mg, 10 wt.% BG, PM with SPS.	HV increased by ~ 3 times, E increased by ~ 65%		DR decreased by ~ 60%, 64 h in PBS.	More live images of MG63 cells.	(Dutta et al. 2019).
AZ80, 1 wt.% TiO$_2$, casting & extrusion. Pure Mg, 2.5 Vol.% TiO$_2$, by DMD.	HV increased by ~ 20%, YS increased by ~ 70%	-	-	Insignificant effect on MC3T3-E1 cells	(Khoshzaban et al. 2014), (Ong et al. 2017).
AZ31, ZrO$_2$, multi-pass FSP.	UTS increased by < 20%, YS increased by ~ 15%.	I_{corr} decreased only by ~ 4%		Insignificant effect on MC3T3-E1 cell viability	(Qiao et al. 2022).
Mg-Zn-Ca, 1 wt.% GNP, by PM.	HV increased by ~ 25%, UCS increased by ~ 140%	CR decreased more than 2 times		MG63 cell viability increased by > 30%.	(Saberi et al. 2020).

Note: PM: powder metallurgy; SPS: spark plasma sintering; UCS: ultimate compressive strength; CR: corrosion rate; DR: degradation rate; YS: yield strength; HV: Vickers hardness; HER: hydrogen evolution rate; I$_{corr}$: corrosion current, UTS: ultimate tensile strength.

REFERENCES

Al-Tamimi, Abdulsalam A., Paulo Rui Alves Fernandes, Chris Peach, Glen Cooper, Carl Diver, and Paulo Jorge Bartolo. 2017. "Metallic Bone Fixation Implants: A Novel Design Approach for Reducing the Stress Shielding Phenomenon." *Virtual and Physical Prototyping* 12 (2): 141–51.

Arora, Gurmeet Singh, Kuldeep K. Saxena, Kahtan A. Mohammed, Chander Prakash, and Saurav Dixit. 2022. "Manufacturing Techniques for Mg-Based Metal Matrix Composite with Different Reinforcements." *Crystals* 12 (7): 945.

Bagheri, Behrouz, Amin Abdollahzadeh, Farzaneh Sharifi, and Mahmoud Abbasi. 2022. "The Role of Vibration and Pass Number on Microstructure and Mechanical Properties of AZ91/SiC Composite Layer During Friction Stir Processing." *Proceedings of the Institution of Mechanical Engineers, Part C: Journal of Mechanical Engineering Science* 236 (5): 2312–26.

Bajakke, Padmakumar A., Vinayak R. Malik, and Anand S. Deshpande. 2019. "Particulate Metal Matrix Composites and Their Fabrication via Friction Stir Processing – a Review." *Materials and Manufacturing Processes* 34 (8): 833–81.

Banijamali, Seyed Masih, Yahya Palizdar, Khan A. Nekouee, Soroush Najafi, and Mahta S. Razavi. 2022. "Effect of B_4C Reinforcement and Hot Rolling on Microstructure and Mechanical Properties of WE43 Magnesium Matrix Composite." *Proceedings of the Institution of Mechanical Engineers, Part L: Journal of Materials: Design and Applications* 236 (9): 1854–68.

Barabás, Réka, Carmen I. Fort, Graziella L. Turdean, and Liliana Bizo. 2021. "Influence of HAP on the Morpho-Structural Properties and Corrosion Resistance of ZrO_2-Based Composites for Biomedical Applications." *Crystals* 11 (2): 202.

Bolokang, A. S., D. E. Motaung, C. J. Arendse, S. T. Camagu, and T. F.G. Muller. 2015. "Structure-Property Analysis of the Mg-TiO_2 and Mg-Sn-TiO_2 Composites Intended for Biomedical Application." *Materials Letters* 161: 328–31.

Bulina, Natalia V., Svetlana V. Makarova, Igor Yu Prosanov, Olga B. Vinokurova, and Nikolay Z. Lyakhov. 2020. "Structure and Thermal Stability of Fluorhydroxyapatite and Fluorapatite Obtained by Mechanochemical Method." *Journal of Solid State Chemistry* 282: 121076.

Butola, Ravi, Deepak Pandit, Chandra Pratap, and Prakash Chandra. 2022. "Two Decades of Friction Stir Processing – a Review of Advancements in Composite Fabrication." *Journal of Adhesion Science and Technology* 36 (8): 795–832.

Carrodeguas, R. G., and S. De Aza. 2011. "α-Tricalcium Phosphate: Synthesis, Properties and Biomedical Applications." *Acta Biomaterialia* 7 (10): 3536–46.

Chen, Yang, Jinhe Dou, Huijun Yu, and Chuanzhong Chen. 2019. "Degradable Magnesium-Based Alloys for Biomedical Applications: The Role of Critical Alloying Elements." *Journal of Biomaterials Applications* 33 (10): 1348–72.

Davis, Rahul, Abhishek Singh, Mark J. Jackson, Reginaldo T. Coelho, Divya Prakash, Charalambos P. Charalambous, Waqar Ahmed, Leonardo R. R. da Silva, and Abner A. Lawrence. 2022. "A Comprehensive Review on Metallic Implant Biomaterials and Their Subtractive Manufacturing." *International Journal of Advanced Manufacturing Technology* 120. Springer London.

Dezfuli, Sina Naddaf, Zhiguang Huan, Arjan Mol, Sander Leeflang, Jiang Chang, and Jie Zhou. 2017. "Advanced Bredigite-Containing Magnesium-Matrix Composites for Biodegradable Bone Implant Applications." *Materials Science and Engineering C* 79: 647–60.

Duan, Huyang, Chuanliang Cao, Xiaolei Wang, Jun Tao, Chen Li, Hongbo Xin, Jing Yang, Yulin Song, and Fanrong Ai. 2020. "Magnesium-Alloy Rods Reinforced Bioglass Bone Cement Composite Scaffolds with Cortical Bone-Matching Mechanical Properties and Excellent Osteoconductivity for Load-Bearing Bone in Vivo Regeneration." *Scientific Reports* 10 (18193).

Dubey, Anshu, Satish Jaiswal, Akshit Garg, Vaibhav Jain, and Debrupa Lahiri. 2021. "Synthesis and Evaluation of Magnesium/Co-Precipitated Hydroxyapatite Based Composite for Biomedical Application." *Journal of the Mechanical Behavior of Biomedical Materials* 118: 104460.

Dutta, Sourav, K. Bavya Devi, Sanjay Gupta, Biswanath Kundu, Vamsi K. Balla, and Mangal Roy. 2019. "Mechanical and in Vitro Degradation Behavior of Magnesium-Bioactive Glass Composites Prepared by SPS for Biomedical Applications." *Journal of Biomedical Materials Research – Part B Applied Biomaterials* 107 (2): 352–65.

Dutta, Sourav, Sanjay Gupta, and Mangal Roy. 2020. "Recent Developments in Magnesium Metal-Matrix Composites for Biomedical Applications: A Review." *ACS Biomaterials Science and Engineering* 6 (9): 4748–73.

Eliaz, Noam. 2019. "Corrosion of Metallic Biomaterials: A Review." *Materials* 12 (3).

Elumalai Pc, Ganesh Radhakrishnan, and Shankar Emarose. 2021. "Investigation into Physical, Microstructural and Mechanical Behaviour of Titanium Dioxide Nanoparticulate Reinforced Magnesium Composite." *Materials Technology* 36 (10): 575–84.

Esmaeilzadeh, O., A. R. Eivani, M. Mehdizade, S. M. A. Boutorabi, and S. M. Masoudpanah. 2022. "An Investigation of Microstructural Background for Improved Corrosion Resistance of WE43 Magnesium-Based Composites with ZnO and Cu/ZnO Additions." *Journal of Alloys and Compounds* 908: 164437.

Feng, Ailing, and Yong Han. 2010. "The Microstructure, Mechanical and Corrosion Properties of Calcium Polyphosphate Reinforced ZK60A Magnesium Alloy Composites." *Journal of Alloys and Compounds* 504 (2): 585–93.

Feng, Ailing, and Yong Han. 2011. "Mechanical and in Vitro Degradation Behavior of Ultrafine Calcium Polyphosphate Reinforced Magnesium-Alloy Composites." *Materials and Design* 32 (5): 2813–20.

Fulmer, Mark T., Ira C. Ison, Christine R. Hankermayer, Brent R. Constantz, and John Ross. 2002. "Measurements of the Solubilities and Dissolution Rates of Several Hydroxyapatites." *Biomaterials* 23 (3): 751–55.

Guo, Yunting, Guangyu Li, Yingchao Xu, Zezhou Xu, Mingqi Gang, Guixun Sun, Zhihui Zhang, et al. 2021. "The Microstructure, Mechanical Properties, Corrosion Performance and Biocompatibility of Hydroxyapatite Reinforced ZK61 Magnesium-Matrix Biological Composite." *Journal of the Mechanical Behavior of Biomedical Materials* 123: 104759.

Gupta, M., M. O. Lai, and D. Saravanaranganathan. 2000. "Synthesis, Microstructure and Properties Characterization of Disintegrated Melt Deposited Mg /SiC Composites." *Journal of Materials Science* 35: 2155–65.

Gupta, M., and W. L. E. Wong. 2015. "Magnesium-Based Nanocomposites: Lightweight Materials of the Future." *Materials Characterization* 105: 30–46.

Hanas, T., T. S. Sampath Kumar, Govindaraj Perumal, Mukesh Doble, and Seeram Ramakrishna. 2018. "Electrospun PCL/HA Coated Friction Stir Processed AZ31/HA Composites for Degradable Implant Applications." *Journal of Materials Processing Technology* 252: 398–406.

Hatt, Luan Phelipe, Keith Thompson, Werner E. G. Müller, Martin James Stoddart, and Angela Rita Armiento. 2019. "Calcium Polyphosphate Nanoparticles Act as an Effective Inorganic Phosphate Source During Osteogenic Differentiation of Human Mesenchymal Stem Cells." *International Journal of Molecular Sciences* 20 (22): 1–15.

Hench, L. L., R. J. Splinter, W. C. Allen, and T. K. Greenlee. 1971. "Bonding Mechanisms at the Interface of Ceramic Prosthetic Materials." *Journal of Biomedical Materials Research* 5 (6): 117–41.

Ho, Yee Hsien, Sameehan S. Joshi, Tso C. Wu, Chu M. Hung, New J. Ho, and Narendra B. Dahotre. 2020. "In-Vitro Bio-Corrosion Behavior of Friction Stir Additively Manufactured AZ31B Magnesium Alloy-Hydroxyapatite Composites." *Materials Science and Engineering C* 109: 110632.

Huang, Yan, Junyi Li, and Li Zhou. 2018. "Mg–3Zn–0.5Zr/HA Nanocomposites Fabricated by High Shear Solidification and Equal Channel Angular Extrusion." *Materials Science and Technology* 34 (15): 1868–79.

Idrisi, Amir Hussain, and Abdel H. Mourad. 2019. "Conventional Stir Casting Versus Ultrasonic Assisted Stir Casting Process: Mechanical and Physical Characteristics of AMCs." *Journal of Alloys and Compounds* 805: 502–8.

Jackson, Lauren E., Benson M. Kariuki, Mark E. Smith, Jake E. Barralet, and Adrian J. Wright. 2005. "Synthesis and Structure of a Calcium Polyphosphate with a Unique Criss-Cross Arrangement of Helical Phosphate Chains." *Chemistry of Materials* 17 (18): 4642–46.

Jahr, Holger, Yageng Li, Jie Zhou, Amir A. Zadpoor, and Kai Uwe Schröder. 2021. "Additively Manufactured Absorbable Porous Metal Implants – Processing, Alloying and Corrosion Behavior." *Frontiers in Materials* 8: 628633.

Jaiswal, Satish, Anshu Dubey, and Debrupa Lahiri. 2020. "The Influence of Bioactive Hydroxyapatite Shape and Size on the Mechanical and Biodegradation Behaviour of Magnesium Based Composite." *Ceramics International* 46 (17): 27205–18.

Jalilvand, Mohammad Mahdi, and Yousef Mazaheri. 2020. "Effect of Mono and Hybrid Ceramic Reinforcement Particles on the Tribological Behavior of the AZ31 Matrix Surface Composites Developed by Friction Stir Processing." *Ceramics International* 46 (12): 20345–56.

James, Johny, A. Raja Annamalai, A. Muthuchamy, and Chun Ping Jen. 2021. "Effect of Wettability and Uniform Distribution of Reinforcement Particle on Mechanical Property (Tensile) in Aluminum Metal Matrix Composite – A Review." *Nanomaterials* 11 (9).

Jeong, Jiwoon, Jung H. Kim, Jung H. Shim, Nathaniel S. Hwang, and Chan Y. Heo. 2019. "Bioactive Calcium Phosphate Materials and Applications in Bone Regeneration." *Biomaterials Research* 23 (4).

Khalili, Vida, Sajjad Moslemi, Benjamin Ruttert, Jan Frenzel, Werner Theisen, and Gunther Eggeler. 2021. "Surface Metal Matrix Nano-Composite of Magnesium/Hydroxyapatite Produced by Stir-Centrifugal Casting." *Surface and Coatings Technology* 406: 126654.

Khodaei, Mohammad, Farahnaz Nejatidanesh, Mohammad Javad Shirani, Alireza Valanezhad, Ikuya Watanabe, and Omid Savabi. 2019. "The Effect of the Nano- Bioglass Reinforcement on Magnesium Based Composite." *Journal of the Mechanical Behavior of Biomedical Materials* 100: 103396.

Khorashadizade, F., S. Abazari, M. Rajabi, H. R. Bakhsheshi-Rad, Ahmad Fauzi Ismail, Safian Sharif, Seeram Ramakrishna, and F. Berto. 2021. "Overview of Magnesium-Ceramic Composites: Mechanical, Corrosion and Biological Properties." *Journal of Materials Research and Technology* 15: 6034–66.

Khoshzaban Khosroshahi, H., F. Fereshteh Saniee, and H. R. Abedi. 2014. "Mechanical Properties Improvement of Cast AZ80 Mg Alloy/Nano-Particles Composite via Thermomechanical Processing." *Materials Science and Engineering A* 595: 284–90.

Kim, J. H., S. H. Kim, H. K. Kim, T. Akaike, and S. C. Kim. 2002. "Synthesis and Characterization of Hydroxyapatite Crystals: A Review Study on the Analytical Methods." *Journal of Biomedical Materials Research* 62 (4): 600–12.

Kumar, Ajit, and Pulak M. Pandey. 2020. "Development of Mg Based Biomaterial with Improved Mechanical and Degradation Properties Using Powder Metallurgy." *Journal of Magnesium and Alloys* 8 (3): 883–98.

Kumar, Anuj, Saeid Kargozar, Francesco Baino, and Sung Soo Han. 2019. "Additive Manufacturing Methods for Producing Hydroxyapatite and Hydroxyapatite-Based Composite Scaffolds: A Review." *Frontiers in Materials* 6 (313).

Kumar, D. Sameer, K. N. S. Suman, and Palash Poddar. 2017. "Effect of Particle Morphology of Ni on the Mechanical Behavior of AZ91E-Ni Coated Nano Al_2O_3 Composites." *Materials Research Express* 4 (6): 066505.

Kumar, Pravir, Ashis Mallick, Milli Suchita Kujur, Khin Sandar Tun, Rajashekhara Shabadi, and Manoj Gupta. 2018. "Strength of Mg–3%Al Alloy in Presence of Graphene Nano-Platelets as Reinforcement." *Materials Science and Technology* 34 (9): 1086–95.

Li, Chunmei, Chengchen Guo, Vincent Fitzpatrick, Ahmed Ibrahim, Myrthe J. Zwierstra, Philip Hanna, Aron Lechtig, Ara Nazarian, Samuel J. Lin, and David L. Kaplan. 2020. "Design of Biodegradable, Implantable Devices towards Clinical Translation." *Nature Reviews Materials* 5 (1): 61–81.

Li, Ling Yu, Zhuang Z. Han, Rong C. Zeng, Wei C. Qi, Xiao F. Zhai, Yi Yang, Yun T. Lou, Tingyue Gu, Dake Xu, and Ji Z. Duan. 2020. "Microbial Ingress and in Vitro Degradation Enhanced by Glucose on Bioabsorbable Mg – Li – Ca Alloy." *Bioactive Materials* 5 (4): 902–16.

Liu, Zhen, Yangchuan Cai, Jie Chen, Jian Han, Zhiyong Mao, and Minfang Chen. 2021. "Fabrication and Characterization of Friction Stir – Processed Mg-Zn-Ca Biomaterials Strengthened with MgO Particles." *International Journal of Advanced Manufacturing Technology* 117 (3–4): 919–32.

Lopes, Debora, Renata B. Soares, Moara M. Castro, Roberto B. Figueiredo, Terence G. Langdon, and Vanessa F. C. Lins. 2020. "Corrosion Behavior in Hank's Solution of a Magnesium – Hydroxyapatite Composite Processed by High-Pressure Torsion." *Advanced Engineering Materials* 22 (12).

Ma, Z. Y. 2008. "Friction Stir Processing Technology: A Review." *Metallurgical and Materials Transactions A: Physical Metallurgy and Materials Science A* 39 (3): 642–58.

Mostafa, Nasser Y., and Paul W. Brown. 2007. "Computer Simulation of Stoichiometric Hydroxyapatite: Structure and Substitutions." *Journal of Physics and Chemistry of Solids* 68 (3): 431–37.

Munir, Khurram, Cuie Wen, and Yuncang Li. 2020. "Graphene Nanoplatelets-Reinforced Magnesium Metal Matrix Nanocomposites with Superior Mechanical and Corrosion Performance for Biomedical Applications." *Journal of Magnesium and Alloys* 8 (1): 269–90.

Narita, Kai, Qiaomu Tian, Ian Johnson, Chaoxing Zhang, Equo Kobayashi, and Huinan Liu. 2019. "Degradation Behaviors and Cytocompatibility of Mg/β-Tricalcium Phosphate Composites Produced by Spark Plasma Sintering." *Journal of Biomedical Materials Research – Part B Applied Biomaterials* 107 (7): 2238–53.

Niu, Xiaomiao, Hongyao Shen, Jianzhong Fu, and Jiawei Feng. 2021. "Effective Control of Microstructure Evolution in AZ91D Magnesium Alloy by SiC Nanoparticles in Laser Powder-Bed Fusion." *Materials and Design* 206: 109787.

Ong, Tiong Hou Damien, Na Yu, Ganesh Kumar Meenashisundaram, Benoit Schaller, and Manoj Gupta. 2017. "Insight into Cytotoxicity of Mg Nanocomposites Using MTT Assay Technique." *Materials Science and Engineering C* 78: 647–52.

Parande, Gururaj, Vyasaraj Manakari, Harshit Gupta, and Manoj Gupta. 2018. "Magnesium-β-Tricalcium Phosphate Composites as a Potential Orthopedic Implant: A Mechanical/Damping/Immersion Perspective." *Metals* 8 (5): 343.

Parande, Gururaj, Vyasaraj Manakari, Somasundaram Prasadh, and Deep Chauhan. 2020. "Strength Retention, Corrosion Control and Biocompatibility of Mg – Zn – Si/HA Nanocomposites." *Journal of the Mechanical Behavior of Biomedical Materials* 103: 103584.

Pasha, Mahammod Babar, R. Narasimha Rao, Syed Ismail, Mutlu Özcan, P. Syam Prasad, and Manoj Gupta. 2022. "Assessing Mg/Si$_3$N$_4$ Biodegradable Nanocomposites for Osteosynthesis Implants with a Focus on Microstructural, Mechanical, in Vitro Corrosion and Bioactivity Aspects." *Journal of Materials Research and Technology* 19: 3803–17.

Qiao, Ke, Ting Zhang, Kuaishe Wang, Shengnan Yuan, Liqiang Wang, Shanyong Chen, Yuhao Wang, Kairui Xue, and Wen Wang. 2022. "Effect of Multi-Pass Friction Stir Processing on the Microstructure Evolution and Corrosion Behavior of ZrO2/AZ31 Magnesium Matrix Composite." *Journal of Materials Research and Technology* 18: 1166–79.

Rahim, Shebeer A., M. A. Joseph, T. S. Sampath Kumar, and T. Hanas. 2022. "Recent Progress in Surface Modification of Mg Alloys for Biodegradable Orthopedic Applications." *Frontiers in Materials* 9: 848980.

Razavi, M., M. H. Fathi, and M. Meratian. 2010a. "Bio-Corrosion Behavior of Magnesium-Fluorapatite Nanocomposite for Biomedical Applications." *Materials Letters* 64 (22): 2487–90.

Razavi, M., M. H. Fathi, and M. Meratian. 2010b. "Microstructure, Mechanical Properties and Bio-Corrosion Evaluation of Biodegradable AZ91-FA Nanocomposites for Biomedical Applications." *Materials Science and Engineering A* 527 (26): 6938–44.

Razavi, Mehdi, and Yan Huang. 2020. "Effect of Hydroxyapatite (HA) Nanoparticles Shape on Biodegradation of Mg/HA Nanocomposites Processed by High Shear Solidification/Equal Channel Angular Extrusion Route." *Materials Letters* 267: 127541.

Razzaghi, Mahmood, Masoud Kasiri-Asgarani, Hamid R. Bakhsheshi-Rad, and Hamid Ghayour. 2020. "Microstructure, Mechanical Properties, and in-Vitro Biocompatibility of Nano- NiTi Reinforced Mg–3Zn-0.5Ag Alloy: Prepared by Mechanical Alloying for Implant Applications." *Composites Part B: Engineering* 190: 107947.

Rey, Christian, Christèle G. Combes, Christophe Drouet, and Hocine Sfihi. 2008. "Fluoride-Based Bioceramics." In *Fluorine and Health*, edited by Alain Tressaud and Günter Haufe, 279–331, Amsterdam, Elsevier B.V.

Ruiz-Aguilar, Criseida, Ulises Olivares-Pinto, Ena A. Aguilar-Reyes, Rigoberto López-Juárez, and Ismeli Alfonso. 2018. "Characterization of β-Tricalcium Phosphate Powders Synthesized by Sol-Gel and Mechanosynthesis." Boletin de La Sociedad Espanola de Ceramica y Vidrio 57 (5): 213–20.

Saberi, A., H. R. Bakhsheshi-Rad, E. Karamian, M. Kasiri-Asgarani, and H. Ghomi. 2020. "Magnesium-Graphene Nano-Platelet Composites: Corrosion Behavior, Mechanical and Biological Properties." *Journal of Alloys and Compounds* 821: 153379.

Shuai, Cijun, Sheng Li, Shuping Peng, Pei Feng, Yuxiao Lai, and Chengde Gao. 2019. "Biodegradable Metallic Bone Implants." *Materials Chemistry Frontiers* 3 (4): 544–62.

Shuai, Cijun, Bing Wang, Shizhen Bin, Shuping Peng, and Chengde Gao. 2020. "Interfacial Strengthening by Reduced Graphene Oxide Coated with MgO in Biodegradable Mg Composites." *Materials and Design* 191: 108612.

Somasundaram, M., Narendra K. Uttamchand, A. Raja Annamalai, and Chun Ping Jen. 2022. "Insights on Spark Plasma Sintering of Magnesium Composites: A Review." *Nanomaterials* 12 (13): 2178.

Su, Jin-long, Jie Teng, Zi-li Xu, and Yuan Li. 2020. "Biodegradable Magnesium-Matrix Composites : A Review." *International Journal of Minerals, Metallurgy and Materials* 27 (6): 724–44.

Sun, Xiaohao, Yue Su, Yan Huang, Minfang Chen, and Debao Liu. 2022. "Microstructure Evolution and Properties of β-TCP/Mg-Zn-Ca Biocomposite Processed by Hot Extrusion Combined with Multi-Pass ECAP." *Metals* 12 (4).

Verma, Vivek, Joy Saha, Abhishek Gautam, and Kaushik Pal. 2022. "Investigation on Microstructure, Mechanical, Biocorrosion and Biocompatibility Behavior of Nano-Sized TiO$_2$@Al$_2$O$_3$ Reinforced Mg-HAp Composites." *Journal of Alloys and Compounds* 910: 164866.

Vidal, Catarina, Patricia Alves, Marta M. Alves, Maria João Carmezim, Maria Helena Fernandes, Liliana Grenho, Patrick L. Inácio, Francisco B. Ferreira, Telmo G. Santos, and Catarina Santos. 2022. "Fabrication of a Biodegradable and Cytocompatible Magnesium/Nanohydroxyapatite/Fluorapatite Composite by Upward Friction Stir Processing for Biomedical Applications." *Journal of the Mechanical Behavior of Biomedical Materials* 129: 105137.

Wang, Jia Li, Jian K. Xu, Chelsea Hopkins, Dick H. K. Chow, and Ling Qin. 2020. "Biodegradable Magnesium-Based Implants in Orthopedics – A General Review and Perspectives." *Advanced Science* 7: 1902443.

Watanabe, Hiroyuki, Naoko Ikeo, and Toshiji Mukai. 2019. "Processing and Mechanical Properties of a Tricalcium Phosphate-Dispersed Magnesium-Based Composite." *Materials Transactions* 60 (1): 105–10.

Zhang, Li Nan, Zeng T. Hou, Xin Ye, Zhao B. Xu, Xue L. Bai, and Peng Shang. 2013. "The Effect of Selected Alloying Element Additions on Properties of Mg-Based Alloy as Bioimplants: A Literature Review." *Frontiers of Materials Science* 7 (3): 227–36.

Zhang, Shuang, Zhen Liu, Yi Xin, Yangchuan Cai, and Jian Han. 2021. "Effect of Equal Channel Angular Pressing on Microstructure and Mechanical Performance of Innovative Nano MgO-Added Mg-Zn-Ca Composite as a Biomaterial." *Materials Letters* 304: 130604.

Zhou, Hongjian, and Jaebeom Lee. 2011. "Nanoscale Hydroxyapatite Particles for Bone Tissue Engineering." *Acta Biomaterialia* 7 (7): 2769–81.

Part III

Protective Coatings

11 Sol-Gel Nanocomposite Coatings

You Zhang, Kai Wei, Dahai Gao,
Zhe Zhang, Yujie Yuan and Liang Wu

CONTENTS

11.1 INTRODUCTION

In this chapter, we present the most recent technologies and advances in sol-gel nanocomposite coatings prepared directly on the surface of magnesium alloys and the use of sol-gel technology in multilayer coatings. First, we introduce the primary process and evolution of sol-gel technology. Subsequently, the practical technologies, including inhibitors, micro-/nanocontainers, and multilayer coatings, are discussed. We also simply display the main application of sol-gel coatings on magnesium alloy substrates.

The sol-gel method is a chemical synthesis method that forms an alkoxide network structure by polycondensation of precursors (alkoxides) in a liquid medium. It is mainly based on metal alkoxides or inorganic compounds as precursors, which are dissolved in organic solutions, usually catalyzed by weak acid or weak base under liquid conditions through hydrolysis, polycondensation, and other chemical processes to form a stable and uniform sol liquid system. The sol will gradually transform to gel with a three-dimensional network structure after a period of aging. The gels generally form porous morphology by removing solvents, which show controllable porosity and specific surface area.

A series of alkoxides was applied as precursors of the sol-gel process, leading to obtaining uniform and highly pure frameworks. Various functional gel materials could be prepared with the help of functionalized alkoxides. For optimizing the sol-gel preparation, some important factors attract the researchers' attention: (a) *hydrolysis temperature* affects the reaction velocity, particle nucleation and colloidal particle size (Omri et al. 2014); (b) *pH value* contributes also to the reaction velocity, as well as the shrinkage of products; (c) *reactant concentration* influence the hydrolysis rate and time, according to the simple principle of chemical equilibrium but also the stability of the reaction system; (d) *aging time* usually decided the microstructure of the prepared gel (Periyasamy et al. 2020).

In 1846, the French chemist J. J. Ebelmen first prepared tetraethoxysilane (TEOS) via the alcoholysis of $SiCl_4$ and found that TEOS slowly hydrolyzed in wet air to become a gel. In the 1930s,

DOI: 10.1201/9781003319856-14

W. Geffcken obtained sol-gel films, which opened a new understanding and development of membrane fabrication. In 1971, H. Dislich from Germany reported the preparation of SiO_2-B_2O-Al_2O_3-Na_2O-K_2O multi-component glass by hydrolysis of metal alkoxides, which marked a new chapter in the study of sol-gel coatings. With the successful preparation of ceramic materials and porous transparent alumina films by B. E. Yoldas and M. Yamane in 1975, the research on sol-gel coatings has gradually been attractive. Since the 1980s, sol-gel technology has been applied in various fields such as glass, oxide coatings, and nuclear functional ceramic powders.

11.2　SOL-GEL ORGANIC-INORGANIC HYBRID COATINGS

Sol-gel coatings have drawn a lot of attention in recent years because of their excellent results and diverse range of uses, especially in the anti-corrosion application. Firstly, the layer formed by the sol-gel process has intrinsically blocked the penetration of the corrosive medium (López et al. 2011). The micromorphologies of this sol-gel coating are displayed in **Figure 11.1**. At first, the obtained coating films were continuous. However, inorganic sol-gel coatings are too brittle to be applied directly into the field of anti-corrosion of magnesium alloys because those films tend to crack under external forces. Instead, hybrid organic/inorganic coating is frequently used to protect the surface of magnesium alloy and restrain the corrosive factors from the surface. For further enhancing the overall properties of hybrid coatings, especially mechanical properties and shielding effect, various organic silanes and polymers are frequently integrated into the sol-gel composition. In most inorganic-organic coating systems, the combination of organic components provides better adjustability; that is, the organic part is responsible for the toughness of coatings, while inorganic frameworks enhance the chemical, thermal, and UV resistance of coating systems (Barroso et al. 2019).

Compared with monophase inorganic coating, organic/inorganic hybrid coating shows (a) low processing temperature; (b) ease of composition control with mixing of various precursors; (c) ease of precursor modification for various products; (d) products with high purity for the sake of easy precursor purification; (e) mild reaction conditions, where the process of preparation is environmentally friendly without much waste. During the coating formed on the magnesium alloy surface, the main adhesion force generated by the sol-gel layer comes from covalent bonding, hydrogen bonding, van der Waals bonding, and physical adsorption interactions. The specific process can be divided into the following three stages. The first stage is the hydrolysis process of precursors such as silane. For instance, alkoxysilane hydrolyzes to yield silanols, which could aggregate and condense into silica sol.

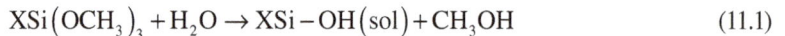

$$XSi(OCH_3)_3 + H_2O \rightarrow XSi-OH(sol) + CH_3OH \qquad (11.1)$$

In the second stage, the Si-OH in the silane sol-gel reacts with the hydroxide of magnesium (Mg-OH) on the surface of the magnesium alloy to form Mg-O-Si covalent bonds:

$$Si-OH(sol) + Mg-OH(Sub) \rightarrow Mg\text{-}O\text{-}Si + H_2O \qquad (11.2)$$

The third stage contains the condensation of the -OH group on the metal surface after drying and heat treatment or the condensation between the deposited Si-OH, which eventually forms the Si-O-Si covalent bond:

$$Si\text{-}OH(sol) + Si\text{-}OH(Sub) \rightarrow Si\text{-}O\text{-}Si + H_2O \qquad (11.3)$$

A covalent bond at the interface is very helpful for the adhesion of the coating. Better adhesion force also allows the sol-gel coating to act as a primer for composite coatings. To summarize, the corrosion resistance of the coating is generated by the shielding effect and good adhesion.

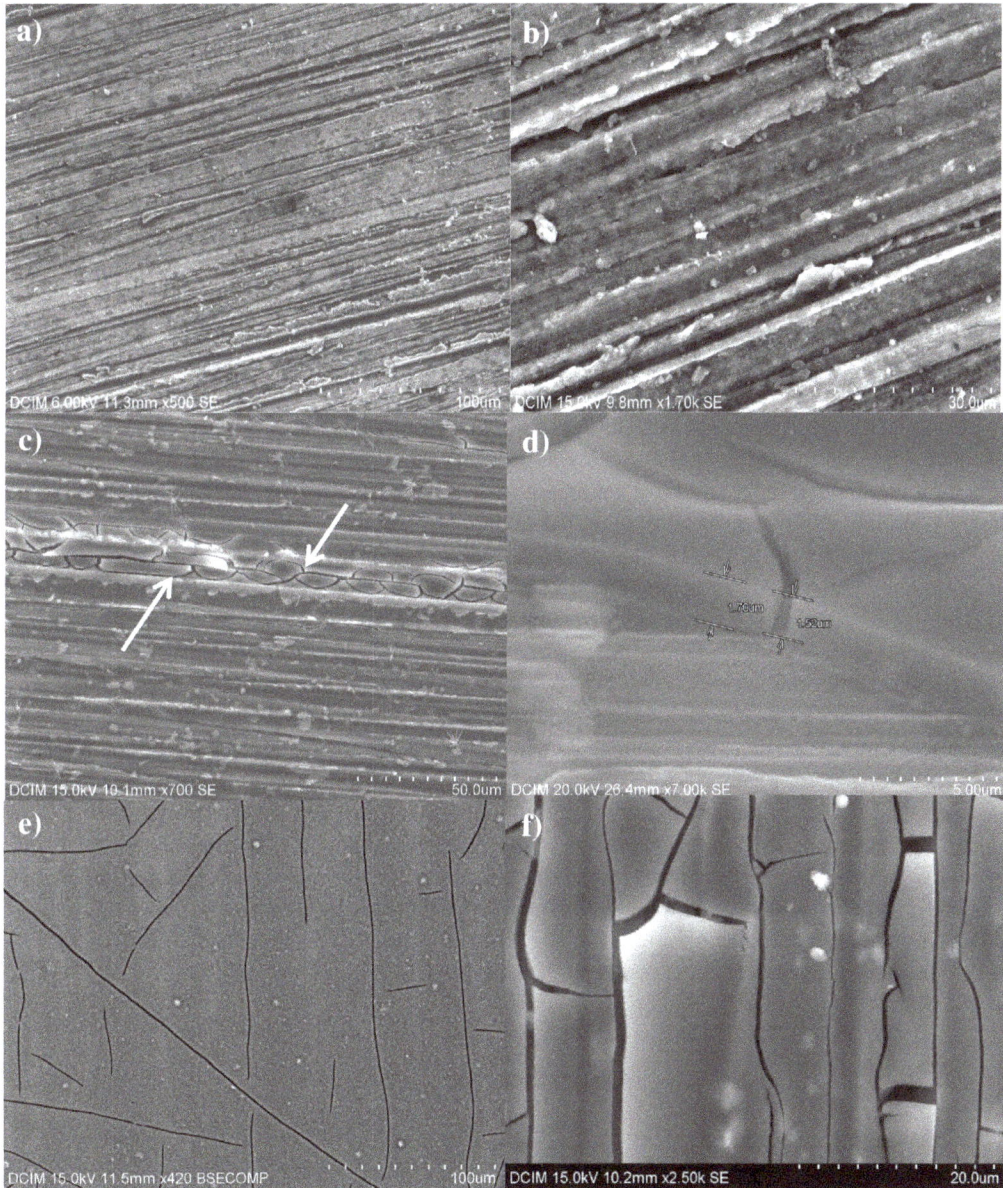

FIGURE 11.1 Surface morphologies of silane films of single layer and three layers after different heat treatment methods: (a) and (b) single layer after 100°C (1 h) and 500°C (1 h) heating; (c) and (d) single layer after 135°C (6 h) heating; (e) and (f) three layers after 100°C (1 h) twice and 135°C (6 h) heating. *Reproduced with permission from (López et al. 2011) © 2011 Elsevier B.V.*

It is a common method to improve the compactness of silica film by introducing other functional groups. The corrosion resistance of silica coating with TEOS and vinyltriethoxysilane (VTEO) as precursors prepared on the surface of AZ91D magnesium alloy is higher than that of a single precursor (Hu et al. 2008). The surface morphology of the film with TEOS and VTEO as precursors was investigated to be denser and less defective than its monosilane counterpart with TEOS only. The main reason is that the presence of vinyl ($CH_2=CH-$) inhibits the formation of aggregated structures in the SiO_2 sol-gel coating. The results suggest that the conventional coatings prepared with single precursors such as tetramethoxysilane (TMOS) and TEOS may result in the formation of pores and channels throughout the coating during condensation or drying. This condition has a significant impact on corrosion resistance, with inorganic silica components increasing material brittleness and decreasing material flexibility. Therefore, single-layer coatings or single-silane systems are unable to meet the needs of the real industry. In turn, the combination of different precursors makes it easier to achieve the desired coating properties. The combination of TEOS, tetrabutoxy titanium (TBOT) or other metal-organic ligands with other types of silanes also has excellent results. 3-glyceropropyltrimethoxysilane (GPTMS) was also used as a precursor to prepare Si-O-Si network films with high crosslinking degrees (Guo et al. 2009).

Circular condensed alkoxysiloxanes have more hydrolyzable Si-O-R groups, which can react sufficiently with the Mg alloy substrate to form a high-density anchoring point. The alkoxysilanes with long alkyl chains tend to transform into dense and continuous films, which positively affects the corrosion-resistance performance (Zucchi et al. 2008).

Metal (Ti, Zr) alkoxides such as TBOT could also transform into a sol-gel coating, which shows sufficient adhesion to magnesium alloy. It was found that the number of coating cycles, heat treatment temperature and time, and other parameters are among the primary determinants of the TiO_2 sol-gel coatings. TBOT can be directly used to prepare sol-gel coating on the surface of AZ31 magnesium alloy (Hu et al. 2011). This coating shows better flatness and cracking resistance than the silica sol-gel coating. The composite layers can be prepared via mixing silica sol and TiO_2 sol as a ratio of 1:1 and showed excellent corrosion potential (-1.359 V vs SCE), low corrosion current density (1.981×10^{-9} A/cm^2), and low corrosion rate (only 0.75 mm/y) (Li et al. 2016).

In these organic-inorganic hybrid coatings, the interface interaction between two components is often a critical factor in determining the performance of the film. Covalent bonds can provide the strongest interaction so that two components can distribute on a molecular scale. Herein, alkoxysilanes with the general formula $RSi(OR)_3$ are frequently imported into the gel networks, in which R is a bridge between organic and inorganic components. To introduce epoxy groups, GPTMS was added to the original ZrO_2 and SiO_2 precursor mixture (Wang et al. 2021). They then used a small amount of 2-methylimidazole to synthesize the polymer matrix by ring-opening addition reaction of epoxy groups at low temperature, creating a new hybrid cross-linking network that significantly reduced the cracks and defects in the film, as shown in **Figure 11.2**. Results demonstrate that the ring-opening addition reaction created cross-linked thick SiO_2-based sol-gel coatings.

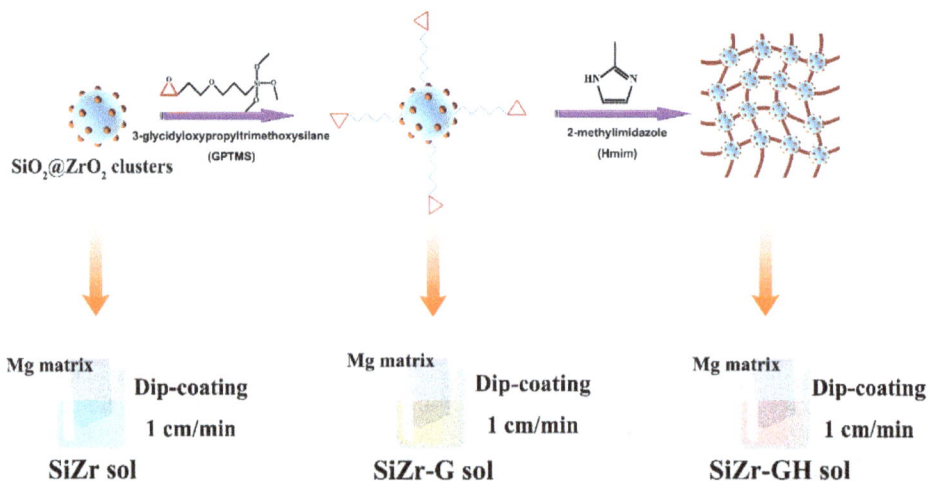

FIGURE 11.2 The schematic of reaction processes and modified sol-gel coating. *Reproduced with permission from (Wang et al. 2021) © 2011 Elsevier B.V.*

Furthermore, nitrates can also be employed as precursors for sol-gel coatings in some cases. For instance, cerium nitrate can serve as a promising substitute for chromate coatings. The acidic solution during the sol-gel process could promote the salt to transform to CeO_2, then forming the layer. But for magnesium alloy, low pH induces severe corrosion on the surface. Herein, researchers used HF to modify the magnesium alloy substrate, generating an immediate layer resisting acid circumstance. As a result, a dense CeO_2 coating grows on the surface (Zhong et al. 2008; Rosero-Navarro et al. 2011). The coatings exhibit excellent corrosion resistance owing to the corrosion inhibition of cerium ions and the barrier effect of the layer.

11.3   SMART SOL-GEL COATINGS WITH MICRO- AND NANOCONTAINERS

The sol-gel coatings show good adhesion and barrier effect, which ensure those can be used in the field of metal anti-corrosion. However, in the long-term process of use, the coating is prone to breakage due to external factors, leading to the loss of the shielding effect of the coating from the corrosive solution. Eventually, it will lead to severe corrosion of the magnesium alloy substrate (Van Ooij et al. 2005). To improve the corrosion resistance of sol-gel coatings, the addition of corrosion inhibitors to achieve the active corrosion resistance of the film is particularly critical. Generally, the corrosion inhibitors should satisfy the following requirements: (a) good corrosion inhibition, (b) compatibility with the sol-gel film, and (c) the amount less than the critical corrosion inhibitor concentration (Yasakau et al. 2008). The corrosion inhibitor/hybrid coating is prepared by adding a certain amount of corrosion inhibitor during the sol-gel process. In some cases, the hybrid coating gradually releases the corrosion inhibitor to improve corrosion resistance of the magnesium alloy.

All inhibitors can be classified into inorganic and organic corrosion inhibitors. Inorganic corrosion inhibitors generally include zirconium salts, cerium salts, zinc nitrate, etc. Cerium salt is a common metal cation corrosion inhibitor used to improve the corrosion resistance of silane film (Correa et al. 2011). The film was prepared with methyltriethoxysilane (MTES) as a precursor, and cerium ions were introduced by adding cerium nitrate during the sol-gel preparation. The corrosion inhibition mechanism of a low concentration of cerium ions is that cerium ions can cross the cathode area and react with hydroxyl groups to form cerium (III) and (IV) hydroxides, which precipitate on the surface of the alloy and form cerium oxide films to prevent further corrosion (Phani et al. 2005). Cerium ion is used as a corrosion inhibitor because it can precipitate in insoluble hydroxide at the cathode position, hindering or even completely blocking the corrosive activity of the mixed coating in some damaged areas and bringing a self-healing effect.

Barranco et al. used TMOS and diethoxydimethylsilane (DEDMS) as precursors to successfully prepare a Ce^{3+} doped sol-gel coating, which was well compatible with the acrylic coating. The hybrid coating showed good corrosion resistance because of the interaction of Ce^{3+} and metal surface (Barranco et al. 2010). Hernández-Barrios et al. investigated the doping of $Ce(NO_3)_3$ in silane sol-gel with TEOS and GPTMS as precursors and acetic acid as a catalyst (Hernández-Barrios et al. 2020). It was shown that the addition of $Ce(NO_3)_3$ reduces the pH of the sol during the mixing process of the sol-gel, which has a protective effect on the magnesium alloy substrate. It also accelerates the hydrolysis-condensation process of the epoxide ring and the pore size. However, the high concentration of $Ce(NO_3)_3$ during the deposition stage will promote the formation of corrosion products on the surface of the AZ31 alloy, as shown in **Figure 11.3**. Other inorganic salt corrosion inhibitors can also be applied in the silane film to achieve improvement in corrosion performance and self-healing properties. For example, the silica film containing zinc nitrate as a corrosion inhibitor (Zhong et al. 2010) was applied as an anti-corrosion coating, generating insoluble zinc hydroxide on the surface to heal the cracks and defects during the corrosion process.

Organic corrosion inhibitors are more widely used in sol-gel coatings for magnesium alloys, including 8-hydroxyquinoline, benzotriazole, phosphonic carboxylic acid, dimethylpiperidine, and other N-containing heterocyclic compounds. 8-Hydroxyquinoline (8-HQ) as a corrosion inhibitor is added to sol-gel films at two synthesis stages before or after hydrolysis of the sol-gel precursor

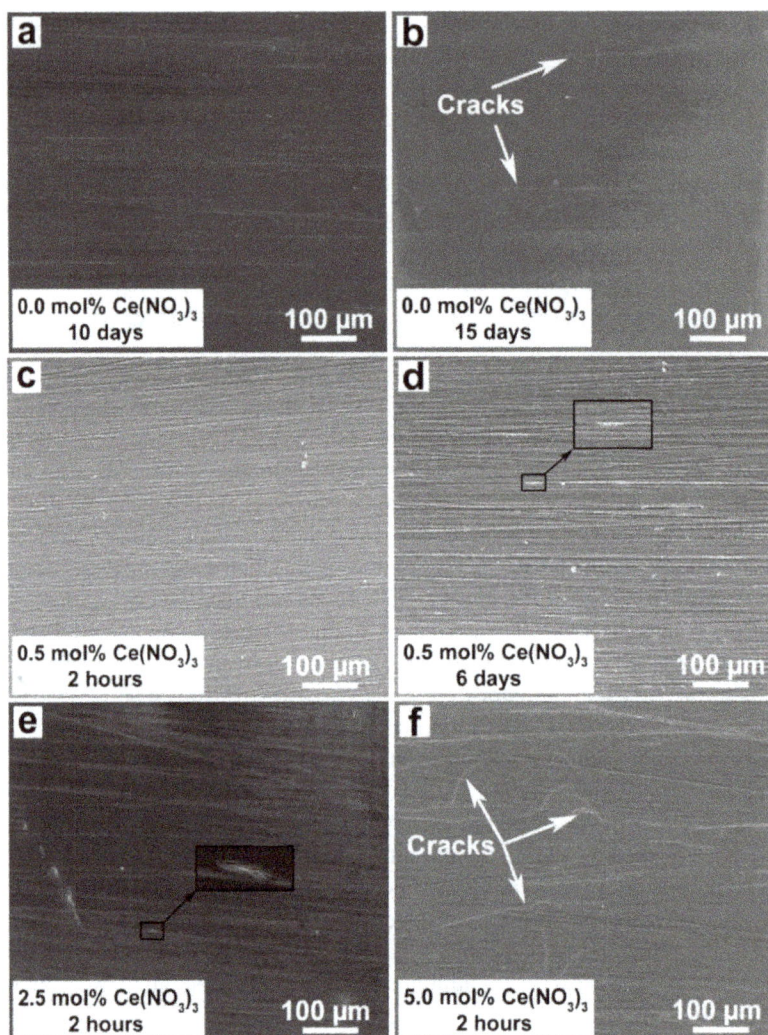

FIGURE 11.3 Cross-sectional BSE-SEM micrographs of coatings obtained from sols with 5.0 vol% CH_3COOH and different $Ce(NO_3)_3$ concentration/aging times: (a) 0.0 mol% $Ce(NO_3)_3$/10 d; (b) 0.0 mol% $Ce(NO_3)_3$/15 d; (c) 0.5 mol% $Ce(NO_3)_3$/2 h; (d) 0.5 mol% $Ce(NO_3)_3$/6 d; (e) 2.5 mol% $Ce(NO_3)_3$/2 h; (f) 5.0 mol% $Ce(NO_3)_3$/2 h. *Reproduced with permission from (Hernández-Barrios et al. 2020) © 2020 Elsevier B.V.*

(Galio et al. 2010). The corrosion resistance of 8-HQ on the surface of magnesium alloy was tested by the scanning vibrating electrode technique (SVET), as shown in **Figure 11.4**. The 0.05% 8-HQ solution treated magnesium alloy shows apparent corrosion inhibition, which is due to the formation of a complex adsorption layer on the surface of magnesium alloy to prevent the penetration of corrosive ions (such as Cl⁻) and protect the substrate.

Some more environmentally friendly inhibitors such as quinaldic acid (QDA), betaine (BET), dopamine hydrochloride (DOP), and diazolidinyl urea (DZU) were investigated in sol-gel films (Upadhyay et al. 2017). Their application in corrosion protection of magnesium alloys is becoming increasingly popular. Ashassi-Sorkhabi et al. (Ashassi-Sorkhabi et al. 2019) prepared an organo-inorganic heterogeneous compound with TEOS and MTES as organic-inorganic hybrid silanes and L -alanine, L-glutamine, L -methionine, and aspartic acid (L-Aspartic) as precursors.

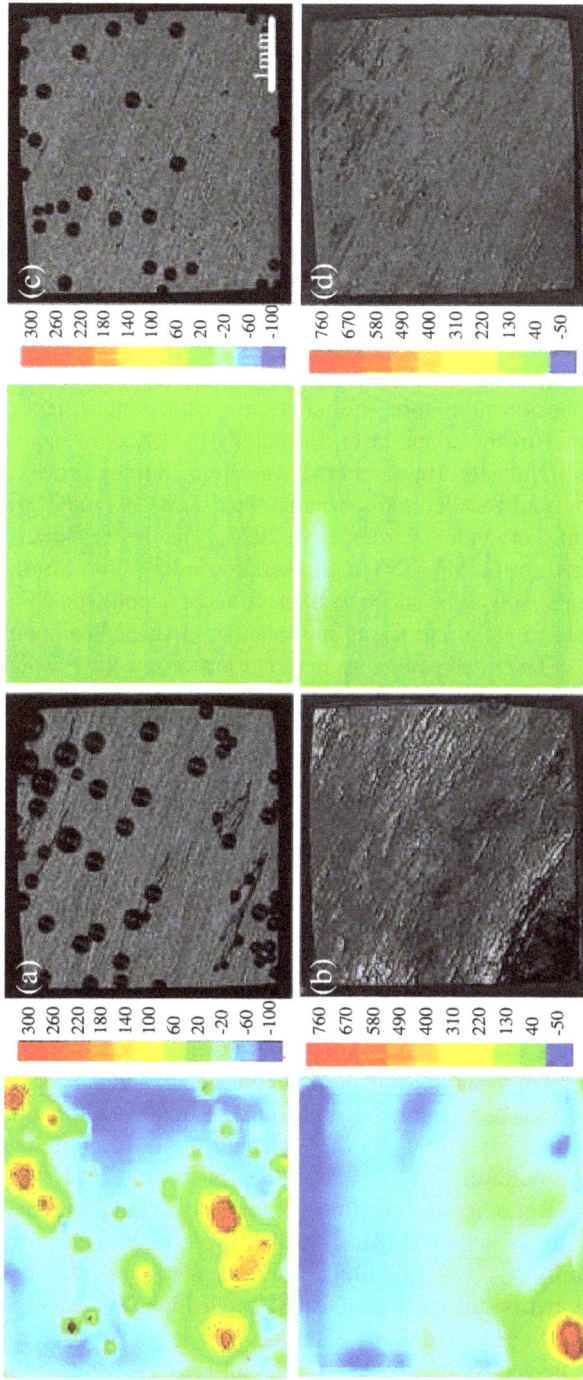

FIGURE 11.4 Optical microphotos and corresponding current density distribution over AZ31 samples in the course of immersion in 0.05 M NaCl: (a) 1 min after beginning of immersion, sample was not treated with inhibitor; (b) the same sample after 17 h of immersion; (c) 1 min after immersion, sample was preliminary treated with alcohol 0.05% solution of 8-HQ; (d) the same sample after 3 days of immersion. The black spots in optical microphotos a and c correspond to hydrogen bubbles. Scale unit in SVET maps is $\mu A/cm^2$. Reproduced with permission from (Galio et al. 2010) © 2009 Elsevier B.V.

However, many inorganic and organic corrosion inhibitors cannot be introduced to the magnesium alloy coating system via direct mixing because of the high activity of the substrate. The controllable release of inhibitors is also an important direction for researching anti-corrosion coatings. Microencapsulated carriers and nanocontainers loading corrosion inhibitors are the main strategies to realize the controlled release of corrosion inhibitors. In addition, the micro-/nanocontainers-loaded inhibitors could overcome the concentration threshold and provide a circumstance for continuous releasing, which is helpful to long-term corrosion resistance.

The common nanocontainers loaded with corrosion inhibitors are mesoporous silica, polyhydric kaolinite, hydrotalcite-like nanocontainers, etc. There are fewer studies on organic nanocontainers loaded with corrosion inhibitors in sol-gel coatings on the magnesium alloy. Halloysite nanotubes (HNTs) are a kind of aluminosilicate with a concentric tubular structure, non-toxic, cheap and easy to prepare. It can be widely used in industry as a material for loading corrosion inhibitors. Cationic corrosion inhibitor Ce^{3+}/Zr^{4+} encapsulated in halloysite nanotubes and montmorillonite clay could substantially reduce the corrosion current density (Adsul et al. 2017; Adsul et al. 2018). Acid modification on the surface of halloysite nanotubes further increases the loading of corrosion inhibitors (Mahmoudi et al. 2019).

Besides the addition of corrosion inhibitors, the addition of functional fillers can also help further improve the performance of the silane layer. Especially the effect of nano-fillers on the performance of silica layers is significant. Researchers have significantly improved the corrosion resistance of coatings by using multi-walled carbon nanotubes (MWCNT) and graphene nanoplatelets (GNPs) in coating film (Fernández-Hernán et al. 2020). The GNPs showed good dispersion in the silane film compared to the MWCNT agglomeration in the film. Similarly, oxidized Fullerene (OF) nanoparticles can likewise be added to silane films as nanofillers. OF is a graphene-based material with a unique shape. Fullerene C_{60} is an icosahedron-shaped compound composed of C atoms placed at the nodes of Hexa and pentagons in a cage lattice structure, which has a pretty low weight and an enormous specific surface area that can be modified in various ways to make it functional. In sol-gel coatings, the compatibility and corrosion resistance of the coating can be improved in this way. OF is activated by nitric acid and dried to form hydroxyl and carboxyl groups on its surface under a strong oxidant. The oxidized OF can quickly react with the hydrolyzed silane precursor to form C-O-Si (Samadianfard et al. 2020). Nanofillers such as graphene nanosheets (Afsharimani et al. 2022) and nanodiamonds (Nezamdoust et al. 2020) were also applied in sol-gel coating systems.

11.4 SOL-GEL COATINGS FOR CONSTRUCTING MULTILAYER COATINGS

As mentioned, a single layer of sol-gel coating probably satisfies the requirement for corrosion resistance. Whereas in some cases, especially for complex circumstances and special requirements, multi-layer coatings should be used, including primer, immediate coat, and top coat. Some conversion films, such as anodic oxidation films, could be applied to the treatment of magnesium alloys, which could be controlled sensitively (Shi et al. 2006). The anodic oxide layer of magnesium alloy is dominated by the ceramic oxide layer $(Mg(OH)_2)$ and the outer oxide layer (MgO). Although the outer layer has a porous surface, other post-treatment technology could overcome that problem (Lkhagvaa et al. 2020). Micro-arc oxidation (MAO), also known as plasma electrolytic oxidation (PEO) or spark anodizing, also forms conversion films, which show many advantages such as efficiency, economy, and ecology and is suitable for industrial applications (Walsh et al. 2009). The corrosion resistance can be improved by optimizing power supply parameters, electrolyte composition, additives, and so on (Farshid and Kharaziha 2021). PEO/sol-gel composite coating ensures the complete coverage of the micropores and microcracks of the PEO coating (Li et al. 2012). Similarly, GPTMS/TEOS silica alkyl sol-gel sealing on the PEO film in the sodium silicate electrolyte system also has good results. As shown in **Figure 11.5**, all the anodic oxide films showed porous structures (Merino et al. 2021). In that system, corrosion resistance can be optimized by adjusting dip coating

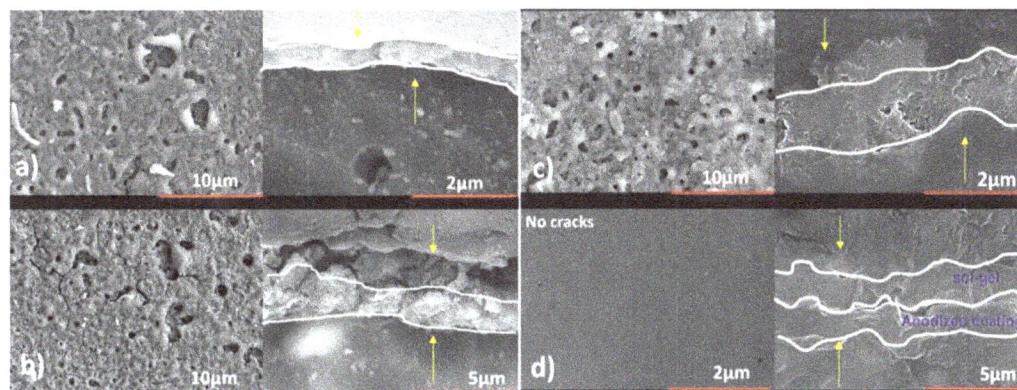

FIGURE 11.5 FESEM surface and cross-sectional morphology of (a–c) anodized AZ31B alloy films obtained at 100 V using the electrolyte of NaOH and various Na_2SiO_3 concentrations (a) 4 g/LNa_2SiO_3, (b) 6 g/L Na_2SiO_3, (c) 8 g/L Na_2SiO_3 and (d) multilayer system (8 g/L Na_2SiO_3+SG). *Reproduced with permission from (Merino et al. 2021) © 2021 The Authors. International Journal of Applied Glass Science published by American Ceramics Society and Wiley Periodicals LLC.*

and immersion times (Dou et al. 2019). Gu et al. prepared a PEO layer on AZ31B magnesium alloy using an electrolyte of $NaAlO_2$, for which the hydroxyapatite (HA) was generated in the presence of $Ca(NO_3)_2$ and P_2O_5 (Gu et al. 2018).

Further improvement of the corrosion resistance of the film can be achieved by constructing a composite film system including fillers and loaded corrosion inhibitors. Inhibitors can also be loaded in the porous oxide conversion film via abundant anchorage points, displaying the controllable releasing performance. A porous oxide layer was impregnated with Ce^{3+} and 8-HQ and sealed by silane film (Lamaka et al. 2009). When immersed in NaCl aqueous solution, inhibitor releasing and physical barrier could have synergistic effects. A PEO/sol-gel composite coating was prepared on the surface of AZ31 magnesium alloy, and three corrosion inhibitors – sodium salts of glycolic, 4-aminosalicylic, and 2,6-pyridinedicarboxylic acids – were added to the porous layer of PEO to yield a film showing good corrosion resistance and self-healing property (Chen et al. 2020). **Figure 11.6** shows the SVET images, indicating that all three coatings successfully released the corrosion inhibitor, which inhibited the corrosion deterioration and protected the magnesium alloy substrate (Chen et al. 2020).

Hydrotalcite-like and layered double hydroxides (LDHs) have promising applications as nanocontainer-loaded corrosion inhibitors. The method of loading corrosion inhibitors in LDH film is simple and has a good corrosion inhibition effect. A composite film prepared by combining 8-HQ intercalated LDH film and the sol-gel layer was prepared to achieve active corrosion protection of AM60B magnesium alloy (Tarzanagh et al. 2022). It is worth noting that the loading of corrosion inhibitors does not affect the structure of the composite film. As the immersion time increases, the modulus of the impedance of the coating could show a part of an ascending curve, which is evidence of the self-healing effect. The loading of corrosion inhibitors in LDH structure can be realized by simple ion exchange. Besides releasing inhibitors, LDH could also catch corrosive anions such as Cl⁻. Researchers used a hydrothermal method to grow an *in situ* nitrate-intercalated LDH film on the magnesium alloy substrate and then completed the preparation of a corrosion inhibitor-doped LDH film by loading F- into the interlayer by the immersion method (Ouyang et al. 2021). The released F- ions can be used with local redeposition of degradation products Mg^{2+} and Ca^{2+} for repairing corrosion pits.

In addition to electrochemical methods, other chemical processes can also be used to prepare a conversion film on the magnesium alloy substrate to enhance corrosion resistance. Among them,

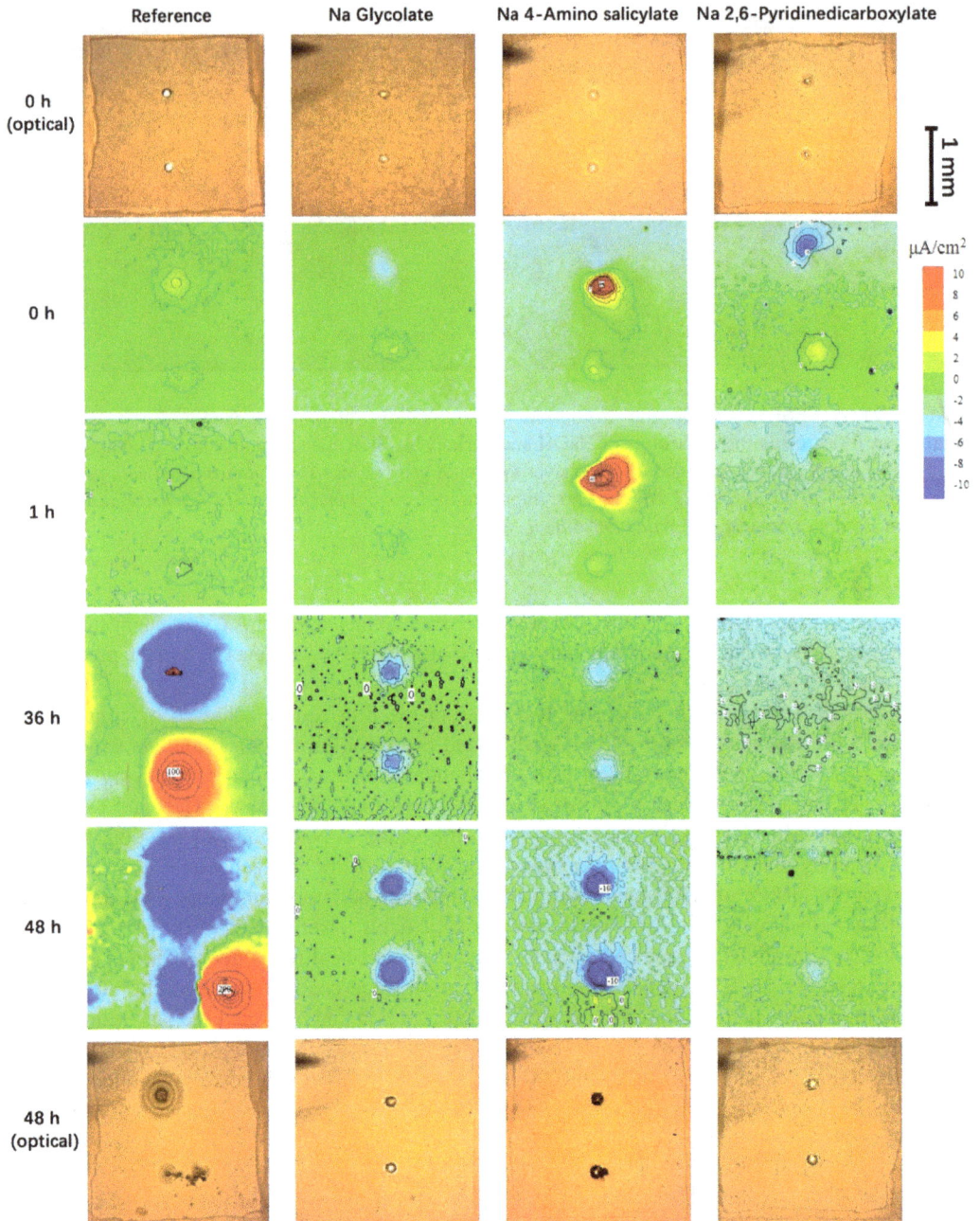

FIGURE 11.6   Optical images of the PEO-SG samples and SVET maps of current recorded above the surface of the PEO-SG samples. The time of exposure and inhibitors are indicated. *Reproduced with permission from (Chen et al. 2020) © 2019 Elsevier B.V.*

fluorination treatment is used to treat the surface of magnesium alloy (Duran et al. 2021). Among them, the long pre-treatment time favored the formation of a hydroxylated magnesium fluoride layer with a high F/O ratio, which significantly contributed to the corrosion resistance of the film. Yue et al. (Yue et al. 2013) prepared a chemical conversion film of about 2 μm on the surface of AZ31 magnesium alloy by pre-treatment, importing Si-O-Mg to the sol-gel film. The alkaline treatment of the magnesium alloy surface is also suitable for loading the silane layer. A stable layer of magnesium hydroxide is formed on the surface of the alloy when the pH value of the alkaline solution was greater than 11 (Liang et al. 2005). The magnesium hydroxide film obtained from the alkali treatment has a strong electrostatic attraction for the silicon hydroxyl groups in the silane sol-gel, attracting them to aggregate and chemically react to form Mg-O-Si chemical bonds (Zhao et al. 2015). Chemical conversion films can also be formed by immersing the surface-treated magnesium alloy substrate in a solution of cerium nitrate and hydrogen peroxide to obtain a cerium conversion film, which is closely combined with sol-gel coating (Pereira et al. 2022). Some other methods of preparing chemical conversion films can also improve the adhesion of sol-gel coatings and prevent the damage of acid gels to magnesium alloy substrates, such as acetic acid (Dalmoro et al. 2019), phytate (Wang et al. 2017), molybdate (Hu et al. 2009), etc.

11.5 APPLICATIONS OF SOL-GEL COMPOSITE COATINGS IN BIOMEDICAL MAGNESIUM ALLOYS

Currently, the application of magnesium alloy sol-gel coating is mainly focused on biocompatibility (Kraus et al. 2012). Unlike traditional biomedical implants such as stainless steel, titanium, and titanium alloys, magnesium-based alloys are biodegradable, avoiding the adverse effects of secondary surgical removal of implants. In addition, the magnesium ion is the fourth most abundant cation in the human body, which is essential for the physiological function of the human body and can promote the growth of bone cells (Witte et al. 2005). However, magnesium alloys have poor corrosion resistance, especially in the physiological environment (human body fluids or plasma). Moreover, a hydrogen evolution reaction will occur with substrate corrosion, limiting its clinical application (Tang et al. 2018).

Nanostructured hydroxyapatite (HA, $Ca_{10}(PO_4)_6(OH)_2$) has good biocompatibility, which is the main inorganic component of human and animal bones. Hydroxyapatite (HA) coating on the surface of metal materials can significantly improve biocompatibility and corrosion resistance of metals (Rojaee et al. 2013). The corrosion resistance and biocompatibility of magnesium alloy can be enhanced by PEO/sol-gel composite film (Zhang et al. 2012). The sealing effect of the HA sol-gel layer on the PEO film improves the stability and biological activity of the film significantly (Tang et al. 2018). The porous ceramic coating formed by PEO can increase the bonding force between the coating and the substrate based on improving the corrosion and wear resistances (Zheng et al. 2018). Compared with the bonding strength between HA coating and magnesium alloy substrate, the bonding strength of PEO/HA composite coating reached 40 MPa, which is attributed to the sealing effect of HA entering the porous layer. PEO/HA can effectively reduce the hydrogen evolution rate and corrosion current density of AZ31 magnesium alloy in SBF, preventing the entry of corrosive ions more effectively (Niu et al. 2016).

The addition of PCTyr (Paeonol condensation tyrosine) Schiff base bio-corrosion inhibitor to silane film with GPTMS and TEOS as precursors can yield a coating with excellent corrosion resistance and biocompatibility on ZE21B alloy (Li et al. 2021). Loading PCTyr-Schiff base in sol-gel can promote cell proliferation and improve the biocompatibility of the sol-gel coating. Phytic acid hybrid sol-gel coating can also enhance the biocompatibility and corrosion resistance of magnesium alloys. It shows strong chelation with various metal ions and can form phytic acid conversion coating on the metal surface (Wang et al. 2017) so that the phytic acid/silane hybrid coating could induce the deposition of calcium phosphate mineralization products (in **Figure 11.7**). The formation of a biocompatible Ta_2O_5 film on the surface of magnesium alloy by the sol-gel method can also improve the corrosion resistance (Gül et al. 2020). Chitosan (CTS) silane composite coating also has development potential in improving the biocompatibility and corrosion resistance of magnesium alloys (Ma et al. 2022).

In addition to biological materials, composite sol-gel coatings can also bring certain antibacterial properties to magnesium alloys. An HA/CuO-TiO$_2$ coating with copper/titanium molar ratio of 2

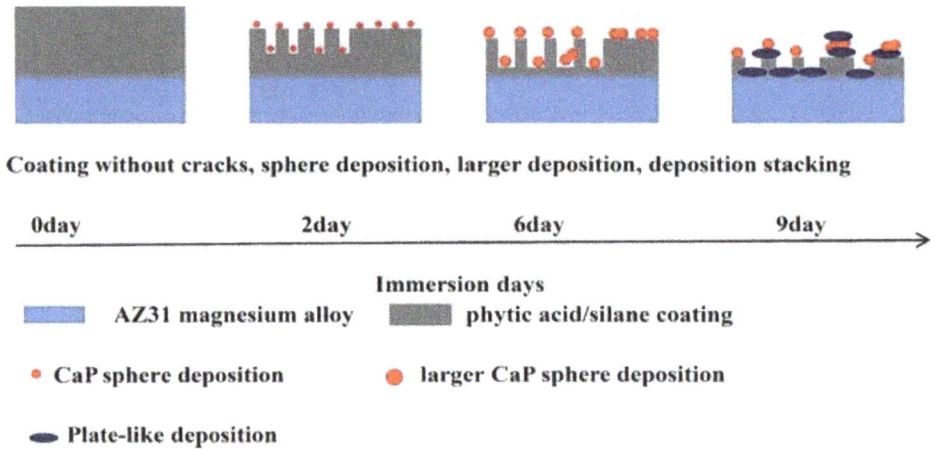

Coating without cracks, sphere deposition, larger deposition, deposition stacking

0day   2day   6day   9day

Immersion days

▬ AZ31 magnesium alloy   ▬ phytic acid/silane coating

• CaP sphere deposition   ● larger CaP sphere deposition

◖ Plate-like deposition

FIGURE 11.7 Schematic illustration of the corrosion process of phytic acid/silane hybrid coating on magnesium alloy in SBF. *Reproduced with permission from (Wang et al. 2017) © 2017, ASM International.*

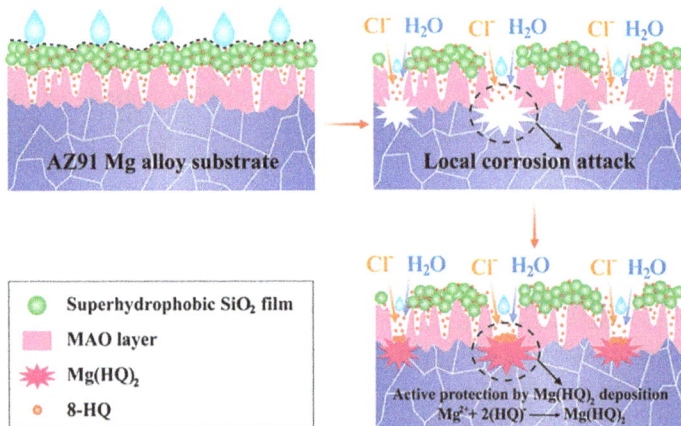

FIGURE 11.8 Schematic showing anti-corrosion mechanism of the bifunctional composite coating. *Reproduced with permission from (Jiang et al. 2019) © 2019, American Chemical Society.*

with flat morphology presented the best corrosion resistance (Xu et al. 2020). The *in vitro* cytotoxicity test and antibacterial performance test showed that the composite coating had good cell compatibility and an obvious antibacterial effect on *Staphylococcus aureus*.

On the other side, manufacturing superhydrophobic coatings can effectively improve the corrosion resistance of magnesium alloys by isolating the substrate from water/humid air (Ishizaki and Saito 2010). The superhydrophobic coating on magnesium alloys can be obtained using MTES as a hydrophobic modifier to modify MAO coating and silane sol-gel composite coating (Wang et al. 2012). Jiang et al. constructed a composite coating with a bottom MAO ceramic layer and an outer sol-gel layer, in which 8-HQ was loaded as a corrosion inhibitor into the sol-gel layer to form an MAO-SHS-8HQ coating (Jiang et al. 2019), as shown in **Figure 11.8**. Self-healing superhydrophobic smart coatings on magnesium alloys is a promising magnesium alloy protection strategy.

11.6   CONCLUSIONS AND OUTLOOK

This chapter describes sol-gel coatings' development and current status on magnesium alloys. The diversity of the development of sol-gel coatings was analyzed from the perspective of precursors. After summarizing some previous studies, it is found that corrosion inhibitors are essential for improving the corrosion resistance of magnesium alloy sol-gel coatings. There will be increasing types and methods of loading corrosion inhibition nanomaterials on magnesium alloy sol-gel coatings. The core lies in the loading of corrosion inhibitors and the construction of nanocontainers. The controlled-release corrosion inhibitors are of great research value for the corrosion resistance of magnesium alloys. The excellent controllability and compatibility of sol-gel technology have shown the development potential of constructing effective multi-layer coating systems. Such composite coating systems could meet the corrosion resistance needs of magnesium alloys under complex corrosion conditions. Summarizing some previous studies, there exist excellent research value and development potential for improving the corrosion resistance of magnesium and its alloys by building a composite film system along with loading corrosion inhibitors. It is expected to replace the conventional but toxic chromate pre-treatment.

However, from the industrial point of view, it is necessary to solve the problems of coating adhesion and cracking to apply the sol-gel coating on the magnesium alloy on a large scale. Meanwhile, a feasible, efficient, and systematic detection scheme and theoretical system are needed to evaluate the performance of the coating better. Moreover, the cost of sol-gel coating is further reduced, breaking the current market of many anticorrosive coatings such as epoxy resin coating monopoly. The advantages of sol-gel coating are pollution-free, easy synthesis, and good corrosion resistance. However, there are still some urgent problems to be solved, such as the protection performance and the long-term durability of the coating in corrosion environments. In addition, it is most important to strengthen the mechanical properties of the sol-gel coating, which will furthermore improve the practical applications of sol-gel coatings.

REFERENCES

Adsul SH, Siva T, Sathiyanarayanan S, et al. Aluminum pillared montmorillonite clay-based self-healing coatings for corrosion protection of magnesium alloy AZ91D. *Surf. Coat. Tech.* 2018, 352, 445–461.

Adsul SH, Siva T, Sathiyanarayanan S, et al. Self-healing ability of nanoclay-based hybrid sol-gel coatings on magnesium alloy AZ91D. *Surf. Coat. Tech.* 2017, 309, 609–620.

Afsharimani N, Talimian A, Merino E, et al. Improving corrosion protection of Mg alloys (AZ31B) using graphene-based hybrid coatings. *Int J Appl Glass Sci.* 2022, 13, 143–150.

Ashassi-Sorkhabi H, Moradi-Alavian S, Esrafili MD, et al. Hybrid sol-gel coatings based on silanes-amino acids for corrosion protection of AZ91 magnesium alloy: Electrochemical and DFT insights – ScienceDirect. *Prog. Org. Coat.* 2019, 131, 191–202.

Barranco V, Carmona N, Galván JC, et al. Electrochemical study of tailored sol – gel thin films as pre-treatment prior to organic coating for AZ91 magnesium alloy. *Prog. Org. Coat.* 2010, 68 (4), 347–355.

Barroso G, Li Q, Bordia RK, et al. Polymeric and ceramic silicon-based coatings-a review. *J. Mater. Chem. A.* 2019, 7, 1936–1963.

Chen Y, Lu X, Lamaka SV, et al. Active protection of Mg alloy by composite PEO coating loaded with corrosion inhibitors. *Appl. Surf. Sci.* 2020, 504, 144462.

Correa PS, Malfatti CF, Azambuja DS. Corrosion behavior study of AZ91 magnesium alloy coated with methyltriethoxysilane doped with cerium ions. *Prog. Org. Coat.* 2011, 72 (4), 739–747.

Dalmoro V, Azambuja DS, Alemán C, et al. Hybrid organophosphonic-silane coating for corrosion protection of magnesium alloy AZ91: The influence of acid and alkali pre-treatments. *Surf. Coat. Tech.* 2019, 357, 728–739.

Dou J, Yu H, Chen C. Preparation and characterization of composite coating on Mg-1.74Zn- 0.55Ca alloy by micro-arc oxidation combined with sol-gel method. *Mater. Lett.* 2019, 255 (15), 126578.

Duran KS, Hernandez-Barrios CA, Coy AE, et al. Effect of fluoride conversion pretreatment time and the microstructure on the corrosion performance of TEOS-GPTMS sol-gel coatings deposited on the WE54 magnesium alloy. *J. Mater. Res. Technol.* 2021, 15, 4220–4242.

Farshid S, Kharaziha M. Micro and nano-enabled approaches to improve the performance of plasma electrolytic oxidation coated magnesium alloys. *J. Magnes. Alloy.* 2021, 9 (5), 1487–1504.

Fernández-Hernán JP, López AJ, Torres B, et al. Silicon oxide multilayer coatings doped with carbon nanotubes and graphene nanoplatelets for corrosion protection of AZ31B magnesium alloy. *Prog. Org. Coat.* 2020, 148, 105836.

Galio AF, Lamaka SV, Zheludkevich ML, et al. Inhibitor-doped sol-gel coatings for corrosion protection of magnesium alloy AZ31. *Surf. Coat. Tech.* 2010, 204, 1479–1486.

Gu Y, Zheng X, Liu Q, et al. Investigating corrosion performance and corrosive wear behavior of sol-gel/ MAO-coated Mg alloy. *Tribol. Lett.* 2018, 66 (3), 101.

Gül C, Albayrak S, Çinici H. Characterization of tantalum oxide Sol-Gel-coated AZ91 Mg alloys. *T. Indian. I. Metals.* 2020, 73 (5), 1249–1256.

Guo X, An M, Yang P, et al. Property characterization and formation mechanism of anticorrosion film coated on AZ31B Mg alloy by SNAP technology. *J. Sol-Gel. Sci. Techn.* 2009, 52 (3), 335–347.

Hernández-Barrios CA, Saavedra JA, Higuera SL, et al. Effect of cerium on the physicochemical and anticorrosive features of TEOS-GPTMS sol-gel coatings deposited on the AZ31 magnesium alloy. *Surf. Interfaces.* 2020, 21, 100617.

Hu J, Li Q, Zhong X, et al. Composite anticorrosion coatings for AZ91D magnesium alloy with molybdate conversion coating and silicon sol-gel coatings. *Prog. Org. Coat.* 2009, 66 (3), 199–205.

Hu J, Li Q, Zhong X, et al. Novel anti-corrosion silicon dioxide coating prepared by sol-gel method for AZ91D magnesium alloy. *Prog. Org. Coat.* 2008, 63 (1), 13–17.

Hu J, Zhang C, Cui B, et al. In vitro degradation of AZ31 magnesium alloy coated with nano TiO_2 film by sol-gel method. *Appl. Surf. Sci.* 2011, 257(21), 8772–8777.

Ishizaki T, Saito N. Rapid formation of a superhydrophobic surface on a magnesium alloy coated with a cerium oxide film by a simple immersion process at room temperature and its chemical stability. *Langmuir.* 2010, 26 (12), 9749–9755.

Jiang D, Xia X, Hou J, et al. Enhanced corrosion barrier of microarc-oxidized Mg alloy by self-Healing superhydrophobic silica coating. *Ind. Eng. Chem. Res.* 2019, 58, 165–178.

Kraus T, Fischerauer SF, Hänzi AC, et al. Magnesium alloys for temporary implants in osteosynthesis: In vivo studies of their degradation and interaction with bone. *Acta. Biomater.* 2012, 8 (3), 1230–1238.

Lamaka SV, Knörnschild G, Snihirova DV, et al. Complex anticorrosion coating for ZK30 magnesium alloy. *Electrochim. Acta.* 2009, 55 (1), 131–141.

Li B, Chen Y, Huang W, et al. In vitro degradation, cytocompatibility and hemolysis tests of CaF_2 doped TiO_2-SiO_2 composite coating on AZ31 alloy. *Appl. Surf. Sci.* 2016, 382, 268–279.

Li W, Su Y, Ma L, et al. Sol-gel coating loaded with inhibitor on ZE21B Mg alloy for improving corrosion resistance and endothelialization aiming at potential cardiovascular application. *Colloids Surf.* 2021, 207, 111993.

Li Z, Jing X, Yi Y, et al. Composite coatings on a Mg-Li alloy prepared by combined plasma electrolytic oxidation and sol-gel techniques. *Corros. Sci.* 2012, 63, 358–366.

Liang J, Guo B, Tian J, et al. Effects of $NaAlO_2$ on structure and corrosion resistance of microarc oxidation coatings formed on AM60B magnesium alloy in phosphate-KOH electrolyte. *Surf. Coat. Tech.* 2005, 199 (2–3), 121–126.

Lkhagvaa T, Rehman ZU, Choi D. Post-anodization methods for improved anticorrosion properties: A review. *J. Coat. Technol. Res.* 2020, 18, 1–17.

López AJ, Rams J, Ureña A. Sol-gel coatings of low sintering temperature for corrosion protection of ZE41 magnesium alloy. *Surf. Coat. Tech.* 2011, 205 (17–18), 4183–4191.

Ma Y, Talha M, Wang Q, et al. Nano-silica/chitosan composite coatings on biodegradable magnesium alloys for enhanced corrosion resistance in simulated body fluid. *Mater. Corros.* 2022, 73 (3), 436–450.

Mahmoudi R, Kardar P, Arabi AM, et al. Acid-modification and praseodymium loading of halloysite nanotubes as a corrosion inhibitor. *Applied Clay Science.* 2019, 184, 105355.

Merino E, Durán A, Castro Y. Integrated corrosion-resistant system for AZ31B Mg alloy via Plasma Electrolytic Oxidation (PEO) and sol-gel processes. *Int. J. Appl. Glass. Sci.* 2021, 12 (4), 519–530.

Nezamdoust S, Seifzadeh D, Habibi-Yangjeh A. Nanodiamond incorporated sol-gel coating for corrosion protection of magnesium alloy. *T. Nonferr. Metal. Soc.* 2020, 30 (6), 1535–1549.

Niu B, Shi P, Shanshan E, et al. Preparation and characterization of HA sol-gel coating on MAO coated AZ31 alloy. *Surf. Coat. Tech.* 2016, 286, 42–48.

Omri K, Najeh I, Dhahri R, et al. Effects of temperature on the optical and electrical properties of ZnO nanoparticles synthesized by sol-gel method. *Microelectron. Eng.* 2014, 128, 53–58.

Ouyang J, Hong X, Gao Y. Retardation and self-repair of erosion pits by a two-stage barrier on bioactive-glass/layered double hydroxide coating of biomedical magnesium alloys. *Surf. Coat. Tech.* 2021, 405 (4), 126562.

Pereira GS, Ramirez OMP, Avila PRT, et al. Cerium conversion coating and sol-gel coating for corrosion protection of the WE43 Mg alloy. *Corros. Sci.* 2022, 206, 110527.

Periyasamy AP, Venkataraman M, Zhao Y. Progress in sol-gel technology for the coatings of *Fabrics. Materials.* 2020, 13 (8), 1838.

Phani AR, Gammel FJ, Hack T, et al. Enhanced corrosion resistance by sol-gel-based ZrO_2-CeO_2 coatings on magnesium alloys. *Mater. Corros.* 2005, 56 (2), 77–82.

Rojaee R, Fathi M, Raeissi K. Controlling the degradation rate of AZ91 magnesium alloy via sol-gel derived nanostructured hydroxyapatite coating. *Mater Sci Eng C Mater Biol Appl.* 2013, 33 (7), 3817–3825.

Rosero-Navarro NC, Curioni M, Castro Y, et al. Glass-like Ce_xO_y sol-gel coatings for corrosion protection of aluminium and magnesium alloys. *Surf. Coat. Tech.* 2011, 206 (2–3), 257–264.

Samadianfard R, Seifzadeh D, Habibi-Yangjeh A, et al. Oxidized fullerene/sol-gel nanocomposite for corrosion protection of AM60B magnesium alloy. *Surf. Coat. Tech.* 2020, 385, 125400.

Shi Z, Song G, Atrens A. Influence of anodising current on the corrosion resistance of anodised AZ91D magnesium alloy. *Corros. Sci.* 2006, 48 (8), 1939–1959.

Tang H, Tao W, Wang C, et al. Fabrication of hydroxyapatite coatings on AZ31 Mg alloy by micro-arc oxidation coupled with sol-gel treatment. *Rsc. Adv.* 2018, 8 (22), 12368–12375.

Tarzanagh YJ, Seifzadeh D, Samadianfard R. Combining the 8-hydroxyquinoline intercalated layered double hydroxide film and sol-gel coating for active corrosion protection of the magnesium alloy. *Int. J. Min. Met. Mater.* 2022, 29 (3), 536–546.

Upadhyay V, Bergseth Z, Kelly B, et al. Silica-based sol-gel coating on magnesium alloy with green inhibitors. *Coatings.* 2017, 7, 86.

Van Ooij WJ, Zhu D, Stacy M, et al. Corrosion protection properties of organofunctional silanes: an overview. *Tsinghua Sci. Technol.* 2005, 10, 639–664.

Walsh FC, Low CTJ, Wood RJK, et al. Plasma electrolytic oxidation (PEO) for production of anodised coatings on lightweight metal (Al, Mg, Ti) alloys. *Transactions of the Imf.* 2009, 87 (3), 122–135.

Wang F, Cai S, Shen S, et al. Preparation of phytic acid/Silane hybrid coating on magnesium alloy and its corrosion resistance in simulated body fluid. *J. Mater. Eng. Perform.* 2017, 26 (9), 4282–4290.

Wang S, Guo X, Xie Y, et al. Preparation of superhydrophobic silica film on Mg-Nd-Zn-Zr magnesium alloy with enhanced corrosion resistance by combining micro-arc oxidation and sol-gel method. *Surf. Coat. Tech.* 2012, 213, 192–201.

Wang W, Yang XN, Wang Y, et al. Endowing magnesium with the corrosion-resistance property through cross-linking polymerized inorganic sol-gel coating. *Rsc. Adv.* 2021, 11 (8), 4365–4372.

Witte F, Kaese V, Haferkamp H, et al. In vivo corrosion of four magnesium alloys and the associated bone response. *Biomaterials.* 2005, 26 (17), 3557–3563.

Xu Y, Wang T, Guo Y, et al. Improvements of corrosion resistance and antibacterial properties of hydroxyapatite/cupric oxide doped titania composite coatings on degradable magnesium alloys. *Langmuir.* 2020, 36 (46), 13937–13948.

Yasakau KA, Zheludkevich ML, Karavai OV, et al. Influence of inhibitor addition on the corrosion protection performance of sol-gel coatings on AA2024. *Prog. Org. Coat.* 2008, 63 (3), 352–361.

Yue YY, Liu ZX, Wan TT, et al. Effect of phosphate-silane pretreatment on the corrosion resistance and adhesive-bonded performance of the AZ31 magnesium alloys. *Prog. Org. Coat.* 2013, 76 (5), 835–843.

Zhang, Y, Bai K, Fu Z, et al. Composite coating prepared by micro-arc oxidation followed by sol-gel process and in vitro degradation properties. *Appl. Surf. Sci.* 2012, 258 (7), 2939–2943.

Zhao H, Cai S, Niu S, et al. The influence of alkali pretreatments of AZ31 magnesium alloys on bonding of bioglass-ceramic coatings and corrosion resistance for biomedical applications. *Ceram. Int.* 2015, 41 (3), 4590–4600.

Zheng X, Liu Q, Ma H, et al. Probing local corrosion performance of sol-gel/MAO composite coating on Mg alloy. *Surf. Coat. Tech.* 2018, 347, 286–296.

Zhong X, Li Q, Hu J, et al. Characterization and corrosion studies of ceria thin film based on fluorinated AZ91D magnesium alloy. *Corros. Sci.* 2008, 50 (8), 2304–2309.

Zhong XK, Li Q, Hu JY, et al. A novel approach to heal the sol-gel coating system on magnesium alloy for corrosion protection. *Electrochimi. Acta.* 2010, 55 (7), 2424–2429.

Zucchi F, Frignani A, Grassi V, et al. Organo-silane coatings for AZ31 magnesium alloy corrosion protection. *Mater. Chem. Phys.* 2008, 110 (2), 263–268.

12 Hydrothermal and Layered Double Hydroxide Coatings

Liang Wu, Yonghua Chen, Wenhui Yao, Jiaxin Li,
Yuhan Zhang, Jiahao Deng and Fusheng Pan

CONTENTS

DOI: 10.1201/9781003319856-15

12.1 INTRODUCTION

Mg/Mg alloys are one of the most abundant light metals on the earth, and their corrosion problem causes great economic losses and environmental problems (Song et al. 2020, 2022). One of the most common ways to protect Mg alloy surfaces from harsh corrosive environments is to prepare protective coating on its surface by surface treatment (Zhang, Peng et al. 2021). However, a single protective coating on the Mg alloy surface cannot prevent the penetration of Cl⁻ and other corrosive substances for a long time. Once the coating is partially damaged, the coating itself can no longer inhibit the corrosion process, as corrosive molecules (such as water and oxygen) can spread to the interface between the alloy and the coating, causing the coating to fail (Li et al. 2018).

Corrosion inhibitors can be added to the coating system to attempt to achieve the self-healing function and extend the protective properties when it is damaged. However, adding corrosion inhibitors directly to the coating can lead to some negative effects, such as uncontrolled release of corrosion inhibitors, rapid degradation of coating properties, and loss of inhibition ability (Zhang, Peng et al. 2021). In recent years, many researchers have studied self-healing coatings on the Mg alloy surface loaded with corrosion inhibitors, which are carried by nanocontainers.

The application of layered double hydroxides (LDHs) as nanocontainers loaded with corrosion inhibitors has been a research hotspot in the field of Mg alloy corrosion protection. LDHs is a lamellar two-dimensional structure with pores. The unique characteristics of interlayer anion exchange make it have more advantages than other porous oxide nanocontainers (Shulha et al. 2022). Due to their extensive advantages, LDHs has been widely used for corrosion protection of Mg alloys, including LDH powder (Salak et al. 2010), LDH conversion coating (Li, Wang et al. 2022), LDH coating loaded with corrosion inhibitors (Ouyang et al. 2020), and LDH composite coatings with MOF (Zhang, Ma, Wang, Sun et al. 2018), MXene (Cai et al. 2021) etc.

Up to now, LDHs have been prepared by a variety of methods (Guo, Wu et al. 2018; Cao et al. 2022). Comparatively speaking, growing LDH coatings directly from the Mg substrate can greatly improve the adhesion and mechanical stability of the coatings produced. Therefore, it is of great significance to explore new methods to prepare directional LDH coatings on the substrate. Among the existing methods of inorganic LDH coating synthesis, hydrothermal synthesis has high flexibility in controlling the structure and morphology. It is also a well-known way to prepare inorganic coatings with the required micro- or nanostructure and controllable crystal orientation.

In conclusion, LDH has broad prospects for developing multifunctional corrosion inhibition technology (Guo, Wu et al. 2018; Bouali et al. 2020; Mir et al. 2020). This chapter summarizes the research progress of LDHs in Mg alloy corrosion protection from the aspects of their structure, preparation methods, growth mechanisms and orientation mechanisms, as well as different corrosion protection effects and prospects of future developments.

12.2 LAYERED DOUBLE HYDROXIDES (LDHS)

12.2.1 STRUCTURE AND PROPERTIES OF LDHS

The general chemical formula of LDHs is given as: $[M^{2+}_{1-x} M^{3+}_x (OH)_2]^{x+} A^n_{x/n} \cdot mH_2O$, where M^{2+} is divalent metal cation, M^{3+} is trivalent metal cation, A^n is interlayer inorganic or organic anion: Cl, NO_3, CO_3^2, SO_4^2, n- is the charge on anion, x is the ratio of M^{2+} to M^{3+}, ranging generally from 0.2 to 0.33 (Kovanda et al. 2010), and m is the number of crystalline H_2O of each LDH. M^{2+} and M^{3+} are generally taken from the third and fourth periods of the periodic table, such as M^{2+}: Mg^{2+}, Cu^{2+}, Zn^{2+}, Co^{2+} (Mochane et al. 2020); M^{3+}: Al^{3+}, Cr^{3+}, Mn^{3+}, Fe^{3+} (Bhavani et al. 2021). In addition to divalent metal cations and trivalent metal cations, LDHs can also contain monovalent (Li^+) (Fogg et al. 2002) and tetravalent (Ti^{4+}) metal cations (Xia et al. 2014) and two or more metal cations.

The structure of LDHs is similar to that of brucite, which is mainly an octahedral structure composed of two parts, the hydroxide layer and the interlayer anion (Li et al. 2012). The two different structures are shown in **Figure 12.1** (Khan et al. 2002). In the hydroxide layer, the metal ion

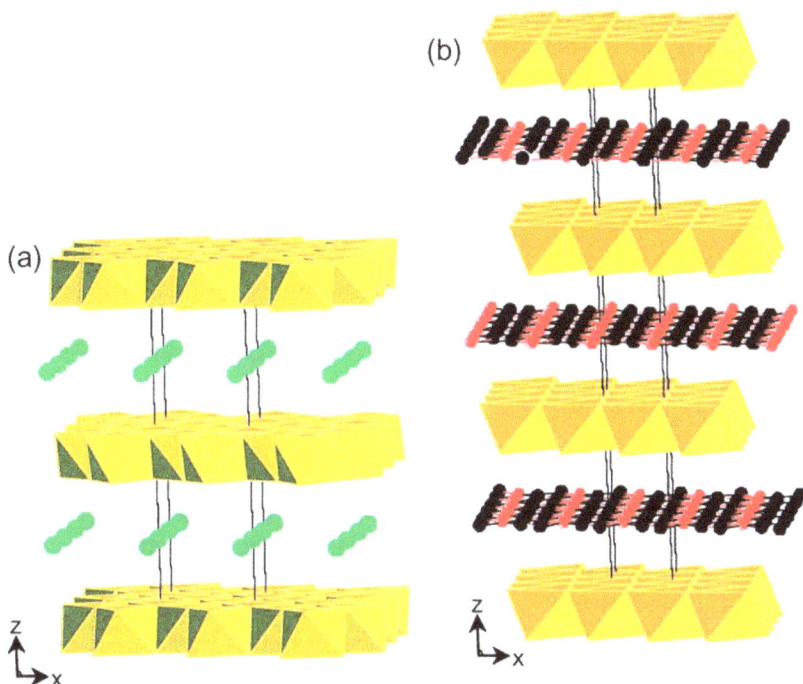

is located in the center of the octahedron and forms coordination bonds with six hydroxide groups. Adjacent octahedra share six edges with each other and thus extend indefinitely.

LDHs have two stacking modes: hexagonal structure and rhombic structure (Mishra et al. 2018; Chen, Hung et al. 2019). The c value of the cell parameter in the hexagonal structure is 2 times the layer spacing, while that in the rhombohedral structure is 3 times the layer spacing. The metal oxide layer in LDHs has a high density of positive charge, and the anions in the layer and the interlayer are mainly electrostatic attraction, with weak van der Waals forces and hydrogen bond. Therefore, anions in the LDH layer can be easily replaced by other anions to form LDHs intercalated with different anions. Thus, LDHs have many of the following different properties.

Acidity and alkalinity are the most basic properties of LDHs. In general, the alkalinity of LDHs is related to M^{2+}. However, the apparent alkalinity of LDHs is very weak because of the limited specific surface area. When the LDH forms bimetallic hydroxides through calcination, it performs strong alkalinity (Taylor 1973). The acidic centers of LDHs are closer to the M^{3+} cations in their composition but are also associated with divalent metal oxides (Inoue et al. 1997). Besides, the acidity of LDHs also depends on the type of anion in the layer. When simple anions are inserted into the LDHs layer, the acidity is weak. However, in the presence of polyacid and heteropoly acid anions, LDHs are more acidic.

The thermal stability of LDHs is closely related to their structure. In the process of heating, LDHs often show some characteristics such as increasing surface area and pore volume (Mishra et al. 2018) The decomposition at lower temperatures ($\leq 600°C$) is reversible, and once the temperature is too high ($> 600°C$), the spinel phase and oxide phase will be formed.

LDHs also has a unique structure memory effect (Yuan et al. 2013). When the composite oxide of LDHs after roasting is put into the solution medium or air containing some anion, it can recrystallize and reabsorb water and anion to restore the original structure.

The most widely used property of LDHs is that their interlayer anions can be replaced by other anions without changing its layered structure (Yu et al. 2017). Through this reversible exchange, new LDH materials with different functions are widely used in biomedicine, environment treatment, corrosion protection, and other fields (Wang and O'Hare 2012).

12.2.2 PREPARATION OF LDHs

12.2.2.1 Preparation of LDH Powders

12.2.2.1.1 Co-Precipitation Method

The co-precipitation method is one of the most common methods to prepare LDHs (Xu et al. 2018). By controlling the relative drop acceleration, the mixed metal salt solution of a certain concentration and the mixed alkali solution of sodium hydroxide and sodium carbonate were slowly added to the container at the same time, and the pH of the reaction system was maintained at a constant value. After fierce stirring, aging, filtering, washing, and drying, the final product was obtained (Aisawa et al. 2001; Chang et al. 2005). Metal salts can be nitrate, sulfate, chloride, and carbonate, etc. Alkali can be sodium hydroxide, potassium hydroxide, ammonia, etc. The main advantage of using the co-precipitation method to prepare LDHs is its wide application range (Ahmed et al. 2012). Firstly, almost all the LDHs formed by M^{2+} and M^{3+} can be prepared by co-precipitation. Secondly, in the same M^{2+} and M^{3+} system, as long as the raw material ratio of M^{2+} and M^{3+} is adjusted, a series of LDHs with different $[M^{2+}]/[M^{3+}]$ ratios can be prepared.

12.2.2.1.2 Anion-Exchange Method

When some LDHs cannot be directly prepared by co-precipitation, the desired anion can be exchanged with the original anion in LDHs prepared by co-precipitation to obtain the target LDHs. In general, the anion with low exchange ability cannot be used to replace the anion with high exchange ability. The higher the charge of the anion and the smaller the radius, the stronger the exchange ability (Costantino et al. 2014).

The priority order of interlayer anion exchange of LDH nanomaterials is as follows: $NO_3^- > Br^- > Cl^- > F^- > OH^- > SO_4^{2-} > CO_3^{2-}$ (Pan et al. 2011). The composition of LDHs also has a certain impact on the anion exchange reaction. Generally, more bound water between layers is conducive to exchange, while more bound water on the surface is not conducive to exchange (Feng et al. 2022).

12.2.2.1.3 Calcination-Reconstruction Method

The calcination-reconstruction method is to put the calcined LDHs in solution or air to recover the layered structure and obtain new materials, making use of the "structure memory effect" of LDHs. The advantage of this method is that it can prepare LDHs with different anions, in which the crystal structure stays the same as the initial structure. In addition, the intercalation competition between inorganic anions of metal salts and organic anions was eliminated by this method, and the anion intercalation ability of the reconstructed LDHs was good and the stability was enhanced (Costa et al. 2008). The calcination temperature of the parent LDHs should be considered when preparing LDHs by the calcination-reconstruction method. According to the different components of the parent LDHs, the original crystal structure can be maintained by choosing the appropriate calcination temperature.

12.2.2.1.4 Hydrothermal Method

As one of the most widely used methods for preparing LDHs, the hydrothermal method can effectively control the grain size, microstructure, crystal structure, and dispersion (Zhang, Li et al. 2022). The hydrothermal synthesis method is to slowly add a certain proportion and concentration of mixed metal salt solution and alkali solution together or directly mix quickly, then immediately transfer them to the reaction kettle, stir for a long time at a certain temperature (usually > 100°C), and finally obtain LDH powder products by centrifugation washing, drying, and grinding. The preparation process of LDHs includes nucleation and crystallization (Gunawan and Xu 2008).

12.2.2.2   Preparation of LDH Coating

12.2.2.2.1   In Situ *Hydrothermal Method*

The *in situ* hydrothermal method is an effective method widely used to prepare inorganic thin coatings (Huang et al. 2012). In the case of *in situ* crystallization, due to the chemical bond between the two states, the adhesion between the substrate and the coating is quite strong. This is clearly advantageous for many applications of such coatings. Various LDH coatings were prepared on the surface of Mg alloys, e.g. Mg-Al LDH (Zhang, Hou et al. 2022), Mg-Fe LDH (Zhang, Zhou et al. 2022), Mg-Al-Co LDH (Zhang, Duan et al. 2021), and Zn-Mg-based LDHs (Ji and Prakash 2021). The Mg substrate was put into metal salt solution as a template, and LDHs were grown on the Mg surface by metal ions. It should be noted that an LDH has a strong adsorption capacity for CO_2. Without autoclave conditions, $LDH\text{-}CO_3^{2-}$, also known as "dead" LDH, will be generated (Chen et al. 2012). This type of LDH does not have ion exchange capability.

12.2.2.2.2   *Co-Precipitation Method*

To prepare LDH coating on Mg alloy by the co-precipitation method, an LDH precursor should be prepared first, and then LDH coating can be formed by interfacial reaction during a hydrothermal treatment. The pH adjustment method is often used in the synthesis of LDH coating. The most critical is that the pH of the precipitate must be higher or at least equal to that of $Mg(OH)_2$. The disadvantage is that the bonding strength between LDH coatings prepared by the co-precipitation method and Mg matrix is weaker than that by the *in situ* hydrothermal method. However, when LDH coatings cannot be directly grown by the *in situ* hydrothermal method, this method can play a role (Zhang et al. 2014; Wu et al. 2021).

12.2.2.2.3   *Anion-Exchange Method*

The anion-exchange method usually introduces different anions into the LDH coating according to the anion-exchange mechanism based on other preparation methods (Grover et al. 2022). The more difficult the anion exchange is, the more stable the LDHs structure is. The high affinity of CO_3^{2-} ions for layered structure makes it difficult to be replaced by other anions, so anion exchange generally occurs between Cl- and other anions (Guo, Zhang et al. 2018; Hou et al. 2019). The LDH coatings prepared by this method can insert different types of ions in the interlayer space and play different functional characteristics (Kaseem et al. 2021).

12.2.2.2.4   *Electrodeposition Method*

The deposition of LDHs mainly consists of the following steps: soak the electrolyte solution containing M^{2+} and M^{3+} ions with specific counter-anions on the cathode surface, and induce hydrogen generation by the action of electric current (Sonoyama et al. 2021). This results in a partial increase in the pH of the electrolyte on the substrate surface due to proton depletion. LDH coatings are deposited on the cathode surface. Oxygen is mainly formed on the counter electrode surface. In the electrodeposition process, LDH coatings with high crystallinity are rarely obtained due to the limited crystal growth time. Compared with other methods, electrodeposition can obtain uniform coatings on uneven surfaces (Zhao et al. 2017).

12.2.2.2.5   *Other Methods*

The spin coating method is an effective method to prepare inorganic coatings with controllable structure and crystal orientation (Frank et al. 1996). In its simplest form, the coatings can be cast directly from aqueous suspensions of precursor crystals without any pre-treatment. However, the bonding strength between the prepared coating and the substrate is not enough, and the equipment required for preparation is relatively complex.

The steam coating method is to prepare LDH coatings on Mg alloys in a closed autoclave. The coating formed by this process is a metal source hydroxide and oxide coating reacting with high-pressure steam at a specific temperature on a Mg substrate. Before steam coating, microwave plasma-enhanced chemical vapour deposition is the oldest method to prepare LDH coatings, especially on large-sized Mg alloys. But the same problem is that this method has certain requirements for experimental equipment.

12.2.3 FORMATION MECHANISM OF LDHs

The most fundamental feature of LDHs is their orderly stacked laminate structure, and the key to the formation mechanism of LDHs lies in the construction process of their laminate structure (Sun et al. 2020). At present, there are three major views on the formation of LHD laminates based on bivalent metal hydroxides and trivalent metal hydroxides and direct topological phase transition mechanisms.

Based on divalent metal hydroxides, Grégoire studied the hydrolysis behavior of mixed metal solutions containing Ni^{2+}-$Fe^{3+}$and Mg^{2+}-Fe^{2+} (Grégoire et al. 2013). The formation mechanism of Mg-Fe/Ni-Fe LDHs was determined by coupling solution chemistry and solid-state analysis. It is a four-step formation process, namely, the formation of hydroxyl oxidation metal phases, condensation and nucleation of adsorbed cations, cation substitution and formation of LDH phase, and the growth and aging of LDHs. In addition, they also found that the precursor can be converted into LDHs, so the growth process of LDHs in mixed metal solution is: Fe^{3+} first formed an iron oxide and hematite phase, and the precursor formed on the iron oxide surface can diffuse through the hematite phase on the surface of Mg alloy, thus forming LDH coating.

Based on trivalent metal hydroxides, Boclair and Braterman (Boclair et al. 1999) titrated NaOH and metal solution(M^{2+}/M^{3+}) to generate LDHs (coprecipitation method) and found that there were two distinct pH variation regions on the titration curve. Generally, the K_{sp} of $M^{(II)}(OH)_2$ is tens of orders of magnitude higher than the K_{sp} of $M^{(III)}(OH)_3$, indicating that $M^{(III)}(OH)_3$ can be formed at low pH. Therefore, it is believed that in the presence of $M^{(III)}$, $M^{(III)}(OH)_3$ is formed in the first stage, and then $M^{(II)}$ precipitates and embeds into $M^{(III)}(OH)_3$ to form LDH. For a given $M^{(II)}$, the lower the solubility of $M^{(II)}(OH)_2$, the lower the pH required to form LDH. For a given $M^{(III)}$, the lower the solubility of $M^{(III)}(OH)_3$, the higher the pH required to form LDH, because $M^{(III)}(OH)_3$ actually acts as a reactant in the system.

The direct topological phase transition process includes electrostatic interaction between the positive laminate and the original anion, the breaking of old bonds such as hydrogen bonds, and the rebuilding of new bonds with new objects, similar to the ion exchange of LDHs (Shulha et al. 2018). The enthalpy of bonding within the LDH layer largely determines the thermal stability of the material, and the effect of the interlayer force is negligible, so that the stability of the interlayer region is low. It allows the embedding/unblocking reactions to occur, which not only maintain the consistency and stability of the parent crystal structure under the condition of interlayer anion changes but also maintain symmetry along a certain selected direction (Acharya et al. 2007).

12.2.4 THE ORIENTATION MECHANISM OF LDH NANOSHEETS

LDH grains are hexagonal sheets; the thickness of their ab-plane scale is greater than the c-oriented scale, so LDHs materials exhibit anisotropic characteristics. The orientation of LDH coatings is related to their formation environment, growth mechanism, and synthesis method. Generally, two different forms of LDHs can be obtained, as shown in **Figure 12.2** (Guo et al. 2010). In Figure 12.2b,

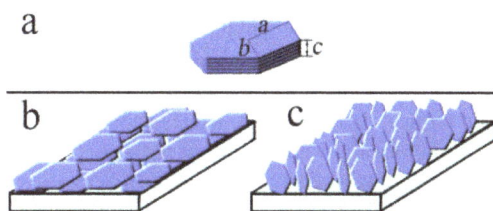

FIGURE 12.2 (a) Single LDH crystal and (b–c) LDH coatings grown in different orientations. *Reproduced with permission from (Guo et al. 2010) © 2010 The Royal Society of Chemistry.*

the LDH nanosheet is parallel to the substrate, while in Figure 12.2c, the LDH nanosheet is perpendicular to the substrate, which is a crystalline form obtained by the *in situ* hydrothermal method (Cao et al. 2018).

The "evolutionary selection" mechanism can well explain the orientation preference of LDH coatings (Jin et al. 2018). At the initial stage, the LDH crystal seed grows in all directions. When two crystals meet, they block each other's growth horizontally, so they continue to grow perpendicular to the substrate and eventually form LDH coatings. However, crystal morphology parallel to the substrate can be observed when the substrate is chemically treated before LDH growth (Pedraza-Chan et al. 2021). Specifically, Wu et al. (Wu, Zhang et al. 2022) prepared reduced graphene oxide (rGO)-MgAl LDH (rGO-LDH) coating on the surface of Mg alloy by the *in situ* hydrothermal growth method, discussed its orientation growth mechanism, and found that at low pH, the nanosheets in the LDH tend to be parallel to the substrate, with good shielding and lubrication capabilities. At high pH, the nanosheets tend to be perpendicular to the substrate. The GO has a strong anchoring effect on the vertically oriented LDH coating, making the coating highly corrosion resistant with good load capacity.

12.3 APPLICATION OF LDHS IN CORROSION PROTECTION

12.3.1 LDH PROTECTIVE COATINGS

LDH coating has a strong chemical bonding between itself and its matrix. The characteristics of LDH are summarized in what follows to prove its wide application value in the field of corrosion protection.

12.3.1.1 Physical Barrier

The layered structure of an LDH itself is a physical barrier (Cao et al. 2021). So far, many researchers have mentioned the good barrier effect of LDHs. Li et al. (Li, Wang et al. 2022) adjusted the pH values of LDH growth solution (9.4, 10.4, 11.2, and 11.4), *in situ* synthesized Mg-Al LDH conversion coatings on Mg alloy AZ91D by the hydrothermal method, and explored the effect of pH on its corrosion resistance. It was proposed that the LDH coating with pH of 11.2 had the best compactness, the best crystallinity, and the lowest corrosion current density (i_{corr}). This was attributed to the synergistic effect of anion exchange reaction of LDH and the anticorrosion physical barrier formed due to the twisted penetration path of staggered LDH nanosheets.

The preparation and application of superhydrophobic LDH anti-corrosion coating could strongly express its barrier effect. The physical barrier effect of the LDH layer, the spatial repulsion effect of the air pockets, and the ion exchange reaction of the interlayer corridor make the prepared LDH functional coating show excellent corrosion resistance to AZ31 matrix in 3.5 wt.% NaCl solution (Huang et al. 2021).

More intuitively, Qu et al. (Qu et al. 2018) conducted a Ca-Al-NO$_3$ LDH rapid chloride ion migration (RCM) test to analyze the relationship between LDH and chloride ion resistance. The decrease of chloride migration coefficient can be attributed to the enhanced barrier effect due to the increase of curvature. In the long-term natural diffusion test, LDHs showed a significantly enhanced barrier effect due to their ability to bind chlorides and improved flexibility.

Although the barrier effect of LDH has been proposed and applied in many existing studies, there is still no clear evidence to prove this view (Hou et al. 2019). Therefore, on the basis of existing research, the detailed structure of LDH coating on Mg alloy surface can be further studied by using advanced technologies (such as synchronous XRD, TEM, and complementary high-resolution SIMS), providing strong evidence for the physical barrier effect of LDH coating on Mg alloy (Cao et al. 2022).

12.3.1.2 Chloride Trapping

Since there is an L-acid center on the surface of LDH laminates, it can use isolated electron pairs to adsorb corrosive chloride ions (Cl$^-$), and the interlayer space can also fix Cl$^-$ (Ke et al. 2019) (Zhou et al. 2015). The chloride trapping effect plays an important role in the corrosion protection of Mg alloys.

Calcined carbonate layered double hydroxides (CLDHs) also have chloride trapping effect. Their structural characteristics, perchlorate adsorption properties, and binding mechanisms are dependent on the metal cation compositions (Mg-Al, Mg-Fe, and Zn-Al) and the molar M^{2+}/M^{3+} ratio (Geng et al. 2021). The binding mechanisms of perchlorate by CLDHs are much more dependent on the unique M^{2+} compared to the type of M^{3+}. The perchlorate adsorption mechanisms of Mg-Al CLDHs and Mg-Fe CLDHs were dominated by the memory effect and hydrogen bonds, which can be differentiated via regulating perchlorate concentrations. To contrast, the perchlorate adsorption by Zn-Al CLDHs was controlled by the memory effect only, as the hydroxyl groups on the hydroxide layers preferred to form strong hydrogen bonds with carbonate rather than perchlorate (Lin et al. 2014).

In addition, ZnAl-LDH-NO$_3$ has a greater chloride binding capacity because of its large interlayer distance. Cl- was controlled between LDH layers to avoid its penetration into the Mg matrix, thereby improving the corrosion resistance of Mg alloy to a certain extent and prolonging the service life (Chen, Zhang et al. 2022).

12.3.1.3 Self-Healing

LDHs can have a certain self-healing ability after proper treatment. The self-healing performance is achieved through the following synergistic effects.

12.3.1.3.1 *Self-Healing Effect Based on Corrosion Inhibitor and Cation Release from LDHs*

After loading corrosion inhibitor in LDH coating, once the coating is corroded, it can form a new material with surface cations and environmental substances (such as Cl$^-$, CO$_3^{2-}$, HCO$_3^-$, etc.) to fill the corroded site. Zhang et al. (Zhang et al. 2016) prepared a crack-free hydroxyapatite/phytic acid (HA/PA) hybrid coating on AZ31 surface and subjected it to hydrothermal treatment in saturated CaO solution. During the hydrothermal treatment, phosphorus and calcium ions diffused to the precursor coating/solution interface to form amorphous CaP compounds, which subsequently crystallized into HA particles and repaired the defects in the precursor coating. By comparison, it was found that after hydrothermal treatment at 120°C for 4 hours, the bonding strength of composite coating and Mg alloy matrix was improved due to the chelation reaction between PA and Mg alloy matrix and the formation of HA to repair defects. At the same time, it had more positive corrosion potential and lower i_{corr}, which had a better protective effect on Mg alloy matrix. Its protection mechanism is shown in **Figure 12.3a**.

Similarly, Li et al. (Li, He et al. 2022) prepared Mg-Al LDH self-healing coating with silicate intercalation on biomedical Mg alloys through the preparation process **of Figure 12.3b**. When

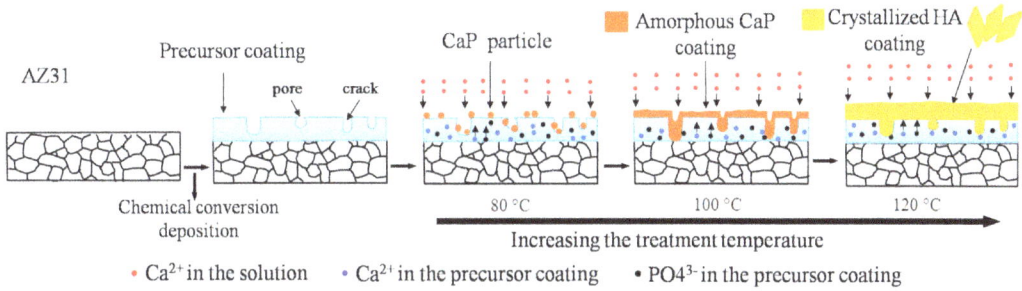

FIGURE 12.3A Protective mechanism of HA/PA hybrid coating on Mg alloy surface. *Reproduced with permission from (Zhang et al. 2015) © 2015 Elsevier B.V.*

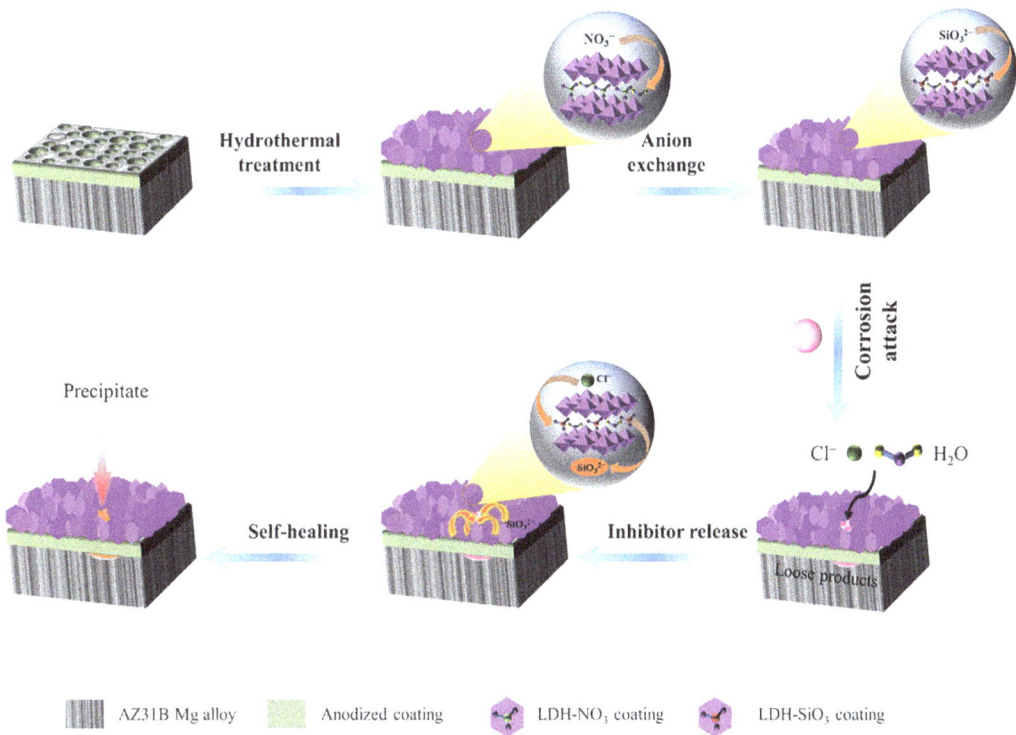

FIGURE 12.3B Preparation process of Mg-Al LDHs self-healing coating. *Reproduced with permission from (Li, He et al. 2022) © 2022 Elsevier B.V.*

corrosion occurs, SiO_3^{2-} ions are released from LDHs and form Mg_2SiO_3 precipitation with Mg^{2+} produced in the natural degradation process, which enhances the cell compatibility of Mg alloy and slows down the corrosion propagation, release of Mg^{2+}, hydrogen precipitation, and pH change. It is proved that the LDHs-SiO_3 coating has obvious long-term protection ability after local corrosion of Mg alloy matrix.

In a word, the release of substances in LDHs can generate new substances by combining/reacting with cations or environmental substances on the surface of Mg alloy to fill the artificially damaged parts in the LDH coating and achieve the self-healing effect. It plays a significant role in the corrosion protection of the alloy.

12.3.1.3.2   Self-Healing Effect Based on LDH Recrystallization

LDH materials have a certain structure memory effect, which also proves that they can be reconstructed through dissolution/recrystallization to achieve self-healing of LDH coatings. The self-healing function of MgAl-LDH (CL)/SiO$_2$@PDMS (SP) coating grown on Mg alloy AZ31 (named CL/SP/Mg) was dominated by the dissolution of the matrix and the recrystallization of LDH (Tang et al. 2023), and it was the same as that of Zn-Al LDH found by Yan et al. for aluminum (Al) alloy (Yan et al. 2013), as shown in **Figure 12.4**.

The artificially scratched specimen was immersed in salt solution, cysteine (cys) in CL interlayer near the Mg exchanged with Cl$^-$ in the scratch area, adsorbed in the scratch area to form a new barrier (Chen, Fang et al. 2019). Similarly, the cys-intercalated Mg-Al LDH grown on the surface of Mg alloy also underwent the same dissolution and recrystallization process. After 5 days of immersion, the recrystallized Mg-Al LDH nanosheets in the scratch area gradually grew up (Zhu et al. 2021). After 10 days, the CL in SP may also be dissolved, and the recrystallized LDH nanosheet formed dense deposition in the scratch area, and the anti-corrosion performance recovered to the best level. However, after prolonged immersion, the lamellar LDH dissolved and combined with cys, forming a dense insoluble precipitation.

The dissolution recrystallization mechanism of Mg alloys shows that when artificial scratches occur, the alloy will undergo anodic dissolution, resulting in changes in pH and ion content at the scratches. At the same time, a large number of Cl- intercalated into LDHs, and some LDH crystals will also be destroyed. The process of recrystallization makes the new LDH crystals filled with scratches, realizing self-healing. Moreover, the self-healing behavior will occur rapidly once the coating is damaged, decreasing the corrosion of Mg dramatically. It has the potential to provide long-term protection for Mg alloy.

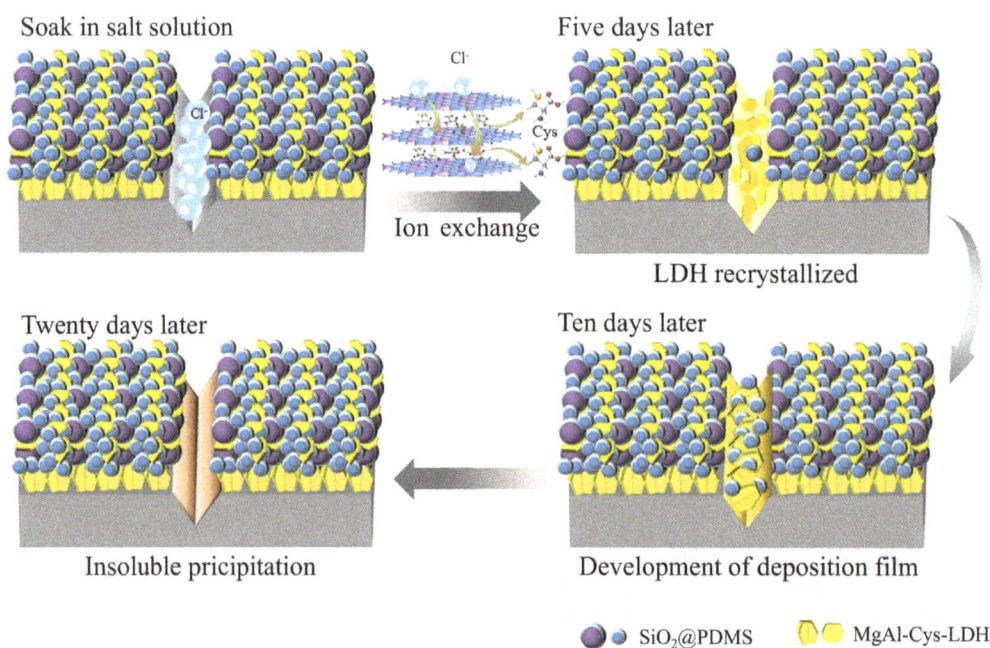

FIGURE 12.4   Self-healing mechanism of CL/SP coating. *Reproduced with permission from (Tang et al. 2023) © 2022 Elsevier B.V.*

12.3.1.3.3   Synergistic Self-Healing Effect

When several self-healing behaviors are present at the same time, a synergistic effect will be produced. The typical situation is that the LDH coating with inhibitor is recrystallized to achieve self-healing. Chen et al. (Chen, Zhang et al. 2019) inserted the corrosion inhibitor aspartic acid (ASP) into Mg-Al LDHs on the surface of AZ31 by the hydrothermal method. Since ASP ion has good corrosion inhibition and has a large specific surface area to capture Cl⁻ and other invasive anions, Mg-Al-ASP-LDHs have better corrosion resistance than Mg-Al-NO$_3$ LDHs. In addition to the inhibition of ASP, Mg-Al LDH coating also undergoes dissolution and recrystallization. The synergistic effect of ASP and LDHs gives the Mg-Al-ASP-LDH coating strong self-healing ability.

This not only makes full use of the existing mature technology but also improves the efficiency of the self-healing, optimizes the corrosion resistance of the self-healing coating, and better protects the substrate.

12.3.1.4   Superhydrophobic LDH Coatings

Superhydrophobic materials can be self-cleaning and can also protect metal surfaces from water corrosion, providing a stronger barrier for Mg alloys. The preparation of superhyrophobic LDH coatings is of great significance for corrosion protection of Mg alloys. Superhydrophobic LDH coating means that the water contact angle (WCA) of the coating surface should be greater than 150°, and the sliding angle (SA) should be less than 10°. For instance, Chen et al. (Chen et al. 2006) prepared oriented Ni-Al LDH coatings by *in situ* crystallization and investigated their hydrophobic properties. The experimental results showed that after low-surface-energy lauric acid modification, the surface showed superhydrophobic phenomenon with WCA up to 163°. Wang et al. (Wang et al. 2020) prepared an anti-corrosion coating by *in situ* growth of molybdate intercalated Mg-Al LDHs on the surface of AZ31 followed by post-sealing with a superhydrophobic coating. Mg-Al LDH modified by molybdate intercalation and lauric acid has excellent corrosion-inhibition performance (99.99%), as shown in **Figure 12.5**.

The main reason was that LDHs with micro/nano binary structures were constructed on the Mg alloy surface during the growth process, which endowed them with corrosion inhibition. The inhibitor carried by anion exchange can give LDH coating a second corrosion inhibition. The superhydrophobic characteristics of LDH modified by low-surface-energy substances provided physical protection against corrosive substances, becoming the third corrosion protection for LDHs. The LDH coatings under the synergistic action showed excellent corrosion resistance and became a functional coating with superhydrophobic property.

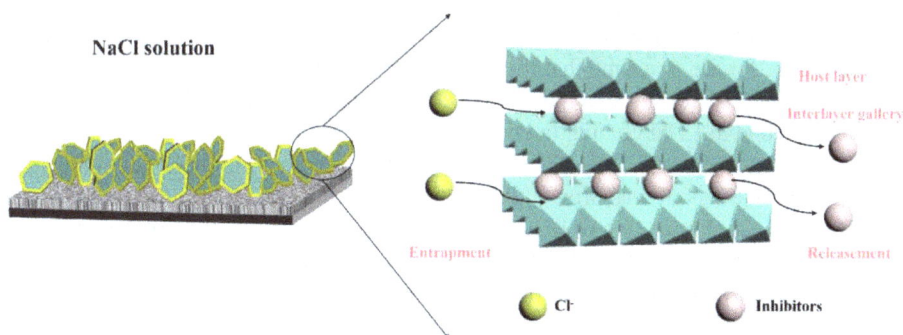

FIGURE 12.5   Corrosion resistance mechanism of superhydrophobic LDH coatings. *Reproduced with permission from (Wang et al. 2020) © 2020 Published by Elsevier B.V. on behalf of Chongqing University.*

12.3.2 LDHs Applied in Different Coatings

Due to the special properties of LDHs such as adjustable laminar composition and adjustable interlayer spacing, many functional assemblies with different physicochemical properties can be derived. Many researchers combine LDHs with other coatings. For example, hydrophobic polymer urea crosslinked polydimethylsiloxane (U-PDMS) was sealed on the surface of tungstate intercalated LDH coating based on AZ31B, and laurate was added to modify LDH powder (La LDH) to prepare WO_4^{2-}-LDH/U-PDMS/La-LDH superhydrophobic self-healing anti-corrosion coating (Ding et al. 2019).

It is also a new trend to prepare hybrid coatings by combining LDH with metal organic frameworks (MOF) (Zhang, Ma, Wang, Sun et al. 2018). Zhou et al. (Zhou et al. 2022) developed a multifunctional nanocontainer using zeolitic imidazolate framework (ZIF)-derived LDHs as gatekeepers for benzotriazole (BTA)-coated mesoporous silica nanoparticles (MSNs-BTA). LDH outer coating and ZIF intermediate layer make MSNs-BTA@ZIF-LDH nanocontainer had good compatibility, dispersibility, and pH responsive controlled-release performance in water-based coating matrix. This excellent corrosion resistance can be attributed to two factors (**Figure 12.6**): (i) Prior to corrosion, MSNs-BTA@ZIF-LDHs acted as a passive nanofiller for corrosion-resistant media, significantly improving the barrier properties of the composite coating. (ii) After corrosion, MSNs-BTA@ZIF-LDHs released preloaded BTA corrosion inhibitor in response to pH changes to prevent further damage to the exposed Mg matrix.

LDHs can be used as powder or coating in various coating systems (sol-gel or epoxy, etc.) (Cao et al. 2022). The protective properties of Mg alloy can be improved by adding appropriate LDH fillers to organic coatings. They can also be improved by growing LDH coatings directly on alloy surfaces. The combination of LDH coatings and other coatings not only retains the properties of LDH itself but also makes full use of the properties of the external coating, making better contributions to the corrosion resistance of the alloy.

12.3.2.1 Labyrinth Effect

The basic concept of labyrinth effect is: in the surface microstate, there are a lot of small voids and pits between the filler and substrate coating, and these voids and pits can form a maze of gases or ions, hindering their transport, so as to achieve the sealing effect. In this way, LDH in the coating can fill in defects and pores, increasing the curvature and difficulty of H_2O molecule penetration. This is known as the labyrinth effect of LDHs, which is suitable for LDH powder and coating.

For LDH powder, it is often used in combination with organic silane. This is because the hydroxyl group of LDH nanoparticles can be condensed with the methoxy group of organic silane so that LDH nanoparticles can be uniformly disperse in the sol-gel coating as a good additive. In addition, the study on the reconstruction characteristics of $Mg_{2-x}M_x/Al_1$ (M=Ca, Sr, Ba) LDHs prepared by sol-gel method proved for the first time that the microstructure of mixed metal oxide reconstructed by LDHs derived from sol-gel showed a memory effect (Valeikiene et al. 2020).

GO was often applied to LDH coating to ensure the barrier performance and labyrinth effect of the coating. Yan et al. (Yan et al. 2019) described the synthesis process and corrosion mechanism of RGO/Zn-Al LDH coatings. It was proved that the LDH coating can be nucleated and grown on the surface of GO. The agglomeration of RGO plate was avoided and the micropores between LDH nanosheet and RGO plate were reduced. This obviously made the transmission path of corrosive medium more tortuous, effectively preventing premature corrosion of the Mg alloy. The combination of RGO plate and LDH coating could yield better composite coatings (Chen et al. 2021; Wu, Chen et al. 2022).

12.3.2.2 Nanocontainer of Corrosion Inhibitor

Most LDH coatings loaded with corrosion inhibitors have self-healing properties due to the controlled release of corrosion inhibitors from the LDH-based nanocontainers. Ouyang et al. (Ouyang

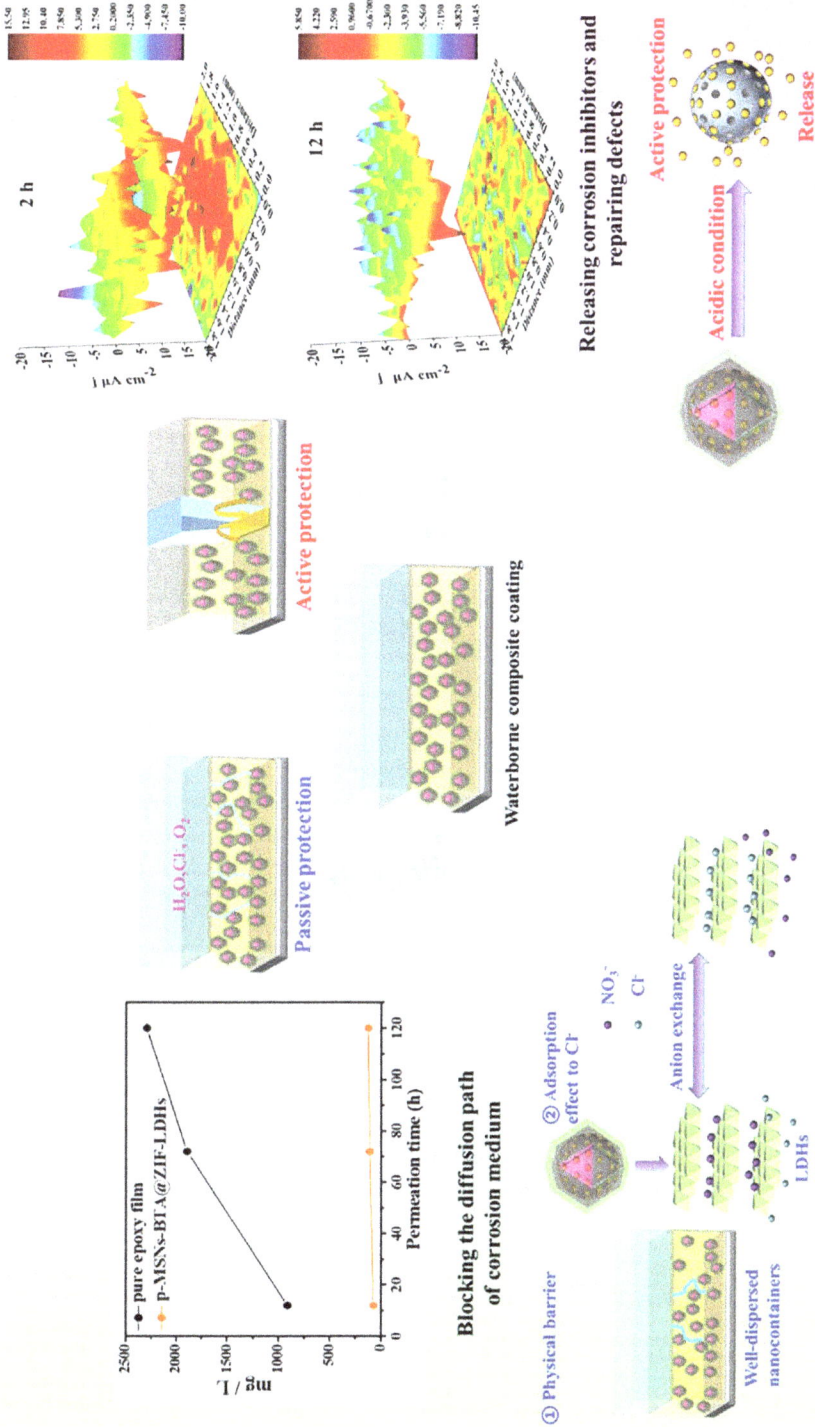

FIGURE 12.6 Dual corrosion resistance and self-repair mechanism of waterborne composite coatings. *Reproduced with permission from (Zhou et al. 2022) © 2022 Elsevier Inc.*

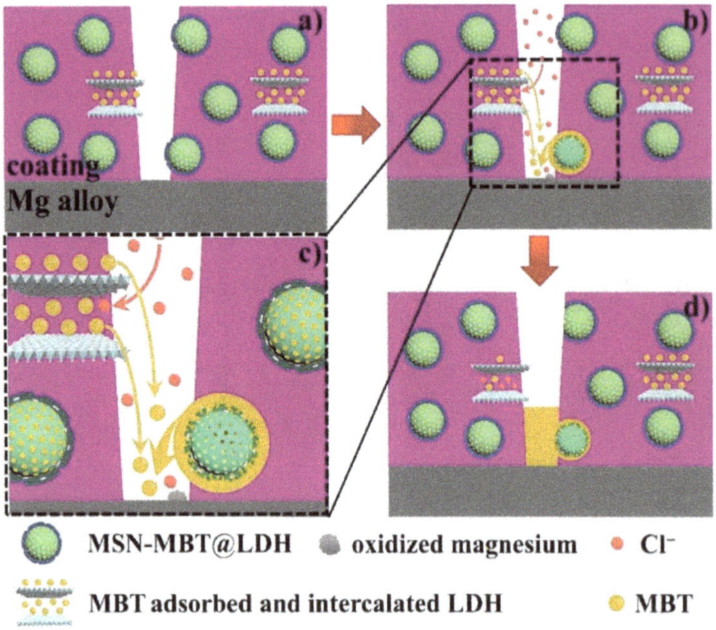

FIGURE 12.7   Self-healing mechanism of SNAP coatings containing MSN-MBT@LDH nanocontainers: (a) SNAP coating for local damage, (b, c) healing process, and (d) SNAP coating after self-repair. *Reproduced with permission from (Ouyang et al. 2020) © 2020 Chongqing University. Publishing services provided by Elsevier B.V. on behalf of KeAi Communications Co. Ltd.*

et al. 2020) constructed a nanocontainer of mesoporous silica nanoparticle (MSN) as the core material and LDHs as the shell material, loaded with corrosion inhibitor MBT. A series of test results proved that LDH nanocontainer had pH-responsive inhibitor release characteristics, and the composite coating had the best barrier properties and corrosion resistance in NaCl corrosion solution. The self-healing process included the desorption of physically adsorbed MBT, the release of intercalated MBT through ion exchange, and the leakage of MBT from the MSN surface through the dissolution of LDH nanoshells, as shown in **Figure 12.7**.

More and more studies have shown that LDH nanocontainers with corrosion inhibitors have excellent anti-corrosion effect in metal coatings (Su et al. 2020). This is mainly due to LDH's excellent Cl- ion capture ability and ion exchange ability, which can reduce the Cl- ion concentration. Secondly, the instability of LDH nanosheets in acid solution makes the coating system have pH response characteristics. When corrosion occurs, the pH at the damaged part of the coating changes (Li, He et al. 2022), and LDH releases the corrosion inhibitor, realizing intelligent corrosion inhibition (Du et al. 2019). The corrosion resistance of the coating has been improved through the blocking effect of the corrosion inhibitor, the sacrificial protection, and the intelligent corrosion inhibition (Xie et al. 2021).

12.3.3 PORE SEALING

Anodic oxide and micro-arc oxidation (MAO) coatings are usually prepared on the surface of Mg alloys to improve the corrosion resistance. However, the surface of such coatings is porous and has limited protective properties against alloys. Therefore, many researchers have found that LDH can grow on these coatings to seal their pores and defects.

Hong et al. (Hong et al. 2022) grew ternary Mg-Al-Y LDHs *in situ* on anodic oxide films of Mg-2Zn-4Y alloy by the hydrothermal method. It has been proved that Mg-2Zn-4Y alloy can provide component cations for ternary LDHs *in situ* and act as a physical barrier to prevent the invasion of corrosive media. Correspondingly, the LDH coating provided self-healing and anti-corrosion properties by sealing the anodic oxide film.

As a matter of fact, Zhang et al. have proposed that the internal source of Mg^{2+} and Al^{3+} cations required for LDH coating growth was attributed to the dissolution of the anodic oxide film and AZ31 matrix after the hydrothermal formation of LDH coating. In other words, the anodic oxide film can be converted into LDH coating. The Stranski-Krastanov growth mode (from 2D to 3D) was quite suitable for describing the transition from anodic oxide film to Mg-Al LDH coating (Zhang, Wu, Tang, Chen et al. 2018). The specific conversion process is shown in **Figure 12.8**.

To sum up, the amorphous $Mg(OH)_2$ in the anodic oxide film gradually aggregated, forming layered structures on the Mg surface. With the increased hydrothermal reaction time, the Mg matrix dissolved, and the pH value increased sharply, leading to the inward growth of the LDH. At the same time, Al^{3+} was doped into $Mg(OH)_2$, resulting in charge imbalance and destruction of inter-layer hydrogen bond (Yang et al. 2012). Therefore, nitrates from the solution were inserted into layers to maintain charge balance. This new structure is a feature of LDHs. Subsequently, LDH grew inward to the substrate and outward to the film/solution interface. Finally, it was worth pointing out that this new method for preparing Mg – Al LDHs has also been used to describe the growth of LDH coatings on MAO coating and to study the effect of MAO properties on the growth of LDHs (Zhang, Wu, Tang, Ma et al. 2018; Zhang, Wu, Tang, Pan et al. 2018).

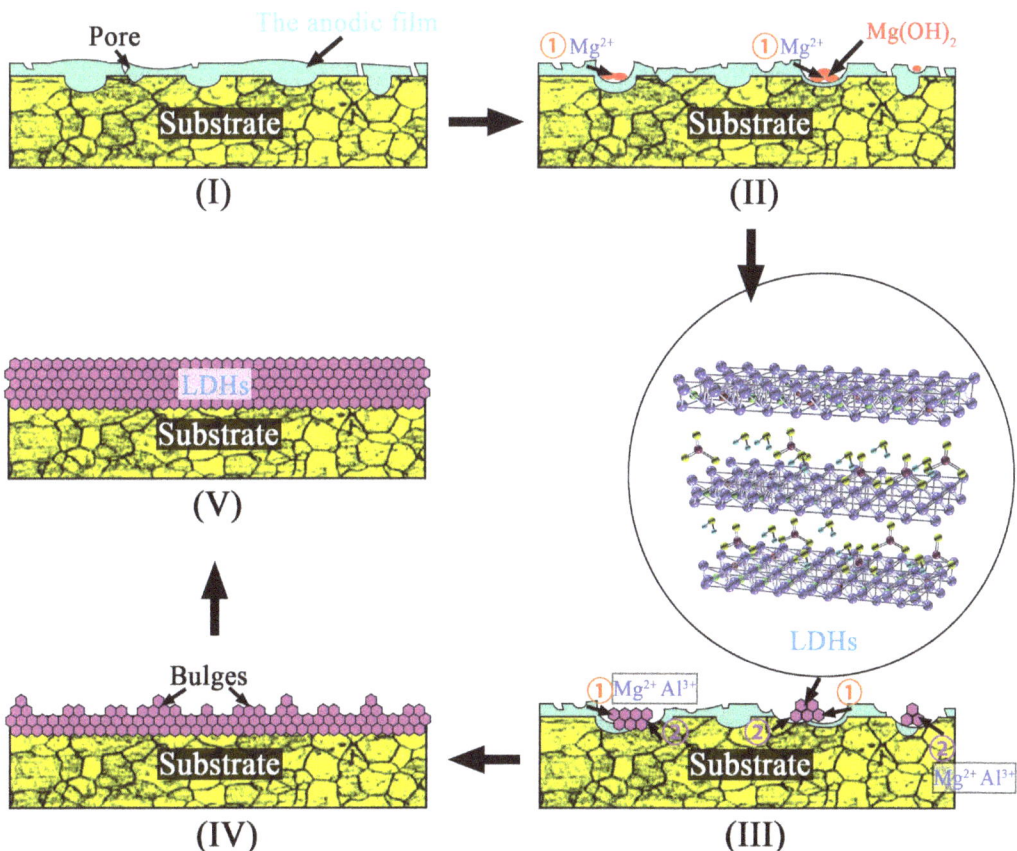

FIGURE 12.8 The process of converting anodic film into LDH coating. *Reproduced with permission from (Zhang, Wu, Tang, Chen et al. 2018) © 2018 Elsevier B.V.*

LDHs can contain rare metal (RE) cations as corrosion inhibitors to seal the defects and micropores of MAO. At the same time, these cations will inhibit the cathodic reaction and protect Mg alloys when released. Additionally, GO has rich negative charges and large specific surface area, which can adsorb a large number of M^{3+} and M^{2+} to promote *in situ* nucleation and provide good support materials for *in situ* growth of LDHs (Chen et al. 2021). Chen et al. prepared RE^{3+}-contained ternary Mg-Al-La LDH/GO (L/G-N) coating by the hydrothermal method based on MAO coating. Then, the ternary L/G-V coating intercalated with vanadate was formed by ion exchange (Chen, Hung et al. 2019). The results showed that MAO could provide endogenous cations (Mg^{2+} and Al^{3+}) for LDH growth. The anticorrosion mechanism of L/G-V coating is as follows (**Figure 12.9**): LDH itself had excellent barrier performance, and the additional GO made LDHs present wavy sheets and curly edges, prolonging the bending degree of the transmission path of corrosive media. The synergistic effect of RE^{3+} cationic and interlayer VO_{3-} anionic corrosion inhibitor in LDHs timely remedied coating defects, realized active/passive dual protection of MAO coating, demonstrated self-healing and anti-corrosion functions, and enhanced the corrosion resistance of Mg alloys.

Therefore, preparing LDH coating on MAO surface can not only seal the pores on MAO surface but also effectively control the redox reactions and coating damages so as to avoid premature failure of the coating.

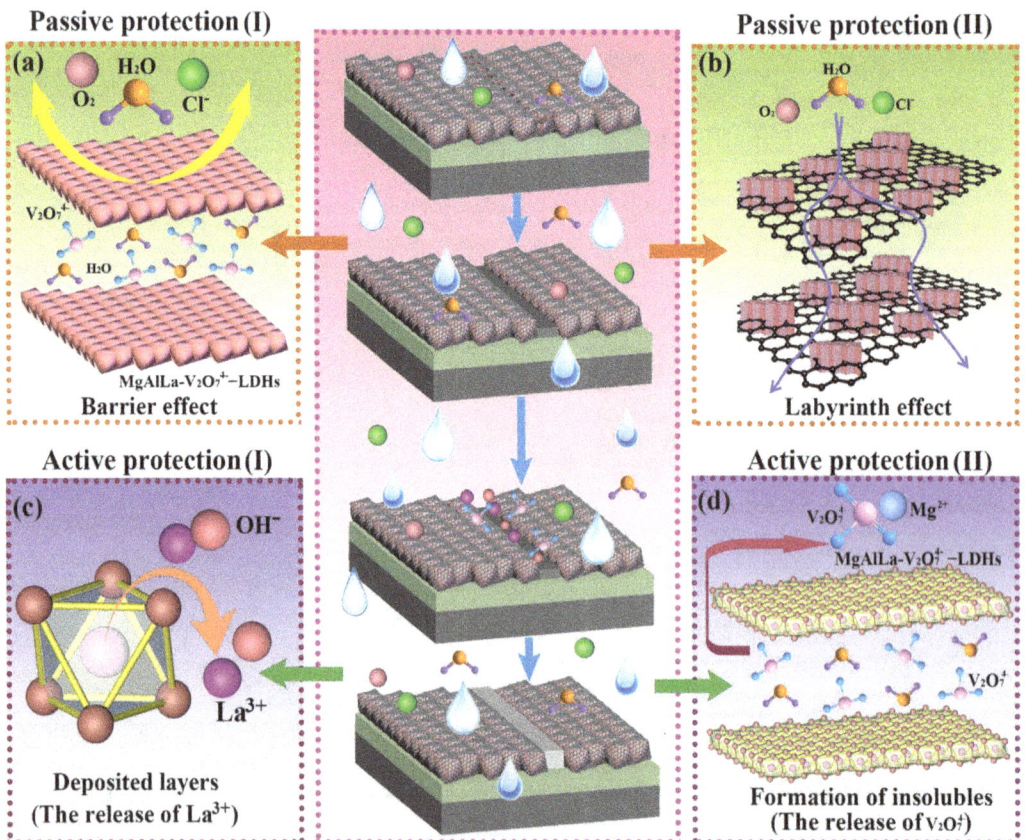

FIGURE 12.9 Anticorrosion mechanism of L/G-V coating. *Reproduced with permission from (Chen, Hung et al. 2019) © 2022 Elsevier B.V.*

12.3.4 DENSIFICATION OF LDH COATINGS

Although LDHs have many excellent properties, the porosity between their lamellar structures is the biggest disadvantage limiting their wide application. Many methods have been developed to seal or fill the pores of nested LDH coatings, such as sol-gel (Tarzanagh et al. 2022), GO modified pore sealing (Wu, Zhang et al. 2020; Asl et al. 2021), nanoparticle deposition filling the pores (Wu et al. 2019), and others (Wu, Chen et al. 2020).

For sol-gel pore sealing, according to the active anti-corrosion of Mg alloy AM60B by 8-hydroxyquinoline (HQ) intercalated LDH coating and sol-gel coating, the alloy surface was completely covered by LDH coating, with typical micro morphology composed of vertically grown nanosheets, and the space in the LDH layer structure was completely sealed by the sol-gel coating (Tarzanagh et al. 2022). The average roughness of the LDH coating was reduced from about 541 nm to 408 nm by sealing the space between LDH sheets with the sol-gel layer. The loading of 8-HQ did not change the morphology of the final composite coating. In LDH/sol-gel or LDH@HQ/sol-gel coating, there was no evidence of cracks, peeling, micropores, or any other low-quality coating between layers. Moreover, the corrosion resistance of LDH/sol-gel composite was much higher than that of LDH coating, which was mainly related to the sealing of solution channel.

For GO modified pore sealing, reduced GO coated Al (G/Al) coating was pre-sprayed on AZ31 to form graphene composite LDH coating (LDH-G/Al) (Wu, Zhang et al. 2020). The graphene in the bottom G/Al coating sealed the LDH defects, making LDH-G/Al have a dense structure. This method provided a corrosion barrier for Mg alloy, and the LDH coating grown *in situ* played a dual role of corrosion inhibitor and anchoring graphene. Combined with the labyrinth effect of graphene, the coating had excellent corrosion resistance.

For nanoparticle deposition filling pore sealing, wear-resistant Al_2O_3 nanoparticles have been used to prepare Mg-Al LDHs/Al_2O_3 composite coatings on the surface of AZ31 (Wu et al. 2019). That is because the pores between LDH nanosheets were used as a "container" for encompassing Al_2O_3 nanoparticles. Conversely, the addition of Al_2O_3 filled the pores of LDH. The synergistic effect of LDH coating and Al_2O_3 nanoparticles on wear resistance and corrosion resistance made the composite coating have excellent corrosion resistance and wear resistance.

Besides, Wu et al. (Wu, Chen et al. 2020) prepared a bottom-layer Mg-Al LDH coating and then modified and sealed it by top layer poly-L-glutamic acid (PGA) coating. The corrosion mechanism is as shown in **Figure 12.10**. A large amount of carboxyl (-COOH) of PGA can be ionized in the

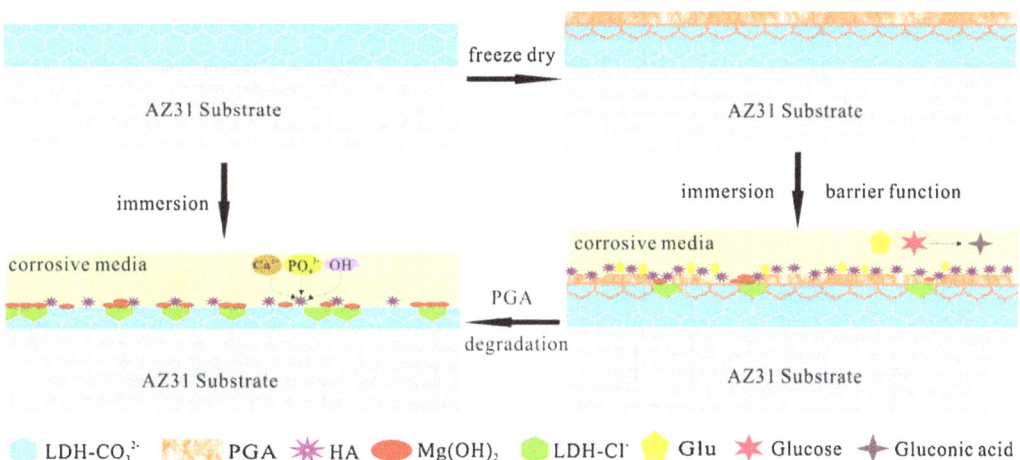

FIGURE 12.10 Corrosion mechanisms of LDH and LDH/PGA-30 coatings. *Reproduced with permission from (Wu, Chen et al. 2020) © 2020 Elsevier B.V.*

solution, showing weak acidity. Consequently, it can react with alkaline LDH in the coating synthesis process and form a composite protective coating on the LDH coating surface through chemical bonding. In addition, a small amount of PGA permeated into the porous LDH and filled and sealed the micropores. Therefore, the composite coating improved the corrosion resistance of AZ31 alloy.

In a word, whether LDH is used to seal the porous coating or other substances seal the LDH coating, the purpose is to make the composite coating more compact. The denser the coating is, the more difficult it is for the corrosion medium to penetrate into the alloy surface, and the better the corrosion-resistance effect will be.

12.3.5 CURRENT LIMITATIONS

In recent years, with the extensive application of LDHs in Mg alloy corrosion and other fields, the focus of researchers has gradually developed from single LDH coating to composite coating. Although much progress has been made, there are still some issues that need further study.

1 *In situ* LDHs is the most widely used synthesis method at present, but the purity and thickness of LDH coating sometimes cannot meet the requirements of industrial applications.
2 LDHs with two-dimensional structure still need post-processing to close the pores in their structure so as to further improve their resistance to corrosive media.
3 As a nanocontainer, the types and quantities of corrosion inhibitors that LDHs can accommodate are limited, and further research is needed to improve the type and capacity of the corrosion inhibitor load.

12.4 CONCLUSIONS AND OUTLOOK

In this chapter, the structure and synthesis methods of LDH coating on Mg alloy surface and its related applications in corrosion protection were covered, and the advantages and disadvantages of various synthesis methods were discussed, with emphasis on the different corrosion-protection effects of LDHs under hydrothermal synthesis methods. The conclusions are as follows:

1 Different synthesis methods have different effects on the growth orientation and crystal structure of LDH nanosheets, which proves the importance of selection of a suitable synthesis method for LDH for a particular application. Hydrothermal preparation of LDH is the most widely used method.
2 Mg alloy coated with LDH coating can inhibit the corrosion of corrosive medium, so it has excellent corrosion resistance. The protective effects of LDH coatings (such as physical barrier, chloride trapping, etc.) are affected by each other. Therefore, the anti-corrosion mechanism of Mg alloy protective coating is the synergistic effect between different protective effects.
3 The growth mechanism and orientation mechanism of LDH have been proposed for a long time, but in specific experimental studies, there are few comparisons between different mechanisms. In addition, the LDH coatings synthesized in different studies are inconsistent, and their corrosion inhibition effects are also different. Therefore, it is difficult to conclude which growth mechanism, orientation mechanism, or coating is better through simple comparison.

In conclusion, LDH protective coating has a good protective effect on Mg alloy and has a broad application background and market prospects in the anticorrosion field, but there are still many problems that need to be improved and solved by researchers.

1 The deep anticorrosion mechanism of the inner dense coating and outer porous coating of LDHs has not been systematically elaborated at present.

2 When LDH coating grows on the Mg matrix, the initial state is heterogeneous growth. Therefore, the growth pattern and *in situ* growth mechanism of LDHs need to be further explored.

3 The combination of corrosion inhibitors or functional species is expected to create multi-functional LDH smart materials. For example, trivalent or macromolecular anionic intercalation LDH is expected to achieve better corrosion resistance or specific functions.

4 Most of the applications reported so far have been carried out in laboratory conditions. In future research, LDH coatings will continue to expand the range of applications, especially in industrial applications. Therefore, translating experimental research into industrial applications will be a great challenge.

REFERENCES

Acharya H, Srivastava SK and Bhowmick AK. Synthesis of partially exfoliated EPDM/LDH nanocomposites by solution intercalation: Structural characterization and properties. *Compos. Sci. Technol.* 2007, 67, 2807–2816.

Ahmed AAA, Talib ZA, bin Hussein MZ and Zakaria A. Zn – Al layered double hydroxide prepared at different molar ratios: Preparation, characterization, optical and dielectric properties. *J. Solid State Chem.* 2012, 191, 271–278.

Aisawa S, Takahashi S, Ogasawara W, Umetsu Y and Narita E. Direct intercalation of amino acids into layered double hydroxides by coprecipitation. *J. Solid State Chem.* 2001, 162, 52–62.

Asl VZ, Chini SF, Zhao J, Palizdar Y, Shaker M and Sadeghi A. Corrosion properties and surface free energy of the ZnAl LDH/rGO coating on MAO pretreated AZ31 magnesium alloy. *Surf. Coat. Tech.* 2021, 426, 127764.

Bhavani AG, Wani TA, Ma'aruf A and Prasad T. Effect of ageing process on crystal morphology of Co-Mg-Al hydrotalcite. *Mater. Today: Proc.* 2021, 44, 2277–2282.

Boclair JW and Braterman PS. Layered double hydroxide stability. 1. Relative stabilities of layered double hydroxides and their simple counterparts. *Chem. Mater.* 1999, 11, 298–302.

Bouali AC, Serdechnova M, Blawert C, Tedim J, Ferreira MGS and Zheludkevich ML. Layered double hydroxides (LDHs) as functional materials for the corrosion protection of aluminum alloys: A review. *Appl. Mater. Today.* 2020, 21, 100857.

Cai M, Fan X, Yan H, Li Y, Song S, Li W, Li H, Lu Z and Zhu M. In situ assemble $Ti_3C_2T_x$ MXene@MgAl-LDH heterostructure towards anticorrosion and antiwear application. *Chem. Eng. J.* 2021, 419, 130050.

Cao K, Yu Z, Zhu L, Yin D, Chen L, Jiang Y and Wang J. Fabrication of superhydrophobic layered double hydroxide composites to enhance the corrosion-resistant performances of epoxy coatings on Mg alloy. *Surf. Coat. Tech.* 2021, 407, 126763.

Cao Y, Zheng D, Li X, Lin J, Wang C, Dong S and Lin C. Enhanced corrosion resistance of superhydrophobic layered double hydroxide films with long-term stability on Al substrate. *ACS Appl. Mater. Interfaces.* 2018, 10, 15150–15162.

Cao Y, Zheng D, Zhang F, Pan J and Lin C. Layered double hydroxide (LDH) for multi-functionalized corrosion protection of metals: A review. *J. Mater. Sci. Technol.* 2022, 102, 232–263.

Chang Z, Evans DG, Duan X, Vial C, Ghanbaja J, Prevot V, De Roy M and Forano C. Synthesis of [Zn – Al – CO_3] layered double hydroxides by a coprecipitation method under steady-state conditions. *J. Solid State Chem.* 2005, 178, 2766–2777.

Chen H, Zhang F, Fu S and Duan X. In situ microstructure control of oriented layered double hydroxide monolayer films with curved hexagonal crystals as superhydrophobic materials. *Adv. Mater.* 2006, 18, 3089–3093.

Chen J, Fang L, Wu F, Xie J, Hu J, Jiang B and Luo H. Corrosion resistance of a self-healing rose-like MgAl-LDH coating intercalated with aspartic acid on AZ31 Mg alloy. *Prog. Org. Coat.* 2019, 136, 105234.

Chen J, Song Y, Shan D and Han E-H. Study of the in situ growth mechanism of Mg – Al hydrotalcite conversion film on AZ31 magnesium alloy. *Corros. Sci.* 2012, 63, 148–158.

Chen R, Hung SF, Zhou D, Gao J, Yang C, Tao H, Yang HB, Zhang L, Zhang L and Xiong Q. Layered structure causes bulk NiFe layered double hydroxide unstable in alkaline oxygen evolution reaction. *Adv. Mater.* 2019, 31, 1903909.

Chen Y, Wu L, Yao W, Chen Y, Zhong Z, Ci W, Wu J, Xie Z, Yuan Y and Pan F. A self-healing corrosion protection coating with graphene oxide carrying 8-hydroxyquinoline doped in layered double hydroxide on a micro-arc oxidation coating. *Corros. Sci.* 2022, 194, 109941.

Chen Y, Wu L, Yao W, Wu J, Yuan Y, Xie Z, Jiang B and Pan F. Synergistic effect of graphene oxide/ternary Mg-Al-La layered double hydroxide for dual self-healing corrosion protection of micro-arc oxide coating of magnesium alloy. *Colloid. Surf. A: Physicochemical and Eng Aspects.* 2022, 130339.

Chen Y, Wu L, Yao W, Zhong Z, Chen Y, Wu J and Pan F. One-step in situ synthesis of graphene oxide/MgAl-layered double hydroxide coating on a micro-arc oxidation coating for enhanced corrosion protection of magnesium alloys. *Surf. Coat. Tech.* 2021, 413, 127083.

Costa FR, Leuteritz A, Wagenknecht U, Jehnichen D, Haeussler L and Heinrich G. Intercalation of Mg – Al layered double hydroxide by anionic surfactants: Preparation and characterization. *Appl. Clay Sci.* 2008, 38, 153–164.

Costantino U, Vivani R, Bastianini M, Costantino F and Nocchetti M. Ion exchange and intercalation properties of layered double hydroxides towards halide anions. *Dalton T.* 2014, 43, 11587–11596.

Ding C, Tai Y, Wang D, Tan L and Fu J. Superhydrophobic composite coating with active corrosion resistance for AZ31B magnesium alloy protection. *Chem. Eng. J.* 2019, 357, 518–532.

Du P, Wang J, Liu G, Zhao H and Wang L. Facile synthesis of intelligent nanocomposites as encapsulation for materials protection. *Mater. Chem. Front.* 2019, 3, 321–330.

Feng X, Long R, Wang L, Liu C, Bai Z and Liu X. A review on heavy metal ions adsorption from water by layered double hydroxide and its composites. *Sep. Purif. Technol.* 2022, 284, 120099.

Fogg AM, Freij AJ and Parkinson GM. Synthesis and anion exchange chemistry of rhombohedral Li/Al layered double hydroxides. *Chem. Mater.* 2002, 14, 232–234.

Frank CW, Rao V, Despotopoulou MM, Pease RFW, Hinsberg WD, Miller RD and Rabolt JF. Structure in thin and ultrathin spin-cast polymer films. *Science.* 1996, 273, 912–915.

Geng J, Pan C, Wang Y, Chen W and Zhu Y. Chloride binding in cement paste with calcined Mg-Al-CO$_3$ LDH (CLDH) under different conditions. *Constr. Build. Mater.* 2021, 273, 121678.

Grégoire B, Ruby C and Carteret C. Hydrolysis of mixed Ni 2$^+$–Fe 3$^+$ and Mg 2$^+$–Fe 3$^+$ solutions and mechanism of formation of layered double hydroxides. *Dalton T.* 2013, 42, 15687–15698.

Grover A, Mohiuddin I, Lee J, Brown RJC, Malik AK, Aulakh JS and Kim K-H. Progress in pre-treatment and extraction of organic and inorganic pollutants by layered double hydroxide for trace-level analysis. *Environ. Res.* 2022, 114166.

Gunawan P and Xu R. Synthesis of unusual coral-like layered double hydroxide microspheres in a nonaqueous polar solvent/surfactant system. *J. Mater. Chem.* 2008, 18, 2112–2120.

Guo L, Wu W, Zhou Y, Zhang F, Zeng R and Zeng J. Layered double hydroxide coatings on magnesium alloys: A review. *J. Mater. Sci. Technol.* 2018, 34, 1455–1466.

Guo L, Zhang F, Lu J-C, Zeng R-C, Li S-Q, Song L and Zeng J-M. A comparison of corrosion inhibition of magnesium aluminum and zinc aluminum vanadate intercalated layered double hydroxides on magnesium alloys. *Front. Mater. Sci.* 2018, 12, 198–206.

Guo X, Zhang F, Evans DG and Duan X. Layered double hydroxide films: Synthesis, properties and applications. *Chem. Commun.* 2010, 46, 5197–5210.

Hong Y, Wu L, Zhang X, Zhan G, Chen Y, Yao W, Dai X, Wu T, Dai X and Xiang J. Effect of Y Concentration on the in situ growth behavior and corrosion protection of the MgAlY-LDH sealing film on the anodized surface of Mg–2Zn–4Y alloy. *Front. Mater.* 2022, 9, 825120.

Hou L, Li Y, Sun J, Zhang SH, Wei H and Wei Y. Enhancement corrosion resistance of MgAl layered double hydroxides films by anion-exchange mechanism on magnesium alloys. *Appl. Surf. Sci.* 2019, 487, 101–108.

Huang M, Lu G, Pu J and Qiang Y. Superhydrophobic and smart MgAl-LDH anti-corrosion coating on AZ31 Mg surface. *J. Ind. Eng. Chem.* 2021, 103, 154–164.

Huang S, Zhu G-N, Zhang C, Tjiu WW, Xia Y-Y and Liu T. Immobilization of Co – Al layered double hydroxides on graphene oxide nanosheets: Growth mechanism and supercapacitor studies. *ACS Appl. Mater. Interfaces.* 2012, 4, 2242–2249.

Inoue M, Kominami H, Kondo Y and Inui T. Organic derivatives of layered inorganics having the second stage structure. *Chem. Mater.* 1997, 9, 1614–1619.

Ji G and Prakash R. Hydrothermal synthesis of Zn-Mg-based layered double hydroxide coatings for the corrosion protection of copper in chloride and hydroxide media. *Int. J. Min. Met. Mater.* 2021, 28, 1991–2000.

Jin H, Yang W and Li Y. *MOF Membranes for Gas Separation. Advanced Material for Membrane Fabrication and Modification,* CRC Press, 2018: 337–368.

Kaseem M, Ramachandraiah K, Hossain S and Dikici B. A review on LDH-smart functionalization of anodic films of Mg alloys. *Nanomaterials.* 2021, 11, 536.

Ke GJ, Zhou Z, Yang PF and Song BX. Preparation of CNTs/MgAl-LDHs composites and their adsorption properties for chloride ions. *Mater Sci Forum, Trans Tech Publications.* 2019, 305–313.

Khan AI and O'hare D. Intercalation chemistry of layered double hydroxides: Recent developments and applications. *J. Mater. Chem.* 2002, 12, 3191–3198.

Kovanda F, Jindová E, Lang K, Kubát P and Sedláková Z. Preparation of layered double hydroxides intercalated with organic anions and their application in LDH/poly (butyl methacrylate) nanocomposites. *Appl. Clay Sci.* 2010, 48, 260–270.

Li J, He N, Li J, Fu Q, Feng M, Jin W, Li W, Xiao Y, Yu Z and Chu PK. A silicate-loaded MgAl LDH self-healing coating on biomedical Mg alloys for corrosion retardation and cytocompatibility enhancement. *Surf. Coat. Tech.* 2022, 439, 128442.

Li LY, Cui LY, Zeng RC, Li SQ, Chen XB, Zheng Y and Kannan MB. Advances in functionalized polymer coatings on biodegradable magnesium alloys – A review. *Acta Biomater.* 2018, 79, 23–36.

Li PB, Wang YX, Shao ZX, Wu BT, Li H, Gao MM, Liu KG and Shi KR. Enhanced corrosion protection of magnesium alloy via in situ Mg – Al LDH coating modified by core – shell structured Zn – Al LDH@ ZIF-8. *Rare Metals.* 2022, 1–14.

Li S, Bai H, Wang J, Jing X, Liu Q, Zhang M, Chen R, Liu L and Jiao C. In situ grown of nano-hydroxyapatite on magnetic CaAl-layered double hydroxides and its application in uranium removal. *Chem. Eng. J.* 2012, 193, 372–380.

Lin Y, Fang Q and Chen B. Metal composition of layered double hydroxides (LDHs) regulating ClO_4^- adsorption to calcined LDHs via the memory effect and hydrogen bonding. *J. Environ. Sci.* 2014, 26, 493–501.

Mir ZM, Bastos A, Höche D and Zheludkevich ML. Recent advances on the application of layered double hydroxides in concrete – a review. *Materials.* 2020, 13, 1426.

Mishra G, Dash B and Pandey S. Layered double hydroxides: A brief review from fundamentals to application as evolving biomaterials. *Appl. Clay Sci.* 2018, 153, 172–186.

Mochane MJ, Magagula SI, Sefadi JS, Sadiku ER and Mokhena TC. Morphology, thermal stability, and flammability properties of polymer-layered double hydroxide (LDH) nanocomposites: A Review. *Crystals.* 2020, 10, 612.

Ouyang Y, Li LX, Xie ZH, Tang L, Wang F and Zhong CJ. A self-healing coating based on facile pH-responsive nanocontainers for corrosion protection of magnesium alloy. *J. Magnes. Alloy.* 2020, 10, 836–849.

Pan GX, Xu MH, Chen HF, Tang PS, Cao F and Ni Z-M. Interlayer structure and ion-exchange properties of hydrotalcite intercalated with CO_3^{2-}, CrO_4^{2-}, SO_4^{2-} and NO_3. *Adv Mat Res, Trans Tech Publ.* 2011, 2102–2105.

Pedraza-Chan MS, Salazar-Kuri U, Sánchez-Zeferino R, Ruiz-López II and Escobedo-Morales A. Emulation of evolutionary selection as the growth mechanism of supported layered double hydroxide frameworks. *Appl. Clay Sci.* 2021, 210, 106159.

Qu ZY, Yu QL and Brouwers HJH. Relationship between the particle size and dosage of LDHs and concrete resistance against chloride ingress. *Cement. Concrete Res.* 2018, 105, 81–90.

Salak AN, Tedim J, Kuznetsova AI, Zheludkevich ML and Ferreira MGS. Anion exchange in Zn – Al layered double hydroxides: in situ X-ray diffraction study. *Chem. Phys. Lett.* 2010, 495, 73–76.

Shulha TN, Serdechnova M, Iuzviuk MH, Zobkalo IA, Karlova P, Scharnagl N, Wieland DCF, Lamaka SV, Yaremchenko AA and Blawert C. In situ formation of LDH-based nanocontainers on the surface of AZ91 magnesium alloy and detailed investigation of their crystal structure. *J. Magnes. Alloy.* 2022, 10, 1268–1285.

Shulha TN, Serdechnova M, Lamaka SV, Wieland DCF, Lapko KN and Zheludkevich ML. Chelating agent-assisted in situ LDH growth on the surface of magnesium alloy. *Sci. Rep.* 2018, 8, 1–10.

Song J, Chen J, Xiong X, Peng X, Chen D and Pan F. Research advances of magnesium and magnesium alloys worldwide in 2021. *J. Magnes. Alloy.* 2022, 10, 863–898.

Song J, She J, Chen D and Pan F. Latest research advances on magnesium and magnesium alloys worldwide. *J. Magnes. Alloy.* 2020, 8, 1–41.

Sonoyama N, Yamada S, Ota T, Inagaki H, Dedetemo PK and Yoshida S. Preparation of layered double hydroxide films using an electrodeposition and subsequent crystal growth method. *Clay Miner.* 2021, 1–8.

Su Y, Qiu S, Yang D, Liu S, Zhao H, Wang L and Xue Q. Active anti-corrosion of epoxy coating by nitrite ions intercalated MgAl LDH. *J. Hazard. Mater.* 2020, 391, 122215.

Sun X, Yao QS, Li YC, Zhang F, Zeng R-C, Zou YH and Li SQ. Biocorrosion resistance and biocompatibility of Mg-Al layered double hydroxide/poly (L-lactic acid) hybrid coating on magnesium alloy AZ31. *Front. Mater. Sci.* 2020, 14, 426–441.

Tang Y, Wu F, Fang L, Ruan H, Hu J, Zeng X, Zhang S, Luo H and Zhou M. Effect of deposition sequence of MgAl-LDH and SiO_2@ PDMS layers on the corrosion resistance of robust superhydrophobic/self-healing multifunctional coatings on magnesium alloy. *Prog. Org. Coat.* 2023, 174, 107299.

Tarzanagh YJ, Seifzadeh D and Samadianfard R. Combining the 8-hydroxyquinoline intercalated layered double hydroxide film and sol – gel coating for active corrosion protection of the magnesium alloy. *Int. J. Min. Met. Mater.* 2022, 29, 536–546.

Taylor HFW. Crystal structures of some double hydroxide minerals. *Mineral. Mag.* 1973, 39, 377–389.

Valeikiene L, Grigoraviciute-Puroniene I and Kareiva A. Alkaline earth metal substitution effects in sol-gel – derived mixed metal oxides and $Mg_{2-x}M_x/Al_1$ (M= Ca, Sr, Ba)–layered double hydroxides. *J. Aust. Ceram. Soc.* 2020, 56, 1531–1541.

Wang Q and O'Hare D. Recent advances in the synthesis and application of layered double hydroxide (LDH) nanosheets. *Chem. Rev.* 2012, 112, 4124–4155.

Wang X, Jing C, Chen Y, Wang X, Zhao G, Zhang X, Wu L, Liu X, Dong B and Zhang Y. Active corrosion protection of super-hydrophobic corrosion inhibitor intercalated Mg – Al layered double hydroxide coating on AZ31 magnesium alloy. *J. Magnes. Alloy.* 2020, 8, 291–300.

Wu H, Shen G, Li R, Chen B, Zhang L, Jie X and Liu G. The growth behavior and properties of orientated LDH film composited with reduced graphene oxide. *Surf. Coat. Tech.* 2022, 436, 128261.

Wu H, Zhang L, Zhang Y, Long S and Jie X. Corrosion behavior of Mg – Al LDH film in-situ assembled with graphene on Mg alloy pre-sprayed Al layer. *J. Alloys Compd.* 2020, 834, 155107.

Wu L, Chen Y, Dai X, Yao W, Wu J, Xie Z, Jiang B, Yuan Y and Pan F. Corrosion resistance of the GO/ZIF-8 hybrid loading benzotriazole as a multifunctional composite filler-modified MgAlY layered double hydroxide coating. *Langmuir.* 2022, 38, 10338–10350.

Wu L, Ding X, Zheng Z, Ma Y, Atrens A, Chen X, Xie Z, Sun D and Pan F. Fabrication and characterization of an actively protective Mg-Al LDHs/Al_2O_3 composite coating on magnesium alloy AZ31. *Appl. Surf. Sci.* 2019, 487, 558–568.

Wu W, Song L, Li YC, Zhang F, Zeng RC, Li SQ and Zou YH. Synthesis of glutamate intercalated Mg-Al layered double hydroxides: Influence of stirring and aging time. *J. Disper. Sci. Technol.* 2021, 42, 2154–2162.

Wu W, Sun X, Zhu C-L, Zhang F, Zeng RC, Zou YH and Li SQ. Biocorrosion resistance and biocompatibility of Mg – Al layered double hydroxide/poly-L-glutamic acid hybrid coating on magnesium alloy AZ31. *Prog. Org. Coat.* 2020, 147, 105746.

Xia SJ, Liu FX, Ni ZM, Shi W, Xue JL and Qian PP. Ti-based layered double hydroxides: Efficient photocatalysts for azo dyes degradation under visible light. *Appl. Catal. B: Environ.* 2014, 144, 570–579.

Xie P, He Y, Zhong F, Zhang C, Chen C, Li H, Liu Y, Bai Y and Chen J. Cu-BTA complexes coated layered double hydroxide for controlled release of corrosion inhibitors in dual self-healing waterborne epoxy coatings. *Prog. Org. Coat.* 2021, 153, 106164.

Xu H, Wu J, Liu J, Chen Y and Fan X. Growth of cobalt – nickel layered double hydroxide on nitrogen-doped graphene by simple co-precipitation method for supercapacitor electrodes. *J. Mater. Sci.: Mater. El.* 2018, 29, 17234–17244.

Yan L, Zhou M, Pang X and Gao K. One-step in situ synthesis of reduced graphene oxide/Zn – Al layered double hydroxide film for enhanced corrosion protection of magnesium alloys. *Langmuir.* 2019, 35, 6312–6320.

Yan T, Xu S, Peng Q, Zhao L, Zhao X, Lei X and Zhang F. Self-healing of layered double hydroxide film by dissolution/recrystallization for corrosion protection of aluminum. *J. Electroch. Soc.* 2013, 160, C480.

Yang Y, Zhao X, Zhu Y and Zhang F. Transformation mechanism of magnesium and aluminum precursor solution into crystallites of layered double hydroxide. *Chem. Mater.* 2012, 24, 81–87.

Yu J, Wang Q, O'Hare D and Sun L. Preparation of two dimensional layered double hydroxide nanosheets and their applications. *Chem. Soc. Rev.* 2017, 46, 5950–5974.

Yuan X, Wang Y, Wang J, Zhou C, Tang Q and Rao X. Calcined graphene/MgAl-layered double hydroxides for enhanced Cr (VI) removal. *Chem. Eng. J.* 2013, 221, 204–213.

Zhang D, Peng F and Liu X. Protection of magnesium alloys: From physical barrier coating to smart self-healing coating. *J. Alloys Compd.* 2021, 853, 157010.

Zhang D, Zhou J, Peng F, Tan J, Zhang X, Qian S, Qiao Y, Zhang Y and Liu X. Mg-Fe LDH sealed PEO coating on magnesium for biodegradation control, antibacteria and osteogenesis. *J. Mater. Sci. Technol.* 2022, 105, 57–67.

Zhang F, Liu ZG, Zeng RC, Li SQ, Cui HZ, Song L and Han E-H. Corrosion resistance of Mg – Al-LDH coating on magnesium alloy AZ31. *Surf. Coat. Tech.* 2014, 258, 1152–1158.

Zhang G, Wu L, Tang A, Chen XB, Ma Y, Long Y, Peng P, Ding X, Pan H and Pan F. Growth behavior of MgAl-layered double hydroxide films by conversion of anodic films on magnesium alloy AZ31 and their corrosion protection. *Appl. Surf. Sci.* 2018, 456, 419–429.

Zhang G, Wu L, Tang A, Ma Y, Song GL, Zheng D, Jiang B, Atrens A and Pan F. Active corrosion protection by a smart coating based on a MgAl-layered double hydroxide on a cerium-modified plasma electrolytic oxidation coating on Mg alloy AZ31. *Corros. Sci.* 2018, 139, 370–382.

Zhang G, Wu L, Tang A, Pan H, Ma Y, Zhan Q, Tan Q, Pan F and Atrens A. Effect of micro-arc oxidation coatings formed at different voltages on the in situ growth of layered double hydroxides and their corrosion protection. *J. Electroch. Soc.* 2018, 165, C317.

Zhang JM, Duan X, Hou AR, Li JC, Wang K and Cai H. In situ preparation of Mg-Al-Co layered double hydroxides on microarc oxidation ceramic coating of LA103Z magnesium-lithium alloy for enhanced corrosion resistance. *J. Mater. Eng. Perform.* 2021, 30, 8490–8499.

Zhang JM, Hou AR, Li JC, Lian DD, Zhang MC and Wang ZH. Enhanced corrosion and wear resistance of LA43M magnesium-lithium alloy with magnesium-aluminum layered double hydroxide coating. *J. Mater. Eng. Perform.* 2022, 1–13.

Zhang JM, Li JC, Hou AR, Duan X, Lian DD, Zhang MC and Zhang T. Investigating the microstructures and properties of Mg-Al LDH film-coated Mg-Li alloy: Effect of hydrothermal temperature. *Int. J. Appl. Ceram. Tec.* 2022, 19, 3062–3071.

Zhang M, Cai S, Shen S, Xu G, Li Y, Ling R and Wu X. In-situ defect repairing in hydroxyapatite/phytic acid hybrid coatings on AZ31 magnesium alloy by hydrothermal treatment. *J. Alloys Compd.* 2016, 658, 649–656.

Zhang M, Ma L, Wang L, Sun Y and Liu Y. Insights into the use of metal – organic framework as high-performance anticorrosion coatings. *ACS Appl. Mater. Interfaces.* 2018, 10, 2259–2263.

Zhao M, Zhao Q, Li B, Xue H, Pang H and Chen C. Recent progress in layered double hydroxide based materials for electrochemical capacitors: design, synthesis and performance. *Nanoscale.* 2017, 9, 15206–15225.

Zhou C, Zhang H, Pan X, Li J, Chen B, Gong W, Yang Q, Luo X, Zeng H and Liu Y. Smart waterborne composite coating with passive/active protective performances using nanocontainers based on metal organic frameworks derived layered double hydroxides. *J. Colloid Interf. Sci.* 2022, 619, 132–147.

Zhou M, Pang X, Wei L and Gao K. Insitu grown superhydrophobic Zn – Al layered double hydroxides films on magnesium alloy to improve corrosion properties. *Appl. Surf. Sci.* 2015, 337, 172–177.

Zhu YX, Song GL and Wu PP. Self-repairing functionality and corrosion resistance of in-situ Mg-Al LDH film on Al-alloyed AZ31 surface. *J. Magnes. Alloy.* 2021, https://doi.org/10.1016/j.jma.2021.11.019.

13 Green Chemicals-Assisted Chemical Conversion Coatings

Thiago Ferreira da Conceição, Carlos Henrique Michelin Beraldo and Augusto Versteg

CONTENTS

13.1 INTRODUCTION

The corrosion protection of Mg and its alloys is one of the most studied topics in corrosion science in the last decade. Hundreds of papers are published each year, reporting new coatings to protect Mg and its alloys from corrosion in a variety of environments (Figure 13.1), with a recent emphasis in biomedical media. The reason for this interest is the combination of good properties, such as light weight and biocompatibility, with the undesired low corrosion resistance of these alloys. Among the many different methods developed to protect Mg from corrosion, the conversion coating stands out as a simple and versatile strategy that allows the buildup of thick and protective layers – comprised of oxides, hydroxides, salts, etc. – on the surface of Mg alloys.

For many years, the standard conversion coating applied to automotive and aerospace Mg alloys was based on Cr (VI) compounds. These Cr (VI) conversion coatings are prepared using solutions of chromium trioxide or dichromate, like the Dow series (e.g. Dow 7 and Dow 19) developed by the Dow Company (Esmaily *et al.* 2017). However, international regulations, such as REACH in the European Union, impose serious restrictions on the Cr (VI) usage. Chromium-free alternatives like Gardobond® X4729, Bonderite® M-NT 5700 and Oxsilan® 0611 – based on titanium/zirconium, silicon and similar chemicals, developed by BASF-Chemetall and Henkel – are commercially available ("Chemetall Group – Gardobond® – High-Efficient Conversion Coatings" n.d.; "Chemetall Group – Oxsilan® – The Eco-Friendly Silan-Based Pretreatment Technology" n.d.; "BONDERITE M-NT 5700" n.d.). Nevertheless, these alternatives are not based on renewable and sustainable chemicals and do not perform adequately in certain conditions. Additionally, these conversion coatings are not intended for biomedical applications, and there is a lack of standard conversion coatings for biomedical Mg implants. For these reasons, the scientific community strives to develop green

DOI: 10.1201/9781003319856-16

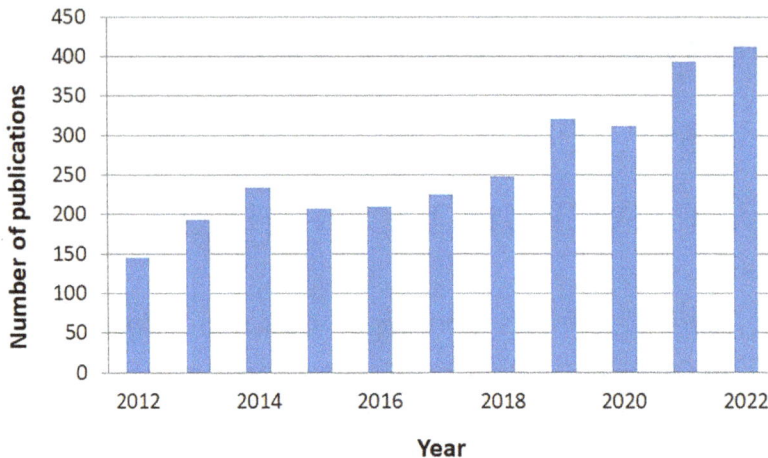

FIGURE 13.1 Number of publications in the Scopus platform found with the keywords "magnesium", "coating" and "corrosion".

and effective conversion coatings for corrosion protection of biomedical and automotive/aerospace magnesium alloys that suits a modern sustainable economy.

In this chapter, selected green environment-friendly conversion coatings are discussed in detail in terms of their preparation procedure, main characteristics and corrosion-protection performance. The selection was based on a strict definition of a "green sustainable process" that let aside conversion coatings based on rare earths (like cerium conversion coatings) and phosphates, which are sometimes considered environment friendly but whose toxicity and sustainability are controversial (Bai *et al.* 2022; Burkholder, Tomasko, and Touchette 2007; Darch *et al.* 2014; Kim *et al.* 2013; Scirè and Palmisano 2020). Additionally, preference was given to methods that use few chemicals, in aqueous solutions, to comply with most of the 12 principles of green chemistry ("12 Principles of Green Chemistry" n.d.).

After screening the literature with these criteria, the following conversion coatings were selected: carbonate coatings, hydrothermal conversion coatings using only water as the conversion medium, steam coating and natural acids conversion coatings. In the next sections, representative publications of each method are discussed, with emphasis on coating morphology, composition and corrosion protection. The chapter concludes with a discussion of the potential of these methods to become standard treatments for Mg components and with perspectives of future developments.

13.2   GREEN CONVERSION COATING METHODS

13.2.1   CARBONATE CONVERSION COATINGS

The first evidence of the protective properties of carbonates on Mg alloys is reported by Al-Abdullat *et al.* in 2001 in a study about the treatment of pure Mg with $NaHCO_3$ and Na_2CO_3 solutions (Al-Abdullat *et al.* 2001), followed by the publications of Lindström *et al.* in 2002 and Lin and Li in 2006, both on the atmospheric corrosion of Mg in the presence of carbon dioxide (Lin and Li 2006; Lindström, Svensson, and Johansson 2002). These studies demonstrate that the presence of magnesium carbonate on the metal surface improves the metal corrosion resistance. Since then, many studies have presented different strategies for the preparation of carbonate coatings on Mg alloys for corrosion protection (Jia *et al.* 2019; Hiromoto *et al.* 2020; Palanisamy, Kulandaivelu, and Nellaiappan 2020). For instance, a $CaCO_3$ coating was developed on a biomedical MgZnCa alloy using a hydrothermal method (Jiang *et al.* 2022), a carbonate conversion coating was prepared on

Mg-Li alloy using low-temperature plasma (Li *et al.* 2022), and an ammonium magnesium carbonate tetrahydrate/calcium carbonate composite coating on AZ91D alloy was reported, prepared by means of a chemical conversion coating (Zhu and Jia 2022), among others. This section focuses on carbonate coatings formed by an environment-friendly chemical conversion process, and the next paragraphs show details of representative studies.

Nam *et al.* produced carbonate conversion films on AZ31 Mg alloys by means of immersion in a 0.5 mol L^{-1} NaOH solution saturated with CO_2 (pH 7.5). Using XRD and XPS, the authors showed that the major product formed on the surface is the magnesium carbonate complex shown in Equations 13.1 and 13.2. Al 2p XPS spectra show evidence of the presence of Al_2O_3, which is probably formed after the dissolution of aluminum during the conversion coating process. In order to establish the optimum immersion time for the treatment, the authors prepared samples treated for 1 minute, 10 minutes, 30 minutes and 60 minutes. The best corrosion protection (evaluated by means of immersion and polarization tests in NaCl 5 wt.%) was obtained with 10 minutes. These samples present less corrosion products in their surface after 180 minutes of immersion in the corrosive solution and a shift in E_{corr} from –1460 mV (untreated sample) to –1239 mV (10 min-treated sample). According to the authors, longer treatment times induced crack formation in the coatings as a result of internal tensions (Nam *et al.* 2018).

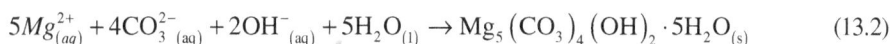

$$5Mg(OH)_{2(s)} + 4HCO_{3(aq)}^- + H_2O_{(l)} \rightarrow Mg_5(CO_3)_4(OH)_2 \cdot 5H_2O_{(s)} + 4OH_{(aq)}^- \quad (13.1)$$

$$5Mg_{(aq)}^{2+} + 4CO_{3(aq)}^{2-} + 2OH_{(aq)}^- + 5H_2O_{(l)} \rightarrow Mg_5(CO_3)_4(OH)_2 \cdot 5H_2O_{(s)} \quad (13.2)$$

The influence of the Mg alloy Al content on the properties of carbonate conversion coatings was investigated by Feliu *et al.* The coatings were prepared on the AZ31 alloy (3 wt.% of Al) and AZ61 alloy (6 wt.% of Al) by immersing the samples in a 9 wt.% $NaHCO_3$ aqueous solution for different times (10 min, 30 min and 60 min). Good results were observed for AZ61 with the treatment times of 10 minutes and 60 minutes, due to the high concentration of aluminum oxide, as shown in the XPS spectra. Interestingly, the treatment time of 30 minutes resulted in the highest carbonate concentration and worst corrosion resistance (tests were performed in NaCl 0.6 mol L^{-1} solution). AZ31 samples showed no significant improvement in corrosion behavior due to the presence of cracks in the coating inner layer (Feliu *et al.* 2013).

A detailed investigation on the formation mechanism of a carbonate conversion coating is reported by Prabhu *et al.* In this study, the authors treated a Mg-4Zn alloy with a 9 wt.% $NaHCO_3$ aqueous solution for different treatment times and recorded the coating morphology using SEM. As shown in Figure 13.2, the coating formation starts with the nucleation of floret-like patches on the surface. Then, these structures spread all over the metal surface within 16 hours of treatment. Interestingly, after this time, the authors observed the formation of a second layer on the top of the first one. The best treatment time is reported to be 24 hours, which leads to the formation of a compact and uniform carbonate layer. Longer immersion times results in cracks in the coating, as shown in Figure 13.2(f). Analysis of linear polarization, in SBF, showed a considerable reduction in i_{corr}, from 72.1 µA cm^{-2} (bare metal) to 1.65 µA cm^{-2} (24-h-treated sample). However, electrochemical impedance spectroscopy (EIS) analyses showed that the protective properties of the coatings significantly decreased after 24 hours of exposure to the corrosive solution. This result was attributed to the permeation of electrolyte through fissures in the coating (Prabhu, Gopalakrishnan, and Ravi 2020).

Another interesting approach is reported by Palanisamy *et al.* whereby the authors used a magnesium carbonate coating as an adhesion promoter for a poly(caprolactone) (PCL) polymer coating. The carbonate layer was prepared on pure Mg substrates by immersion in a sodium hydrogen carbonate solution (90 g L^{-1}), at pH 8.4 and 27°C, for different times (3, 6, 12 and 24 h). The morphology of the formed $MgCO_3$ coating is very similar to the ones reported by Prabhu *et al.* (Prabhu, Gopalakrishnan, and Ravi 2020). The carbonate layer reduced the corrosion rate from 1.27 mm

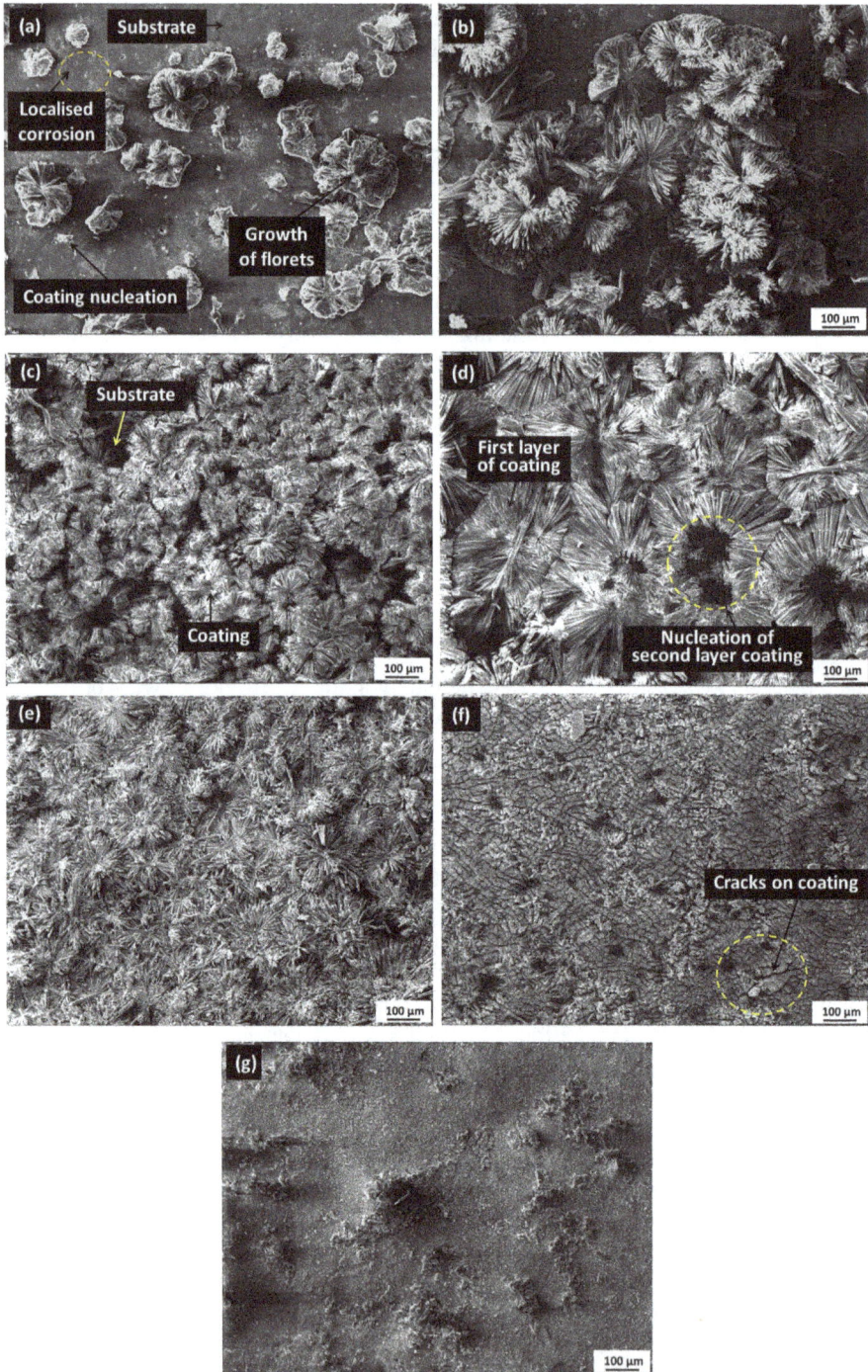

FIGURE 13.2   SEM images of magnesium surface after (a) 1 h, (b) 2 h, (c) 8 h, (d) 16 h, (e) 24 h, (f) 48 h and (g) 96 h of immersion in $NaHCO_3$ 9 wt.%. *Reproduced with permission from (Prabhu, Gopalakrishnan, and Ravi 2020) © 2020 Elsevier B.V.*

year[1] (untreated alloy) to 0.14 mm year[1], and the authors observed good adhesion between it and the polymer coating (Palanisamy, Kulandaivelu, and Nellaiappan 2020).

13.2.2 HYDROTHERMAL CONVERSION COATINGS

The term "hydrothermal" was used for the first time by the British geologist Sir Roderick Muchinson in the 19th century to describe the formation of rocks and minerals by the action of water, at high pressure and temperature, on the earth's crust (Thankachan and Balakrishnan 2018). The term was quickly adopted by mineralogists to describe their synthesis of minerals, such as quartz and feldspar, in laboratory conditions that resemble those found in the earth's crust (that is, high pressure and temperature). Nowadays, "hydrothermal process" can be defined as a heterogeneous process that uses water at high temperature and pressure – in an appropriate container, such as autoclaves – to dissolve and recrystallize materials, usually leading to the coverage of a substrate with a crystalline solid (Byrappa and Yoshimura 2001).

In the field of coatings for corrosion protection of Mg alloys, hydrothermal processes are extensively used for the preparation of coatings comprised of hydroxyapatite, Mg-Al layered double hydroxides (LDH), hydrotalcite, brucite etc. For instance, Wen et al. formed a hydroxyapatite coating on the surface of a medical Mg alloy for corrosion protection (Wen et al. 2021). In a recent review, Liu et al. described the use of hydrothermal processes to form LDH coatings to protect different Mg alloys from corrosion (Liu et al. 2022). Ali et al. used a hydrothermal process to deposit a calcium phosphate layer on the surface of AZ91–3Ca Mg alloy, aiming for an increase in biocompatibility and corrosion resistance (Ali et al. 2019).

Different chemicals have been used for the hydrothermal treatment of magnesium alloys, with emphasis on hydroxides and phosphates. However, from a green chemistry point of view, it is desirable to perform the hydrothermal process using distilled or deionized water alone. Thus, in the next paragraphs, focus is given to conversion coatings formed by hydrothermal treatment of magnesium alloy using only water as the conversion medium.

Gupta et al. performed a hydrothermal treatment on pure Mg disks, using deionized water in a reaction vessel at 160°C, with different treatment times (1 h, 3 h, 4 h, 5 h, 7 h). The authors described the obtention of uniform, compact and adherent $Mg(OH)_2$ coatings. They also observed that the thickness of the coatings increases with the treatment time. EIS analysis showed that the samples treated for 3, 5 and 7 hours reached |Z| in order of $10^4 \ \Omega \ cm^2$, while the untreated sample and the one treated for 1 hour presented |Z| values in the order of $10^3 \ \Omega \ cm^2$. The potentiodynamic polarization essays showed a reduction in i_{corr} of 97% for the 5-hour-treated sample, in comparison to the untreated one. The corrosion tests were performed in phosphate buffered saline solution (PBS) (Gupta, Mensah-Darkwa, and Kumar 2014).

Song et al. applied a hydrothermal treatment on a self-made Mg-2Zn-Mn-Ca-Ce alloy. Samples were placed in a steel autoclave with deionized water at 160°C for different times (3, 4.5 and 6 h). This process resulted in coatings consisting of $Mg(OH)_2$ hexagonal flakes, with a thickness ranging from 14 to 20 μm, depending on the treatment time. According to the authors, by increasing the treatment time, the coatings become denser and practically free of pores. Corrosion tests showed that the treated samples present significantly lower hydrogen evolution rate and i_{corr}, more positive open circuit potential (OCP) and higher impedances (all tests were performed in Hanks' solution). The best corrosion protection was observed for the sample treated for 6 hours (Song et al. 2016).

In another investigation, Song et al. conducted a hydrothermal treatment on a Mg-9Li alloy using deionized water in an autoclave for 2 hours in different temperatures (120, 130, 140 and 150°C) and evaluated the corrosion protection in 0.1 mol L^{-1} NaCl media. SEM/EDS measurements indicated that the coating is composed of $Mg(OH)_2$ and LiOH. The authors also confirmed that the coating thickness increases with the temperature (1.2 μm to 9.2 μm). Furthermore, it was observed a uniform and dense coating until 140 °C. For the 150°C-treated sample, a significant change in the coating morphology was observed, becoming severely damaged, as shown in Figure 13.3. EIS tests

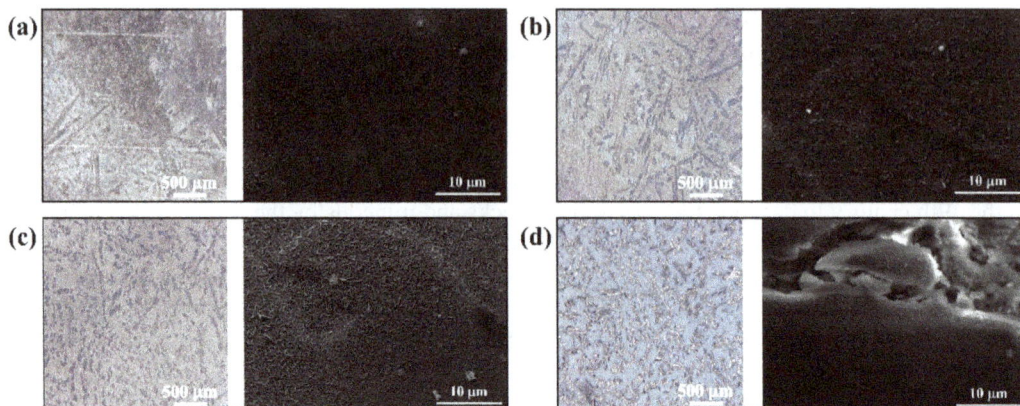

FIGURE 13.3 SEM images of the surfaces of the Mg-9Li samples treated at (a) 120°C, (b) 130°C, (c) 140°C and (d) 150°C. *Reproduced with permission from (Song* et al. *2021) © 2021 by the authors. Licensee MDPI, Basel, Switzerland, under the terms and conditions of the Creative Commons Attribution (CC BY) license.*

revealed an increase in corrosion protection. The |Z| values increased from *ca.* 2000 Ω cm^2, for the untreated sample, to around 4000 Ω cm^2 for the sample treated at 140°C. The authors attributed this improvement to the presence of a dual-layer coating, composed of an external stacking structure and an inner denser structure (Song *et al.* 2021).

In the study of Ren *et al.*, plates of AZ31 Mg alloy were treated by a hydrothermal process using distilled water or an alkaline aqueous solution containing $Al(NO_3)_3$ and sodium hydroxide. The hydrothermal process consisted in placing the samples in the selected aqueous medium, inside a hydrothermal reaction kettle, at 150°C for 18 hours. Both solutions formed a flake-like coating with a thickness of *ca.* 40 µm, composed mainly of $Mg(OH)_2$. For the samples treated with the $Al(NO_3)_3$ containing solution, a significant amount of LDH was also observed. Both coatings improved the surface hardness and the wear resistance of the alloy surface. Additionally, electrochemical polarization tests, performed in 3.5 wt.% NaCl solution show that both coatings reduced i_{corr} and shifted E_{corr} to more positive values. The coatings prepared with the $Al(NO_3)_3$-containing solution present the best results (Ren *et al.* 2022).

13.2.3 STEAM COATINGS

Steam coating is a coating method developed by Ishizaki et al. in 2013, similar to the hydrothermal process in the sense that it uses water at high temperature and pressure inside an autoclave or a similar vessel. However, whereas in the hydrothermal process, the sample is placed inside the liquid, in the steam coating method, the sample is placed above the liquid and gets in contact only with its vapour, at high temperature and pressure. Furthermore, the method forms thick and compact coatings using pure water as the conversion medium. Figure 13.4 shows a general scheme usually adopted for the preparation of steam coatings.

In the first publication of Ishizaki *et al.*, the steam coating process was applied to panels of AZ31 Mg alloy, using deionized water in an autoclave at 150°C, for 6 hours. The treatment resulted in the formation of an 80-µm-thick coating, composed mainly of brucite-type $Mg(OH)_2$ and lesser amounts of Mg-Al LDH containing CO_3^{2-}. Electrochemical polarization tests, performed after 30 minutes of exposure to a 5 wt.% NaCl solution, showed that the coating shifted E_{corr} from −1460 mV (untreated alloy) to −360 mV (treated alloy) and that i_{corr} is 4 orders of magnitude lower for the coated samples (Ishizaki, Chiba, and Suzuki 2013). In another publication in the same year, Ishizaki *et al.* evaluated the influence of temperature (150°C to 180°C) and treatment time (1 h to 6 h) on

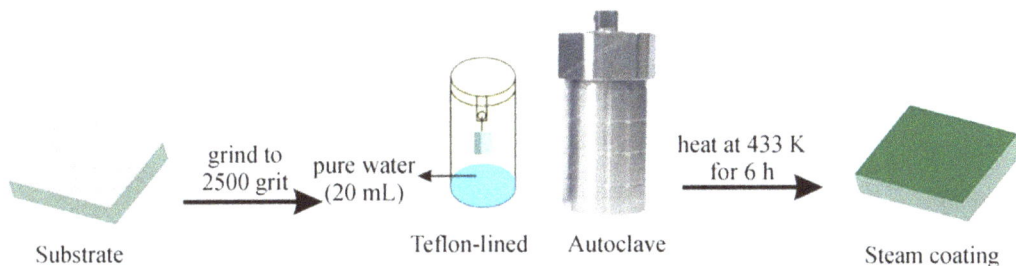

FIGURE 13.4 General procedure for the steam coating process. *Reproduced with permission from (Qiu, Zhang et al. 2020) © 2020, Higher Education Press and Springer-Verlag GmbH Germany, part of Springer Nature.*

the properties of steam coatings on AZ31 Mg alloy and concluded that the treatment at 160°C for 6 hours resulted in the best corrosion protection (Ishizaki, Chiba *et al.* 2013). In the subsequent years, the group of Ishizaki has applied the steam coating method to the Mg alloys AMCa602 (Ishizaki *et al.* 2015), AZCa602 (Nakamura *et al.* 2017) and AZ61 (Nakamura *et al.* 2018).

Recently, the steam coating process has been explored in combination with other methods. For instance, Qiu *et al.* applied the steam coating process (160°C for 6 h) on AZ31 Mg alloy, followed by a treatment with myristic acid. From the SEM/EDS measurements, the authors concluded that the coating was comprised of $Mg(OH)_2$ and Mg-Al LDH, with CO_3^{2-} in the LDH galleries. EIS analysis, performed in 3.5 wt.% NaCl solution, showed that |Z| was 1 order of magnitude higher for the treated sample ($10^4 \, \Omega \, cm^2$) in comparison to the untreated one ($10^3 \, \Omega \, cm^2$) (Qiu, Zhang *et al.* 2020). In another study, Qiu *et al.* used the steam coating method on a AZ31 Mg alloy, followed by a treatment with bis-[triethoxysilylpropyl]tetrasulfide silane (BTESPT) and $Ce(NO_3)_3$. The steam coating process resulted in a coating of $Mg(OH)_2$ and LDH (CO_3^{2-} in the LDH galleries) with a thickness of 30 μm, that reduced i_{corr} from $1.3 \times 10^{-5} \, A \, cm^{-2}$ to $1.9 \times 10^{-7} \, A \, cm^{-2}$. The composite coating (obtained after the BTESPT and $Ce(NO_3)_3$ treatment) reduced the i_{corr} to $4,0 \times 10^{-8} \, A \, cm^{-2}$. Impedance tests showed that the steam coating improved the impedance from $10^2 \, \Omega \, cm^2$ (untreated alloy) to $10^4 \, \Omega \, cm^2$ (steam coating), whereas the composite coating further improved the impedances to $10^5 \, \Omega \, cm^2$ (Qiu, Zeng *et al.* 2020).

Wang *et al.* have applied the steam coating method (160°C, 6 h) on AZ80 sheets previously treated with citric acid (CA) for different times (10 s, 20 s and 30 s). According to the authors, the treatment resulted in coatings mainly constituted of $Mg(OH)_2$ and Mg-Al LDH (CO_3^{2-} in the LDH galleries) with thicknesses ranging from 30 μm to 40 μm. Electrochemical tests showed that the sample previously treated with CA for 30 seconds had the best corrosion resistance, with i_{corr} three orders of magnitude lower than that of the untreated alloy and impedances in the order of $10^5 \, \Omega \, cm^2$. According to the authors, by treating the alloy with CA for 30 seconds, it is possible to remove the oxide/hydroxide layer that covers the alloy secondary phases (Al_8Mn_5 and $Mg_{17}Al_{12}$), leaving them more exposed to the water vapour. This higher exposure enhances the conversion coating deposition rate and results in a coating with higher thickness (Wang, Sun *et al.* 2022).

To clarify this process, the authors presented a detailed investigation on the coating formation mechanism by following its properties after different treatment times (0.5 h, 1.5 h, 3.0 h). The authors observed that after the initial removal of the natural oxide layer, $Mg(OH)_2$ starts to deposit on the secondary phases, which act as cathodes. Then, Al^{3+} ions (from the secondary phases) gradually replaced some Mg^{2+} ions in the deposited $Mg(OH)_2$ film, turning it into a LDH layer. The presence of CO_2 in the atmosphere renders the CO_3^{2-} ions that, together with the OH^- ions, neutralize the charge in the LDH film. Figure 13.5 presents a schematic representation of this mechanism.

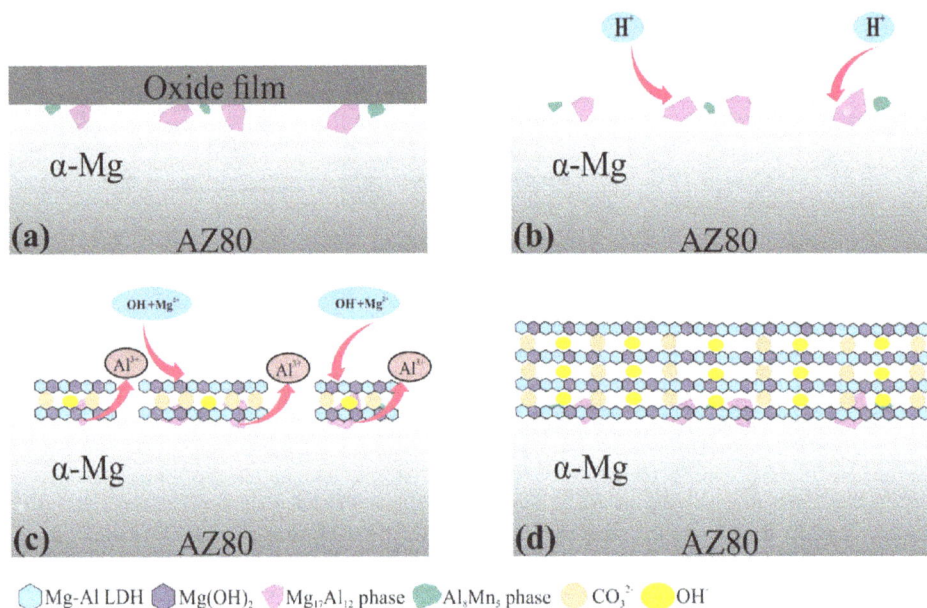

Mg-Al LDH ⬡ Mg(OH)₂ ⬣ Mg₁₇Al₁₂ phase Al₈Mn₅ phase CO₃²⁻ OH

FIGURE 13.5 Schematic mechanism of the coating formation and LDH charge neutralization. *Reproduced with permission from (Wang, Sun et al. 2022) © 2022 Chongqing University. Publishing services provided by Elsevier B.V. on behalf of KeAi Communications Co. Ltd.*

13.2.4 NATURAL ACIDS CONVERSION COATINGS

Some of the most explored methods used for the preparation of conversion coatings on Mg alloys are based on reactions with acids. The treatment of Mg alloys with HF and H_3PO_4, for example, can be considered now as classical ones (Hafeez *et al.* 2020; Wang, Zhang *et al.* 2022). The easiness of forming protective coatings on Mg alloys by simply reacting them with acids has stimulated many investigations on natural acids as green and effective chemicals for corrosion protection.

Recently, great attention has been given to natural acids containing phenol groups, such as gallic acid (GA), tannic acid (TA) and vanilic acid (VA). GA is usually discussed together with TA, since both chemicals present polyphenol units that chelate Mg^{2+} ions, and that GA can be obtained from the hydrolysis of TA. However, GA differs from TA in the chemical structure, as it presents a carboxylic acid group besides the polyphenol units. Despite the apparent interest in this acid as a suitable chemical to prepare conversion coatings for Mg alloys, the number of studies in the literature about GA conversion coatings is very low. In fact, there are two studies from the group of Chen that describe the properties of GA coatings, in combination with hexamethylenediamine (HD) (a system described as GAHD coating) on MZM magnesium alloys (Chen *et al.* 2015; Chen *et al.* 2019). This coating was developed to improve the corrosion resistance of the alloy and to enhance its surface functionalization with biomedical molecules. The authors studied the influence of the GA/HD ratio over the coating's corrosion protection and concluded that by increasing the GA amount, the anti-corrosion behavior improves due to a higher number of chelates that enhances the coating barrier properties.

Apart from these studies, one can find the publication of Lin *et al.*, in which the authors prepared a sandwich-like coating consisting of a layer of GA between layers of poly(D,L-lactide-co-glycolide) (PLGA) on a ZK60 Mg alloy. It is shown that the presence of GA on the PLGA coatings reduced i_{corr} from 1.79×10^{-6} A cm⁻² (PLGA) to 0.01×10^{-6} A cm⁻² (sandwich coating). This coating is not a conversion coating but is another example of the unexplored potential of GA in protecting

Mg from corrosion (Lin, Lee, and Yeh 2020). Nevertheless, it can be concluded that the use of GA to protect Mg alloys from corrosion is still in its initial stage of development, and more investigations are required to evaluate its true potential to form protective conversion coatings.

On the other hand, TA has been much more explored for the preparation of conversion coatings on Mg alloys than the GA. The first reports on the use of a TA-containing solution for the preparation of conversion coatings on Mg alloys were published in 2008 (Chen *et al.* 2008) and 2009 (Chen *et al.* 2009) by Chen *et al.*, whereby the authors treated an AZ91D Mg alloy with a solution containing TA and many other compounds, such as NH_4VO_3 and K_2ZrF_6. According to the authors, the coating consisted of Al_2O_3, MgF_2 and a penta-hydroxy benzamide Mg complex, formed by the hydrolysis of TA.

Recently, Asgari *et al.* prepared a metal-phenolic network (MPN) coating composed of Mg^{2+} ions chelated with TA on a AZ31 Mg alloy pre-coated with $Mg(OH)_2$. The process consisted in immersing the alloy in a 1 mg mL^{-1} TA solution containing different amounts of $MgCl_2$ and NaOH (pH = 10), in 3 cycles of 30 minutes each. The results demonstrated that the corrosion resistance of the formed coatings improved with the Mg^{2+} concentration. In the best conditions, electrochemical analysis showed an improvement of 180% in the corrosion resistance of the coated alloys, in comparison to the uncoated ones (Asgari *et al.* 2019).

Zhang *et al.* prepared a tea stain–like conversion coating of TA on AZ31 Mg alloys, which the authors called "bionic coating". The coatings were prepared by immersing the alloy in solutions containing TA, with or without $MgSO_4$, for 15 minutes at 80°C, in a total of 5 cycles. The treatment resulted in coatings with a thickness around 200 μm that considerably improved the alloy corrosion resistance, as shown by polarization tests, EIS and hydrogen evolution measurements. The superior performance was observed for the coating prepared with $MgSO_4$ due to a higher amount of formed chelates (Zhang, Yao, Li, Li, Yang *et al.* 2019).

TA has also been tested in combination with other coating methods, such as HF treatment (Wang *et al.* 2019) and hydroxyapatite (Zhu *et al.* 2017). The interesting results obtained has stimulated research on similar compounds such as epigallocatechin gallate, a polyphenol present in green tea that is under investigation for the preparation of conversion coatings on Mg alloys (Zhang, Yao, Li, Li, Wang *et al.* 2019; Zhang, Yao *et al.* 2021).

VA is another carboxylic acid that presents a phenol group, but contrary to GA and TA, it has only one phenol group on its chemical structure; thus, it cannot form chelates. Nevertheless, the literature shows good potential for this acid to form protective conversion coatings. Abatti *et al.* prepared a magnesium vanillate coating by dipping a pre-treated AZ31 alloy (pre-coated with $Mg(OH)_2$) in a 1 mmol L^{-1} VA aqueous solution for 24 hours. Electrochemical results showed a decrease in 98 folds in corrosion current density, confirming the protectiveness of the coating. Additionally, the authors observed that the vanillate conversion coatings can be used as a primer to improve the adhesion of organic coatings (Abatti *et al.* 2018). Similar results are presented by Li *et al.* in their study on VA conversion coating on AZ31 Mg alloy (Li *et al.* 2020).

Fatty acids – another group of naturally occurring carboxylic acids – have been extensively investigated in recent years as suitable chemicals for the preparation of superhydrophobic conversion coatings. Due to their amphiphilic behavior, they have been proposed as a way to enhance hydrophobicity of metal alloy surfaces. The most explored compound in this application is stearic acid (SA) (Cao *et al.* 2018; Cui *et al.* 2015; Jin *et al.* 2019). In these studies, generally, a pre-layer of $Mg(OH)_2$ is formed on the Mg alloy surface, and it is then modified with the fatty acid by the reaction shown in Equation 13.3:

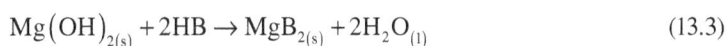

$$Mg(OH)_{2(s)} + 2HB \rightarrow MgB_{2(s)} + 2H_2O_{(l)} \qquad (13.3)$$

in which HB represents a generic fatty acid.

As an example of the fatty acid approach, Jin *et al.* modified a layer of hydrothermally formed $Mg(OH)_2$ on the surface of an AZ31B alloy with SA by soaking the samples in a SA solution in n-hexane. The authors observed the influence of SA concentration and soaking time on the hydro-phobicity and corrosion resistance of the formed coating. In the best conditions (7 mmol L^{-1} of SA for 4 h), the surfaces showed a very hydrophobic behavior, with water contact angle of 159°. The electrochemical tests demonstrated the improvement of corrosion resistance, with the SA treated samples showing a 2 orders of magnitude reduction in i_{corr} (1.62×10^{-5} A cm^{-2} for untreated samples and 1.73×10^{-7} A cm^{-2} for the coated samples) (Jin *et al.* 2019). In more recent papers, SA have been used to modify the surface of other coatings, such as chitosan-based (Wang, Xiao *et al.* 2022) and fluoride-based (Zhang, Zhang *et al.* 2021) coatings to improve the surfaces' hydrophobicity.

Finally, another natural acid that has caught the attention of the scientific community in the context of greener chemistry is the phytic acid (PA). PA is a natural inositol derivative comprised of six monophosphate ester groups linked to an inositol unit (Higuchi 2014). The first publication about PA conversion coatings on Mg alloys is from Liu *et al.*, in which the authors report the treatment of a AZ91D Mg alloy with PA solutions (concentration ranging from 0.3 to 2.5%) at different temperatures, pH and treatment time. The authors reported that the treatment resulted in cracked coatings that, however, provided corrosion protection similar to that observed for chromate conversion coatings (Liu, Guo, and Huang 2006).

Significant contribution to the development of PA coatings was given by Cui *et al.* (Cui, Li, Li, Wang *et al.* 2008; Cui, Li, Li, Jin et al. 2008, 2010, 2012), which published many studies of PA conversion coatings, mainly on AZ91D alloy. In the subsequent years, many different groups have developed methods for the preparation of PA conversion coatings on different Mg alloys, as shown in details by a recent review in the literature (Xu *et al.* 2022). In general, the protectiveness of PA conversion coatings is associated with the amount of chelates in the coating, in a similar way as discussed for TA conversion coatings. However, PA conversion coatings are usually cracked and thin and do not offer adequate protection when used alone. For that reason, many research groups have investigated the influence of pre-treatments and post-treatments on PA conversion coatings to improve their protectiveness *(Zhang et al. 2014*; Liu, Guo, and Huang 2012).

13.3 COMPARATIVE PERFORMANCE

Among the processes described for the preparation of green conversion coatings, the one that uses gaseous CO_2 dissolved in water (by Equations 13.4 and 13.5) to form carbonate conversion coatings has the greatest environmental appeal (Feliu *et al.* 2013; Nam *et al.* 2018):

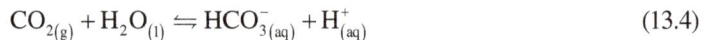

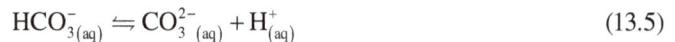

$$CO_{2(g)} + H_2O_{(l)} \rightleftharpoons HCO_3^-{}_{(aq)} + H^+{}_{(aq)} \tag{13.4}$$

$$HCO_3^-{}_{(aq)} \rightleftharpoons CO_3^{2-}{}_{(aq)} + H^+{}_{(aq)} \tag{13.5}$$

CO_2 is a toxic gas emitted by the exhausting system of vehicles and is one of the main factors responsible for the air pollution (US EPA 2015). The development of a method that effectively converts toxic CO_2 gas into a harmless carbonate coating is of great relevance. Nevertheless, despite this environmental appeal, this strategy has been investigated by few researchers. Furthermore, the results presented in the preceding sections indicate that carbonate coatings have a limited efficiency in protecting Mg alloys from corrosion. Similar observation is made by Zhu and Jia on a study whereby they propose the preparation of a composite carbonate coating to overcome the limitations of neat carbonate conversion coating (Zhu and Jia 2022).

The hydrothermal and steam coating methods are both interesting from the environmental point of view and show similarities in the experimental setup. However, significant differences are observed in the properties of coatings prepared with both methods. The hydrothermal treatment of Mg alloys has received a great deal of attention in recent years, especially in the context of biomedical alloys. The formed coatings usually present a rough surface, thickness ranging from 1 to 40 μm, and moderate impedance (10^3 to 10^4 Ω cm^2). On the other hand, the literature indicates that the steam coating method forms thicker (up to 80 μm thick) and denser coatings with slightly superior corrosion protection (10^4 to 10^5 Ω cm^2). Interestingly, this method has received less attention in comparison to the hydrothermal treatment.

In the context of conversion coatings based on natural acids, there is a great interest in TA, GA and PA, as can be seen in recent reviews and book chapters (Xu *et al.* 2022; Luo *et al.* 2022; Conceição, Abatti, and Beraldo 2022). These three acids have in common the ability to form chelates with Mg^{2+} ions, a characteristic that is beneficial to the coating's protective properties. However, whereas there are a significant number of studies about TA and PA conversion coatings on Mg alloys, the same is not true for GA. Many more studies on the influence of different parameters on the properties of GA conversion coatings are required to evaluate the potential of this acid to protect Mg from corrosion. For TA and PA, it is observed that the formed coatings are usually thin and offer limited corrosion protection. The optimal conditions for the preparation of these coatings need to be determined in future studies.

In the case of fatty acids, there are plenty of possibilities for the development of hydrophobic conversion coatings, since most of the investigations deal with stearic acid and myristic acid, despite the existence of many different fatty acids (Conceição, Abatti, and Beraldo 2022). In particular, fatty acids show a great potential as a final coating step to convert a previously prepared layer (like Mg(OH)$_2$) into a superhydrophobic one.

13.4 CONCLUSIONS AND OUTLOOK

The preceding sections demonstrate that there is a great variety of green conversion coatings for corrosion protection of Mg alloys. Additionally, the discussed methods demonstrate the possibility of improving the corrosion performance of Mg alloys and, at the same time, complying with many of the 12 principles of green chemistry, such as atom economy, safer solvents and auxiliaries, and use of renewable feedstock. Nevertheless, two important aspects need special attention in order to turn some of these methods suitable for commercial and industrial applications.

In the first place, more comparisons between the proposed green alternative and commercially available products are required, as well as more standardized evaluation of the corrosion performance. The majority of the studies in the literature rely on inner comparisons – comparisons between samples in the same study, prepared with different conditions – without referring to commercially available methods. Furthermore, standard corrosion tests, such as those in a salt-spray chamber, could be more applied to green conversion coatings to allow a straightforward analysis of their protective performance.

Another important aspect to point out is that in many cases, the experimental conditions for coating preparation are randomly chosen and do not follow a systematic order to allow a conclusion about the optimal conditions. In these cases, the true potential of a coating may be hidden by the unfavorable conditions used in the preparation process. Thus, systematic studies on the influence of parameters such as treatment time, temperature, concentration and pH, on the coating's properties are required for a comprehensive understanding of the potential of these coatings.

In conclusion, the great diversity of green methods used for the preparation of corrosion protective coatings on Mg alloys, presented in this chapter, shows the relevance of this topic. The limitations of these methods, however, demonstrate that much more research is required to effectively protect Mg from corrosion with sustainable and green chemicals. Thus, it is expected that this research field will keep growing in the coming years, maintaining its position on the top of the most investigated topics in corrosion science.

REFERENCES

"12 Principles of Green Chemistry." N.d. American Chemical Society. Accessed November 16, 2022. www.acs.org/content/acs/en/greenchemistry/principles/12-principles-of-green-chemistry.html.

Abatti, G. P., A. T. N. Pires, A. Spinelli, N. Scharnagl, and T. F. da Conceição. 2018. "Conversion Coating on Magnesium Alloy Sheet (AZ31) by Vanillic Acid Treatment: Preparation, Characterization and Corrosion Behavior." *Journal of Alloys and Compounds* 738 (March): 224–32. https://doi.org/10.1016/j.jallcom.2017.12.115.

Al-Abdullat, Y., S. Tsutsumi, N. Nakajima, M. Ohta, H. Kuwahara, and K. Ikeuchi. 2001. "Surface Modification of Magnesium by NaHCO3 and Corrosion Behavior in Hank's Solution for New Biomaterial Applications." *Materials Transactions* 42 (8): 1777–80. https://doi.org/10.2320/matertrans.42.1777.

Ali, A., F. Iqbal, A. Ahmad, F. Ikram, A. Nawaz, A. A. Chaudhry, S. A. Siddiqi, and I. Rehman. 2019. "Hydrothermal Deposition of High Strength Calcium Phosphate Coatings on Magnesium Alloy for Biomedical Applications." *Surface and Coatings Technology* 357 (January): 716–27. https://doi.org/10.1016/j.surfcoat.2018.09.016.

Asgari, M., Y. Yang, S. Yang, Z. Yu, P. K. D. V. Yarlagadda, Y. Xiao, and Z. Li. 2019. "Mg – Phenolic Network Strategy for Enhancing Corrosion Resistance and Osteocompatibility of Degradable Magnesium Alloys." *ACS Omega* 4 (26): 21931–44. https://doi.org/10.1021/acsomega.9b02976.

Bai, J., X. Xu, Y. Duan, G. Zhang, Z. Wang, L. Wang, and C. Zheng. 2022. "Evaluation of Resource and Environmental Carrying Capacity in Rare Earth Mining Areas in China." *Scientific Reports* 12 (1): 6105. https://doi.org/10.1038/s41598-022-10105-2.

"BONDERITE M-NT 5700." N.d. Accessed November 16, 2022. www.henkel-adhesives.com/ca/en/product/conversion-coatings/bonderite_m-nt_5700.html.

Burkholder, J. M., D. A. Tomasko, and B. W. Touchette. 2007. "Seagrasses and Eutrophication." *Journal of Experimental Marine Biology and Ecology, the Biology and Ecology of Seagrasses* 350 (1): 46–72. https://doi.org/10.1016/j.jembe.2007.06.024.

Byrappa, K., and M. Yoshimura. 2001. "2 – History of Hydrothermal Technology." In *Handbook of Hydrothermal Technology*, edited by K. Byrappa and Masahiro Yoshimura, 53–81. Norwich, NY: William Andrew Publishing. https://doi.org/10.1016/B978-081551445-9.50003-9.

Cao, Z., F. Lu, P. Qiu, F. Yang, G. Liu, S. Wang, and H. Zhong. 2018. "Formation of a Hydrophobic and Corrosion Resistant Coating on Manganese Surface via Stearic Acid and Oleic Acid Diethanolamide." *Colloids and Surfaces A: Physicochemical and Engineering Aspects* 555 (October): 372–80. https://doi.org/10.1016/j.colsurfa.2018.07.020.

"Chemetall Group – Gardobond® – High-Efficient Conversion Coatings." N.d. Accessed November 16, 2022. www.chemetall.com/Products/Trademarks/Gardobond/index.jsp.

"Chemetall Group – Oxsilan® – The Eco-Friendly Silan-Based Pretreatment Technology." N.d. Accessed November 16, 2022. www.chemetall.com/oxsilan/.

Chen, S., J. Zhang, Y. Chen, S. Zhao, M. Chen, X. Li, M. F. Maitz, J. Wang, and N. Huang. 2015. "Application of Phenol/Amine Copolymerized Film Modified Magnesium Alloys: Anticorrosion and Surface Biofunctionalization." *ACS Applied Materials & Interfaces* 7 (44): 24510–22. https://doi.org/10.1021/acsami.5b05851.

Chen, S., S. Zhao, M. Chen, X. Zhang, J. Zhang, X. Li, H. Zhang, X. Shen, J. Wang, and N. Huang. 2019. "The Anticorrosion Mechanism of Phenolic Conversion Coating Applied on Magnesium Implants." *Applied Surface Science* 463 (January): 953–67. https://doi.org/10.1016/j.apsusc.2018.08.261.

Chen, X., G. Li, J. Lian, and Q. Jiang. 2008. "An Organic Chromium-Free Conversion Coating on AZ91D Magnesium Alloy." *Applied Surface Science* 255 (5, Part 1): 2322–8. https://doi.org/10.1016/j.apsusc.2008.07.092.

Chen, X., G. Li, J. Lian, and Q. Jiang. 2009. "Study of the Formation and Growth of Tannic Acid Based Conversion Coating on AZ91D Magnesium Alloy." *Surface and Coatings Technology* 204 (5): 736–47. https://doi.org/10.1016/j.surfcoat.2009.09.022.

Conceição, T. F. da, G. P. Abatti, and C. H. M. Beraldo. 2022. "Hydroxy BenzeneHydroxy Benzene/Phenolic AcidsPhenolic Acids and Carboxylic/Fatty AcidFatty Acids Conversion Coatings." In *Conversion Coatings for Magnesium and Its Alloys*, edited by Viswanathan S. Saji, T. S. N. Sankara Narayanan, and Xiaobo Chen, 279–96. Cham: Springer International Publishing. https://doi.org/10.1007/978-3-030-89976-9_13.

Cui, X., G. Jin, Q. Li, Y. Yang, Y. Li, and F. Wang. 2010. "Electroless Ni – P Plating with a Phytic Acid Pretreatment on AZ91D Magnesium Alloy." *Materials Chemistry and Physics* 121 (1): 308–13. https://doi.org/10.1016/j.matchemphys.2010.01.042.

Cui, X., G. Jin, E. Liu, M. Ding, Q. Li, and F. Wang. 2012. "Influence of Substrate Composition on the Formation of Phytic Acid Conversion Coatings." *Materials and Corrosion* 63 (3): 215–22. https://doi.org/10.1002/maco.201005686.

Cui, X., Q. Li, Y. Li, F. Wang, G. Jin, and M. Ding. 2008. "Microstructure and Corrosion Resistance of Phytic Acid Conversion Coatings for Magnesium Alloy." *Applied Surface Science* 255 (5, Part 1): 2098–103. https://doi.org/10.1016/j.apsusc.2008.06.199.

Cui, X., Y. Li, Q. Li, G. Jin, M. Ding, and F. Wang. 2008. "Influence of Phytic Acid Concentration on Performance of Phytic Acid Conversion Coatings on the AZ91D Magnesium Alloy." *Materials Chemistry and Physics* 111 (2): 503–7. https://doi.org/10.1016/j.matchemphys.2008.05.009.

Cui, X., X. Lin, C. Liu, R. Yang, X. Zheng, and M. Gong. 2015. "Fabrication and Corrosion Resistance of a Hydrophobic Micro-Arc Oxidation Coating on AZ31 Mg Alloy." *Corrosion Science* 90 (January): 402–12. https://doi.org/10.1016/j.corsci.2014.10.041.

Darch, T., M. S. A. Blackwell, J. M. B. Hawkins, P. M. Haygarth, and D. Chadwick. 2014. "A Meta-Analysis of Organic and Inorganic Phosphorus in Organic Fertilizers, Soils, and Water: Implications for Water Quality." *Critical Reviews in Environmental Science and Technology* 44 (19): 2172–202. https://doi.org/10.1080/10643389.2013.790752.

Esmaily, M., J. E. Svensson, S. Fajardo, N. Birbilis, G. S. Frankel, S. Virtanen, R. Arrabal, S. Thomas, and L. G. Johansson. 2017. "Fundamentals and Advances in Magnesium Alloy Corrosion." *Progress in Materials Science* 89 (August): 92–193. https://doi.org/10.1016/j.pmatsci.2017.04.011.

Feliu, S., A. Samaniego, A. A. El-Hadad, and I. Llorente. 2013. "The Effect of NaHCO3 Treatment Time on the Corrosion Resistance of Commercial Magnesium Alloys AZ31 and AZ61 in 0.6M NaCl Solution." *Corrosion Science* 67 (February): 204–16. https://doi.org/10.1016/j.corsci.2012.10.020.

Gupta, R. K., K. Mensah-Darkwa, and D. Kumar. 2014. "Corrosion Protective Conversion Coatings on Magnesium Disks Using a Hydrothermal Technique." *Journal of Materials Science and Technology* 30 (1): 47–53. https://doi.org/10.1016/j.jmst.2013.07.012.

Hafeez, M. A., A. Farooq, A. Zang, A. Saleem, and K. M. Deen. 2020. "Phosphate Chemical Conversion Coatings for Magnesium Alloys: A Review." *Journal of Coatings Technology and Research* 17 (4): 827–49. https://doi.org/10.1007/s11998-020-00335-2.

Higuchi, M. 2014. "Chapter 15 – Antioxidant Properties of Wheat Bran against Oxidative Stress." In *Wheat and Rice in Disease Prevention and Health*, edited by Ronald Ross Watson, Victor R. Preedy, and Sherma Zibadi, 181–99. San Diego: Academic Press. https://doi.org/10.1016/B978-0-12-401716-0.00015-5.

Hiromoto, S., S. Itoh, N. Noda, T. Yamazaki, H. Katayama, and T. Akashi. 2020. "Osteoclast and Osteoblast Responsive Carbonate Apatite Coatings for Biodegradable Magnesium Alloys." *Science and Technology of Advanced Materials* 21 (1): 346–58. https://doi.org/10.1080/14686996.2020.1761237.

Ishizaki, T., S. Chiba, and H. Suzuki. 2013. "In Situ Formation of Anticorrosive Mg – Al Layered Double Hydroxide-Containing Magnesium Hydroxide Film on Magnesium Alloy by Steam Coating." *ECS Electrochemistry Letters* 2 (5): C15. https://doi.org/10.1149/2.006305eel.

Ishizaki, T., S. Chiba, K. Watanabe, and H. Suzuki. 2013. "Corrosion Resistance of Mg – Al Layered Double Hydroxide Container-Containing Magnesium Hydroxide Films Formed Directly on Magnesium Alloy by Chemical-Free Steam Coating." *Journal of Materials Chemistry A* 1 (31): 8968–77. https://doi.org/10.1039/C3TA11015J.

Ishizaki, T., N. Kamiyama, K. Watanabe, and A. Serizawa. 2015. "Corrosion Resistance of Mg(OH)2/Mg – Al Layered Double Hydroxide Composite Film Formed Directly on Combustion-Resistant Magnesium Alloy AMCa602 by Steam Coating." *Corrosion Science* 92 (March): 76–84. https://doi.org/10.1016/j.corsci.2014.11.031.

Jia, S., Y. Guo, W. Zai, Y. Su, S. Yuan, X. Yu, Y. Xu, and G. Li. 2019. "Preparation and Characterization of a Composite Coating Composed of Polycaprolactone (PCL) and Amorphous Calcium Carbonate (ACC) Particles for Enhancing Corrosion Resistance of Magnesium Implants." *Progress in Organic Coatings* 136 (November): 105225. https://doi.org/10.1016/j.porgcoat.2019.105225.

Jiang, P., R. Hou, S. Zhu, and S. Guan. 2022. "A Robust Calcium Carbonate (CaCO3) Coating on Biomedical MgZnCa Alloy for Promising Corrosion Protection." *Corrosion Science* 198 (April): 110124. https://doi.org/10.1016/j.corsci.2022.110124.

Jin, Q., G. Tian, J. Li, Y. Zhao, and H. Yan. 2019. "The Study on Corrosion Resistance of Superhydrophobic Magnesium Hydroxide Coating on AZ31B Magnesium Alloy." *Colloids and Surfaces A: Physicochemical and Engineering Aspects* 577 (September): 8–16. https://doi.org/10.1016/j.colsurfa.2019.05.060.

Kim, E., S. Yoo, H. Ro, H. Han, Y. Baek, I. Eom, H. Kim, P. Kim, and K. Choi. 2013. "Aquatic Toxicity Assessment of Phosphate Compounds." *Environmental Health and Toxicology* 28 (February): e2013002. https://doi.org/10.5620/eht.2013.28.e2013002.

Li, S., L. Yi, T. Liu, H. Deng, B. Ji, K. Zhang, and L. Zhou. 2020. "Formation of a Protective Layer against Corrosion on Mg Alloy via Alkali Pretreatment Followed by Vanillic Acid Treatment." *Materials and Corrosion* 71 (8): 1330–8. https://doi.org/10.1002/maco.201911488.

Li, Y., Z. Kang, X. Zhang, J. Pan, Y. Ren, and G. Zhou. 2022. "Fabricating an Anti-Corrosion Carbonate Coating on MgLi Alloy by Low-Temperature Plasma." *Surface and Coatings Technology* 439 (June): 128418. https://doi.org/10.1016/j.surfcoat.2022.128418.

Lin, C., and X. Li. 2006. "Role of CO_2 in the Initial Stage of Atmospheric Corrosion of AZ91 Magnesium Alloy in the Presence of NaCl." *Rare Metals* 25 (2): 190–6. https://doi.org/10.1016/S1001-0521(06)60038-7.

Lin, L., H. Lee, and M. Yeh. 2020. "Characterization of a Sandwich PLGA-Gallic Acid-PLGA Coating on Mg Alloy ZK60 for Bioresorbable Coronary Artery Stents." *Materials* 13 (23): 5538. https://doi.org/10.3390/ma13235538.

Lindström, R., J.-E. Svensson, and L.-G. Johansson. 2002. "The Influence of Carbon Dioxide on the Atmospheric Corrosion of Some Magnesium Alloys in the Presence of NaCl." *Journal of The Electrochemical Society* 149 (4): B103. https://doi.org/10.1149/1.1452115.

Liu, L., Q. Deng, P. White, S. Dong, I. S. Cole, J. Dong, and X. Chen. 2022. "Hydrothermally Prepared Layered Double Hydroxide Coatings for Corrosion Protection of Mg Alloys – A Critical Review." *Corrosion Communications* (September). https://doi.org/10.1016/j.corcom.2022.07.001.

Liu, J., Y. Guo, and W. Huang. 2006. "Study on the Corrosion Resistance of Phytic Acid Conversion Coating for Magnesium Alloys." *Surface and Coatings Technology* 201 (3): 1536–41. https://doi.org/10.1016/j.surfcoat.2006.02.020.

Liu, J., Y. Guo, and W. Huang. 2012. "Phytic Acid Conversion Coatings on Magnesium Surface Treatment with Cerium Chloride Solution." *Protection of Metals and Physical Chemistry of Surfaces* 48 (2): 233–7. https://doi.org/10.1134/S2070205112020116.

Luo, R., B. Zhang, H. Zhang, L. Yang, and Y. Wang. 2022. "Tannic and Gallic AcidGallic Acids (GA) Conversion Coatings." In *Conversion Coatings for Magnesium and Its Alloys*, edited by Viswanathan S. Saji, T. S. N. Sankara Narayanan, and Xiaobo Chen, 261–77. Cham: Springer International Publishing. https://doi.org/10.1007/978-3-030-89976-9_12.

Nakamura, K., Y. Shimada, T. Miyashita, A. Serizawa, and T. Ishizaki. 2018. "Effect of Vapor Pressure During the Steam Coating Treatment on Structure and Corrosion Resistance of the Mg(OH)2/Mg-Al LDH Composite Film Formed on Mg Alloy AZ61." *Materials* 11 (9): 1659. https://doi.org/10.3390/ma11091659.

Nakamura, K., M. Tsunakawa, Y. Shimada, A. Serizawa, and T. Ishizaki. 2017. "Formation Mechanism of Mg-Al Layered Double Hydroxide-Containing Magnesium Hydroxide Films Prepared on Ca-Added Flame-Resistant Magnesium Alloy by Steam Coating." *Surface and Coatings Technology* 328 (November): 436–43. https://doi.org/10.1016/j.surfcoat.2017.08.060.

Nam, D., D. Lim, S. Kim, D. Seo, S. E. Shim, and S. Baeck. 2018. "The Fabrication of a Conversion Film on AZ31 Containing Carbonate Product and Evaluation of Its Corrosion Resistance." *Journal of Alloys and Compounds* 737 (March): 597–602. https://doi.org/10.1016/j.jallcom.2017.12.061.

Palanisamy, M. S., R. Kulandaivelu, and Narayanan T. S. Nellaiappan. 2020. "Improving the Corrosion Resistance and Bioactivity of Magnesium by a Carbonate Conversion-Polycaprolactone Duplex Coating Approach." *New Journal of Chemistry* 44 (12): 4772–85. https://doi.org/10.1039/C9NJ06030H.

Prabhu, D. B., P. Gopalakrishnan, and K. R. Ravi. 2020. "Morphological Studies on the Development of Chemical Conversion Coating on Surface of Mg–4Zn Alloy and Its Corrosion and Bio Mineralisation Behaviour in Simulated Body Fluid." *Journal of Alloys and Compounds* 812 (January): 152146. https://doi.org/10.1016/j.jallcom.2019.152146.

Qiu, Z. M., R. C. Zeng, F. Zhang, L. Song, and S. Q. Li. 2020. "Corrosion Resistance of Mg–Al LDH/Mg(OH)2/silane–Ce Hybrid Coating on Magnesium Alloy AZ31." *Transactions of Nonferrous Metals Society of China* 30 (11): 2967–79. https://doi.org/10.1016/S1003-6326(20)65435-8.

Qiu, Z. M., F. Zhang, J. T. Chu, Y. C. Li, and L. Song. 2020. "Corrosion Resistance and Hydrophobicity of Myristic Acid Modified Mg-Al LDH/Mg(OH)2 Steam Coating on Magnesium Alloy AZ31." *Frontiers of Materials Science* 14 (1): 96–107. https://doi.org/10.1007/s11706-020-0492-x.

Ren, L., S. Gao, Z. Chen, D. Jiang, and H. Huang. 2022. "Facile Preparation of Wear-Resistant and Anti-Corrosion Films on Magnesium Alloy." *Surface Engineering* 38 (1): 22–9. https://doi.org/10.1080/02670844.2021.2025312.

Scirè, S., and L. Palmisano. 2020. "1 – Cerium and Cerium Oxide: A Brief Introduction." In *Cerium Oxide (CeO_2): Synthesis, Properties and Applications*, edited by Salvatore Scirè and Leonardo Palmisano, 1–12. Metal Oxides. Elsevier. https://doi.org/10.1016/B978-0-12-815661-2.00001-3.

Song, D., G. Guo, J. Jiang, L. Zhang, A. Ma, X. Ma, J. Chen, and Z. Cheng. 2016. "Hydrothermal Synthesis and Corrosion Behavior of the Protective Coating on Mg-2Zn-Mn-Ca-Ce Alloy." *Progress in Natural Science: Materials International* 26 (6): 590– 9. https://doi.org/10.1016/j.pnsc.2016.11.002.

Song, D., B. Lian, Y. Fu, G. Wang, Y. Qiao, E. E. Klu, X. Gong, and J. Jiang. 2021. "Dual-Layer Corrosion-Resistant Conversion Coatings on Mg-9li Alloy via Hydrothermal Synthesis in Deionized Water." *Metals* 11 (9). https://doi.org/10.3390/met11091396.

Thankachan, R. M., and R. Balakrishnan. 2018. "Chapter 8 – Synthesis Strategies of Single-Phase and Composite Multiferroic Nanostructures." In *Synthesis of Inorganic Nanomaterials*, edited by Sneha Mohan Bhagyaraj, Oluwatobi Samuel Oluwafemi, Nandakumar Kalarikkal, and Sabu Thomas, 185–211. Micro and Nano Technologies. Woodhead Publishing. https://doi.org/10.1016/B978-0-08-101975-7.00008-7.

US EPA, OAR. 2015. "Overview of Greenhouse Gases." *Overviews and Factsheets*. December 23, 2015. www.epa.gov/ghgemissions/overview-greenhouse-gases.

Wang, J., X. Sun, L. Song, M. B. Kannan, F. Zhang, L. Cui, Y. Zou, S. Li, and R. Zeng. 2022. "Corrosion Resistance of Mg-Al-LDH Steam Coating on AZ80 Mg Alloy: Effects of Citric Acid Pretreatment and Intermetallic Compounds." *Journal of Magnesium and Alloys* (March). https://doi.org/10.1016/j.jma.2022.01.004.

Wang, L., X. Xiao, X. Yin, J. Wang, G. Zhu, S. Yu, E. Liu, B. Wang, and X. Yang. 2022. "Preparation of Robust, Self-Cleaning and Anti-Corrosion Superhydrophobic Ca-P/Chitosan (CS) Composite Coating on AZ31 Magnesium Alloy." *Surface and Coatings Technology* 432 (February): 128074. https://doi.org/10.1016/j.surfcoat.2021.128074.

Wang, P., J. Liu, X. Luo, P. Xiong, S. Gao, J. Yan, Y. Li, Y. Cheng, and T. Xi. 2019. "A Tannic Acid-Modified Fluoride Pre-Treated Mg – Zn – Y – Nd Alloy with Antioxidant and Platelet-Repellent Functionalities for Vascular Stent Application." *Journal of Materials Chemistry B* 7 (46): 7314–25. https://doi.org/10.1039/C9TB01587F.

Wang, X., Z. Zhang, S. Li, M. B. Kannan, and R. Zeng. 2022. "Chemical Conversion Coatings: Fundamentals and Recent Advances." In *Conversion Coatings for Magnesium and Its Alloys*, edited by Viswanathan S. Saji, T. S. N. Sankara Narayanan, and Xiaobo Chen, 3–28. Cham: Springer International Publishing. https://doi.org/10.1007/978-3-030-89976-9_1.

Wen, S., X. Liu, J. Ding, Y. Liu, Z. Lan, Z. Zhang, and G. Chen. 2021. "Hydrothermal Synthesis of Hydroxyapatite Coating on the Surface of Medical Magnesium Alloy and Its Corrosion Resistance." *Progress in Natural Science: Materials International* 31 (2): 324–33. https://doi.org/10.1016/j.pnsc.2020.12.013.

Xu, Y., Y. Guo, G. Li, and J. Lian. 2022. "Biodegradable Phytic Acid Conversion Coatings on Magnesium Alloy for Temporary Orthopedic Implant: A Review." *Progress in Organic Coatings* 169 (August): 106920. https://doi.org/10.1016/j.porgcoat.2022.106920.

Zhang, B., R. Yao, L. Li, M. Li, L. Yang, Z. Liang, H. Yu, Hao Zhang, Rifang Luo, and Yunbing Wang. 2019. "Bionic Tea Stain – Like, All-Nanoparticle Coating for Biocompatible Corrosion Protection." *Advanced Materials Interfaces* 6 (20): 1900899. https://doi.org/10.1002/admi.201900899.

Zhang, B., R. Yao, L. Li, Y. Wang, R. Luo, L. Yang, and Y. Wang. 2019. "Green Tea Polyphenol Induced Mg2+-Rich Multilayer Conversion Coating: Toward Enhanced Corrosion Resistance and Promoted in Situ Endothelialization of AZ31 for Potential Cardiovascular Applications." *ACS Applied Materials & Interfaces* 11 (44): 41165–77. https://doi.org/10.1021/acsami.9b17221.

Zhang, B., R. Yao, M. F. Maitz, G. Mao, Z. Hou, H. Yu, R. Luo, and Y. Wang. 2021. "Poly (Dimethyl Diallyl Ammonium Chloride) Incorporated Multilayer Coating on Biodegradable AZ31 Magnesium Alloy with Enhanced Resistance to Chloride Corrosion and Promoted Endothelialization." *Chemical Engineering Journal* 421 (October): 127724. https://doi.org/10.1016/j.cej.2020.127724.

Zhang, C., S. Zhang, D. Sun, J. Lin, F. Meng, and H. Liu. 2021. "Superhydrophobic Fluoride Conversion Coating on Bioresorbable Magnesium Alloy – Fabrication, Characterization, Degradation and Cytocompatibility with BMSCs." *Journal of Magnesium and Alloys* 9 (4): 1246–60. https://doi.org/10.1016/j.jma.2020.05.017.

Zhang, R., S. Cai, G. Xu, H. Zhao, Y. Li, X. Wang, K. Huang, M. Ren, and X. Wu. 2014. "Crack Self-Healing of Phytic Acid Conversion Coating on AZ31 Magnesium Alloy by Heat Treatment and the Corrosion Resistance." *Applied Surface Science* 313 (September): 896–904. https://doi.org/10.1016/j.apsusc.2014.06.104.

Zhu, B., S. Wang, L. Wang, Y. Yang, J. Liang, and B. Cao. 2017. "Preparation of Hydroxyapatite/Tannic Acid Coating to Enhance the Corrosion Resistance and Cytocompatibility of AZ31 Magnesium Alloys." *Coatings* 7 (7): 105. https://doi.org/10.3390/coatings7070105.

Zhu, J., and C. Jia. 2022. "Electrochemical Studies on Ammonium Magnesium Carbonate Tetrahydrate/Calcium Carbonate Composite Coating on AZ91D Magnesium Alloy." *Materials Chemistry and Physics* 292 (December): 126787. https://doi.org/10.1016/j.matchemphys.2022.126787.

14 Electrochemical Anodic Oxidation/Micro-Arc Oxidation-Based Composite Coatings

T.S.N. Sankara Narayanan and K. Ravichandran

CONTENTS

14.1 INTRODUCTION

Magnesium (Mg) and its alloys have received considerable attention in the automotive and aerospace industries due to their unique attributes, which include light weight, high strength-to-weight ratio and amenability for machining and casting (Tan and Ramakrishna, 2021). Adopting Mg and its alloys in automotive and aerospace industries would help improve fuel efficiency and reduce

CO_2 emissions. In addition, the biocompatibility, biodegradability and lower elastic modulus make Mg and some of its alloys as a material of choice for the development of absorbable stents (Azadani et al., 2022). However, the high reactivity of Mg and its alloys limit their widespread acceptance, particularly for biomedical applications. Alloying and surface modification are considered as viable options to improve the corrosion resistance of Mg and its alloys. Among the various surface modification methods hitherto available, electrochemical anodizing and microarc oxidation (MAO) are considered simple, cost-effective and easy to scale up.

Anodizing is an electrochemical oxidation process in which an oxide film is developed on the surface of the metal/alloy at relatively lower voltages. MAO is a plasma-assisted anodic oxidation process in which the oxide layer is generated at higher voltages. The first stage of the MAO process is anodizing. However, the application of a high voltage leads to the dielectric breakdown of the oxide layer. The electrical discharges generated during the MAO process lead to the development of a porous coating. A detailed account of the discharge characteristics during MAO of various metals is provided by Clyne et al. (2019). The partial short-term melting of the metal/alloy as well as the already-formed oxide layer establishes a metallurgical bonding between the metal/alloy and the oxide coating, resulting in the formation of a highly adherent MAO coating.

The MAO coating consists of a porous outer layer with large pores, a relatively less porous inner layer and a pore-free compact barrier layer. The recent progress in surface modification of metals by MAO, focusing on the principle, structure, and performance is reviewed by Kaseem et al. (2020). The mechanism, properties and applications of MAO coatings formed on Mg and its alloys, in particular, are addressed by Darband et al. (2017). The porous outer layer would facilitate mechanical interlocking and promote cell adhesion and growth. However, permeation of the corrosive medium through the pores to the base metal could undermine the corrosion resistance and mechanical integrity of Mg and its alloys. The strategies adopted to improve the corrosion resistance of MAO-coated Mg alloys for degradable implants are reviewed by Sankara Narayanan et al. (2014). Among them, incorporation of particles in the MAO coating to seal the pores and cracks is considered a simple and interesting strategy. Besides improving the corrosion resistance, particle incorporation in the MAO coating also improved the desired characteristics such as hardness and wear resistance as well as impart useful functionalities such as bioactivity and antibacterial properties.

The type of particles incorporated in MAO coatings formed on Mg/Mg alloys include silicon carbide (SiC) (Yang and Wu, 2010; Yu et al., 2015; Wang et al., 2015), tungsten carbide (WC) (NasiriVatan et al., 2016), Si_3N_4 (silicon nitride) (Lu et al., 2013; Lou et al., 2017), titanium nitride (TiN) (Mashtalyar et al., 2017), titanium dioxde (TiO_2) (Liang et al., 2007; Song et al., 2011; Li, Tang et al., 2012), silica (SiO_2) (Gnedenkov et al., 2015; Lu, Blawert, Huang et al., 2016; Zoubi et al., 2017), alumina (Al_2O_3) (Laleh et al., 2010; Li and Luan, 2012; Zhang et al., 2013; Wang et al., 2014; Asgari et al., 2017), ceria (CeO_2) (Lim et al., 2012; Mohedano et al., 2015; Atapour et al., 2019), zirconia (ZrO_2) (Arrabal, Matykina, Skeldon et al., 2008; Arrabal, Matykina, Viejo et al. 2008; Lee et al., 2011; Tang et al., 2011; Gnedenkov et al., 2015; Zoubi et al., 2017), molybdenum disulphide (MoS_2) (Lou et al., 2018), carbon nanotube (CNT) (Hwang and Chung, 2018), clay (Blawert et al., 2012), graphene oxide (GO) (Askarnia et al., 2021), polytetrafluoroethylene (PTFE) (Wang et al., 2022), zinc oxide (ZnO) (Bordbar-Khiabani et al., 2019), copper (Cu) (Chen et al., 2019; Liang et al., 2021), silver (Ag) (Necula et al., 2009; Ryu and Hong, 2010; Chen et al., 2021), hydroxyapatite (HA) (Lin et al., 2014), biphasic calcium phosphate (Seyfoori et al., 2013) and Ag-doped HA (Yazici et al., 2017). The effect of particle addition on the properties of MAO coatings formed on Mg and its alloys is addressed by Fattah-alhosseini, Chaharmahali et al. (2020).

The present chapter provides a detailed account of MAO-based composite coatings with a focus on particle dispersion, mechanisms and modes of particle incorporation, location of particles, influence of electrical parameters on particle incorporation and influence of particle incorporation on the characteristic properties of MAO composite coatings.

14.2   DISPERSION OF PARTICLES IN THE ELECTROLYTE

For preparing MAO composite coatings, it is a common practice to add particles as either powders or sols in the electrolyte. As particles in the electrolyte tend to settle down by sedimentation, it is imperative that they should be kept in suspension. This is usually accomplished by ultrasonication during the electrolyte preparation stage followed by stirring or agitation of the electrolyte during the MAO process. Achieving uniform dispersion of particles in the electrolyte is one of the major challenges of preparing MAO composite coatings. The dispersion stability of the particles in suspension is determined by the repulsion between adjacent particles having a similar surface charge (Lu, Mohedano et al., 2016; Hwang and Chung, 2018). Since alkaline electrolytes are commonly used for preparing MAO coatings on Mg/Mg alloys, particles with a negative zeta potential such as carbon nanotubes (CNT) (–66.8 mV) will easily move towards the anode by electrophoresis (Hwang and Chung, 2018). With the onset of potential/current, rate of movement of the particles with a higher negative potential will be increased. Surfactants are used as additives to impart a desirable surface charge for particles. Addition of sodium dodecyl sulphate (SDS) in the electrolyte imparts a negative zeta potential to ZrO_2, SiO_2, TiN and Si_3N_4 nanoparticles (NPs) that helps to effectively disperse them in the electrolyte (Gnedenkov et al., 2015; Lou et al., 2017; Mashtalyar et al., 2017). A polymeric surfactant has been used to disperse Al_2O_3 NPs in the electrolyte (Zhang et al., 2013). Asgari et al. (2020) have used urea and SDS as surfactants to disperse Al_2O_3 NPs in an alkaline phosphate electrolyte, and among them, urea enables a higher level of particle incorporation. A non-ionic surfactant is used to disperse PTFE NPs in the electrolyte (Wang et al., 2022).

14.3   EFFECT OF PARTICLE ADDITION IN THE ELECTROLYTE ON THE VOLTAGE-TIME TRANSIENTS

Addition of NPs in the electrolyte, depending on their electrical conductivity, could alter the breakdown potential as well as the time to reach it, and this can be inferred well from the voltage-time transients recorded during the MAO process. Addition of Ag, WC, MoS_2, Si_3N_4, HA and CNT has been shown to reduce the breakdown potential (Necula et al., 2009; NasiriVatan et al., 2016; Yazici et al., 2017; Lou et al., 2017, 2018; Hwang and Chung, 2018). The extent of decrease in breakdown voltage and time to reach it appears to be a function of concentration of the NPs. The breakdown voltage is decreased from 302 V to 267 V when the concentration of Si_3N_4 NPs is increased from 0 to 4 g/L, while the time to reach the breakdown potential is decreased from 28.5 seconds to 17.0 seconds (Lou et al., 2017). The effect of varying concentrations of the CNT is clearly reflected in the voltage-time transients (Figure 14.1). Irrespective of the concentration of CNT (0–10 g/L), there is no significant change in the voltage-time curves during the first stage of the MAO process. Nevertheless, the breakdown voltage, the working voltage and the time to reach them are considerably reduced with an increase in concentration of CNT from 0 to 10 g/L (Hwang and Chung, 2018). This is mainly due to the high electrical conductivity of the CNT. The absence of any significant change in voltage during the first stage of the MAO process is also observed by Necula et al. (2009) for particle-free electrolyte as well as those containing 1, 3 and 5 g/L of Ag NPs. They have also reported that the anodic forming voltage is decreased from 280 V (for particle-free electrolyte and the one containing 1 g/L of Ag NPs) to 220 V and 170 V for the electrolytes modified with 3 and 5 g/L Ag NPs. According to NasiriVatan et al. (2016), the voltage-transient recorded during the formation of MAO coating on AZ31B Mg alloy using a phosphate-based electrolyte containing 5 g/L of WC nanopowder is relatively smooth with less fluctuations than those recorded from a particle-free electrolyte. Hwang and Chung (2018) have shown the dependence of voltage fluctuations on the concentration of CNT; the higher the concentration of the CNT particles, the larger is the voltage fluctuation (Figure 14.1).

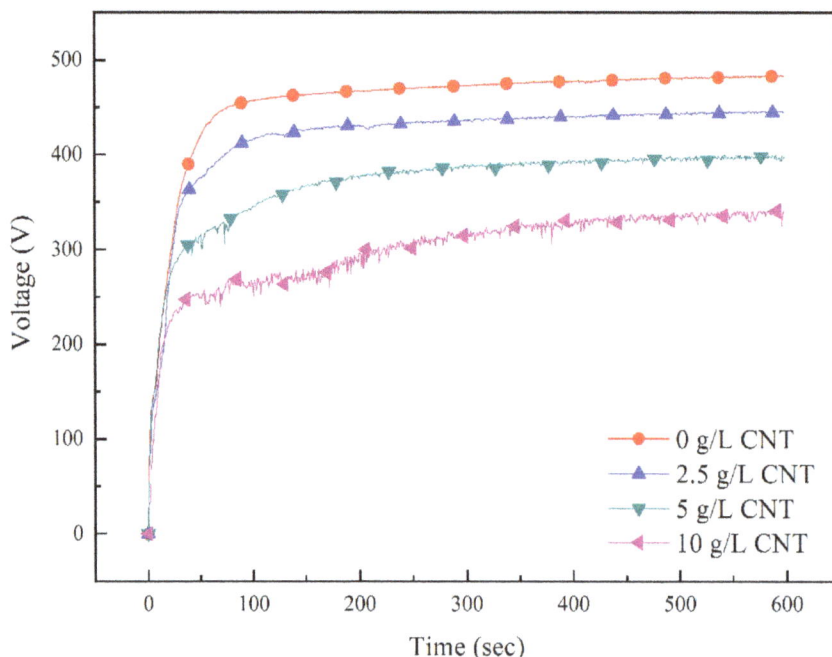

FIGURE 14.1 Voltage-time transients recorded during the MAO process using an alkaline silicate electrolyte containing varying concentrations of carbon nanotube. *Reproduced with permission from (Hwang and Chung, 2018) © 2018, under Creative Commons Attribution 4.0 License.*

14.4 MECHANISMS GOVERNING PARTICLE INCORPORATION IN MAO COATINGS

MAO of Mg and its alloys involves four stages, which include anodic oxidation of Mg/Mg alloy, breakdown of the oxide layer, microarc oxidation and arc oxidation. During the anodic oxidation stage, a passive oxide layer is formed on the surface of the Mg/Mg alloy with a surge in voltage. The presence of particles in the electrolyte has not caused any significant effect during this stage (Wang et al., 2014). Electrophoretic movement of particles and their adsorption on the surface of the Mg/Mg alloy occurs during this stage. When the voltage is increased to breakdown potential, electrical discharge occurs all over the surface. The high temperature (~10^3 to 10^4 K) and pressure (~ 100 MPa) generated in the discharge channels melt the Mg/Mg alloy as well as the previously formed oxide layer. Subsequently, the molten mass is pushed out through the discharge channels and meets the cold electrolyte-containing particles. The particles attached with molten mass get incorporated in the pores. The incorporated particles tend to fill smaller-size pores and partially block bigger-size pores. Yet another possibility of particle incorporation is also suggested. Accordingly, the electrolyte and particles fill the discharge channels, and during sparking, the electrolyte gets evaporated, leaving behind the particles (Yazici et al., 2017). The distribution of particles not only at the surface but also at the inner layers justifies the later possibility. Indeed, the discharge channels, pores and cracks serve as transport paths through which the electrolyte along with the particles reach the inner layers (Arrabal, Matykina, Skeldon et al., 2008).

14.5 MODES OF PARTICLE INCORPORATION

It is certain that particles or sols added to the electrolyte get incorporated in the MAO coating, and their mode of incorporation could be classified as "inert", "partially reactive" and "reactive". The exact mode by which particles get incorporated is a function of the applied potential/current, the strength of electrical discharges, the temperature and pressure experienced at the discharge site, the type, size and melting point of the particles, type of electrolyte and the reactivity of the particles with the electrolyte (Lou et al., 2017, 2018). Incorporation of particles and absence of formation of any new phases suggest "inert" incorporation. Melting of the particles and formation of new phases suggest "partially reactive" or "reactive" incorporation. Development of a dense microstructure is possible with both inert and reactive modes. Nevertheless, reactive incorporation makes a significant change in the microstructure of the MAO coating. In terms of particles, the intrinsic chemical stability and melting point of the particles during the electrical discharges decide their mode of incorporation. A higher potential/current facilitates reactive mode, while lower potential/current favours inert mode. MAO coating prepared with the addition of ZrO_2 NPs at 50 mA/cm^2 indicates inert incorporation, while that formed at 200 mA/cm^2 leads to the formation of a $Mg_2Zr_5O_{12}$ phase, suggesting reactive incorporation (Arrabal, Matykina, Skeldon et al., 2008; Lee et al., 2011).

Micron-sized particles, in general, are incorporated by inert mode since the short-lived electrical discharges could not facilitate their melting. Nano-sized particles, due to their low melting point, tend to show preference for reactive incorporation. Particles such as SiC, Si_3N_4, CeO_2, SiO_2, TiN, CNT and PTFE exhibit inert incorporation. HA follows partially reactive incorporation, while Al_2O_3 shows reactive incorporation. It has been reported that during the formation of MAO coating on AZ31 Mg alloy using an alkaline silicate-fluoride-based electrolyte, a part of the HA NPs incorporated in the coating gets dissociated to β-tricalcium phosphate (β-TCP), resulting in the formation of biphasic calcium phosphate coating (Seyfoori et al., 2013). Al_2O_3 particles added to an aluminate-based electrolyte displayed reactive incorporation during the MAO process, which is evidenced by the absence of Al_2O_3 in the coating, formation of $MgAl_2O_4$ phase and an increase in Al/Mg as well as $MgAl_2O_4$/MgO ratios (Li and Luan, 2012). This inference is further substantiated by Wang et al. (2014) by the increase in volume fraction of the $MgAl_2O_4$ phase during the formation of MAO coating using the electrolyte containing Al_2O_3 NPs. In spite of the preferential reactive incorporation, mechanical trapping of the particles cannot be ignored (Li and Luan, 2012).

14.6   LOCATION OF THE PARTICLES IN THE MAO COATING

During the MAO process, depending on the type of electrolyte and electrical parameters, the pores generated on the surface of the MAO coating could vary from 1 to 10 μm (Wang et al., 2014). Since the size of the pores is bigger than the size of the particles, transport of particles through the pores is highly amenable. According to Necula et al. (2009), Ag NPs are embedded across the entire thickness of the oxide layer. However, CeO_2 and ZrO_2 NPs are located throughout the MAO coating except the inner barrier layer (Arrabal, Matykina, Skeldon et al., 2008; Arrabal, Matykina, Viejo et al. 2008; Lim et al., 2012; Mohedano et al., 2015). The preferential location of the particles in the MAO coating has been identified as the pores and cracks, particularly at the surface or outer layer (Mashtalyar et al., 2017; Hwang and Chung, 2018). Even within the pores, the locations of the particles are varied. The fraction of the WC NPs at bottom of the pores is found to be much higher than those on the walls of the pores (NasiriVatan et al., 2016). The distribution of HA NPs either localized in the pores or over the entire surface shows a dependence on its concentration. At a lower concentration of 1 g/L of HA, the HA particles are identified only at the pores, whereas they are found to be distributed all over the surface when the concentration of HA is increased from 1 to 10 g/L (Yazici et al., 2017).

The preferential location of the incorporated particles as pores and cracks poses the following questions: (i) Whether particle incorporation occurs only after the microdischarge and not during the early anodizing stage? (ii) What is the limiting factor? (iii) Is it the size of the particles? Lim et al. (2012) have located CeO_2 particles during the early stages of MAO process, i.e. from 15 seconds onwards. This leads to the question, if particle inclusion occurs during the early stages of the MAO process, why is their presence not quite evident in the inner barrier layer? This can be explained based on the mechanism of particle incorporation. The particles added to the electrolyte might have been included in the MAO coating right from the beginning, as pointed out by Lim et al. (2012). Once the breakdown potential is reached, the onset of sparking increases the temperature and pressure in the discharge channels that facilitates melting of the Mg/Mg alloy and the already-formed oxide layer. Subsequently, the molten mass is pushed outside, during which the particles incorporated during the first stage might have also moved out to the surface. How the pores at the surface or outer layer become the preferential location of the particles can also be explained based on the mechanism of the MAO process. When the molten mass is pushed outside through the discharge channels (pores), it comes in contact with the cold electrolyte containing suspended particles, leading to trapping of the particles within the pores. The other possible mechanism is that the discharge channels are filled with electrolyte containing particles, and the high temperature promotes evaporation of the electrolyte, leaving behind the particles (Yazici et al., 2017). The preferential location of the nano-sized SiO_2 particle is the pores and cracks while micron sized SiO_2 particles mainly adhere to the molten mass (Lu et al., 2015).

14.7 LEVEL OF PARTICLE INCORPORATION

14.7.1 Nano- vs Micron-Sized Particles

Lu et al. (2015) have compared MAO composite coatings using an alkaline phosphate-based electrolyte containing 5 g/L SiO_2 particles with nano-sized (12 nm) and micron-sized (1–5 μm) particles. For MAO coating prepared using nano-sized SiO_2 particles, an intense Si signal is observed over the entire surface of the coating. Besides, the presence of Si is identified between the inner and outer layers of the MAO coating. Conversely, for the MAO coating prepared using micron-sized SiO_2 particles, the Si signal is rather localized, and it becomes weak in the regions surrounding the particles. The higher Si content suggests that it is easy to incorporate nano-sized than micron-size SiO_2 particles. This is not surprising, as the nano-sized particles could easily pass through the pores and cracks. The mode of incorporation of nano-sized SiO_2 particles is reactive, which is evidenced by a large decrease in the volume fraction of MgO phase accompanied with the formation of $MgSiO_3$ phase. On the contrary, the volume fraction of the MgO phase remains unaltered for MAO coating prepared using micron-sized SiO_2 particles. Since the melting point of nano-sized particles is lesser than that of the micron-sized particles, melting and chemical reaction with other phases, resulting in the formation of a new phase, is possible for nano-sized particles. Hence, it is evident that particle size plays a key role in determining the amount of particles as well as their mode of incorporation.

14.7.2 Addition of Sols Instead of Particles

It is interesting to note that MAO-based composite coatings on Mg and its alloys can also be formed with the addition of sols instead of particles in the electrolyte (Liang et al., 2007; Laleh et al., 2010; Tang et al., 2011). Similar to the particles, movement of the sols is also governed by electrophoresis and/or adsorption during the initial stages of MAO and their inclusion in the MAO coating by transport through the discharge channels and by sticking to the molten metal/already-formed coated layer. Addition of sols decreased the breakdown voltage and increased the growth rate (Tang et al., 2011).

14.7.3 Influence of Electrical Parameters

The effect of duty ratio and frequency on the uptake of particles in MAO coating formed on AM50 Mg alloy using an alkaline phosphate electrolyte containing 5 g/L SiO_2 (1–5 μm) was studied by Lu, Blawert, Mohedano et al. (2016). The choice of a higher duty cycle and lower frequency has enabled a higher level of particle incorporation in the MAO coating. Irrespective of the pulse-off time (1.8, 3.6 and 7.2 ms), the Si content of the MAO composite coating incorporated with SiO_2 particles is ~ 6.1 at. %, suggesting that the pulse-off time has no major effect on the level of particle incorporation in the MAO coating. For a given duty ratio of 10%, frequency of 250 Hz and a pulse-off time of 3.6 ms, an increase in pulse-on time from 0.2 ms to 0.4 and 0.8 ms has increased the Si content of the coating from 5 at. % to 6 and 7.2 at. %, suggesting that pulse-on time is an important parameter in determining the level of particle incorporation (Lu, Blawert, Mohedano et al., 2016). Yang and Wu (2010) have studied the effect of current frequency (500, 700 and 900 Hz) on the extent of incorporation of SiC NPs in the MAO coating formed on AZ91D Mg alloy using an alkaline silicate electrolyte containing 2 g/L of SiC NPs. A higher level of incorporation of SiC NPs is observed for MAO composite coatings prepared at 900 Hz.

14.8   INFLUENCE OF PARTICLE INCORPORATION ON THE CHARACTERISTICS OF MAO COATINGS

14.8.1   COLOUR

Incorporation of certain particles like TiN, TiO_2, CNT and Ag-doped HA changed the colour of the MAO coating, and the extent of change in colour is found to be dependent on the concentration of the particles added to the electrolyte. MAO coatings formed on Mg/Mg alloys from silicate-, phosphate- and aluminate-based electrolytes are in general white or grayish white in colour. MAO coating prepared using a silicate-based electrolyte containing TiN particles (black in colour) is changed from white to gray, and the colour shade is changed from light gray to dark gray when the concentration of TiN particles is increased from 1 to 4 g/L (Mashtalyar et al., 2017). An increase in concentration of Ag-doped HA from 1 to 10 g/L in the phosphate-based electrolyte has changed the colour of the MAO coating from light yellow to tan (Yazici et al., 2017). A change in colour from grayish white to blue is observed for MAO coating prepared using a phosphate-based electrolyte containing 0 to 6 g/L of TiO_2 particles, and the colour shade becomes darker with an increase in concentration of the TiO_2 particles (Figure 14.2) (Li, Tang et al., 2012).

The colour of the MAO coating is changed from light gray to dark gray when the concentration of TiO_2 particles in an aluminate-based electrolyte is increased from 1.6 g/L to 4.8 g/L (Song et al., 2011). An increase in concentration of CNT in silicate-based electrolyte from 0 to 10 g/L has changed the colour of the MAO coating from white to black (Hwang and Chung, 2018). The MAO coating formed on Mg–Zn–Zr–RE alloy using a Keronite electrolyte is white-grey in colour. Addition of 1.0, 3.0 and 5.0 g/L of colloidal Ag NPs has changed the colour of the MAO coating to

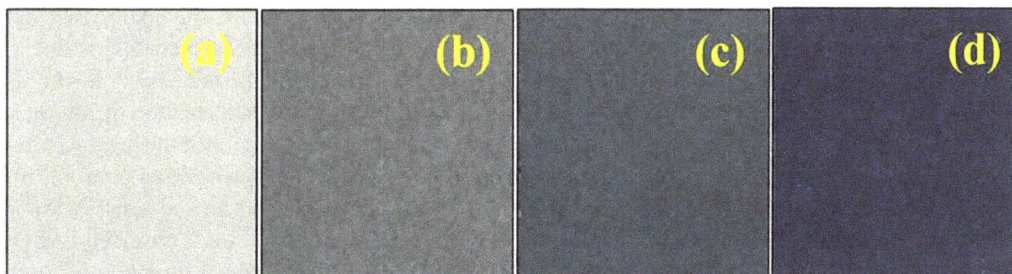

FIGURE 14.2   Colour of the MAO coating obtained using: (a) the base electrolyte (BE) (6.0 g/L $(NaPO_3)_6$, 3.0 g/L NaOH and 10.0 mL/L triethanolamine (TEA); (b) BE + 2 g/L TiO_2; (c) BE + 4 g/L TiO_2; and (d) BE + 6 g/L TiO_2. *Reproduced with permission from (Li, Tang et al., 2012) © 2012 Elsevier B.V.*

shiny light yellow, dark yellow and matte dark brown, respectively (Necula et al., 2009). The colour of the MAO coating formed on AZ31 Mg alloy using an alkaline silicate electrolyte is changed from light gray to dark yellow when the electrolyte is modified with the addition of 0.5 g/L AgNO$_3$. An increase in concentration of AgNO$_3$ or treatment time has changed the colour of the MAO coating from dark yellow to brown (Ryu and Hong, 2010).

14.8.2   Uniformity and Compactness

MAO composite coatings prepared using a phosphate-based electrolyte containing 5 g/L of WC nanopowder has been shown to be uniform and compact (NasiriVatan et al., 2016). Addition of 2.5 g/L of MoS$_2$ NPs to a aluminate-phosphate-based electrolyte has enabled the formation of a uniform and compact MAO composite coating. However, when the concentration of MoS$_2$ NPs is increased to 10 g/L, the extent of adsorption of NPs is increased. This, in turn, promotes a higher level of segregation of MoS$_2$ NPs at the MAO coating/AZ31 Mg alloy interface, which deleteriously influence the uniformity and compactness of the MAO composite coating (Lou et al., 2018). The composite coatings prepared using TiO$_2$, Al$_2$O$_3$ and ZrO$_2$ sols are more uniform and compact, with fewer defects (Liang et al., 2007; Laleh et al., 2010; Tang et al., 2011).

14.8.3   Thickness

Addition of NPs in the electrolyte alters the thickness of the MAO composite coating. An increase or decrease in thickness is largely determined by the particle size, concentration of the particles and how the NPs influence the voltage-transient (Lu, Mohedano et al., 2016; Lou et al., 2017). Lu et al. (2013) have prepared MAO composite coatings with the addition of Si$_3$N$_4$ particles with three different sizes, viz., 0.02 μm, 0.1–0.8 μm and 1–5 μm. Addition of 0.1–0.8 μm- and 1–5 μm-sized particles decreased the thickness of the MAO coating from 40 μm to 25 μm and 10 μm, respectively. The thickness of the MAO composite coating is found to increase from 13.1 μm to 17.8 μm with an increase in concentration of Si$_3$N$_4$ NPs from 0 to 3 g/L. However, a further increase in its concentration to 4 g/L has led to a decrease in thickness from 17.8 μm to 16.8 μm (Lou et al., 2017). Addition of 2.5 g/L MoS$_2$ has increased the thickness of the MAO coating from 13.4 μm to 20.1 μm. However, a further increase in its concentration from 2.5 g/L to 5.0, 7.5 and 10 g/L has led to a decrease in coating thickness from 20.1 μm to 18.2 μm, 16.2 μm and 14.8 μm (Lou et al., 2018). The average thickness of the MAO composite coating is decreased from 19.3 μm to 14.2 μm, 12.5 μm and 8.0 μm with an increase in concentration of CNT from 0 to 2.5, 5 and 10 g/L, respectively. With an increase in concentration of CNT from 0 to 10 g/L, a steady decrease in the breakdown and working voltage is observed in the voltage-time transient (Figure 14.1) (Hwang and Chung, 2018). Wang et al. (2014) have reported that the thickness of the MAO coating prepared using an alkaline aluminate-based electrolyte containing 0, 5, 10 and 15 g/L of Al$_2$O$_3$ NPs are 23.4, 33.5, 36.9 and 40.4 μm, respectively. According to them, the increase in thickness of the MAO coating is due to the reactive incorporation of Al$_2$O$_3$ NPs. Lou et al. (2017) have correlated the increase in thickness of the MAO coating prepared with the addition of 1 to 3 g/L of Si$_3$N$_4$ to the formation of Mg$_2$SiO$_4$ phase following the reaction between the Si$_3$N$_4$ NPs and the molten mass generated during the MAO process. The growth rate determines the thickness of the MAO coating. NasiriVatan et al. (2016) have shown that MAO coating incorporated with WC NPs has lower resistance, and hence, a lower energy is required for its formation. This results in an increase in growth rate. Conversely, Al$_2$O$_3$ NPs reacts with MgO and promotes the formation of MgAl$_2$O$_4$ phase. The consumption of MgO phase has led to a decrease in growth rate (Li and Luan, 2012).

Lu, Blawert, Mohedano et al. (2016) have evaluated the effect of duty ratio and frequency on the thickness of the MAO coatings formed on AM50 Mg alloy using an alkaline phosphate electrolyte containing 5 g/L SiO$_2$ (1–5 μm). A combination of lower frequency and higher duty ratio facilitates the formation of a thicker coating, with lesser number of pores having a larger area. For a given duty ratio of 10%, the coating thickness is decreased from 84 μm to 45 and 35 μm with an increase in frequency from 50 Hz to 250 and 500 Hz. The choice of a lower frequency facilitates a higher growth rate, which results in the formation of MAO coating with a higher thickness. Yang and Wu (2010) have studied the effect of current frequency on the thickness of MAO coating on AZ91D Mg alloy using an alkaline silicate electrolyte containing 2 g/L of SiC NPs. For a given treatment time of 45 minutes, the thickness of the MAO composite coating has increased from 12 μm to 20 and 24 μm with an increase in frequency from 500 Hz to 700 and 900 Hz. The long-lived discharges generated at 900 Hz increased the growth rate, which results in the formation of a thicker coating. Lu, Blawert, Mohedano et al. (2016) have also evaluated the effect of pulse-on time and pulse-off time on the thickness of the MAO composite coating incorporated with SiO_2 particles. For a given duty ratio of 10% and frequency of 250 Hz, decreasing the pulse-off time from 3.6 ms to 1.8 ms decreased the coating thickness from 41 μm to 20 μm and makes the MAO coating uneven. However, under similar conditions, an increase in pulse off-time from 3.6 ms to 7.2 ms has caused no significant effect on coating thickness. For a given duty ratio of 10%, frequency of 250 Hz and a pulse-off time of 3.6 ms, an increase in pulse-on time from 0.2 ms to 0.4 and 0.8 ms has increased the thickness of the MAO coating from 28 μm to 41 and 49 μm.

14.8.4 SURFACE MORPHOLOGY

The surface morphology of the MAO coating reveals the presence of pores and cracks from particle-free as well as particle-containing electrolytes. The pores are emanated due to gas evolution, while the cracks are originated due to thermal stresses generated due to the rapid cooling of the molten mass by the electrolyte. In terms of particles, the morphology of MAO coating is not changed except when the mode of incorporation becomes "partially reactive" or "reactive" (Arrabal, Matykina, Skeldon et al., 2008). An increase in concentration of HA particles in an alkaline phosphate-based electrolyte from 1 to 10 g/L has altered the surface morphology of the MAO coating (partially reactive incorporation) following an increase in the number of discharge channels and a decrease in the size of the discharge channels (Yazici et al., 2017).

14.8.5 POROSITY

Addition of 5 g/L of WC NPs (< 80 nm) to an alkaline phosphate-based electrolyte has reduced the surface porosity of the MAO coating from 10.59% to 9.81% and the mean pore diameter from 8.42 μm to 7.19 μm (NasiriVatan et al., 2016). The porosity of MAO coating prepared using a silicate-fluoride-based electrolyte is decreased with the addition of TiN NPs (20 nm) and the extent of decrease in porosity is increased with an increase in concentration of TiN NPs from 1 to 4 g/L (Mashtalyar et al., 2017). The decrease in porosity and average pore diameter are due to the filling of the pores by WC and TiN NPs. In spite of whether the NPs are incorporated as individual particles or agglomerates, they could be able to fill the pores. The NPs are identified at the bottom of the pores as well as on the walls, though the fraction appears to be relatively higher at the bottom than on the walls (NasiriVatan et al., 2016). Hwang and Chung (2018) have also observed the dependence of porosity on the concentration of CNT in an alkaline silicate-fluoride electrolyte; the higher the concentration of CNT (10 g/L), the lower is the porosity. According to them, the high electrical conductivity of CNT enables a reduction in both the breakdown and working potential, which results in the generation of fine microarc discharges and smaller pores. With an increase in concentration of CNT from 0 to 10 g/L, the porosity is considerably reduced. Lou et al. (2017, 2018) have observed an increase in surface porosity of MAO coatings formed with the addition of MoS_2 and Si_3N_4 NPs in aluminate-phosphate electrolyte. The surface porosity is increased from 4.31% to 7.35% when the concentration of MoS_2 in the electrolyte is increased from 0 to 10 g/L. An increase in concentration of Si_3N_4 NPs from 0 to 4 g/L has increased the surface porosity from 4.31% to 8.40%.

The effect of duty ratio and frequency on the porosity of MAO coating formed on AM50 Mg alloy using an alkaline phosphate electrolyte containing 5 g/L SiO_2 (1–5 μm) was studied by Lu, Mohedano et al. (2016). Irrespective of the duty ratio, MAO coating formed at 50 Hz consists of larger pores than those prepared at 250 and 500 Hz. An increase in frequency facilitates the formation of MAO coating with smaller pores that are uniformly distributed over the entire surface of the coating. For a given duty ratio of 10%, the number of pores is increased from 90 to 242 and 378 with an increase in frequency from 50 Hz to 250 and 500 Hz, and this trend is also followed for other duty ratios. However, the total area of the pores is decreased from 17696 $μm^2$ to 14508 and 13889 $μm^2$ when the frequency is increased from 50 Hz to 250 and 500 Hz. For a given frequency, an increase in duty ratio leads to the formation of MAO coatings with larger pores. The effect of current frequency on the porosity of MAO coating formed on AZ91D Mg alloy using an alkaline silicate electrolyte containing 2 g/L of SiC NPs was studied by Yang and Wu (2010). Although the number of pores is increased, the mean pore diameter is decreased with an increase in frequency from 500 to 900 Hz. Although an increase in pulse-on time from 0.2 ms to 0.4 and 0.8 ms has increased the level of incorporation of SiO_2 particles in the MAO coating, such an increase in pulse-on time makes the MAO coating more porous and uneven (Lu, Mohedano et al., 2016). Addition of TiO_2, Al_2O_3 and ZrO_2 sols in the electrolyte considerably changed the surface morphology of the composite coatings, as there was observed to be lesser number of smaller pores and cracks. The pores appeared to be blocked internally (Liang et al., 2007).

14.8.6 Phase Transformation and Formation of New Phases

MgO is identified as one of the phases of MAO coatings formed on Mg and its alloys. The diffusion of anions present in the electrolyte through the pores of the MAO coating enables formation of additional phases, such as magnesium silicate, magnesium phosphate and magnesium aluminate from silicate-, phosphate- and aluminate-based electrolytes, respectively (Yang and Wu, 2010; Lee et al., 2011). The particles added to the electrolyte might also react with the MgO, resulting in the formation of a new phase. Arrabal, Matykina, Skeldon et al. (2008) have studied the formation of MAO composite coatings on pure Mg using both alkaline silicate and alkaline phosphate electrolytes with the addition of 2 g/L of monoclinic ZrO_2 under direct current conditions at 3 A/dm^2 for 2400 seconds. For the MAO coating prepared using the phosphate electrolyte, the microdischarges induce sufficient heating, which enabled transformation of monoclinic to tetragonal ZrO_2 phase. In addition, the formation of $Mg_2Zr_5O_{12}$ phase is also promoted. The monoclinic ZrO_2 phase identified in the MAO composite coating could either be due to the untransformed ZrO_2 particles or due to the transformation of tetragonal to monoclinic phase when the surface comes in contact with the cold electrolyte. The formation of $Mg_2Zr_5O_{12}$ phase is not promoted for coatings prepared using silicate electrolyte. This is due to the weak microdischarges generated from the silicate electrolyte. Arrabal, Matykina, Viejo et al. (2008) have also prepared MAO composite coatings using an alkaline silicate electrolyte containing 10 g/L monoclinic ZrO_2 NPs under alternating-current conditions at 20 A/dm^2 for 2400 seconds. The microdischarges generated during the MAO process tends to induce sufficient heating, which facilitates the reaction of monoclinic ZrO_2 NPs with the molten metal/oxide coating, resulting in the formation of $Mg_2Zr_5O_{12}$ phase. Nevertheless, the transformation of monoclinic to tetragonal ZrO_2 is not observed.

The presence of anatase and rutile TiO_2 phase is identified in MAO coating formed on AM60B Mg alloy using an alkaline phosphate electrolyte containing 4 vol. % of TiO_2 sol. The anatase phase is formed during the early stages of the MAO process as the electrical discharge energy is relatively lower. With an increase in treatment time, the surge in temperature and pressure experienced at the discharge channels has enabled transformation of the metastable anatase phase to stable rutile phase. Nevertheless, the external surface layer that is in contact with the electrolyte gets cooled rapidly, which prevents transformation of the anatase to rutile phase (Liang et al., 2007). Lou et al. (2017, 2018) have reported that MAO coating formed using an aluminate-phosphate-based electrolyte without any particle addition consists of $MgAl_2O_4$ as the major phase and MgO as the minor phase. Addition of 2.5, 5.0, 7.5 and 10 g/L of MoS_2 particles to this electrolyte has promoted the formation of MgO phase at the expense of $MgAl_2O_4$ phase (Lou et al., 2018). Addition of 0 to 10 g/L CNT to a silicate-based electrolyte fails to show any phase transition or formation of new phases during the MAO process (Hwang and Chung, 2018). The effect of current frequency on the phase contents of the MAO coating on AZ91D Mg alloy using an alkaline aluminate-based electrolyte containing 2 g/L of SiC NPs was evaluated by Yang and Wu (2010). MgO and $MgAl_2O_4$ phases are identified as the major phases, and the volume fraction of these phases is not changed with the current frequency. Addition of sols increased the volume fraction of phases such as $MgAl_2O_3$ and $Mg_2Zr_5O_{12}$ at the expense of MgO, suggesting a better reactivity with the molten mass (Liang et al., 2007; Laleh et al., 2010; Tang et al., 2011).

14.8.7 SURFACE ROUGHNESS

Addition of MoS_2 NPs in an aluminate phosphate electrolyte has increased the average surface roughness (R_a) of the MAO coating. Addition of 2.5 and 5.0 g/L of MoS_2 NPs has increased the R_a from 1.04 μm (particle-free electrolyte) to 1.99 μm and 2.05 μm, respectively. However, the R_a is increased to 3.56 μm and 4.88 μm when the concentration of the MoS_2 NPs in the electrolyte is increased to 7.5 and 10.0 g/L (Lou et al., 2018). A similar behaviour is also observed with the addition of 1 to 4 g/L of Si_3N_4 NPs to an aluminate-phosphate-based electrolyte. Addition of 1 and 2 g/L of Si_3N_4 NPs has increased the R_a from 1.04 μm (particle-free electrolyte) to 1.63 μm and 1.76 μm, respectively, while the R_a is increased to 2.59 μm and 2.93 μm when the content of Si_3N_4 NPs in the electrolyte is increased to 3 and 4 g/L (Lou et al., 2017). The R_a of MAO coating is increased from 1.21 to 1.56, 1.83, 2.00 and 2.13 when an alkaline silicate electrolyte is modified with the addition of 0, 1, 2, 3 and 4 g/L of TiN NPs (Mashtalyar et al., 2017). Asgari et al. (2017) have observed a steady decrease in R_a from 3.70 μm to 1.50 μm when the concentration of Al_2O_3 NPs in an alkaline phosphate-based electrolyte is increased from 0 to 30 g/L. However, the trend is reversed at 40 g/L of Al_2O_3 NPs, resulting in a R_a of 2.75 μm. Hence, it is clear that addition of NPs in the electrolyte, in general, increases the R_a of the MAO coating, and the extent of increase in R_a is a function of the concentration of the NPs in the electrolyte.

Atapour et al. (2019) have reported that the R_a of MAO composite coating incorporated with CeO_2 NPs (< 5 nm) is 0.91 μm, which is relatively lower than those prepared using particle-free electrolyte with R_a of 1.4 μm. This inference is contrary to the findings of Mohedano et al. (2015). They have reported a higher R_a for MAO coatings prepared using micron-sized CeO_2 particles (< 5 μm). These observations can be rationalized based on how the CeO_2 particles are incorporated in the coating. The micron-sized CeO_2 particles partially filled the pores of the MAO coating, whereas a fraction of them might have protruded outside the pores, which might have caused an increase in R_a. The findings of Mohedano et al. (2015) and Atapour et al. (2019) suggest that particle size and how they are incorporated in the coating decides the R_a of the resultant coating. Yang and Wu (2010) have studied the effect of current frequency on the R_a of MAO coating on AZ91D Mg alloy using an alkaline silicate electrolyte containing 2 g/L of SiC NPs. The R_a is found to decrease with an increase in current frequency, and the MAO composite coating formed at 900 Hz exhibits a finer microstructure with a lower surface roughness. Unlike the MAO composite coatings prepared with the addition of particles, the R_a of coatings prepared with the addition of sols is considerably lower (Liang et al., 2007; Laleh et al., 2010; Tang et al., 2011).

Yu et al. (2015) have prepared MAO coatings on AZ31 Mg alloy using a silicate-hexametaphosphate-based electrolyte with and without the addition of 2 g/L of SiC NPs at high and low current densities for different durations of time up to 1200 seconds. Irrespective of the current density, both R_a and R_z are increased with an increase in treatment time for MAO coatings prepared from particle-free electrolyte as well as the one with 2 g/L SiC NPs. At lower treatment times up to 300 seconds, both R_a and R_z values of MAO coating prepared using the electrolyte having SiC

NPs are relatively higher than those prepared from particle-free electrolyte, and this trend is similar for both high and low current densities. Beyond 300 seconds and up to 1200 seconds, a decrease in R_a and R_z is observed for MAO coatings prepared at high current density using the electrolyte having 2 g/L of SiC NPs. For MAO coatings prepared at low current density, there is no considerable difference in the R_a and R_z values of MAO coatings obtained from particle-free and SiC NPs containing electrolytes. The higher R_a and R_z values of MAO coating prepared using the electrolyte having 2 g/L SiC NPs up to 300 seconds is in line with the inferences made by other researchers. The decrease in R_a and R_z values observed beyond 300 seconds to 1200 seconds for MAO coatings prepared from SiC NPs containing electrolyte at high and low current densities could be due to the filling up of the pores by the SiC NPs.

14.8.8 ADHESION

In general, the adhesive strength of MAO coatings formed on Mg/Mg alloys is higher due to the metallurgical bonding of the oxide layer with the base metal. It is important to ensure how particle incorporation could influence the adhesive strength of the MAO coating. Evaluation of the coating adhesion is commonly determined by a scratch test, and the critical load (L_{C2}) required for delamination or disruption of the MAO coating gives a measure of the adhesive strength. The L_{C2} of MAO coating prepared using a silicate-fluoride-based electrolyte is 8.4 N. Addition of TiN NPs in the electrolyte up to 3 g/L enables an increase in the L_{C2} value of the MAO coating up to 10.5 N, beyond which the trend is reversed. Agglomeration of the TiN NPs as well as formation of a brittle coating are considered responsible for the lower L_{C2} value (9.3 N) of the MAO coating prepared using 4 g/L of TiN NPs (Mashtalyar et al., 2017).

MAO coating prepared using an aluminate-phosphate-based electrolyte exhibits a L_{C2} value of 30.3 N. Addition of 2 and 3 g/L of Si_3N_4 NPs in the electrolyte has enabled an increase in L_{C2} value from 30.3 N to 33 and 32.3 N, respectively. Nevertheless, the L_{C2} value is decreased to 23.9 N when the concentration of the Si_3N_4 NPs in the electrolyte is increased to 4 g/L, suggesting a decrease in adhesive strength of the MAO coating when compared to those prepared using 2 and 3 g/L of Si_3N_4 NPs. The increase in porosity is considered responsible for the poor adhesive strength of the MAO coating prepared using 4 g/L of Si_3N_4 NPs (Lou et al., 2017). For MAO coatings formed using the same electrolyte containing 0 to 10 g/L MoS_2 NPs, the L_{C2} value is found to be 50 N for the coating prepared using 2.5 g/L of MoS_2 NPs, which is higher than 33.5 N for coating prepared using particle-free electrolyte. When the concentration of MoS_2 NPs in the electrolyte is increased beyond 2.5 g/L, there is observed to be a considerable decrease in the L_{C2} value, suggesting a large decrease in the adhesive strength of these MAO coatings (Lou et al., 2018). High porosity, uneven distribution of the MoS_2 NPs and the lack of formation of a continuous inner layer are considered the major reasons for the poor adhesion of the MAO coatings prepared using 5, 7.5 and 10 g/L of MoS_2 NPs.

Askarnia et al. (2021) have evaluated the scratch resistance of MAO coatings formed on AZ91D Mg alloy using an alkaline metasilicate electrolyte containing 0, 10, 20 and 30 mg/L GO (100 nm) nanosheets at a load of 10 N. The width of the damage zone after the scratch test gives a measure of the adhesive strength and continuity of the MAO coating. Scanning electron micrographs, acquired at the damage zone of MAO coatings prepared in the absence and presence of GO nanosheets, are shown in Figure 14.3. Cracking and delamination are prevalent on the damaged zone for MAO coating prepared in the absence of GO nanosheets (Figure 14.3(a)). For MAO coatings prepared using 10, 20 and 30 mg/L GO nanosheets, the width of the damaged zone is considerably reduced, and the extent of cracking and delamination is rather limited (Figures. 14.3(b), 14.3(c) and 14.3(d)). Among them, the width of the damaged zone is relatively less for MAO coating prepared using 20 mg/L of GO nanosheets (Figure 14.3(c)). Hence, it is clear that incorporation of particles is likely to increase the adhesive strength of the MAO coating as long as formation of a dense layer is promoted. If particle incorporation makes the coating more porous or highly brittle or it affects the formation of a continuous inner layer, then the coating adhesion will become poor.

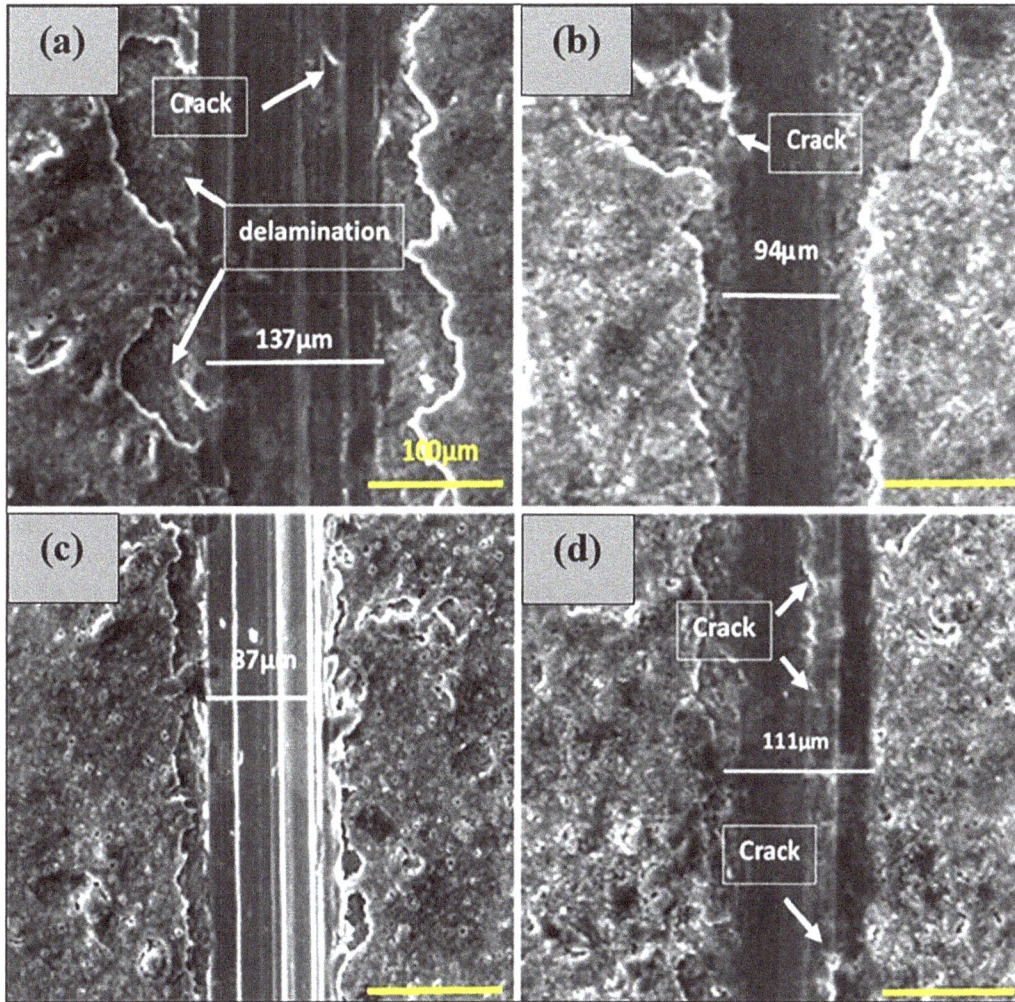

FIGURE 14.3 Scanning electron micrographs acquired at the damage zone of MAO composite–coated AZ91D Mg alloy prepared using varying concentrations of GO nanosheets after scratch test under a load of 10 N: (a) 0 g/L GO; (b) 10 g/L GO (b); (c) 20 g/L GO; and (d) 30 g/L GO. *Reproduced with permission from (Askarnia et al., 2021) © 2021 Elsevier B.V.*

14.8.9 HARDNESS

Addition of 5 g/L of WC NPs in a phosphate-based electrolyte has increased the hardness of the MAO coating from 408 to 422 HV (NasiriVatan et al., 2016). The hardness of the MAO coating formed on AZ31 Mg alloy using an aluminate-phosphate-based electrolyte is increased from 1142 to 1693 HV with the addition of 10 g/L of MoS_2 in the electrolyte (Lou et al., 2018). Addition of 3 g/L of Si_3N_4 NPs in the same electrolyte has enabled an increase in hardness from 1142 to 1672 HV. However, a further increase in concentration of the Si_3N_4 NPs to 4 g/L has resulted in a decrease in hardness of the MAO coating to 1234 HV. This is due to the higher porosity of the MAO coating obtained using 4 g/L of Si_3N_4 NPs (Lou et al., 2017). MAO coating formed on AZ91D Mg alloy using a silicate-aluminate-based electrolyte containing 4 g/L SiC NPs exhibits a hardness of 445.3 HV, while it is 372.3 HV for the one prepared using the particle-free electrolyte (Wang et al., 2015).

Lu, Blawert, Mohedano et al. (2016) have compared the hardness of MAO composite coatings formed on AM50 Mg alloy using an alkaline phosphate electrolyte containing both nano- (12 nm) and micron- (1–5 μm) sized SiO_2 particles. Incorporation of the 1–5 μm SiO_2 particles has caused only a slight increase in hardness of the MAO coating from 327 HV to 338 HV, whereas the hardness of the MAO coating is increased from 327 HV to 396 HV when SiO_2 NPs are added to the electrolyte. Addition of TiN particles in a silicate-fluoride electrolyte has shown a strong dependence of hardness of the MAO coating on the concentration of the particles. With the addition of 1 g/L of TiN particles in the electrolyte, only a marginal increase in hardness from 214 to 224 HV is observed. With an increase in concentration of TiN from 1 to 2 g/L, a large increase in hardness from 214 to 459 HV is noticed. However, further increase in concentration of TiN particles to 3 and 4 g/L has decreased hardness of the MAO coating to 428 and 377 HV (Mashtalyar et al., 2017). MAO coating formed on AM50 Mg alloy using an alkaline aluminate-phosphate-based electrolyte exhibits a hardness of 329 HV, while addition of 5 g/L CeO_2 NPs in that electrolyte has increased the hardness to 456 HV. The increase in hardness has been attributed to the incorporation of CeO_2 NPs in the MAO coating as well as due to the lower porosity of the composite coating (Atapour et al., 2019).

Wang et al. (2014) have evaluated the hardness of MAO coatings prepared using an alkaline aluminate-based electrolyte containing 5, 10 and 15 g/L Al_2O_3 NPs. The hardness of the MAO coating obtained from the particle-free electrolyte is 156.3 HV. Addition of 15 g/L of Al_2O_3 NPs in the electrolyte has increased the hardness of the MAO coating from 156.3 to 291.5 HV. Nevertheless, the extent of increase in hardness with an increase in concentration of the Al_2O_3 NPs is not linear. The increase in hardness of the MAO coating has been attributed not only due to the incorporation of Al_2O_3 NPs but also due to the higher volume fraction of the $MgAl_2O_4$ phase formed due to the reactive incorporation of the Al_2O_3 NPs. MAO coating formed on AZ31 Mg alloy using an alkaline silicate electrolyte containing 10 g/L of Al_2O_3 has been shown to possess a hardness of 358 HV as opposed to a hardness of 130 HV for the MAO coating prepared using particle-free electrolyte. The increase in hardness of the MAO composite coating is ascribed to the ceramic nature of the coating as well as filling of the pores by the Al_2O_3 particles, which helped to strengthen the MAO coating (Zhang et al., 2013). It is evident that particle addition in the electrolyte has enabled an increase in hardness of the MAO coating, which would enable an increase in wear resistance. The extent of increase in hardness of the MAO coating is determined by the type of electrolyte, type of particles, particle size, particle concentration, porosity of the MAO coating, ability of the particles to fill the pores of the MAO coating, ceramic nature of the MAO coating and volume fraction of new phases that could contribute to the hardness.

14.8.10 CORROSION RESISTANCE

MAO composite coatings, in general, increase the corrosion resistance of Mg/Mg alloys when compared to their particle-free counterparts. Addition of Al_2O_3, CNT and CeO_2 in the electrolyte has enabled an improvement in the corrosion resistance of the corresponding MAO composite coatings (Wang et al., 2014; Hwang and Chung, 2018; Atapour et al., 2019). In order to achieve an improvement in corrosion resistance, it is imperative that the incorporated particles must seal the pores and cracks as well as decrease the R_a of the MAO composite coating. Formation of a new phase such as $MgAl_2O_4$ following the reaction between MgO and an aluminate-based electrolyte or Al_2O_3 NPs is considered to be beneficial, as it would promote the corrosion resistance (Wang et al., 2014). A combination of formation of a new phase and incorporation of NPs throughout the coating offer significant improvement in corrosion resistance of MAO composite coatings. Higher chemical stability of the particles, better compactness of the coating, absence of pores and cracks, which are the pathways for the diffusion of aggressive Cl- ions, formation of a new phase with better chemical stability, formation of an intact inner layer that could serve as a barrier between the coating and the substrate are the major factors that contribute to the higher corrosion resistance of MAO composite coatings.

The corrosion resistance of MAO composite coatings prepared using an aluminate-phosphate-based electrolyte containing 1, 2, 3 and 4 g/L of Si_3N_4 NPs in 3.5% NaCl was evaluated by Lou et al. (2017). When compared to the coating prepared from a particle-free electrolyte, there observed to be a slight reduction in the corrosion current density (i_{corr}) value of coatings prepared using 1 and 2 g/L of Si_3N_4 NPs. However, the trend is reversed when the concentration of Si_3N_4 NPs is increased to 3 and 4 g/L. To account for the corrosion behaviour of these coatings, the thickness, surface porosity and R_a should be considered carefully. When compared to the coating obtained from particle-free electrolyte (13.1 μm), the coating thickness is increased with an increase in concentration of Si_3N_4 NPs from 1 to 3 g/L (16.0 to 17.8 μm), while it is decreased at 4 g/L (16.8 μm). The R_a is increased from 1.04 μm to 2.93 μm when the concentration of Si_3N_4 NPs is increased from 0 to 4 g/L. The surface porosity is increased from 4.31% to 8.40% when the concentration of Si_3N_4 NPs is increased from 0 to 4 g/L. Hence, it is clear that the beneficial effect of higher coating thickness on corrosion resistance is outweighed by the higher R_a and higher surface porosity for coatings prepared using 4 g/L of Si_3N_4 NPs.

Lou et al. (2018) have evaluated the corrosion resistance of MAO composite coatings prepared using an aluminate-phosphate-based electrolyte containing 2.5, 5.0, 7.5 and 10.0 g/L MoS_2 NPs in 3.55% NaCl. Potentiodynamic polarization tests performed after 0.5 hours of immersion in 3.5% NaCl reveals that when compared to the MAO coating prepared from particle-free electrolyte, the i_{corr} of MAO composite coating prepared using 2.5, 5.0 and 7.5 g/L of MoS_2 NPs is decreased from 2.96 μA/cm^2 to 0.83, 1.22 and 1.73 μA/cm^2, respectively. However, an increase in concentration of MoS_2 NPs to 10 g/L has increased the i_{corr} to 4.88 μA/cm^2. However, after 6 hours of immersion in 3.5% NaCl, the i_{corr} value of MAO coating prepared from particle-free electrolyte becomes 1.35 μA/cm^2, whereas for those prepared using 2.5, 5.0, 7.5 and 10.0 g/L of MoS_2 NPs, the i_{corr} is increased to 2.86, 2.80, 5.40 and 282.3 μA/cm^2, respectively. The decrease in i_{corr} value of MAO coating prepared using particle-free electrolyte from 2.96 to 1.35 μA/cm^2 suggests development of corrosion products that could have blocked the pores of the coating. Higher surface porosity, higher R_a, lack of uniform distribution of MoS_2 NPs in the oxide layer and the lack of formation of a continuous and dense inner layer are the major reasons for the poor corrosion resistance of MAO composite coatings prepared using 10 g/L of MoS_2 NPs after 0.5 of immersion in 3.5% NaCl as well as for the MAO composite coatings incorporated with MoS_2 NPs after 6 hours of immersion in 3.5% NaCl.

The surface morphology of MAO coatings prepared using an alkaline phosphate-based electrolyte containing 0, 1.5, 3.0 and 4.5 g/L of ZnO NPs is shown in Figure 14.4 (a–d). It is evident that the level of incorporation of ZnO NPs is increased with an increase in its concentration. Accumulation of the ZnO NPs inside the pores is quite evident, and the porosity of the coatings is decreased from 36.4% to 14.6% with an increase in concentration of ZnO in the electrolyte from 1 to 4 g/L. The potentiodynamic polarization curves of uncoated AZ91 Mg alloy and MAO-coated alloys formed with and without addition of ZnO in the electrolyte in simulate body fluid (SBF) at 37°C are shown in Figure 14.4(e), while the corresponding Nyquist plots are shown in Figure 14.4(f). Addition of 1.5, 3.0 and 4.5 g/L of ZnO in the electrolyte has decreased the i_{corr} of the MAO coating from 6.17 μA/cm^2 to 0.56, 0.21 and 0.063 μA/cm^2, respectively. This corresponds to an increase in R_p from 9.26 kΩ.cm^2 to 85.72, 220.8 and 683.2 kΩ.cm^2, respectively. MAO coatings prepared with the addition of ZnO NPs have enabled an increase in the resistance of both the outer porous layer and inner barrier layer. The total resistance of MAO coating is increased from 38.7 kΩ.cm^2 (prepared from particle-free electrolyte) to 385.1 kΩ.cm^2 for the one prepared using 4.5 g/L ZnO NPs in the electrolyte. It is obvious that the decrease in porosity and filling up of the pores by the ZnO NPs is the main reason for the observed in improvement in corrosion resistance of the MAO composite coatings (Bordbar-Khiabani et al., 2019).

Tang et al. (2011) have compared the corrosion resistance of MAO coatings formed on AZ91D Mg alloy using an alkaline silicate-based electrolyte containing 5 vol. % ZrO_2 sol in 3.5% NaCl. The surface morphology of the MAO coating prepared without the addition of the ZrO_2 sol reveals the presence of many pores and cracks with a rough surface (Figure 14.5(a)). Conversely, the MAO

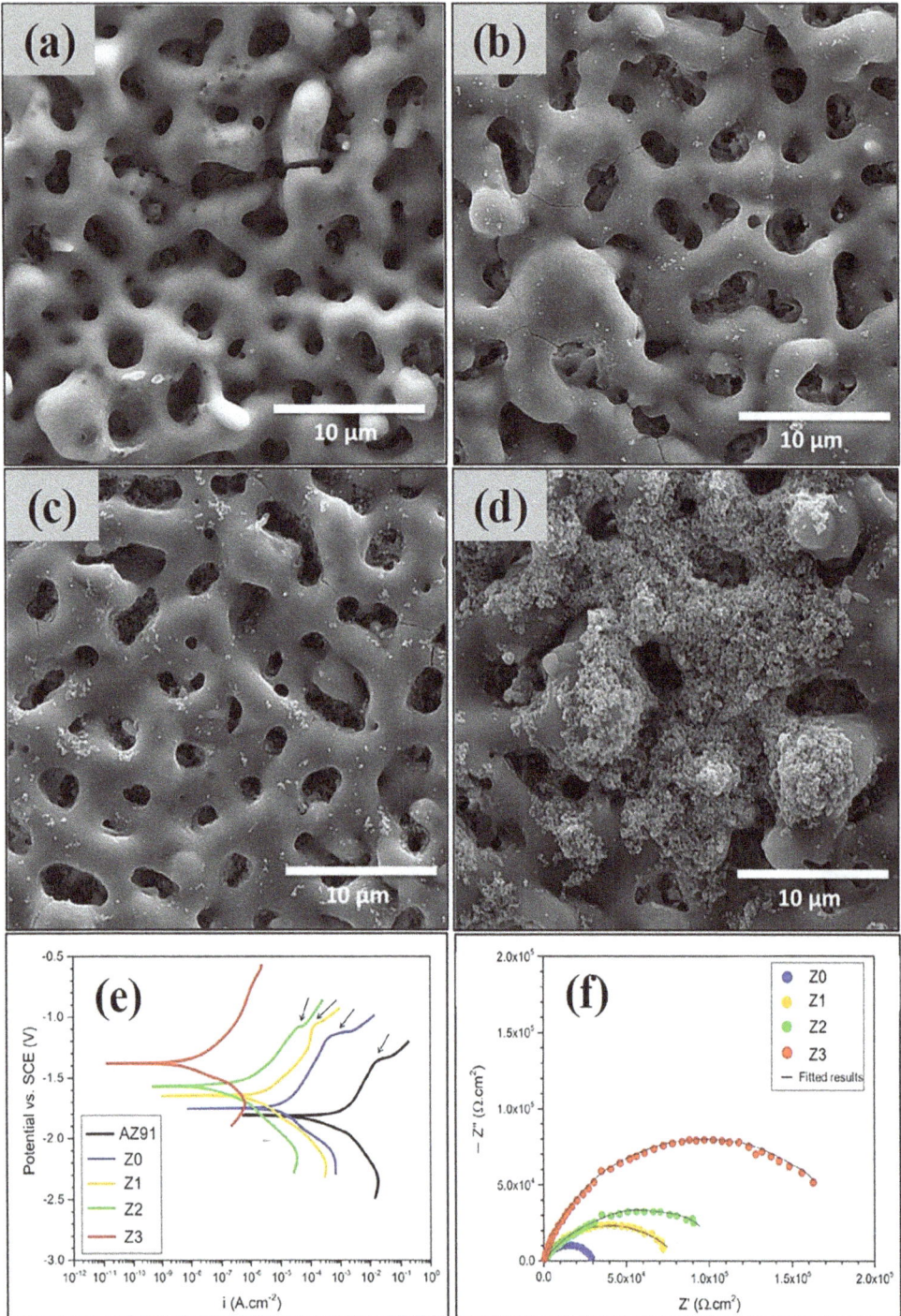

FIGURE 14.4 (a–d) Surface morphology of MAO coatings prepared using an alkaline phosphate-based electrolyte containing (a) 0 g/L ZnO (Z0); (b) 1.5 g/L ZnO (Z1); (c) 3.0 g/L ZnO (Z2); and (d) 4.5 g/L ZnO (Z3); (e) potentiodynamic polarization curves; and (f) Nyquist plots of the uncoated AZ91 Mg alloy and MAO-coated alloys formed with and without addition of ZnO in the electrolyte. *Reproduced with permission from (Bordbar-Khiabani et al., 2019) @ 2019 Elsevier B.V.*

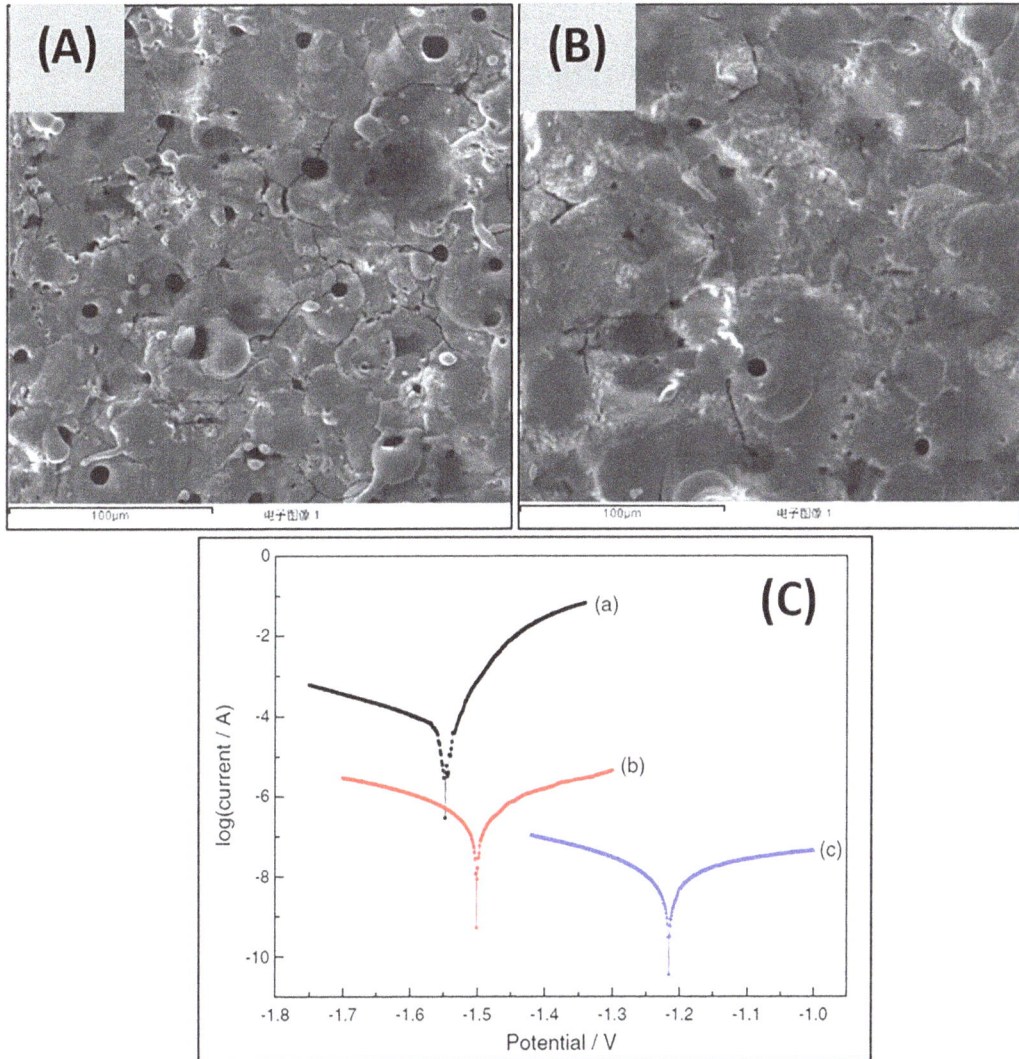

FIGURE 14.5 (A, B) Surface morphology of the MAO coatings formed on AZ91D Mg alloy using (a) an alkaline silicate electrolyte without the addition of ZrO_2 sol; (b) with the addition of 5 vol. % of ZrO_2 sol in the electrolyte; and (C) polarization curves of uncoated AZ91D Mg alloy (curve "a"), MAO-coated alloy formed without the addition of ZrO_2 sol (curve "b"); and MAO-coated alloy formed with the addition of ZrO_2 sol in the electrolyte (curve "c"). *Reproduced with permission from (Tang et al., 2011) @ 2011 Elsevier B.V.*

composite coating prepared with the addition of 5 vol. % ZrO_2 sol in the electrolyte is relatively smooth, with much better uniformity and lesser amount of pores and cracks (Figure 14.5(b)). The potentiodynamic polarization curves of uncoated AZ91D Mg alloy, MAO-coated alloy using the alkaline silicate electrolyte and MAO-coated alloy using the ZrO_2 sol are shown in Figure 14.5(c). The i_{corr} of MAO coating is decreased from 5.32×10^{-7} mA/cm^2 (sol-free electrolyte) to 1.44×10^{-8} mA/cm^2 for the one prepared using ZrO_2 sol in the electrolyte.

MAO coatings incorporated with Ag-HA have been shown to decrease the corrosion resistance in simulated body fluid (SBF). An increase in concentration of Ag-HA has led to an increase in its level of incorporation. Nevertheless, the high reactivity makes the coating more porous and decreased the compactness of the coating, which accounts for its poor corrosion resistance (Yazici et al., 2017). Incorporation of particles with high conductivity such as TiN NPs has led to a decrease in corrosion resistance of MAO composite coatings (Mashtalyar et al., 2017).

Immersion of MAO-coated AM50B and AM60B Mg alloys in 12% potassium dihydrogen phosphate and 5% sodium silicate solutions has enabled sealing of the pores and decreased the R_a. For MAO-coated AM50B Mg alloy, the alkaline phosphate and silicate treatment has reduced the i_{corr} from 0.145 µA/cm² to 0.026 and 0.024 µA/cm², respectively. Similarly, for AM60B Mg alloy, a decrease in i_{corr} from 0.104 µA/cm² to 0.032 and 0.022 µA/cm² is observed following the alkaline phosphate and silicate sealing treatment, respectively (Malayoglu et al., 2010). The effectiveness of an organic sealing agent to seal the pores of MAO coating formed on AZ91D Mg alloy using an alkaline silicate electrolyte was evaluated by Duan et al. (2006). The ability of the organic sealing agent to permeate through the pores and cracks of the MAO coating helps to seal them effectively and improves the corrosion-protective ability in 3.5% NaCl. However, the organic sealing agent fails to provide long-term corrosion protection, which is evidenced from the decrease in |Z| with an increase in immersion time from 28 to 288 hours. Dip coating of poly-L-Lactic acid (PLLA) has been shown to effectively seal the pores of the MAO coating formed on WE42 Mg alloy, which is evidenced by the absence of any signs of corrosion after 4 weeks of immersion in simulated body fluid (SBF) (Lu et al., 2011; Guo et al., 2011). Deposition of sol-gel TiO_2 and SiO_2 coatings has been suggested as a viable approach to seal the pores of MAO coatings on Mg and its alloys and to improve their corrosion resistance (Shi et al., 2009; Shang et al., 2009; Laleh et al., 2011; Li, Jing et al., 2012). The thickness of a single layer is insufficient to offer reasonable corrosion protection, which warrants dip coating for several cycles. Development of a thicker coating, though possible by an increase in the number of cycles, cracking of the coating limits the number of cycles. In addition, subsequent heat-treatment has increased the R_a. Development of cracks as and an increase in R_a, are detrimental to corrosion protection. High-intensity pulsed ion beam (HIPIB) radiation and steam treatment have been explored as post-treatments to seal the pores of the MAO coating (Han et al., 2011; Hiromoto and Yamamoto, 2010). These sealing treatments can also be applied to MAO composite coatings to further enhance their corrosion protective ability.

14.8.11   WEAR RESISTANCE

Yu et al. (2015) have studied the tribological behaviour of MAO coatings formed on AZ31 Mg alloy using an alkaline silicate-hexametaphosphate electrolyte containing 2 g/L SiC NPs (~50 nm) at low and high current densities using a ball-on-flat contact configuration with a SAE 521000 Cr steel ball (diameter: 9.5 mm; hardness: 62 HRC) at normal loads of 10 and 20 N. The coefficient of friction (COF) vs. sliding time curves of MAO coatings prepared at low and high current densities exhibit a sharp increase in COF followed by a steady state after a few seconds of sliding distance. The steady-state value of COF after 1800 seconds varies between 0.72 and 0.80 for coatings prepared at low current density, while for those prepared at high current density, it varies between 0.53 and 0.59. At a normal load of 10 N, the wear depth of MAO composite coatings incorporated with SiC particles, prepared at high and low current densities, is 101 and 68 µm, respectively. With a further increase in load to 20 N, no significant difference in wear depth is observed for MAO composite coatings prepared at high and low current densities, but the width of the wear track is increased. When compared to the MAO coating prepared using particle-free electrolyte, the MAO composite coating prepared using SiC NPs offers a better wear resistance, and this trend is observed for coatings prepared at both low and high current densities. The difference in wear behaviour is due to the higher thickness, lower surface roughness of the MAO composite coatings incorporated with SiC NPs. Coatings prepared at high current density exhibit a higher wear rate than those prepared at low current density. The difference in wear behaviour between the coatings prepared at low and high current densities is due to the compactness of the coating, which is much better for those prepared at low current density.

The wear behaviour of MAO coatings formed on AM50 Mg alloy using an alkaline phosphate electrolyte containing 5 g/L of micron- and nano-sized SiO_2 particles using a ball-on-disc contact configuration with a 6 mm diameter AISI 52100 steel ball as the counterbody under a normal load of 5 N was studied by Lu, Blawert, Huang et al. (2016). For MAO coating prepared using a particle-free electrolyte, the COF is increased to 0.87 after a sliding distance of 2 meters, whereas for the same sliding distance, the COF is about 0.6 for coatings prepared using both micron- and nano-sized SiO_2 particles. A large fluctuation in COF, ranging from 0.46 to 0.78, from 2 to 12 meters of sliding distance, is observed for MAO coating prepared using a particle-free electrolyte. Conversely, the fluctuations in COF are rather limited for MAO coatings prepared using both micron- and nano-sized SiO_2 particles, and they exhibit a steady-state value of 0.66 and 0.72, respectively, after 12 meters of sliding. The specific wear rate of MAO coating prepared using a particle-free electrolyte is 3.7×10^{-3} mm^3/N/m, while it is decreased to 4.0×10^{-4} mm^3/N/m and 7.3×10^{-4} mm^3/N/m for MAO coatings prepared using nano- and micron-sized SiO_2 particles, respectively. The decrease in specific wear rate observed for MAO composite coatings incorporated with nano- and micron-sized SiO_2 particles correlates well with their hardness. The hardness of MAO coating prepared using particle-free electrolyte and those containing micron- and nano-sized SiO_2 particles are 327 HV, 338 HV and 396 HV, respectively. The wear track pattern indicates complete removal of the MAO coating prepared using particle-free electrolyte. The width and depth of the wear track are 987 μm and 55 μm, respectively. In contrast, MAO coatings prepared using micron- and nano-sized SiO_2 particles are intact. The steel ball counterbody has encountered severe abrasive damage and exhibits a high volume loss when mated against MAO coatings prepared using particle-free electrolyte. This is due to the removal of the MAO coating, entrapment of wear debris between the mating couples and involvement of a three-body abrasive wear mechanism. The extent of wear of the steel ball is rather limited when mated against MAO coatings prepared using micron- and nano-sized SiO_2 particles, and in this respect, the later coating performed much better.

Mashtalyar et al. (2017) have evaluated the tribological behaviour of MAO coatings formed on MA8 Mg alloy using an alkaline silicate-fluoride-based electrolyte containing 1 to 4 g/L TiN NPs (20 nm) using a ball-on-plate configuration with a 10-mm-diameter Si_3N_4 ball as the counterbody at a normal load of 10 N. The wear rate is decreased from 1.12×10^{-5} mm^3/N/m (for particle-free coating) to 4.97×10^{-6} mm^3/N/m for MAO coating prepared using 3 g/L of TiN NPs. The increase in wear resistance can be correlated to the increase in hardness from 2.1 ± 0.3 GPa (for particle-free coating) to 4.2 ± 0.5 GPa for the MAO coating prepared using 3 g/L of TiN NPs. However, a further increase in concentration of TiN NPs to 4 g/L has increased the wear rate to 6.50×10^{-6} mm^3/N/m. The decrease in hardness to 4.2 ± 0.3 GPa substantiates the lower wear resistance. Besides, the MAO coating formed using 4 g/L of TiN NPs is brittle and consist of many cracks, which could not provide sufficient load-bearing capacity.

The wear resistance of MAO coatings on AZ31 Mg alloy prepared using an alkaline phosphate electrolyte containing 10 to 40 g/L Al_2O_3 NPs using a pin-on-disc contact configuration with a polymer pin at a normal load of 45 N was evaluated by Asgari et al. (2017). MAO coatings prepared using Al_2O_3 NPs exhibit a lower wear rate than those formed using a particle-free electrolyte, which is attributed due to the filling of the pores by the NPs. The COF vs. sliding distance curve shows an upward trend of an increase in COF with sliding distance for MAO coatings prepared using 20, 30 and 40 g/L of Al_2O_3 NPs. However, a downward trend of decrease in COF with sliding distance is observed for MAO coatings prepared using particle-free electrolyte and those containing 10 g/L of Al_2O_3 NPs. The fluctuations in COF observed for all samples indicate removal of the coating from the AZ31 Mg alloy, and the debris is in contact with the counterbody. MAO coating prepared using 30 g/L of Al_2O_3 NPs offered a better wear resistance.

The wear characteristics of MAO coatings prepared using a particle-free aluminate-phosphate-based electrolyte and those prepared using the same electrolyte containing 5 g/L of CeO_2 NPs (< 5 nm) was studied by Atapour et al. (2019). A ball-on disc contact configuration with a 6-mm-diameter AISI 52100 steel ball as the counterbody at three different normal loads, viz., 2, 5 and 10 N was used to evaluate the wear resistance. At 2 N, MAO coatings prepared using particle-free and particle-containing electrolyte remained intact. When the normal load is increased to 5 N, coating prepared using particle-free electrolyte failed, whereas the one prepared using the electrolyte containing CeO_2 NPs offered good wear resistance. At 10 N, complete removal of the coating is observed for the one prepared using particle-free electrolyte, while an excellent wear resistance is offered by the coating prepared using the electrolyte containing CeO_2 NPs. Transfer of the wear debris to the steel counterbody is very high for MAO coating prepared using particle-free electrolyte when compared to the one prepared using CeO_2 NPs. The development of a deep score on the steel ball mated with the MAO coating formed using particle-free electrolyte indicates the involvement of a three-body abrasive wear mechanism. The improvement in wear resistance observed for MAO composite coatings incorporated with CeO_2 NPs is due to their higher hardness, lower surface roughness and lower porosity. The large fluctuations in the COF of the coating prepared without particles points out its low load-bearing capacity. For the coating prepared with the addition of CeO_2 NPs, the extent of fluctuation is relatively lower. Similar inferences of the ability of MAO composite coatings prepared using WC, Si_3N_4 and MoS_2 to offer excellent wear resistance when compared to their particle-free counterparts has also been reported elsewhere (NasiriVatan et al., 2016; Lou et al., 2017, 2018).

14.8.12 BIOACTIVITY

Bioactivity is an important property of materials used for implant applications. MAO coatings formed on Mg and its alloys lack bioactivity. In order to impart better bioactivity, addition of CaP-based compounds in the electrolyte has been attempted. However, lack of solubility of many Ca-based salts such as $Ca(OH)_2$, CaF_2, $CaCO_3$ etc., and preferential diffusion of OH- ion when compared to the PO_4^{3-} ion pose a major challenge (Chaharmahali et al., 2020). Addition of pure and doped HA particles is a viable option to incorporate bioactive materials in the MAO coating. Yazici et al. (2017) have prepared MAO coatings on Mg-Sr-Ca ternary alloy using an alkaline phosphate-based electrolyte containing 0, 1 and 10 g/L Ag-doped HA particles (AgHA-0, AgHA-1 and AgHA-10). The level of incorporation of the Ag-HA particles in the MAO coating is increased with an increase in its concentration in the electrolyte. When subjected to immersion in SBF, the extent of formation of apatite crystals is appreciable for MAO coating prepared using AgHA-10, while no significant apatite growth could be observed on MAO coatings prepared using AgHA-0 and AgHA-1 (Figure 14.6 (a–c)). MAO coating prepared using AgHA-10 is capable of promoting apatite growth as early as 5 days of immersion in SBF, and the extent of growth of apatite becomes much appreciable covering the entire surface after 15 days of immersion in SBF (Figure 14.6 (d–f)). For MAO coatings formed on AZ91D Mg alloy, an increase in Ca/P ratio from 1.58 to 1.75 is observed when the concentration of GO nanosheets in the alkaline metasilicate electrolyte is increased from 0 to 30 g/L, suggesting the ability of MAO composite coating to exhibit better bioactivity (Askarnia et al., 2021).

14.8.13 ANTIBACTERIAL ACTIVITY

Ren et al. (2011) have evaluated the antibacterial activity of pure Mg, AZ31 Mg alloy, MAO-coated Mg and F- and Si-coated Mg prepared by chemical conversion method against *Escherichia coli (E.coli)* and *Staphylococcus aureus (S. aureus)*. Both pure Mg and AZ31 Mg alloy exhibits a strong inhibiting effect against the growth of *E.coli* and *S. aureus*. MAO-coated Mg also shows a reasonable inhibition, whereas pure Mg and AZ31 Mg alloy coated with F and Si by chemical conversion method fails to reveal any antibacterial activity. This is due to the large increase in pH following the dissolution of pure Mg and AZ31 Mg alloy in the culture medium (pH: 7.40), reaching 8.50 and 8.20, respectively, after 6 hours of incubation. The pH is increased further to 10.0 and 9.50, respectively, after 24 hours of incubation.

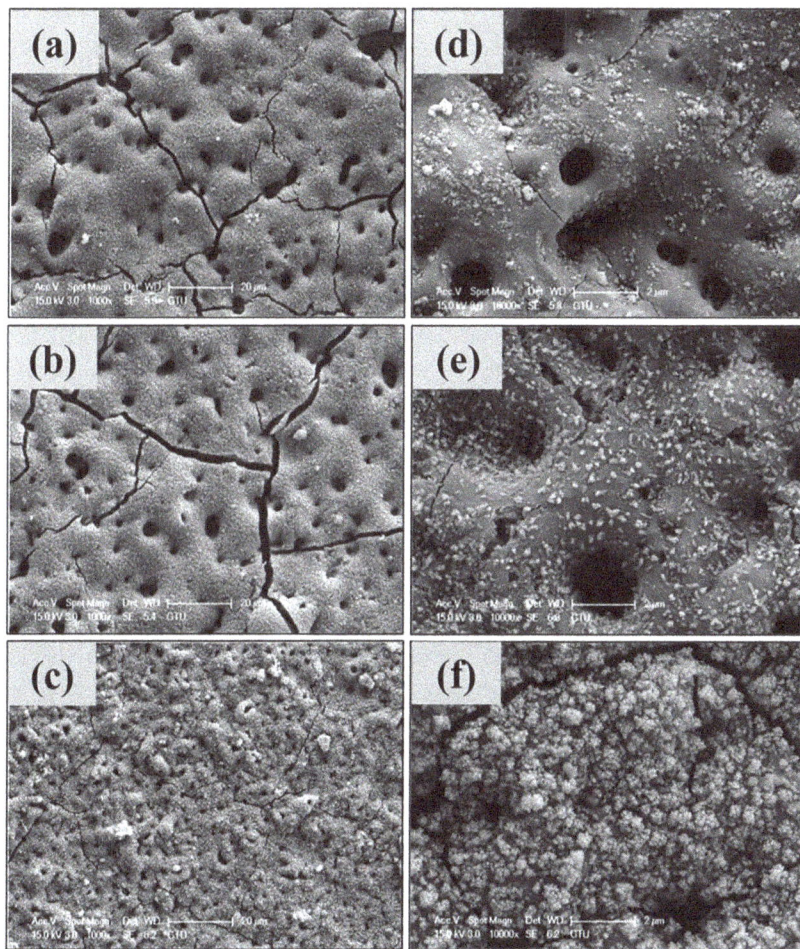

FIGURE 14.6 (a–c) Surface morphology of MAO coatings prepared using varying concentrations of Ag-doped HA after 15 days of immersion in SBF: (a) AgHA-0; (b) AgHA-1; and (c) AgHA-10; (d–f) surface morphology of MAO coating prepared using AgHA-10 before and after immersion in SBF: (d) before immersion; (e) after 5 days of immersion; and (f) after 15 days of immersion. *Source: Reproduced with permission from (Yazici et al., 2017) © 2017 Elsevier B.V.*

According to Robinson et al. (2010), the bacteria could survive only if the pH of the environment is in the range of 6.0 to 8.0. At pHs higher than 8.0, the cytoplasmic proteins could lose their structural integrity. Hence, the antibacterial effect observed for pure Mg and AZ31 Mg alloy is due to the large increase in pH of the culture medium following dissolution of Mg, which makes survival of the bacteria impossible. Due to the lack of formation of a compact MAO coating on Mg as well as its porous structure, the pH is increased to 8.00, 8.50 and 8.80 after 6, 12 and 24 hours of incubation, respectively. The increase in pH beyond the threshold is the main reason for its antibacterial activity. The inability of Mg and AZ31 Mg alloy coated with F and Si by the conversion coating method to exhibit any antibacterial activity is due to the formation of a uniform and dense coating that effectively prevents dissolution of Mg/AZ31 Mg alloy, which is evidenced by the absence of any significant change in the pH of the culture medium even after 24 hours of incubation.

The pores and cracks present in the MAO coating allow permeation of the culture medium, leading to dissolution of Mg/Mg alloy accompanied by an increase in the pH of the medium. However, the intactness and thickness of the inner barrier layer determines the extent of increase in pH. MAO composite coatings clearly reveal that the preferential locations of the NPs are the pores and cracks, and such conditions would not facilitate corrosion of Mg/Mg alloy and limit the extent of increase in pH. This leads to the question whether MAO composite coatings could offer antibacterial activity, and if not, what are the strategies to impart antibacterial activity for them? Antibacterial activity is an important attribute of MAO composite coatings formed on Mg and its alloys, particularly for those meant for biomedical applications. The possible ways of enhancing the antibacterial activity of MAO composite coatings incorporated with particles is recently reviewed by Fattah-alhosseini, Molaei et al. (2020). The choice of a Mg alloy containing an antibacterial element such as Mg-Ag or Mg-Cu alloy, addition of Ag- and Cu-containing compounds in the electrolyte such as Ag-doped HA, silver acetate, copper sulphate, etc., and deposition of another layer over the MAO coating with compounds that possess antibacterial effect such as tannic acid, ciprofloxacin etc., are suggested as viable options to impact antibacterial activity for the MAO coating (Ryu and Hong, 2010; Yazici et al., 2017; Shao et al., 2020; Yi et al., 2021; Liang et al., 2021; Lin et al., 2021; Wang et al., 2021; Xue et al., 2021).

The antibacterial activity of MAO coatings prepared using a silicate-fluoride-based electrolyte containing 0 and 0.5 g/L of $AgNO_3$ against the growth of *E. coli* and *S. aureus* is evaluated by Ryu and Hong (2010). For the MAO coating prepared without the addition of $AgNO_3$, the colony forming unit (CFU) of *E.coli* after 24 hours of incubation is increased from 1.5×10^5 to 2.4×10^7 cells/mL, while that of the *S. aureus* is increased from 2.2×10^5 to 3.9×10^5 cells/mL. For MAO coatings prepared using 0.5 g/L of $AgNO_3$, the CFU of *E.coli* and *S. aureus* is decreased to < 10 cells/mL, amounting to an antibacterial efficiency of 99.9%. Yazici et al. (2017) have prepared MAO coatings on Mg-3Sr-6Ca alloy using an alkaline phosphate-based electrolyte containing 0, 1 and 10 g/L of Ag-doped HA (AgHA-0, AgHA-1 and AgHA-10) and evaluated their antibacterial activity against the growth of *E. coli*. The viability for the growth of *E.coli* is 49%, 20% and 0% for AgHA-0, AgHA-1 and AgHA-10, respectively. Chen et al. (2021) have prepared MAO coatings on Mg-3Zn-0.5Sr alloy using an alkaline phosphate-based electrolyte with the addition of 0, 1, 2 and 3 g/L of silver acetate (CH_3COOAg). MAO coatings prepared using 0 and 1 g/L of silver acetate fail to exhibit any antibacterial activity against the growth of *E. coli*. However, addition of 2 and 3 g/L of silver acetate in the electrolyte has enabled the formation of MAO coatings with excellent antibacterial activity. The antibacterial efficacy (ABE) of MAO coating prepared using 2 g/L of silver acetate is 90.89%. The morphological features revealing the extent of growth of *E, coli* on the surface of MAO coatings prepared using 0, 1, 2 and 3 g/L of silver acetate is shown in Figure 14.5. A large number of *E. coli* growth could be observed on MAO coatings prepared using 0 and 1 g/L of silver acetate (Figure 14.7 (a, a1) and (b, b1)). The extent of growth of *E. coli* is minimal on MAO coatings formed using 2 and 3 g/L of silver acetate (Figure 14.7 (c, c1) and (d, d1)).

Yan et al. (2019) have compared the antibacterial activity of as-cast and as-solution treated Mg-0.06Cu alloy as well as pure Mg before and after MAO treatment. The *S. aureus* colony count after 6 hours of incubation is found to be 1×10^5, 4×10^4 and 6×10^4 CFU/mL for pure Mg, as-cast and as-solution treated Mg-0.06Cu alloy, respectively. Conversely, there is no bacterial growth on MAO-coated Mg and Mg-0.06Cu alloys after 6 hours of incubation. Chen et al. (2019) have prepared MAO coating Mg-2Zn-1Gd-0.5Zr alloy using a fluoride-phosphate-based electrolyte containing 1 g/L nano CuO and evaluated its antibacterial activity against the growth of *S. aureus*. The antibacterial rate of the MAO composite coating is found to be 93% and 98% after 6 and 12 hours of incubation, while it is > 99% after 24 hours.

Copper sulphate (0.1, 0.5 and 1.0 g/L) was added to an alkaline silicate-fluoride-based electrolyte to prepare MAO composite coatings with antibacterial property. MAO coating prepared using 0.1 g/L of $CuSO_4$ inhibits the growth of *S. aureus* only by 10%, whereas those prepared using 0.5 and 1 g/L of $CuSO_4$ exhibit up to 50% inhibition (Liang et al., 2021). The release of Ag^+ and Cu^{2+} ions

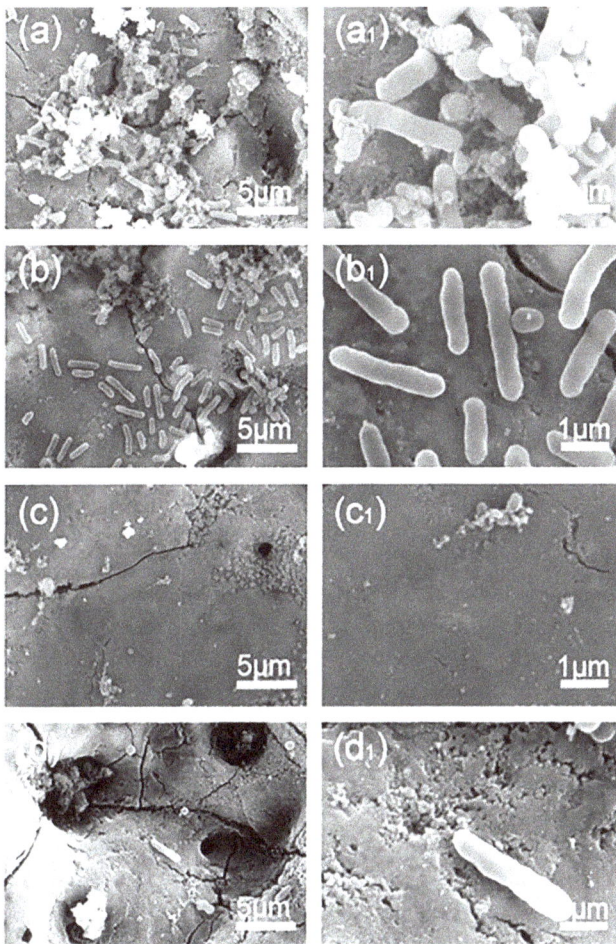

FIGURE 14.7 Surface morphology of MAO coatings after culturing E.coli for 24 h: (a, a1) AgAC-0, (b, b1) AgAC-1, (c, c1) AgAC-2, (d, d1) AgAC-3. *Reproduced with permission from (Chen et al., 2021) © 2021 Taylor & Francis.*

from the MAO coating is considered responsible for the excellent antibacterial activity against the growth of both *E. coli* and *S. aureus*.

MAO coatings formed on AZ91D Mg alloy using an alkaline metasilicate electrolyte containing 10 to 30 mg/L of GO nanosheets exhibits good antibacterial activity against the growth of *E. coli* and *S. aureus*, and the extent of inhibition is much better for coatings prepared using 30 mg/L of GO nanosheets. MAO coating prepared using 20 mg/L of GO nanosheets exhibits a preference to inhibit the growth of *E. coli* than *S. aureus*, whereas the one prepared using 30 mg/L of GO nanosheets is capable of inhibiting the growth of both *E. coli* and *S. aureus* (Askarnia et al., 2021).

The strong attraction between the Ag^+ ions and the thiol group of the enzyme proteins makes the bacteria lose its activity and eventually die. The binding of Ag^+ ions with the RNA and DNA prevents reproduction of the bacteria. The Ag^+ ions initially bind with the proteins present in the cell walls then reach the cytoplasm and change the structure of the bacteria and, finally, cause the cell death. Adsorption on the surface of the bacteria, damage of the cell membrane and solidification of the structure of proteins is the mechanism by which the Cu^{2+} ions exhibit antibacterial activity.

Attack of the cell membrane by the sharp edge of GO nanosheets affects the viability of the bacterial growth.

14.9   CONCLUSIONS AND OUTLOOK

Effective dispersion of the particles in the electrolyte without agglomeration is one of the important requirements of preparing MAO composite coatings. Particles with a negative zeta potential will easily move towards the anode by electrophoresis. Additives and/or surfactants can be used to impart desirable surface charge for the particles. Movement of the particles by electrophoresis during the anodic oxidation stage, followed by transport of the particles through the discharge channels, pores and cracks and attachment of particles with the molten mass, are the key steps in determining incorporation of particles in the MAO coating. The mode of incorporation of particles can be "inert", "partially reactive" and "reactive". Micron-sized particles with a higher melting point are incorporated by inert mode, while nano-sized particles with a lower melting point tend to show preference for reactive incorporation. The preferential location of the nano-sized particles is the pores and cracks, while micron-sized particles mainly adhere to the molten mass. Particle size plays a key role in determining the amount of particles as well as their mode of incorporation.

MAO composite coatings are uniform and compact as long as agglomeration of the particles in the electrolyte and their segregation at the MAO coating/Mg alloy interface is avoided. Particle incorporation, in general, reduced the surface porosity and the mean pore diameter of MAO composite coatings if the electrical discharges are controlled by a careful choice of experimental conditions. Addition of particles in the electrolyte, in general, increases the R_a of the MAO coating. An increase in treatment time also leads to an increase in the R_a. Particle incorporation increases the adhesive strength of the MAO coating as long as formation of a dense layer is promoted. However, if particle incorporation makes the coating more porous and highly brittle and affects the formation of a continuous inner layer then the coating adhesion will become poor. Particle incorporation, in general, increases the hardness of the MAO composite coatings as long as it is not increasing the porosity of the coating and makes the coating less compact and brittle. Besides the incorporated particles, the formation of new phases would also contribute to the hardness of the MAO composite coating. An increase in hardness is likely to increase the wear resistance. MAO composite coatings with a uniform and compact layer, higher thickness and lower surface roughness offer better wear resistance.

Higher chemical stability and lower electrical conductivity of the particles, formation of a compact coating, sealing of pores and cracks and formation of new phases with better chemical stability, formation of an intact inner layer would contribute to an improvement in corrosion resistance of MAO composite coatings. Conversely, a higher surface porosity, higher R_a, uneven distribution of particles and lack of formation of a continuous and dense inner layer could deleteriously influence the corrosion resistance of MAO composite coatings. With an appropriate choice of particles such as Ag-HA and GO, it would be possible to impart better bioactivity for MAO composite coatings. Similarly, the antibacterial activity can be imparted for the MAO composite coating by choosing a Mg alloy containing an antibacterial element such as Mg-Ag or Mg-Cu alloy, addition of Ag- and Cu-containing compounds in the electrolyte such as Ag-doped HA, silver acetate, copper sulphate, etc. and deposition of another layer over the MAO coating with compounds that possess antibacterial effect such as tannic acid, ciprofloxacin, etc.

Nano-sized particles appear to be a preferred choice, as it is easier to incorporate them than micron-sized particles. However, agglomeration of NPs in the electrolyte should be avoided by a suitable choice of additives and/or surfactants. As nano-sized particles could show preference for reactive incorporation with a suitable combination of type of electrolyte and electrical parameters, the formation of new phases would facilitate an increase in hardness and improve the corrosion resistance. Addition of sols instead of particles has promoted the formation of a uniform and compact coating with fewer defects, lower porosity, lower R_a and higher volume fraction of new phases. Hence, sols appear to be a preferred choice to particles for preparing MAO composite coatings. However, studies that compare the characteristics of MAO composite coatings prepared using sols and particles are rather limited. Also, studies on the hardness and wear resistance of MAO-based composite coatings prepared using sols are scarce. Currently, the ways of imparting suitable functionalities for MAO composite coatings and development of duplex and multi-layer coating systems involving MAO composite coating have received considerable attention, and more developments like this will also be made in the near future.

REFERENCES

Arrabal, R., Matykina, E., Skeldon, P., Thompson, GE. Incorporation of zirconia particles into coatings formed on magnesium by plasma electrolytic oxidation. *J. Mater. Sci.* 2008, 43, 1532–1538.

Arrabal, R., Matykina, E., Viejo, F., Skeldon, P., Thompson, GE., Merino, MC. AC plasma electrolytic oxidation of magnesium with zirconia nanoparticles. *Appl. Surf. Sci.* 2008, 254, 6937–6942.

Asgari, M., Aliofkhazraei, M., Barati Darband, GH., Rouhaghdam, AS. Evaluation of alumina nanoparticles concentration and stirring rate on wear and corrosion behavior of nanocomposite PEO coating on AZ31 magnesium alloy. *Surf. Coat. Technol.* 2017, 309, 124–135.

Asgari, M., Aliofkhazraei, M., Barati Darband, GH., Sabour Rouhaghdam, A. How nanoparticles and submicron particles adsorb inside coating during plasma electrolytic oxidation of magnesium? *Surf. Coat. Technol.* 2020, 383, 125252.

Askarnia, R., Roueini Fardi, S., Sobhani, M., Staji, H., Aghamohammadi, H. Effect of graphene oxide on properties of AZ91 magnesium alloys coating developed by micro-arc oxidation process, *J. Alloys Compd.* 2021, 892, 162106.

Atapour, M., Blawert, C., Zheludkevich, ML. The wear characteristics of CeO_2 containing nanocomposite coating made by aluminate-based PEO on AM 50 magnesium alloy. *Surf. Coat. Technol.* 2019, 357, 626–637.

Azadani, MN., Zahedi, A., Bowoto, AK., Oladapo, BI. A review of current challenges and prospects of magnesium and its alloy for bone implant applications. *Prog. Biomater.* 2022, 11, 1–26.

Blawert, C., Sah, SP., Liang, J., Huang, Y., Höche, D. Role of sintering and clay particle additions on coating formation during PEO processing of AM50 magnesium alloy. *Surf. Coat. Technol.* 2012, 213, 48–58.

Bordbar-Khiabani, A., Yarmand, B., Mozafari, M. Enhanced corrosion resistance and in-vitro biodegradation of plasma electrolytic oxidation coatings prepared on AZ91 Mg alloy using ZnO nanoparticles-incorporated electrolyte. *Surf. Coat. Technol.* 2019, 360, 153–171.

Chaharmahali, R., Fattah-alhosseini, A., Babaei, K. Surface characterization and corrosion behavior of calcium phosphate (Ca-P) base composite layer on Mg and its alloys using plasma electrolytic oxidation (PEO): A review. *J. Magnes. Alloy.* 2020, 9, 21–40.

Chen, J., Zhang, Y., Ibrahim, M., Etim, IP., Tan, L., Yang, K. In vitro degradation and antibacterial property of a copper-containing micro-arc oxidation coating on Mg-2Zn-1Gd-0.5Zr alloy. *Colloids Surf. B Biointerfaces.* 2019, 179, 77–86.

Chen, Y., Dou, J., Pang, Z., Zheng, Z., Yu, H., Chen, C. Ag-containing antibacterial self-healing micro-arc oxidation coatings on Mg – Zn – Sr alloys. *Surf. Eng.* 2021, 37, 926–941.

Clyne, TW., Troughton, SC. A review of recent work on discharge characteristics during plasma electrolytic oxidation of various metals. *Int. Mater. Rev.* 2019, 64, 127–162.

Darband, GHB., Aliofkhazraei, M., Hamghalam, P., Valizade, N. Plasma electrolytic oxidation of magnesium and its alloys: Mechanism, properties and applications. *J. Magnes. Alloy.* 2017, 5, 74–132.

Duan, H., Du, K., Yan, C., Wang, F. Electrochemical corrosion behavior of composite coatings of sealed MAO film on magnesium alloy AZ91D. *Electrochim Acta.* 2006, 51, 2898–2908.

Fattah-alhosseini, A., Chaharmahali, R., Babaei, K. Effect of particles addition to solution of plasma electrolytic oxidation (PEO) on the properties of PEO coatings formed on magnesium and its alloys: A review. *J. Magnes. Alloy.* 2020, 8, 799–818.

Fattah-alhosseini, A., Molaei, M., Attarzadeh, N., Babaei, K., Attarzadeh, F. On the enhanced antibacterial activity of plasma electrolytic oxidation (PEO) coatings that incorporate particles: A review. *Ceram. Int.* 2020, 46, 20587–20607.

Gnedenkov, SV., Sinebryukhov, SL., Mashtalyar, DV., Imshinetskiy, IM., Samokhin, AV., Tsvetkov, YV. Fabrication of coatings on the surface of magnesium alloy by plasma electrolytic oxidation using ZrO_2 and SiO_2 nanoparticles. *J. Nanomater.* 2015, Article ID 154298.

Guo, M., Cao, L., Lu, P., Liu, Y., Xu, X. Anticorrosion and cytocompatibility behavior of MAO/PLLA modified magnesium alloy WE42. *J. Mater. Sci.: Mater. Med.* 2011, 22, 1735–1740.

Han, XG., Zhu, XP., Lei, MK. Electrochemical properties of microarc oxidation films on a magnesium alloy modified by high intensity pulsed ion beam. *Surf. Coat. Technol.* 2011, 206, 874–878.

Hiromoto, S., Yamamoto, A. Control of degradation rate of bioabsorbable magnesium by anodization and steam treatment. *Mater. Sci. Eng. C.* 2010, 30, 1085–1093.

Hwang M, Chung W. Effects of a carbon nanotube additive on the corrosion-resistance and heat-dissipation properties of plasma electrolytic oxidation on AZ31 magnesium alloy. *Materials.* 2018, 11, 2438.

Kaseem, M., Fatimah, S., Nashrah, N., Gun Ko, Y. Recent progress in surface modification of metals coated by plasma electrolytic oxidation: Principle, structure, and performance. *Prog. Mater. Sci.* 2020, 117, 100735.

Laleh, M., Kargar, F., Sabour Rouhaghdam, A. Improvement in corrosion resistance of micro arc oxidation coating formed on AZ91D magnesium alloy via applying a nano-crystalline sol – gel layer. *J. Sol-Gel Sci. Technol.* 2011, 59, 297–303.

Laleh, M., Rouhaghdam, AS., Shahrabi, T., Shanghi, A. Effect of alumina sol addition to micro-arc oxidation electrolyte on the properties of MAO coatings formed on magnesium alloy AZ91D. *J. Alloys Compd.* 2010, 496, 548–552.

Lee, KM., Shin, KR., Namgung, S., Yoo, B., Shin, DH. Electrochemical response of ZrO_2-incorporated oxide layer on AZ91 Mg alloy processed by plasma electrolytic oxidation. *Surf. Coat. Technol.* 2011, 205, 3779–3784.

Li, W., Tang, M., Zhu, L., Liu, H. Formation of microarc oxidation coatings on magnesium alloy with photo-catalytic performance. *Appl. Surf. Sci.* 2012, 258, 10017–10021.

Li, X., Luan, BL. Discovery of Al_2O_3 particles incorporation mechanism in plasma electrolytic oxidation of AM60B magnesium alloy. *Mater. Lett.* 2012, 86, 88–91.

Li, Z., Jing, X., Yuan, Y., Zhang, M. Composite coatings on a Mg – Li alloy prepared by combined plasma electrolytic oxidation and sol – gel techniques. *Corros. Sci.* 2012, 63, 358–366.

Liang, D., Liang, P., Yi, Q., Sha, S., Shi, J., Chang, Q. Copper coating formed by micro-arc oxidation on pure Mg improved antibacterial activity, osteogenesis, and angiogenesis in vivo and in vitro. *Biomed. Microdevices.* 2021, 23, 39.

Liang, J., Hu, L., Hao, J. Preparation and characterization of oxide films containing crystalline TiO_2 on magnesium alloy by plasma electrolytic oxidation. *Electrochim. Acta.* 2007, 52, 4836–4840.

Lim, TS., Ryu, HS., Hong, SH. Electrochemical corrosion properties of CeO_2-containing coatings on AZ31 magnesium alloys prepared by plasma electrolytic oxidation. *Corros. Sci.* 2012, 62, 104–111.

Lin, X., Wang, X., Tan, L., Wan, P., Yu, X., Li, Q., Yang, K. Effect of preparation parameters on the properties of hydroxyapatite containing micro-arc oxidation coating on biodegradable ZK60 magnesium alloy. *Ceram. Int.* 2014, 40, 10043–10051.

Lin, Z., Wang, T., Yu, X., Sun, X., Yang, H. Functionalization treatment of micro-arc oxidation coatings on magnesium alloys: A review. *J. Alloys Compd.* 2021, 879, 160453.

Lou, BS., Lee, JW., Tseng, CM., Lin, YY., Yen, CA. Mechanical property and corrosion resistance evaluation of AZ31 magnesium alloys by plasma electrolytic oxidation treatment: Effect of MoS_2 particle addition. *Surf. Coat. Technol.* 2018, 350, 813–822.

Lou, BS., Lin, YY., Tseng, CM., Lu, YC., Duh, JG., Lee, JW. Plasma electrolytic oxidation coatings on AZ31 magnesium alloys with Si_3N_4 nanoparticle additives. *Surf. Coat. Technol.* 2017, 332, 358–367.

Lu, P., Cao, L., Lin, Y., Xu, X., Wu, X. Evaluation of magnesium ions release, biocorrosion, and hemocompatibility of MAO/PLLA modified magnesium alloy WE42. *J. Biomed. Mater. Res. B.* 2011, 96B, 101–109.

Lu, X., Blawert, C., Huang, Y., Ovri, H., Zheludkevich, ML., Kainer, KU. Plasma electrolytic oxidation coatings on Mg alloy with addition of SiO_2 particles. *Electrochim. Acta.* 2016, 187, 20–33.

Lu, X., Blawert, C., Mohedano, M., Scharnagl, N., Zheludkevich, ML., Kainer, KU. Influence of electrical parameters on particle uptake during plasma electrolytic oxidation processing of AM50 Mg alloy. *Surf. Coat. Technol.* 2016, 289, 179–185.

Lu, X., Blawert, C., Scharnagl, N., Kainer, KU. Influence of incorporating Si_3N_4 particles into the oxide layer produced by plasma electrolytic oxidation on AM50 Mg alloy on coating morphology and corrosion properties. *J. Magnes. Alloy.* 2013, 1, 267–274.

Lu, X., Blawert, C., Zheludkevich, ML., Kainer, KU. Insights into plasma electrolytic oxidation treatment with particle addition. *Corros. Sci.* 2015, 101, 201–207.

Lu, X., Mohedano, M., Blawert, C., Matykina, E., Arrabal, R., Kainer, KU., Zheludkevich, ML. Plasma electrolytic oxidation coatings with particle additions – A review. *Surf. Coat. Technol.* 2016, 307, 1165–1182.

Malayoglu, U., Tekin, KC., Shrestha, S. Influence of post-treatment on the corrosion resistance of PEO coated AM50B and AM60B Mg alloys. *Surf. Coat. Technol.* 2010, 205, 1793–1798.

Mashtalyar, DV., Gnedenkov, SV., Sinebryukhov, SL., Imshinetskiy, IM., Puz', AV. Plasma electrolytic oxidation of the magnesium alloy MA8 in electrolytes containing TiN nanoparticles. *J. Mater. Sci. Technol.* 2017, 33, 461–468.

Mohedano, M., Blawert, C., Zheludkevich, ML. Silicate-based plasma electrolytic oxidation (PEO) coatings with incorporated CeO_2 particles on AM50 magnesium alloy. *Mater. Des.* 2015, 86, 735–744.

NasiriVatan, H., Ebrahimi-Kahrizsangi, R., Asgarani, MK. Tribological performance of PEO-WC nano-composite coating on Mg alloys deposited by plasma electrolytic oxidation. *Tribol. Int.* 2016, 98, 253–260.

Necula, BS., Fratila-Apachitei, LE., Berkani, A., Apachitei I., Duszczyk, J. Enrichment of anodic MgO layers with Ag nanoparticles for biomedical applications. *J. Mater. Sci.: Mater. Med.* 2009, 20, 339–345.

Ren, L., Lin, X., Tan, L., Yang, K. Effect of surface coating on antibacterial behavior of magnesium based metals. *Mater. Lett.* 2011, 65, 3509–3511.

Robinson, DA., Griffith, RW., Shechtman, D., Evans, RB., Conzemius, MG. In vitro antibacterial properties of magnesium metal against Escherichia coli, Pseudomonas aeruginosa and Staphylococcus aureus. *Acta Biomater.* 2010, 6, 1869–1877.

Ryu, HS., Hong, SH. Corrosion resistance and antibacterial properties of Ag-containing MAO coatings on AZ31 magnesium alloy formed by microarc oxidation. *J. Electrochem. Soc.* 2010, 157, C131–C136.

Sankara Narayanan, TSN., Park, IS., Lee, MH. Strategies to improve the corrosion resistance of microarc oxidation (MAO) coated magnesium alloys for degradable implants: Prospects and challenges. *Prog. Mater. Sci.* 2014, 60, 1–71.

Seyfoori, A., Mirdamadi, S., Seyedraoufi, ZS., Khavandi, A., Aliofkhazraei, M. Synthesis of biphasic calcium phosphate containing nanostructured films by micro arc oxidation on magnesium alloy. *Mater. Chem. Phys.* 2013, 142, 87–94.

Shang, W., Chen, B., Shi, X., Chen, Y., Xiao X. Electrochemical corrosion behavior of composite MAO/sol – gel coatings on magnesium alloy AZ91D using combined micro-arc oxidation and sol – gel technique. *J. Alloys Compd.* 2009, 474, 541–545.

Shao, Y., Zeng, RC., Li, SQ., Cui, LY., Zou, YH., Guan, SK., Zheng, YF. Advance in antibacterial magnesium alloys and surface coatings on magnesium alloys: A review. *Acta Metall. Sin. (Engl. Lett.).* 2020, 33, 615–629.

Shi, P., Ng, WF., Wong, MH., Cheng, FT. Improvement of corrosion resistance of pure magnesium in Hanks' solution by microarc oxidation with sol – gel TiO_2 sealing. *J. Alloys Compd.* 2009, 469, 286–292.

Song, YL., Sun, XY., Liu, YH. Effect of TiO_2 nanoparticles on the microstructure and corrosion behavior of MAO coatings on magnesium alloy. *Mater. Corros.* 2011, 62, No. 9999.

Tan, J., Ramakrishna, S. Applications of magnesium and its alloys: A review. *Appl. Sci.* 2021, 11, 6861.

Tang, M., Liu, H., Li, W., Zhu, L. Effect of zirconia sol in electrolyte on the characteristics of microarc oxidation coating on AZ91D magnesium. *Mater. Lett.* 2011, 65, 413–415.

Wang, S., Si, N., Xia, Y., Liu, L. Influence of nano-SiC on microstructure and property of MAO coating formed on AZ91D magnesium alloy. *Trans. Nonferrous Met. Soc. China.* 2015, 25, 1926–1934.

Wang, S., Wen, L., Wang, Y., Cheng, Y., Cheng, YL., Zou, Y., Zhu, Y., Chen, G., Ouyang, J., Jia, D., Zhou, Y. One-step fabrication of double-layer nanocomposite coating by plasma electrolytic oxidation with particle addition. *Appl. Surf. Sci.* 2022, 592, 153043.

Wang, X., Yan, H., Hang, R., Shi, H., Wang, L., Ma, J., Liu, X. Yao, X. Enhanced anticorrosive and antibacterial performances of silver nanoparticles/polyethyleneimine/MAO composite coating on magnesium alloys. *J. Mater. Res. Technol.* 2021, 11, 2354–2364.

Wang, Y., Wei, D., Yu, J., Di, S. Effects of Al_2O_3 nano-additive on performance of micro-arc oxidation coatings formed on AZ91D Mg alloy. *J. Mater. Sci. Tech.* 2014, 30, 984–990.

Xue, K., Liang, LX., Cheng, SC., Liu, HP., Cui, LY., Zeng, RC., Li, SQ., Wang, ZL. Corrosion resistance, antibacterial activity and drug release of ciprofloxacin-loaded micro-arc oxidation/silane coating on magnesium alloy AZ31. *Prog. Org. Coat.* 2021, 158, 106357.

Yan, X., Zhao, MC., Yang, Y., Tan, L., Zhao, YC., Yin, DF., Yang, K., Atrens, A. Improvement of biodegradable and antibacterial properties by solution treatment and micro-arc oxidation (MAO) of a magnesium alloy with a trace of copper. *Corros. Sci.* 2019, 156, 125–138.

Yang, Y., Wu, H. Effects of current frequency on the microstructure and wear resistance of ceramic coatings embedded with SiC nano-particles produced by micro-arc oxidation on AZ91D magnesium alloy. *J. Mater. Sci. Tech.* 2010, 26, 865–871.

Yazici, M., Gulec, AE., Gurbuz, M., Gencer, Y., Tarakci, M. Biodegradability and antibacterial properties of MAO coatings formed on Mg-Sr-Ca alloys in an electrolyte containing Ag doped hydroxyapatite. *Thin Solid Films*. 2017, 644, 92–98.

Yi, Q., Liang, P., Liang, D., Shi, J., Sha, S., Chang, Q. Multifunction Sr doped microporous coating on pure magnesium of antibacterial, osteogenic and angiogenic activities. *Ceram. Int.* 2021, 47, 8133–8141.

Yu, L., Cao, J., Cheng, Y. An improvement of the wear and corrosion resistances of AZ31 magnesium alloy by plasma electrolytic oxidation in a silicate – hexametaphosphate electrolyte with the suspension of SiC nanoparticles. *Surf. Coat. Technol.* 2015, 276, 266–278.

Zhang, D., Gou, Y., Liu, Y., Guo, X. A composite anodizing coating containing superfine Al_2O_3 particles on AZ31 magnesium alloy. *Surf. Coat. Technol.* 2013, 236, 52–57.

Zoubi, WA., Kamil, MP., Ko, YG. Synergistic influence of inorganic oxides (ZrO_2 and SiO_2) with N_2H_4 to protect composite coatings obtained via plasma electrolyte oxidation on Mg alloy. *Phys. Chem. Chem. Phys.* 2017, 19, 2372–2382.

15 Electrodeposited Metallic/Nanocomposite Coatings

Liju Elias and A.C. Hegde

CONTENTS

15.1 INTRODUCTION

Magnesium (Mg) alloys have emerged as some of the most attractive engineering materials of the twenty first century in the fields of transportation (Vijaya Ramnath et al. 2022; Thomas et al. 2022; Song et al. 2022), electronics (Landkof 2000), aerospace (Vijaya Ramnath et al. 2022; Landkof 2000), biomedical (Jana et al. 2022; Seetharaman et al. 2022), etc. The significantly low density of Mg (1.7 g cm^{-3}) (Wang, Yuan et al. 2022), compared to the other main engineering materials such as Al and Fe, makes Mg and its alloys as more attractive lightweight materials (Vijaya Ramnath et al. 2022; Wang, Guo et al. 2022). However, their high intrinsic reactivity (Polmear 1994) and poor corrosion resistance impedes their wide applications (Song and Atrens 1999; Friedrich and Mordike 2006; Atrens et al. 2011). Unlike in Al (Mott 1939), the inherent oxide film formed on the surface of Mg is less effective to prevent corrosion due to its porous nature (Atrens et al. 2011; Shih et al. 2007). Hence, it is necessary to adopt suitable corrosion protection strategies.

Generally, the anticorrosion properties of Mg alloys can be improved either by optimising the manufacturing processes according to the requirement (Xin et al. 2009; Hort et al. 2010) or by adopting suitable surface treatment strategies (Gray and Luan 2002; Hornberger et al. 2012; Hu et al. 2012; Zheng et al. 2017). Mg alloys of desired composition and microstructure (Xin et al. 2009), and thereby enhanced corrosion resistance, can be obtained through suitable selection of raw materials or manufacturing process (Song et al. 2022; Hornberger et al. 2012). On the other hand, the surface treatments involve the development of a suitable protective layer on the alloy surface to prevent the direct contact between the substrate and the corrosive environment (Gray and Luan 2002; Yao et al. 2022). Surface coatings may also offer added advantages like wear resistance (Gray and Luan 2002), antibacterial activity (Huang et al. 2022), hydrophobicity (Li, Xue et al. 2022), decorative appeal (Gray and Luan 2002), etc., depending on the type of the coatings developed. The coatings can be based on organic compounds (Hu et al. 2012), metals (Gray and Luan 2002), alloys (Hornberger et al. 2012; Yao et al. 2022), composites (Gray and Luan 2002; Hornberger et al.

DOI: 10.1201/9781003319856-18

2012), metal oxides (Yao et al. 2022), and functionalised (hydrophobic, antibacterial, etc.) (Huang et al. 2022; Li, Yin et al. 2022) according to the intended application and working environment. However, among these coatings, metal/alloy/composite coatings are reported to have wide practical application to enhance the corrosion resistance of Mg alloys due to their long-term stability and post-treatment possibilities (Yao et al. 2022). The commonly used surface treatment techniques for the development of corrosion-resistant coatings on Mg alloys are fluoride treatment (Olugbade et al. 2022), microarc oxidation (Yao et al. 2022), dip/spray coatings (Peng et al. 2021), anodisation (Shi et al. 2006), and electrodeposition (Liu et al. 2017) (**Figure 15.1**). Among all these techniques, electrodeposition is emerged as a most reliable and cost-effective technique to develop protective coatings for Mg alloy substrates with application-specific characteristics.

Apart from the conventional techniques, many advanced techniques are also available for the surface modification of Mg alloys through the development of metal or alloy coatings, like thermal spraying (Parco et al. 2006), chemical vapour deposition (Christoglou et al. 2004), laser cladding (Volovitch et al. 2008), etc. However, these methods can lead to microstructural changes in Mg alloy substrate. Electrodeposition is attractive here, as it will not typically affect the bulk characteristics of Mg alloys (Zhang, Yan et al. 2009; Nasirpouri 2017; Zheng et al. 2021; Di Girolamo and Dini 2022). It can also offer the development of coatings with controlled thickness and tuned characteristics by optimising the parameters such as deposition time, current density, and plating bath composition (Brenner 1963; Parthasaradhy 1989). By considering the practical significance of electrodeposition in modifying the Mg alloy surface for various applications, this chapter presents a concise description of the different types of electrodeposited metallic/nanocomposite coatings for the corrosion protection of Mg alloys.

15.2   ELECTRODEPOSITION ON Mg ALLOY SUBSTRATES

The electroplating or electrodeposition process involves the deposition of metal ions from the plating solution through electrochemical reduction at the cathode surface, and the electrochemistry involved in the process is governed by Faraday's laws of electrolysis (Brenner 1963; Parthasaradhy 1989). Electrodeposition can be performed with the help of a simple experimental setup comprising a plating bath, electrodes (anode and cathode), and a direct-current (DC) power source. The requirement of only simple salts as metal source, cheap chemicals as additive, and water as solvent for this process making it more attractive and cost effective (Elias and Hegde 2017a). The optimisation of a plating bath for the development of a desired coating is the crucial step or most important requirement for electrodeposition (Brenner 1963; Elias and Hegde 2016a). During the electrodeposition process, the Mg alloy substrate is used as the cathode, and the dissolved metal ions from the plating bath get deposited on the surface to form a protective coating. The coating characteristics can be tuned according to the requirement by controlling the process parameters such as deposition current density, bath composition, and deposition time (Brenner 1963; Parthasaradhy 1989). Other than conventional electrodeposition (Elias and Hegde 2016a), modern techniques such as composition modulated multilayer deposition (Elias and Hegde 2015; Elias, Bhat et al. 2016), magneto-electrodeposition (Elias, Cao et al. 2016; Elias and Hegde 2017b), sono-electrodeposition (Costa and de Almeida Neto 2020), sol-enhanced electrodeposition (Elias and Hegde 2016b), and composite electrodeposition (Elias and Hegde 2016c) are also attaining much attention owing to their potential in the development of efficient materials for corrosion protection (Elias and Hegde 2015; Elias, Bhat et al. 2016; Elias and Hegde 2017b) and energy-related applications (Elias, Cao et al. 2016; Elias and Hegde 2017c; Elias et al. 2022).

Although electroplating is a simple and effective technique for the development of coatings, the quality of the developed coating varies from substrate to substrate (Parthasaradhy 1989). As there is a possibility for the formation of an oxide or hydroxide ($MgO/Mg(OH)_2$) layer on the surface of Mg substrate when it comes in contact with air or electroplating solution (Huang et al. 2010; Yang et al. 2014), the process of electroplating is not so easy to develop uniform and adherent coating.

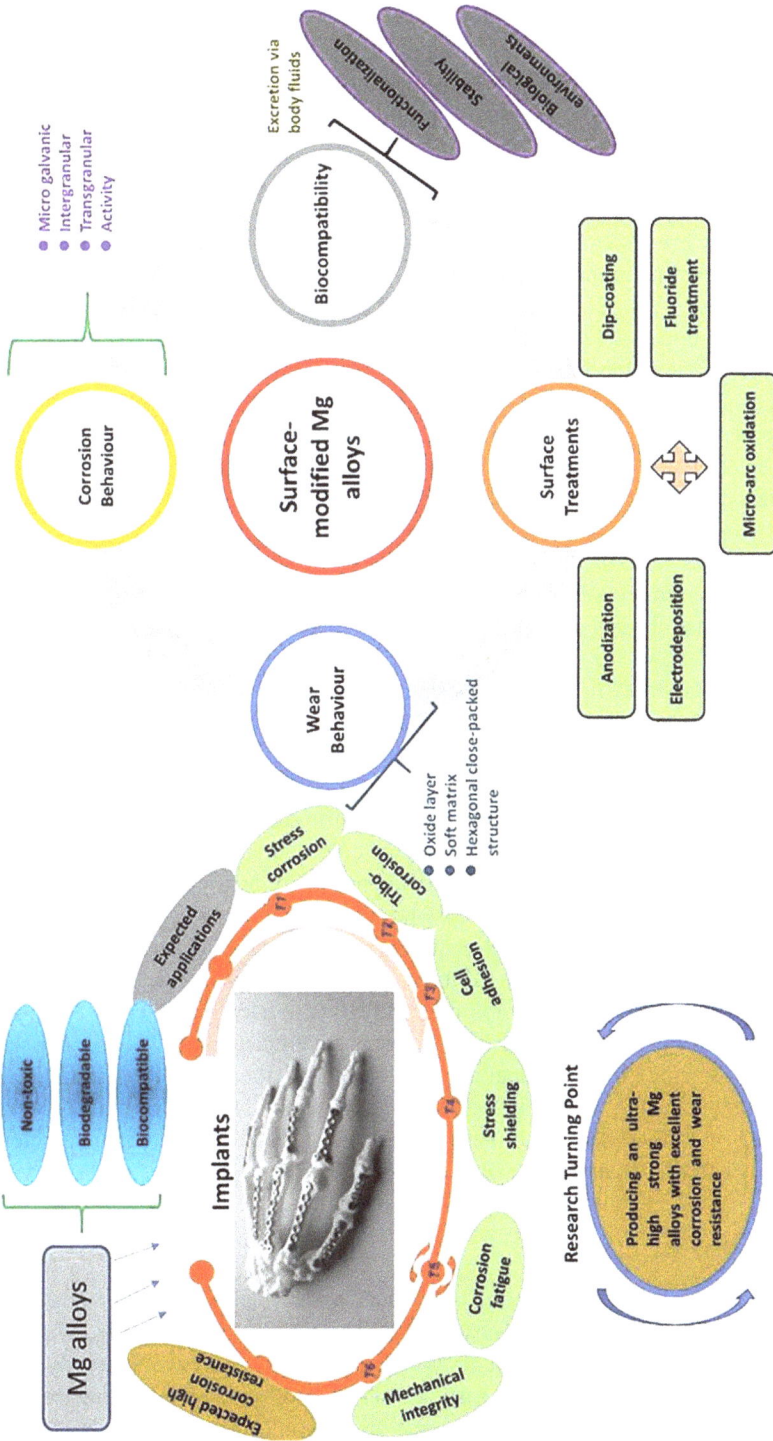

FIGURE 15.1 Illustration of the importance and applications of Mg alloys along with the commonly used surface treatment techniques for the development of corrosion-resistant coatings. *Reproduced with permission from (Olugbade et al. 2022) © 2021, ASM International.*

The main two problems that can adversely affect the electrodeposition process on Mg alloy are their extremely high chemical reactivity at pH below 10 and galvanic corrosion of the substrate (Shih et al. 2007; Yang et al. 2014). As there are various heterogeneous phases having different electrode potentials in Mg alloy, the substrate can undergo galvanic corrosion when it comes in contact with the aqueous plating bath and thereby lead to its destruction (Wu et al. 2010; Zhao et al. 2010). Hence, it is necessary to adopt effective pre-treatment techniques to achieve electrodeposited coatings on Mg alloys by preventing the corrosive dissolution of the substrate.

15.2.1 Pre-Treatments

The general steps in the pre-treatment of Mg alloys to ensure good-quality coating involve degreasing, pickling, activation, zinc immersion, and cyanide copper plating followed by the final electroplating of the metal/alloy (Yang et al. 2014, 2021). The formation of zinc film on the surface of Mg alloy through the zinc immersion step can help to prevent the re-oxidation of Mg and to reduce the potential difference between the intrinsically active Mg alloy substrate and the finally deposited coating (Zhao et al. 2010; Zhang, Yan et al. 2009). In another route, the pre-treatment can be performed by replacing the zinc immersion and cyanide Cu plating steps with electroless Ni plating prior to the final electroplating (Yang et al. 2021; Zhang et al. 2011). Also, to attain electrochemical surface homogeneity, Mg alloys are pre-treated through immersion processes to achieve selective dissolution under suitable pH ranges (Zhang, Yan et al. 2009).

Zhu et al. (2010) reported a detailed study on the effectiveness of different pre-treatment techniques for electrodeposition on AZ91D alloy and the electrochemical behaviour of the substrate in various pre-treatment media such as pickling, activation, and Zn immersion solutions. Pickling of the substrate helps to remove the oxides/hydroxides film on the alloy and leads to the formation of a coarse surface to facilitate improved adhesion of the deposit (Zhao et al. 2010; Zhang et al. 2011). The combinations of HNO_3 and CrO_3 or H_3PO_4 are the commonly used pickling solutions for Mg alloy substrates to achieve uniform corrosion of the different phases (Zhu et al. 2010; Zhang et al. 2011). The activation process is performed to remove the corrosion products formed on the substrate after pickling and also to form a new activation film on the surface to reduce their further corrosion in the plating solution. HF and $K_4P_2O_7$ are the commonly used activation solutions for Mg alloys, and they result in the formation of thin activation films of MgF_2 and $Mg(OH)_2$, respectively (Zhang, Yu et al. 2009). Further, $KMnO_4$ can also be used as a pickling-activation solution, but the oxide film (Mn_xO_y) obtained after this treatment may contain cracks. However, $KMnO_4$ pickling-activation and $K_4P_2O_7$ activation are reported to be more suitable for the development of primary Zn layer on the substrate through immersion than the HF activation (Zhao et al. 2010; Zhu et al. 2010). The MgF_2 film formed after HF activation reduces the efficacy of zinc immersion, leading to the formation of a poor zinc film as compared to the other activation procedures. But the substrate surface having a compact MgF_2 layer achieved through HF activation can facilitate the direct electrochemical deposition of Ni-P alloy onto the surface without the development of any intermediate Zn layer (Xiang et al. 2000; Zhang et al. 2011).

15.2.2 Metal Coatings

The electrodeposited metal coatings such as Zn, Ni, and Cu are widely explored as a base layer coating and a physical barrier to prevent the substrate's corrosion during top coating (Wu et al. 2010; Zhao et al. 2010; Zhang et al. 2011; Yang et al. 2021). The metal coatings can not only offer corrosion protection but also enhance the wear resistance of Mg alloys without affecting their mechanical characteristics (Wu et al. 2010). With Al being a primary alloy element for Mg, Al or Al alloy coatings are widely explored (Zhang, Yu et al. 2009). The ability to form stable oxide film (Al_2O_3) having high corrosion resistance and the lesser possibility for galvanic corrosion (because of the close electrode potential to that of Mg) make Al/Al alloy coatings the best choice for Mg alloys

(Zhang, Yan et al. 2009; Mordike and Ebert 2001). As high temperature technique such as hot-dip aluminising is not viable for Mg alloy substrates due to their low melting point, electrodeposition is considered the most promising for the deposition of Al-based coatings on Mg alloys (Mordike and Ebert 2001).

Further, Zhang, Yu et al. (2009) studied the effect of fluoride in Ni plating bath on the coating characteristics and corrosion protection of AZ91D alloy. The study revealed that the fluoride in the Ni plating bath can significantly reduce the extent of Mg alloy substrate corrosion during the plating process by serving as an activator of Ni anodic dissolution. For the development of Ni coating on the surface of Mg alloy, the homogenisation treatment of the substrate can also be carried out through Zn immersion followed by Cu plating (Yang et al. 2014). The electrodeposited nanocrystalline Ni coatings are reported to be more effective towards the corrosion protection of Mg alloys than the electroless Ni coatings (Gu et al. 2006).

As Mg alloy substrates are highly reactive when exposed to air or water, electroplating from non-aqueous electrolyte solvent such as ionic liquids is attaining much attention to overcome the inherent problems of instability/dissolution of the deposit during electroplating from aqueous solutions (Bakkar and Neubert 2007; Gao et al. 2021). Akin to aqueous electrodeposition, there are many reports on the electroplating of Zn onto different Mg alloys from choline chloride-based electrolytes (Bakkar and Neubert 2007; Chang et al. 2008). Chang et al. (2008) reported the electrodeposition of Al on AZ91D alloy from aluminum chloride-1-ethyl-3-methylimidazolium chloride ($AlCl_3$–EMIC) ionic liquid under galvanostatic conditions. The study revealed that the corrosion-protection characteristics of the coating is greatly influenced by the deposition current density. The coating developed at lower current density is reported to have more uniform and compact Al layer and thereby the best corrosion-resistance characteristics. Recently, Chen et al. (2022) reported a detailed theoretical simulation study on the factors influencing the electrodeposition of Al on complex Mg alloy geometries from ionic liquids. This study gives a detailed insight on the thickness distribution of the coatings and the influence of various electrochemical parameters on the coating process.

Similarly, Singh et al. (2021) reported the development of Ni coatings on AZ91 alloy from a newly optimised water-soluble eutectic-based ionic liquid. The coating was observed to exhibit attractive corrosion resistance with a very low corrosion current density (0.9 μA cm^{-2}) and a longer stability over 10 hours under salt spray exposure. The obtained results of a salt spray exposure (**Figure 15.2**) test indicates that the multilayered coating developed on pre-treated substrate ($Ce(NO_3)_3$ treatment) followed by post heat treatment at 90°C exhibits attractive stability without any failure. The obtained electrochemical corrosion results are also in compliance with the salt spray analysis results.

15.2.3 Metal Oxide Coatings

Metal oxide coatings are also reported to be effective towards the corrosion protection of Mg alloys. Generally, in electroplating, the metal oxide coatings are developed through anodic electrodeposition by using the substrate as anode during deposition process rather than using it as cathode in conventional or cathodic electrodeposition. Lei et al. (2010) reported a simple strategy for the development of corrosion-resistant MgO coatings on die-cast homemade Mg-Zn-Ca alloy through an anodic electrodeposition in concentrated KOH solution followed by heat treatment in the presence of air. Although plasma electrolytic oxidation (PEO) techniques are widely used for the development of oxides or ceramic coatings for Mg alloys (Cai et al. 2006; Zhang and Zhang 2009), this simple anodic electrodeposition technique has attracted much attention, as it can overcome one of the major issues of micropores/micro-cracks generation in PEO coatings. Unlike the naturally formed MgO film on Mg alloys when exposed to the atmosphere, a thick MgO coating with improved stability and lower corrosion rate was obtained through this electrodeposition route.

Yang et al. (2021) reported a very interesting indirect corrosion-protection strategy for AZ31B alloy through photoelectrochemical cathodic protection by the development of an electrodeposited

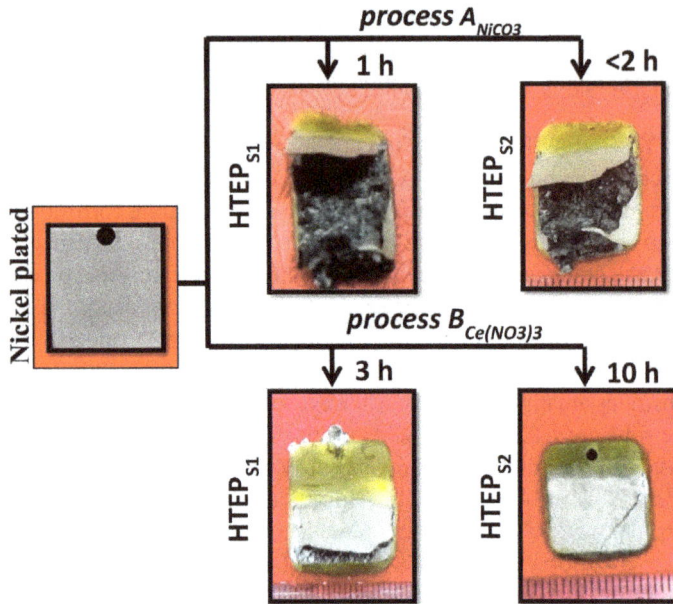

FIGURE 15.2 Comparison of the performance of Ni coatings developed on AZ91 alloy (pretreated through different methods NiCO$_3$ and Ce(NO$_3$)$_3$ treatments) from a newly optimised water-soluble eutectic based ionic liquid after salt spray exposure test. *Reproduced with permission from (Singh et al. 2021) © 2021 Elsevier B.V.*

Cu$_2$O semiconductor layer over the electroless plated Ni layer on the substrate. The Ni interlayer having higher corrosion potential than the Mg alloy substrate can be easily polarised by receiving photogenerated electrons from the Cu$_2$O layer under visible light illumination and thereby can offer a prolonged indirect cathodic protection to the substrate (Zhang, Rahman et al. 2018; Yang et al. 2021). A schematic of the mechanism of indirect corrosion protection to AZ31B alloy by electrodeposited Cu$_2$O semiconductor through photoelectrochemical cathodic protection is given in **Figure 15.3(a)**. The experimental data of OCP vs. time curve of the electroless Ni coated Mg alloy with and without Cu$_2$O layer, as shown in **Figure 15.3(b)**, indicates the change in corrosion-protection efficacy of the Mg alloy after the development of electrodeposited Cu$_2$O semiconductor layer over the electroless plated Ni layer on the substrate (Yang et al. 2021).

15.2.4 ALLOY COATINGS

Among the different types of electrodeposited alloy coatings reported for the corrosion protection of Mg alloys, phosphorous-based alloy coatings are attaining much attention owing to their tunable structural characteristics from amorphous to microcrystalline or nanocrystalline with change in P contents (Hu and Bai 2003; Zhang et al. 2011). Ni-P alloy coatings are observed to have more enhanced corrosion-protection efficacy than Ni coatings. Zhang et al. (2011) reported the electroplating of Ni-P coating on AZ91D alloy after suitable pre-treatment involving the development of a protective underlayer of Cu. The developed Ni-P coatings exhibited more amorphous character and enhanced corrosion resistance with increase of P content. The surface morphology of the AZ91D alloy after different types of pre-treatments is shown in **Figure 15.4**.

The SEM image shown in Figure 15.4a clearly indicates the formation of a smooth and slightly non-uniform substrate surface with α (Mg) and β (Mg$_{17}$Al$_{12}$) phases having pits and cracks after HF pickling-activation treatment. Further, the substrate surface is observed to be covered by Zn through Zn immersion treatment (Figure 15.4b) that results in reduced difference between the different

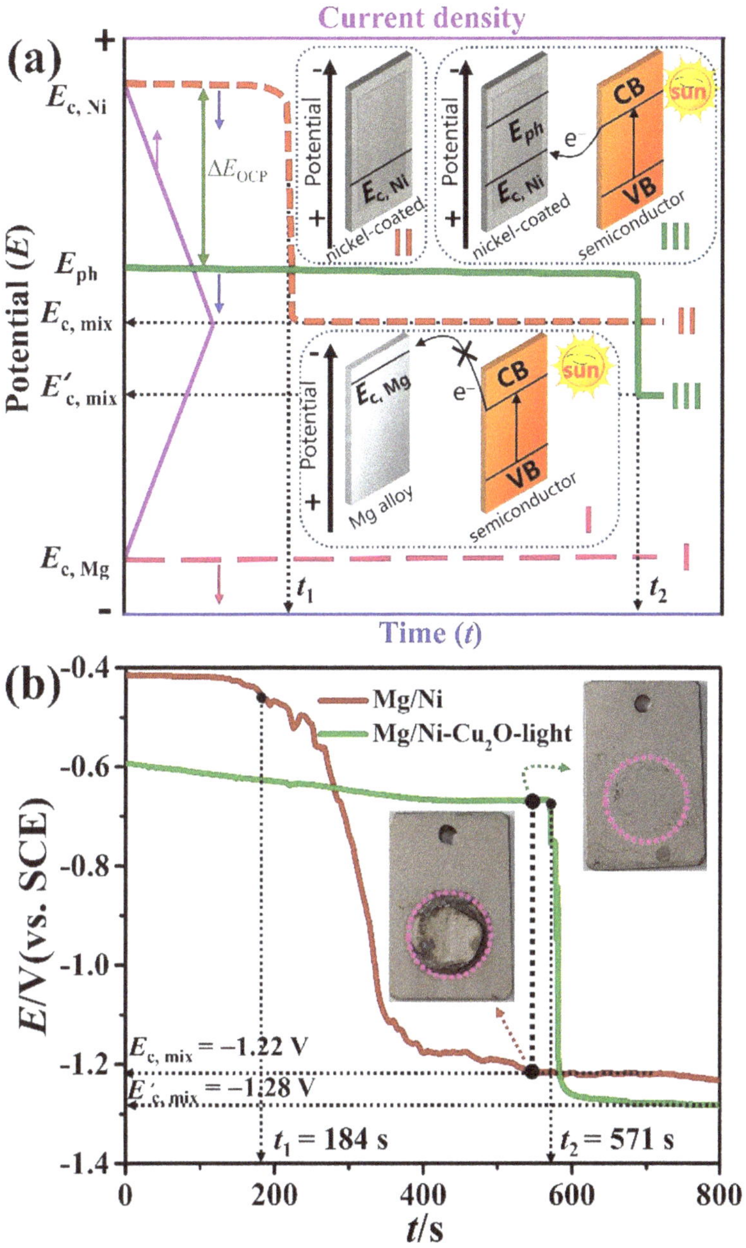

FIGURE 15.3 (a) Schematic of the mechanism of indirect corrosion protection to AZ31B alloy substrate by electrodeposited Cu_2O semiconductor layer through photoelectrochemical cathodic protection, and (b) OCP vs. time curves of the electroless Ni coated Mg alloy with and without Cu_2O layer. *Reproduced with permission from (Yang et al. 2021) © 2021 Elsevier B.V.*

phases on the substrate. This process further reduces the possible potential difference between the substrate and the intended metal or alloy layer. Figure 15.4C clearly depicts the formation of a uniform and crack-free Ni-P coating with nodular morphology on Cu pre-plated AZ91D alloy. However, the electroless Ni-P coating developed directly on the substrate (Figure 15.4D), in the absence of a Cu underlayer, is observed to be rough, loose, and porous, with a large number of

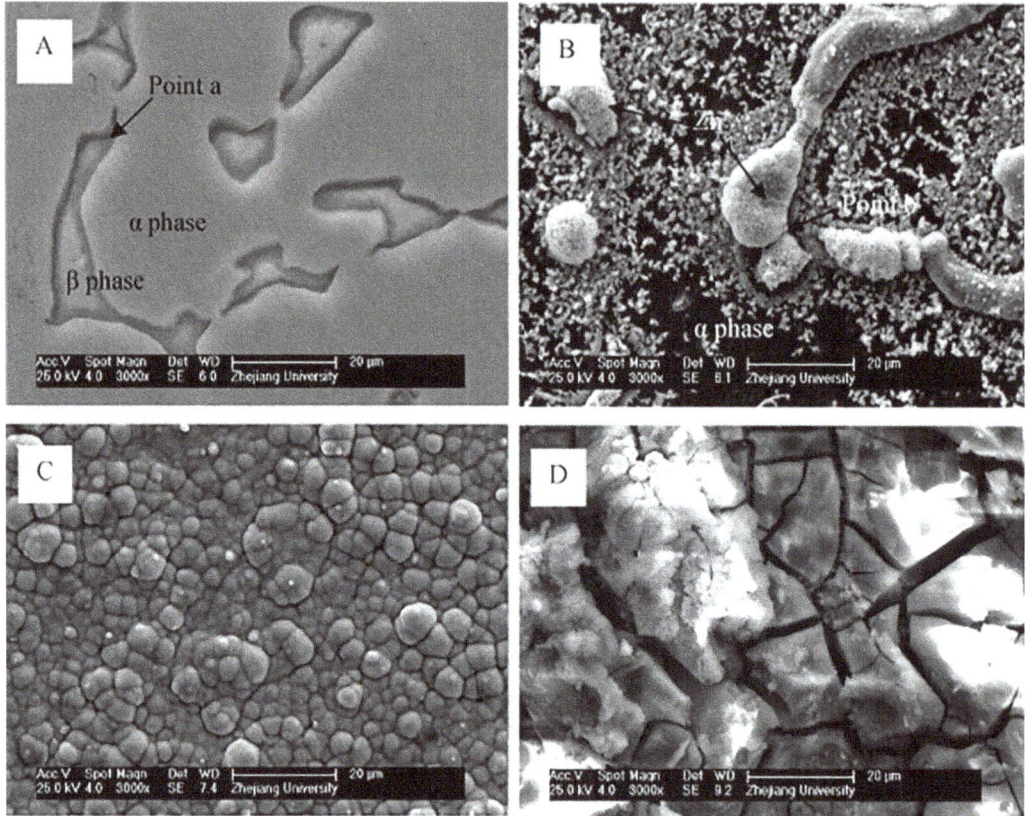

FIGURE 15.4 Scanning electron microscopy images of the AZ91D alloy after different types of pre-treatments: (A) HF pickling-activation, (B) Zn immersion, (C) Ni-P coating after copper plating, and (D) Ni-P deposited without the underlayer of Cu. *Reproduced with permission from (Zhang et al. 2011) © 2011 Elsevier B.V.*

cracks on the surface (Zhang et al. 2011). The formation of cracks on the coating surface is responsible for the reduced corrosion resistance of the Ni-P coated substrate without a Cu underlayer as compared to the Ni-P coating developed on Cu pre-plated AZ91D alloy (Figure 15.5).

Further, Zhang, Yu et al. (2009) reported the deposition of Al-Mn alloy on AZ31B alloy from $AlCl_3$-NaCl-KCl-$MnCl_2$ molten salts at 170°C after pre-treatment involving the deposition of a thin underlayer of Zn. The Mn content and thereby the phase structure of the Al-Mn alloy coating was tuned by changing the composition of $MnCl_2$ in the molten plating bath. The morphology and phase structure of the developed Al-Mn coatings were observed to have great influence on their corrosion protection characteristics in 3.5% NaCl medium.

15.2.5 COMPOSITE COATINGS

Composite coatings can have enhanced mechanical and corrosion-resistance characteristics as compared to their metal or alloy counterparts. Electrodeposited Ni-SiC nanocomposite coating on AZ91 alloy substrate is reported to enhance the corrosion resistance of the substrate significantly (Fini and Amadeh 2013). The potentiodynamic polarisation curves of the composite-coated substrate are observed to exhibit manifold decrease in corrosion current density as compared to the bare substrate in 3.5 wt.% NaCl medium. Similarly, Ni-SiO_2 coating is reported to enhance the corrosion resistance and wear resistance of the AZ91HP alloy (Yan et al. 2011). Further, chitosan/graphene

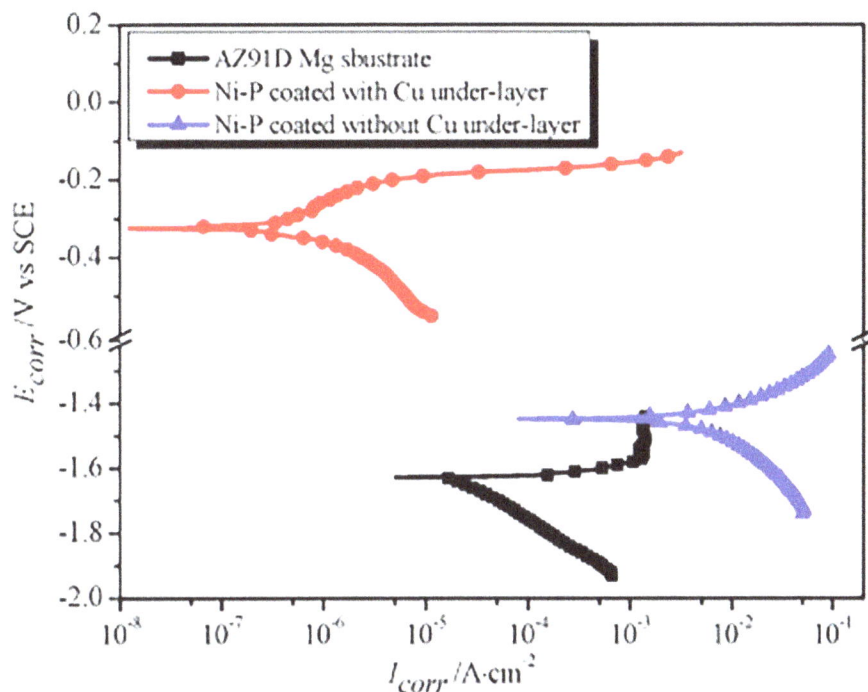

FIGURE 15.5 Comparison of the potentiodynamic polarisation responses of the bare and Ni-P-coated AZ91D alloy with and without Cu underlayer in 3.5 wt.% NaCl medium. *Reproduced with permission from (Zhang et al. 2011) © 2011 Elsevier B.V.*

oxide nanocomposite coating developed through pulse electrodeposition is observed to enhance the mechanical and corrosion-resistance characteristics of a Mg-Zn alloy in simulated body fluid (Hamghavandi et al. 2021).

Cui et al. (2017) reported the development of TiO_2 layer-by-layer assembled composite coating on as-extruded Mg-1Li-1Ca alloy by electrodeposition followed by surface silanization. The electrochemical impedance spectroscopy (EIS) responses of the bare, alkali-treated, TiO_2-coated, and polymethyltrimethoxysilane (PMTMS)/TiO_2-coated alloys indicate an enhanced corrosion resistance for the composite coating with large charge transfer resistance and a large low-frequency impedance modulus, obtained respectively from the Nyquist and Bode plots, as shown in **Figure 15.6**.

Among the various composite coatings, layered double hydroxides (LDH) with nanoparticles incorporated between the interlayers as nanocontainers can offer enhanced corrosion protection (Tan et al. 2022). The protection can be offered in two ways: either by facilitating the release of inhibitors retained between the layers by weak electrostatic force of attraction or by trapping the corrosive anions (e.g. Cl–) into the interlayer space. Tan et al. (2022) reported such a composite coating of Mg-Fe LDH intercalated with 8-hydroxyquinoline and modified with citric acid on WE43 alloy by combining electrodeposition, hydrothermal, and anion exchange synthesis routes. The obtained coating is observed to have a self-healing characteristic suitable for biomedical applications.

Huang et al. (2022) reported the electrodeposition of chitosan/boron nitride composite coating with enhanced corrosion resistance and antibacterial properties on Mg-Zn-Y-Nd-Zr alloy. The results clearly indicate that the chitosan/boron nitride composite coating acts as a physical barrier against the corrosion of the underlying substrate and exhibits antibacterial activity against *E. coli* and *S. aureus*. The corrosion resistance characteristics of many such composite coatings are discussed in the following sections.

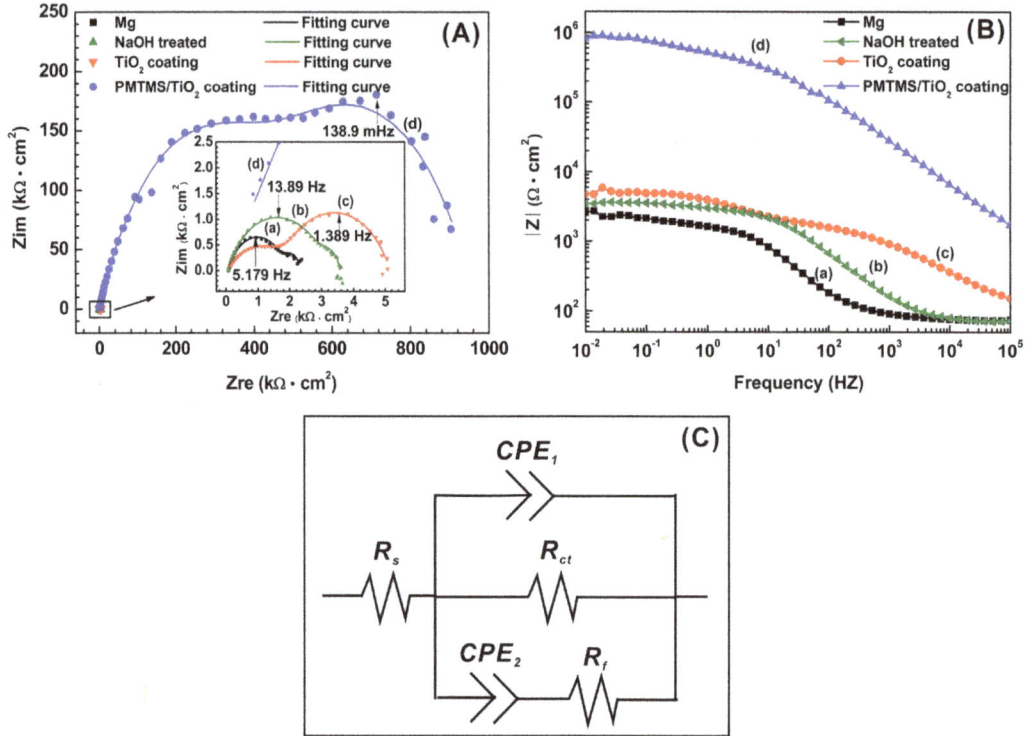

FIGURE 15.6 EIS responses of the bare, alkali-treated, TiO_2-coated, and PMTMS/TiO_2-coated alloys: (A) Nyquist plot, (B) Bode plot, and (C) electrical equivalent circuit used for fitting the Nyquist data. *Reproduced with permission from (Cui et al. 2017) © 2017 Elsevier B.V.*

15.2.6 BIOMEDICAL COATINGS

The biomedical application of Mg alloys, especially, as implantable medical devices (e.g. orthopedic implants, bone nails etc.), utilises their poor corrosion resistance for the development of biodegradable materials (Liu et al. 2017; Jana et al. 2022; Seetharaman et al. 2022). Although stainless steels and Ti alloys are the conventionally used materials as biomedical implants, these materials can become toxic and adversely affect human health after long-term usage (Song et al. 2008; Liu et al. 2017). Also, such implants mandate the requirement of a second surgery for the removal of the same after healing. Hence, biodegradable implants that exhibits similar mechanical properties of natural bone and that can be gradually dissolved, absorbed, consumed, or excreted after a desired time required for bone tissue healing are attaining much attention (Song et al. 2008).

The recent trend in biomedical research is focused on coordinating the biodegradability and corrosion resistance to reduce the discrepancy between service life and expectancy life of Mg alloy-based implants in practical applications (Liu et al. 2017; Witte 2010; Wang et al. 2015). In this context, different types of coatings have been explored for the surface modification and thereby to improve the corrosion resistance.

As the fast corrosion of Mg alloy implants can significantly affect the osteointegration due to the possible hydrogen gas entrapment around the implant (Song et al. 2010), the development of coatings based on the constituent elements in the bone minerals such as Ca and P have high significance. Among the different Ca-P coatings, hydroxyapatite (HA) and tricalcium phosphate coatings are reported to be the most effective coatings for biodegradable implants to enhance the corrosion resistance without affecting biocompatibility (Song et al. 2010). Although there are advanced methods like hot spraying (Parco et al. 2006), laser cladding (Volovitch et al. 2008), etc., electrodeposition is considered the most reliable technique to develop coatings having comparable characteristics of bone minerals at ambient temperature.

Aboudzadeh et al. (2019) presented a detailed study on the influence of pulse electrodeposition parameters on the morphology and characteristics of Si-hydroxyapatite coating. The study indicates that the deposition current density, duty cycle, and pH of the medium play a significant role in the formation of uniform and adherent coatings. Similarly, Song et al. (2010) reported the development of Ca-P coating on Mg-Zn alloy by electrodeposition from a solution containing $Ca(NO_3)_2 \cdot 4H_2O$ and $NH_4H_2PO_4$. The deposited Ca-P coatings were then converted to hydroxyapatite by alkali treatment, and a fluoridated hydroxyapatite coating was developed through direct deposition by adding $NaNO_3$ and NaF into the plating solution. The developed hydroxyapatite coating is observed to make a significant reduction in degradation rate of the Mg alloy substrate in simulated body fluid. Compared with the hydroxyapatite coating, the fluoridated hydroxyapatite coating is observed to exhibit better stability and corrosion resistance and more influence on the nucleation of osteoconductive minerals.

Song et al. (2008) reported the electrodeposition of hydroxyapatite on AZ91D alloy. The developed coating is observed to exhibit enhanced corrosion protection and thereby modifies the biodegradation rate in simulated body fluid. A wide range of modified hydroxyapatite-based multilayer coatings like SiC-SiC nanowire-Si doped HA coating (Zhang, Li et al. 2018), pyrolytic carbon-SiC-fluoridated-HA-HA coating (Zhang et al. 2019), carbon/SiC nanowire/Na-doped carbonated HA (Leilei et al. 2014), etc. are attaining much attention as bioactive coatings with enhanced stability.

The protective coating becomes more attractive if it can address the issue of post-surgical infection that leads to the early failure of the implants (Song et al. 2008; Witte 2010). In this context, Tian et al. (2022) reported the development of a hydroxyapatite-based composite coating with bacterial adhesion resistance and bacteria-killing properties on AZ31 alloy. The surface of substrate was tuned to become superhydrophilic by the development of hydroxyapatite/hydroxypropyltrimethyl ammonium chloride chitosan composite coating by combining hydrothermal and electrodeposition techniques. The combined effect of superhydrophilicity that results in reduced bacterial adhesion and the bactericidal characteristics of hydroxypropyltrimethyl ammonium chloride chitosan in the coating ensures effective protection of the substrate from microbial infection. The composite coating is observed to have bacteria-killing efficiency of 99.9% and 94.7% against *S. aureus* and *E. coli*, respectively, along with good anticorrosion performance (Tian et al. 2022). The mechanism of bacterial adhesion resistance of hydroxyapatite coating and bacteria-killing properties of hydroxypropyltrimethyl ammonium chloride chitosan composite coatings are shown in **Figure 15.7**.

15.2.7 SUPERHYDROPHOBIC COATINGS

Superhydrophobic coatings are an attractive choice for surface modification of Mg alloys to serve as a barrier coating to reduce the contact between the medium and the substrate (Zhang et al. 2016; Guo et al. 2020; Zhang et al. 2016) (Yao et al. 2020). The poor mechanical stability is one of the major issues of superhydrophobic coatings that significantly affects their wide application. Generally, the development of a superhydrophobic surface involves the microstructure tuning and surface energy modification (Tuteja et al. 2007), and these can be achieved either in a single step or in two steps. The multi-step fabrication of a superhydrophobic surface can either involve a pre-treatment followed by a post-modification route or pre-modification followed by a post-texturing route (**Figure 15.8**) through the use of different material synthesis strategies. Among the different methods available for the fabrication of superhydrophobic coatings, electrodeposition is considered the simple and effective technique for the direct deposition of superhydrophobic materials onto the substrate surface. Other than following a complicated multistep processing, electrodeposition enables a single-step fabrication of superhydrophobic coatings without the requirement of any sophisticated instrumentation or costly chemicals. There are many reports on the development of superhydrophobic coatings on Mg alloys by electrodeposition.

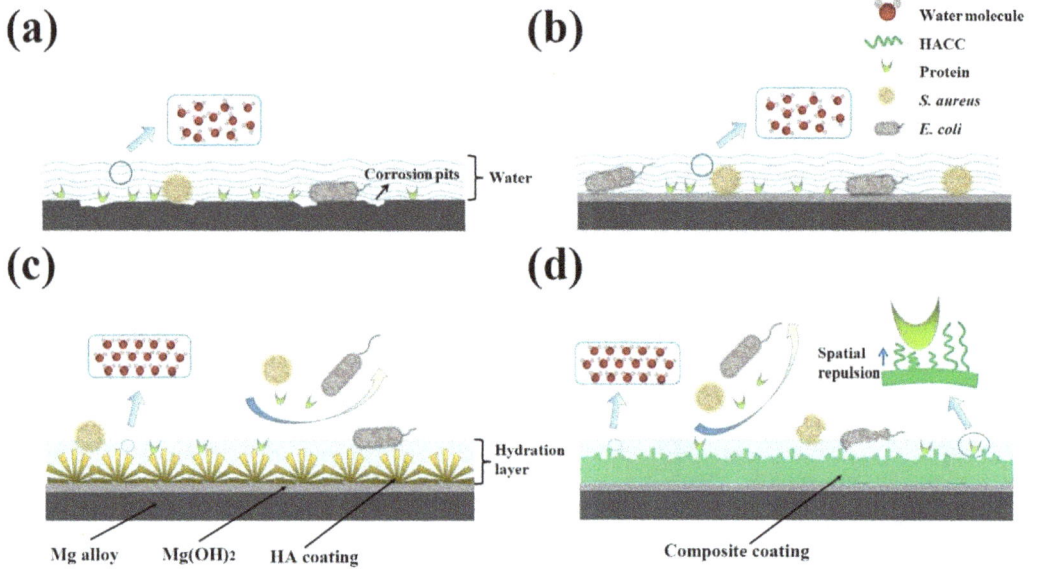

FIGURE 15.7 Schematic of the bacterial interaction/adhesion on different surfaces of AZ31 alloy substrate: (a) bare Mg alloy, (b) Mg alloy with Mg(OH)$_2$ layer, (c) Mg alloy with Mg(OH)$_2$ and hydroxyapatite coating, and (d) Mg alloy with hydroxypropyltrimethyl ammonium chloride chitosan composite coating. *Reproduced with permission from (Tian et al. 2022) © 2022 Elsevier B.V.*

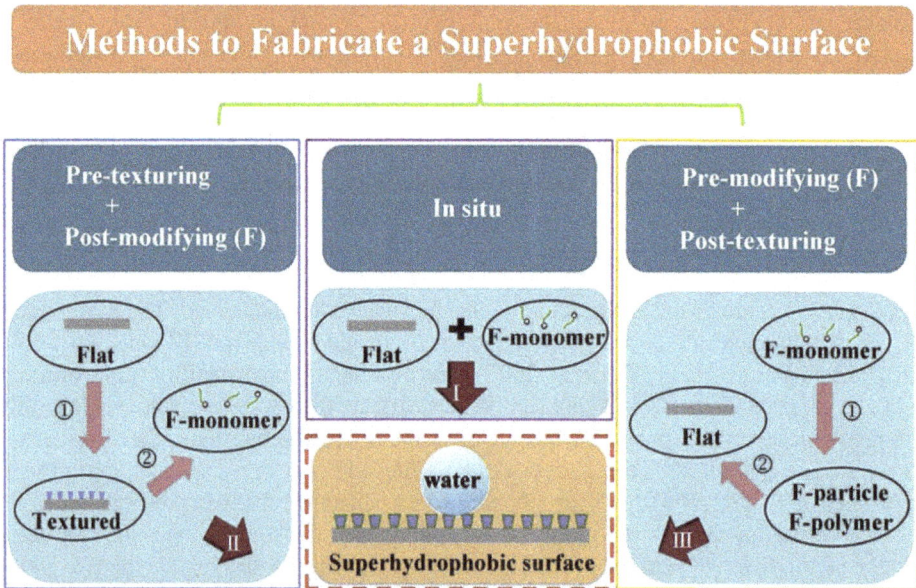

FIGURE 15.8 Schematic of the common strategies for the development of hydrophobic surfaces. *Reproduced with permission from (Yao et al. 2020) © 2020 Elsevier.*

Liu and Kang (2014) reported the development of a superhydrophobic coating on MB8 alloy by electrodeposition from ethanol solution containing cerium nitrate and myristic acid. The developed coating with hierarchical micro/nanoparticles of cerium myristate displayed enhanced superhydrophobicity and corrosion resistance. In a similar route, Kang and Li (2017) and Zheng et al. (2019) also reported the development of superhydrophobic coating on AZ31 alloy by a single-step electrodeposition process from ethanol solution containing cerium nitrate and stearic acid. The obtained coating was observed to have improved anticorrosion and self-cleaning characteristics. This superhydrophobic coating was then modified appropriately for the biomedical application of TZ51 alloy, and the coating was observed to exhibit enhanced corrosion resistance in simulated body fluid (Liu et al. 2017). Further, Liu et al. (2014) reported the development of superhydrophobic coating on AZ91D alloy by suitably controlling the Ni deposition process to achieve micro/nanoclusters grown in cauliflower-like structures followed by modification with stearic acid. The coating was able to trap large amount of air on the surface due to its peculiar structural/morphological features and thereby exhibit enhanced corrosion resistance.

Then, Li, Yin et al. (2022) reported the development of a graphene oxide (GO) and metal organic framework (MOF)-based superhydrophobic dual-layer coating on AZ31B alloy by combining electrodeposition and spray coating techniques (**Figure 15.9**). The structure of the coating comprising the bottom layer of GO and the upper superhydrophobic polypyrrole-zeolitic imidazolate framework-8 layer is observed to be effective in preventing the penetration of corrosive ions into the substrate surface and also to improve the mechanical characteristics of the substrate (Li, Yin et al. 2022).

Syu et al. (2013) reported the development of an optically transparent Li-Al-CO$_3$ LDH coating on AZ31 alloy. The electrodeposition strategy is observed to be simple for the development of LDH coating under ambient operating conditions. The corrosion resistance of the developed LDH is observed to be varied with coating thickness and the hydrophobic characteristics of the coating. As

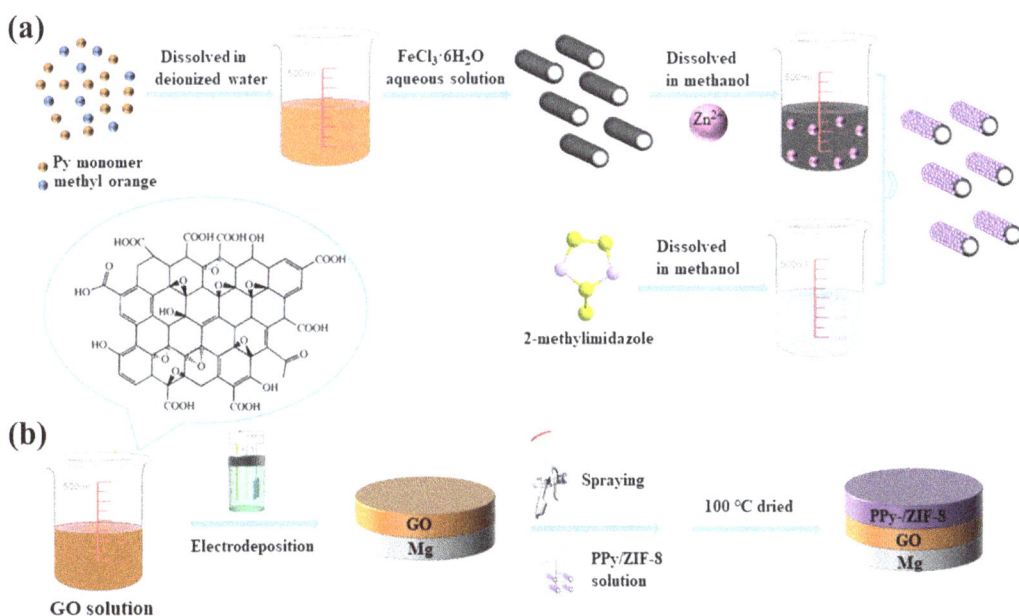

FIGURE 15.9 Schematic of: (a) the synthesis of polypyrrole-zeolitic imidazolate framework-8 and (b) development of superhydrophobic graphene oxide polypyrrole-zeolitic imidazolate framework-8 dual layer coating on AZ31B alloy. *Reproduced with permission from (Li, Yin et al. 2022) © 2022 Elsevier.*

the dissociation of LDH coating can lead to the formation of a highly alkaline environment through the release of OH– ions, coatings with larger thickness can only ensure corrosion protection under these conditions as compared to the thin LDH films (Syu et al. 2013).

Wang, Guo et al. (2022) reported the development of hexadecyltrimethoxysilane (HDTMS)-modified diatomite/epoxy resin composite coating on AZ31B alloy by combining electrodeposition and spray coating techniques. The superhydrophobic HDTMS-modified diatomite powder was prepared through electrodeposition by using HDTMS and tetraethyl orthosilicate (TEOS) in ethanol. Then it was spray coated on the substrate by using E44 epoxy resin as the adhesive to obtain the composite coating. The composite coating is observed to have hydrophobicity, corrosion resistance, and anti-icing performance.

Zhang and Lin (2019) reported the development of a superhydrophobic coating on anodised AZ21 alloy by pulse electrodeposition from a plating bath containing calcium nitrate and steric acid. The pulse deposition duty cycle is observed to have great influence on the superhydrophobic characteristics and corrosion resistance of the coatings. The coating developed under pulse mode with 50% duty cycle is obtained as the best coating with ideal coating characteristics and corrosion resistance. Furthermore, Zhang et al. (2021) reported the development of a superhydrophobic coating of dodecyltrimethoxysilane on AZ31 alloy by electrodeposition from a solution containing ethanol, potassium nitrate, and dodecyltrimethoxysilane. The coating is observed to have low wettability, excellent self-cleaning characteristics, and anticorrosion performance.

A triple-layer superhydrophobic composite coating of nickel-phosphorus/nickel/fluorinated polysiloxane on AZ31 alloy, reported by Fang et al. (2022), is attaining much attention owing to the attractive coating characteristics achieved through the innovative design to utilise the synergism between different layers. The conversion of hydrophilic nickel electrodeposited on pre-treated Mg alloy to a superhydrophobic surface was achieved through direct silane modification. The triple-layer coating is observed to have enhanced superhydrophobicity and corrosion resistance (Fang et al. 2022). The mechanism of formation of the superhydrophobic composite coating on AZ31 alloy by the initial electroless deposition of Ni (ELNi) followed by the electrochemical deposition of Ni (ECNi) and the final immersion step to form the fluorinated polysiloxane layer through the dehydration-condensation reaction is schematically shown in **Figure 15.10**.

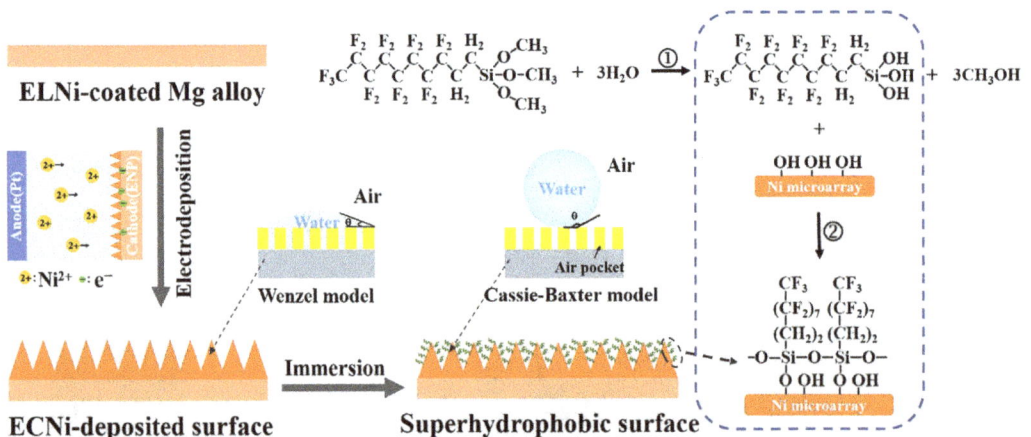

FIGURE 15.10 Schematic of the mechanism of formation of the superhydrophobic composite coating on AZ31 alloy by the initial electroless deposition of Ni (ELNi) followed by the electrochemical deposition of Ni (ECNi) and the final immersion step to form the fluorinated polysiloxane layer through the dehydration-condensation reaction. *Reproduced with permission from (Fang et al. 2022) © 2022 Elsevier B.V.*

Apart from superhydrophobicity, superoleophobicity is also attaining much attention, and the development of superamphiphobic coatings that can exhibit both these characteristics are of great demand in many practical applications. Liu et al. (2016) reported such a superamphiphobic coating on AZ91D alloy by adopting nickel plating in combination with a low-surface-energy perfluoro-caprylic acid modification. The coating was observed to have stability over a wide pH range and enhanced corrosion resistance. The contact angles of 160.2° and 152.4°, obtained for the coating, respectively, in water and oil confirm the superamphiphobic nature.

15.3 CONCLUSIONS AND OUTLOOK

Electrodeposition has emerged as one of the efficient methods for the development of protective metal/alloy/composite coatings on Mg alloys with tuned characteristics, great degree of reproducibility, long-term stability, and post-treatment possibilities for different practical applications. Apart from the other advanced surface coating strategies for Mg alloys like thermal spraying, chemical vapour deposition, laser cladding, etc., electrodeposition is considered as the simple and suitable technique for the development of compact and uniform coatings without affecting the microstructural characteristics of Mg alloy substrates. The development of coatings with controlled thickness and tuned characteristics can be achieved through electrodeposition by optimising the parameters such as deposition time, current density, and plating bath composition. Also, other than following a complicated multi-step process, electrodeposition enables a single step fabrication of coatings with corrosion resistance, wear resistance, antibacterial activity, hydrophobicity, biocompatibility, and decorative appeal without the requirement of any sophisticated instrumentation or costly chemicals. The suitable optimisation and use of advanced electrodeposition techniques such as magneto-electrodeposition and sono-electrodeposition can open up new possibilities for the practical application of Mg alloys in the fields of electronics, aerospace, biomedicine, etc.

REFERENCES

Aboudzadeh, N, C Dehghanian, MA Shokrgozar, Effect of electrodeposition parameters and substrate on morphology of Si-HA coating, Surface and Coatings Technology 375 (2019) 341–351.

Atrens, A, M Liu, NI Zainal Abidin, GL Song, Corrosion of magnesium (Mg) alloys and metallurgical influence, in Woodhead Publishing Series in Metals and Surface Engineering, Corrosion of Magnesium Alloys, Woodhead Publishing, Sawston, 2011: pp. 117–165.

Bakkar, A, V Neubert, Electrodeposition onto magnesium in air and water stable ionic liquids: From corrosion to successful plating, Electrochemistry Communications 9 (2007) 2428–2435.

Brenner, A, Electrodeposition of Alloys: Principles and Practice, Academic Press, Massachusetts, 1963.

Cai, Q, L Wang, B Wei, Q Liu, Electrochemical performance of microarc oxidation films formed on AZ91D magnesium alloy in silicate and phosphate electrolytes, Surface and Coatings Technology 200 (2006) 3727–3733.

Chang, JK, SY Chen, WT Tsai, MJ Deng, IW Sun, Improved corrosion resistance of magnesium alloy with a surface aluminum coating electrodeposited in ionic liquid, Journal of the Electrochemical Society 155 (2008) C112.

Chen, L, B Liang, M Cao, Y Yang, D Wang, Q Yang, Z Han, G Wang, Y Wang, M Zhang, Simulation of aluminium electrodeposition on complex geometry: Influence of input parameters and electrodeposition conditions, Surfaces and Interfaces 33 (2022) 102208.

Christoglou, CH, N Voudouris, GN Angelopoulos, M Pant, W Dahl, Deposition of aluminium on magnesium by a CVD process, Surface and Coatings Technology 184 (2004) 149–155.

Costa, JM, AF de Almeida Neto, Ultrasound-assisted electrodeposition and synthesis of alloys and composite materials: A review, Ultrasonics Sonochemistry 68 (2020) 105193.

Cui, LY, PH Qin, XL Huang, ZZ Yin, RC Zeng, SQ Li, EH Han, ZL Wang, Electrodeposition of TiO_2 layer-by-layer assembled composite coating and silane treatment on Mg alloy for corrosion resistance, Surface and Coatings Technology 324 (2017) 560–568.

Di Girolamo, D, D Dini, Electrodeposition as a versatile preparative tool for perovskite photovoltaics: Aspects of metallization and selective contacts/active layer formation, Solar RRL 6 (2022) 2100993.

Elias, L, M Bhar, S Ghosh, SK Martha, Effect of alloying on the electrochemical performance of Sb and Sn deposits as an anode material for lithium-ion and sodium-ion batteries, Ionics 28 (2022) 2759–2768.

Elias, L, KU Bhat, AC Hegde, Development of nanolaminated multilayer Ni – P alloy coatings for better corrosion protection, RSC Advances 6 (2016) 34005–34013.

Elias, L, P Cao, AC Hegde, Magnetoelectrodeposition of Ni – W alloy coatings for enhanced hydrogen evolution reaction, RSC Advances 6 (2016) 111358–111365.

Elias, L, AC Hegde, Electrodeposition of laminar coatings of Ni – W alloy and their corrosion behaviour, Surface and Coatings Technology 283 (2015) 61–69.

Elias, L, AC Hegde, Development of Ni-P alloy coatings for better corrosion protection using glycerol as additive, Analytical & Bioanalytical Electrochemistry 8 (2016a) 629–643.

Elias, L, AC Hegde, Synthesis and characterization of Ni-P-Ag composite coating as efficient electrocatalyst for alkaline hydrogen evolution reaction, Electrochimica Acta 219 (2016b) 377–385.

Elias, L, AC Hegde, Modification of Ni – P alloy coatings for better hydrogen production by electrochemical dissolution and TiO_2 nanoparticles, RSC Advances 6 (2016c) 66204–66214.

Elias, L, AC Hegde, Alloy Coatings for Corrosion Protection and Electrocatalysis, LAP LAMBERT Academic Publishing, Mauritius, 2017a.

Elias, L, AC Hegde, Effect of magnetic field on corrosion protection efficacy of Ni-W alloy coatings, Journal of Alloys and Compounds 712 (2017b) 618–626.

Elias, L, AC Hegde, Effect of magnetic field on HER of water electrolysis on Ni – W Alloy, Electrocatalysis 8 (2017c) 375–382.

Fang, R, R Liu, ZH Xie, L Wu, Y Ouyang, M Li, Corrosion-resistant and superhydrophobic nickel-phosphorus/nickel/PFDTMS triple-layer coating on magnesium alloy, Surface and Coatings Technology 432 (2022) 128054.

Fini, MH, A Amadeh, Improvement of wear and corrosion resistance of AZ91 magnesium alloy by applying Ni – SiC nanocomposite coating via pulse electrodeposition, Transactions of Nonferrous Metals Society of China 23 (2013) 2914–2922.

Friedrich, HE, BL Mordike, eds., Corrosion and surface protections, in: Magnesium Technology: Metallurgy, Design Data, Applications, Springer, Berlin, Heidelberg, 2006: pp. 431–497.

Gao, X, Q Huang, D Ma, Y Jiang, T Ren, X Guo, J Zhang, L Guo, Improving environmental adaptability and long-term corrosion resistance of Mg alloys by pyrazole ionic liquids: Experimental and theoretical studies, Journal of Molecular Liquids 333 (2021) 115964.

Gray, J, B Luan, Protective coatings on magnesium and its alloys – a critical review, Journal of Alloys and Compounds 336 (2002) 88–113.

Gu, C, J Lian, J He, Z Jiang, Q Jiang, High corrosion-resistance nanocrystalline Ni coating on AZ91D magnesium alloy, Surface and Coatings Technology 200 (2006) 5413–5418.

Guo, L, C Gu, J Feng, Y Guo, Y Jin, J Tu, Hydrophobic epoxy resin coating with ionic liquid conversion pretreatment on magnesium alloy for promoting corrosion resistance, Journal of Materials Science & Technology 37 (2020) 9–18.

Hamghavandi, MR, A Montazeri, AA Daryakenari, M Pishvaei, Preparation and characterization of chitosan/graphene oxide nanocomposite coatings on Mg–2 wt% Zn scaffold by pulse electrodeposition process, Biomedical Materials 16 (2021) 065005.

Hornberger, H, S Virtanen, AR Boccaccini, Biomedical coatings on magnesium alloys – a review, Acta Biomaterialia 8 (2012) 2442–2455.

Hort, N, Y Huang, D Fechner, M Störmer, C Blawert, F Witte, C Vogt, H Drücker, R Willumeit, KU Kainer, Magnesium alloys as implant materials – Principles of property design for Mg – RE alloys, Acta Biomaterialia 6 (2010) 1714–1725.

Hu, CC, A Bai, Influences of the phosphorus content on physicochemical properties of nickel – phosphorus deposits, Materials Chemistry and Physics 77 (2003) 215–225.

Hu, RG, S Zhang, JF Bu, CJ Lin, GL Song, Recent progress in corrosion protection of magnesium alloys by organic coatings, Progress in Organic Coatings 73 (2012) 129–141.

Huang, CA, CK Lin, YH Yeh, Corrosion behavior of Cr/Cu-coated Mg alloy (AZ91D) in 0.1 M H_2SO_4 with different concentrations of NaCl, Corrosion Science 52 (2010) 1326–1332.

Huang, W, D Mei, H Qin, J Li, L Wang, X Ma, S Zhu, S Guan, Electrophoretic deposited boron nitride nanosheets-containing chitosan-based coating on Mg alloy for better corrosion resistance, biocompatibility and antibacterial properties, Colloids and Surfaces A: Physicochemical and Engineering Aspects 638 (2022) 128303.

Jana, A, M Das, VK Balla, In vitro and in vivo degradation assessment and preventive measures of biodegradable Mg alloys for biomedical applications, Journal of Biomedical Materials Research Part A 110 (2022) 462–487.

Kang, Z, W Li, Facile and fast fabrication of superhydrophobic surface on magnesium alloy by one-step electrodeposition method, Journal of Industrial and Engineering Chemistry 50 (2017) 50–56.

Landkof, B, Magnesium applications in aerospace and electronic industries, Magnesium Alloys and Their Applications (2000) 168–172.

Lei, T, C Ouyang, W Tang, LF Li, LS Zhou, Enhanced corrosion protection of MgO coatings on magnesium alloy deposited by an anodic electrodeposition process, Corrosion Science 52 (2010) 3504–3508.

Leilei, Z, L Hejun, L Kezhi, Z Shouyang, F Qiangang, Z Yulei, L Jinhua, L Wei, Preparation and characterization of carbon/SiC nanowire/Na-doped carbonated hydroxyapatite multilayer coating for carbon/carbon composites, Applied Surface Science 313 (2014) 85–92.

Li, B, S Xue, P Mu, J Li, Robust self-healing graphene oxide-based superhydrophobic coatings for efficient corrosion protection of magnesium alloys, ACS Applied Materials & Interfaces 14 (2022) 30192–30204.

Li, B, X Yin, S Xue, P Mu, J Li, Facile fabrication of graphene oxide and MOF-based superhydrophobic dual-layer coatings for enhanced corrosion protection on magnesium alloy, Applied Surface Science 580 (2022) 152305.

Liu, Q, Z Kang, One-step electrodeposition process to fabricate superhydrophobic surface with improved anticorrosion property on magnesium alloy, Materials Letters 137 (2014) 210–213.

Liu, Y, S Li, Y Wang, H Wang, K Gao, Z Han, L Ren, Superhydrophobic and superoleophobic surface by electrodeposition on magnesium alloy substrate: Wettability and corrosion inhibition, Journal of Colloid and Interface Science 478 (2016) 164–171.

Liu, Y, J Xue, D Luo, H Wang, X Gong, Z Han, L Ren, One-step fabrication of biomimetic superhydrophobic surface by electrodeposition on magnesium alloy and its corrosion inhibition, Journal of Colloid and Interface Science 491 (2017) 313–320.

Liu, Y, X Yin, J Zhang, S Yu, Z Han, L Ren, A electro-deposition process for fabrication of biomimetic superhydrophobic surface and its corrosion resistance on magnesium alloy, Electrochimica Acta 125 (2014) 395–403.

Mordike, BL, T Ebert, Magnesium: Properties – applications – potential, Materials Science and Engineering: A 302 (2001) 37–45.

Mott, NF, A theory of the formation of protective oxide films on metals, Transactions of the Faraday Society 35 (1939) 1175–1177.

Nasirpouri, F, Fundamentals and principles of electrode-position, in: Electrodeposition of Nanostructured Materials, Springer, Switzerland, 2017: pp. 75–121.

Olugbade, TO, BO Omiyale, OT Ojo, Corrosion, corrosion fatigue, and protection of magnesium alloys: Mechanisms, measurements, and mitigation, Journal of Materials Engineering and Performance 31 (2022) 1707–1727.

Parco, M, L Zhao, J Zwick, K Bobzin, E Lugscheider, Investigation of HVOF spraying on magnesium alloys, Surface and Coatings Technology 201 (2006) 3269–3274.

Parthasaradhy, N, Practical Electroplating Handbook, Prentice-Hall, New Jersey, 1989: p. 1444.

Peng, F, D Zhang, X Liu, Y Zhang, Recent progress in superhydrophobic coating on Mg alloys: A general review, Journal of Magnesium and Alloys 9 (2021) 1471–1486.

Polmear, IJ, Magnesium alloys and applications, Materials Science and Technology 10 (1994) 1–16.

Seetharaman, S, S Jayalakshmi, R Arvind Singh, M Gupta, The potential of magnesium-based materials for engineering and biomedical applications, Journal of the Indian Institute of Science 102 (2022) 421–437.

Shi, Z, G Song, A Atrens, The corrosion performance of omogeni magnesium alloys, Corrosion Science 48 (2006) 3531–3546.

Shih, TS, JB Liu, PS Wei, Oxide films on magnesium and magnesium alloys, Materials Chemistry and Physics 104 (2007) 497–504.

Singh, C, SK Tiwari, R Singh, Development of corrosion-resistant electroplating on AZ91 Mg alloy by employing air and water-stable eutectic based ionic liquid bath, Surface and Coatings Technology 428 (2021) 127881.

Song, GL, A Atrens, Corrosion mechanisms of magnesium alloys, Advanced Engineering Materials 1 (1999) 11–33.

Song, J, J Chen, X Xiong, X Peng, D Chen, F Pan, Research advances of magnesium and magnesium alloys worldwide in 2021, Journal of Magnesium and Alloys 10 (2022) 863–898.

Song, Y, S Zhang, J Li, C Zhao, X Zhang, Electrodeposition of Ca – P coatings on biodegradable Mg alloy: in vitro biomineralization behavior, Acta Biomaterialia 6 (2010) 1736–1742.

Song, YW, DY Shan, EH Han, Electrodeposition of hydroxyapatite coating on AZ91D magnesium alloy for biomaterial application, Materials Letters 62 (2008) 3276–3279.

Syu, JH, JY Uan, MC Lin, ZY Lin, Optically transparent Li – Al – CO$_3$ layered double hydroxide thin films on an AZ31 Mg alloy formed by electrochemical deposition and their corrosion resistance in a dilute chloride environment, Corrosion Science 68 (2013) 238–248.

Tan, JK, N Birbilis, S Choudhary, S Thomas, P Balan, Corrosion protection enhancement of Mg alloy WE43 by in-situ synthesis of MgFe LDH/citric acid composite coating intercalated with 8HQ, Corrosion Science 205 (2022) 110444.

Thomas, P, CW Lai, MR Johan, Prospective of magnesium and alloy-based composites for lightweight railway rolling stocks, Applied Science and Engineering Progress 15 (2022) 5716–5716.

Tian, M, S Cai, L Ling, Y Zuo, Z Wang, P Liu, X Bao, G Xu, Superhydrophilic hydroxyapatite/hydroxypropyl-trimethyl ammonium chloride chitosan composite coating for enhancing the antibacterial and corrosion resistance of magnesium alloy, Progress in Organic Coatings 165 (2022) 106745.

Tuteja, A, W Choi, M Ma, JM Mabry, SA Mazzella, GC Rutledge, GH McKinley, RE Cohen, Designing superoleophobic surfaces, Science 318 (2007) 1618–1622.

Vijaya Ramnath, B, D Kumaran, J Melvin Antony, M Rama Subramanian, S Venkatram, Studies on magnesium alloy: Composites for aerospace structural applications, in: N Mazlan, SM Sapuan, RA Ilyas (Eds.), Advanced Composites in Aerospace Engineering Applications, Springer International Publishing, Cham, 2022: pp. 163–175.

Volovitch, P, JE Masse, A Fabre, L Barrallier, W Saikaly, Microstructure and corrosion resistance of magnesium alloy ZE41 with laser surface cladding by Al – Si powder, Surface and Coatings Technology 202 (2008) 4901–4914.

Wang, J, Y Yuan, T Chen, L Wu, X Chen, B Jiang, J Wang, F Pan, Multi-solute solid solution behavior and its effect on the properties of magnesium alloys, Journal of Magnesium and Alloys 10 (2022) 1786–1820.

Wang, T, Q Guo, TC Zhang, YX, Zhang, S Yuan, Large-scale prepared superhydrophobic HDTMS-modified diatomite/epoxy resin composite coatings for high-performance corrosion protection of magnesium alloys, Progress in Organic Coatings 170 (2022) 106999.

Wang, Z, Y Su, Q Li, Y Liu, Z She, F Chen, L Li, X Zhang, P Zhang, Researching a highly anti-corrosion superhydrophobic film fabricated on AZ91D magnesium alloy and its anti-bacteria adhesion effect, Materials Characterization 99 (2015) 200–209.

Witte, F, The history of biodegradable magnesium implants: A review, Acta Biomaterialia 6 (2010) 1680–1692.

Wu, L, J Zhao, Y Xie, ZD Yang, Progress of electroplating and electroless plating on magnesium alloy, Transactions of Nonferrous Metals Society of China 20 (2010) s630–s637.

Xiang, YH, WB Hu, B Shen, Initial deposition mechanism of direct electroless nickel plating on magnesium alloy, Journal-Shanghai Jiaotong University-Chinese Edition 34 (2000) 1638–1640.

Xin, RL, MY Wang, JC Gao, P Liu, Q Liu, Effect of microstructure and texture on corrosion resistance of magnesium alloy, in: Materials Science Forum, Trans Tech Publications, Stafa-Zurich, 2009: pp. 1160–1163.

Yan, L, S Yu, JD Liu, ZW Han, D Yuan, Microstructure and wear resistance of electrodeposited Ni-SiO2 nano-composite coatings on AZ91HP magnesium alloy substrate, Transactions of Nonferrous Metals Society of China 21 (2011) s483–s488.

Yang, GL, Y Ouyang, ZH Xie, Y Liu, W Dai, L Wu, Nickel interlayer enables indirect corrosion protection of magnesium alloy by photoelectrochemical cathodic protection, Applied Surface Science 558 (2021) 149840.

Yang, H, X Guo, X Chen, N Birbilis, A homogenization pre-treatment for adherent and corrosion-resistant Ni electroplated coatings on Mg-alloy AZ91D, Corrosion Science 79 (2014) 41–49.

Yao, W, W Liang, G Huang, B Jiang, A Atrens, F Pan, Superhydrophobic coatings for corrosion protection of magnesium alloys, Journal of Materials Science & Technology 52 (2020) 100–118.

Yao, W, L Wu, J Wang, B Jiang, D Zhang, M Serdechnova, T Shulha, C Blawert, ML Zheludkevich, F Pan, Micro-arc oxidation of magnesium alloys: A review, Journal of Materials Science & Technology 118 (2022) 158–180.

Zhang, J, C Gu, Y Tong, W Yan, J Tu, A smart superhydrophobic coating on AZ31B magnesium alloy with self-healing effect, Advanced Materials Interfaces 3 (2016) 1500694.

Zhang, J, ZU Rahman, Y Zheng, C Zhu, M Tian, D Wang, Nanoflower like SnO$_2$-TiO$_2$ nanotubes composite photoelectrode for efficient photocathodic protection of 304 stainless steel, Applied Surface Science 457 (2018) 516–521.

Zhang, J, C Yan, F Wang, Electrodeposition of Al – Mn alloy on AZ31B magnesium alloy in molten salts, Applied Surface Science 255 (2009) 4926–4932.

Zhang, L, S Li, H Li, L Pei, Bioactive surface modification of carbon/carbon composites with multilayer SiC-SiC nanowire-Si doped hydroxyapatite coating, Journal of Alloys and Compounds 740 (2018) 109–117.

Zhang, L, L Pei, H Li, F Zhu, Design and fabrication of pyrolytic carbon-SiC-fluoridated hydroxyapatite-hydroxyapatite multilayered coating on carbon fibers, Applied Surface Science 473 (2019) 571–577.

Zhang, RF, SF Zhang, Formation of micro-arc oxidation coatings on AZ91HP magnesium alloys, Corrosion Science 51 (2009) 2820–2825.

Zhang, S, F Cao, L Chang, J Zheng, Z Zhang, J Zhang, C Cao, Electrodeposition of high corrosion resistance Cu/Ni – P coating on AZ91D magnesium alloy, Applied Surface Science 257 (2011) 9213–9220.

Zhang, SJ, DL Cao, LK Xu, JK Tang, RQ Meng, HD Li, Corrosion resistance of a superhydrophobic dodecyltrimethoxysilane coating on magnesium alloy AZ31 fabricated by one-step electrodeposition, New Journal of Chemistry 45 (2021) 14665–14676.

Zhang, Y, T Lin, Influence of duty cycle on properties of the superhydrophobic coating on an anodized magnesium alloy fabricated by pulse electrodeposition, Colloids and Surfaces A: Physicochemical and Engineering Aspects 568 (2019) 43–50.

Zhang, Z, G Yu, Y Ouyang, X He, B Hu, J Zhang, Z Wu, Studies on influence of zinc immersion and fluoride on nickel electroplating on magnesium alloy AZ91D, Applied Surface Science 255 (2009) 7773–7779.

Zhao, MJ, C Cai, L Wang, Z Zhang, JQ Zhang, Effect of zinc immersion pretreatment on the electro-deposition of Ni onto AZ91D magnesium alloy, Surface and Coatings Technology 205 (2010) 2160–2166.

Zheng, J, DC Bock, T Tang, Q Zhao, J Yin, KR Tallman, G Wheeler, X Liu, Y Deng, S Jin, Regulating electrodeposition morphology in high-capacity aluminium and zinc battery anodes using interfacial metal – substrate bonding, Nature Energy 6 (2021) 398–406.

Zheng, T, Y Hu, F Pan, Y Zhang, A Tang, Fabrication of corrosion-resistant superhydrophobic coating on magnesium alloy by one-step electrodeposition method, Journal of Magnesium and Alloys 7 (2019) 193–202.

Zheng, T, Y Hu, Y Zhang, F Pan, Formation of a hydrophobic and corrosion resistant coating on magnesium alloy via a one-step hydrothermal method, Journal of Colloid and Interface Science 505 (2017) 87–95.

Zhu, Y, G Yu, B Hu, X Lei, H Yi, J Zhang, Electrochemical behaviors of the magnesium alloy substrates in various pretreatment solutions, Applied Surface Science 256 (2010) 2988–2994.

16 Electroless Nanocomposite Coatings

T.S.N. Sankara Narayanan and K. Ravichandran

CONTENTS

16.1 INTRODUCTION

Electroless (EL) plating involves reduction of metal ions using chemical reducing agents such as sodium hypophosphite or sodium borohydride. It is an autocatalytic process. Hence, the metal that is being plated should also be catalytically active, which would otherwise require activation. In spite of its resemblance to electroplating, EL plating possess has some unique attributes. Unlike electroplated coatings, EL-plated coatings are uniform irrespective of the geometry of the component being plated, and it is a metal-metalloid alloy coating. The major limitations of EL plating are low bath stability and poor efficiency (Balaraju et al., 2003; Sankara Narayanan and Seshadri, 2012; Sudagar et al., 2013; Sankara Narayanan et al., 2016).

Among the various metals that can be plated by EL deposition, EL nickel (Ni) assumes significance due to its importance in many engineering applications. Sodium hypophosphite (Ni-P), sodium borohydride (Ni-B), dimethylamine borane (Ni-B) and hydrazine (pure Ni) can be used as reducing agents for EL Ni plating. Among them, EL Ni-P coating has offered an excellent corrosion resistance, while EL Ni-B coating has imparted a high hardness and superior wear resistance. EL-plated pure Ni coatings prepared using hydrazine as a reducing agent has been used mainly for semiconductor applications. They are not popular due to the high cost as well as the health hazards associated with hydrazine. Hypophosphite-reduced EL Ni plating has received commercial success due to the ease of control and cost-effectiveness (Balaraju et al., 2003; Sankara Narayanan and Seshadri, 2012; Sudagar et al., 2013; Sankara Narayanan et al., 2016).

The low density, high specific strength, good thermal conductivity and better electromagnetic shielding are some of the important characteristics of magnesium (Mg) alloys. However, Mg alloys are highly active and exhibit poor corrosion resistance. Several surface modification methods such as conversion coatings, anodizing and microarc oxidation have been explored to impart the desirable characteristics for Mg alloys. Although each of these methods has improved some characteristic properties of Mg alloys, they fail to improve certain other properties. Materials used for aerospace industries warrant good electrical conductivity, low magnetism, high corrosion resistance and better wear resistance. Chemical conversion coatings formed on Mg alloys are non-magnetic. In addition, they could not offer good wear resistance. Anodic oxidation and microarc oxidation based coatings on Mg alloys increase the hardness and improve the wear resistance. However, they lack electrical conductivity. The porous nature of microarc oxidation coatings allows permeation of the corrosive medium through them and deleteriously influence the corrosion resistance of the Mg alloys. EL Ni plating could endow the Mg alloys with better wear resistance, higher corrosion resistance, electrical conductivity and magnetic property. Being a metal-metalloid alloy coating, by suitably controlling the P and B content of the Ni-P and Ni-B coatings, it would be possible to manipulate the magnetic properties. Addition of second-phase particles in EL Ni plating baths would enable formation of EL Ni-P and Ni-B based composite coatings with better corrosion resistance and wear resistance.

Electroless (EL) nanocomposite coatings are prepared by dispersing nanoparticles (NPs) in the plating bath. Inclusion of NPs has been attempted both in hypophosphite-reduced (Ni-P) and borohydride-reduced (Ni-B) EL nickel plating baths. A variety of NPs are incorporated in EL Ni-P and Ni-B matrices, which includes carbides such as silicon carbide (SiC), boron carbide (B_4C) and tungsten carbide (WC), nitrides such as silicon nitride (Si_3N_4) and titanium nitride (TiN), oxides such as alumina (Al_2O_3), zirconia (ZrO_2), titania (TiO_2) and silica (SiO_2), sulphides such as molybdenum disulphide (MoS_2) and tungsten disulphide (WS_2), carbon nanostructures (diamond, carbon black, carbon nanotubes (CNTs)), metal organic frameworks (MOF), montmorillonite (MMT), graphitic carbon nitride (GCN), etc. The key requirements of particles to be incorporated in EL Ni-P and Ni-B matrices are (i) they should be chemically inert and (ii) they should remain in suspension. However, NPs tend to agglomerate due to their higher surface energy, which would affect the homogeneity of the coating and deleteriously influence the mechanical properties. Many reviews have addressed the developments hitherto made in electroless nanocomposite coatings (Balaraju et al., 2003; Sankara Narayanan and Seshadri, 2012; Sudagar et al., 2013; Sankara Narayanan et al., 2016; Zhang, 2016; Pancrecious et al., 2018; Barati and Hadavi, 2020). The present chapter highlights some of the salient features and recent developments in this area of research.

16.2 IMPORTANCE OF PRE-TREATMENT AND ACTIVATION

Deposition of EL Ni-P coating directly on Mg and its alloy is complicated and really challenging. Zinc immersion treatment (followed by direct EL Ni plating) and acid pickling in CrO_3 followed by activation in HF are the conventionally used pre-treatment processes for the deposition of EL Ni-P coating on Mg and its alloys. Acid pickling is necessary to remove the oxide film and any metallic impurities that are present on the surface of Mg and its alloys. Besides, pickling imparts a rough surface profile, which is an important requirement for the initiation of EL Ni plating. However, the methods for generating a rough surface profile on the surface of Mg and its alloys should be chosen wisely. Grit blasting has been used to prepare the surface of Mg before EL Ni-B coating. However, grit blasting of Mg using alumina (150 μm) has been shown to be detrimental, as it results in the formation of an irregular Mg/EL Ni-B coating interface (Correa et al., 2013). Activation is essential to prevent rapid corrosion of Mg and its alloys in the EL Ni plating bath. However, the toxicity issues associated with the use of CrO_3 and HF have warranted the development of eco-friendly pre-treatment processes for EL Ni plating. Many alternatives have been explored. The use of H_3PO_4 instead of CrO_3 for pickling has led to a poor adhesion of the EL Ni-P coating on AZ31 Mg alloy. Addition of HF along with H_3PO_4 has also not improved the adhesion. However, pickling in H_3PO_4 followed by activation by NH_4HF_2 has enabled an excellent adhesion of the EL Ni-P coating on

AZ31 Mg alloy. However, the volume of the H_3PO_4 solution used for pickling should be < 400 ml/L (Xie et al., 2015). NH_4HF_2 has been suggested as an alternative to HF for activating Mg and its alloys before EL Ni plating. This leads to the question whether NH_4HF_2 is less toxic than HF? In terms of toxicity, HF and NH_4HF_2 are on par with each other. However, it is easier to handle NH_4HF_2, as it is less volatile than HF (Zuleta et al., 2012). Pre-plating using an alkaline hypophosphite-reduced EL plating bath is considered a viable choice. However, the P content of the EL Ni-P coating will be low. Many types of conversion coatings have been explored as pre-treatment. However, the use of $NaNO_2$ as an accelerator in forming them is carcinogenic.

Pickling of AZ31 Mg alloy using a solution mixture of 30 ml/L H_3PO_4 (85% V/V), 5 g/L $Mn(H_2PO_4)_2$, 50 ml/L HNO_3(80%V/V) and 150 ml/L acetic acid at 20°C for 5 minutes has led to the formation of a Mg-Mn phosphate coating, which helped to increase the adhesion, hardness and corrosion resistance of the EL Ni-P coating subsequently deposited over it (Liu et al., 2012). Correa et al. (2012) have suggested a pre-treatment process for the deposition of EL Ni-B coating on Mg and AZ91D Mg alloy. The surface of the Mg and Mg alloy was prepared by mechanical polishing using SiC-coated abrasive paper (grit size: 100) followed by grit blasting using alumina (150 μm). They were subjected to alkaline cleaning using 37 g/L NaOH + 10 g/L Na_3PO_4 at 65°C for 10 minutes. The cleaned surfaces were activated using 200 g/L NH_4F_2 at 25°C for 10 minutes. In addition to fluoride activation, up to 10 g/L of NH_4F_2 was added to the alkaline borohydride-reduced EL Ni plating bath. This sequence has enabled the formation of a uniform and compact EL Ni-B coating on both Mg as well as AZ91D Mg alloy. The presence of NH_4F_2 in the plating bath has enabled surface passivation, accelerated the initiation of the EL Ni-B coating and reduced the number of defects. Coatings prepared using 5 g/L of NH_4F_2 offered good corrosion resistance. Tran et al. (2012) have compared the characteristics of EL Ni-P coating formed on Mg alloy prepared by two different combinations of picking and activation. Pickling using a mixture of HNO_3 and H_3PO_4 followed by two-stage activation, first using a mixture of $K_4P_2O_7$, Na_2CO_3 and KF and then using a mixture of NH_4HF_2 and H_3PO_4, has improved the adhesion of the EL Ni-P coating. In contrast, pickling using a mixture of CrO_3 and HNO_3 followed by activation using HF has offered only a limited improvement in adhesion of the EL Ni-P coating.

Xu et al. (2014) have compared the effect of two different pickling solutions on the characteristics of EL Ni-P coatings deposited on Mg-7.5Li-2Zn-1Y alloy. The surface preparation, pickling and activation treatments include surface grinding using SiC-coated abrasive papers (grit size: 200, 600, 1000 and 2000), ultrasonic cleaning using acetone, alkaline cleaning using 50 g/L NaOH and 10 g/L $Na_{12}PO_4$.12H_2O at 65°C for 10 minutes, pickling using either (i) 180 g/L CrO_3 and 1 g/L KF or (ii) 125 g/L CrO_3 and 110 ml/L HNO_3 and activation using 40% HF at 25°C for 10 minutes. Pickling of the Mg alloy using 125 g/L CrO_3 and 110 ml/L HNO_3 has enabled the formation of a uniform, pore free and compact coating with a higher P content, excellent adhesion and superior corrosion resistance. In contrast, the Mg alloy pickled using 180 g/L CrO_3 and 1 g/L KF has resulted in the formation of a relatively porous coating with weak adhesion, lower P content and relatively poor corrosion resistance.

Strontium phosphate (Sr-P) conversion coating has been suggested as a cost-effective and eco-friendly pre-treatment for the deposition of EL Ni-P coating on AM60B Mg alloy (Rajabalizadeh and Seifzadeh, 2016). The Sr-P conversion coating is porous. Nucleation of Ni occurs at these pores during the initial stages of EL Ni-P plating. Subsequently, the growth occurs in X, Y and Z directions, resulting in the formation of a uniform EL Ni-P coating. The treatment time employed for Sr-P conversion coating has exerted a strong influence on the quality of the EL Ni-P coating. AM60B Mg alloy pretreated with Sr-P conversion coating for 2 and 5 minutes has enabled the formation of a uniform and pore-free EL Ni-P coating with compact grain boundaries, higher P content and better corrosion resistance. Conversely, an increase in the pre-treatment time to 15 minutes has led to the development of a granular structure with more intergranular spacing, lower P content and inferior corrosion resistance.

Immersion in EL Ni solution followed by immersion in an alkaline $NaBH_4$ solution has been shown to activate MAO-coated AZ31 Mg alloy with the formation of Ni^0 clusters so that it can be subsequently plated using EL plating bath (Ezhilselvi et al., 2017). The NaBH$_4$ activation treatment increased the number of nucleation sites and enabled the deposition of a uniform and homogeneous EL Ni-P coating over the MAO-coated AZ31 Mg alloy. It would be difficult to use the conventional chromic acid etching and HF activation treatments for activating MAO-coated Mg alloys, as the MAO coating will also be etched and/or removed. In this perspective, activation using alkaline NaBH$_4$ solution appears to be a viable option.

Pre-treatment in the form of surface preparation, acid pickling and activation is an essential prerequisite for the development of EL Ni-P and EL Ni-B composite coatings with desired characteristics.

16.3   DISPERSION OF NANOPARTICLES IN ELECTROLESS NICKEL-BASED PLATING BATHS

The deposition of EL nanocomposite coatings involves dispersion of NPs in the EL plating bath, formation of an ionic cloud over the NPs, transport of ion-adsorbed NPs towards the substrate by convection and diffusion and incorporation of NPs in the EL deposited coating without any bonding between them and the metal matrix. The common protocol employed to achieve an effective dispersion of the NPs is to mix them with a small portion of the plating bath by ultrasonication for 30 minutes. Subsequently, this solution is mixed with the plating bath. Since the plating rate of EL Ni-P and Ni-B is of the order of 12 to 15 µm/h and 8–10 µm/h, respectively, it is essential that the plating bath should be mechanically agitated to keep the NPs in suspension.

Ball milling can be used to reduce the size of the NPs. Pre-treatment is required for certain NPs. CNTs should be pre-treated with acid to remove impurities such as SiO$_2$ and amorphous carbon before dispersing them in the plating bath. In addition, the length of the CNTs should be reduced by ball milling, and their surface should be functionalized to achieve an effective dispersion. The presence of Ni^{2+}, Na$^+$, K$^+$ and NH$_4^+$ ions increases the ionic strength of EL Ni plating baths. A higher ionic strength would compress the double layer around the particles, which would enable agglomeration of the particles. Hence, conditions to achieve dispersion of NPs in deionized water are not applicable for dispersion of the same particles in EL Ni plating baths and warrant careful optimization.

Addition of surfactants in the plating bath enables an effective dispersion of the NPs. Surfactants, by virtue of their adsorption, impart a surface charge on the NPs. The steric hindrance between the surfactant molecules helps to improve the stability of NPs suspension. In addition, surfactants facilitate an easy removal of the H$_2$ bubbles by reducing the surface tension. Different types of surfactants, viz., cationic, anionic, non-ionic and zwitterionic, have been used in EL plating baths (Mafi and Dehghanian, 2011; Zielińska et al., 2012; Ansari and Thakur, 2017; Fayyad et al., 2019). Nwosu et al. (2012) have suggested the selection criteria for surfactants to be used in EL Ni plating baths. Surfactants, in general, possess good adhesion tension and leveling power. Nevertheless, compatibility with the plating bath, stability and reactivity are the key factors in determining the choice of the surfactants. The excellent compatibility between the Ni^{2+} ions in the EL plating bath makes cationic surfactants a suitable choice. Nonionic surfactants don't have a charge on them, but they are tolerant with the Ni^{2+} ions. Fluorosurfactants possess excellent stability in acidic EL Ni-P and alkaline EL Ni-B plating baths that operate at 80°C to 98°C. They are capable of reducing the surface tension much better than hydrocarbon based surfactants. Anionic surfactants, due to their negative charge, are considered suitable for the incorporation of positively charged particles (Abdoli and Sabour Rouhaghdam, 2013).

Addition of surfactants is required to achieve an effective dispersion of particles such as CNT, MoS$_2$, PTFE, etc. (Mafi and Dehghanian, 2011; Zarebidaki and Allahkaram, 2011). Zeta potential provides an idea about the particle charge. Addition of surfactants to the EL plating bath could alter the surface charge of the particle, which can be identified by the zeta potential (Shen et al., 2021). For instance, addition of fluorinated alkyl quaternary ammonium iodide has been shown to change the zeta potential of PTFE particles dispersed in EL Ni plating bath from –25 mV to + 40 mV (Ger and Hwang, 2002). With an appropriate choice of surfactants, it would be possible to impart the desired charge on the particles so as to achieve an effective dispersion and maximize the level of incorporation of particles.

16.4   STABILITY OF THE PLATING BATHS

It is imperative to eliminate impurities in EL plating baths to prevent their decomposition. Addition of NPs with a high surface area is likely to increase the chance of decomposition of the plating bath. This is particularly important for borohydride-reduced EL Ni plating baths. Hence, utmost care should be exercised in preparing EL nanocomposite coatings. A palladium stability test is commonly used to evaluate the stability of EL plating baths. Accordingly, the plating solution will be heated to 60°C, and a 100 mg/L $PdCl_2$ will be slowly added from 1 to 50 ml to the plating bath. The time for decomposition of the plating bath will be recorded; the lower the time, the poorer is the stability of the plating bath and vice versa (Sun et al., 2017).

16.5   RATE OF DEPOSITION OF ELECTROLESS NICKEL-BASED NANOCOMPOSITE COATINGS

The rate of deposition of EL Ni-P and Ni-B composite coatings is likely to increase if the NPs increase the number of nucleation centers. However, adsorption of NPs and their surface coverage also play a vital role in determining the rate of deposition. For EL Ni-W-P-CeO_2 composite coating, the rate of deposition is increased with an increase concentration of CeO_2 particles from 0 to 8 mg/L. However, a further increase in the concentration of the CeO_2 particles from 8 mg/L to 10 and 12 mg/L has led to a decrease in plating rate (Sun et al., 2017). Addition of GO nanosheets with a large surface area has been shown to reduce the rate of deposition of EL Ni-P-GO nanocomposite coating (Li et al., 2022). This is due to the adsorption of GO NPs on the surface of Mg-Li alloy as well as to the decrease in diffusion and adsorption of Ni^{2+} ions during plating. The dependence of the rate of deposition of EL Ni-P-TiO_2 nanocomposite coatings formed on AZ91D Mg alloy on the size as well as the concentration of the TiO_2 NPs is shown by Carrillo et al. (2019). According to them, irrespective of the size (5 nm and 100 nm) and concentration (0.5 g/L and 2 g/L) of the TiO_2 NPs, the rate of deposition of EL Ni-P-TiO_2 nanocomposite coatings (1.64 to 3.97 µm) is lower than that of the plain EL Ni-P coating (4.62 µm). For a given size of 5 nm, the rate of deposition is decreased from 3.97 µm to 2.24 µm when the concentration of the TiO_2 NPs is increased from 0.5 to 2.0 g/L. For a given concentration of 2 g/L, the rate of deposition is decreased from 2.24 µm to 1.64 µm when the size of the TiO_2 NPs is increased from 5 nm to 100 nm. The decrease in plating rate of the composite coatings is due to the adsorption of the TiO_2 NPs on the catalytic surface, which becomes prominent with the increase in size and concentration of the NPs. To achieve an optimum rate of deposition, the size and concentration of the NPs should be carefully selected.

16.6   LEVEL OF INCORPORATION OF NANOPARTICLES IN THE DEPOSITED MATRIX

The level of incorporation of NPs in the EL Ni-P and Ni-B matrices is determined by particle flux at the interface, i.e. the amount of particles available at the vicinity of the Mg alloy and residency time, i.e. for how long the particle remains on the surface of the Mg alloy before it gets engulfed by the growing EL Ni-P matrix. Particle flux at the interface is mainly determined by the concentration of the NPs in the plating bath. Hence, the level of incorporation would increase with concentration of the NPs in the plating bath. However, beyond a threshold level, the extent of incorporation will reach saturation due to the leveling of the particle flux. Wang et al. (2013) have made a similar inference for EL Ni-P/SiC gradient composite coating in which the level of incorporation is increased from 0 to 8 wt.% when the concentration of SiC particles is increased from 0 to 4 g/L. However, a further increase in concentration of SiC particles to 6 and 8 g/L has led to only a small increase in level of incorporation of SiC particles from 8 to 9 wt.%. At higher concentrations, agglomeration of the NPs tends to decrease the level of incorporation. The level of incorporation of particles is commonly expressed in wt. % or vol. %. However, Sarret et al. (2006) have recommended the use of number density of particles instead of expressing them in wt. % or vol. %.

Since the plating bath is agitated to keep the particles in suspension, the flow conditions need to be optimized to achieve a maximum level of particle incorporation. Laminar flow condition would decrease the particle flux at the interface. Turbulent flow condition is likely to throw away the particles and decrease the residency time. In addition, collision of particles that are thrown away with the incoming particles greatly reduces the particle flux at the interface. Hence, both laminar and turbulent flow conditions are not suitable to achieve a higher level of particle incorporation. Laminar-to-turbulent transition is considered to be a suitable condition to maximize the level of incorporation, and as this varies with the size of the particles, this flow condition should be carefully optimized (Lajevardi et al. 2011). Excessive hydrogen evolution during plating would decrease the residency time of the particles and decrease their chance of incorporation. This is particularly important in alkaline borohydride-reduced EL Ni plating baths where the extent of hydrogen evolution is much higher.

16.7   SURFACE MORPHOLOGY OF THE NANOCOMPOSITE COATINGS

Both plain EL Ni-P and EL Ni-P nanocomposite coatings exhibit a nodular structure with typical cauliflower morphology. In spite of their similar morphological features, many researchers have reported that particle incorporation has led to refinement of the nodules. According to Song et al. (2007), the nodules of EL Ni-P coating are mostly round in shape, and their size is in the range of a few micrometers. However, the nodules of EL Ni-P-ZrO$_2$ nanocomposite coatings have different dimensions, with their size ranging from a few micrometers to nanometers. Wang et al. (2013) have compared the morphological features of EL Ni-P coating and EL Ni-P/SiC gradient composite coating (**Figure 16.1**). The EL Ni-P coating exhibits a spherical nodular structure, while the EL Ni-P/SiC gradient composite coating shows a coarse nodular structure. The EL Ni-P/SiC gradient composite coating has smaller nodules than the plain EL Ni-P coating. This is due to the ability of the SiC NPs to increase the number of catalytically active sites. A decrease in size of the nodules is also observed for EL Ni-P-TiN nanocomposite coating following incorporation of TiN NPs in the EL Ni-P matrix (Yu et al., 2011). Li et al. (2022) have compared the morphological features of EL Ni-P, EL Ni-P-GO and EL Ni-P-CeO$_2$@GO composite coatings (**Figure 16.2**). A considerable refinement of the nodules is observed for the composite coatings. The average size of the nodules is decreased from 34.04 µm (for EL Ni-P coating) to 11.71 µm following the incorporation of CeO$_2$@GO composite particles in the EL Ni-P matrix. The decrease in the size of the nodule of the composite coatings is due to the increase in number of nucleation centers and the ability of the incorporated NPs to inhibit the growth of the nodules. The surface morphology of EL Ni-Mo-P-CNT nanocomposite coatings prepared using 0.1 to 0.5 g/L of CNTs exhibits a typical cauliflower-like morphology (Xiao et al., 2022). Large protrusion of particles could be observed for coatings prepared using 0.1 g/L of CNTs. However, the protrusions in the composite coatings are considerably reduced when the concentration of CNTs in the plating bath is increased up to 0.4 g/L. The size of the nodules is also reduced with an increase in concentration of CNTs up to 0.4 g/L. However, the trend is reversed for the coating prepared using 0.5 g/L CNTs wherein the protrusion of the particles and the size of the nodules are increased. This is due to agglomeration of the CNTs.

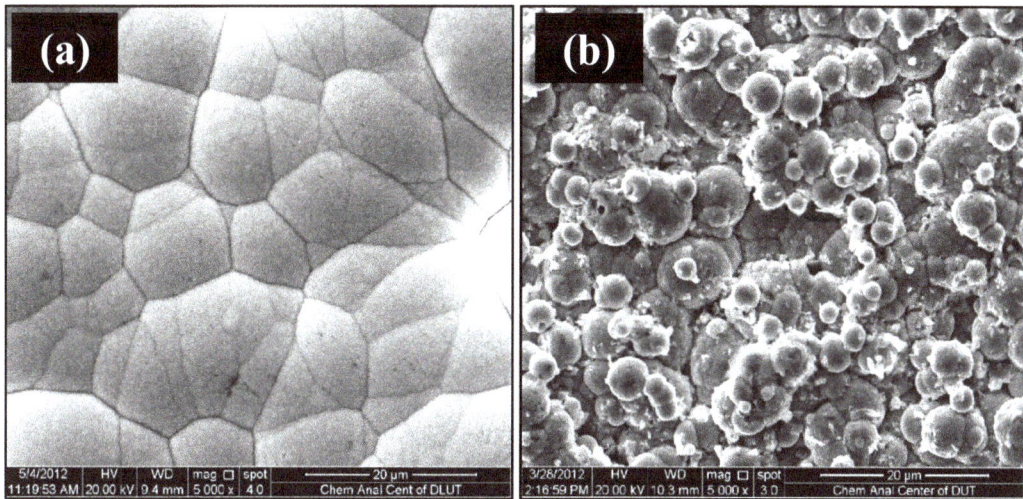

FIGURE 16.1   Surface morphology of (a) EL Ni-P and (b) EL Ni-P-SiC gradient composite coating formed on AZ91D Mg alloy. *Reproduced with permission from (Wang et al., 2013) © 2013 Elsevier B.V.*

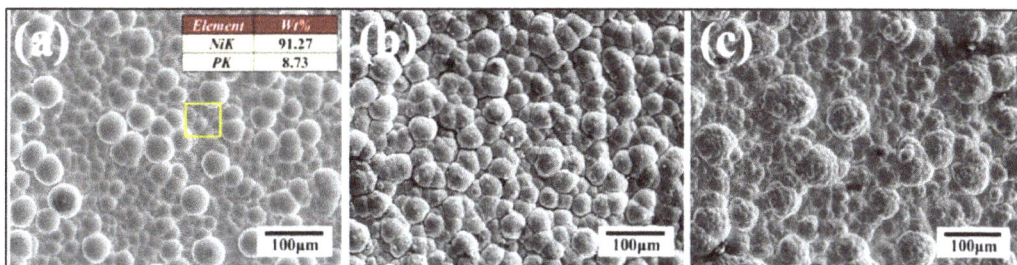

FIGURE 16.2   Surface morphologies of (a) EL Ni-P coating; (b) EL Ni-P-GO composite coating; and (c) EL Ni-P-CeO$_2$@GO composite coating. *Reproduced with permission from (Li et al., 2022) © 2022 Elsevier B.V.*

16.8   SURFACE ROUGHNESS OF THE NANOCOMPOSITE COATINGS

Incorporation of NPs in the EL Ni-P and Ni-B matrices alters the surface roughness of the composite coatings. Li et al. (2022) have reported that the average surface roughness (R_a) of EL Ni-P, EL GO-Ni-P and EL CeO$_2$@GO-Ni-P composite coating are 602 nm, 428 nm and 367 nm, respectively. The lower R_a of the EL Ni-P nanocomposite coatings when compared to the plain EL Ni-P coating is due to the filling up of the pores by the GO nanosheets, which increased the uniformity of the composite coatings. Agglomeration of the NPs, particularly at higher concentration, has been identified as the major reason for the increase in R_a (Carrillo et al., 2019; Sayyad and Senanayake, 2021a, 2021b; Xiao et al., 2022). A higher R_a is not desirable, as it would increase the coefficient of friction (COF) (Carrillo et al., 2019).

16.9   COATING ADHESION

The adhesive strength of EL composite coatings can be evaluated by the thermal shock method as per ASTM B733–04. Accordingly, the composite-coated Mg alloy will be subjected to thermal stresses (24–200°C) for 20 cycles. The absence of any blisters, cracks, delamination and spalling of the coated layer after the test indicates a good adhesion of the composite coating on the Mg alloy

(Hu et al., 2018; Rajabalizadeh et al., 2022). Since the thermal expansion coefficient of the Mg alloy and the coating are different, the thermal stresses generated during the thermal shock test could undermine the adhesion of the coated layer. If the coating is weakly adherent, then it would easily detach from the Mg alloy. Development of cracks and blisters is also classified as possible modes of adhesive failure. The cross-sectional morphology of the composite coating could also be used to evaluate the adhesion of the EL composite coatings. The absence of any apparent defect between the composite coating and the Mg alloy indicate a good adhesion between them.

A gradation in the SiC content from the AZ91D Mg alloy to the top surface of EL Ni-P/SiC gradient composite coating has been shown to decrease the thermal stress in the coated layer and helps to increase the adhesion of the composite coating to the Mg alloy (Wang et al., 2013). The absence of blisters and cracks in the coated layer substantiates the excellent adhesion of the EL Ni-P/SiC gradient composite coating. Hu et al. (2018) have shown that EL Ni-P-Al$_2$O$_3$ nanocomposite coating exhibits a good adhesion on AZ91D die-cast alloy. The absence of any defects at the Mg alloy-composite coating interface points out that the incorporation of Al$_2$O$_3$ NPs in the EL Ni-P matrix has caused no deleterious effect on coating adhesion. Similarly, no cracks, blisters and detachment of the coated layer are observed for both EL Ni-P and EL Ni-P MIL-53 MOF composite coating deposited on AM60B Mg alloy, suggesting an excellent adhesion of these coatings on AM60B alloy (Rajabalizadeh et al., 2022) (**Figure 16.3**). After 20 cycles of heating and cooling, both EL Ni-B and EL Ni-B-MMT composite coating did not show the development of any cracks and blisters, suggesting that both of them are adherent to the AM60B Mg alloy (Miri et al., 2021).

Lakavat et al. (2022) have evaluated the adhesion of EL Ni-P-B-Al$_2$O$_3$ composite coating on AZ91D Mg alloy as per VDI 3198 standard. According to this standard, development of microcracks with and without a smaller extent of delamination at the circumference of the imprint points out the existence of a strong interfacial adhesion between the coating and the substrate. Conversely, an extended delamination over a wider region of the circumference of the imprint suggests poor adhesion of the coating. EL Ni-P-B-Al$_2$O$_3$ composite coating fails to show any delamination. The development of microcracks on the circumference of the imprint indicates a strong adhesion between the EL Ni-P-B-Al$_2$O$_3$ composite coatings and the AZ91D Mg alloy. Dhakal et al. (2022) have suggested depositing a thin layer of plain EL Ni-P coating before the EL Ni-P-Si$_3$N$_4$ composite coating as a strategy to improve the adhesion of the later. Although they have realized the

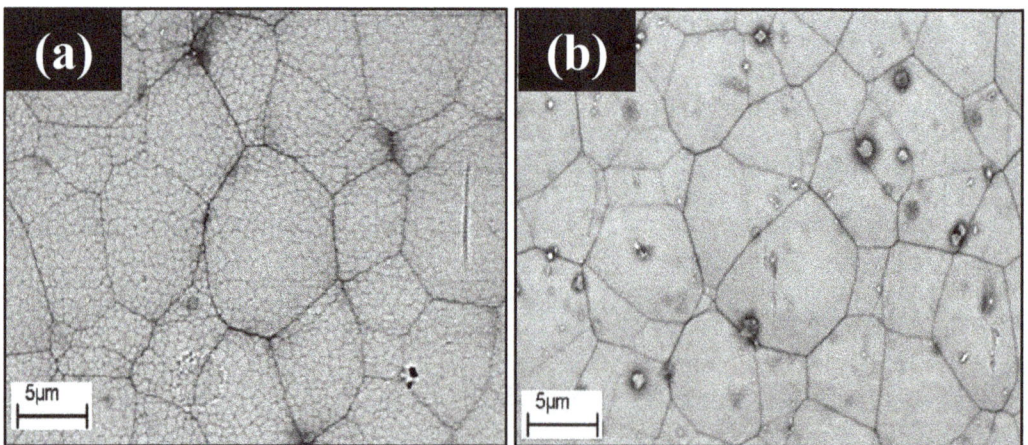

FIGURE 16.3 Surface morphology of (a) EL Ni-P and (b) EL Ni-P-MOF composite coatings after 20 cycles (24–200°C) of thermal shock test. *Reproduced from (Rajabalizadeh et al., 2022) © 2021 Chongqing University. Publishing services provided by Elsevier B.V. on behalf of KeAi Communications Co. Ltd.*

improvement of adhesion for EL Ni-P-Si$_3$N$_4$ composite coating on steel, the strategy would work for Mg and its alloys. In fact, this approach is followed for preparing EL Ni-P and EL Ni-B based composite coatings in which the plain EL Ni-P coating served as a pre-treatment of the Mg alloy to prevent excessive dissolution of the alloy during EL Ni plating.

16.10   COATING HARDNESS

The hardness of EL Ni composite coatings is an important attribute, and coatings with a higher hardness are likely to offer better wear resistance. An increase in hardness following incorporation of B$_4$C, SiC, ZrO$_2$, SiO$_2$, Al$_2$O$_3$, TiO$_2$, TiN and CNTs particles in the EL Ni matrix has been observed by many researchers. The increment in hardness is due to dispersion strengthening effect caused by the incorporated particles in the EL Ni matrix. The microhardness of EL Ni-P-B$_4$C and Ni-W-P-B$_4$C composite coatings was evaluated by Araghi and Paydar (2010, 2013, 2014). Incorporation of B$_4$C particles has led to an increase in hardness of the EL Ni-P matrix from 700 MPa to 1200 MPa. Similarly, the hardness is increased from 800 MPa to 1290 MPa following the incorporation of B$_4$C particles in the EL Ni-W-P matrix. The observed increase in hardness for the composite coatings is due to the ability of the B$_4$C particles to hinder the movement of dislocations. The relatively higher hardness of EL Ni-W-P and EL Ni-W-P-B$_4$C composite coatings when compared to their EL Ni-P counterparts is due to the formation of a Ni-W-P solid solution, which also contributes to the hardness by a solid solution hardening mechanism.

Gou et al. (2010) have studied the variation in the hardness of EL Ni-P-SiC composite coatings as a function of pH (4.40 to 6.00) of the hypophosphite-reduced EL nickel plating bath. According to them, the pH of the plating bath could alter the grain size of the composite coating and hence the hardness. The EL Ni-P-SiC composite coating exhibits a steady increase in hardness with the pH of the bath from 4.40 to 5.20. However, a further increase in pH to 6.00 has led to a rapid decrease in hardness. EL Ni-P-SiC composite coating obtained at a pH of 5.20 exhibits a higher hardness of ~650 HV due to its fine-grained structure. Incorporation of SiC in the EL Ni-P matrix has increased the hardness of EL Ni-P/SiC gradient composite coating from 560 HV (for plain EL Ni-P coating) to 620 HV. The gradation in the SiC content among the layers hindered the movement of dislocations and enabled an increase in hardness (Wang et al., 2013).

The microhardness of EL Ni-P-SiO$_2$ nanocomposite coatings prepared using varying concentrations (7.5, 12.5 and 17.5 g/L) of SiO$_2$ NPs in the hypophosphite-reduced EL nickel plating bath was determined by Sadreddini et al. (2015). Incorporation of 2.74 wt.% of SiO$_2$ NPs in the EL Ni-P matrix (for 7.5 g/L SiO$_2$ NPs) has enabled an increase in hardness from 342 HV (for EL Ni-P coating) to 386 HV. With an increase in the level of incorporation of SiO$_2$ NPs to 4.62 wt.% (for 12.5 g/L SiO$_2$ NPs), the hardness reached a maximum of 429 HV. For the EL Ni-P-SiC nanocomposite coating prepared using 17.5 g/L of SiO$_2$ NPs, the hardness is decreased to 403 HV as the level of particle incorporation is also decreased from 4.62 wt.% to 3.97 wt.%. Hence, it is evident that the dispersion strengthening mechanism becomes prevalent with the level of incorporation of the second-phase particles. This is further substantiated by the marginal increase in hardness of EL Ni-B coating from 1021 HV to 1034 HV following the incorporation of a smaller amount of MMT NPs (Miri et al., 2021).

The variation in hardness with the level of incorporation of Al$_2$O$_3$ particles in the EL Ni-P matrix was studied by Hu et al. (2018). With an incorporation of 1.7 wt.% Al$_2$O$_3$ particles in the EL Ni-P matrix, only a marginal increase in hardness up to 520 HV is observed. With a further increase in the level of incorporation of Al$_2$O$_3$ particles to 3.6 wt.%, the hardness is increased to 638 HV. However, the hardness is decreased to 576 HV when the level of incorporation of Al$_2$O$_3$ particles is increased to 4.2 wt.%. This is due to the unfavourable effect caused by the Al$_2$O$_3$ particles on the crystal structure of EL Ni-P coating. Hence, it is clear that the hardness of the EL Ni-P composite coatings is not totally a function of the level of incorporation of the particles, and other factors also exert a strong influence on the hardness.

The variation in hardness of EL Ni-P-B-Al$_2$O$_3$ nanocomposite coatings prepared using varying concentrations of Al$_2$O$_3$ NPs (1, 2 and 3 wt.%) in the EL plating bath was studied by Lakavat et al. (2022). Composite coating obtained at a bath loading of 1 wt.% of Al$_2$O$_3$ NPs exhibits a higher hardness of 950 HV when compared to the EL Ni-P-B coating with a hardness of 870 HV. The hardness of the composite coating is increased further to 986 HV when the bath loading of Al$_2$O$_3$ NPs in the plating bath is increased to 2 wt.%. However, a reversal in trend in hardness is observed for a bath loading of 3 wt.%, as evidenced by a decrease in hardness to 967 HV. The uniform distribution of Al$_2$O$_3$ NPs in the EL Ni-P-B matrix, filling up of the pores and gaps and formation of a dense coating account for the higher hardness observed for the EL Ni-P-B-Al$_2$O$_3$ nanocomposite coatings prepared at a bath loading of 1 and 2 wt.% Al$_2$O$_3$ NPs. Agglomeration of the particles at a bath loading of 3 wt.% Al$_2$O$_3$ NPs has increased the porosity of the resultant composite coating. The decrease in rate of deposition and an increase in R_a validate such an occurrence.

Xiao et al. (2022) have studied the effect of loading of carbon nanotubes (CNTs) in the EL plating bath on the hardness of EL Ni-Mo-P-CNT nanocomposite coatings. With an increase in bath loading of CNTs from 0.1 g/L to 0.4 g/L, the hardness is increased from 988.77 HV to 1263.54 HV, whereas it decreased to 1126.03 HV at 0.5 g/L. The increase in hardness of the nanocomposite coating observed up to a bath loading of 0.4 g/L is due to the reinforcement of CNTs in the EL Ni-Mo-P-CNT matrix. Dispersion strengthening and grain refinement have contributed to the increase in hardness. Agglomeration of the CNTs at 0.5 g/L has led to the generation of a defective composite coating with many pores and cracks, which accounts for the decrease in hardness.

Carrillo et al. (2019) have evaluated the hardness of EL Ni-P-TiO$_2$ nanocomposite coatings deposited on AZ91D Mg alloy by dispersing 5-nm- and 100-nm-sized TiO$_2$ NPs in hypophosphite-reduced EL nickel plating bath at two different particle loadings of 0.5 g/L and 2.0 g/L. EL Ni-P-TiO$_2$ nanocomposite coatings prepared using 5-nm-sized TiO$_2$ NPs did not show any significant variation in hardness from EL Ni-P coating irrespective of bath loadings, and all of them lie in the range of 14 GPa. However, an increase in hardness (19.68 GPa) is observed for EL Ni-P-TiO$_2$ nanocomposite coatings prepared using 0.5 g/L of 5-nm-sized TiO$_2$ NPs. Nevertheless, when the bath loading of TiO$_2$ NPs is increased from 0.5 g/L to 2.0 g/L, the hardness is decreased to 16.87 GPa. The decrease in hardness is due to the increase in porosity of the coating caused by agglomeration of the TiO$_2$ NPs at a higher bath loading of 2 g/L. Hence, it is clear that an appropriate particle size and bath loading should be made to achieve a higher hardness of EL Ni-P composite coatings.

The effect of heat treatment at various temperatures of EL Ni-P composite coatings deposited on Mg alloys was addressed by Song et al. (2007), Yu et al. (2011), Ghavidel et al. (2020) and Rajabalizadeh et al. (2022). Only a slight increase in hardness is observed for EL Ni-P-MIL-53 MOF composite coating after heat treating it at 200°C, which can be attributed to the relaxation of internal stress. However, a reasonable increase in hardness is observed after heat treating the composite coating at 400°C, which can be rationalized due to the formation of hard intermetallic Ni$_3$P phase (Rajabalizadeh et al., 2022). Yu et al. (2011) have observed a maximum hardness of 855 HV for EL Ni-P coated AZ31 Mg alloy after a heat treatment at 400°C for 1 hour. The nanocomposite coating exhibits a hardness of 932 HV when the heat treatment is conducted at 350°C for 1 hour. According to Song et al. (2007), the hardness of EL Ni-P-ZrO$_2$ nanocomposite coatings is increased with an increase in heat treatment temperature up to 350°C reaching a maximum of 1400 HV, whereas it decreased at 400°C. Solution strengthening and precipitation strengthening of the matrix as well as dispersion strengthening of the ZrO$_2$ NPs incorporated in the EL Ni-P matrix accounts for the increase in hardness observed up to 350°C. Precipitation of the hard intermetallic Ni$_3$P phase following crystallization of the amorphous N-P matrix enables an increase in hardness up to 350°C. However, aggregation and coarsening of the Ni$_3$P phase at 400°C has led to softening of the matrix, which results in a decrease in hardness (Song et al., 2007).

Ghavidel et al. (2020) have evaluated the variation in hardness of EL Ni-P and EL Ni-P-SiC nanocomposite coatings deposited on AZ31 Mg alloy as a function of temperature (150, 230, 300,

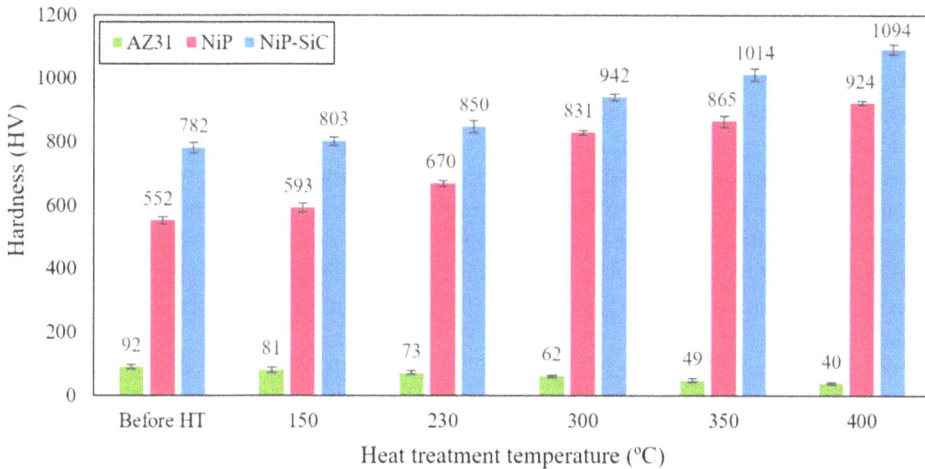

FIGURE 16.4 Variation in hardness of uncoated AZ31 Mg alloy, EL Ni-P coated alloy and EL Ni-P-SiC composite-coated alloy as a function of heat treatment temperature. *Reproduced with permission from (Ghavidel et al., 2020) © 2020 Elsevier B.V.*

350 and 400°C) employed for heat treatment. Only a marginal increment in hardness is observed at 150°C. The hardness starts to increase at 230°C (850 HV) following the precipitation of the hard intermetallic Ni_3P phase that reaches a maximum of 1094 HV at 400°C (**Figure 16.4**). The transformation of the amorphous phase of the as-plated coating to crystalline Ni and Ni_3P phases accounts for the improvement in hardness after heat treatment. Nevertheless, the hardness of the AZ31 Mg alloy is decreased with an increase in heat treatment temperature from 150°C to 400°C due to grain growth. Hence, it is recommended to adopt a suitable temperature for heat treatment without compromising the characteristics of the Mg alloy. Ghavidel et al. (2020) have suggested that the EL Ni-P-SiC nanocomposite coating deposited on AZ31 Mg alloy can be heat treated at 300°C. Since the heat-treatment temperature is limited to 300°C for Mg and its alloys, ternary alloys such as Ni-W-P and their composites can be considered as a good choice as they could offer a much better hardness in their as-plated condition itself.

16.11 TRIBOLOGICAL BEHAVIOUR

Lakavat et al. (2022) have evaluated the tribological behaviour of EL Ni-P-B and EL Ni-P-B-Al_2O_3 nanocomposite coatings prepared using varying concentrations (1, 2 and 3 wt.%) of Al_2O_3 NPs. A pin-on-disc contact configuration was used to evaluate the wear behaviour in which the coated AZ91D Mg pin was mated against an EN18 steel disc. The applied force, sliding speed and sliding distance were 20 N, 0.5 m/s and 1000 m, respectively. The specific wear rate and average coefficient of friction (COF) of EL Ni-P-B coating are 4.8×10^{-10} kg/N/m and 0.20, respectively. For EL Ni-P-B-Al_2O_3 nanocomposite coatings prepared using 1, 2 and 3 wt.% Al_2O_3 NPs, the specific wear rate is decreased to 2.1×10^{-10} kg/N/m, 1.2×10^{-10} kg/N/m and 1.9×10^{-10} kg/N/m, respectively. The COF is decreased to 0.12, 0.08 and 1.1, respectively. The lower specific wear rate and COF observed for EL Ni-P-B-Al_2O_3 nanocomposite coatings when compared to EL Ni-P-B coatings is due to their higher hardness and lower R_a. The increase in specific wear rate and COF observed for EL Ni-P-B-Al_2O_3 nanocomposite coating prepared using 3 wt.% Al_2O_3 NPs when compared to those prepared using 2 wt.% Al_2O_3 NPs is due to the lower hardness and higher R_a. Among the coatings evaluated EL Ni-P-B-Al_2O_3 nanocomposite coating prepared using 2 wt.% Al_2O_3 NPs with a higher hardness of 986 HV and lower R_a of 0.42 μm offered a higher wear resistance.

The tribological behaviour of EL Ni-P and EL Ni-P-MIL-53 MOF composite coatings formed on AM60B Mg alloy, both in as-plated and heat-treated conditions, was evaluated by Rajabalizadeh et al. (2022). A pin-on-disc contact configuration was used to ascertain the wear behaviour in which the coated Mg alloy was mated against a 6-mm-diameter AISI 52100 steel pin (60 HR). The applied normal load, sliding speed and sliding distance were 3 N, 0.5 m/s and 300 m, respectively. The specific wear rate of as-plated EL Ni-P coating is decreased from 3.77×10^{-9} kg/N/m to 3.44×10^{-9} kg/N/m following the inclusion of MIL-53 MOF in the EL Ni-P matrix. This is due to the higher hardness of the as-plated composite coating (564 HV) when compared to its plain counterpart (518 HV). However, the COF of the as-plated composite coating is found to be higher (0.87) than that of the plain EL Ni-P coating (0.60). This is due to the higher contact area of the as-plated composite coating with the steel pin.

The wear behaviour of heat-treated EL Ni-P and EL Ni-P-MIL-53 MOF composite coatings exhibit some interesting features. Heat treatment of both EL Ni-P and EL Ni-P-MIL-53 MOF composite coatings at 400°C has led to considerable decrease in the specific wear rate when compared to their as-plated counter parts. This is due to an increase in microhardness from 518 HV to 861 HV for EL Ni-P coating and from 564 HV to 907 HV for EL Ni-P-MIL-53 MOF composite coating due to the precipitation of the hard intermetallic Ni_3P phase. Although a decrease in specific wear rate following a marginal increase in hardness is observed for EL Ni-P-MIL-53 MOF composite coating when compared to EL Ni-P coating after heat treatment at 200°C, their specific wear rate is higher than their as-plated counterparts. This is due to the absence of any phase transformation after heat treating them at 200°C. Nevertheless, the development of cracks in the coating has led to the generation of wear debris and increase the mass loss due to wear. Though such cracks could also form after heat treatment at 400°C, the increase in hardness becomes the predominant factor in determining the wear behaviour (Rajabalizadeh et al., 2022). Adhesive wear is found to be the governing mechanism for both as-plated and heat-treated EL Ni-P and EL Ni-P-MIL-53 MOF composite coatings (**Figure 16.5**).

Araghi and Paydar (2010, 2014) have evaluated the tribological behaviour of EL Ni-P, EL Ni-W-P, EL Ni-P-B_4C and EL Ni-W-P-B_4C composite coatings formed on AZ91D Mg alloy. A pin-on-disk contact configuration was employed to ascertain the wear resistance in which the coated Mg alloy disk was mated against a hardened high-carbon steel pin. The applied normal load, sliding speed and sliding distance were 10 N, 0.06 m/s and 500 m, respectively. EL Ni-P-B_4C and EL Ni-W-P-B_4C composite coatings offered better wear resistance than their plain counterparts, which is due to the higher hardness of the coatings following incorporation of B_4C particles in the EL Ni-P matrix. The COF of EL Ni-P-B_4C and EL Ni-W-P-B_4C composite coatings are higher than their plain counterparts. This is due to the detachment of B_4C particles that subsequently get trapped between the mating couples. Adhesive wear is the main mechanism for EL Ni-P coating. Both adhesive and abrasive wear mechanisms are operative for EL Ni-W-P coating. The brittle nature of EL Ni-W-P coating makes two-body abrasive wear as the predominant one. The wear mechanism of EL Ni-P-B_4C composite coating is governed by a combination of both adhesive and three-body abrasive wear in which trapping of the detached B_4C particles contribute to the later mechanism. For EL Ni-W-P-B_4C composite coating, abrasive and three-body abrasive wear are the main wear mechanisms.

The tribological behavior of the as-plated EL Ni-Mo-P and EL Ni-Mo-P-CNT nanocomposite coatings prepared using 0.1–0.5 g/L of CNTs was studied by Xiao et al. (2022). The wear resistance of the coatings was evaluated using a reciprocating friction and wear tester in which the EL Ni-Mo-P and EL Ni-Mo-P-CNT nanocomposite-coated AZ91D Mg alloys were rubbed against a GCr15 steel counterface at a load of 20 N for 10 minutes. Among the composite coatings, the one prepared using 0.4 g/L of CNT exhibits a low COF of 0.20. This is due to the uniform distribution of CNT in the EL Ni-P matrix, which reduces the extent of plastic deformation and the change in friction mode from sliding to rolling. Agglomeration of CNTs at 0.5 g/L deleteriously influenced the quality of the composite coating, and that has resulted in poor wear resistance (Xiao et al., 2022).

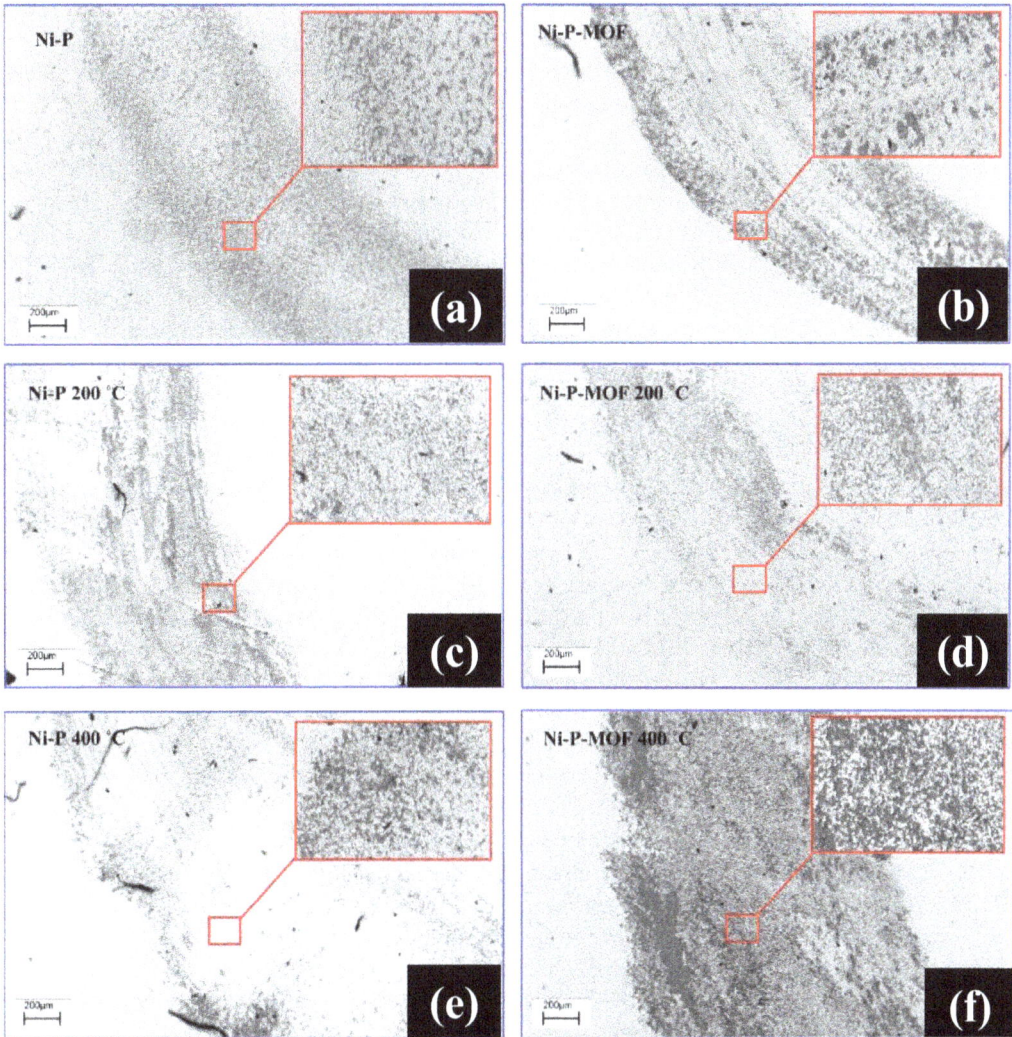

FIGURE 16.5 SEM images of the wear tracks and worn surfaces of: (a) as-plated EL Ni-P coating; (b) as-plated EL Ni-P-MOF composite coating; (c) EL Ni-P coating heat-treated at 200°C; (d) EL Ni-P-MOF composite coating heat treated at 200°C; (e) EL Ni-P coating heat-treated at 400°C; and (f) EL Ni-P-MOF composite coating heat treated at 400°C. *Reproduced from (Rajabalizadeh et al., 2022) © 2021 Chongqing University. Publishing services provided by Elsevier B.V. on behalf of KeAi Communications Co. Ltd.*

Carrillo et al. (2019) have assessed the tribological properties of uncoated AZ91D Mg alloy, EL Ni-P and EL Ni-P-TiO$_2$ nanocomposite-coated Mg alloy using 5-nm and 100-nm-sized TiO$_2$ NPs at two different bath loadings of 0.5 and 2 g/L. A ball-on-disk contact configuration was employed to study the wear behaviour in which a 6-mm-diameter alumina ball was used as the counterface material to rub against the uncoated and coated Mg alloy. The applied force, sliding speed and sliding distance were 10 N, 10.5 mm/s and 56.6 m, respectively. The specific wear rate and COF of EL Ni-P coated alloy is much lower than that of the uncoated alloy. For EL Ni-P-TiO$_2$ nanocomposite coatings, the particle size and particle loading in the plating bath exert a strong influence on the wear behaviour. The composite coating prepared using 5-nm-sized TiO$_2$ NPs at a particle load of 0.5 g/L exhibits a lower specific wear rate and COF than the plain EL Ni-P coating. However, for the same particle loading, the composite coating prepared using 100-nm-sized TiO_2 NPs shows a slightly higher specific wear rate and COF than the one prepared using 5-nm-sized TiO_2 NPs. When the particle load is increased from 0.5 g/L to 2 g/L, a low specific wear rate is observed only for the composite coatings prepared using 5-nm-sized TiO_2 NPs. For the same particle loading of 2 g/L, the specific wear rate becomes much higher than plain EL Ni-P coating when the size of the particles is increased to 100 nm. At a particle load of 2 g/L, irrespective of the size of the NPs, the COF becomes much higher than the plain EL Ni-P coating as well as the uncoated AZ91 Mg alloy. This is due to the change in wear mechanism to three-body abrasive wear for coatings prepared at higher particle loading. EL Ni-P-TiO_2 nanocomposite coating prepared using 5-nm-sized TiO_2 NPs at a particle load of 0.5 g/L exhibits a lower specific wear rate of 0.92×10^{-11} kg/N/m and a low COF of 0.21. Flattening of the nodules and accumulation of wear debris between the mating couples are observed for EL Ni-P coating. Abrasive wear and three-body abrasive wear govern the wear process of EL Ni-P-TiO_2 nanocomposite coatings. For composite coatings prepared using 0.5 g/L of 100-nm-sized TiO_2 NPs, the extent of severe plastic deformation is less. However, for composite coatings prepared using 2 g/L of 100-nm-sized TiO_2 NPs, the three-body abrasive wear mechanism becomes prevalent, which is evidenced by the grooves and microcuts in the wear track pattern.

Li et al. (2022) have studied the wear behaviour of uncoated Mg-Li alloy, EL Ni-P, EL Ni-P-GO and EL Ni-P-CeO_2@GO composite coating under dry sliding conditions. A ball-on-disk contact configuration was used to assess the wear resistance in which the uncoated and coated Mg-Li alloy disks were mated against a GCr15 steel ball. The normal load, frequency, friction radius and test duration were 2 N, 5 Hz, 2 mm and 40 minutes, respectively. The COF values of the uncoated Mg-Li alloy is decreased from 0.548 to 0.513, 0.399 and 0.331 after the deposition of EL Ni-P, EL Ni-P-GO and EL Ni-P-CeO_2@GO composite coatings, respectively. The width of the wear track of the uncoated Mg-Li alloy is also decreased from 1535 μm to 512, 437 and 346 μm after the deposition of EL Ni-P, EL Ni-P-GO and EL Ni-P-CeO_2@GO composite coatings, respectively. The higher extent of wear debris generated for EL Ni-P coating when compared to EL Ni-P-GO and EL Ni-P-CeO_2@GO composite coatings signify a higher load-bearing capacity of the composite coatings. The dispersion strengthening and grain refinement caused by the incorporation of GO and CeO_2@GO composites enables a better strength for the composite coatings. Abrasive wear and oxidative wear are the main wear mechanisms for EL Ni-P coating. Although there is no change in the wear mechanisms for EL Ni-P-GO and EL Ni-P-CeO_2@GO composite coatings, the presence of lamellar GO and CeO_2@GO composites limits the extent of heat generated during the sliding wear and reduces the degree of oxidation. The synergistic effect of load-bearing capacity of the CeO_2 NPs and the ability of GO to provide self-lubrication account for the excellent wear resistance of EL Ni-P-CeO_2@GO composite coating.

The wear resistance of EL Ni-P and EL Ni-P-SiC composite coatings deposited on AZ31 Mg alloy both in as-plated and heat-treated (150, 230, 300, 350 and 400°C for 1 hour) conditions was evaluated by Ghavidel et al. (2020). A pin-on-disk contact configuration was employed to study the wear behaviour in which the coated Mg alloy disk was mated against a AISI 52100/100Cr6 steel pin (6 mm diameter and 12 mm length; HRC 64) under dry sliding conditions. The normal load, sliding speed and sliding distance were 4 N, 0.1 m/s and 500 m, respectively. The EL Ni-P-SiC composite coating exhibits a better wear resistance than EL Ni-P coating in the as-plated condition. This is due to uniform distribution of the SiC NPs in the EL Ni-P matrix and an increase in hardness of the composite coating. During the initial periods of wear, the COF of as-plated EL Ni-P coating lies in the range of 0.4, and it reaches to 0.9 after a sliding distance of 100 m, which remained stable up to 500 m. The COF of as-plated EL Ni-P-SiC composite coating is increased from 0.30 to 0.55 up to a sliding distance of 100 m and remained stable up to 500 m. The incorporation of SiC NPs in the EL Ni-P matrix reduced the wear scars, and the worn surface is relatively smooth with a few shallow grooves. The higher hardness, lower surface roughness and excellent compatibility of the SiC NPs with the EL Ni-P matrix reduced the extent of wear debris for the as-plated EL Ni-P-SiC composite coating. This is reflected by the lower depth and width of the wear scar and a lower COF of the EL Ni-P-SiC composite coating when compared to that of the EL Ni-P coating.

After heat treatment, the wear mechanism of EL Ni-P-SiC nanocomposite coatings is completely changed. The stress buildup during heat treatment makes the composite coating brittle, resulting in the formation of many cracks. During wear, the generation of more wear debris and detachment of the SiC NPs from the matrix has increased the mass loss due to wear and wear rate to 12.2×10^{-5} mm³/N/m. Such an occurrence also changed the wear mechanism from a two-body abrasive wear to three-body abrasive wear and increased the COF to 0.48 (Ghavidel et al., 2020). Two different factors could influence the wear rate of heat-treated EL Ni-P nanocomposite coatings. The increase in hardness following the precipitation of hard intermetallic Ni_3P phase is likely to decrease the wear rate. Conversely, the development of cracks during heat treatment promotes the extent of debris formation and increases the loss due to wear.

Incorporation of ZrO_2, SiC and TiN particles in the EL Ni-P matrix has been shown to improve the wear resistance (Song et al., 2007; Gou et al., 2010; Yu et al., 2011) following the dispersion-strengthening effect and load-bearing support provided by the particles.

16.12   CORROSION RESISTANCE

Song et al. (2007) have prepared EL Ni-P-ZrO_2 nanocomposite coatings on AZ91D die-cast alloy with the addition of 2 to 6 g/L of ZrO_2 NPs in the electrolyte and evaluated their corrosion resistance in 3.5% NaCl. Incorporation of ZrO_2 NPs in the EL Ni-P matrix has enabled the formation of a uniform, compact, highly adherent and pore-free composite coating with excellent corrosion resistance when compared to its particle-free counterpart. Seifzadeh and Rahimzadeh Hollagh (2014) have prepared EL Ni-Co-P-SiO_2 nanocomposite coating on AZ91D Mg alloy and compared its corrosion resistance with EL Ni-Co-P coating in 3.5% NaCl. Addition of 2 g/L of SiO_2 NPs in the hypophosphite-reduced EL plating bath has enabled the formation of a uniform, compact and pore-free composite coating that decreased the corrosion current density (i_{corr}) from 2.0 µA/cm² (for EL Ni-Co-P coating) to 0.6 µA/cm². Electrochemical impedance spectroscopy (EIS) studies revealed that the charge transfer resistance (R_{ct}) of both EL Ni-Co-P and EL Ni-Co-P-SiO_2 nanocomposite coating is decreased with an increase in immersion time from 2 to 12 hours. However, for any given duration of time, the EL Ni-Co-P-SiO_2 nanocomposite coating exhibits a higher R_{ct} and a lower double layer capacitance (C_{dl}) than the EL Ni-Co-P coating. The higher corrosion resistance of EL Ni-Co-P coated AZ91D Mg alloy is due to its relatively higher thickness and the development of tortuous nodule boundaries (Seifzadeh and Rahimzadeh Hollagh, 2014).

Wang et al. (2013) have studied the corrosion behaviour of EL Ni-P/SiC gradient composite coatings deposited on AZ91D Mg alloy. The gradation in the SiC content in the composite coating was achieved by a continuous drop-wise addition of 0.5- to 0.7-µm-sized SiC particles to the plating bath over a period of 1.5 hours during EL plating. The corrosion resistance of the EL Ni-P/SiC gradient composite coatings in 3.5% NaCl was evaluated by potentiodynamic polarization and EIS studies. The decrease in effective metallic area following the incorporation of SiC particles in the EL Ni-P matrix has restrained the anodic dissolution reaction. Among the EL Ni-P/SiC gradient composite coatings, those prepared using 4 g/L of SiC particles in the plating bath offered better corrosion resistance. A further increase in concentration of the SiC particles in the bath beyond 4 g/L has led to a decrease in corrosion resistance, as they destroy the structure of EL Ni-P/SiC gradient composite coating. Heat-treatment of EL Ni-P/SiC gradient composite coating at 400°C for 1 hour has changed the crystal structure, enabled the formation of Ni_3P and Ni_5P_2 phases and increased the corrosion resistance when compared to its as-plated counterpart.

The corrosion behaviour of EL Ni-P binary and Ni-W-P ternary alloy composite-coated AZ91D Mg alloy incorporated with B_4C particles in 3 wt.% NaCl was evaluated by Araghi and Paydar (2010, 2013, 2014). According to them, incorporation of B_4C particles in EL Ni-P as well as EL Ni-W-P matrices has led to a decrease in corrosion resistance. Deposition of EL Ni-P coating has caused a shift in corrosion potential (E_{corr}) of the AZ91D Mg alloy from -1.636 V to -0.808 V (vs. Ag/AgCl), and a corresponding decrease in i_{corr} from 350 μA/cm² to 7 μA/cm². EL Ni-W-P coating offered relatively better corrosion resistance than EL Ni-P coating, which is reflected by a positive shift in E_{corr} to -0.776 V and a lower i_{corr} of 5 μA/cm². Nevertheless, both Ni-P-B$_4$C and Ni-W-P-B$_4$C composite coatings exhibit inferior corrosion resistance when compared to their particle-free counterparts. The E_{corr} of Ni-P-B$_4$C and Ni-W-P-B$_4$C composite coatings were -1.031 V and -0.942 V, and the corresponding i_{corr} were 84 μA/cm² and 18.5 μA/cm², respectively. Based on the corrosion protective ability, the coatings can be ranked as follows: EL Ni-W-P > EL Ni-P > EL Ni-W-P-B$_4$C > EL Ni-P-B$_4$C. The generation of microcracks and the development of local stress in the coated layer following the incorporation of B$_4$C particles in the EL Ni-P matrix are considered the major reasons for the poor corrosion-protective ability of EL Ni-W-P-B$_4$C and EL Ni-P-B$_4$C composite coatings. The findings of Araghi and Paydar (2010, 2013, 2014) have revealed that it is important to avoid the buildup of local stress in the composite coatings following particle incorporation, which would otherwise affect the corrosion resistance.

The corrosion resistance of EL Ni-P-SiO$_2$ nanocomposite coatings deposited on AZ91HP Mg alloy in 3.5% NaCl was evaluated by Sadreddini et al. (2015). The concentration of the SiO$_2$ NPs added to the plating bath played a key role in determining the extent of corrosion protection of the EL Ni-P-SiO$_2$ coating. Addition of 7.5 and 12.5 g/L of SiO$_2$ NPs in the hypophosphite-reduced EL plating bath has led to a positive shift in E_{corr} from -0.44 V (vs. SCE) to -0.40 V and -0.29 V and a decrease in i_{corr} from 4 μA/cm² to 2.2 μA/cm² and 1.3 μA/cm², respectively. However, a further increase in concentration of SiO$_2$ NPs to 17.5 g/L has led to a reversal in trend, which is evidenced by a higher i_{corr} of 1.9 μA/cm². Addition of SiO$_2$ NPs in the EL plating up to 12.5 g/L has led to the development of a uniform and compact coating, which accounts for the observed improvement in corrosion resistance. At a concentration of 17.5 g/L, agglomeration of the SiO$_2$ NPs becomes prevalent, which decreased the phosphorous content as well as the level of incorporation of SiO$_2$ NPs. EIS studies further confirm that the R_{ct} of EL Ni-P-SiO$_2$ nanocomposite coating prepared using 12.5 g/L of SiO$_2$ NPs is much higher (5072 Ω.cm²) than other coatings. The Bode impedance and Bode phase angle plots of uncoated AZ91HP Mg alloy, EL Ni-P coated alloy and EL Ni-P-SiO$_2$ composite-coated alloy incorporated with varying levels of SiO$_2$ NPs are shown in **Figure 16.6**. It is evident that the EL Ni-P-SiO$_2$ composite-coated alloy prepared using 12.5 g/L of SiO$_2$ NPs with a

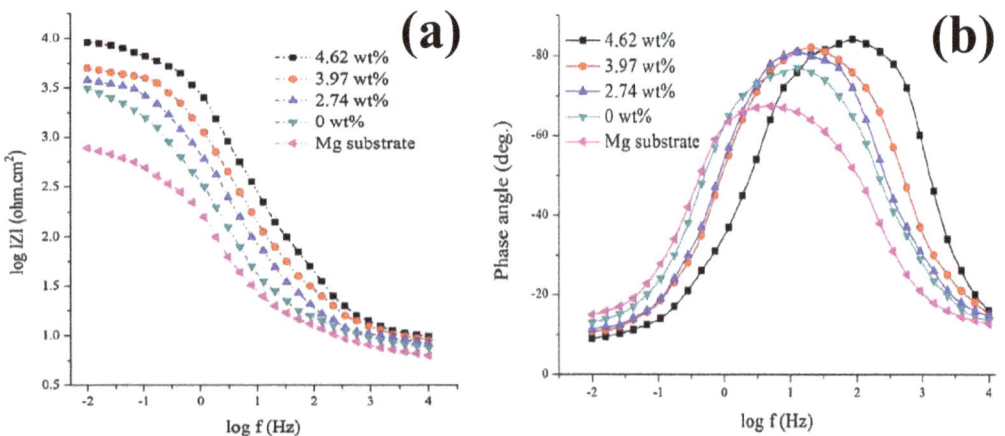

FIGURE 16.6 (a) Bode impedance and (b) Bode phase angle plots of uncoated AZ91HP Mg alloy, EL Ni-P coated alloy and EL Ni-P-SiO$_2$ composite-coated alloy incorporated with varying levels of SiO$_2$ NPs. *Reproduced with permission from (Sadreddini et al., 2015) © 2015 Elsevier B.V.*

level of incorporation of 4.62 wt.% SiO_2 NPs in the coating has a higher |Z| and a higher phase angle maximum when compared to other composite coatings.

Li et al. (2022) have evaluated the corrosion resistance of uncoated, EL Ni-P coated, EL Ni-P-GO composite-coated and EL Ni-P-CeO_2@GO composite-coated Mg-Li alloys in 3.5% NaCl by potentiodynamic polarization and EIS. All three types of coatings have offered excellent corrosion resistance for the Mg-Li alloy, which is evidenced by a large shift in E_{corr} towards the noble direction (from -163 V to -0.35 V (vs. SCE)) and a twofold decrease in the i_{corr} (from 5.90×10^{-4} A/cm^2 to 1.41×10^{-6} A/cm^2). Incorporation of GO and CeO_2@GO in the EL Ni-P matrix has enabled a shift in E_{corr} from -0.46 V (for EL Ni-P coated Mg-Li alloy) to -0.42 V and -0.35 V, respectively. The i_{corr} is decreased from 7.44×10^{-6} A/cm^2 (for EL Ni-P coated Mg-Li alloy) to 3.86×10^{-6} A/cm^2 and 1.41×10^{-6} A/cm^2 following the incorporation of GO and CeO_2@GO in the EL Ni-P coating. EIS studies further validate the inference made by the polarisation tests, which is reflected by the large increase in resistance accompanied by a strong decrease in the capacitance of the EL Ni-P-GO and EL Ni-P-CeO_2@GO composite-coated Mg-Li alloys. Among the two types of composite coatings, the EL Ni-P-CeO_2@GO composite coating has an edge over EL Ni-P-GO composite coating in offering a better corrosion resistance. This is due to the synergistic effect of the barrier property of GO nanosheets toward the permeation of NaCl solution as well as the corrosion inhibiting effect of CeO_2 NPs.

The corrosion resistance of EL Ni-W-P-CeO_2 composite coatings deposited on AZ91D Mg alloy in 3.5% NaCl was evaluated by Sun et al. (2017). The corrosion protective ability of the composite coatings varies with the concentration of CeO_2 particles in the plating bath. Addition of 8 mg/L of CeO_2 particles has enabled a positive shift in E_{corr} from -0.53 V (for EL Ni-W-P coating) to -0.31 V (vs. SCE) with a corresponding decrease in i_{corr} from 4.35 μA/cm^2 to 0.45 μA/cm^2. Refinement of microstructure and development of a smooth and compact coating following incorporation of CeO_2 particles in EL Ni-W-P matrix are considered the major reasons for the observed improvement in corrosion resistance. An increase in concentration of CeO_2 particles beyond 8 mg/L has led to a reversal in that trend. The EIS studies further validate the inferences made from polarization studies with a relatively high R_{ct} value of 2.45×10^4 Ω.cm^2 for EL Ni-W-P-CeO_2 composite coating prepared using 8 mg/L of CeO_2 particles. After heat treatment, the corrosion resistance of EL Ni-W-P-CeO_2 composite coating is decreased. In spite of an improvement in the uniformity, compactness and the development of smooth surface, precipitation of some crystalline phases, viz., Ni_3P, Ni_4W etc., has deleteriously influenced the corrosion resistance.

Hu et al. (2018) have evaluated the corrosion resistance of uncoated AZ91D die-cast Mg alloy, EL Ni-P coated alloy and EL Ni-P-Al_2O_3 composite-coated alloy incorporated with 1.7, 3.6 and 4.2 wt.% of Al_2O_3 particles in 3.5% NaCl. Deposition of EL Ni-P coating has enabled a positive shift in E_{corr} from -1.47 V (for AZ91D Mg alloy) to -0.51 V (vs. SCE) and a corresponding decrease in i_{corr} from 1.4×10^{-4} A/cm^2 to 3.1×10^{-6} A/cm^2. For EL Ni-P-Al_2O_3 composite coating incorporated with 3.6 wt.% Al_2O_3 particles, the E_{corr} is further shifted in the positive direction to -0.35 V, while the i_{corr} is decreased to 4.5×10^{-7} A/cm^2. The i_{corr} of composite coatings incorporated with 1.7 wt.% and 4.2 wt.% Al_2O_3 particles are relatively higher than that of the one having 3.6 wt.% of Al_2O_3 particles. The difference in corrosion resistance observed among the composite coatings is due to the influence of Al_2O_3 particles on the grain size and porosity. At a lower level of particle incorporation (1.7 wt.%), the lack of grain refinement limits the improvement in corrosion resistance. At a higher level of particle incorporation (4.2 wt.%), the increase in porosity accounts for the relatively poor corrosion resistance.

The corrosion resistance of EL Ni-P-SiC nanocomposite coatings, prepared using varying concentrations of SiC NPs (0.5, 1.0, 2.0 and 4.0 g/L) in the plating bath, in 3.5% NaCl, was evaluated by Ghavidel et al. (2020). The quality of the composite coatings varied with the amount of SiC NPs added to the plating bath, and that is also reflected in their corrosion behaviour. Composite coating prepared using 0.5 g/L SiC NPs fails to show any significant change in the morphological features when compared to EL Ni-P coating. Hence, their corrosion behaviour was quite similar. When the concentration of SiC NPs in the plating bath was increased to 1 g/L, a uniform and compact coating was formed, which accounted for the positive shift in E_{corr} from -0.387 V (for EL Ni-P coating) to -0.308 V (vs. Ag/AgCl) and a corresponding decrease in i_{corr} from 1.7 μA/cm² to 1.2 μA/cm². However, a reversal in trend was observed for composite coatings prepared using 2 and 4 g/L of SiC NPs. Aggregation of SiC NPs has affected the quality of the coatings, making them more porous and less protective. Annealing of the EL Ni-P-SiC coating up to 300°C has enabled an increase in corrosion resistance, which is evidenced by the positive shift in E_{corr} from -0.336 V (for as-plated EL Ni-P-SiC coating) to -0.262 V and a decrease in i_{corr} from 0.9 μA/cm² (for as-plated EL Ni-P-SiC coating) to 0.6 μA/cm². A further increase in annealing temperature beyond 300°C has deteriorated the quality of the coating and increased the i_{corr}. Annealing at 400°C has led to cracking of the coated layer due to the difference in thermal expansion coefficient between the Mg alloy and the coating. Detachment of particles could worsen the scenario.

The corrosion behaviour of EL Ni-B and EL MMT nanocomposite-coated AM60B Mg alloy in 3.5% NaCl was evaluated by Miri et al. (2021). Deposition of EL Ni-B coating has led to a positive shift in E_{corr} from -1.54 V (for uncoated AM60B Mg alloy) to -0.33 V (vs. Ag/AgCl) and a corresponding decrease in i_{corr} from 46.89 μA/cm² to 0.12 μA/cm². Incorporation of MMT NPs in the EL Ni-B matrix has led to a further decrease in i_{corr} to 0.06 μA/cm². The barrier effect provided by the MMT NPs, filling up of the pores and grain boundaries were considered responsible for the observed improvement in corrosion resistance. EIS studies further substantiate the higher corrosion resistance offered by the EL Ni-B-MMT nanocomposite coating when compared to EL Ni-B coating, which is evidenced by an increase in polarisation resistance (R_p) from 240 kΩ cm² to 343 kΩ cm².

Rajabalizadeh et al. (2022) have incorporated an Al-based MIL-53 MOF nanostructure in the EL Ni-P coating on AM60B Mg alloy and evaluated the corrosion resistance of the composite coatings in 3.5% NaCl by EIS. Incorporation of MIL-53 MOF in the EL Ni-B matrix has led to an increase in R_{ct} from 19.65 kΩ cm² (for EL Ni-B coating) to 29.79 kΩ cm². The capacitance (Q_{dl}) value is decreased from 26.095 μs n/Ω/cm² (for EL Ni-B coating) to 23.84 μs n/Ω/cm². Since the Q_{dl} value can be related to the porosity of the coating, it is obvious that incorporation of MIL-53 MOF nanostructure has decreased the porosity of the EL Ni-P- MIL-53 MOF composite coating. The decrease in permeation of 3.5% NaCl through the composite coating has contributed to the improvement in corrosion resistance. The presence of MIL-53 MOF nanostructure in the coating has slowed down the approach of the corrosive medium and delayed the onset of corrosion of AM60B Mg alloy.

Lakavat et al. (2022) have prepared EL Ni-P-B and EL Ni-P-B-Al₂O₃ nanocomposite coatings on AZ91D Mg alloy and evaluated their corrosion behaviour in 3.5% NaCl. EL Ni-P-B-Al₂O₃ coatings prepared using 1 and 2 wt.% Al₂O₃ NPs in the electrolyte has led to a decrease in i_{corr} from 1.1×10^{-6} A/cm² to 5.01×10^{-7} A/cm² and 1.47×10^{-8} A/cm², respectively. However, a further increase in concentration of the Al₂O₃ NPs to 3 wt.% has increased the i_{corr} to 6.3×10^{-7} A/cm². The R_{ct} value is increased from 5823 (for EL Ni-P-B coating) to 8769 Ω cm² for the composite coating prepared using 2 wt.% Al₂O₃ NPs. The C_{dl} is decreased from 11.43 μF (for EL Ni-P-B coating) to 7.23 μF for the composite coating prepared using 2 wt.% Al₂O₃ NPs. The improvement in corrosion resistance of EL Ni-P-B-Al₂O₃ nanocomposite coatings prepared using 1 and 2 wt.% Al₂O₃ NPs is due to the decrease in available metallic area for corrosion. In addition, the presence of Al₂O₃ NPs might promote the formation of a passive layer. Agglomeration of Al₂O₃ NPs becomes prevalent at 3 wt.%, which increased the viscosity of the plating bath and decreased the level of incorporation of particles in the EL Ni-P matrix.

16.13   DUPLEX LAYERED ELECTROLESS NANOCOMPOSITE COATINGS

Georgiza et al. (2013) have developed a duplex EL Ni coating on AZ31 Mg alloy, which consists of a mid-phosphorus coating as the first layer over which a high-phosphorus coating incorporated with TiO₂ and ZrO₂ particles is deposited as the second layer. The amorphous nature of the duplex layered composite coatings offered an excellent corrosion resistance for the AZ31 Mg alloy in 3.5% NaCl, as it eliminates the diffusion paths among the grain boundaries. This is evidenced by the positive shift in E_{corr} from -0.6 V (for as-plated plain EL Ni-P coating) to -0.4 V (vs. SCE) for the duplex layered EL Ni-mid P/Ni-high P-ZrO_2 composite coatings with a decrease in i_{corr} from 12.4 $\mu A/cm^2$ (for as-plated plain EL Ni-P coating) to 2.2 $\mu A/cm^2$. In general, heat treatment tends to induce cracks in the EL-plated coating and decrease its corrosion resistance. This is evidenced by an increase in i_{corr} from 12.4 $\mu A/cm^2$ (for as plated plain EL Ni-P coating) to 203.5 $\mu A/cm^2$ after heat treatment at 200°C for 2 hours. However, for the duplex layered composite coating, only a slight increase in i_{corr} from 2.2 $\mu A/cm^2$ to 3.8 $\mu A/cm^2$ is observed after a heat treatment at 200°C for 2 hours.

A double-layered EL Ni-high P/Ni-low P-ZrO_2 composite coating on AZ31 Mg alloy is developed by Shu et al. (2015). The composite layer is prepared by the addition of ZrO_2 sol in the plating bath. The EL Ni-high P inner layer prevents the diffusion of the aggressive chloride ions, while the sol-modified EL Ni-low P-ZrO_2 composite outer layer improves the mechanical properties. The double-layered EL Ni-high P/Ni-low P-ZrO_2 composite coating could withstand salt spray test for > 480 hours. This is due to the amorphous nature of the EL Ni-high P coating underlayer. The incorporation of ZrO_2 particles in the EL Ni-low P layer has increased the hardness from 640 HV to 820 HV. A combination of better corrosion resistance and mechanical properties makes the coating suitable for many engineering applications.

16.14 CONCLUSIONS AND OUTLOOK

This chapter highlighted the key aspects of electroless deposited composite coatings on Mg alloys. Unlike other surface modification methods, EL Ni plating could endow the Mg alloys with better wear resistance, higher corrosion resistance, electrical conductivity and magnetic property. In addition, by a careful choice of experimental conditions, it would be possible to manipulate the P and B content of the Ni-P and Ni-B coatings and impart the desirable characteristics such as higher corrosion resistance, higher hardness, better wear resistance, electrical conductivity and magnetic properties. Addition of second-phase particles in EL Ni plating baths would enable formation of EL Ni-P and Ni-B based composite coatings with better corrosion resistance and wear resistance. EL Ni-P/Ni-B-based nanocomposite coatings can be easily prepared by dispersing the NPs in the plating bath, which would enable incorporation of a portion of the NPs in the EL Ni-P/Ni-B matrix. A variety of NPs can be incorporated in EL Ni-P and Ni-B matrices. The key requirements of particles to be incorporated in EL Ni-P and Ni-B matrices are (i) they should be chemically inert and (ii) they should remain in suspension. Agglomeration of nanoparticles in the EL Ni plating bath is a major problem in achieving an effective dispersion of the NPs; surfactants play a critical role in dispersing the particles. Choosing the right type of surfactants is the key to achieving maximum level of incorporation of particles in the EL N-P matrix. Incorporation of NPs in the EL Ni-P and Ni-B matrices has enabled an increase in hardness, wear resistance and corrosion resistance. Incorporation of hard particles such as SiC, Si_3N_4, diamond, etc. increased the hardness, while soft particles tend to decrease the hardness. Incorporation of new types of particles such as GO, graphitic carbon nitride, etc. has been shown to be very effective. Duplex and multi-layered nanocomposite coatings are currently emerging, and they would continue to occupy the key position in future developments in this field of research.

REFERENCES

Abdoli, M., Sabour Rouhaghdam, A. Preparation and characterization of Ni – P/nanodiamond coatings: Effects of surfactants. *Diam. Relat. Mater.* 2013, 31, 30–37.

Ansari, MI., Thakur, DG. Influence of surfactant: Using electroless ternary nanocomposite coatings to enhance the surface properties on AZ91 magnesium alloy. *Surf. Interfaces.* 2017, 7, 20–28.

Araghi, A., Paydar, MH. Electroless deposition of Ni – P – B_4C composite coating on AZ91D magnesium alloy and investigation on its wear and corrosion resistance. *Mater. Des.* 2010, 31, 3095–3099.

Araghi, A., Paydar, MH. Electroless deposition of Ni – W – P – B_4C nanocomposite coating on AZ91D magnesium alloy and investigation on its properties. *Vacuum.* 2013, 89, 67–70.

Araghi, A., Paydar, MH. Wear and corrosion characteristics of electroless Ni – W – P – B_4C and Ni – P – B_4C coatings. *Tribol. – Mater. Surf. Interfaces.* 2014, 8, 146–153.

Balaraju, JN., Sankara Narayanan, TSN., Seshadri, SK. Electroless Ni – P composite coatings. *J. Appl. Electrochem.* 2003, 33, 807–816.

Barati, Q., Hadavi, SMM. Electroless Ni-B and composite coatings: A critical review on formation mechanism, properties, applications and future trends. *Surf. Interfaces.* 2020, 21, 100702.

Carrillo, DF., Santa, AC., Valencia-Escobar, A., Zapata, A., Echeverría, F., Gómez, MA., Zuleta, AA., Castaño, JG. Tribological behavior of electroless Ni – P/Ni – P – TiO_2 coatings obtained on AZ91D magnesium alloy by a chromium-free process. *Int. J. Adv. Manuf. Technol.* 2019, 105, 1745–1756.

Correa, E., Zuleta, AA., Guerra, L., Castaño, JG., Echeverría, F., Baron-Wiecheć, A., Skeldon, P., Thompson, GE. Formation of electroless Ni-B on bifluoride-activated magnesium and AZ91D alloy. *J. Electrochem. Soc.* 2013, 160, D327–D336.

Correa, E., Zuleta, AA., Sepúlveda, M., Guerra, L., Castaño, JG., Echeverría, F., Liu, H., Skeldon, P., Thompson, GE. Nickel – boron plating on magnesium and AZ91D alloy by a chromium-free electroless process. *Surf. Coat. Technol.* 2012, 206, 3088–3093.

Dhakal, DR., Kshetri, YK., Chaudhary, B., Kim, TH., Lee, SW., Kim, BS., Song, Y., Kim, HS., Kim, HH. Particle-size-dependent anticorrosion performance of the Si_3N_4 – nanoparticle -incorporated electroless Ni-P coating. *Coatings.* 2022, 12, 9.

Ezhilselvi, V., Balaraju, JN., Subramanian, S. Chromate and HF free pretreatment for MAO/electroless nickel coating on AZ31B magnesium alloy. *Surf. Coat. Technol.* 2017, 325, 270–276.

Fayyad, EM., Abdullah, AM., Hassan, MK., Mohamed, A., Jarjoura, G., Farhat, Z. Effect of electroless bath composition on the mechanical, chemical, and electrochemical properties of new NiP – C_3N_4 nanocomposite coatings. *Surf. Coat. Technol.* 2019, 362, 239–251.

Georgiza, E., Novakovic, J., Vassiliou, P. Characterization and corrosion resistance of duplex electroless Ni-P composite coatings on magnesium alloy. *Surf. Coat. Technol.* 2013, 232, 432–439.

Ger, MD., Hwang, BJ. Effect of surfactants on codeposition of PTFE particles with electroless Ni-P coating. *Mater. Chem. Phys.* 2002, 76, 38–45.

Ghavidel, N., Allahkaram, SR., Naderi, R., Barzegar, M., Bakhshandeh, H. Corrosion and wear behavior of an electroless Ni-P/nano-SiC coating on AZ31 Mg alloy obtained through environmentally-friendly conversion coating. *Surf. Coat. Technol.* 2020, 382, 125156.

Gou, Y., Huang, W., Zeng, R., Zhu, Y. Influence of pH values on electroless Ni-P-SiC plating on AZ91D magnesium alloy. *T. Nonferr. Metal Soc. China.* 2010, 20, s674–s678.

Hu, R., Su, Y., Liu, Y., Liu, H., Chen, Y., Cao, C., Ni, H. Deposition process and properties of electroless Ni-P-Al_2O_3 composite coatings on magnesium alloy. *Nanoscale Res. Lett.* 2018, 13, 198.

Lajevardi, SA., Shahrabi, T., Hasannaeimi, V. Synthesis and mechanical properties of nickel-titania composite coatings. *Mater. Corros.* 2011, 62, 29–34.

Lakavat, M., Bhaumik, A., Gandi, S., Parne, SR. Electroless Ni – P – B coatings on magnesium alloy AZ91D: Influence of nano Al_2O_3 on corrosion, wear, and hardness behaviour. *Surf. Topogr.: Metrol. Prop.* 2022, 10, 025021.

Li, D., Cui, X., Wen, X., Feng, L., Hu, Y., Jin, G., Liu, E., Zheng, W. Effect of CeO_2 nanoparticles modified graphene oxide on electroless Ni-P coating for Mg-Li alloys. *Appl. Surf. Sci.* 2022, 593, 153381.

Liu, H., Bi, S., Cao, L., Bai, Q., Teng, X., Yu, Y. The deposition process and the properties of direct electroless nickel-phosphorous coating with chromium-free phosphate pickling pretreatment on AZ31 magnesium alloy. *Int. J. Electrochem. Sci.* 2012, 7, 8337–8355.

Mafi, IR., Dehghanian, C. Comparison of the coating properties and corrosion rates in electroless Ni – P/PTFE composites prepared by different types of surfactants. *Appl. Surf. Sci.* 2011, 257, 8653–8658.

Miri, T., Seifzadeh, D., Rajabalizadeh, Z. Electroless Ni – B – MMT nanocomposite on magnesium alloy. *Surf. Eng.* 2021, 37, 1194–1205.

Nwosu, N., Davidson, A., Hindle, C., Barker, M. On the influence of surfactant incorporation during electroless nickel plating. *Ind. Eng. Chem. Res.* 2012, 51, 5635–5644.

Pancrecious, JK., Ulaeto, SB., Ramya, R., Rajan, TPD., Pai, BC. Metallic composite coatings by electroless technique – a critical review. *Int. Mater. Rev.* 2018, 63, 488–512.

Rajabalizadeh, Z., Seifzadeh, D. Strontium phosphate conversion coating as an economical and environmentally-friendly pretreatment for electroless plating on AM60B magnesium alloy. *Surf. Coat. Technol.* 2016, 304, 450–458.

Rajabalizadeh, Z., Seifzadeh, D., Khodayari, A., Sohrabnezhad, S. Corrosion protection and mechanical properties of the electroless Ni-P-MOF nanocomposite coating on AM60B magnesium alloy. *J. Magnes. Alloys*. 2022, 10, 2280–2295.

Sadreddini, S., Salehi, Z., Rassaie, H. Characterization of Ni – P – SiO_2 nano-composite coating on magnesium. *Appl. Surf. Sci*. 2015, 324, 393–398.

Sankara Narayanan, TSN., Seshadri, SK. Electro- and electroless deposited composite coatings: Preparation, characteristics, and applications. In: Nicolais, L., Borzacchiello, A. (Eds.), *Wiley Encyclopedia of Composites*. 2nd Edition. John Wiley & Sons, Inc., Hoboken, NJ. 2012, pp. 777–791.

Sankara Narayanan, TSN., Seshadri, SK., Park, IS., Lee, MH. Electroless nanocomposite coatings: Synthesis, characteristics, and applications. In: Aliofkhazraei, M., Makhlouf, A. (Eds.), *Handbook of Nanoelectrochemistry*. Springer, Cham, Switzerland. 2016, pp. 389–416.

Sarret, M., Muller, C., Amell, A. Electroless NiP micro- and nano-composite coatings. *Surf. Coat. Technol*. 2006, 201, 389–395.

Sayyad, F., Senanayake, R. Experimental investigation on surface roughness of electroless Ni – B – TiO_2 nanocomposite coatings. *Sadhana*. 2021a, 46, 61.

Sayyad, F., Senanayake, R. Improving the deposit efficiency of nano composite deposits on AZ91 magnesium alloy by using a suitable bath composition and operating conditions. *Mater. Today: Proc*. 2021b, 47, 2990–2993.

Seifzadeh, D., Rahimzadeh Hollagh, A. Corrosion resistance enhancement of AZ91D magnesium alloy by electroless Ni-Co-P coating and Ni-Co-P-SiO_2 nanocomposite. *J. Mater. Eng. Perform*. 2014, 23, 4109–4121.

Shen, X., Wang, J., Xin, G. Effect of the zeta potential on the corrosion resistance of electroless nickel and PVDF composite layers using surfactants. *ACS Omega* 2021, 6, 33122–33129.

Shu, X., Wang, Y., Liu, C., Aljaafari, A., Gao, W. Double-layered Ni-P/Ni-P-ZrO_2 electroless coatings on AZ31 magnesium alloy with improved corrosion resistance. *Surf. Coat. Technol*. 2015, 261, 161–166.

Song, YW., Shan, DY., Chen, RS., Han, EH. Study on electroless Ni – P – ZrO_2 composite coatings on AZ91D magnesium alloys. *Surf. Eng*. 2007, 23, 334–338.

Sudagar, J., Lian, J., Sha, W. Electroless nickel, alloy, composite and nano coatings – a critical review. *J. Alloys Compd*. 2013, 571, 183–204.

Sun, W., Xu, JM., Wang, Y., Guo, F., Jia, ZW. Effect of cerium oxide on morphologies and electrochemical properties of Ni-W-P coating on AZ91D magnesium. *J. Mater. Eng. Perform*. 2017, 26, 5753–5759.

Tran, TN., Yu, G., Hu, BN., Xie, ZH., Tang, R., Zhang, XY. Effects of pretreatments of magnesium alloys on direct electroless nickel plating. *Trans. Inst. Met. Finish*. 2012, 90, 209–214.

Wang, HL., Liu, LY., Dou, Y., Zhang, WZ., Jiang, WF. Preparation and corrosion resistance of electroless Ni-P/SiC functionally gradient coatings on AZ91D magnesium alloy. *Appl. Surf. Sci*. 2013, 286, 319–327.

Xiao, Y., Sun, WC., Jia, YP., Liu, YW., Tian, SS. Electroless of Ni – Mo – P/CNTs composite coatings on AZ91D magnesium alloy: Corrosion and wear resistance. *Trans. Indian Inst. Met*. 2022, 75, 2117–2127.

Xie, ZH., Chen, F., Xiang, SR., Zhou, JL., Song, ZW., Yu, G. Studies of several pickling and activation processes for electroless Ni-P plating on AZ31 magnesium alloy. *J. Electrochem. Soc*. 2015, 162, D115–D123.

Xu, C., Chen, L., Yu, L., Zhang, J., Zhang, Z., Wang, J. Effect of pickling processes on the microstructure and properties of electroless Ni – P coating on Mg–7.5Li–2Zn–1Y alloy. *Prog. Nat. Sci*. 2014, 24, 655–662.

Yu, L., Huang, W., Zhao, X. Preparation and characterization of Ni – P – nano TiN electroless composite coatings. *J. Alloys Compd*. 2011, 509, 4154–4159.

Zarebidaki, A., Allahkaram, SR. Effect of surfactant on the fabrication and characterization of Ni-P-CNT composite coatings. *J. Alloys Compd*. 2011, 509, 1836–1840.

Zhang, B. Electroless composite plating. In: *Amorphous and Nano Alloys Electroless Depositions Technology, Composition, Structure and Theory*. 1st Edition, Chapter 3. Elsevier, Amsterdam. 2016, pp. 107–140.

Zielińska, K., Stankiewicz, A., Szczygieł, I. Electroless deposition of Ni – P – nano-ZrO_2 composite coatings in the presence of various types of surfactants. *J. Colloid Interface Sci*. 2012, 377, 362–367.

Zuleta, AA., Correa, E., Sepúlveda, M., Guerra, L., Castaño, JG., Echeverría, F., Skeldon, P., Thompson, GE. Effect of NH_4HF_2 on deposition of alkaline electroless Ni – P coatings as a chromium-free pre-treatment for magnesium. *Corros. Sci*. 2012, 55, 194–200.

17 Electrophoretic Coatings

Viswanathan S. Saji

CONTENTS

17.1 INTRODUCTION

Electrophoretic deposition (EPD) is a colloidal process which involves the deposition of charged particles/polymer macromolecules via electric field-assisted migration (electrophoresis) to attain a dense film formation (maximum thickness ~ 100 μm) on an oppositely charged working electrode (**Figure 17.1**). EPD can be used for the deposition of metals, metal oxides, ceramics, polymers, and composites for various applications, such as solar cells, sensors, energy storage devices, catalysis, electronics, biomedical, and anti-corrosion. EPD is a cost-effective, fast, and scalable approach capable of depositing dense and high-purity coatings on complex-shaped objects at room temperature. The technique is desirable for biomedical implant applications due to its high deposit purity (Sarkar and Nicholson 1996) (Boccaccini, Keim et al. 2010) (Ammam 2012).

EPD is usually conducted using the typical two-electrode cell setup (**Figure 17.1a**), characteristic of conventional electroplating. The significant difference is that while EPD uses a colloidal suspension of nanoparticles, conventional electroplating uses ionic species from a salt solution via Faradaic processes. EPD can be cathodic (positively charged particles deposit on the cathode) or anodic (negatively charged particles deposit on the anode). The applied electric field can be either direct or modulated (alternating or pulsed-direct current). The deposition can be performed in organic or aqueous suspensions. Despite the several advantages of organic suspensions, they typically suffer from high cost, toxicity, volatility, flammability, and high electric field strength requirement. Aqueous suspensions are appropriate due to their low cost, eco-friendliness, and low electric field strength norm; nevertheless, potential water electrolysis and bubble evolution could happen. High-temperature sintering is usually required to make an as-deposited EPD coating denser (Sarkar and Nicholson 1996) (Boccaccini, Keim et al. 2010) (Ammam 2012) (Seuss and Boccaccini 2013) (Diba et al. 2016) (Hu et al. 2020) (Saji 2018, 2021a, 2021b).

The coating formation is determined by both (i) deposition-related and (ii) suspension-related parameters. The suspension-related parameters include particle size, conductivity, viscosity and concentration of the suspension, the solvent's dielectric constant, co-depositing particles, charging agents, zeta (ζ) potential, and colloidal stability. The major deposition-related parameters are the substrate conductivity and the deposition voltage and time. Typically, to avoid particle aggregation, a high

DOI: 10.1201/9781003319856-20

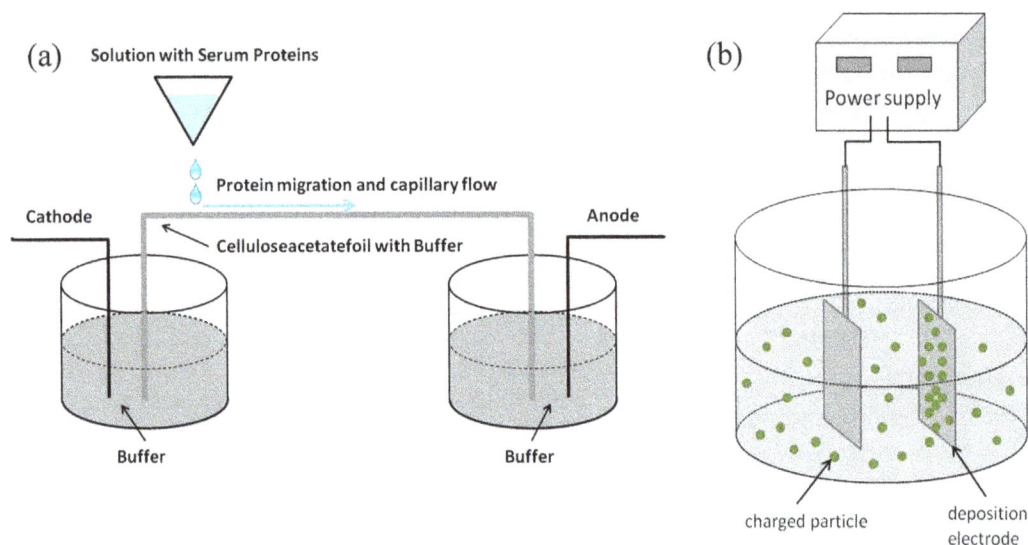

FIGURE 17.1   Schematics of typical (a) EPD two-electrode cell set-up and (b) Conventional electrophoresis cell. *Reproduced with permission from (Seuss and Boccaccini 2013) © 2013 American Chemical Society.*

ζ-potential is obligatory to generate a high electrostatic repulsion, which could be attained via charging agents (Sarkar and Nicholson 1996) (Boccaccini, Keim et al. 2010) (Ammam 2012) (Van der Biest et al. 1999) (Besra and Liu 2007) (Diba et al. 2016) (Avcu et al. 2019). The EPD mechanism was explained by several theories, including flocculation by particle accumulation (Hamaker 1940; Hamaker and Verwey 1940), particle charge neutralization (Grillon et al. 1992), electrochemical particle coagulation (Koelmans 1995), electrical double layer (EDL) distortion and thinning (Sarkar and Nicholson 1996), and pH localization (De and Nicholson 1999). The most widely accepted mechanism among these is EDL distortion and thinning (**Figure 17.2**). Clear fundamentals of EPD are described elsewhere (Besra and Liu 2007) (Ammam 2012) (Ata et al. 2014) (Diba et al. 2016) (Avcu et al. 2019) (Obregón et al. 2019) (Hu et al. 2020).

EPD has been widely explored for depositing anti-corrosive and bioactive coatings on Mg alloys. Optimizing process parameters such as deposition voltage, time, and nanoparticle concentration is critical in making a homogeneous and dense EPD coating with increased adhesion and reduced surface cracks/pores. A recent comprehensive review of the topic is available (Saji 2021a). This chapter provides the most recent updates on EPD coatings reported for Mg alloys, emphasizing EPD composite coatings.

17.2   EPD COATINGS ON Mg ALLOYS

17.2.1   Hydroxyapatite, Chitosan, and Bioactive Glass

EPD coatings based on hydroxyapatite (HA), chitosan (CS), and bioactive glasses (BG) are the primarily investigated types for Mg alloys. This is indeed be attributed to the excellent biomedical properties, eco-friendliness, and economic feasibility of HA, CS, and BG. Their composite coatings with pertinent ceramic oxides, carbon nanomaterials, and/or polymers could enhance the bioactivity, mechanical properties, and corrosion/wear resistances of Mg implants while controlling their uncontrollable degradation. Mg-based implants are excellent candidates for biodegradable bone regeneration owing to their compatibility with human bone and favourable mechanical properties (Saji 2021a).

DIFFUSE DOUBLE-LAYER DISTORTION BY EPD

LOCAL DIFFUSE DOUBLE-LAYER THINNING

COAGULATION

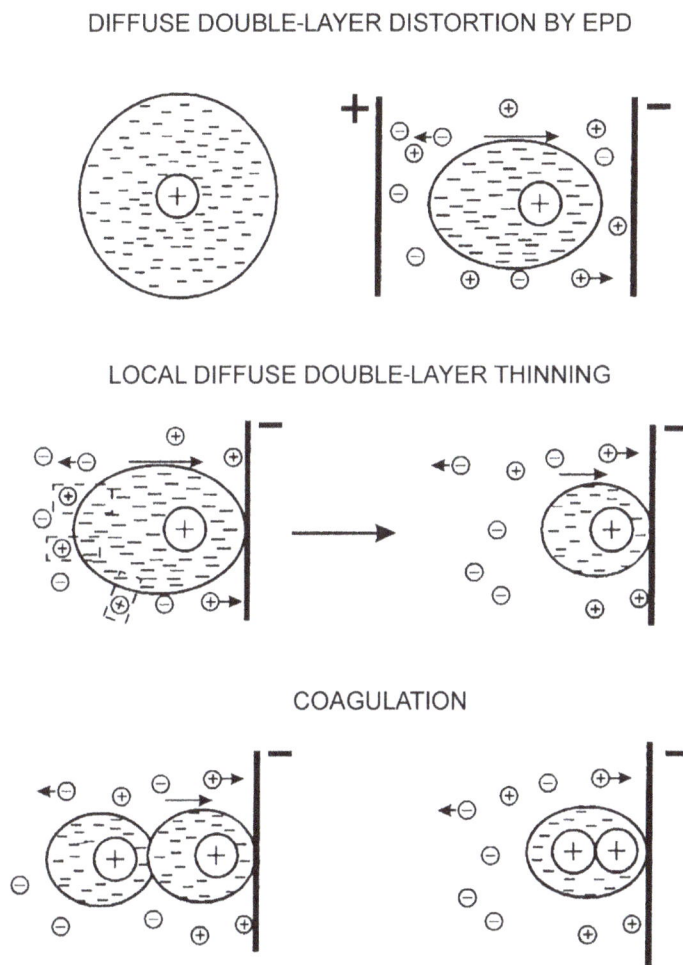

FIGURE 17.2   EDL distortion and thinning mechanism. *Reproduced with permission from (Ammam 2012)* © *2012 The Royal Society of Chemistry.*

Several reports are available on HA-only coatings (Kaabi Falahieh Asl et al. 2014) (Manoj Kumar et al. 2016) (Sankar et al. 2017) (Tayyaba et al. 2019) (Saadati et al. 2020). Most of the reported works focused on composite coatings of HA with CS (Zhang et al. 2013) (Singh et al. 2021a) (Singh et al. 2019) (Zhang, Wen et al. 2016; Zhang, Zhu 2016) and BG (Singh et al. 2020, 2021b) with or without the incorporation of metal oxides (Singh et al. 2020, 2021b), carbon nanomaterials (Zhang, Wen et al. 2016; Zhang, Zhu 2016), and/or polymers (Chen et al. 2016), (Tian et al. 2017). All the studies on HA-only coatings displayed substantially improved corrosion protection for the base Mg substrate. The corrosion current density (i_{corr}) recorded for 30 V HA-coated and equivalent uncoated samples in simulated body fluid (SBF) were 7.1×10^{-6} A/cm^2 and 1.2×10^{-4} A/cm^2, respectively (Kaabi Falahieh Asl et al. 2014). Potentiodynamic polarization studies showed that the degradation rate of a 15 μm thick HA-coated and 400°C-annealed Mg-6Zn-0.5Zr alloy in Ringer's solution was ~ 5 times lower than the bare counterpart (Tayyaba et al. 2019). A 300°C-annealed HA coating presented ~ 25 times increased corrosion resistance for a Mg-Zn alloy in SBF (Manoj Kumar et al. 2016).

SBF immersion studies undeniably supported the intrinsic greater bioactivity (inducing calcium phosphate's precipitation/mineralization) of EPD HA-coated Mg alloys. Razavi et al. disclosed that the L-929 fibroblast cell's viability within 7 days on bare, micro-arc oxidized (MAO), and MAO/EPD-fluoridated-HA coated samples were ~50–58, 70–85 and 160–175%, respectively (Razavi et al. 2015).

Several works explored combined MAO-EPD processes. The combined procedure is very efficient for reduced degradation, enhanced biocompatibility/bioactivity, and other desired functionalities for Mg alloys. MAO coatings are advantageous in having superior adhesion and barrier protection and can be a suitable matrix for composite coatings. An MAO layer typically comprises an exterior porous layer and an interior compact layer. Most reports in this area utilized MAO to deposit a thicker oxide protective layer on the Mg sample and a succeeding EPD to seal the porous MAO layer as well as to deliver the chosen functionalities. Several works explored concurrent MAO and EPD to make the composite coating in one step where the preferred EPD nanoparticles were incorporated in the MAO deposition bath. The MAO-EPD coupled process could achieve a higher coating thickness and improved corrosion resistance. Electrochemical impedance spectroscopy studies on nanostructured fluoridated-HA (FHA) coated MAO/Mg-alloy presented the polarization resistance (R_p) in the order: FHA/MAO/Mg-alloy (9325.8 Ω), MAO/Mg-alloy (957.2 Ω), and bare alloy (305.5 Ω). The weight loss data after SBF immersion studies showed that all the samples exhibited a fast degradation after 72 hours of immersion; thereafter, the weight loss increment was comparatively slow (**Figure 17.3a**). The weight loss of MAO/FHA-coated alloy was the lowermost. The weight losses of the three samples after 672 hours of immersion were 67, 42, and 16 mg, respectively. The associated variation of the pH of the SBF and the Mg-ion concentration supported the enhanced corrosion resistance of MAO/FHA-coated alloy (**Figure 17.3b,c**) (Razavi et al. 2015).

Many recent studies focused on CS-based composite coatings of HA, such as CS-HA (Wang et al. 2022), CS-BG (Heise et al. 2017) (Alaei et al. 2020), CS-gelatin (GL) (Qi et al. 2019), and CS-GL-BG (Akram et al. 2020). The positively charged CS due to protonation in CH_3COOH could endorse HA-CS composite deposition (Zhang et al. 2012). Wang et al. fabricated calcium phosphate (Ca-P)-CS composite coating on AZ31 alloy via EPD and chemical conversion. The coating was further modified with stearic acid to make it superhydrophobic (water contact angle ~158.6°). The composite superhydrophobic coated alloy displayed excellent self-cleaning properties with two orders of lower i_{corr} (1.6135 × 10^{-7} A/cm^2) than the bare alloy. After 216 hours of immersion in Hanks' solution, the coated alloy remained protected (Wang et al. 2022). A GL-CS coating was deposited on WE43 alloy with simvastatin drug loading into GL. The coating minimized the substrate degradation and enhanced osteogenic differentiation concurrently (Qi et al. 2019).

BG coatings are known for their immediate hydroxycarbonate apatite formation in SBF (Heise et al. 2017) (Witecka et al. 2021). The 45S5 BG is commonly used owing to its capability to develop robust bone contact. Rojaee et al. reported an EPD BG-only coating for biodegradable implant engineering (Rojaee et al. 2014). However, BG-only coating is typically not preferred due to its high brittleness and less coating adhesion, and it is generally used as a component in composite formulations. SBF immersion studies of CS-BG coated alloy disclosed copious HA formation compared to the bare alloy. The best bioactivity recorded for a CS-0.4 BG coated alloy was attributed to its lower corrosion and biodegradability (Alaei et al. 2020). Heise et al. investigated CS-BG coating for WE43 alloy. The ζ-potential of −41 mV (pH ~10) supports the suspension stability of BG in water. The ζ-potential decreased close to the isoelectric point on further adding ethanol. Upon CS addition, the ζ-potential was moved to +42 mV (**Figure 17.4a**), favouring cathodic EPD. The cathodic EPD is desirable for Mg alloys, as it could evade the expected initial metal dissolution in anodic EPD. The EPD mass variations with deposition time and current are shown in **Figure 17.4b**. The interpretation of slopes suggested that the deposition current has a more noteworthy effect than the time (Heise et al. 2017).

Partially replacing BG particles with mesoporous BG in CS-BG composite coating increased the surface roughness and hydrophobicity and induced a positive cell response. The study showed that it is key to reduce the deposition time to minimize the adverse impact of the CS-based acidic solution on the Mg substrate (Witecka et al. 2021). Höehlinger et al. studied the influence of various chemical

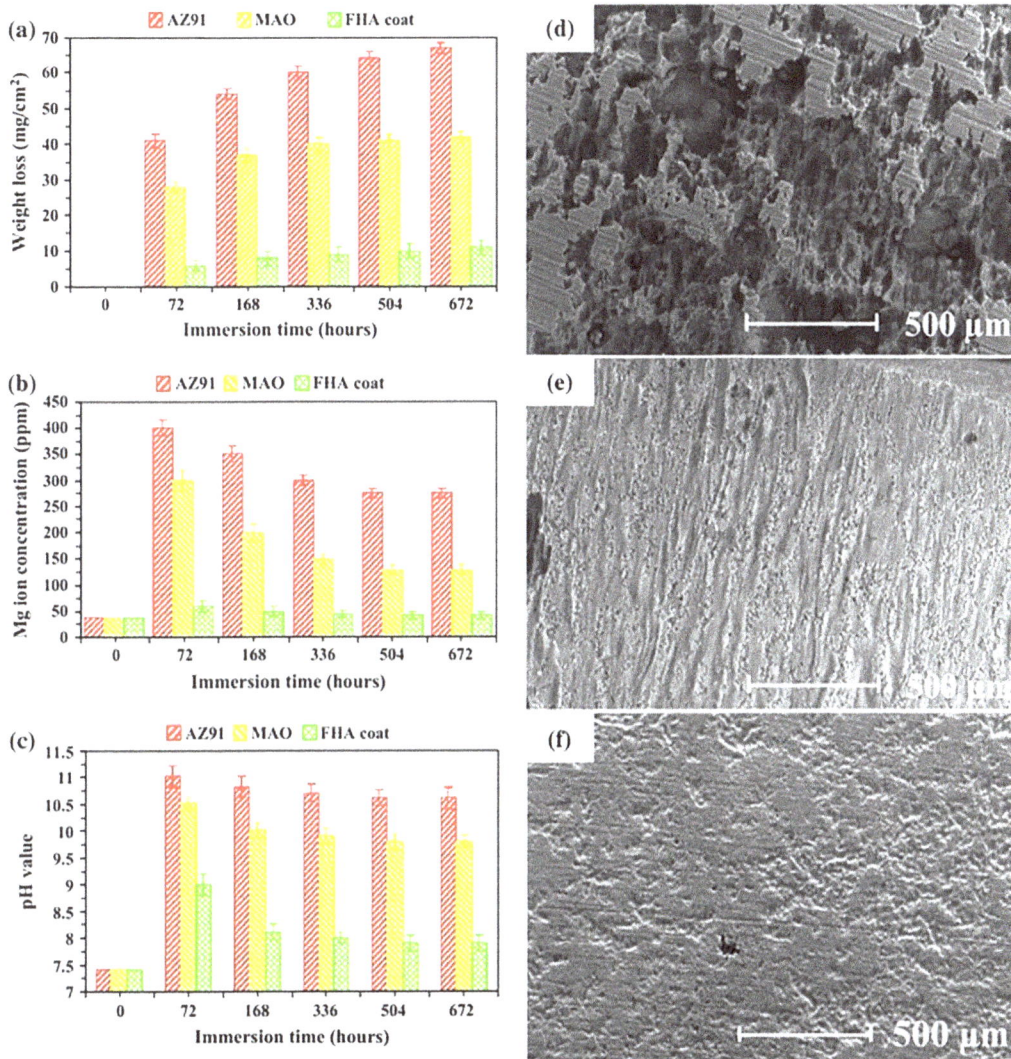

FIGURE 17.3   Variation of (a) weight loss of bare, MAO-coated, and MAO/FHA-coated alloys as a function of immersion time in SBF, (b, c) corresponding variation of (b) Mg ion concentration, and (c) pH of SBF after 672 h of immersion. (d–f) Surface morphologies after 672 h of immersion (after cleaning the corrosion products) of bare, MAO-coated, and MAO/FHA-coated alloys. *Reproduced with permission from (Razavi et al. 2015) © 2014, The Minerals, Metals & Materials Society and ASM International.*

pre-treatments, including Dulbecco's modified eagle's medium (DME), CaP, HF and NaOH. The EPD suspension was prepared with 1 g/L 45S5 BG and 0.5 g/L CS in 1 vol.% CH_3COOH + 20 vol.% deionized water + 79 vol.% ethanol. The ethanol addition reduced the hydrogen evolution during the EPD performed at 50 V for 1 minute. The DME and HF pre-treatments had more positive effects for the better anti-corrosion properties of the coated sample (Höehlinger et al. 2017). Sol-gel produced 63S BG nanoparticles (0.4–1.2 g/L) in 0.5 g/L of 0.2 vol % CH_3COOH solution of CS was employed for cathodic EPD at 10 V for 5 min. Reducing CH_3COOH concentration from 1 to 0.2 vol.% was beneficial in lowering hydrogen reduction and forming a coating with fewer cracks. The deposition yield was directly proportional to the BG concentration in the bath and the time of deposition. Their scanning electron microscopy studies disclosed that a lower concentration of BG (0.4 g/L) assisted

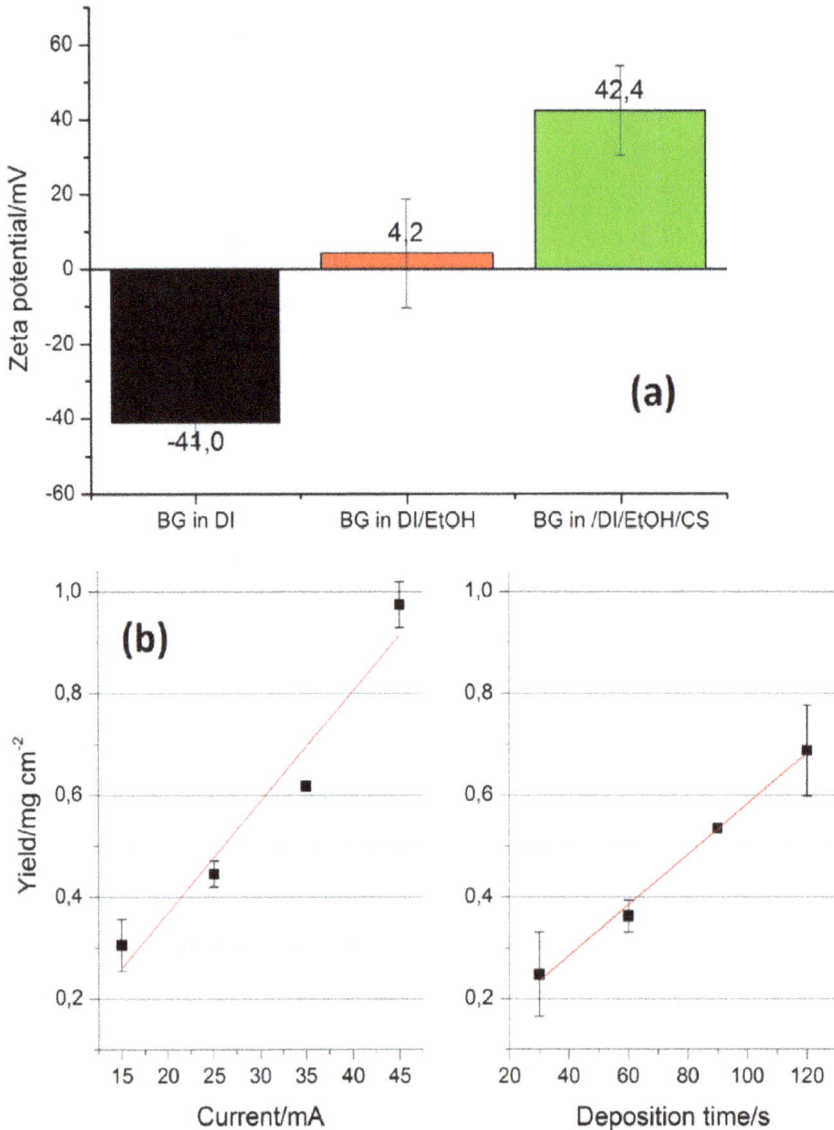

FIGURE 17.4   (a) ζ-potentials of BG in deionized water, water-ethanol, and water-ethanol-CS suspensions. (b) Variation of EPD yield with deposition current and time. *Reproduced with permission from (Heise et al. 2017) © 2017 Elsevier Ltd.*

in forming a thicker (3–4 μm), uniform, and crackless coating, attributed to a lower rate of deposition. The study also suggested that turbulence associated with higher applied voltages can affect the deposition quality. The i_{corr} of CS-0.4 BG coated and bare samples were ~ 2 and 40 μA/cm², respectively (Alaei et al. 2020).

AC EPD was employed to fabricate a CS-GL-BG coating on Mg-Si-Sr alloy using two types of BG: melt-derived and sol-gel-derived. The EPD suspension was made using 0.8 g/L each of GL and CS in aqueous CH_3COOH, to which 1 g/L BG in ethanol was added. The BG addition decreased the ζ-potential of CS-GL suspension from +46.28 mV to +39.90 mV (pH 5.5). At a higher pH (6.5–7.5), a thicker coating was deposited but with deep cracks. The corrosion rate of the EPD-coated and the uncoated alloys were 0.08 and 0.69 mm/year, respectively (Akram et al. 2020).

17.2.1.1 Incorporation of Ceramic Oxides, Carbon Nanostructures, and Polymers

Integrating selected ceramic oxides improves the mechanical properties and wear resistance of HA-, CS-, or BG-based EPD coatings. To fabricate a CS-BG-SiO$_2$ coating, Heise et al. prepared a BG-SiO$_2$ suspension using ~2 µm 45S5 BG and ~0.43 µm SiO$_2$ particles in ethanol-tetraethyl ortho-silicate-ammonia solvent blend, to which 0.5 g/L CS in a water-CH$_3$COOH-ethanol mix solvent was added. The optimized EPD current and time were 35–40 mA and 90 seconds, respectively (Heise et al. 2018). Singh et al. have shown that incorporating Fe$_3$O$_4$ nanoparticles in the HA-CS matrix improved the bioactive and anti-microbial properties (Singh et al. 2021a). The Fe$_3$O$_4$ particle's clustering resulted in a porous composite coating; the porosity augmented from 2.7% to 7.3% as Fe$_3$O$_4$ concentration in the deposition bath varied from 1 to 5 wt.% (Singh et al. 2019). Immersion studies of HA-CA-Fe$_3$O$_4$ coated AZ91 alloy in Ringer's solution for 21 days disclosed that the corrosion resistance of 1 wt.% Fe$_3$O$_4$-added HA-CS coating was 6–14% higher than 3–5 wt.% Fe$_3$O$_4$-added coating. Fe$_3$O$_4$ nanoparticle incorporation was useful in increasing the surface roughness (Singh et al. 2019). Improved corrosion protection was also noted with a HA-BG-Fe$_3$O$_4$-CS coating (Singh et al. 2021b).

Graphene (GR) assimilated HA-composite coatings demonstrated better corrosion resistance (Santos et al. 2015) (Ahangari et al. 2021). The i_{corr} was the lowermost (0.846 µA) for the HA-carboxymethyl cellulose (CMC)-GR coated alloy, compared to HA-CMC (2.96 µA), HA-GR (3.06 µA), and HA (6.31 µA) coated alloys. The i_{corr} of the uncoated alloy was 15.35 µA. The assimilation of CMC was advantageous in enhancing the adhesion strength (Ahangari et al. 2021). Analogous results were also documented for CaP-CS-GR coating (Zhang Zhu et al. 2016). Related works on composite coatings of ceramic oxides, carbon nanomaterials, and polymers are described in the following sections.

17.2.2 Ceramic Oxides, Metals, Mg Compounds

EPD of SiO$_2$ or TiO$_2$ nanoparticles (Fukuda and Matsumoto 2005) could significantly enhance Mg alloy's corrosion and wear resistance. Feil et al. fabricated EPD SiO$_2$ coating from a suspension of ethanol-SiO$_2$ containing boron/sodium oxide as a sintering additive and polydimethylsiloxane as an adhesion promoter (Feil et al. 2008). Heise et al. have shown that the addition of SiO$_2$ to CS-BG coating enhanced the hardness and scratch resistance and reduced creep/stress relaxation and elastic recovery. The study showed that partially replacing BG with SiO$_2$ could provide a topography suitable for HA deposition in SBF (Heise et al. 2018). Impedance spectroscopy studies of TiO$_2$/alginate composite-coated alloy revealed 3- to 7-times greater corrosion resistance than the uncoated alloy (Cordero-Arias et al. 2015). A supplementary interlayer could provide a more desirable microstructure for the outer EPD layer. A double-layer coating of TiO$_2$/hardystonite was comprised of a hybrid ion beam deposited TiO$_2$ interlayer layer (1–1.5 µm) and the EPD made hardystonite outer layer (5–6 µm). The TiO$_2$ inner layer assisted in reducing the porosity of the composite coating and significantly enhanced corrosion resistance (Bakhsheshi-Rad et al. 2016).

An Al$_2$O$_3$-based composite coating was deposited on layered double hydroxides (LDHs)-covered AZ31 alloy by EPD at 150 V from an ethanol suspension of Al$_2$O$_3$ and Al(NO$_3$)$_3$. The EPD coating efficiently shielded the cracks and voids on the LDHs layer. Wear studies showed superior performance for the composite coating deposited from a bath having 2.5 g/L of Al$_2$O$_3$ nanoparticles (Wu et al. 2019). The assimilation of graphene oxide (GO) has been revealed to augment the wear resistance of Al$_2$O$_3$ coating, which was ascribed to the better bonding between the Al$_2$O$_3$ nanoparticles and the GO nanoplatelets (Askarnia et al. 2020). As discussed, several works explored Fe$_3$O$_4$ nanoparticles, particularly with HA and CS. Incorporating Fe$_3$O$_4$ nanoparticles was shown to be effective in suppressing microorganisms' growth (Singh et al. 2019, 2021a). Reports are also available on BG-CS-Fe$_3$O$_4$ (Singh et al. 2020) and HA-BG-CS-Fe$_3$O$_4$ (Singh et al. 2021b) coatings.

EPD ZrO$_2$ protective coatings can be a good choice. Amiri et al. used an aqueous suspension comprising cetyltrimethylammonium bromide (CTAB) as a dispersant/charging agent and poly(vinyl alcohol) (PVA) as a binder. The deposition yield and coating thickness varied with the ZrO_2 concentration in the bath and the deposition current density and time (**Figure 17.5**). The coating thickness was reduced after an optimum current density of 3 mA/cm^2, owing to the increased hydrogen gas generation. A more homogenous coating was formed at a lower ZrO_2 concentration of 10 g/L than 30 g/L. The optimum deposition time was 50 minutes, forming a dense layer with fewer pores. Corrosion studies of the EPD-coated alloys in SBF displayed that the sample made using a 20 g/L ZrO_2 bath had the maximum corrosion resistance (Amiri et al. 2017). An additional laser texturing of a MAO/ZrO_2-EPD coating enhanced corrosion resistance and fibroblast cell proliferation. The i_{corr} values recorded for MAO-EPD/Laser, MAO/Laser, MAO-EPD, MAO, and uncoated alloys were 1.73 ×10^{-8}, 1.47 ×10^{-7}, 2.36 ×10^{-5}, 2.04 ×10^{-4}, and 0.121 mA/cm^2, respectively (Sampatirao et al. 2021). Zhu et al. investigated EPD of ZrO_2 along with silanes and electrophoretic paint (E-paint) (Zhu et al. 2013).

Yttria-stabilized zirconia (YSZ) coating was fabricated using ethanol-acetyl acetone mixed solvent with iodine dispersant. The ethanol addition aided in reducing the ζ-potential owing to the increased positive ion's formation via the acetylacetone-iodine reaction. Dense and even coating formation happened in 1: 1 acetylacetone: ethanol (50 vol.% each). The optimized deposition parameters were 40 V for 3 minutes (Behrangi and Aghajani 2018). A study on the effect of Al interlayer before EPD YSZ coating showed that the interlayer considerably densified the EPD coating via Al-oxide formation. The double-layer coating was sintered at 400°C (Shahriari and Aghajani 2016, 2017).

Several works studied the EPD of ZnO nanoparticles on Mg alloys. An EPD ZnO coating made with an ethanol suspension of 0.001 M $Zn(NO_3)_2$ and 5 g/L ZnO was subsequently subjected to a sintering at 400°C (Qu et al. 2019). EPD ZnO coating can be used as an effective pore-sealing layer on MAO-coated alloys (Bordbar-Khiabani et al. 2020). A recent study reported ZnO-BG coating on MAO/Mg alloy. The composite nanoparticles used for EPD consisted of ZnO cores (259 ± 9 nm) covered by BG shells (26 ± 3 nm). The EPD bath also consisted of dopamine surfactant and polyethyleneimine binder. The rough surfaces of the nanoparticles were helpful for coating adhesion. Impedance spectroscopy studies of the coated alloy in phosphate-buffered saline demonstrated noteworthy enhancement in the barrier protection. The composite coating resulted in higher Ca-P deposition after 7 days of SBF immersion compared to the ZnO coating (Aghili et al. 2022). A combined PVD-EPD process was used to fabricate ZnO/$Ca_3ZrSi_2O_9$ composite coating, where a 10-μm-thick EPD $Ca_3ZrSi_2O_9$ layer was deposited over a PVD-made ZnO underlayer (~900 nm). Anti-bacterial studies showed that the composite coating had better inhibition against *E. coli*, *S. dysenteriae* and *K. pneumonia* (Bakhsheshi-Rad et al. 2017).

A few reports employed rare-earth (RE) oxides. Cathodic EPD at 10 mA/cm^2 for 15 minutes was used to deposit gadolinium oxide on anodized Mg alloy using a bath comprised of gadolinium nitrate (3 mg/mL) and ethylene glycol, and subsequently, the coated alloy was heat-treated at 250°C. The R_p and i_{corr} of the Gd-coated alloy were 80 kΩ and 0.08 μA, respectively. The corresponding values of the bare alloy were 985 Ω and 31.1 μA (Saranya et al. 2020).

The EPD of metallic nanoparticles was studied by Shahriari and Aghajani (2016, 2017). A few works investigated EPD of MgO nanoparticles (Cortez Alcaraz et al. 2019) (Zhou et al. 2021) (Lin et al. 2020). Many reports addressed EPD of Mg compounds such as $MgSiO_3$ (Bakhsheshi-Rad et al. 2020), $CaMgSiO_4$ (Bakhsheshi-Rad et al. 2019), $Ca_7MgSi_4O_{16}$ (Razavi et al. 2013), $CaMgSi_2O_6$ (Razavi, Faithi, Savabi, Beni et al. 2014), and $Ca_2MgSi_2O_7$ (Razavi, Faithi, Savabi et al. 2014). These coatings revealed enhanced bioactivity and corrosion resistance for Mg alloys. SBF immersion studies of MAO/$CaMgSi_2O_6$ coated alloy demonstrated that the corrosion rate continuously augmented in the first 72 hours of immersion and then reduced. The corrosion rates recorded for MAO/$CaMgSi_2O_6$-coated, MAO-coated, and bare alloys after 72 hours of immersion were 0.14, 0.38, and 0.57 mg/cm^2/h, respectively (Razavi, Faithi, Savabi, Beni et al. 2014). The higher diameters of the Nyquist spectra's capacitive loops (**Figure 17.6a**) confirm the good protection capability

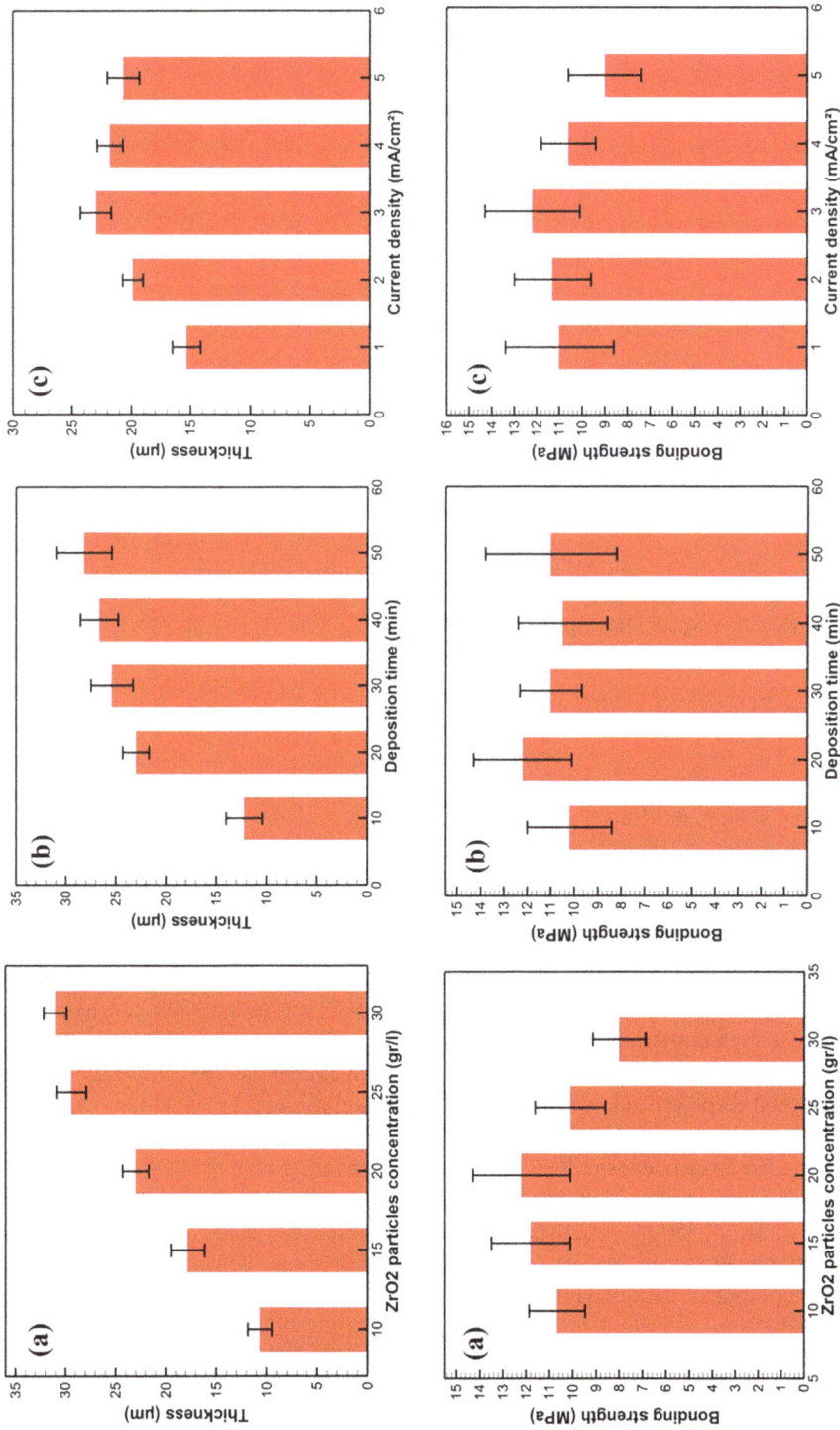

FIGURE 17.5 Effect of solution and deposition parameters on (top) thickness and (bottom) bonding strength of EPD coatings: (a) ZrO_2 particles concentration, (b) EPD time, and (c) EPD current density. *Reproduced with permission from (Amiri et al. 2017) © 2017 Elsevier B.V.*

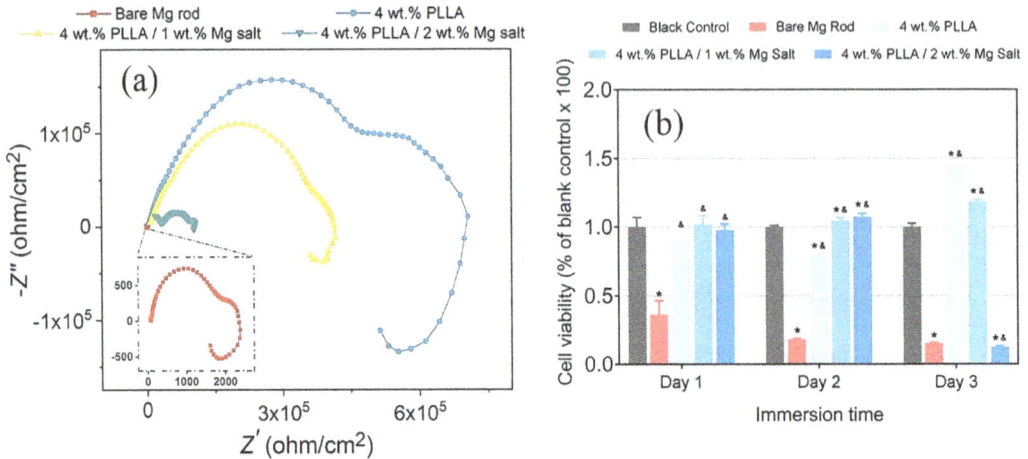

FIGURE 17.6 (a) Nyquist plots in D-Hank's buffer solution at 37°C. (b) Viability of MC3T3-E1 cells through MTT assay after 24 h of cultivation. *Reproduced with permission from (Zhu et al. 2020) © 2020 American Chemical Society.*

of the composite coated alloy. The inductive loops indicate persisting corrosive attack but with a lower grade. Results of the MTT assay on MC3T3-E1 cell viability revealed far better cytocompatibility for the composite coated alloy, as revealed by **Figure 17.6b** (Zhu et al. 2020).

17.2.3   Carbon Nanostructures

Incorporating carbon nanostructures, including carbon nanotubes (CNT) and GR, is a better approach to augmenting EPD coating's hydrophobicity and durability. Immersion studies of EPD CaP-CS-CNT coated alloy in SBF showed significantly enhanced phosphate crystal growth at optimum CNT concentrations (0.05–0.125 g). The composite coating continued undamaged under 70 N load. Nevertheless, a higher or lower CNT concentration was unsuitable for coating integrity (Zhang, Wen et al. 2016). Similar reports are also available on CaP-CS-GR (Zhang, Zhu et al. 2016) and GO-HA (Santos et al. 2015) coatings. EPD of GO was shown to be advantageous in efficiently covering the porous MAO layer and thereby enhancing corrosion resistance (Khiabani et al. 2018). A cathodic EPD at 80 V, followed by aqueous reduction at 60°C, was used to fabricate a reduced GO coating with excellent photothermal effect (Liu et al. 2022). Another work used anodic EPD at a low deposition voltage of 3 V to make a thin GO coating from an aqueous suspension. The thin coating (0.7–1 nm thickness) displayed relatively crack-free surface morphology. GO was partly reduced during the deposition due to peripheral carboxylate groups' reduction (Maqsood et al. 2020).

    A combined pulse-electrodeposited/EPD-made Ni/GR coating provided excellent wear resistance. The GR EPD was conducted at 140 V in a bath comprising Al(NO$_3$)$_3$ (0.07 g), GR (0.07 g), and 2-propanol (7 mL) in 350 mL of ethanol. Above the optimum voltage (140 V), the turbulence in the bath resulted in a reduced coating deposition. Also, inhomogeneous GR deposition has happened above the optimum level of GR in the bath (0.2 g/L). Ni was electrodeposited over the EPD GR (Dong et al. 2019).

    Several works on GR-assimilated composite coatings are available, including HA-GO (Santos et al. 2015), HA-CMC-GR (Ahangari et al. 2021), and CaP-CS-GR (Zhang, Zhu et al. 2016). A HA-CMC-GR coating was deposited by two-step EPD involving depositions at 50 and 30 V, followed by sintering at 450°C. The addition of GR assisted in minimizing the extent of microcracks, attributed to the better HA-GR interaction (Ahangari et al. 2021). Studies on Al$_2$O$_3$-GO coating disclosed that ~2 wt.% of GO yielded a nearly crack-free coating (Askarnia et al. 2020).

The mechanical properties of EPD coatings could be significantly enhanced by the GO incorporation, as revealed by a study on HA-CS-GO coating on AZ91D alloy. GO also improved the HA growth during immersion studies in SBF for 24 days. New HA grains with leaf-like morphologies were developed at higher GO concentrations (~2 wt.%). The corrosion rate of the coated alloy was decreased from 4.3 to 0.2 mpy, which was correlated to the reduced surface cracks. The GO addition was beneficial in enhancing the antibacterial properties (Askarnia et al. 2021).

Instead of carbon nanostructures, a recent work employed BN where a composite coating of CS and BN nanosheets (BNNS) was developed on Mg-Zn-Y-Nd-Zr alloy. Firstly, the Mg sample was subjected to MAO and then to EPD in 1 g/L BNNS dispersed in a CS solution. The composite coating formed a continuous casing layer on the MAO surface with evenly embedded BNNS nanoparticles. The 3D AFM images and roughness (R_a) values of the coated alloy are provided in **Figure 17.7a**. The R_a of the composite coating was significantly higher, attributed to the exposure of BNNS on the coating surface. The composite formation greatly enhanced the coating adhesion (by scratch test), surface hydrophobicity (**Figure 17.7b**), and corrosion resistance (**Figure 17.7c,d**). Their cell culture studies showed no negating influence on the mouse osteoblasts adhesion (Huang et al. 2022).

17.2.4 POLYMERS, PROTEINS, E-PAINTS

EPD is suitable for fabricating polymers, E-paints, and protein coatings on Mg alloys. A composite coating of dimethylaminoethyl methacrylate (charging agent), poly isobornyl acrylate (anti-inflammatory), and tannic acid (functional assembly unit) fabricated by EPD significantly reduced the degradation rate of Mg during 12 weeks of implantation. The tannic acid concentration was decisive for a stable colloidal suspension. Particle aggregation at higher tannic acid concentrations was attributed to the hydrogen binding of polyphenols (Sun et al. 2016). Similarly, a composite coating of tetraaniline, dextran, and caffeic acid enhanced Mg alloy's cytocompatibility and corrosion resistance. The EPD was performed in 0.10 mg/mL of the composite at 150 V for 5 minutes. The coated sample was UV irradiated for photo-crosslinking via caffeic acid's cinnamoyl groups (Li et al. 2020). The corrosion rates of HA-only and HA-poly(lactic-co-glycolic acid) coated samples in SBF were 104 ± 22 and 50 ± 16 mm/yr, respectively (Tian and Liu 2015). Recently, Dai et al. reported an EPD composite coating of amphiphilic electroactive tetraaniline (TAN)-conjugated poly(γ-glutamic acid) (PGA). The superior corrosion of the γ-PGA-TAN coated Mg alloy was mainly attributed to the electroactivity and the rapid passive film formation. The coated alloy efficiently regulated the release of Mg ions during 40 days of immersion. The coated sample also presented improved cytocompatibility for L929 cells (Dai et al. 2022).

Several other polymer-based composite coatings were reported, including HA-PGA-g-7-amino-4-methylcoumarin (AMC) (Sun et al. 2015) (Chen et al. 2016), HA-poly(lactic-co-glycolic acid) (Tian et al. 2017), and TiO$_2$-alginate (Cordero-Arias et al. 2015). Lee et al. deposited drug-loaded poly(lactic-co-glycolic acid) on a silane-pretreated sample (Lee et al. 2016). Gnedenkov et al. studied EPD of polytetrafluoroethylene (PTFE) on MAO-coated alloy (Gnedenkov et al. 2015) (Gnedenkov et al. 2021). *In vivo* studies on HA/PGA-AMC coated samples yielded reduced degradation with better histocompatibility and osteoinductivity (Chen et al. 2016).

A few works explored proteins-based EPD coatings such as Zein/45S5 BG (Rehman, 2020) (Ahmed et al. 2020). These coatings were effective in reducing degradation and enhancing bioactivity. The ζ-potential of the Zein-BG suspension was ~ +30 mV (pH 3). BG addition was beneficial in depositing a uniform and dense coating (Ahmed et al. 2020).

Pan et al. subjected EPD aliphatic polycarbonate coating to photo-crosslinking to enhance corrosion resistance. A cationic polycarbonate (PMDMT) was synthesized and assembled with a crosslinker pentaerythritol tetrakis(3-mercaptopropionate) (PETMP); their colloidal particles were subsequently coated on AZ31 sheets and stents by cathodic EPD. The coated alloy was further subjected to UV irradiation for crosslinking (Mg-CP-SH). The coating thickness altered with the

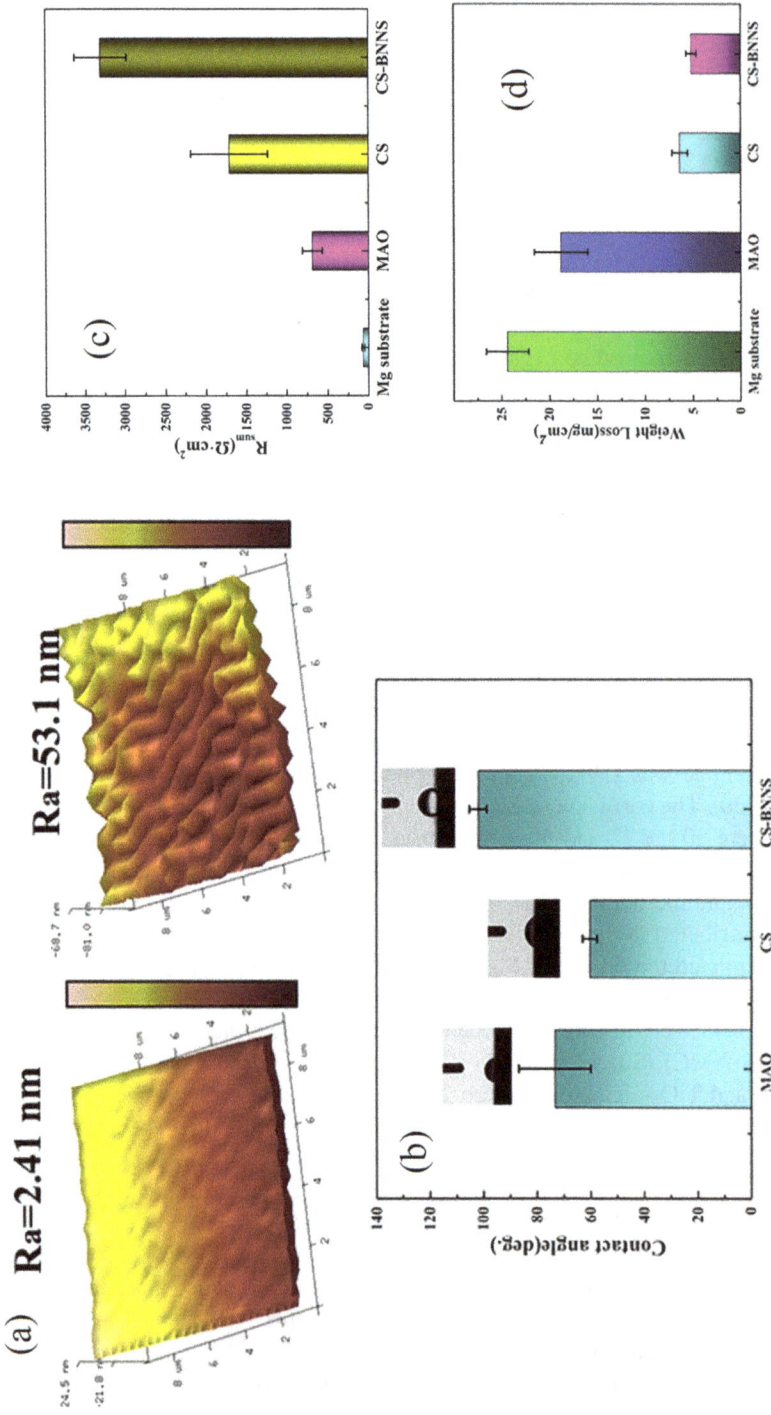

FIGURE 17.7 (a) 3D surface morphologies of CS and CS-BNNS, (b) Water contact angles, (c) Coating resistance (R_{sum}) from impedance spectroscopy, and (d) Weight loss data after 7 days of SBF immersion. *Reproduced with permission from (Huang et al. 2022) © 2022 Elsevier B.V.*

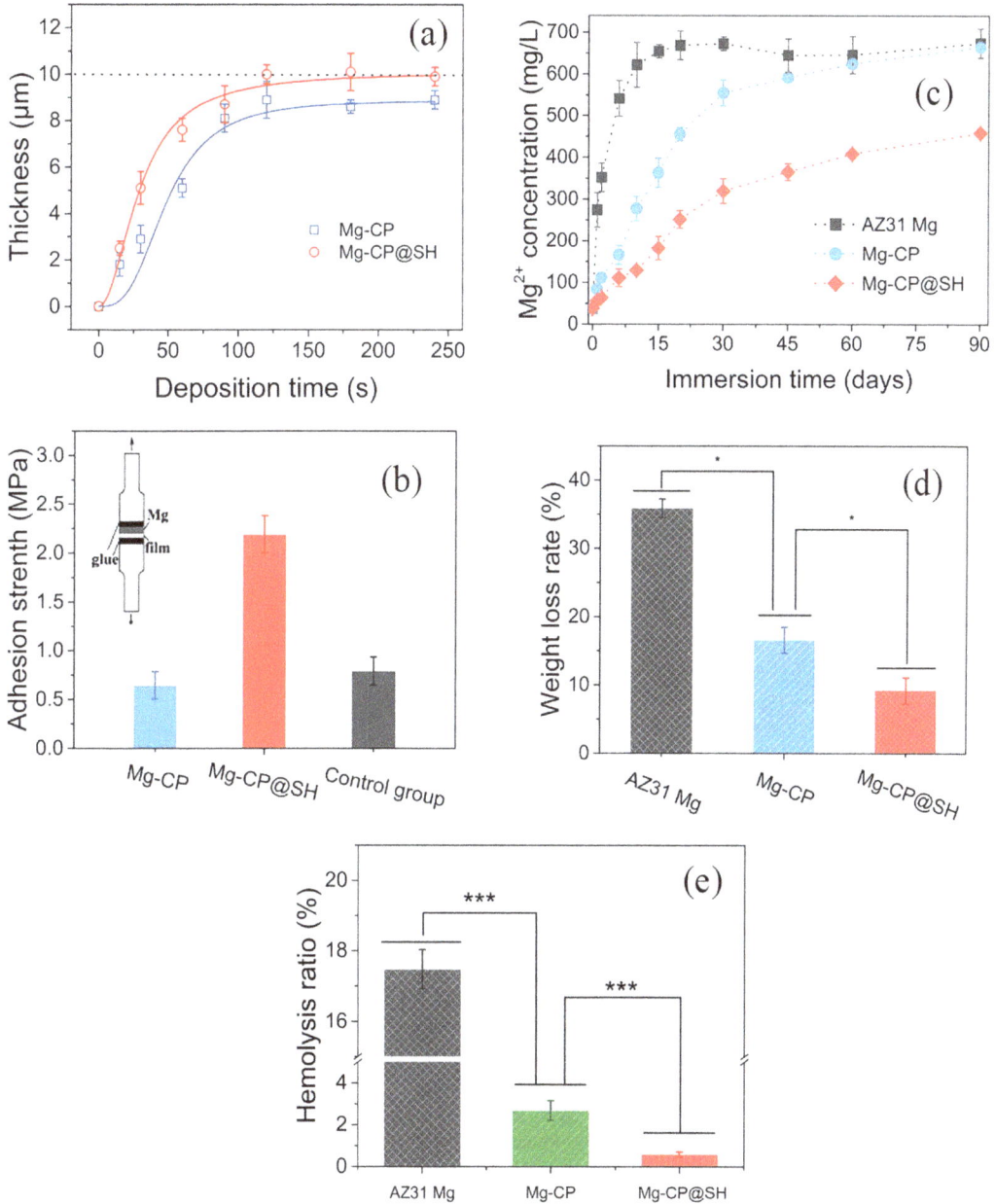

FIGURE 17.8   (a) Coating thickness vs. EPD time (voltage – 40 V, colloid concentration – 15 mg/mL) (Mg-CP – coating without PMDMT). (b) Adhesion strength of different coated samples (reference – ISO 4624–2016). (c, d) Results of long-term degradation tests: variation of (c) Mg²⁺ concentration, and (d) mass loss during 90 days of SBF immersion. (e) Hemolysis rates. *Reproduced with permission from (Pan et al. 2022) © 2021 Elsevier B.V.*

deposition time (**Figure 17.8a**). The composite alloy presented excellent adhesion (**Figure 17.8b**), corrosion resistance (**Figure 17.8c,d**), and hemocompatibility (**Figure 17.8e**). The polar sulfide groups enhanced the coating/substrate interaction, and the crosslinking improved the cohesive force within the coating. The low hemolysis rate of the composite coating was attributed to the coated sample's reduced degradation and ion release (Pan et al. 2022).

Several studies used commercial E-paints for EPD (Jiang et al. 2011) (Tang et al. 2013) (Phoung and Moon 2015) (Phuong et al. 2017). These EPD coatings have been shown to have improved mechanical interlocking with a base MAO layer. Luo et al. deposited cathodic-type epoxy resin on

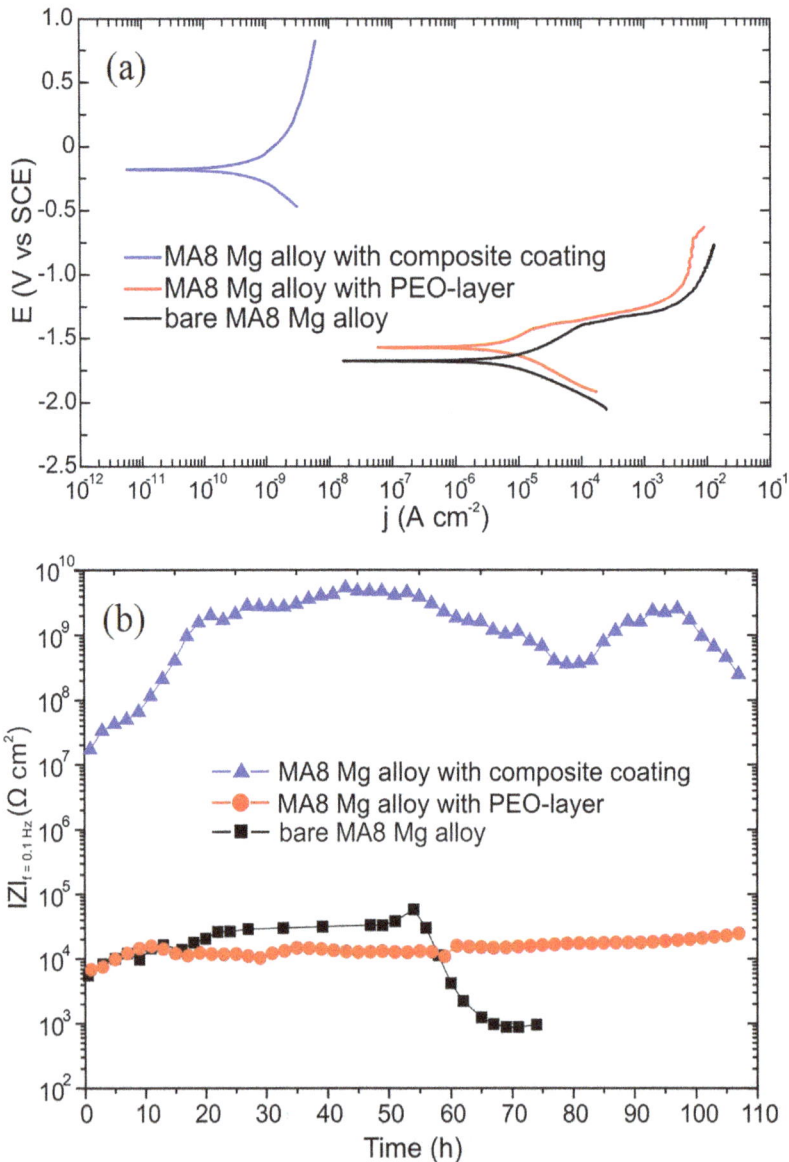

FIGURE 17.9 (a) Potentiodynamic polarization and (b) impedance modulus (at 0.1 Hz) vs. time plots in minimum essential medium. PEO corresponds to MAO. *Reproduced with permission from (Gnedenkov et al. 2021) © 2021 Elsevier Ltd.*

MAO/Mg-alloy at 185 V. A curing step at 170°C enhanced the weak bonding of the polymer (Luo et al. 2014). Gnedenkov et al. disclosed considerably improved corrosion and wear resistance for an EPD polymer coating on MAO/Mg-alloy (Gnedenkov et al. 2015). Wu et al. conducted cathodic EPD of an aqueous HG-91 E-paint on an MAO-coated alloy, and the sample after EPD was water-washed and heated at 220°C. An EPD at 100 V yielded coatings with ~ 20 μm thickness (Wu et al. 2020). In a former study, Zhu et al. explored ZrO_2/E-paint coating using silane-modified nano ZrO_2 and epoxy-modified polyurethane resin, which was fabricated at 140 V and the coating cured at 160°C (Zhu et al. 2013). **Figure 17.9a** shows potentiodynamic polarization curves of PTFE/HA-MAO coated MA8 alloy. From the i_{corr} and E_{corr} values, it is evident that the barrier protection was only marginally increased by the MAO-only coating. The i_{corr} of the bare and MAO-coated alloys was 9.2×10^{-6} A/cm^2 and 5.4×10^{-6} A/cm^2, respectively. Instead, the composite coated alloy demonstrated excellent protection with a low i_{corr} of 7.6×10^{-10} A/cm^2. A typical anodic passivation behaviour is evident in the potentiodynamic polarization plot of the composite coated alloy. Their time-dependent impedance spectroscopy studies supported the enhanced corrosion resistance (**Figure 17.9b**). The impedance of the composite coated sample was at ~ 5.4×10^9 Ω/cm^2 (Gnedenkov et al. 2021).

17.3   CONCLUSIONS AND OUTLOOK

EPD is a well-established coating method for Mg alloys. The technique is simple and eco-friendly. By integrating appropriate nanoparticles in the deposition bath, coatings with improved corrosion resistance and required functionalities such as anti-wear, anti-bacterial, and bioactivity could be straightforwardly fabricated. The nature and thickness of the layer can be simply tuned by optimizing the deposition voltage, time, and concentration of the colloidal particles.

The chapter briefly describes different EPD coatings reported for Mg and its alloys, emphasizing recent reports on EPD composite coatings. The sections are classified based on the type of nanoparticles used for EPD, viz. (i) hydroxyapatite, chitosan, bioactive glass, (ii) ceramic oxides, metals, Mg compounds, (iii) carbon nanomaterials and (iv) polymers, proteins, E-paints. The most studied coating is hydroxyapatite-based. Biologically relevant materials such as bioglass and chitosan were extensively used as a component in the coating formulations. Several reports addressed EPD nanocomposite coatings with assimilated ceramic oxides and carbon nanostructures. The additions of SiO_2, Al_2O_3, ZrO_2, carbon nanotubes, and graphene boost the scratch-resistance and hardness of the EPD coatings. EPD composite polymer coatings have high significance for Mg alloys. Most works utilized post-high-temperature sintering to develop dense and adhesive coatings.

Several prospects exist to explore novel nanocomposite EPD coatings of polymers, ceramics, and metals to better protect Mg and its alloys. An effective strategy is EPD/MAO combined coatings. An optimized EPD coating on the MAO layer could substantially boost the corrosion resistance.

REFERENCES

Aghili F., Hoomehr B., Saidi R., Raeissi K., Synthesis and electrophoretic deposition of zinc oxide and zinc oxide-bioactive glass composite nanoparticles on AZ31 Mg Alloy for biomedical applications, *Ceram. Int.*, 48, 2022, 34013–34024.

Ahangari M., Johar M. H., Saremi M., Hydroxyapatite-carboxymethyl cellulose-graphene composite coating development on AZ31 magnesium alloy: Corrosion behavior and mechanical properties, *Ceram. Int.*, 47, 2021, 3529–3539.

Ahmed Y., Nawaz A., Virk R. S., Wadood A., Rehman M. A. U., Fabrication and characterization of zein/bioactive glass deposited on pretreated magnesium via electrophoretic deposition, *Int. J. Ceram. Eng. Sci.*, 2, 2020, 254–263.

Akram M., Arshad N., Aktan M. K., Braem A., Alternating current electrophoretic deposition of chitosan-gelatin-bioactive glass on Mg-Si-Sr alloy for corrosion protection, *ACS Appl. Bio Mater.*, 3, 2020, 7052–7060.

Alaei M., Atapour M., Labbaf S., Electrophoretic deposition of chitosan-bioactive glass nanocomposite coatings on AZ91 Mg alloy for biomedical applications, *Prog. Org. Coat.*, 147, 2020, 105803.

Ammam M., Electrophoretic deposition under modulated electric fields: A review, *RSC Adv.*, 2, 2012, 7633–7646.

Amiri H., Mohammadi I., Afshar A., Electrophoretic deposition of nano-zirconia coating on AZ91D magnesium alloy for bio-corrosion control purposes, *Surf. Coat. Technol.*, 311, 2017, 182–190.

Askarnia R., Ghasemi B., Fardi S. R., Adabifiroozjaei E., Improvement of tribological, mechanical and chemical properties of Mg alloy (AZ91) by electrophoretic deposition of alumina/GO coating, *Surf. Coat. Technol.*, 403, 2020, 126410.

Askarnia R., Roueini Fardi S., Sobhani M., Staji H., Ternary hydroxyapatite/chitosan/graphene oxide composite coating on AZ91D magnesium alloy by electrophoretic deposition, *Ceram. Int.*, 47, 2021, 27071–27081.

Ata M. S., Liu Y., Zhitomirsky I., A review of new methods of surface chemical modification, dispersion and electrophoretic deposition of metal oxide particles, *RSC Adv.*, 4, 2014, 22716–22732.

Avcu E., Baştan F. E., Abdullah H. Z., Ur Rehman M. A., Avcu Y. Y., Boccaccini A. R., Electrophoretic deposition of chitosan-based composite coatings for biomedical applications: A review, *Prog. Mater. Sci.*, 103, 2019, 69–108.

Bakhsheshi-Rad H. R., Hamzah E., Ismail A. F., Aziz M., Kasiri-Asgarani M., Akbari E., Jabbarzare S., QNajafinezhad A., Hadisi Z., Synthesis of a novel nanostructured zinc oxide/baghdadite coating on Mg alloy for biomedical application: in-vitro degradation behavior and antibacterial activities, *Ceram. Int.*, 43, 2017, 14842–14850.

Bakhsheshi-Rad H. R., Hamzah E., Ismail A. F., Aziz M., Najafinezhad A., Daroonparvar M., Synthesis and in-vitro performance of nanostructured monticellite coating on magnesium alloy for biomedical applications, *J. Alloys Compnds*, 773, 2019, 180–193.

Bakhsheshi-Rad H. R., Hamzah E., Kasiri-Asgarani M., Jabbarzare S., Daroonparvar M., Najafinezhad A., Fabrication, degradation behavior and cytotoxicity of nanostructured hardystonite and titania/hardystonite coatings on Mg alloys, *Vacuum.*, 129, 2016, 9–12.

Bakhsheshi-Rad H. R., Najafinezhad A., Hamzah E., Ismail A. F., Berto F., Chen X. B., Clinoenstatite/tantalum coating for enhancement of biocompatibility and corrosion protection of Mg alloy, *J. Funct. Biomater.*, 11, 2020, 26.

Behrangi S., Aghajani H., Nano 3YSZ electrophoretic deposition from acetylacetone + ethanol solvent on the surface of AZ91 magnesium alloy, *Micro Nano Lett.*, 13, 2018, 611–616.

Besra L., Liu M., A review on fundamentals and applications of electrophoretic deposition (EPD), *Prog. Mater. Sci.*, 52, 2007, 1–61.

Boccaccini A. R., Keim S., Ma R., Li Y., Zhitomirsky I., Electrophoretic deposition of biomaterials, *J. R. Soc. Interface.*, 7, 2010, S581–S613.

Bordbar-Khiabani A., Yarmand B., Mozafari M., Effect of ZnO pore-sealing layer on anti-corrosion and in-vitro bioactivity behavior of plasma electrolytic oxidized AZ91 magnesium alloy, *Mater. Lett.*, 258, 2020, 126779.

Chen P., Sun J., Zhu Y., Yu X., Meng L., Li Y., Liu X., Corrosion resistance of biodegradable Mg with a composite polymer coating, *J. Biomater. Sci. Polym. Ed.*, 27, 2016, 1763–1774.

Cordero-Arias L., Boccaccini A. R., Virtanen S., Electrochemical behavior of nanostructured TiO_2/alginate composite coating on magnesium alloy AZ91D via electrophoretic deposition, *Surf. Coat. Technol.*, 265, 2015, 212–217.

Cortez Alcaraz M. C., Cipriano A. F., Lin J., Soria Jr P., Tian Q., Liu H., Electrophoretic deposition of magnesium oxide nanoparticles on magnesium: Processing parameters, microstructures, degradation, and cytocompatibility, *ACS Appl. Bio Mater.*, 2, 2019, 5634–5652.

Dai M., Li S., Cui Y., Zhang W., Shi H., Pan K., Wei W., Liu X., Li X., Fabrication of electroactive poly(γ-glutamic acid) coating for improving corrosion resistance and cytocompatibility of magnesium alloy, *Polym Int.*, 71, 2022, 1278–1286.

De D., Nicholson P. S., Role of ionic depletion in deposition during electrophoretic deposition, *J. Am. Ceram. Soc.*, 82, 1999, 3031–3036.

Diba M., Fam D. W. H., Boccaccini A. R., Shaffer M. S. P., Electrophoretic deposition of graphene-related materials: A review of the fundamentals, *Prog. Mater. Sci.*, 82, 2016, 83–117.

Dong Y., Sun W., Liu X., Ma M., Zhang Y., Liu Y., Electrophoretic-deposition of graphene and microstructure and friction behavior of Ni-graphene composite coatings, *Adv. Eng. Mater.*, 21, 2019, 1900327.

Feil F., Füerbeth W., Schütze M., Nanoparticles based inorganic coatings for corrosion protection of magnesium alloys, *Surf. Eng.*, 24, 2008, 198–203.

Fukuda H., Matsumoto Y., Formation of Ti-Si composite oxide films on Mg-Al-Zn alloy by electrophoretic deposition and anodization, *Electrochim. Acta.*, 50, 2005, 5329–5333.

Gnedenkov A. S., Lamaka S. V., Sinebryukhov S. L., Mashtalyar D. V., Egorkin V. S., Imshinetskiy I. M., Zheludkevich M. L., Gnedenkov S. V., Control of the Mg alloy biodegradation via PEO and polymer-containing coatings, *Corros. Sci.*, 182, 2021, 109254.

Gnedenkov S. V., Sinebryukhov S. L., Mashtalyar D. V., Imshinetskiy I. M., Composite fluoropolymer coatings on Mg alloys formed by plasma electrolytic oxidation in combination with electrophoretic deposition, *Surf. Coat. Technol.*, 283, 2015, 347–352.

Grillon F., Fayeulle D., Jeandin M., Quantitative image analysis of electrophoretic coatings, *J. Mater. Sci. Lett.*, 11, 1992, 272–275.

Hamaker H. C., Formation of a deposit by electrophoresis, *Trans. Farad. Soc.*, 35, 1940, 279–287.

Hamaker H. C., Verwey E. J. W., Part II.-(C) Colloid stability. The role of the forces between the particles in electrodeposition and other phenomena, *Trans. Farad. Soc.*, 1940, 35, 180–185.

Heise S., Höehlinger M., Hernández Y. T., Palacio J. J. P., Ortiz J. A. R., Wagener V., Virtanen S., Boccaccini A. R., Electrophoretic deposition and characterization of chitosan/bioactive glass composite coatings on Mg alloy substrates, *Electrochim. Acta.*, 232, 2017, 456–464.

Heise S., Wirth T., Höehlinger M., Hernández Y. T., Ortiz J. A. R., Wagener V., Virtanen S., Boccaccini A. R., Electrophoretic deposition of chitosan/bioactive glass/silica coatings on stainless steel and WE43 Mg alloy substrates, *Surf. Coat. Technol.*, 344, 2018, 553–563.

Höehlinger M., Heise S., Wagener V., Boccaccini A. R., Virtanen S., Developing surface pre-treatments for electrophoretic deposition of biofunctional chitosan-bioactive glass coatings on a WE43 magnesium alloy, *Appl. Surf. Sci.*, 405, 2017, 441–448.

Hu S., Li W., Finklea H., Liu X., A review of electrophoretic deposition of metal oxides and its application in solid oxide fuel cells, *Adv. Colloid Interface Sci.*, 276, 2020, 102102.

Huang W., Mei D., Qin H., Li J., Wang L., Ma X., Zhu S., Guan S., Electrophoretic deposited boron nitride nanosheets-containing chitosan-based coating on Mg alloy for better corrosion resistance, biocompatibility and antibacterial properties, *Colloid. Surf. A.*, 638, 2022, 128303.

Jiang Y., Yang H., Bao Y., Zhang Y., Characterization of composite coatings produced by plasma electrolytic oxidation plus cathodic electrophoretic deposition on magnesium alloy, *Adv. Mater. Res.*, 146–147, 2011, 1126–1131.

Kaabi Falahieh Asl S., Nemeth S., Tan M. J., Electrophoretic deposition of hydroxyapatite coatings on AZ31 magnesium substrate for biodegradable implant application, *Prog. Crys. Growth Charact. Mater.*, 60, 2014, 74–79.

Khiabani A. B., Rahimi S., Yarmand B., Mozafari M., Electrophoretic deposition of graphene oxide on plasma electrolytic oxidized-magnesium implants for bone tissue engineering applications, *Mater. Today Proc.*, 5, 2018, 15603–15612.

Koelmans H., Suspensions in non-aqueous media, *Philips Res. Rep.*, 10, 1995, 161–193.

Lee W. S., Park M., Kim M. H., Park C. G., Huh B. K., Seok H. K., Choy Y. B., Nanoparticle coating on the silane modified surface of magnesium for local drug delivery and controlled corrosion, *J. Biomater. Appl.*, 30, 2016, 651–661.

Li X., Shi H., Cui Y., Pan K., Wei W., Liu X., Dextran-caffeic acid/tetraaniline composite coatings for simultaneous improvement of cytocompatibility and corrosion resistance of magnesium alloy, *Prog. Org. Coat.*, 149, 2020, 105928.

Lin J., Nguyen N. T., Zhang C., Ha A., Liu H. H., Antimicrobial properties of MgO nanostructures on magnesium substrates, *ACS Omega.*, 5, 2020, 24613–24627.

Liu L., Peng F., Zhang D., Li M., Hunag J., Liu X., A tightly bonded reduced graphene oxide coating on magnesium alloy with photothermal effect for tumor therapy, *J. Magnes. Alloys*, 10, 2022, 3031–3040.

Luo Z., Hao Z., Jiang B., Ge Y., Zheng X., Micro arc oxidation and electrophoretic deposition effect on damping and sound transmission characteristics of AZ31B magnesium alloy, *J. Central South Univ.*, 21, 2014, 3419–3425.

Manoj Kumar R., Kuntal K. K., Singh S., Gupta P., Bhushan B., Gopinath P., Lahiri D., Electrophoretic deposition of hydroxyapatite coating on Mg-3Zn alloy for orthopaedic application, *Surf. Coat. Technol.*, 287, 2016, 82–92.

Maqsood M. F., Raza M. A., Ghauri F. A., Rehman Z. U., Ilyas M. T., Corrosion study of graphene oxide coatings on AZ31B magnesium alloy, *J. Coat. Technol. Res.*, 17, 2020, 1321–1329.

Obregón S., Amor G., Vázquez A., Electrophoretic deposition of photocatalytic materials, *Adv. Colloid Interface Sci.*, 269, 2019, 236–255.

Pan K., Li X., Shi H., Dai M., Yang Z., Chen M., Wei W., Liu X., Zheng Y., Preparation of photo-crosslinked aliphatic polycarbonate coatings with predictable degradation behavior on magnesium-alloy stents by electrophoretic deposition, *Chem. Eng. J.*, 427, 2022, 131596.

Phoung N. V., Fazal, B. R., Moon S., Electrophoretic painting on AZ31 Mg alloy pretreated in cerium conversion coating solutions prepared in ethanol-water mixtures, *Metals Mater. Int.*, 23, 2017, 106–114.

Phoung N. V., Moon S., Deposition and characterization of E-paint on magnesium alloys, *Prog. Org. Coat.*, 89, 2015, 91–99.

Qi H., Heise S., Zhou J., Schuhladen K., Yang Y., Cui N., Dong R., Virtanen S., Chen Q., Boccaccini A. R., Lu T., Electrophoretic deposition of bioadaptive drug delivery coatings on magnesium alloy for bone repair, *ACS Appl. Mater. Interface.*, 11, 2019, 8625–8634.

Qu J. E., Ascencio M., Jiang L. M., Omanovic S., Yang L. X., Improvement in corrosion resistance of WE43 magnesium alloy by the electrophoretic formation of a ZnO surface coating, *J. Coat. Technol. Res.*, 16, 2019, 1559–1570.

Razavi M., Faithi M., Savabi O., Beni B. H., Razavi S. M., Vashaee D., Tayebi L., Coating of biodegradable magnesium alloy bone implants using nanostructured diopside (CaMgSi$_2$O$_6$), *Appl. Surf. Sci.*, 288, 2014, 130–137.

Razavi M., Faithi M., Savabi O., Razavi S. M., Beni B. H., Vashaee D., Tayebi L., Surface modification of magnesium alloy implants by nanostructured bredigite coating, *Mater. Lett.*, 113, 2013, 174–178.

Razavi M., Faithi M., Savabi O., Razavi S. M., Beni B. H., Vashaee D., Tayebi L., Controlling the degradation rate of bioactive magnesium implants by electrophoretic deposition of akermanite coating, *Ceram. Int.*, 40, 2014, 3865–3872.

Razavi M., Fathi M., Savabi O., Vashaee D., Tayebi L., Biodegradable magnesium alloy coated by fluoridated hydroxyapatite using MAO/EPD technique, *Surf. Eng.*, 30, 2014, 545–551.

Razavi M., Fathi M., Savabi O., Vashaee D., Tayebi L., In vitro analysis of electrophoretic deposited fluoridated hydroxyapatite coating on micro-arc oxidized AZ91 magnesium alloy for biomaterials applications, *Metall. Mater. Trans. A.*, 46, 2015, 1394–1404.

Rehman M. A. U., Zein/bioactive glass coatings with controlled degradation of magnesium under physiological conditions: Designed for orthopedic implants, *Prosthesis.*, 2, 2020, 211–224.

Rojaee R., Fathi M., Raeissi K., Taherian M., Electrophoretic deposition of bioactive glass nanopowders on magnesium based alloy for biomedical applications, *Ceram. Int.*, 40, 2014, 7879–7888.

Saadati A., Hesarikia H., Nourani M. R., Taheri R. A., Electrophoretic deposition of hydroxyapatite coating on biodegradable Mg-4Zn-4Sn-0.6Ca-0.5Mn alloy, *Surf. Eng.*, 36, 2020, 908–918.

Saji V. S., Electrodeposition in bulk metallic glasses, *Materialia.*, 3, 2018, 1–11.

Saji V. S., Electrophoretic (EPD) coatings for magnesium alloys, *J. Ind. Eng. Chem.*, 103, 2021a, 358–372.

Saji V. S., Electrophoretic-deposited superhydrophobic coatings, *Chem. Asian J.*, 16, 2021b, 474–491.

Sampatirao H., Amruthaluru S., Chennampalli P., Lingamaneni R. K., Nagumothu R., Fabrication of ceramic coatings on the biodegradable ZM21 magnesium alloy by PEO coupled EPD followed by laser texturing process, *J. Magnes. Alloys.*, 9, 2021, 910–926.

Sankar M., Suwas S., Balasubramanian S., Manivasagam G., Comparison of electrochemical behavior of hydroxyapatite coated onto WE43 Mg alloy by electrophoretic and pulsed laser deposition, *Surf. Coat. Technol.*, 309, 2017, 840–848.

Santos C., Piedade C., Uggowitzer P. J., Montemor M. F., Carmezim M. J., Parallel nano-assembling of a multifunctional GO/Hap nanoparticle coating on ultrahigh-purity magnesium for biodegradable implants, *Appl. Surf. Sci.*, 345, 2015, 387–393.

Saranya K., Bhuvaneswari S., Chatterjee S., Rajendran N., Biocompatible gadolinium-coated magnesium alloy for biomedical applications, *J. Mater. Sci.*, 55, 2020, 11582–11596.

Sarkar P, Nicholson P. S., Electrophoretic deposition (EPD): Mechanisms, kinetics, and application to ceramics, *J. Am. Ceram. Soc.*, 79, 1996, 1987–2002.

Seuss S., Boccaccini A. R., Electrophoretic deposition of biological macromolecules, drugs, and cells, *Biomacromolecules.*, 14, 2013, 3355–3369.

Shahriari A., Aghajani H., Electrophoretic deposition of 3YSZ coating on AZ91D alloy using Al and Ni-P interlayers, *J. Mater. Eng. Perform.*, 25, 2016, 4369–4382.

Shahriari A., Aghajani H., Electrophoretic deposition of 3YSZ coating on AZ91D using an aluminum interlayer, *Prot. Metal. Phys. Chem. Surf.*, 53, 2017, 518–526.

Singh S., Singh G., Bala N., Corrosion behavior and characterization of HA/Fe$_3$O$_4$/CS composite coatings on AZ91 Mg alloy by electrophoretic deposition, *Mater. Chem. Phys.*, 237, 2019, 121884.

Singh S., Singh G., Bala N., Synthesis and characterization of iron oxide- hydroxyapatite-chitosan composite coating and its biological assessment for biomedical applications, *Prog. Org. Coat.*, 150, 2021a, 106011.

Singh S., Singh G., Bala N., Electrophoretic deposition of Fe$_3$O$_4$ nanoparticles incorporated HA-bioglass-chitosan nanocomposite coating on AZ91 Mg alloy, *Mater. Today Commun.*, 26, 2021b, 101870.

Singh S., Singh G., Bala N., Aggarwal K., Characterization and preparation of Fe$_3$O$_4$ nanoparticles loaded bioglass-chitosan nanocomposite coating on Mg alloy and in vitro bioactivity assessment, *Int. J. Biol. Macromol.*, 151, 2020, 519–528.

Sun J., Zhu Y., Meng L., Chen P., Shi T., Liu X., Zheng Y., Electrophoretic deposition of colloidal particles on Mg with cytocompatibility, antibacterial performance, and corrosion resistance, *Acta Biomater.*, 45, 2016, 387–398.

Sun J., Zhu Y., Meng L., Shi T., Liu X., Zheng Y., A biodegradable coating based on self-assembled hybrid nanoparticles to control the performance of magnesium, *Macromol. Chem. Phys.*, 216, 2015, 1952–1962.

Tang J., Wang J., Xie X., Zhang P., Lai Y., Li Y., Qin L., Surface coating reduces degradation rate of magnesium alloy developed for orthopaedic applications, *J. Orthopaed. Trans.*, 1, 2013, 41–48.

Tayyaba Q., Shahzad M., Butt A. Q., Din R. U., Khan M., Qureshi A. H., The influence of electrophoretic deposition of hydroxyapatite on Mg-Zn-Zr alloy on its in-vitro degradation behaviour in the Ringer's solution, *Surf. Coat. Technol.*, 375, 2019, 197–204.

Tian Q., Liu H., Electrophoretic deposition and characterization of nanocomposites and nanoparticles on magnesium substrates, *Nanotechnology.*, 26, 2015, 175102.

Tian Q., Rivera-Castaneda L., Liu H., Optimization of nano-hydroxyapatite/poly(lactic-co-glycolic acid) coatings on magnesium substrates using one-step electrophoretic deposition, *Mater. Lett.*, 186, 2017, 12–16.

Van der Biest O. O., Vandeperre L. J., Electrophoretic deposition of materials, *Ann. Rev. Mater. Sci.*, 29, 1999, 327–352.

Wang L., Xiao X., Yin X., Wang J., Zhu G., Yu S., Liu E., Wang B., Yang X., Preparation of robust, self-cleaning and anti-corrosion superhydrophobic Ca-P/chitosan (CS) composite coating on AZ31 magnesium alloy, *Surf. Coat. Technol.*, 432, 2022, 128074.

Witecka A., Valet S., Basista M., Boccaccini A. R., Electrophoretically deposited high molecular weight chitosan/bioactive glass composite coatings on WE43 magnesium alloy, *Surf. Coat. Technol.*, 418, 2021, 127232.

Wu L., Ding X. X., Zheng Z. C., Ma Y. L., Atrens A., Chen X. B., Xie Z. H., Sun D., Pan F., Fabrication and characterization of an actively protective Mg-Al LDHs/Al$_2$O$_3$ composite coating on magnesium alloy AZ31, *Appl. Surf. Sci.*, 487, 2019, 558–568.

Wu M., Guo Y., Xu G., Cui Y., Effects of deposition thickness on electrochemical behaviors of AZ31B magnesium alloy with composite coatings prepared by micro-arc oxidation and electrophoretic deposition, *Int. J. Electrochem. Sci.*, 15, 2020, 1378–1390.

Zhang J., Dai C. S., Wei J., Wen Z. H., Study on the bonding strength between calcium phosphate/chitosan composite coatings and a Mg alloy substrate, *Appl. Surf. Sci.*, 261, 2012, 276–286.

Zhang J., Dai C. S., Wei J., Wen Z. H., Zhang S., Chen C., Degradable behavior and bioactivity of micro-arc oxidized AZ91D Mg alloy with calcium phosphate/chitosan composite coating in m-SBF, *Colloid Surf. B*, 111, 2013, 179–187.

Zhang J., Wen Z., Zhao M., Li G., Dai C., Effect of the addition CNTs on performance of CaP/chitosan/coating deposited on magnesium alloy by electrophoretic deposition, *Mater. Sci. Eng. C*, 58, 2016, 992–1000.

Zhang J., Zhu F., Zhang Y., Hu M., Chi Y., Zhang X., Guo X., In vitro bioactivity, degradation property and cell viability of the CaP/chitosan/graphene coating on magnesium alloy in m-SBF, *Int. J. Electrochem. Sci.*, 11, 2016, 9326–9339.

Zhou W., Yan J., Li Y., Wang L., Jing L., Li M., Yu S., Cheng Y., Zheng Y., Based on the synergistic effect of Mg^{2+} and antibacterial peptides to improve the corrosion resistance, antibacterial ability and osteogenic activity of magnesium-based degradable metals, *Biomater. Sci.*, 9, 2021, 807–825.

Zhu R., Zhang J., Chang C., Gao S., Ni N., Effect of silane and zirconia on the thermal property of cathodic electrophoretic coating on AZ31 magnesium alloy, *J. Magnes. Alloys.*, 1, 2013, 235–241.

Zhu Y., Zheng L., Liu W., Qin L., Ngai T., Poly(L-lactic acid) (PLLA)/MgSO$_4$·7H$_2$O composite coating on magnesium substrates for corrosion protection and cytocompatibility promotion, *ACS Appl. Bio Mater.*, 3, 2020, 1364–1373.

18 Physical and Chemical Vapour Deposition Coatings

Akeem Yusuf Adesina and Nasirudeen Ogunlakin

CONTENTS

18.1 INTRODUCTION

Magnesium and its alloys are promising lightweight materials that are highly desirable in many applications, especially where energy consumption and conservation are crucial. Magnesium has many unique properties, besides its easy castability, excellent hot formability and recyclability, and good machinability, its low density of $1.74 g/cm^3$, which is about two-thirds the density of aluminum, two-fifth the density of titanium, and one-fourth the density of steel (Manoj and Nai Mui, 2011), makes it very attractive to the industries. In addition, it has a high strength-to-weight ratio, sound-damping capabilities, and good thermal and electrical conductivity (Hassan et al., 2015). It is a good choice for applications where weight is a concern, such as in the aerospace and automotive industries. Furthermore, magnesium is abundant in both seawater and the earth's crust. In seawater, it is the third-most-dissolved mineral with $1.1 kg/m^3$ availability, and it is the sixth-most-abundant element with 2% mass density in the earth's crust (Neite et al., 2006).

Since its commercial exploitation began in 1886, magnesium has found applications in a variety of industries. One of the main applications of magnesium in the automotive industry is in the production of vehicle parts. Magnesium is used in the manufacturing of vehicle/car parts, such as

DOI: 10.1201/9781003319856-21

the wheels, steering wheels, gearbox components, etc. Magnesium is favored for the production of these parts because of the metal's low weight and high strength-to-weight ratio, which is important in the reduction of the overall weight of the vehicle and thereby improves fuel efficiency and performance (Kainer and von Buch, 2003; Kim and Han, 2008). In the aerospace industry, magnesium is widely used in the thrust reversals, gearboxes, engines, and transmission casings of both military and civilian helicopters and aircraft (Friedrich and Mordike, 2006). In the electronics industry, it is used in a variety of applications; one of the main uses of magnesium in electronics is in the production of cell phones, laptops, and tablet casings. Magnesium alloys are often used in the casing of these devices because they are strong, light, and able to dissipate heat effectively (Ogunlakin, 2016). Magnesium alloys are used in the production of medical implants, such as pacemakers and stents, because of their biocompatibility and ability to dissolve gradually in the body (Farraro et al., 2014; Hassan et al., 2017). In addition, Mg is also a potential energy storage material because of its high energy density and low cost. It can store hydrogen with a significant capability of reversibility (Liu et al., 2014).

Despite the huge benefits and potentials of magnesium and its alloys, its wider adoption and application have been limited due to some of its undesirable inherent properties, such as limited strength, fatigue and creep resistance, low stiffness, and limited ductility. Furthermore, magnesium is relatively reactive, which can make it susceptible to corrosion and degradation when exposed to certain chemicals or aqueous environments (Olalekan et al., 2022). Moreover, magnesium alloys demonstrate poor surface properties such as high chemical reactivity, low hardness, and wear resistance. It exhibits high corrosion and wear rates in several environments, making its wide adoption and utilization in many applications very challenging.

The attractive potentials and huge advantages of magnesium have thus led to several efforts and investigations toward enhancing its corrosion and wear resistance properties. Two main approaches are often considered to improve the corrosion and wear resistance of magnesium alloys: alloying/strengthening and surface modification. Developing new magnesium alloys that contain elements such as aluminum, zinc, manganese, and rare earth metals can improve the corrosion and wear resistance (Westengen, 2001), However, major alloying elements further deteriorate some intrinsic properties of magnesium such as ductility, creep, and impact strength (Somekawa, 2019). Surface modification, on the other hand, involves the application of a high corrosion resistance material to magnesium to serve as a protective barrier between the magnesium substrate and the environment.

Two main approaches are usually employed in the surface modification of magnesium: ion implantation and coating. Ion implantation involves the acceleration of another element of higher corrosion resistance towards the surface of magnesium up to a depth of approximately 10 nanometres. Success has been recorded in improving the corrosion resistance of magnesium through titanium ion implantation (Wu et al., 2020). However, the corrosion resistance cannot be increased significantly because of the presence of active magnesium on the surface. The corrosion and wear resistance properties of Mg and its alloys can be improved significantly by the use of protective coatings that possess high corrosion- and wear-resistance properties without affecting the intrinsic properties of magnesium. Coatings on magnesium are a potentially effective way to significantly improve surface properties and widen their application. The coating provides a barrier between the magnesium and its environment; hence, the Mg is not exposed to the environment and is protected from corrosion, wear, and other damages. However, it is of utmost importance that the coating material is deliberately designed and applied to ensure its full compatibility with the magnesium substrate.

18.2　COATING SELECTION CRITERIA FOR MAGNESIUM ALLOYS

The requirements for coatings on magnesium alloys are shown in Figure 18.1, and these requirements will depend on the intended application of the magnesium alloy and the desired properties.

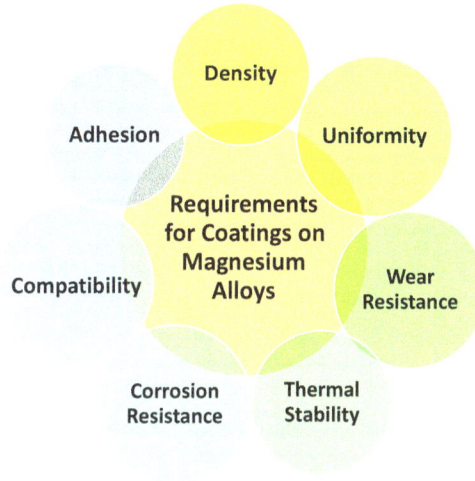

FIGURE 18.1 Requirements for coatings on magnesium and its alloys.

Generally, for improved surface properties, effective protection, and enhanced performance, the coating on magnesium substrate should meet the following requirements:

i **Uniformity:** The uniformity of a coating on magnesium is important for several reasons. A uniform coating can provide better protection and performance for the magnesium alloy, as the entire surface is covered with a consistent layer of coating material. A non-uniform coating leaves areas of the magnesium alloy exposed to the environment, thereby reducing the effectiveness of the coating. To achieve a uniform coating on magnesium, it is important to carefully control the coating process and optimize the deposition parameters. This can involve using a coating method that is suitable for the intended coating material and the properties of the magnesium alloy, as well as carefully controlling the coating temperature, pressure, and power. It may also be necessary to preheat and optimize the surface preparation and cleaning methods to ensure that the coating adheres uniformly to the magnesium alloy surface (Fan et al., 2016).

ii **Compatibility with magnesium:** The coating material should be compatible with magnesium and should not react with the magnesium alloy or degrade its properties. This can be achieved by using coating materials that have a similar crystal structure and chemical composition to magnesium or by using inert materials that do not react with magnesium (Gray and Luan, 2002).

iii **Adhesion:** Adhesion is an interfacial property that enables the coating to remain attached to the substrate material even after the application of tangential or normal load, and it is an important property of coatings on magnesium alloys. Good adhesion of a coating to a magnesium alloy substrate can improve the performance and durability of the coated magnesium alloy, while poor adhesion can lead to delamination or peeling of the coating, reducing the performance and service life of the coated magnesium alloy (Lu et al., 2012). Several factors can affect the adhesion of a coating to a magnesium alloy substrate. One of the key factors is the compatibility of the coating material with the magnesium alloy. The coating material should be able to form a strong metallurgical and/or physical bond with the magnesium alloy to ensure that the coating remains attached to the substrate. Another important factor is the surface preparation of the magnesium alloy before coating. The surface of the magnesium alloy should be clean and free of contaminants, such as oil, dirt, or oxide layers, to ensure that the coating adheres properly to the substrate. It may also be necessary to use a surface treatment, such as etching or roughening, to improve the adhesion of the coating to the magnesium alloy (Fan et al., 2016). The coating process and deposition parameters can also affect the adhesion of a coating to a magnesium alloy substrate. The coating method should be suitable for the intended coating material and the properties of the magnesium alloy, and the coating temperature, pressure, and power should be carefully controlled to ensure that the coating adheres uniformly to the substrate. Overall, good adhesion of a coating to a magnesium alloy substrate is important for the performance and durability of the coated magnesium alloy. Proper selection of the coating material, surface preparation, coating process, and deposition parameters can help to ensure good adhesion of the coating to the magnesium alloy.

iv **Denseness:** The coating should be dense and not porous. Porosity in coatings on magnesium alloys can have several negative effects. Firstly, porosity can reduce the strength and durability of the coating, as the voids or pores can act as stress concentration points and reduce the overall strength of the coating. This can reduce the performance and service life of the coated magnesium alloy in applications where the coating is subjected to mechanical stresses or impacts. Secondly, porosity in coatings on magnesium alloys can reduce the corrosion and wear resistance of the coating, as the voids or pores can provide pathways for the ingress of corrosive agents into the coating and the magnesium alloy substrate. This can accelerate the corrosion of the coated magnesium alloy, reducing its performance and durability in corrosive environments. Thirdly, porosity in coatings on magnesium alloys can affect the adhesion of the coating to the magnesium alloy substrate, as the voids or pores can provide points of weakness at the interface between the coating and the substrate. This can lead to delamination or peeling of the coating, reducing the performance and durability of the coated magnesium alloy.

To minimize porosity in coatings on magnesium alloys, it is important to carefully control the coating process and optimize the deposition parameters. This can involve using an appropriate coating deposition method and controlling deposition parameters, such as the coating temperature, pressure, and power. It may also be necessary to optimize the surface preparation and cleaning methods to ensure that the coating adheres uniformly to the magnesium alloy surface, with minimal voids or pores.

v **Corrosion resistance and galvanic couple:** The coating should provide corrosion resistance to the magnesium alloy and should prevent the ingress of corrosive agents, such as moisture and chemicals. This can be achieved by using corrosion-resistant materials, such as silicon carbide or aluminum oxide, or by applying a multi-layer coating with different corrosion-resistant layers. Galvanic corrosion can occur in coatings on magnesium alloys if the coating material is different from the magnesium alloy substrate and if the coating and the substrate are in contact with an electrolyte. The coating and the substrate will have different electrical potentials, and the electrolyte will facilitate the transfer of ions between the two materials. Magnesium, being the least noble metal, will corrode severely in galvanic connection with most metal coatings. To prevent galvanic corrosion of coatings on magnesium alloys, the coating material should be compatible with the magnesium alloy substrate. The coating material should have a similar electrical potential to the magnesium alloy to minimize the difference in potential between the coating and magnesium (Zhang et al., 2016). It is also important to prevent the ingress of electrolytes into the interface between the coating and the substrate to prevent the transfer of ions between the two materials. In some cases, it may be necessary to use a protective layer or barrier between the coating and the magnesium alloy substrate to prevent galvanic corrosion. The protective layer should have a low electrical conductivity and a similar electrical potential to the coating and the substrate to minimize the potential difference between the three materials. This can effectively isolate the coating and the substrate from each other, preventing the transfer of ions and the occurrence of galvanic corrosion.

vi **Wear resistance:** The coatings on magnesium alloys should induce surface with to improve mechanical and tribological properties beyond that of the magnesium substrate, especially for wear and friction reduction applications. The coatings should exhibit improved hardness and lower coefficient of friction to enhance the wear and friction resistance, respectively. Besides tailoring the process conditions, these properties can be achieved by systematic selection of the coating material as well as the coating design strategy. Multi-layer coatings with alternating functionality, graded coatings, hierarchical structure coatings, and nanocomposite coating designs are typical examples that have been implemented on magnesium (Hoche et al., 2007, 2014).

vii **Thermal stability:** The coating should have good thermal stability and should not degrade or lose its properties when exposed to high temperatures. This is particularly important for magnesium alloys that are used in high-temperature applications, such as automotive or aerospace components. The coating material should have a high melting point and a similar thermal expansion coefficient as that of the substrate to avoid thermal mismatch and should be able to withstand the thermal cycling and thermal shock encountered in these applications.

18.3 VACUUM DEPOSITION TECHNIQUES

Vacuum deposition (VD) is a group of processes that are used to deposit thin films or layers of materials (elements, alloys, oxides, etc.) via atom-by-atom or molecule-by-molecule on surfaces at low pressure (vacuum). The thickness ranges from one atom up to millimeters, and as a result, vacuum deposition is also referred to as thin film deposition. Several technologies have been proposed for the application of coating materials on magnesium alloys, e.g. conversion coating, electrochemical plating, thermal spraying, etc. (Altun and Sen, 2007). However, VD techniques stand out in that the vacuum environment reduces contaminants, increases the mean free path for collisions of atoms and high-energy ions for better deposition properties, and with the flexibility to control the process parameters such as gas and vapour composition. Furthermore, with VD techniques, thin films and coatings with high purity, tailored properties, thermal stability, and that are economical, can be deposited. Most VD processes are also sustainable and environmentally friendly, they can deposit a wide variety of inorganic materials, and some organic materials can also be deposited on a diverse and wide group of surfaces and substrates. Generally, the VD techniques are divided into two (2) main categories depending on the vapour and/or plasma source. In physical vapour deposition (PVD), the source of the vapour is liquid or solid, while chemical vapour deposition (CVD) uses chemical vapour. Figure 18.2 shows a typical classification of the various type of variants of physical and chemical vapour deposition techniques.

18.4 PHYSICAL VAPOUR DEPOSITION (PVD) COATINGS FOR Mg ALLOYS

Physical vapour deposition is a technique for modifying the surface of a substrate by applying a thin film on it. It involves the vapourization of a solid material in a vacuum or low-pressure gas and the subsequent deposition of the same on the substrate. PVD can provide pure and high-performance coatings compared to other methods. Coating thickness in the range of angstroms to millimeters can be achieved with the PVD technique. PVD coatings are commonly used on magnesium alloys for various applications, including automotive, aerospace, and medical industries. The specific materials and methods used for PVD coating on magnesium may vary depending on the desired properties and the intended application.

Coating deposition on magnesium substrate can be achieved through several PVD techniques such as electron-beam deposition, cathodic-arc deposition, and magnetron sputtering. These techniques vary in the method by which the coating target is vapourized and the method by which the vapour is delivered to the magnesium substrate.

Thin Film Deposition Techniques

PHYSICAL

CHEMICAL

Sputtering	Evaporation	Gas Phase	Liquid Phase
Glow discharge DC sputtering	Vacuum Evaporation	Chemical vapour Deposition	Electro-deposition
Triode sputtering	Resistive heating Evaporation	Laser Chemical Vapour deposition	Chemical bath deposition (CBD) / Arrested Precipitation Technique (APT)
Getter sputtering	Flash Evaporation	Photo-chemical vapour deposition	Electro less deposition
Radio Frequency sputtering	Electron beam Evaporation	Plasma enhanced vapour deposition	Anodisation / Molecular Beam Epitaxy / Liquid phase Epitaxy
Magnetron & Ion Beam sputtering	Laser Evaporation	Metal-Organo Chemical Vapour Deposition (MOCVD)	Sol- gel / Spin Coating / Spray-pyrolysis technique (SPT)
A.C. Sputtering	Arc & R. F. Heating	Atomic Layer Epitaxy (ALE)	Ultrasonic (SPT) / Polymer assisted deposition (PAD)

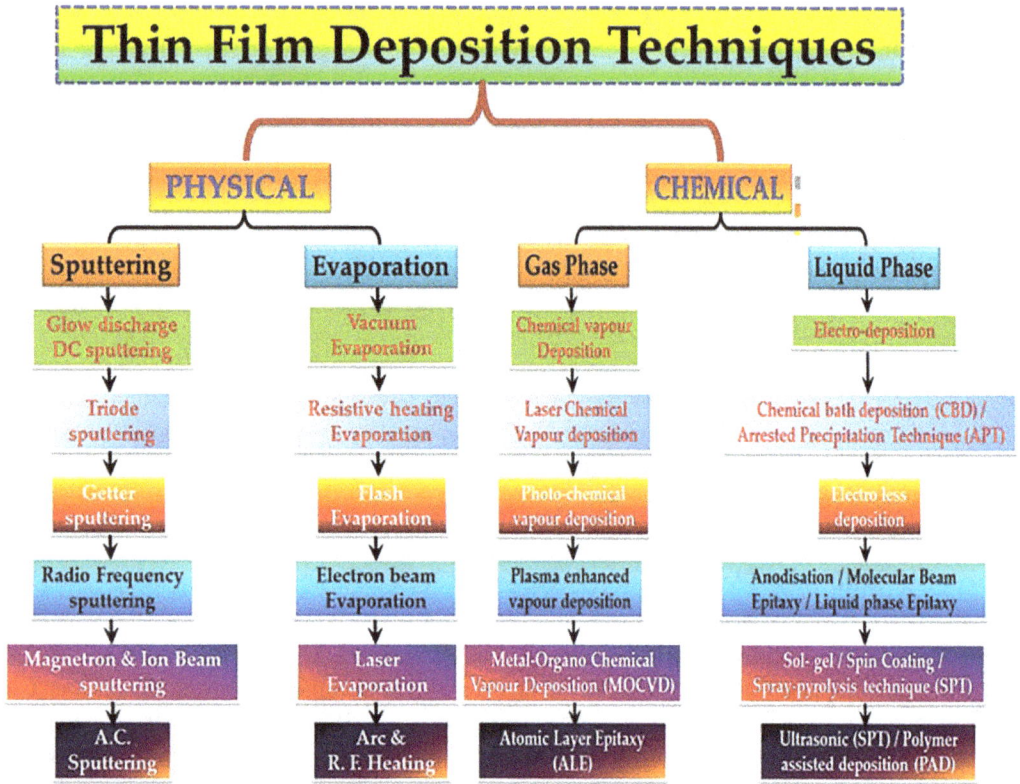

FIGURE 18.2 Classification of vapour (thin film) deposition techniques (Das and Sahoo, 2020). *Reproduced with permission from the authors.*

18.4.1 PVD TECHNIQUES

18.4.1.1 Advanced Ion Beam–Assisted Electron Beam PVD

Advanced assisted electron beam physical vapour deposition (AAEB-PVD) is a PVD process that utilizes a beam of ions to assist in the deposition of thin layers of material onto a surface. In an AAEB-PVD process, a beam of ions from an ion source (typically Ar) is directed to the substrate. The ion beams arrive alongside the desired coating material from an electron beam evaporator, as shown in Figure 18.3. The ions impart energy to the atoms of the coating material, thereby increasing their mobility on the surface of the substrate. The adhesion, density, uniformity, and grain structure of the deposited coating are greatly improved with the increased energy and surface mobility of the atoms. The beam can also be incorporated with oxygen ions and nitrogen ions when depositing oxides or nitrides. AAEB-PVD is commonly used to deposit dielectric materials, such as silicon dioxide and aluminum oxide, onto semiconductor wafers and other substrates. It is also used to deposit other materials, such as metals and ceramics, onto various surfaces. Carbon nitride (CN) coatings have been deposited successfully on magnesium alloy substrate using ion beam–assisted deposition, the coating led to an increase in the hardness and elastic modulus of the magnesium substrate by 90% and 80% respectively (Yang et al., 2008). AAEB-PVD is known for its high precision and control, allowing for the deposition of highly uniform and conformal coatings.

18.4.1.2 Plasma-Assisted Electron Beam PVD

Plasma-assisted electron beam physical vapour deposition (PAEB-PVD) is a type of physical vapour deposition (PVD) process that uses both a plasma and an electron beam to deposit thin layers of

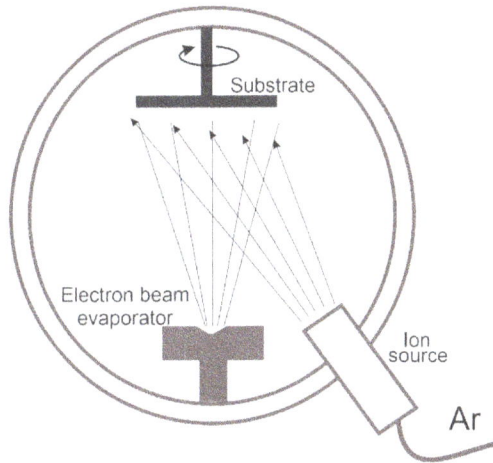

FIGURE 18.3 Schematic diagram of AAEB-PVD deposition.

material onto a surface. In a PAEB-PVD process, a gas or vapour containing the desired material is introduced into a reaction chamber, where it is ionized by a plasma to form reactive species. These species are then bombarded by an electron beam, which helps to improve the adhesion and uniformity of the deposited coating. The resulting plasma and electron beam help to deposit a thin and uniform layer of the desired material onto the surface of the substrate. In high-temperature applications, PAEB-PVD is frequently used to produce thermal barrier coatings to shield the substrate from overheating and creep. PAEB-PVD coatings have improved oxidation and erosion resistance (Prater and Courtright, 1987) and improved adhesion (Liebert and Stepka, 1977) and can produce coatings with columnar microstructures with increased strain tolerance (Liebert and Stepka, 1976).

18.4.1.3 Electromagnetically Steered Cathodic Arc Evaporation

In cathodic arc deposition (CAE), by applying a high-current, low-voltage arc on the surface of a target material a high-velocity jet of the vapourized cathode material is produced. The process involves the creation of a cathodic arc by passing a high-voltage electric current between a cathode and an anode in a reaction chamber. The resulting plasma is then directed onto the surface of the substrate by an electromagnet, which helps to deposit a thin and uniform layer of the desired coating material. The electromagnet also allows for precise control over the deposition process, allowing for the creation of complex and custom-shaped coatings. The performance of the CAE process is enhanced by the introduction of a steering variable magnetic field to steer the arc spot and prevent arc discharge at high current density.

Electromagnetically steered cathodic arc evaporation (ESCAE) as shown in Figure 18.4 involves the introduction of a steering variable magnetic field into the CAE process: the variable magnetic field steers the arc spot and prevents the arc from discharging at high current density (Lindfors et al., 1986; Sanders, 1989). ESCAE is commonly used to deposit metals, ceramics, and diamond-like carbon onto various surfaces. The process is usually associated with microdroplets, which, if they adhere to the growing films, will lead to the degradation of the film (Anders et al., 1993; Boxman and Goldsmith, 1989). The droplets can be prevented by placing a superconductor shield on the substrate to protect it from the incoming droplets while the depositing ions are transported behind the shield (Aksenov et al., 1999; Takikawa et al., 2003). The cathodic arc PVD process has been successfully used to deposit TiN coatings on AZ91 magnesium alloys for improved corrosion resistance (Altun and Sinici, 2008).

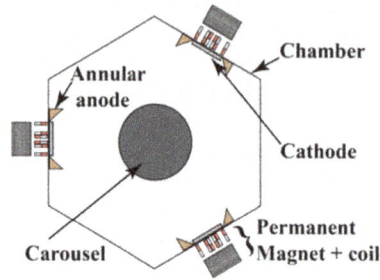

FIGURE 18.4 Top-view schematic representation (not to scale) of the ESCAE deposition setup. *Reproduced with permission from (Chaar et al., 2019); © 2019 by the authors. Licensee MDPI, Basel, Switzerland.*

18.4.1.4 Closed-Field Unbalanced-Magnetron Sputtering

Low ionization efficiency, high heat output, and low deposition rate are the main characteristics of the conventional sputtering process. The development of the magnetron configuration has overcome these limitations through effective control of the secondary electrons to enhance the plasma densification, which subsequently led to a high deposition rate (Sproul et al., 1994).

The magnetron sputtering process consists of a cathode made up of a magnetic material and a non-magnetic material placed in a reaction chamber, along with the substrate to be coated as shown in Figure 18.5. A high-voltage electric current is passed between the cathode (target) and the substrate, creating a plasma in the space between them. The plasma then sputters or ejects, atoms of the cathode material onto the surface of the substrate, depositing a thin layer of the desired material. The magnetic field in the cathode helps to improve the uniformity and adhesion of the deposited coating, making it smoother and more durable. In conventional or balanced magnetron sputtering, the limitation in the control of the plasma makes it difficult to produce fully dense films on large complex substrates (Musil and Kadlec, 1990). The unbalanced magneton system produces a more uniform film due to the enhanced control and directionality of the magnetic field (Savvides and Window, 1986). Multiple magnetron systems have been introduced to further enhance the deposition rate. In multiple magnetron systems, the field lines between adjacent magnetrons can be linked by configuring the magnetic arrays with opposing magnetic polarities, creating a close field where the substrate is covered in a high-density plasma region (Liu, 2017). Closed-field unbalanced-magnetron sputtering (CFUBMS) is an extremely versatile technique for producing well-adhered, high-quality films. Each target can be of different materials to achieve multi-component coatings or coatings with a gradient layer.

18.4.2 PVD Coating Selection Criteria

The selection of an appropriate PVD coating for magnesium alloys depends on the specific properties and performance requirements of the coating, as well as the intended application of the magnesium alloy. Some common factors to consider when selecting a PVD coating for magnesium alloys include the following:

i **The type of material to be used in the coating:** Different materials have different properties and functions, so it is important to choose a material that is suitable for the intended application. Generally, oxides, nitrides, borides, and carbides of transition metals are employed as monolithic coatings in conventional PVD coatings to enhance corrosion and wear resistance (Holmberg and Matthews, 2009). Producing ever-harder coatings has long been a top priority for many tribologists. However, these hard coatings are not suitable for magnesium alloys because of the high mismatch in the elastic modulus between the

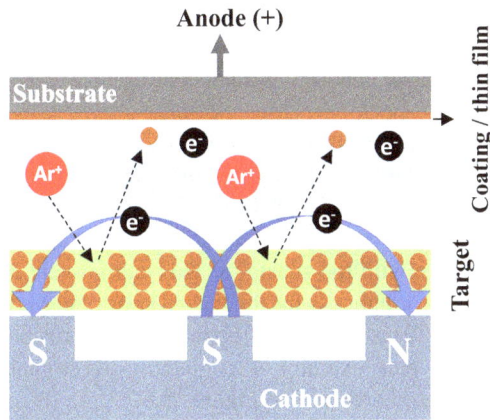

FIGURE 18.5 Schematic diagram of the closed-field (unbalanced) magnetron sputtering setup.

coating and the magnesium substrate. Magnesium has low strength and modulus compared with these coating materials; hence, when coated with hard materials, the magnesium will be under excessive tension that can lead to the failure of the coating in tribological applications. The importance of employing the hardness-to-elastic modulus (H/E) ratio as a measure of coating mechanical durability was brought up by Leyland and Matthews (Leyland and Matthews, 2000), who also proposed the idea of metallic nanocomposite coatings (Leyland and Matthews, 2004), where the issue associated with the modulus mismatch with magnesium substrate can be addressed. The properties of the metallic nanocomposite coatings can be tuned to achieve the moderately high hardness and low elastic modulus as desired for use with magnesium-based substrates.

ii **The thickness of the coating:** The thickness of a coating on magnesium can affect the properties and performance of the coated magnesium alloy. In general, thicker coatings can provide better protection against corrosion and wear, as they have a greater ability to absorb and dissipate the forces that can damage the surface of the magnesium alloy. However, thicker coatings can also increase the weight and cost of the coated magnesium alloy, and they can potentially impair the mechanical properties of the alloy, such as its strength and ductility. Therefore, it is important to carefully consider the desired thickness of a coating on magnesium to balance the benefits and drawbacks of the coating. The optimal coating thickness will depend on the specific application and requirements of the coated magnesium alloy.

iii **The adhesion of the coating:** The adhesion of a coating can affect the performance and durability of the coated magnesium alloy. A coating with good adhesion will remain firmly attached to the surface of the magnesium alloy, providing effective protection against corrosion and wear. However, a coating with poor adhesion may peel, flake, or otherwise detach from the surface of the magnesium alloy, reducing its effectiveness and potentially causing damage to the alloy. Therefore, it is important to select a coating with good adhesion to ensure the long-term performance and reliability of the coated magnesium alloy. The adhesion of a coating can be influenced by several factors, such as the composition and surface roughness of the coating, the surface preparation of the magnesium substrate, and the curing conditions of the coating.

18.4.3 DESIGN CONCEPTS OF NANOCOMPOSITE PVD COATINGS

Nanocomposite PVD coatings are coatings that are composed of multiple materials at the nanoscale to provide improved properties and performance for the coated magnesium substrate. The design of nanocomposite PVD coatings involves selecting the materials to be used and determining their proportions and arrangement in the coating. Some common design concepts for nanocomposite PVD coatings include multi-phase or layer coatings, graded coatings, and hierarchical coatings.

i **Multi-phase or multi-layer coatings** involve the use of multiple phases or layers in the coating, with each phase or layer providing specific properties or functions (Stueber et al., 2009). For example, a nanocomposite PVD coating can have a hard and wear-resistant layer on the surface, followed by a corrosion-resistant layer underneath, and a ductile or elastic layer at the interface with the substrate. This can provide improved wear and corrosion resistance, as well as toughness for the coated substrate. Multilayer coatings of AlN and TiN on magnesium AZ91 alloys led to the increased corrosion resistance of the alloy (Altun and Sen, 2006). A typical example of toughening and strengthening mechanisms on ceramic multilayer coatings is provided in Figure 18.6.

ii **Graded coatings** have a gradual variation in the composition of the material, allowing for a smooth transition between different properties (Figure 18.7). There is a gradual change in the composition or structure from the surface to the interface with the substrate (Kamalan Kirubaharan and Kuppusami, 2020). Graded coatings on magnesium alloys can provide improved properties and performance for the coated magnesium alloy, compared to uniform coatings with a single composition or structure (Yue and Li, 2008).

iii **Hierarchical coatings** are a type of coating that is arranged in multiple levels, each with its unique microstructure and function. These coatings can be applied to magnesium alloys to

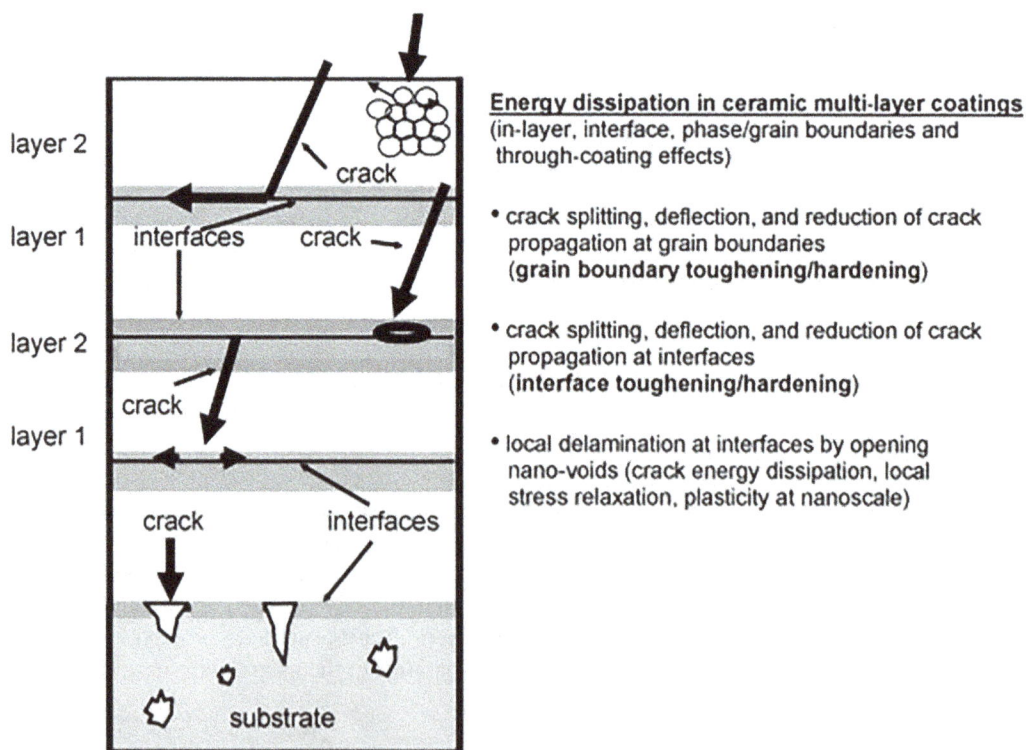

FIGURE 18.6 Schematics of toughening and strengthening mechanisms in ceramic multi-layer coatings (Holleck and Schier, 1995; Stueber et al., 2009). *Reproduced with permission from (Stueber et al., 2009) © 2008 Elsevier B.V.*

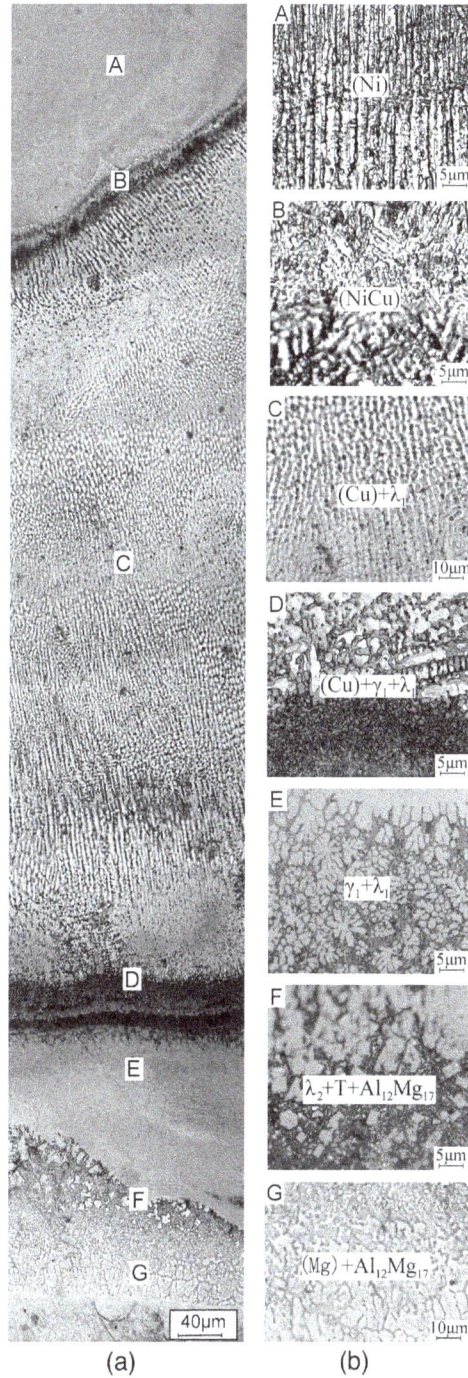

FIGURE 18.7 Microstructure of a graded coating on magnesium alloy. *Reproduced with permission from (Yue and Li, 2008) © 2007 Elsevier B.V.*

provide multiple levels of protection and performance. For example, a hierarchical coating on magnesium may consist of a base layer of a corrosion-resistant material, such as aluminum oxide, followed by a second layer of a harder material, such as titanium carbide, and a third layer of a lubricious material, such as diamond-like carbon as shown in Figure 18.8. This multi-layered structure allows the hierarchical coating to provide corrosion resistance,

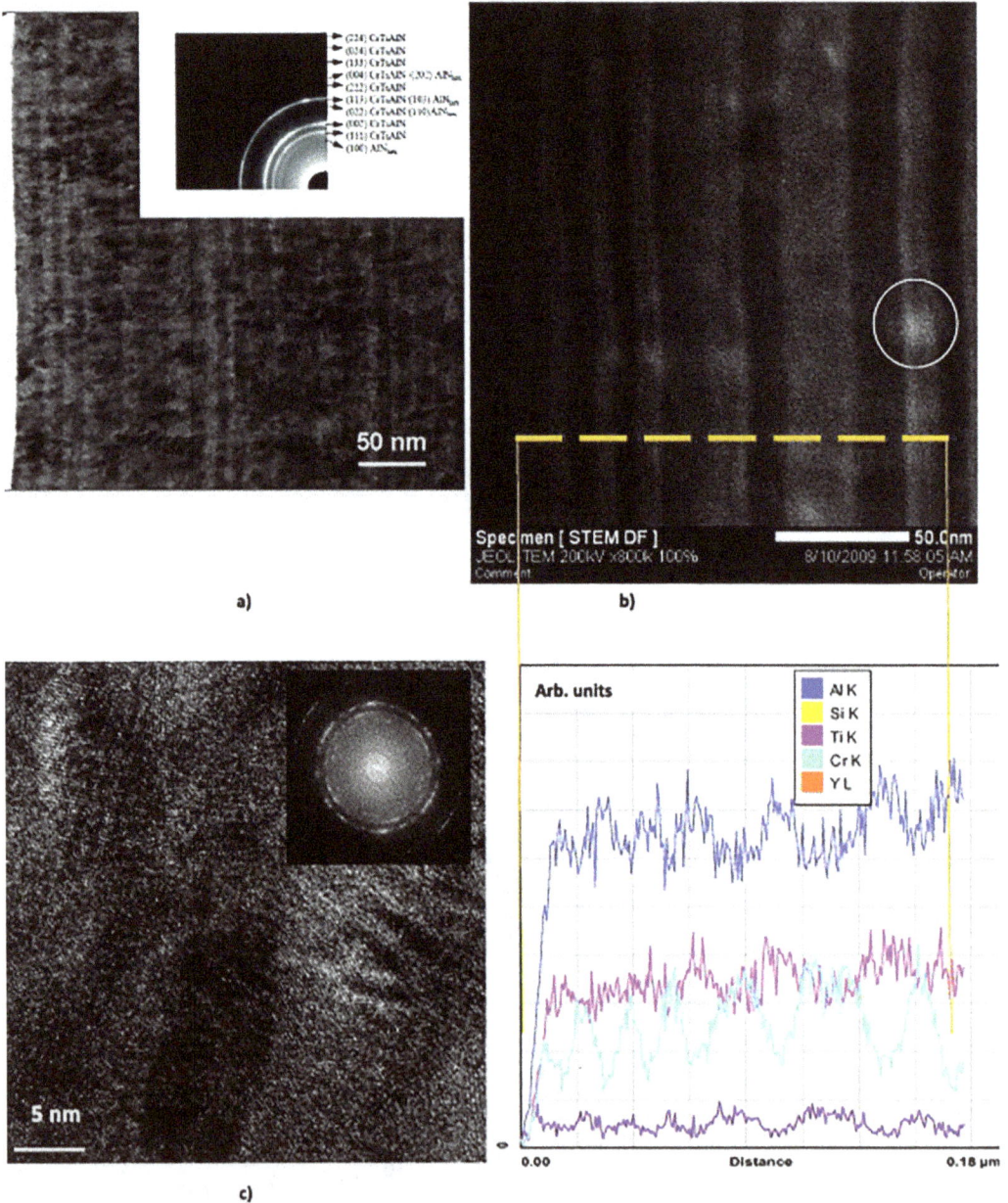

FIGURE 18.8 (a) TEM image of the $Ti_{0.2}Al_{0.55}Cr_{0.2}Si_{0.03}Y_{0.02}N/Ti_{0.25}Al_{0.65}Cr_{0.1}N$ coating (cross-sectional view) with SAED pattern inset, (b) HAADF-STEM image and energy-dispersive spectroscopy profile of the coating, and (c) TEM image with SAED pattern inset. *Reproduced with permission from (Fox-Rabinovich et al., 2012) © 2012 National Institute for Materials Science.*

wear resistance, and low friction, all in a single coating (Fox-Rabinovich et al., 2012). These coatings can provide improved performance and functionality compared to single-layer coatings, making them a promising option for various applications.

18.4.4 Diamond-Like Carbon

Diamond-like carbon (DLC) films are a type of thin film coating made of carbon-based materials that exhibit some of the properties of diamonds. They possess excellent mechanical and tribological properties (Zia, 2020). DLCs are also chemically inert and biocompatible, making them a good prospect for magnesium in biomedical applications. they can be deposited on metals and non-metal substrates. They can be produced using PVD processes such as ion beams (Puzikov and Semenov, 1991), cathodic arc (Vetter, 2014), and sputtering (Yeldose and Ramamoorthy, 2008). The coating is formed when ionized and decomposed carbon or hydrocarbon species land on the surface of a substrate with energy in the range of 10 to 300eV. DLC films are known for their high hardness, low friction, and excellent wear resistance, making them suitable for a variety of applications.

Some common properties of DLC films include:

- **High hardness:** DLC films are typically harder than most other PVD coatings, with a hardness on the order of 10 to 20 GPa. This makes them ideal for applications in which wear resistance and durability is important, such as in cutting tools, automotive parts, and medical implants.
- **Low friction:** DLC films have a low coefficient of friction, typically in the range of 0.05 to 0.2. This makes them useful for reducing friction and wear in sliding or rotating components, such as bearings, gears, and seals.
- **Excellent wear resistance:** DLC films have excellent wear resistance, with a low wear rate and high resistance to scratching, abrasion, and other forms of mechanical damage. This makes them suitable for applications in which the coated component will be subjected to high levels of wear and tear, such as in the aerospace and automotive industries.
- **High thermal stability:** DLC films are highly stable at high temperatures and can maintain their properties and performance up to temperatures of 600 to 1000°C. This makes them suitable for applications in which the coated component will be exposed to high temperatures, such as in engine parts, exhaust systems, and heat exchanger tubes.

Despite the many desirous properties of DLC coatings, their wider application is limited by their high residual stress and low toughness (Zia et al., 2019). Hence, the need to further improve the properties of DLC to meet specification applications. Several elements have been doped or deposited simultaneously with DLCs to achieve desired features. Cr, Ni, Mo, and Ta improve the tribological performance of DLC (Khun et al., 2016; Tagami et al., 2016; Tang et al., 2014); Cu and Ti for improved adhesion (Ma et al., 2012), N and Si for improved thermal stability (Lee et al., 2016), and Ag and Fl to boost their biocompatibility (Bociaga et al., 2015).

DLC coatings on magnesium alloys have been achieved using a low-temperature plasma-assisted PVD technique (Iriyama and Yoshihar, 2011), the result of which led to an improved formability of the Magnesium alloy. Similarly, the corrosion resistance of AZ31 magnesium alloy has been improved with the deposition of Si-doped DLC film using the sputtering technique (Choi et al., 2007).

18.4.5 Future Trends

The future trend in PVD coatings for magnesium alloys is likely to focus on the development of new coating materials and processes that can improve the performance and durability of magnesium alloys in a wide range of applications. This may include the development of new coating materials with improved corrosion resistance, wear resistance, and other desirable properties, as well as the development of new PVD coating processes that can provide more precise and controllable coatings. In addition, there is likely to be a focus on reducing the cost and environmental impact of PVD coating processes, as well as improving the sustainability and recyclability of coated magnesium alloys. Overall, the future trend in PVD coatings for magnesium alloys is likely to be driven by the need for advanced materials and technologies that can support the growing demand for lightweight and high-performance materials in various industries.

18.5 CHEMICAL VAPOUR DEPOSITION (CVD) COATINGS FOR Mg ALLOYS

The CVD process involves reacting a gaseous phase with a heated substrate to create a coating. It involves exposing the substrate to a more volatile precursor, which decomposes on it to form a thin coating. The method can be used to apply coatings of various compositions consisting of metallic, ceramic, intermetallic, or refractory materials (Sun et al., 2021). CVD offers the advantage of depositing coatings of a variety of compositions at or near theoretical density and chemical compositions. The primary distinction between CVD and PVD is that the deposition process in CVD is multidirectional, whereas in PVD, it is a line-of-site impingement kind of deposition (Behera et al., 2020). The process of CVD has been widely employed to create coatings of excellent quality. The underlying procedures have great levels of flexibility and can produce coatings with uniform thickness and little porosity. It is possible to meet the requirements for quality and properties of a coating (chemical, physical, tribological, etc.) for a variety of applications by properly selecting the material and process parameters, which include the composition and mixture of the reaction gas, the substrate material, the temperature and pressure of the chamber, and the flow of gas. They are also used in the production of cutting tools, medical implants, and other industrial and consumer products. In the conventional CVD process, heat energy is used to provide the high temperature required for the decomposition of the gas molecules, which limits the coating material and substrates that can be coated (Ohring, 2002). Alternative sources of energy such as plasma and photons have been widely used to widen the application of the CVD process. Several CVD techniques have been employed to deposit coatings on magnesium and its alloys. These technologies can vary in their requirements for temperature, pressure, and other process conditions, as well as the materials and surfaces they can be used with. They include low pressure, fluidized bed, plasma enhanced, and atomic-layer CVD.

18.5.1 Low-Pressure CVD

Low-pressure CVD, or LPCVD, is a type of CVD process that can be used to deposit thin layers of material onto the surface of magnesium alloys. In a LPCVD, a gas or vapour containing the desired material is introduced into a reaction chamber, where it is heated and allowed to react with the surface of the magnesium alloy. The low pressure in the chamber allows the gas or vapour to expand and cover a larger area of the substrate, resulting in a uniform and smooth coating. LPCVD is commonly used to deposit silicon-based materials, such as silicon dioxide and silicon nitride, onto magnesium alloys. These coatings can improve the corrosion resistance and wear resistance of the magnesium alloys, making them more durable and longer-lasting. LPCVD is also used to deposit other materials, such as metals and ceramics, onto magnesium alloys for various applications.

18.5.2 Fluidized Bed CVD

Fluidized bed chemical vapour deposition (FBCVD) is a type of CVD process in which a gaseous precursor material is coated on a powder, particles, or solid substrate. In an FBCVD process, the substrate is suspended in a fluidized bed, which is typically made up of a gas (such as air) and small particles (such as beads or granules) that are agitated to create a fluid-like state. This allows the gaseous precursor to permeate and react to form a thin film of nanomaterials that are evenly deposited on the substrate. Al_2O_3 (Baiocco et al., 2018) and Al (Christoglou et al., 2004) are examples of materials that have been successfully coated on magnesium using the FBCVD.

18.5.3 PLASMA-ENHANCED CVD

Plasma-enhanced chemical vapour deposition (PECVD) uses plasma to enhance the deposition of thin layers of material onto the surface of magnesium alloys. In the plasma-enhanced or plasma-assisted CVD process, plasma is utilized where the ionized gases are generated by electrical discharges, thereby reducing the process temperature. Deposition of layers that cannot resist high temperatures and coatings on temperature-sensitive substrates are made possible by PECVD's utilization of lower temperatures. Additionally, because the precursor activated by the plasma is more reactive and the rate of arrival of the ionizing precursor may be controlled by a bias voltage, the deposition rate in PECVD is typically higher, and coating characteristics are simpler to control (Förch et al., 2007). PECVD coatings have good adhesion to a variety of substrates, including metals. It can produce a variety of void-free layers with thicknesses ranging from 10 to 1 um. PECVD can be used to create films with a variety of microstructures and compositions that can be used to create functional coatings (Martinu et al., 2010) and robust protective coatings for biocompatible magnesium alloys (Deng, 2019). Many protective coatings, such as organic coatings, silicon carbide, silicon nitride, silicon dioxide, amorphous silicon, and diamond-like carbon, have been deposited using PECVD.

18.5.4 ATOMIC LAYER CVD

Atomic layer chemical vapour deposition (ALCVD) can be used to deposit thin layers of material onto the surface of magnesium alloys. In an ALCVD process, two or more gases are introduced into a reaction chamber, where they react with the surface of the magnesium alloy in a sequence of alternating pulses. Each pulse provides a monolayer of a material, which is then allowed to react and form a stable layer on the surface. By repeating this process, multiple layers of material can be deposited onto the surface, resulting in a thin and uniform coating. ALCVD is commonly used to deposit dielectric materials, such as silicon dioxide and aluminum oxide, onto magnesium alloys. These coatings can improve the corrosion resistance and wear resistance of the magnesium alloy, making it more durable and longer-lasting. ALCVD is also used to deposit other materials, such as metals and ceramics, onto magnesium alloys for various applications. ALCVD is known for its high precision and control, allowing for the deposition of highly uniform and conformal coatings. Coating of TiO_2 nanofilms on magnesium-zinc alloy for biomedical applications has been made possible via ALCVD (Yang et al., 2019).

18.6 CONCLUSIONS AND OUTLOOK

The prospects for PVD and CVD coatings on magnesium alloys are promising, with a growing demand for high-performance materials and coatings in various industries. Several research directions and recommendations can be considered to further advance the development and application of PVD and CVD coatings on magnesium alloys. Some potential research directions and recommendations include:

- Developing new coating materials: One of the key challenges in PVD and CVD coatings on magnesium alloys is the limited availability of coating materials that are compatible with magnesium and can provide the desired properties and performance. Research can be focused on developing new coating materials that can be deposited using PVD and CVD techniques, and that can provide improved corrosion resistance, wear resistance, and biocompatibility for magnesium alloys.

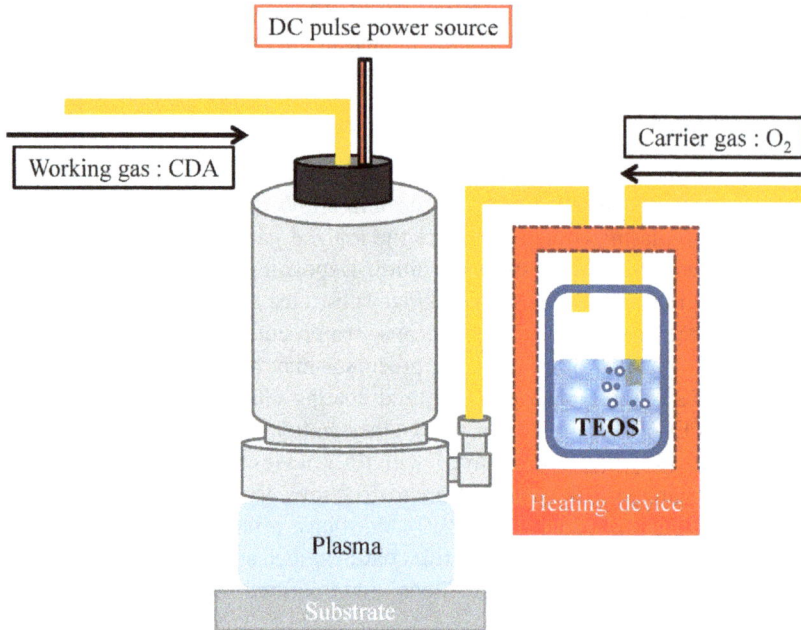

FIGURE 18.9   Schematic diagram of the atmospheric pressure plasma enhanced chemical vapour deposition system. *Reproduced with permission from (Kuo and Chang, 2015) © 2015 Elsevier B.V.*

- Optimizing coating processes: Another important research direction is to optimize the PVD and CVD coating processes to improve the quality and performance of the coated magnesium alloys. This can involve optimizing the deposition parameters, such as the coating temperature, pressure, and power, as well as the substrate preparation and cleaning methods. Research can also be focused on developing new coating methods, such as hybrid PVD/CVD or nanostructured coatings, that can provide improved properties and performance for magnesium alloys.
- Evaluating the long-term performance of coated magnesium alloys: PVD- and CVD-coated magnesium alloys have great potential for use in high-stress and high-temperature applications, where the coating and the substrate will be subjected to severe conditions. Research can be focused on evaluating the long-term performance of coated magnesium alloys, including their corrosion resistance, wear resistance, and mechanical properties, under various environmental and operational conditions. This can provide valuable information on the durability and reliability of coated magnesium alloys in different applications.
- Investigating the environmental impact of PVD and CVD coatings: PVD and CVD coatings on magnesium alloys can potentially have an impact on the environment due to the release of coating materials and byproducts during the coating process. Research can be focused on investigating the environmental impact of PVD and CVD coatings and on developing sustainable coating methods that can minimize the environmental impact of coated magnesium alloys.

Overall, PVD and CVD coatings on magnesium alloys have a bright future, with a growing demand for high-performance materials and coatings in various industries. Further research and development can help to advance the capabilities and applications of PVD and CVD coatings on magnesium alloys.

REFERENCES

Aksenov, I.I., Belous, V.A., Vasil'ev, V.V., Volkov, Y.Y., Strel'nitskij, V.E., 1999. A rectilinear plasma filtering system for vacuum-arc deposition of diamond-like carbon coatings. Diam. Relat. Mater. 8, 468–471. https://doi.org/10.1016/S0925-9635(98)00402-6.

Altun, H., Sen, S., 2006. The effect of PVD coatings on the corrosion behavior of AZ91 magnesium alloy. Mater. Des. 27, 1174–1179. https://doi.org/10.1016/j.matdes.2005.02.004.

Altun, H., Sen, S., 2007. The effect of PVD coatings on the wear behavior of magnesium alloys. Mater. Charact. 58, 917–921. https://doi.org/10.1016/j.matchar.2006.09.007.

Altun, H., Sinici, H., 2008. Corrosion behaviour of magnesium alloys coated with TiN by cathodic arc deposition in NaCl and Na_2SO_4 solutions. Mater. Charact. 59, 266–270. https://doi.org/10.1016/j.matchar.2007.01.004.

Anders, S., Anders, A., Yu, K.M., Yao, X.Y., Brown, I.G., 1993. On the macroparticle flux from vacuum arc cathode spots. IEEE Trans. Plasma Sci. 21, 440–446. https://doi.org/10.1109/27.249623.

Baiocco, G., Rubino, G., Tagliaferri, V., Ucciardello, N., 2018. Al2O3 coatings on magnesium alloy deposited by the fluidized bed (FB) technique. Materials. 11, 94. https://doi.org/10.3390/ma11010094.

Behera, A., Mallick, P., Mohapatra, S.S., 2020. Chapter 13 – Nanocoatings for anticorrosion: An introduction, in: Rajendran, S., Nguyen, T.A., Kakooei, S., Yeganeh, M., Li, Y. (Eds.), Corrosion Protection at the Nanoscale, Micro and Nano Technologies. Elsevier, pp. 227–243. https://doi.org/10.1016/B978-0-12-819359-4.00013-1.

Bociaga, D., Komorowski, P., Batory, D., Szymanski, W., Olejnik, A., Jastrzebski, K., Jakubowski, W., 2015. Silver-doped nanocomposite carbon coatings (Ag-DLC) for biomedical applications – Physiochemical and biological evaluation. Appl. Surf. Sci. 355, 388–397. https://doi.org/10.1016/j.apsusc.2015.07.117.

Boxman, R.L., Goldsmith, S., 1989. Principles and applications of vacuum arc coatings. IEEE Trans. Plasma Sci. 17, 705–712. https://doi.org/10.1109/27.41186.

Chaar, A.B.B., Syed, B., Hsu, T.-W., Johansson-Jöesaar, M., Andersson, J.M., Henrion, G., Johnson, L.J.S., Mücklich, F., Odén, M., 2019. The effect of cathodic arc guiding magnetic field on the growth of (Ti0.36Al0.64)N coatings. Coatings. 9, 660. https://doi.org/10.3390/coatings9100660.

Choi, J., Nakao, S., Kim, J., Ikeyama, M., Kato, T., 2007. Corrosion protection of DLC coatings on magnesium alloy. Diam. Relat. Mater., Proceedings of Diamond 2006, the 17th European Conference on Diamond, Diamond-Like Materials, Carbon Nanotubes, Nitrides and Silicon Carbide 16, 1361–1364. https://doi.org/10.1016/j.diamond.2006.11.088.

Christoglou, C., Voudouris, N., Angelopoulos, G.N., Pant, M., Dahl, W., 2004. Deposition of aluminium on magnesium by a CVD process. Surf. Coat. Technol. 184, 149–155. https://doi.org/10.1016/j.surfcoat.2003.10.065.

Das, B.K., Sahoo, S., 2020. Growth of ZnO Thin Film on Silicon and Quartz Substrate by Pulsed Laser Deposition Technique (MSc Thesis). Utkal University, India. https://doi.org/10.13140/RG.2.2.18125.67042/1.

Deng, Z., 2019. PECVD-formed Polymer Coatings for the Corrosion Protection of Biocompatible Magnesium Alloy WE43 (MEng Thesis). McGill University, Canada.

Fan, L., Xu, C., Yuekun, G., Chao, M., Xiaopeng, H., Yida, D., Wenbin, H., Zhong, C., 2016. Effect of pretreatment and annealing on aluminum coating prepared by physical vapor deposition on AZ91D magnesium alloys. Int. J. Electrochem. Sci. 11, 5655–5668. https://doi.org/10.20964/2016.07.57.

Farraro, K.F., Kim, K.E., Woo, S.L.-Y., Flowers, J.R., McCullough, M.B., 2014. Revolutionizing orthopaedic biomaterials: The potential of biodegradable and bioresorbable magnesium-based materials for functional tissue engineering. J. Biomech. 47, 1979–1986. https://doi.org/10.1016/j.jbiomech.2013.12.003.

Förch, R., Chifen, A.N., Bousquet, A., Khor, H.L., Jungblut, M., Chu, L.-Q., Zhang, Z., Osey-Mensah, I., Sinner, E.-K., Knoll, W., 2007. Recent and expected roles of plasma-polymerized films for biomedical applications. Chem. Vap. Depos. 13, 280–294. https://doi.org/10.1002/cvde.200604035.

Fox-Rabinovich, G.S., Yamamoto, K., Beake, B.D., Gershman, I.S., Kovalev, A.I., Veldhuis, S.C., Aguirre, M.H., Dosbaeva, G., Endrino, J.L., 2012. Hierarchical adaptive nanostructured PVD coatings for extreme tribological applications: The quest for nonequilibrium states and emergent behavior. Sci. Technol. Adv. Mater. 13, 043001. https://doi.org/10.1088/1468-6996/13/4/043001.

Friedrich, H.E., Mordike, B.L., 2006. Magnesium Technology: Metallurgy, Design Data, Applications. Springer, Berlin, Heidelberg. https://doi.org/10.1007/3-540-30812-1.

Gray, J.E., Luan, B., 2002. Protective coatings on magnesium and its alloys – a critical review. J. Alloys Compd. 336, 88–113. https://doi.org/10.1016/S0925-8388(01)01899-0.

Hassan, S.F., Nasirudeen, O.O., Al-Aqeeli, N., Saheb, N., Patel, F., Baig, M.M.A., 2017. Processing, microstructure and mechanical properties of a TiO_2 nanoparticles reinforced magnesium for biocompatible application. Metall. Res. Technol. 114, 214. https://doi.org/10.1051/metal/2017015.

Hassan, S.F., Ogunlakin, N.O., Al-Aqeeli, N., Saheb, N., Patel, F., Baig, M.M.A., 2015. Magnesium – nickel composite: Preparation, microstructure and mechanical properties. J. Alloys Compd. 646, 333–338. https://doi.org/10.1016/j.jallcom.2015.06.099.

Hoche, H., Allebrandt, D., Scheerer, H., Broszeit, E., Berger, C., 2007. Design of wear and corrosion resistant PVD-coatings for magnesium alloys. Mater. Werkst. 38, 365–371. https://doi.org/10.1002/mawe.200700143.

Hoche, H., Groß, S., Oechsner, M., 2014. Development of new PVD coatings for magnesium alloys with improved corrosion properties. Surf. Coat. Technol. 259, 102–108. https://doi.org/10.1016/j.surfcoat.2014.04.038.

Holleck, H., Schier, V., 1995. Multilayer PVD coatings for wear protection. Surf. Coat. Technol. 76–77, 328–336. https://doi.org/10.1016/0257-8972(95)02555-3.

Holmberg, K., Matthews, A., 2009. Coatings Tribology: Properties, Mechanisms, Techniques and Applications in Surface Engineering (Second Edition). Elsevier Science, Amsterdam Heidelberg.

Iriyama, Y., Yoshihar, S., 2011. DLC coating on magnesium alloy sheet by low-temperature plasma for better formability, in: Czerwinski, F. (Ed.), Magnesium Alloys – Corrosion and Surface Treatments. InTech. https://doi.org/10.5772/13352.

Kainer, K.U., von Buch, F., 2003. The current state of technology and potential for further development of magnesium applications, in: Magnesium – Alloys and Technology. John Wiley & Sons, Ltd, pp. 1–22. https://doi.org/10.1002/3527602046.ch1.

Kamalan Kirubaharan, A.M., Kuppusami, P., 2020. Chapter 16 – Corrosion behavior of ceramic nanocomposite coatings at nanoscale, in: Rajendran, S., Nguyen, T.A., Kakooei, S., Yeganeh, M., Li, Y. (Eds.), Corrosion Protection at the Nanoscale, Micro and Nano Technologies. Elsevier, pp. 295–314. https://doi.org/10.1016/B978-0-12-819359-4.00016-7.

Khun, N.W., Lee, P.M., Toh, W.Q., Liu, E., 2016. Tribological behavior of nickel-doped diamond-like carbon thin films prepared on silicon substrates via magnetron sputtering deposition. Tribol. Trans. 59, 845–855. https://doi.org/10.1080/10402004.2015.1110864.

Kim, J., Han, D., 2008. Recent development and applications of magnesium alloys in the Hyundai and kia motors corporation. Mater. Trans. 49, 894–897. https://doi.org/10.2320/matertrans.MC200731.

Kuo, Y.-L., Chang, K.-H., 2015. Atmospheric pressure plasma enhanced chemical vapor deposition of SiOx films for improved corrosion resistant properties of AZ31 magnesium alloys. Surf. Coat. Technol. 283, 194–200. https://doi.org/10.1016/j.surfcoat.2015.11.004.

Lee, J., Choi, B.H., Yun, J.-H., Park, Y.S., 2016. Characteristics of nitrogen doped diamond-like carbon films prepared by unbalanced magnetron sputtering for electronic devices. J. Nanosci. Nanotechnol. 16, 4893–4896. https://doi.org/10.1166/jnn.2016.12225.

Leyland, A., Matthews, A., 2000. On the significance of the H/E ratio in wear control: A nanocomposite coating approach to optimised tribological behaviour. Wear. 246, 1–11. https://doi.org/10.1016/S0043-1648(00)00488-9.

Leyland, A., Matthews, A., 2004. Design criteria for wear-resistant nanostructured and glassy-metal coatings. Surf. Coat. Technol., Proceedings of the 30th International Conference on Metallurgical Coatings and Thin Films 177–178, 317–324. https://doi.org/10.1016/j.surfcoat.2003.09.011.

Liebert, C.H., Stepka, F.S., 1976. Ceramic thermal-barrier coatings for cooled Turbines. Presented at the Propulsion Conf., Palo Alto, CA.

Liebert, C.H., Stepka, F.S., 1977. Ceramic thermal barrier coatings for cooled turbines. J. Aircr. 14, 487–493. https://doi.org/10.2514/3.58805.

Lindfors, P.A., Mularie, W.M., Wehner, G.K., 1986. Cathodic arc deposition technology. Surf. Coat. Technol. 29, 275–290. https://doi.org/10.1016/0257-8972(86)90001-0.

Liu, L., 2017. Development of Novel Nanocomposite PVD Coatings to Improve Wear and Corrosion Resistance of Magnesium Alloys (PhD Thesis). University of Sheffield.

Liu, T., Chen, C., Qin, C., Li, X., 2014. Improved hydrogen storage properties of Mg-based nanocomposite by addition of LaNi5 nanoparticles. Int. J. Hydrog. Energy. 39, 18273–18279. https://doi.org/10.1016/j.ijhydene.2014.03.041.

Lu, X., Zuo, Y., Zhao, X., Tang, Y., 2012. The improved performance of a Mg-rich epoxy coating on AZ91D magnesium alloy by silane pretreatment. Corros. Sci. 60, 165–172. https://doi.org/10.1016/j.corsci.2012.03.041.

Ma, G., Gong, S., Lin, G., Zhang, L., Sun, G., 2012. A study of structure and properties of Ti-doped DLC film by reactive magnetron sputtering with ion implantation. Appl. Surf. Sci. 258, 3045–3050. https://doi.org/10.1016/j.apsusc.2011.11.034.

Manoj, G., Nai Mui, L.S., 2011. Magnesium, Magnesium Alloys, and Magnesium Composites. John Wiley & Sons, Ltd. https://doi.org/10.1002/9780470905098.

Martinu, L., Zabeida, O., Klemberg-Sapieha, J.E., 2010. Plasma-enhanced chemical vapor deposition of functional coatings, in: Martin, P.M. (Ed.), Handbook of Deposition Technologies for Films and Coatings (Third Edition). William Andrew Publishing, Boston, pp. 392–465. https://doi.org/10.1016/B978-0-8155-2031-3.00009-0.

Musil, J., Kadlec, S., 1990. Reactive sputtering of TiN films at large substrate to target distances. Vacuum. 40, 435–444. https://doi.org/10.1016/0042-207X(90)90241-P.

Neite, G., Kubota, K., Higashi, K., Hehmann, F., 2006. Magnesium-based alloys, in: Materials Science and Technology. John Wiley & Sons, Ltd. https://doi.org/10.1002/9783527603978.mst0082.

Ogunlakin, N.O., 2016. Development of Magnesium Based Hybrid Nanocomposite (MSc Thesis). King Fahd University of Petroleum and Minerals, Saudi Arabia.

Ohring, M., 2002. Chemical vapor deposition, in: Ohring, M. (Ed.), Materials Science of Thin Films (Second Edition). Academic Press, San Diego, pp. 277–355. https://doi.org/10.1016/B978-012524975-1/50009-4.

Olalekan, O.N., Abdul Samad, M., Hassan, S.F., Elhady, M.M.I., 2022. Tribological evaluations of spark plasma sintered Mg – Ni composite. Tribol. – Mater. Surf. Interfaces. 16, 110–118. https://doi.org/10.1080/17515831.2021.1898898.

Prater, J.T., Courtright, E.L., 1987. Ceramic thermal barrier coatings with improved corrosion resistance. Surf. Coat. Technol. 32, 389–397. https://doi.org/10.1016/0257-8972(87)90122-8.

Puzikov, V.M., Semenov, A.V., 1991. Ion beam deposition of diamond-like carbon films. Surf. Coat. Technol. 47, 445–454. https://doi.org/10.1016/0257-8972(91)90310-S.

Sanders, D.M., 1989. Review of ion-based coating processes derived from the cathodic arc. J. Vac. Sci. Technol. U. S. 7, 3. https://doi.org/10.1116/1.575939.

Savvides, N., Window, B., 1986. Unbalanced magnetron ion-assisted deposition and property modification of thin films. J. Vac. Sci. Technol. Vac. Surf. Films. 4, 504–508. https://doi.org/10.1116/1.573869.

Somekawa, H., 2019. Effect of alloying elements on toughness and ductility of magnesium. J. Jpn. Inst. Met. Mater. 83. https://doi.org/10.2320/jinstmet.J2018067.

Sproul, W.D., Legg, K.O., National Institute of Standards and Technology (Eds.), 1994. Advanced Surface Engineering, Opportunities for Innovation. Technomic Publishing, Lancaster, PA.

Stueber, M., Holleck, H., Leiste, H., Seemann, K., Ulrich, S., Ziebert, C., 2009. Concepts for the design of advanced nanoscale PVD multilayer protective thin films. J. Alloys Compd., 14th International Symposium on Metastable and Nano-Materials (ISMANAM-2007) 483, 321–333. https://doi.org/10.1016/j.jallcom.2008.08.133.

Sun, L., Yuan, G., Gao, L., Yang, J., Chhowalla, M., Gharahcheshmeh, M.H., Gleason, K.K., Choi, Y.S., Hong, B.H., Liu, Z., 2021. Chemical vapour deposition. Nat. Rev. Methods Primer. 1, 1–20. https://doi.org/10.1038/s43586-020-00005-y.

Tagami, Y., Umehara, N., Kousaka, H., Xingrui, D., 2016. The effect of tantalum on the friction and wear properties of DLC deposited by filtered arc deposition. The Proceedings of the Machine Design and Tribology Division Meeting. JSME 16, A1–A5.

Takikawa, H., Miyakawa, N., Sakakibara, T., 2003. Development of shielded cathodic arc deposition with a superconductor shield. Surf. Coat. Technol., Proceedings from the Joint International Symposia of the 6th APCPST, 15th SPSM, 4th International Conference on Open Magnetic Systems for Plasma Confinement and 11th KAPRA 171, 162–166. https://doi.org/10.1016/S0257-8972(03)00262-7.

Tang, X.S., Wang, H.J., Feng, L., Shao, L.X., Zou, C.W., 2014. Mo doped DLC nanocomposite coatings with improved mechanical and blood compatibility properties. Appl. Surf. Sci. 311, 758–762. https://doi.org/10.1016/j.apsusc.2014.05.155.

Vetter, J., 2014. 60years of DLC coatings: Historical highlights and technical review of cathodic arc processes to synthesize various DLC types, and their evolution for industrial applications. Surf. Coat. Technol., 25 years of TiAlN hard coatings in research and industry 257, 213–240. https://doi.org/10.1016/j.surfcoat.2014.08.017.

Westengen, H., 2001. Magnesium: Alloying, in: Buschow, K.H.J., Cahn, R.W., Flemings, M.C., Ilschner, B., Kramer, E.J., Mahajan, S., Veyssière, P. (Eds.), Encyclopedia of Materials: Science and Technology. Elsevier, Oxford, pp. 4739–4743. https://doi.org/10.1016/B0-08-043152-6/00825-1.

Wu, H., Xi, K., Xiao, S., Qasim, A.M., Fu, R.K.Y., Shi, K., Ding, K., Chen, G., Wu, G., Chu, P.K., 2020. Formation of self-layered hydrothermal coating on magnesium aided by titanium ion implantation: Synergistic control of corrosion resistance and cytocompatibility. Surf. Coat. Technol. 401, 126251. https://doi.org/10.1016/j.surfcoat.2020.126251.

Yang, F., Chang, R., Webster, T.J., 2019. Atomic layer deposition coating of TiO_2 nano-thin films on magnesium-zinc alloys to enhance cytocompatibility for bioresorbable vascular stents. Int. J. Nanomed. 14, 9955–9970. https://doi.org/10.2147/IJN.S199093.

Yang, J.X., Cui, F.Z., Lee, I.-S., Jiao, Y.P., Yin, Q.S., Zhang, Y., 2008. Ion-beam assisted deposited C – N coating on magnesium alloys. Surf. Coat. Technol. 202, 5737–5741. https://doi.org/10.1016/j.surfcoat.2008.06.116.

Yeldose, B.C., Ramamoorthy, B., 2008. Characterization of DC magnetron sputtered diamond-like carbon (DLC) nano coating. Int. J. Adv. Manuf. Technol. 38, 705–717. https://doi.org/10.1007/s00170-007-1131-8.

Yue, T.M., Li, T., 2008. Laser cladding of Ni/Cu/Al functionally graded coating on magnesium substrate. Surf. Coat. Technol. 202, 3043–3049. https://doi.org/10.1016/j.surfcoat.2007.11.007.

Zhang, D., Wei, B., Wu, Z., Qi, Z., Wang, Z., 2016. A comparative study on the corrosion behaviour of Al, Ti, Zr and Hf metallic coatings deposited on AZ91D magnesium alloys. Surf. Coat. Technol., 2015 International Thin Films Conference [TACT 2015] 303, 94–102. https://doi.org/10.1016/j.surfcoat.2016.03.079.

Zia, A.W., 2020. New generation carbon particles embedded diamond-like carbon coatings for transportation industry, in: Advances in Smart Coatings and Thin Films for Future Industrial and Biomedical Engineering Applications. Elsevier, pp. 307–332. https://doi.org/10.1016/B978-0-12-849870-5.00004-5.

Zia, A.W., Zhou, Z., Li, L.K.-Y., 2019. Structural, mechanical, and tribological characteristics of diamond-like carbon coatings, in: Nanomaterials-Based Coatings. Elsevier, pp. 171–194. https://doi.org/10.1016/B978-0-12-815884-5.00007-7.

19 Thermal and Cold Spray Coatings

Mohamed Abdrabou Hussein

CONTENTS

19.1 INTRODUCTION

Magnesium (Mg) is the lightest structural metal, weighing approximately 35% less than Al and 78% less than steel, and it possesses remarkable rigidity and damping capability (Kleiner et al. 2003). Mg alloy's low density and excellent castability make them ideal for mass-produced and lightweight applications, such as vehicle and aviation components (Salonitis et al. 2009). However, their widespread application was limited due to their weak corrosion resistance. Surface engineering is one of the potential ways to avoid these restrictions, as it can provide sufficient corrosion protection without changing the microstructure and characteristics of the base material. There are various surface protection processes reported for Mg alloys, such as anodizing, physical/chemical vapour deposition conversion coating, electroplating, and thermal spray (TS) and cold spray (CS) coating.

Cold spray (CS) operates at low temperatures and shows the potential to be applied for thermally sensitive materials such as Mg alloys. The CS coating is generated by the extreme plastic deformation of solid particles. CS has the benefit of minimizing thermal defects, such as oxidation, porosity, and phase evolution, when compared to TS systems that use partially or entirely molten particles during coating deposition (Salonitis et al. 2009). Therefore, CS is a promising coating for materials that are sensitive to heat or oxidation such as Mg (Vilardell et al. 2015).

19.2 THERMAL SPRAY COATINGS

19.2.1 THERMAL SPRAY PROCESS

Different TS processes include high-velocity oxy-fuel spray (HVOF), plasma spray, arc spray, and flame spray. Though the particulars of these different spraying techniques vary, the essential

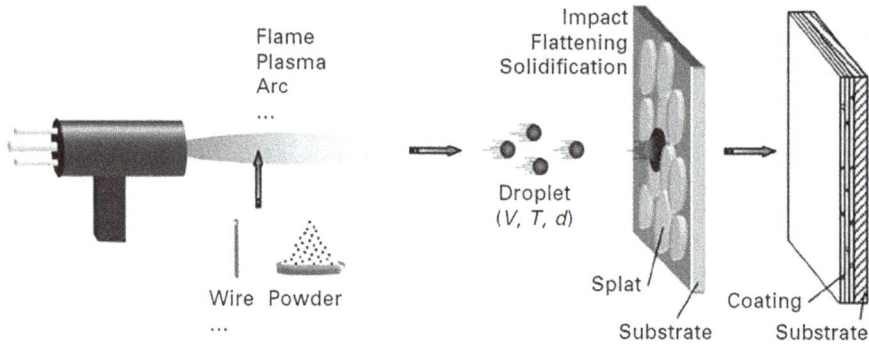

FIGURE 19.1 Schematic representation of thermal spray coating technique. *Reproduced with permission from (Li 2010) © 2010 Woodhead Publishing Limited.*

concepts underpinning TS systems are comparable. TS systems utilize feedstock materials either in a powder or wire form. The particles that are molten or partially molten are sprayed onto the substrate surface and solidified, generating a coating, a lamellar, layered structure. Depending on the energy input, particles are directed toward the surface of the coated material in a variety of ways: it may be electrical (such as plasma spray and electric wire arc techniques), chemical (like combustion), or a mix of the two (flame spray and HVOF spray techniques) (Herman et al. 2000). Coating deposition processes are influenced by the properties of the droplets, such as their temperature, velocity, and size, which are governed by the spraying methods and conditions. Depending on the required thickness of the deposit, a single or numerous torch passes can be performed in each particular region. A TS coating that is properly sprayed will give effective protection against wear, corrosion, and high-temperature conditions. Each spray technique and spray process impart a unique amount of thermal and kinetic energy to the sprayed particles. The thermal and kinetic energy of the spray particles can be directly determined from the velocity and temperature of the flame (spray jet). In contrast, the HVOF approach utilizes a high amount of kinetic energy and a comparatively moderate amount of heat energy.

In a fundamental TS technique (**Figure 19.1**) (Li 2010), chemical or electrical energy is utilized to transform powder, wire, or rod material into small molten or semi-molten droplets with a diameter of 10 to 100 μm. These droplets are forced onto a substrate by a subsonic or supersonic gas stream at rates ranging from a few tens to about 1000 meters per second (3280 feet per second). Upon impact, each droplet disperses and immediately hardens at cooling rates on the order of 104 to 108 K/s, forming the fundamental microstructural unit of spray-deposited material, known as a "splat." A spatter often solidifies before the following molten droplet impacts on top of it due to the quick cooling. The usual microstructure of TS coatings consists of a lamellar buildup of melted, partially melted, and un-melted particles, as well as oxides and inevitable pores. It is interesting to observe that nearly all materials can be sprayed using TS methods. It is anticipated that the sprayed coatings created by such technologies will have higher oxygen and porosity levels. Plasma spraying is commonly used to apply ceramic coatings (Champagne 2007). HVOF (Pawlowski 2008) generally applies to metallic and cermet (hard metal) coatings (Pawlowski 2008).

19.2.2 CORROSION PROTECTION WITH THERMAL SPRAY COATINGS

By applying various coating materials via thermal spray techniques, the corrosion protection of Mg alloy engineering components can be greatly enhanced. The substrate should be suitably prepared before the TS coating for better surface adhesion of the coating. After washing and degreasing, the substrate surface should be roughened by grit blasting or another means. Additionally, this might enhance the surface area of the coated surface. Mismatch oxidation and the coefficient of thermal expansion among coated Mg and the coating are the two most essential aspects that could eventually lead to the spallation of the coating. The creation of a "thermally grown oxide" layer on the Mg alloy during the TS method can greatly improve adhesion at the interface between the deposited coating and the base Mg alloy (Maev and Leshchynsky 2009). As a result, a lower temperature of Mg substrate and/or a metallic bond coat were suggested for reducing the difficulties related to Mg alloy during the TS method.

Post-processes showed a potential to enhance the corrosion protection of TS-coated Mg alloys. Due to Mg's high affinity for oxygen, the substrate surface oxidizes rapidly in air environment. Typically, a hydroxyl layer forms on the surface within a few minutes. The presence of this film can drastically impair the adherence of TS coatings to the substrate (Lugscheider et al. 2003). Some post-processing is required due to the insufficient adhesion between the TS coatings and the Mg substrate. An AZ61 alloy with a NiCrAl intermediate layer and an outer layer of Fe-based amorphous coating was deposited using HVOF. The intermediate metallic layer overcomes the typical mismatch between Fe-based materials and Mg-based alloys, resulting in the formation of strong metallurgical bonds at the interface regions. The results showed that the corrosion protection of AZ61 alloy was doubled with Fe coating (Guo et al. 2016). In another study, Al coatings were applied to AZ91D, AZ31, and AZ80 alloys using the TS process (Arrabal et al. 2009). The corrosion protection was studied in 3.5% NaCl solution. Their result showed that the TS Al coating developed interconnecting pores, which caused galvanic corrosion of the coated substrates.

Using the HVOF process at different conditions, stainless steel (SS) was deposited on ZE41 alloy to enhance its corrosion protection (García-Rodríguez et al. 2016). It was found that coatings with thickness ranging from 42 to 478 μm were consistent, crack-free, and dense. The coated samples exhibited comparable corrosion protection in 3.5% NaCl to that of bulk SS sample. The anodic – cathodic polarization plots are depicted in **Figure 19.2**. The open circuit potential (OCP) of the SS-1 coated sample (1.3 V) was extremely near to that of the uncoated Mg sample, indicating that the substrate is playing a role in the corrosion behavior of the sample and the limited corrosion protection of this coating within 1 hour after application. After 168 hours of immersion, the polarization resistances of SS-2 and SS-3 coatings were 3 and 17 times greater than the untreated sample, respectively (García-Rodríguez et al. 2016).

In 3.5 wt.% NaCl, the corrosion resistance of Al- and Al-11Si-coated AZ31-, AZ80-, and AZ91D Mg-based alloys was examined. Their results showed that all the Mg-based substrates displayed galvanic corrosion at the substrate-coating layer contact due to the coating's porous structure (Pardo et al. 2011).

The TS technique was also employed for biomedical applications involving Mg alloys. To improve the corrosion protection, high-velocity suspension flame spray (HVSFS) was utilized to coat a hydroxyapatite (HAp)/Mg composite coating on AZ91D Mg alloy. The finding indicated that the HAp/Mg composite coating consisted primarily of Mg and HAp, with a trace quantity of MgO. The HAp/Mg composite coating significantly improved the corrosion protection of the AZ91D alloy in Hanks' Balanced Salt Solution (HBSS) (**Figure 19.3**) (Yao et al. 2019).

TABLE 19.1

TS Coating Conditions for Different Samples. *Reproduced with permission from (García-Rodríguez et al. 2016) © 2015 Elsevier B.V.*

Specimen	Spraying distance (mm)	Number of layers	Gun speed (mm/s)	Feedstock speed rotation (%vol.)
SS-1	200	1	350	50
SS-2	400	3	350	40
SS-3	300	4	250	40

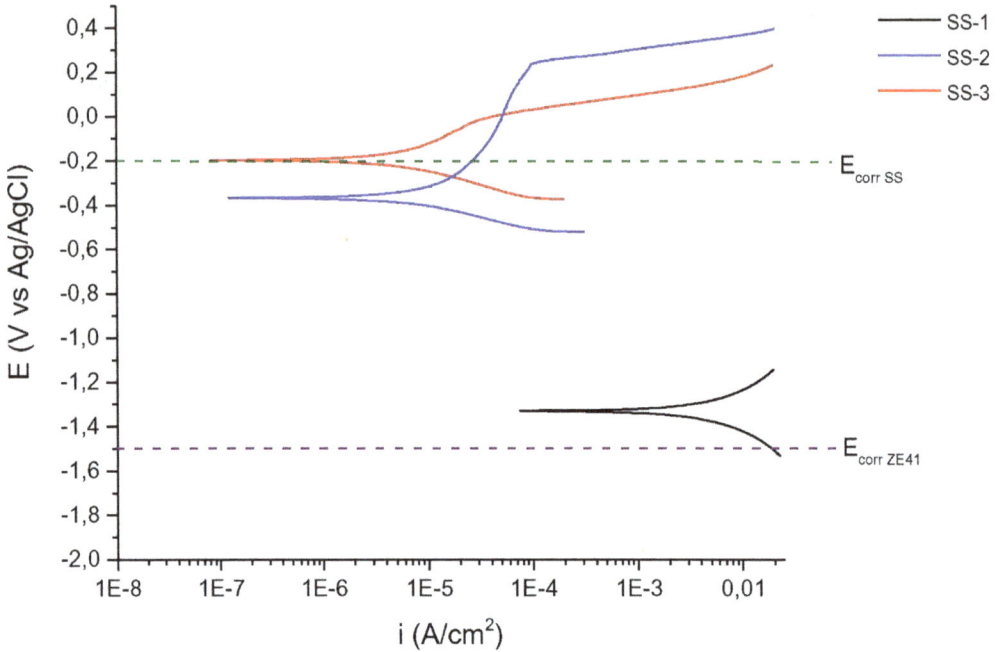

FIGURE 19.2 Potentiodynamic polarization plots (PDP) plots after 1 hour of immersion in 3.5% NaCl medium. *Reproduced with permission from (García-Rodríguez et al. 2016) © 2015 Elsevier B.V.*

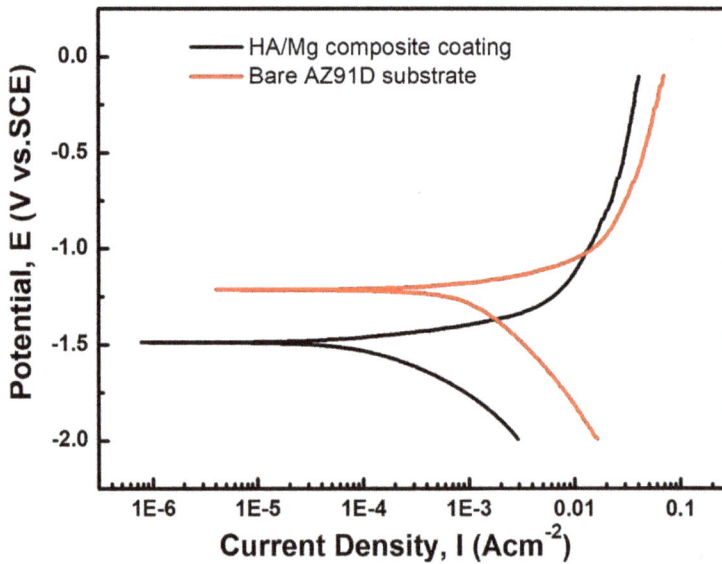

FIGURE 19.3 PDP plots of coated and uncoated substrates. *Reproduced with permission from (Yao et al. 2019) © 2018, ASM International.*

19.3   COLD SPRAY COATINGS

19.3.1   GENERAL FEATURES OF COLD SPRAYING

CS is a solid-state technique in which solid powder particles are propelled to speed between 300 to 1200 ms^{-1} over a de Laval nozzle, impact, and adhere to a sample surface, hence achieving a high coating deposition rate. During deposition, the powder feedstock remains considerably below its melting point. Therefore, CS is categorized as a solid-state deposition technique, and it offers significant advantages over other usual TS techniques (Poza and Garrido-Maneiro 2022). The physical and mechanical characteristics of a coating-substrate system are significantly impacted by particle-substrate contact upon high-velocity impact and subsequent bonding. Adhesion happens when powder particles surpass a critical impact velocity, which is essential for determining the optimal spraying settings and reducing production costs by enhancing deposition efficiency. The reported literature showed that the critical velocity is dependent on spray material characteristics, particle temperature, particle size distribution, surface oxidation, and the type and properties of the substrate. Critical and erosion velocities determine the impact velocity window within which powders must be accelerated for the cold spray technique to be successful (Li et al. 2006). It is crucial to optimize deposition settings and deposition procedures to enhance coating performance and deposition efficiency (Hassani-Gangaraj et al. 2015).

19.3.2   TYPES OF COLD SPRAY TECHNIQUES

19.3.2.1   Low-Pressure Cold Spray (LPCS)

In LPCS, the accelerating gas, which is common air or N_2, is pressured at relatively low pressure (5–10 bar), preheated (up to 600°C) to enhance its aerodynamic properties, and then propelled through a DeLaval nozzle (Schmidt et al. 2009). On the diverging side of the nozzle, the velocity of the hot gas is between 300 and 600 ms^{-1}. In the LPCS system, powder particles are delivered radially downstream of the throat of the supersonic nozzle and driven toward the substrate, as shown in **Figure 19.4**. Keeping the static pressure within the nozzle below that of the surrounding atmosphere enables the Venturi effect to successfully suck feedstock particles from the powder feeder.

19.3.2.2   High-Pressure Cold Spray (HPCS)

Before passing through a converging-diverging DeLaval nozzle (Singh et al. 2012), the accelerating He or N_2 gas under pressure of (25–30 bar) is heated (up to 1000°C) to maximize its aerodynamic characteristics in HPCS. The expansion of the gas at the nozzle promotes the conversion of enthalpy to kinetic energy, which speeds the gas flow to the supersonic domain (1,200 ms^{-1}) while decreasing its temperature. As depicted in **Figure 19.5**, the solid particles impact the substrate with sufficient kinetic energy (between 600 and 1200 ms^{-1}) to promote mechanical and/or metallic bonding. The spray efficiency of this HPCS system is up to 90%, compared to 50% for the LPCS system.

19.3.3   CORROSION PROTECTION WITH COLD SPRAY COATINGS

Al is one of the most commonly utilized spray feedstocks on Mg alloys in the CS process due to its light weight, high ductility, and superior corrosion resistance. Al has higher general corrosion protection related to other metals due to its passivation characteristics. On the other hand, CS is a cost-effective method for forming thick metallic Al coatings on Mg alloys with minimal surface preparation and without degrading the substrate's mechanical or thermal properties (**Figure 19.6**) (Singh et al. 2012).

On the Mg alloy surface, the corrosion performance of CS Al coatings was comparable to that of bulk-cast Al. Due to galvanic coupling, porosity within the CS coatings on Mg substrates could accelerate corrosion. Consequently, it is crucial to control the porosity of CS coatings (Spencer

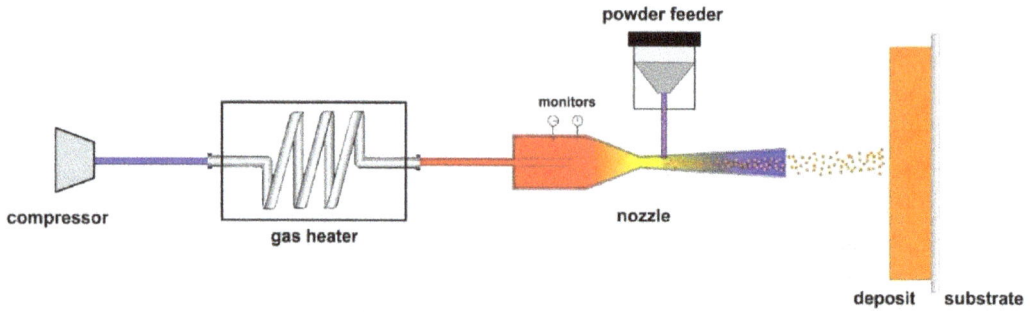

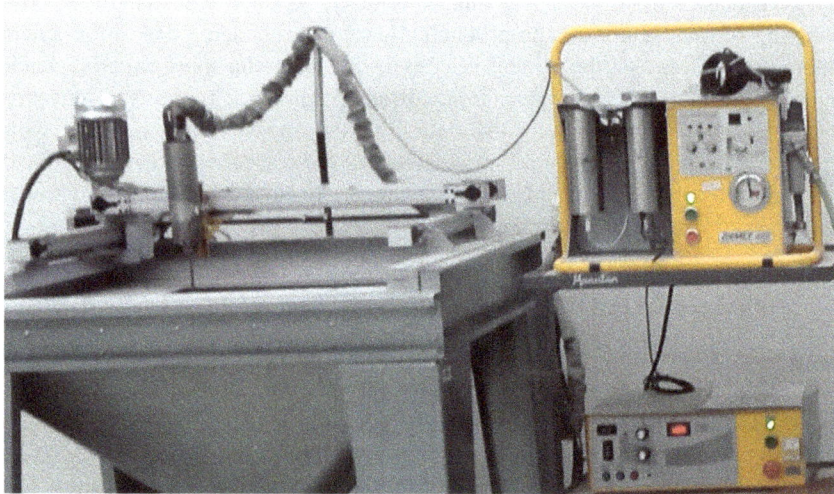

FIGURE 19.4 (a) Schematic diagram of LPCS and (b) photograph showing a real LPCS with control system. *Reproduced (a) with permission from (Yin et al. 2018) © 2018 Elsevier B.V.*

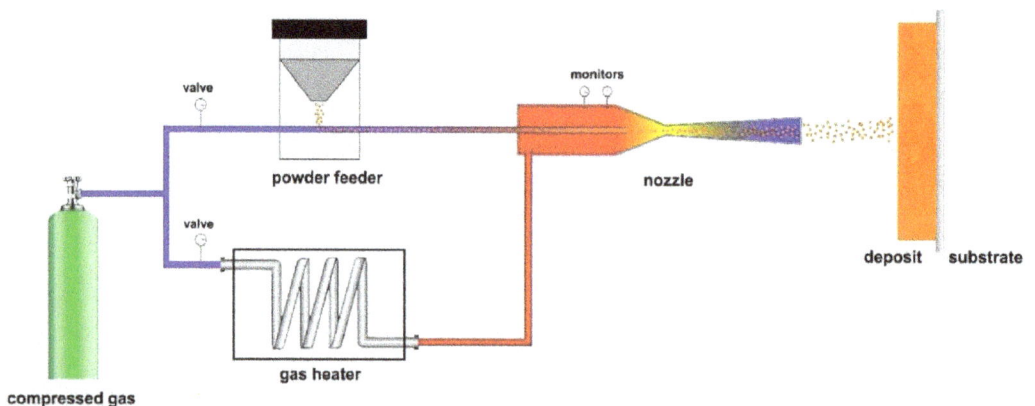

FIGURE 19.5 Schematic diagram of HPCS. *Reproduced with permission from (Yin et al. 2018) © 2018 Elsevier B.V.*

and Zhang 2008). CS coatings with increased density can enhance corrosion protection. It is common knowledge that the addition of hard ceramic particles to CS coatings increases not just the coatings' wear characteristics and hardness but also their density. As the creation of a diffusion layer during thermal treatment enhances bond strength at the interface and makes delamination more difficult, it is commonly believed that post-spray thermal treatment could further improve

the corrosion resistance of CS coatings. In addition, the thermal treatment technique reduces the residual stress that occurred at the interface of the coating during spraying, hence minimizing the risk of delamination.

The pure Al coating on AZ91 alloy exhibited greater resistance to pitting corrosion than bulk Al in NaCl medium (Tao et al. 2010). After 1 day of immersion, the smooth surface of a large quantity of Al transformed into highly corroded portions in isolated spots. The coated surface characterized by fish scales and holes did not change significantly after one day of immersion. After 10 days, fish scales have corroded, and there are signs of localized corrosion. However, Mg was not identified as the primary component of corrosion products, indicating that the Al covering provides significant corrosion protection. In contrast, after 10 days of immersion, deeper corrosion pits and local fissures in the bulk material showed intensified localized corrosion (**Figure 19.7**) (Tao et al. 2010).

A high-purity Al coating achieved the lowest galvanic corrosion rate when compared to commercially pure Al and Al–5Mg applied powders (DeForce et al. 2011). In terms of overall performance, including hardness, adhesion strength, and corrosion resistance, Al–5Mg particles appeared to be the most effective barrier coating for Mg alloys. The corrosion performance of Al and 6061 alloys mixed with Al_2O_3 composite on AZ91E substrates deposited with cold spray was studied. Even though 1% to 1.5% porosity was present in the 6061Al coating, the Al_2O_3 composite coating has a dense and uniform distribution.

After depositing an Al60Ni40 wt.% coating with a thickness of 240 μm using LPCS (Chakradhar et al. 2021), the corrosion resistance of the AZ31B alloy was investigated. The corrosion protection of coated samples was greater than that of uncoated samples, and corrosion protection increased with immersion time in the NaCl medium due to the formation of protective oxide layers. The Al60Ni40 coating sample exhibited a potential shift toward the negative side, possibly due to Al oxidation. Due to the growth of an oxide layer, a positive shift was observed that increased with immersion time (**Figure 19.8** and Table 19.2).

The corrosion properties of Ti and Al coatings deposited on AZ31B alloy using the HPCS process were studied (Daroonparvar et al. 2021). The result showed that both Al and Ti coating enhanced the corrosion protection of the substrate. However, Ti coating's performance was superior to Al coating (Table 19.3). The Ti coating showed remarkable resistance to pitting corrosion in NaCl medium compared to Al coating.

CS deposition of a CoCrFeNi high-entropy alloy (HEA) for corrosion protection of AZ91 in a NaCl medium was reported recently (Zhu et al. 2022). The results demonstrated that a homogeneous microstructure of the coating and the stability of the multilayer passive film improved the corrosion resistance of HEA coating. The corrosion current density of HEA coating was approximately

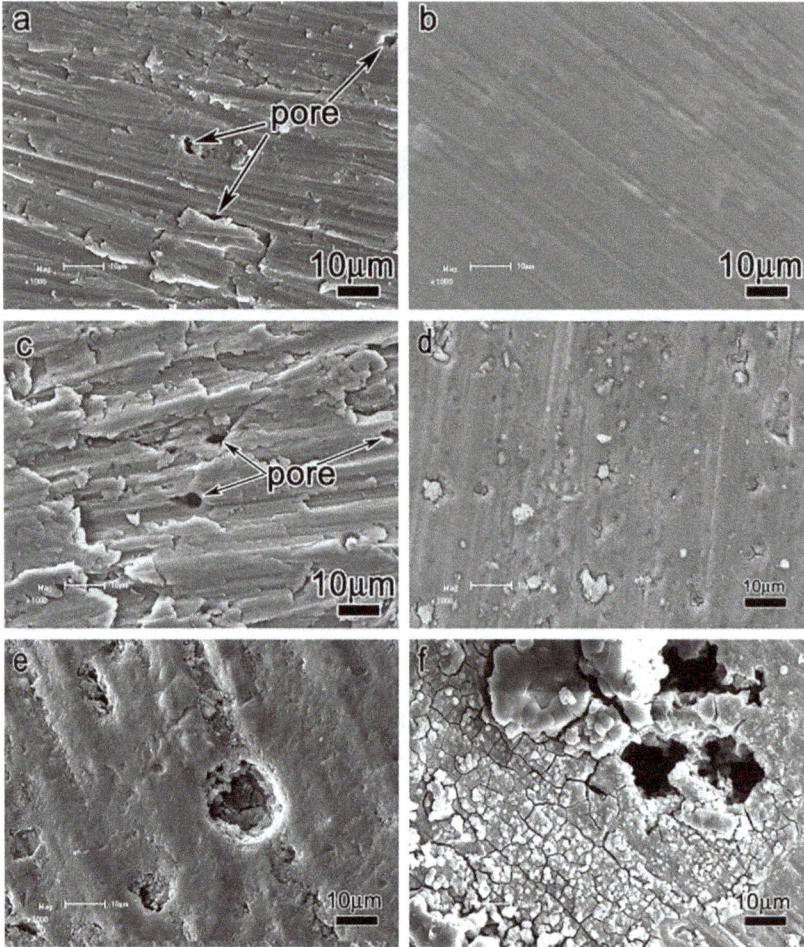

FIGURE 19.7   SEM of CS pure Al coating and bulk pure Al (a, b) before the test, (c, d) after 1 day, and (e, f) after 10 days of immersion. *Reproduced with permission from (Tao et al. 2010) © 2010 Elsevier Ltd.*

5 times lower than that of uncoated AZ91 alloy (**Figure 19.9**). The weight-loss test results for four weeks indicated that the coated sample lost less weight than the uncoated substrate (**Figure 19.10**).

19.4   COLD SPRAY VERSUS THERMAL SPRAY

High temperatures accompanied by the conventional TS process cause negative thermal effects. However, in CS, the likelihood of high-temperature reactions and oxidation is significantly reduced. Some of the shortcomings of typical TS techniques are tensile residual stresses resulting from thermal contraction during solidification and cooling, which promote delamination and border the maximum achievable thickness. This issue is exacerbated when the substrate material and coating differ in terms of hardness, density, and thermal expansion coefficient. However, there is a relatively slight thermally induced dimensional change with CS because particle consolidation occurs in the solid state. Additionally, the high-impact velocities of the solid particles are extremely effective at peening the underlying material and generating deposits that are normally under compressive stress. Compared to standard TS processes, CS often produces less porous materials with higher hardness, conductivity, and lower oxide content.

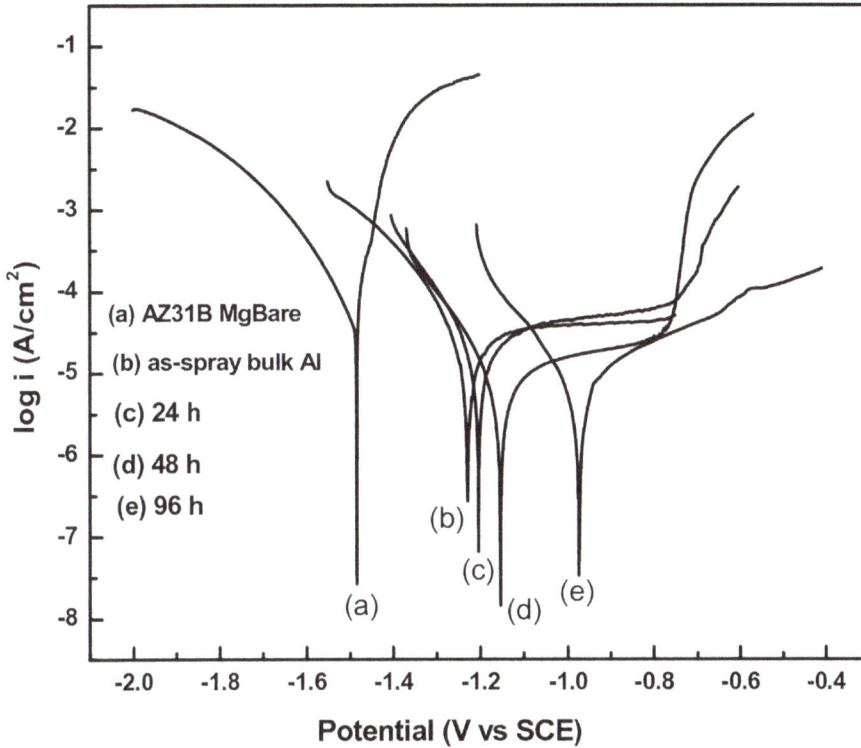

FIGURE 19.8   PDP of (a) uncoated and (b) Al coated AZ31B Mg alloys at different immersion times (c) 24 h, (d) 48 h, and (e) 96 h. *Reproduced with permission from (Chakradhar et al. 2021) © 2021, ASM International.*

TABLE 19.2

PDP Data of Uncoated and Al-Coated Alloys at Different Immersion Times. *Reproduced with permission from (Chakradhar et al. 2021) © 2021, ASM International.*

Samples	E_{corr} (V)	I_{corr} ($\mu A/cm^2$)	R_p ($\Omega.cm^2$)
Bare AZ31B	−1.48	833.5	777
As-sprayed bulk Al coating	−1.22	5.574	2536
24 h	−1.20	5.465	2835
48 h	−1.15	5.455	2935
96 h	−0.97	4.105	4073

Several properties of CS in comparison to TS are of special importance for the development of a surface layer resistant to corrosion. CS is coated with feedstock particles in a solid state, and no inherent flaws are produced as a result of the liquefaction and solidification processes. This characteristic produces dense, low-porosity coatings that are desirable for corrosion protection. This is especially significant in the case of galvanic protection, when it is preferable to have the surface layer consist of pure and oxidizable components. Furthermore, as the coating buildup is accompanied by the peening effect of incoming particles, compressive residual stresses can form in both the coating and the substrate during cold spraying. In contrast, TS introduces tensile

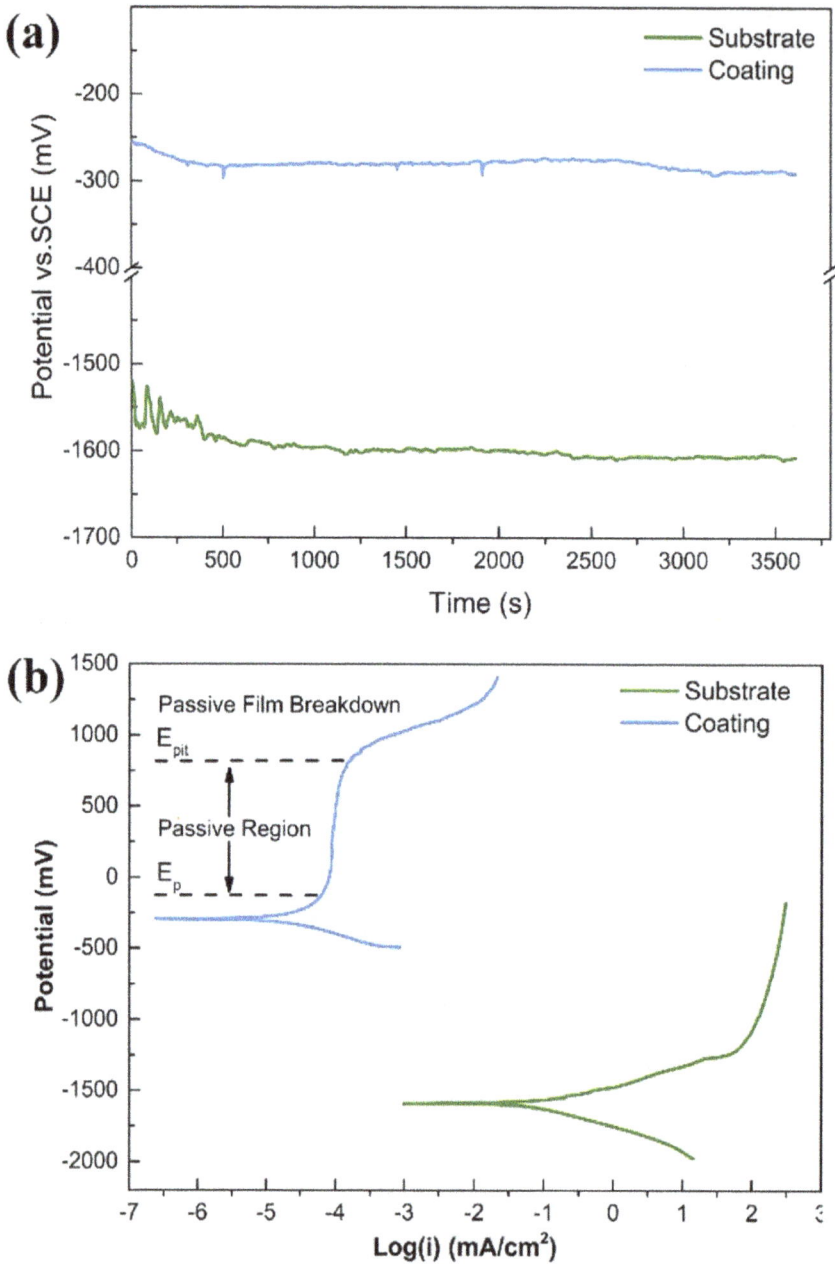

FIGURE 19.9 OCP (a) and PDP (b) results of the uncoated and coated samples in NaCl solution. *Reproduced with permission from (Zhu et al. 2022) © 2022 Elsevier B.V.*

residual stresses; as a result, the likelihood of delamination of the coating from the substrate increases with coating thickness. The CS has some downsides as well. Among these include the immobility of many cold spray units, powder handling to reduce the risk of combustion, and the cost-effectiveness of the operation. The CS provides improved corrosion prevention; however, when oxide formation is the major protective mechanism, TS can improve performance (Hassani-Gangaraj et al. 2015).

TABLE 19.3

The Electrochemical Corrosion Parameters of Al and Ti Coating Compared to the Uncoated AZ31B Alloy. *Reproduced with permission from (Daroonparvar et al. 2021) © 2021 by the authors. Licensee MDPI, Basel, Switzerland.*

Type of Sample	E_{corr} (mV$_{SCE}$)	β_a (mV/dec)	β_c (mV/dec)	i_{corr} (μA cm^{-2})
AZ31B	−1453.86	115.3	67.6	2.504
Al-coated AZ31B	−804.59	75.5	114	0.092
Ti-coated AZ31B	−387.299	159.9	99.7	0.049

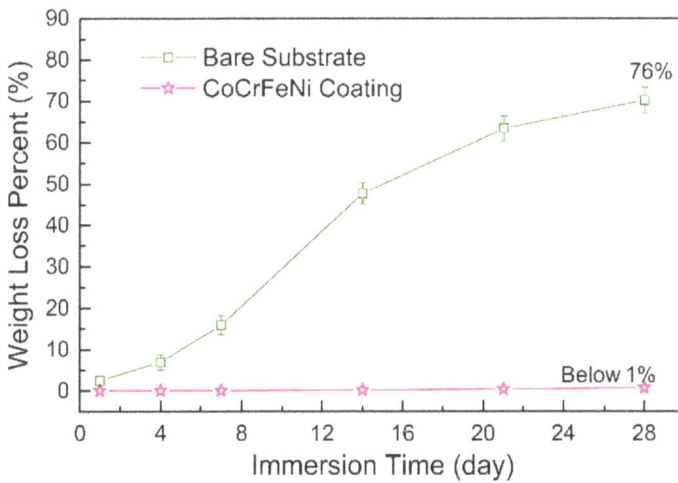

FIGURE 19.10   The weight loss results for coated and uncoated samples. *Reproduced with permission from (Zhu et al. 2022) © 2022 Elsevier B.V.*

19.5   CONCLUSIONS AND OUTLOOK

Mg and its alloys have garnered attention due to their potential industrial applications. However, their lower resistance to corrosion limits their applications. In this chapter, the TS and CS processes and case studies for their applications for surface engineering of Mg alloys for corrosion protection were discussed. The CS technique has developed from an emerging technology to a viable choice for the corrosion protection of Mg alloys. It has become a potential, economical, and environmentally friendly method for protecting Mg components from corrosion compared to TS. In an aggressive medium, CS coatings created under optimum circumstances may exhibit very high adhesion, minimal porosity, and enhanced long-term corrosion resistance. To properly prepare coatings with strong corrosion resistance, it is necessary to optimize all spraying parameters. Future research should involve systematic and long-term comparisons and assessments to establish the benefits and drawbacks of each coating (Daroonparvar et al. 2022). Galvanic corrosion is the greatest issue that noble barrier coatings face while protecting Mg alloys. Therefore, there is a need for more durable coatings with minimum porosity levels, the development of coating materials with a reduced standard reduction potential mismatch with Mg alloy, a low corrosion rate, and strong resistance to pitting corrosion. Moreover, HEA coating using cold spray process would be a promising direction to be explored for corrosion protection of Mg alloys.

ACKNOWLEDGEMENTS

The author would like to thank the Interdisciplinary Research Center for Advanced Materials (IRC-AM), King Fahd University of Petroleum & Minerals (KFUPM), for their support.

REFERENCES

Arrabal, R., A. Pardo, M. C. Merino, S. Merino, P. Casajús, M. Mohedano, and P. Rodrigo, Corrosion behavior of mg-al alloys with aluminum thermal spray coatings in humid and saline environments, *Corrosion* 65 (2009) 817–830.

Chakradhar, R. P. S., G. Chandra Mouli, Harish Barshilia, and Meenu Srivastava, Improved corrosion protection of magnesium alloys AZ31B and AZ91 by cold-sprayed aluminum coatings, *Journal of Thermal Spray Technology* 30 (2021) 371–384.

Champagne, Victor K. *The cold spray materials deposition process*. Vol. 187. Elsevier Science, 2007.

Daroonparvar, M., Hamid Reza Bakhsheshi-Rad, Abbas Saberi, Mahmood Razzaghi, Ashish K. Kasar, Seeram Ramakrishna, Pradeep L. Menezes, Manoranjan Misrad, Ahmad Fauzi Ismail, Safian Shariff, and Filippo Berto, Surface modification of magnesium alloys using thermal and solid-state cold spray processes: Challenges and latest progresses, *Journal of Magnesium and Alloys* 10 (2022) 2025–2061.

Daroonparvar, Mohammadreza, Ashish K. Kasar, Mohammad Umar Farooq Khan, Pradeep L. Menezes, Charles M. Kay, Manoranjan Misra, and Rajeev K. Gupta, Improvement of wear, pitting corrosion resistance and repassivation ability of mg-based alloys using high pressure cold sprayed (HPCS) commercially pure-titanium coatings, *Coatings* 11 (2021) 57.

DeForce, Brian S., Timothy J. Eden, and John K. Potter, Cold spray Al-5% Mg coatings for the corrosion protection of magnesium alloys, *Journal of Thermal Spray Technology* 20 (2011) 1352–1358.

García-Rodríguez, S., A. J. López, B. Torres, and J. Rams, 316L stainless steel coatings on ZE41 magnesium alloy using HVOF thermal spray for corrosion protection, *Surface and Coatings Technology* 287 (2016) 9–19.

Guo, S. F., F. S. Pan, H. J. Zhang, D. F. Zhang, J. F. Wang, J. Miao, C. Su, and C. Zhang, Fe-based amorphous coating for corrosion protection of magnesium alloy, *Materials & Design* 108 (2016) 624–631.

Hassani-Gangaraj, Seyyed Mostafa, A. T. I. E. H. Moridi, and M. A. R. I. O. Guagliano, Critical review of corrosion protection by cold spray coatings. *Surface Engineering* 31 (2015) 803–815.

Herman, Herbert, Sanjay Sampath, and Robert McCune, Thermal spray: Current status and future trends, *MRS Bulletin* 25 (2000) 17–25.

Kleiner, Matthias, Manuel Geiger, A. Klaus, Manufacturing of lightweight components by metal forming. *CIRP Annals* 52 (2003) 521–542.

Li, C. J., Thermal spraying of light alloys. In *Surface engineering of light alloys*. Woodhead Publishing, 2010, pp. 184–241.

Li, Chang-Jiu, Wen-Ya Li, and Hanlin Liao, Examination of the critical velocity for deposition of particles in cold spraying, *Journal of Thermal Spray Technology* 15 (2006) 212–222.

Lugscheider, Erich, Maria Parco, K. U. Kainer, and N. Hort, Thermal spraying of magnesium alloys for corrosion and wear protection. In *Magnesium: Proceedings of the 6th international conference magnesium alloys and their applications*, pp. 860–868. Weinheim, Wiley-VCH, 2003.

Maev, Roman Gr, and Volf Leshchynsky, *Introduction to Low Pressure Gas Dynamic Spray: Physics and Technology*. John Wiley & Sons, 2009.

Pardo, A., M. C. Merino, R. Arrabal, P. Casajús, M. Mohedano, S. Feliu Jr, and S. Merino, Corrosion behavior of AZ magnesium alloys with Al and Al-11Si thermal spray coatings, *Corrosion* 67 (2011) 025003–1.

Pawlowski, Lech, *The science and engineering of thermal spray coatings*. John Wiley & Sons, 2008.

Poza, Pedro, and Miguel Ángel Garrido-Maneiro, Cold-sprayed coatings: Microstructure, mechanical properties, and wear behaviour, *Progress in Materials Science* 123 (2022) 100839.

Salonitis, Konstantinos, John Pandremenos, John Paralikas, and George Chryssolouris, Multifunctional materials used in automotive industry: A critical review, *Engineering Against Fracture* (2009) 59–70. https://doi.org/10.1007/978-1-4020-9402-6_5.

Schmidt, Tobias, Hamid Assadi, Frank Gärtner, Horst Richter, Thorsten Stoltenhoff, Heinrich Kreye, and Thomas Klassen, From particle acceleration to impact and bonding in cold spraying, *Journal of Thermal Spray Technology* 18 (2009) 794–808.

Singh, Harminder, T. S. Sidhu, and S. B. S. Kalsi, Cold spray technology: Future of coating deposition processes, *Frattura ed Integrità Strutturale* 6 (2012) 69–84.

Spencer, Kevin, and Ming Xing Zhang, The emergence of cold spray as a tool for surface modification, *Key Engineering Materials* 384 (2008) 61–74.

Tao, Yongshan, Tianying Xiong, Chao Sun, Lingyan Kong, Xinyu Cui, Tiefan Li, and Guang-Ling Song, Microstructure and corrosion performance of a cold sprayed aluminium coating on AZ91D magnesium alloy, *Corrosion Science* 52 (2010) 3191–3197.

Vilardell, A. Martín, Nuria Cinca, Amadeu Concustell, S. Dosta, I. Garcia Cano, and José María Guilemany, Cold spray as an emerging technology for biocompatible and antibacterial coatings: State of art, *Journal of Materials Science* 50 (2015) 4441–4462.

Villafuerte, Julio, and Wenyue Zheng, Corrosion protection of magnesium alloys by cold spray. In Czerwinski F (Ed.), *Magnesium alloys – corrosion and surface treatments*. Intechopen, Croatia, 2011, pp.185–194. ISBN 978-953-307-972-1.

Yao, Hai-Long, Xiao-Zhen Hu, Hong-Tao Wang, Qing-Yu Chen, Xiao-Bo Bai, Meng-Xian Zhang, and Gang-Chang Ji, Microstructure and corrosion behavior of thermal-sprayed hydroxyapatite/magnesium composite coating on the surface of AZ91D magnesium alloy, Journal of Thermal Spray Technology 28 (2019) 495–503.

Yin, Shuo, Pasquale Cavaliere, Barry Aldwell, Richard Jenkins, Hanlin Liao, Wenya Li, and Rocco Lupoi, Cold spray additive manufacturing and repair: Fundamentals and applications, *Additive Manufacturing* 21 (2018) 628–650.

Zhu, Jian, Xiang Cheng, Longmei Zhang, Xidong Hui, Yongling Wu, Hongyu Zheng, Zhiqiang Ren et al., Microstructures, wear resistance and corrosion resistance of CoCrFeNi high entropy alloys coating on AZ91 Mg alloy prepared by cold spray, *Journal of Alloys and Compounds* 925 (2022) 166698.

20 Polymer Nanocomposite Coatings

A. Madhan Kumar

CONTENTS

20.1 INTRODUCTION

Protective coatings are the most familiar and lucrative corrosion mitigation strategy for magnesium (Mg) and its alloys, which are more prone to environmental degradation, and they are the foremost choice for protecting Mg materials against corrosion in aggressive environments. In general, protective coatings can generate an external film, performing as a barrier surface to obstruct the permeation of exterior constituents such as moisture, oxygen, and aggressive ions. Polymers have extraordinary barrier characteristics in a wide variation in coating constituents and validate excellent adherence to metallic surfaces, if well designed, ascribed to the numerous functional groups and adsorption spots in the polymeric chain. To illuminate the importance of utilizing polymeric coatings in commercial applications, it has been reported that around $2 billion have been yearly spent on polymeric coatings in the USA (Pourhashem, Saba et al. 2020). However, these coatings could not deliver longstanding surface protection because of their brittle nature and hydrolytic degradation in aggressive conditions. Furthermore, microdefects are formed in polymeric film through the curing/crosslinking reactions due to release of vapourized solvents and exterior environmental stresses. The formed microdefects cause the permeation path for the aggressive species through coatings to metallic surfaces (Ramezanzadeh et al. 2015). To overcome these shortcomings, functional additives/fillers have been reinforced into polymeric matrix to improve the barrier performance. In this way, about 60 vol.% of microfillers are reinforced in polymeric film to attain the desired protective features. However, the higher loading of micro fillers exhibits an unfavorable effect on the chemical interactions between polymer/filler and the increase in the composite density (Azeez et al. 2013). In contrast, polymeric composite coatings with outstanding performance have been obtained through the use of fillers in nanometre scale with portions less than 2 vol.% in polymeric film (Ramanathan et al. 2008).

DOI: 10.1201/9781003319856-23

FIGURE 20.1 Advantages of polymer nanocomposites coatings.

A promising strategy for enhancing the overall performance of polymeric coatings and compensating for their undesirable features is to fabricate the polymer nanocomposite with the appropriate reinforcements of functional nanomaterials (**Figure 20.1**). The essential components of a simple polymer nanocomposite coating are polymeric resins and nanomaterial fillers. Most utilized nanomaterials as fillers in polymeric coatings, including ZnO, TiO_2, SiO_2, Al_3O_2, graphene, carbon nanotubes, and nanoclays. Generally, the nanofillers are broadly categorized into three main types based on the shapes such as zero-dimensional (0D), one-dimensional (1D), and two-dimensional (2D) (Nazari et al. 2022). The 0D nanoparticles (NPs) mostly used are sphere-shaped metal oxides (i.e. TiO_2, Al_2O_3, SiO_2).

The 1D nanomaterials are nanotubes, nanowires, nanofibers, and nanorods, which have a diameter of less than 100 nm and length up to few hundred nanometres. Further, clay and graphene are renowned 2D nanomaterials with a sheet-like assembly. Numerous researches have revealed that these nanomaterials exhibit distinctive characteristics of higher surface area, homogeneous surfaces, and regulatable length-to-width ratio that can extend the permeation routes of aggressive vapours and ions and hence improve the surface protective performance of the polymeric films (Shi et al. 2009). In addition, as UV-resistant materials, the reinforcement of nano ZnO and TiO_2 in the polymeric matrix enhances weathering and anti-bacterial resistance of the coatings. A wide variety of polymer nanocomposite coatings (PNCs) have been reported for Mg alloys with different

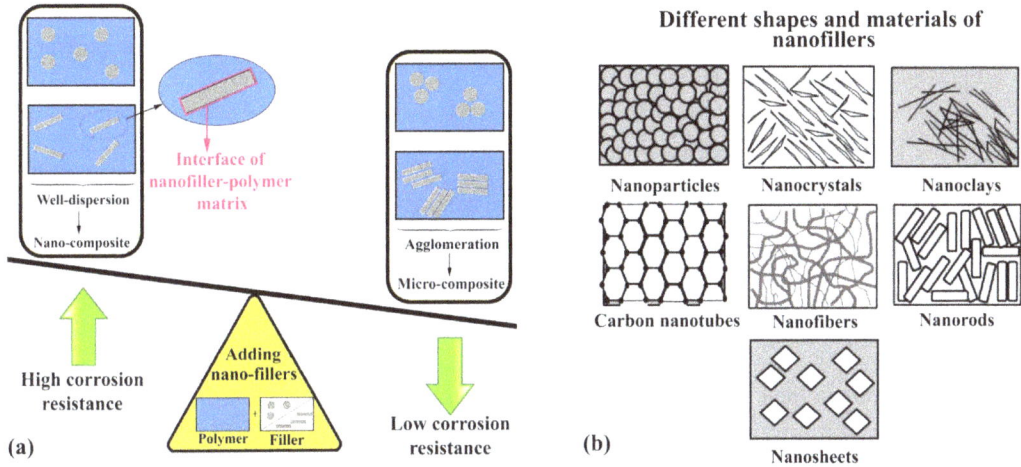

FIGURE 20.2 Schematic representation of (a) dispersion degree of nanofillers in the polymeric matrix and (b) different shape of the nanomaterials. *Reproduced with permission from (Pourhashem, Saba et al. 2020) © 2020 The Korean Society of Industrial and Engineering Chemistry. Published by Elsevier B.V.*

characteristics, such as superhydrophobic, self-cleaning, UV resistant and enhanced mechanical durability, that can be prepared with a moderately low methodical effort.

The distribution of reinforced nanomaterials inside polymeric matrix is an important feature in governing the overall characteristics and performance of prepared PNCs. As presented in **Figure 20.2**, the nano-reinforcements can be either dispersed well or poorly dispersed inside the polymeric film and it is significant to ensure the uniform distribution of nanomaterials within the polymeric matrix in order to deliver high protection performance (Gray and Luan 2002; Pourhashem, Saba et al. 2020).

The agglomeration of nanomaterials is the main issue in reinforcing them into the polymer matrix. Several strategies have been already established to overcome this issue by preventing the nanofiller's agglomeration. In this respect, nanomaterials can be reinforced homogeneously through suitable synthesis routes, using shear mixers, and chemically altering the nanofiller's surface to improve the chemical interactions between nanofillers and polymeric chains (Rong et al. 2006). Further, the overall performance of PNCs is strongly governed by the surface morphology and shape of the reinforced nanofillers. In addition, the configurations of nanofillers and their volume fractions are another vital factor in the desired characteristics of PNCs. Uniformly distributed nano-materials inside polymeric matrix with improved interfacial interactions can form a continuous assembly with transformed polymeric chains; delivering novel high-performance PNCs with the desired characteristics.

20.2 POLYMER MATRIXES

Protecting a reactive Mg surface by covering it with a polymeric protective coating is a keen route to attain the corrosion mitigation and acquire desired surface features without compromising the mechanical characteristics. The selection of appropriate polymeric matrix is the main concern, which is majorly dependent on the operating conditions of the base metallic materials. PNCs have been mainly prepared through three methods, namely solution exfoliation, melt intercalation, and *in situ* polymerization (Zhang et al. 2020). Recently, importing of covalent and non-covalent groups on the surface of the nanofillers have been efficiently utilized to improve the chemical interactions between polymer and nanofillers and also bring more compatibility with polymeric molecules in their chain, which further leads to attaining effective distribution of nanofillers in the polymeric coatings (Shahryari et al. 2021).

20.2.1 CONDUCTING POLYMERS

In developing multifunctional and high-performance coatings, researchers explored novel appropriate polymeric materials that are lucrative, easy to process, durable, and possess notable features. In this juncture, conducting polymers (CPs) possessing high electrical conductivity, light weight, good thermal and environmental stability, and low-cost processing routes are an absolute match (Namsheer and Rout 2021). These attracting characteristics attracted significant attention from the coating's scientists. CPs have also labelled as conjugated polymers since they have alternative single and double bonds in their polymeric chain, and their conductivity is governed by the delocalized configurations.

Generally, the intrinsic conductivity of CPs is ranging from 10^{-5} to 10^{-16} S/cm (**Figure 20.3**). The polyaniline (PANI), polypyrrole (PPy), polythiophene (PT), and poly (3,4 dimethoxy thiophene) (PEDOT) are the commonly utilized CPs in anticorrosion applications (**Figure 20.3**). Undoped structure of CPs usually exhibits insulating features, whereas doped structure provides electrical conductivity. The corrosion-protection mechanism using the CP is not completely elucidated so far, and a few important mechanisms were proposed, as follows (Deshpande et al. 2014):

1 the CP could perform as a physical barrier film
2 the CP could act as a passivation film

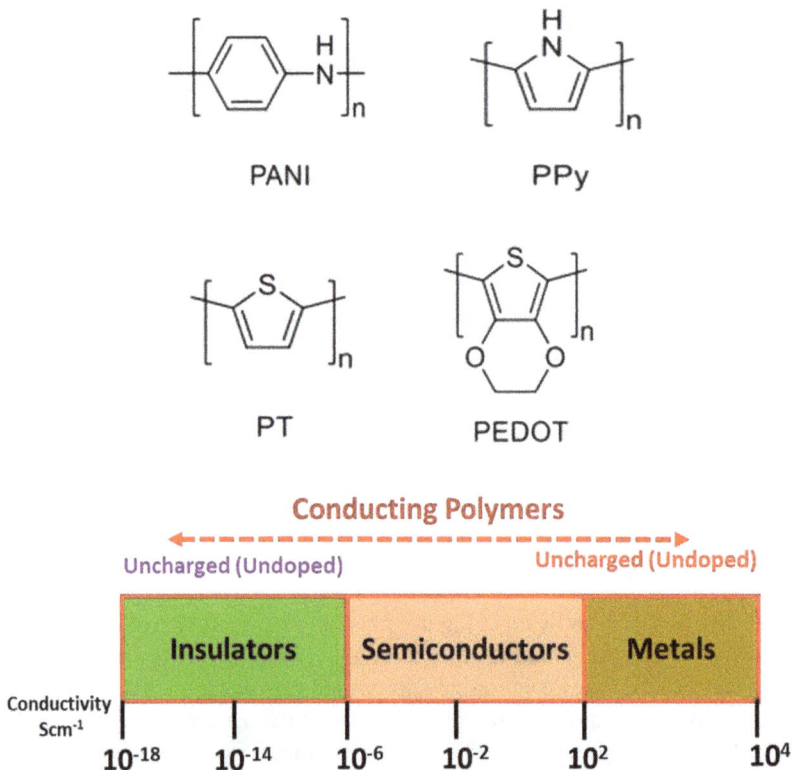

FIGURE 20.3 Examples of conducting polymers and their conductivity range.

3 the CP could cathodically protect the metallic surface
4 the CP could function as a binder or reservoir for inhibiting species and discharge them during the change in their redox state

CPs can also be utilized to protect the Mg alloys from corrosion. Truong et al. first utilized PPy as additives in an acrylic coating, and their obtained results revealed that the surface protective action of acrylic coatings is improved with the addition of PPy particles (Truong et al. 2000). Sathiyanarayanan et al. investigated the effect of PANI in an epoxy coating on Mg alloy and compared their results with that of strontium chromate–containing epoxy coating and revealed that PANI-containing coating exhibited improved corrosion protection than the chromate reinforced coating (Sathiyanarayanan et al. 2006). In another investigation (Sathiyanarayanan et al. 2007), PANI and PANI–TiO_2 nanocomposite reinforced with an acrylic coating was studied, and the results indicated that the PANI–TiO_2 composite containing coating showed higher corrosion protection than that of a coating containing PANI alone. These results are possibly accompanied by the homogeneous distribution of the reinforced nanocomposites in the coating, which enables effective protection for the Mg alloy.

Jothi et al. have performed the anodization technique on AZ31 Mg alloy (**Figure 20.4**) and subsequently deposited a PPy coating with the inclusion of different gelatine content, and their obtained results revealed that the hydrogen evolution from the Mg alloy was decreased pointedly in SBF medium in the synchronized existence of the anodized film and gelatin included PPy coating (Jothi, Adesina, Madhan Kumar et al. 2020). In spite of the accomplishments achieved so far in conducting polymer coatings, still, enormous efforts are needed to utilize it as an effective substitute for conventional coatings. Progressive research is essential to exactly elucidate the corrosion protective phenomenon of CPs.

20.2.2 HIGH-PERFORMANCE POLYMERS

High-performance polymeric coatings were utilized mainly to protect Mg components in industrial applications and mostly consist of different components including a polymeric resin binder, inhibitive pigments, and several functional additives including dryers, dispersion, stabilizing and hardening agents. This coating could act as a physical barrier and also a corrosion-inhibiting layer to

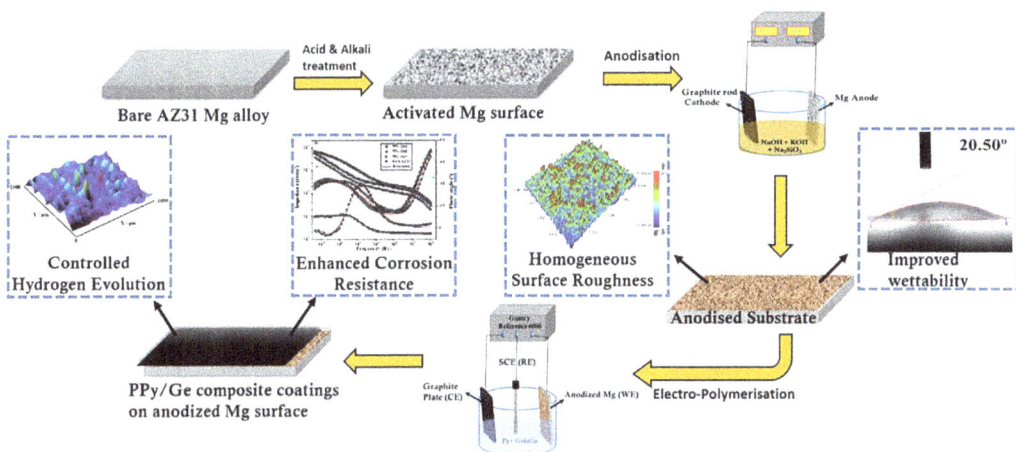

FIGURE 20.4 Schematic representation of fabrication and characteristics of polypyrrole/gelatin composite coating on anodized AZ31 Mg alloy. *Reproduced with permission from (Jothi, Adesina, Madhan Kumar et al. 2020) © 2019 Elsevier B.V.*

protect the Mg surface from aggressive environments. These coatings can be employed as a single coat or one of the layers in a multiple-coating system. The primary step in developing the high-performance coatings is to select a proper polymeric alkali-resistant resin including epoxy, polyurethane, polyvinyl butyral, and acrylic derivatives. To further enhance the performance and add other characteristics, pigments and additives can be added in the coating. Polymeric resins based on polyesters, acrylates, and polyurethane, which are generally utilized for steel or aluminum, are not completely effective for Mg and its alloys (Hu et al. 2012). This is possibly due to the low adhesion, low UV resistance, low stability in alkaline environment, and brittleness of these coatings on the Mg surface, etc.

Mg-rich primers for protecting Mg alloys have been discovered to enhance the anticorrosion performance of epoxy coatings, similar to zinc-rich primers for steels (Lu et al. 2011). Shen and Zuo prepared a Mg-rich epoxy primer with the incorporation of ZnO nanoparticles on AZ91D Mg alloy and reported that the incorporated ZnO particles enhanced the crosslinking density of the epoxy chain, diminished pores and microdefects, and improved the adhesion and barrier property of the coating (Shen and Zuo 2014). Jothi et al. deposited polyurethane coatings on anodized AZ31 Mg substrates and evaluated the corrosion protection performance in NaCl solution. The achieved data illustrated that the adhesion and corrosion-protective action of PU coatings was enhanced considerably in the presence of the anodized film on Mg alloy surface (Jothi, Adesina, Rahman et al. 2020). Xie et al. fabricated a superhydrophobic coating based on polydimethyl siloxane (PDMS) and different sized SiO_2 NPs (40 nm and 50–250 nm) on a Mg alloy and reported significantly enhanced surface hydrophobicity. Corrosion and wear performance was depended on the size and extent of SiO_2 NPs addition (Xie et al. 2018).

20.2.3 BIOPOLYMERS

For the past decade, numerous research works have focused on using Mg and its alloys as biodegradable implants in biomedical applications (Sarian et al. 2022). However, achieving the controlled degradation of Mg in a physiological medium is a challenging task for the researchers. In this juncture, biocompatible and biodegradable polymeric coatings have been utilized on Mg implants to account for a slow degradation that can be easily regulated (Li et al. 2018).

Recently, coatings made of biodegradable polymers have undergone substantial research to enhance Mg implants' biocompatibility and/or healing characteristics. Preferably, the biodegradable polymer used should be simple to metabolize and excrete via the physiological route, degrade to non-toxic compounds, and not cause an inflammatory reaction *in vivo*. Based on the preparation route, biodegradable polymeric coatings are generally categorized into natural and synthetic polymers (Tipan et al. 2022). Collagen, cellulose, chitosan, and pectin are well-known examples for natural polymers. In recent years, silk fibroin derived from polymeric structural protein of silk fibers has also been explored to protect Mg alloys in the physiological medium. On the other hand, the most studied synthetic polymers are polycaprolactone (PCL), poly (lactic acid) (PLA), poly (glycolic acid) (PGA), and poly(lactic-co-glycolic acid) (PLGA) (**Figure 20.5**).

These polymers can also be used for diagnostic imaging for healing evaluations since they have appropriate mechanical characteristics and are degradable in human tissue (Ibrahim et al. 2015; Tian et al. 2016). In addition, the available heteroatoms such as nitrogen and oxygen in the polymeric chain can behave as adsorption spots to inhibit the corrosion on Mg surface. Degradation time for synthetic polymers usually ranges from 1 month to 1 year, and the degradation rate can be governed by glass transition temperature, pH hydrophilicity, and molecular weight. Most natural polymers are soluble in water and can be crosslinked to produce a water-insoluble polymeric chain. The crosslinking tendency influences the drug-release profile from the host matrix fabricated through these polymers. The degradation occurs through an enzymatic process, and its rate depends on purity, crosslinking tendency, molecular weight, and the availability of enzymes near bone tissues. These circumstances further control the drug-release rate from the host polymer matrix (Kravanja

CLASSIFICATION OF BIOPOLYMER	ADVANTAGES	LIMITATIONS
NATURALS  CHITOSAN DEXTRAN 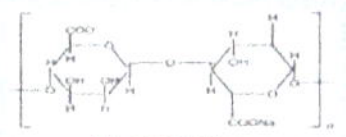 ALGINATE	✓ Hydrophilic, biocompatible, ✓ Cell/tissue specific binding affinity, safe ✓ Readily available	➤ Possible immunogenicity, ➤ Require purification ➤ Less controlled degradation ➤ Short release profile
SYNTHETIC 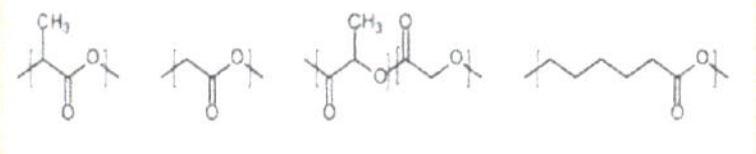 PLA PGA PLGA PCL	✓ Design desired features ✓ Easy to add functional groups ✓ Precise controlled release ✓ No immunogenicity ✓ Control of mechanical and physical features	➤ Required ligands attached to achieve cell/tissue specific binding affinity ➤ Require synthesis ➤ Scale up challenges ➤ Hydrophobic

FIGURE 20.5 Advantages and limitations of natural and synthetic biopolymers.

and Finšgar 2022). However, natural polymers lack a reproducible degradation rate and often possess an uncontrolled drug release.

In recent years, several research works have focused much attention to enhance the controlled degradation of the biopolymer coatings via reinforcing them with TiO_2, Ag, MgO, graphene oxide, and metal-organic frameworks (MOFs) (Zhou et al. 2022; Zhao et al. 2020; Ge et al. 2022). Jia et al. reinforced amorphous calcium carbonate particles into PCL coating and reported that a layer containing Mg phosphate was formed by the dissolution in SBF medium, which further improved the corrosion resistance of the Mg implant. Moreover, the surface analyses after immersing in SBF revealed a higher extent of Mg-modified amorphous calcium carbonate and apatite precipitation on the Mg surface, indicating its excellent biomineralization tendency (Jia et al. 2019). Abdelrahman et al. prepared a PCL nanocomposite coating with the inclusion of hydroxyapatite (HA) NPs reinforced with simvastatin on AZ31 alloy to expand the invitro corrosion resistance and osteocompatibility (Rezk et al. 2019).

20.3 NANOREINFORCEMENTS

The utilization of nanotechnologies in coatings is strategic and permits coatings to generate multifunctional, novel functionalities like superhydrophobic, self-cleaning, UV resistant and enhanced mechanical durability are the emerging demands for highly precise applications. Nanocomposite coatings developed by reinforcing functional nanofillers in polymeric matrix have been considered as novel corrosion-mitigation materials that exhibit improved surface protective action and mechanical strength compared to their pure polymeric coatings. In general, nanomaterials such as metal and metal oxide nanoparticles, nanolayered materials, carbides, and nitrides offer substantial potential for improving the barrier performance of polymeric coatings. The inherent features of nanofillers as well as their surface morphology, functional groups, dimensions, and loading fractions can affect the overall performance of PNCs.

20.3.1 METAL OXIDE NANOPARTICLES

Various nanometal oxides, including TiO_2, SiO_2, ZnO, Al_2O_3, CeO_2, Fe_2O_3, Fe_3O_4, and ZrO_2, have been utilized as corrosion-inhibitive nanofillers. These metal oxide nanofillers can cover the pores and microdefects of polymeric coatings and improve the barrier features of polymeric film against the penetration of aggressive species (Sahoo et al. 2010). They can also deliver additional advantages; for example, ZnO NPs is one of the most commonly employed nanofillers in polymeric coatings to accomplish hydrophobicity, higher hardness, lower refractive index, and improved barrier performance (Ramezanzadeh et al. 2011). Among the different crystalline structures of TiO_2, rutile TiO_2 NPs are generally employed as inorganic pigment due to their exclusive features such as eco-friendliness, corrosion and chemical resistance, good photocatalytic and electrical features, capability to absorb UV radiation, etc. (Peng et al. 2020). SiO_2 as a nanofiller enhances the hydrophobicity, mechanical durability, and chemical and thermal stability of polymeric resins. ZrO_2 displays intrinsic characteristics such as effective strength, high hardness, fracture toughness, and wear and chemical resistance, making them a suitable nanofiller for PNCs (Behzadnasab et al. 2011). The rare earth metal oxide, CeO_2, is also explored due to its characteristics such as low water solubility, suitable optical and electrical properties, and high corrosion resistance. Nazeer et al. prepared a poly(butyl methacrylate) nanocomposite coating with the addition of graphene oxide (GO) and TiO_2 on AZ31 Mg alloy. Their obtained results indicated that the reinforcement of GO in the PNC significantly reduced the permeation of aggressive species. The reinforced TiO_2 disclosed charge transfer obstruction at the metal/solution interface (Nazeer et al. 2019).

20.3.2 CARBON NANOSTRUCTURES

Due to the high specific surface area and favorable mechanical, electrical, and thermal characteristics of carbon nanostructures, the PNCs reinforced with the carbon nanostructures can significantly enhance the overall performance. Among the different carbon nanostructures (**Figure 20.6**), carbon nanotubes (CNTs) and graphene (GPE) are the most commonly utilized carbon nanomaterials in protective coatings due to their high aspect ratio, good electrical conductivity, and thermal and mechanical stability (Pourhashem, Ghasemy et al. 2020). Raja et al., in an earlier work, prepared a PNC based on polyurethane (PU) and acrylic resins reinforced with the graphite, and their findings revealed that the reinforcements of graphite in acrylic coating pointedly reduced the electrical resistance. Besides, the coating did not exhibit adequate corrosion resistance due to formation of galvanic cells between Mg surface and graphite. In contrast, PU coating hinders the formation of galvanic cells and thus noticeably enhances the protective action (Raja et al. 2010).

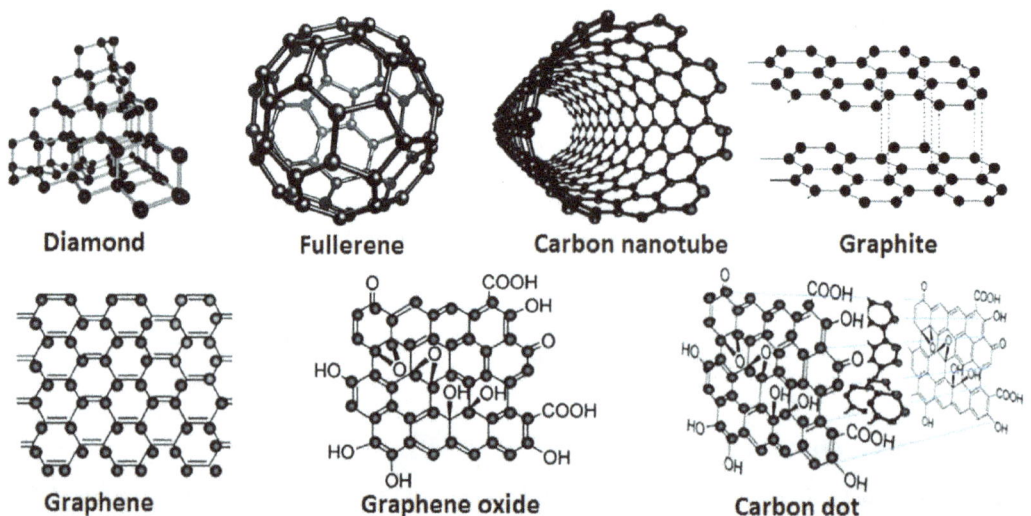

FIGURE 20.6 Schematic representation of different carbon nanostructures. *Reproduced with permission from (Yan et al. 2016) © Under CC-BY, The Royal Society of Chemistry.*

It is well known that the Van der Waals radius of carbon atoms on GPE is around 0.11 nm, and their presence in the coating can effectively impede the permeation of most aggressive ions. However, the insoluble nature of GPE in water or other organic solvents is a major concern, leading to minimal dispersion in the coating matrix (Cui et al. 2021). Hence, the derivatives of GPE, including graphene oxide (GO) and reduced graphene oxide (rGO), have been introduced as nanofillers in polymeric coatings. Unlike GPE, its derivatives have functional groups containing hydrophilic oxygen, such as carboxyl, hydroxyl, and epoxy groups, on their surface, which permit them not only to improve the solubility but also to perform as functional polymers, providing bonding sites for attaching other nanomaterials (Cui et al. 2019). Hence, both the compatibility and distribution of GPE inside the polymeric matrix can be significantly enhanced, and numerous research works have preferred GPE derivatives as nanofillers in PNCs.

Alternatives to graphene materials, graphitic carbon nitrides (GCN) and hexagonal boron nitrides (h-BN), can also be utilized. GCN exhibits a stacked configuration containing π-conjugated graphitic planes generated by sp2 hybridisation of the C and N atoms (Yang et al. 2019). The layered h-BN possess alternating boron and nitrogen atoms in a honeycomb structure and weak Van der Waals interactions present between the h-BN planes. Researchers have focused on using these materials to substitute the GPE based nanofillers (Teijido et al. 2022; Hu et al. 2012).

20.3.3 Emerging Materials

In recent years, metal organic framework (MOFs)-based PNCs have revealed prodigious capacity in developing protective coatings for Mg with effective and strong corrosion-protective performance. MOFs are a kind of nanoporous materials configured by linking organic ligands into inorganic metal clusters or ion-based nodes into stretched networks. The tuneable porous assembly, controllable composition, many active sites, and flexible release features of MOFs make them a perfect material for emerging high-performance PNCs with multiple functionalities, including self-healing, superhydrophobicity, and physical barrier effects (Jiang et al. 2022). Contrasting to conventional nanofillers acting only as the passive fillers hindering the permeation route of aggressive species into polymer matrix, the adaptability of MOFs provides a remarkable potential to construct novel candidates in protecting the Mg surface against diverse aggressive environments. Several kinds of MOFs have good compatibility with many polymeric matrixes owing to their functional groups (active sites) in organic ligands of MOFs, which can generate non-covalent and/or covalent bond interactions with the polymeric molecules. Further, 2D MOF nanofillers with lamellar morphology, high active sites, and exclusive chemical structure could instantaneously provide good imperviousness and compatibility to the polymeric matrix (Zhou et al. 2022).

Ren et al. developed an epoxy nanocomposite coating reinforced with the zinc gluconate (ZnG)-encapsulated ZIF-8 nanocontainers (ZnG@ZIF-8), and their obtained results revealed that the prepared nanocomposite coatings exhibited significant improvement in corrosion-protection performance compared to that of pure epoxy coating. They further proposed that this improvement was ascribed with the homogeneous distribution of reinforced ZnG@ZIF-8 nanocontainers in the epoxy matrix (**Figure 20.7**) (Ren, Chen et al. 2020).

In general, small molecular parts or monomers can frequently be introduced easily into the nanochannels of MOFs. Then, these small molecular constituents undergo *in situ* polymerization in the nanochannels of MOFs to achieve an effective nanocomposite. For instance, the small aniline monomers can be *in situ* polymerized on the pores of MIL-101 MOF, and the controlled structure of PANI covers the nanochannels of the MIL-101 (**Figure 20.8**) that further circumvent the loading issue of PANI in epoxy coatings. It is well known that the pure epoxy coating is easily attacked by aggressive ions due to its defects and pores formed during curing, which enable the formation of corrosion channels. However, epoxy coating reinforced with PANI@MIL-101 has made the coating compact and denser, which aids in covering the defects and pores and then enhancing the anti-permeation capacity. In addition, PANI is an effective corrosion inhibitor for metallic surfaces, and

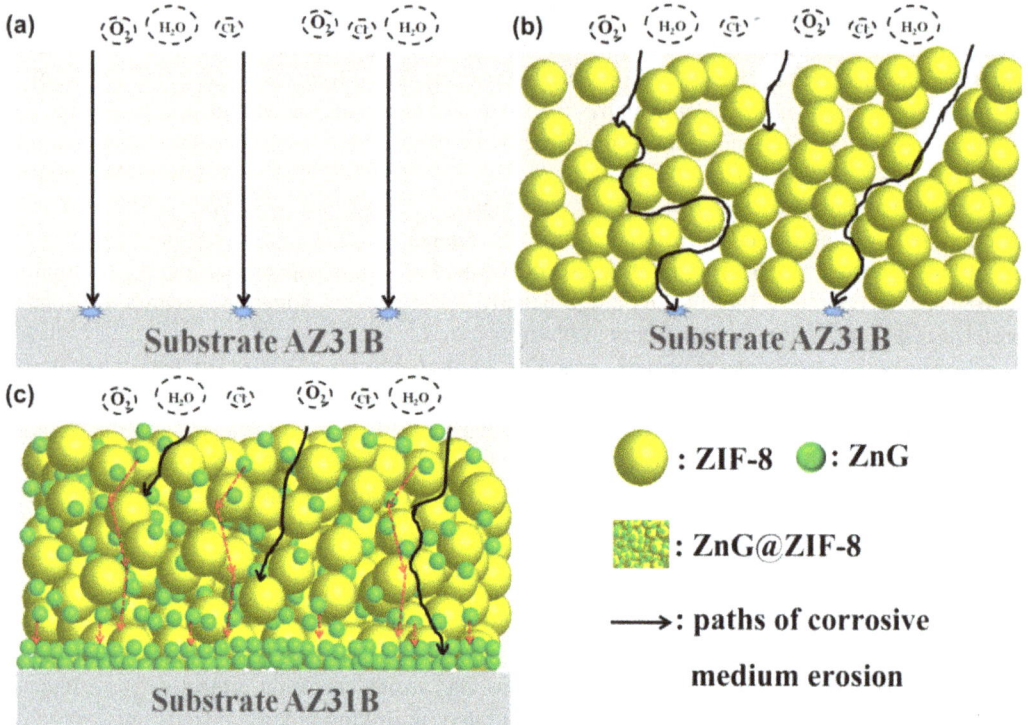

FIGURE 20.7   Schematic diagram of proposed corrosion protection mechanism by (a) pure epoxy coating, (b) epoxy/ZIF-8, and (c) Epoxy/ZIF-8/ZnG coating. *Reproduced with permission from (Ren, Chen et al. 2020) © 2019 Elsevier B.V.*

self-polymerization of aniline into pores of MIL-101 not only enhanced the loading of PANI but also regulated the length of PANI chains (Ren, Li et al. 2020).

Nanofillers based on the MOFs have been utilized as both passive and active nanofillers. Passive nanofillers based on MOFs can further improve the barrier performance of composite coatings in addition to reducing the number of micropores in the coating matrix. Abundant functional groups existed in the organic linkers of MOFs can improve the distribution and compatibility of nanofillers in the composite coatings (Zhou et al. 2022). On the other hand, active nanofillers can inevitably self-heal the microdefects in the polymeric films and reinstate the protecting effect of the coatings to some extent. In general, active nanofillers based on MOFs are categorized into two kinds: the MOFs nanocontainers and composite nanofillers using MOFs as triggers to have controlled release of healing agents. In the former case, healing agents are directly incorporated into the cavity and pores of MOFs, and MOFs instantaneously function as nanocarriers with controlled release. For the latter, the healing agents are encapsulated in other porous materials, and MOFs only perform as the nanovalves on the surface of nanocomposites to achieve the controlled release. The *in situ* preparation approach, as one of the main effective strategies, can directly load healing agents into the MOF nanochannels during the synthesis of MOFs or healing agents from small components to larger molecules in the cavity of as-prepared MOFs (Kitao et al. 2017). In comparison with the passive nanofillers that chiefly fill the micro-defects and decrease the dispersion pathways of aggressive species in the polymeric matrix, the active nanofillers generally start to act after the local failure arises in the polymeric coating. When the polymeric coating is degraded/destroyed, the healing agents from active MOFs can vigorously cover the defects and extend the operation life of the coatings (Wu et al. 2021).

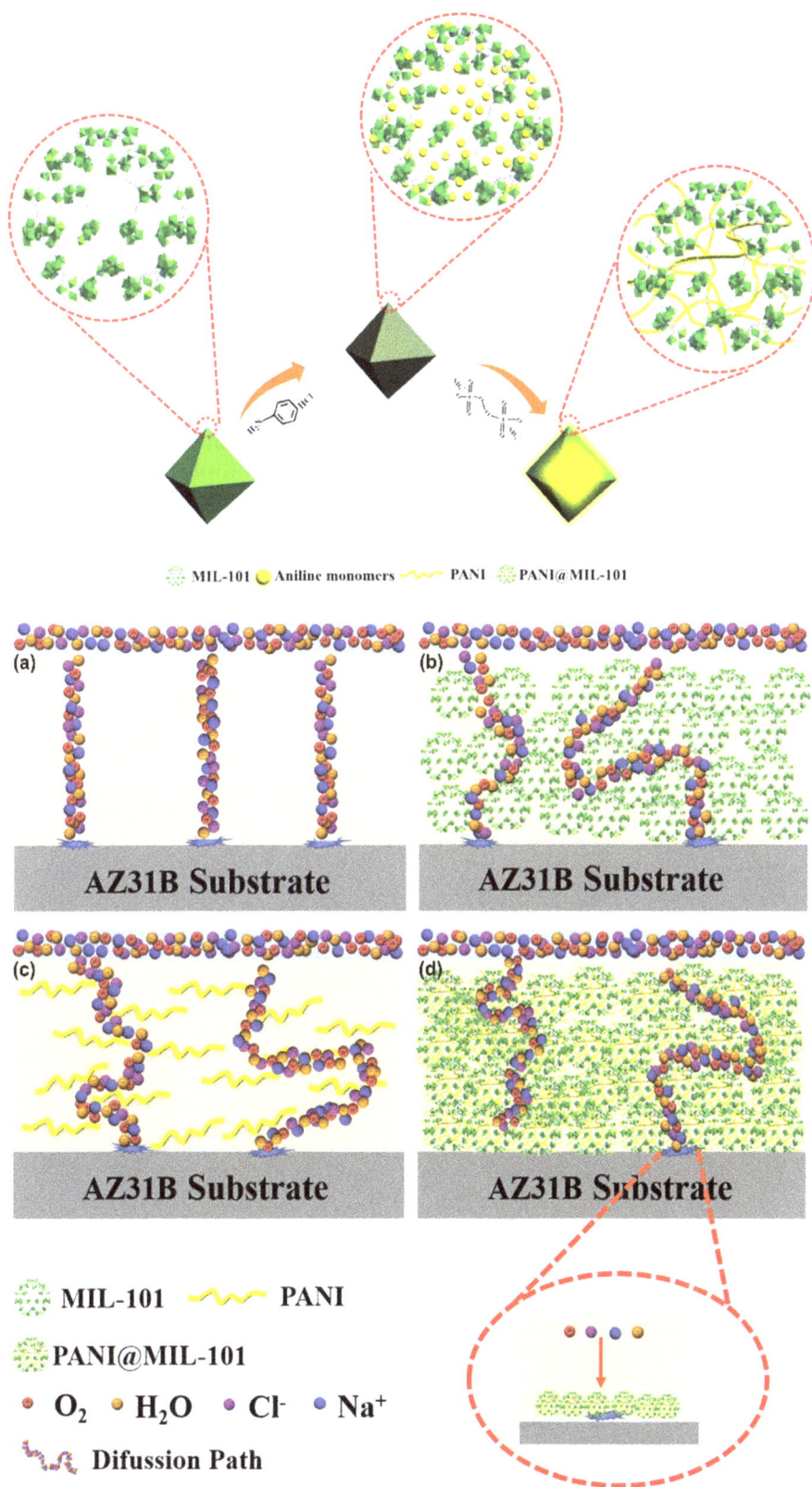

MIL-101　○ **Aniline monomers**　～～ **PANI**　**PANI@MIL-101**

MIL-101　～～ **PANI**

PANI@MIL-101

● **O₂**　● **H₂O**　● **Cl⁻**　● **Na⁺**

Difussion Path

FIGURE 20.8 (A) Schematic representation of preparation of PANI@MIL-101 and (B) their corrosion protection mechanism on AZ31 Mg alloys, (a) Epoxy coating, (b) Epoxy/MIL-101, (c) Epoxy/PANI and (d) Epoxy/PANI@MIL-101 coatings. *Reproduced with permission from (Ren, Li et al. 2020) © 2020 Elsevier Inc.*

20.4   KEY PARAMETERS IN POLYMER NANOCOMPOSITE COATING FABRICATION

One of the main factors regulating the dispersion of nanofillers in the polymeric matrix is the preparation method. PNCs have been obtained mainly using the three chief strategies: *in situ* polymerization (ISP), solution exfoliation (SE), and melt intercalation (MI) (**Figure 20.9**). In the SE route, the polymer and nanofillers were separately dispersed in suitable solvents and then mixed together using the high-shear mechanical stirring or ultrasonic methods. During this, the polymer molecules were attached on the nanofiller's surface by intercalating and displacing the solvent between them. Finally, the unused solvent is eliminated through precipitation or vapourization. In the MI approach, the polymeric resin is heated at above its softening temperature, and the nanomaterials are directly inserted into the melted polymer matrix under continuous shear mixing. However, the range of temperature should be carefully monitored to stop the thermal degradation of the polymer. In the ISP route, the nanofillers are added in monomer solution when the polymerization is obtained by the application of heat, diffused initiator, and irradiation, etc. The distribution capacity of nanofillers mainly depends on rate of the polymerization, and it is frequently utilized for polymers with inadequate solubility in solvents.

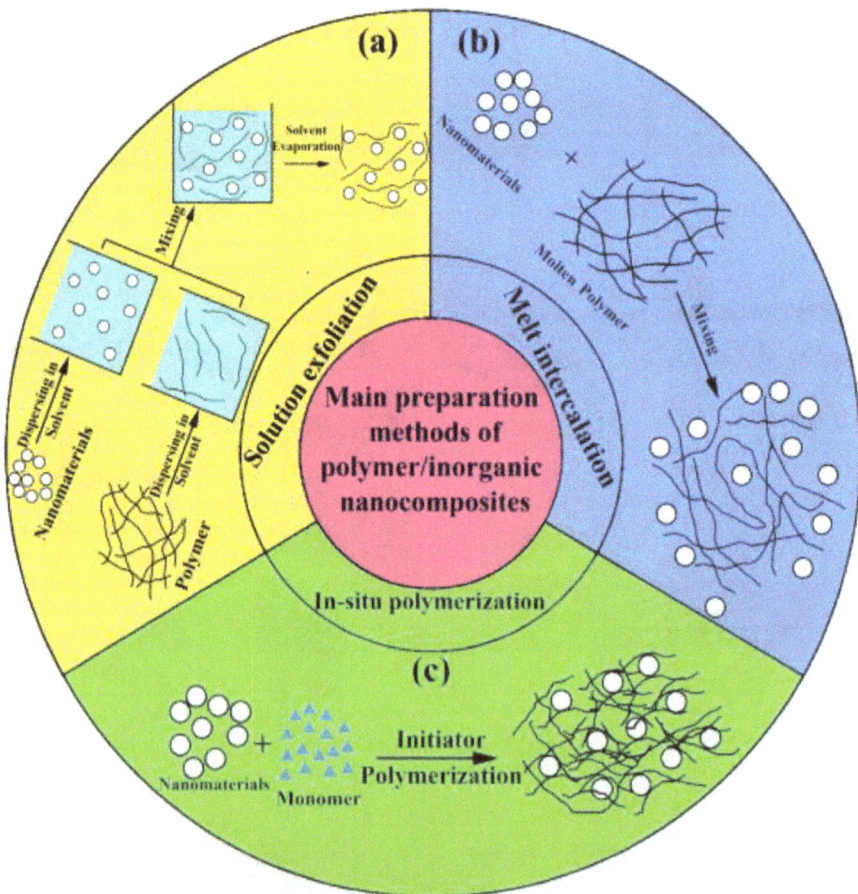

FIGURE 20.9   Important approaches to develop polymer nanocomposite coatings: (a) solution exfoliation, (b) melt intercalation, and (c) *in situ* polymerization. *Reproduced with permission from (Pourhashem, Saba et al. 2020) © 2020 The Korean Society of Industrial and Engineering Chemistry. Published by Elsevier B.V.*

The MI and SE techniques are the widely employed routes to prepare PNCs and these approaches form Van der Waals interactions between the polymeric matrix, and the reinforced nanofillers. In contrast, ISP route forms covalent bond interactions between polymeric matrix and reinforced nanofillers. Further, mechanical dispersion approaches using high-speed mixing, impact, or shearing forces improve the chemical interaction between nanofillers and matrix to consistently disperse them, whereas ultrasonic treatments can reduce the interactions between nanofillers and avoid their agglomeration. Hence, the type and nature of the methods used for developing PNCs have an enormous influence on the dispersion degree of nanofillers inside polymeric coatings, the interfacial chemical interactions at nanofiller/polymeric matrixes, and the subsequent surface protective efficiency of PNCs.

20.5   CONCLUSIONS AND OUTLOOK

Continuous progress in nanotechnology permitted scientists to design and develop novel PNCs linking the desired characteristics of polymers and nanofiller materials providing more than one surface protective mechanism in a single coat. A survey of literature obviously suggested that salient features of the polymer and the functional nanofillers can be effectively optimised to develop high-performance PNCs for protecting the Mg and its alloys in aggressive environments. In comparison with the conventional fillers, nanofiller materials possess numerous merits such as reducing the amount of fillers in polymer matrix and enhancing the strength-to-weight ratio, improving the interfacial interactions between the polymer chain and nanofiller, achieving lengthy permeation routes for penetrating aggressive species, and improving the corrosion-protection efficiency. In addition, multifunctionalities such as UV resistance, mechanical durability, self-cleaning, and thermal and mechanical stability can also be achieved via the intrinsic features of appropriate nanofillers.

The important facts determined from this chapter that can upcoming explorations to develop high-performance multifunctional PNCs:

1 The quality of a nanofiller's distribution inside the polymer matrix has considered a significant factor in attaining the effective barrier protection of the resultant coatings. The amount of reinforced nanofillers in the polymeric matrix, the preparation techniques used (such as *in situ* polymerization, solution intercalation, mechanical stirring, and melt processing), and the functional groups and chemical structure of the nanofillers are the major parameters in influencing their homogeneous dispersion inside the coating matrix.

2 Agglomeration of nanofillers is the main concern during the preparation of PNCs, and numerous approaches related to the mechanical, physical and chemical treatments have been adopted to solve it. Further, surface treatments have been broadly employed to form effective covalent bond or chemical interaction between the polymeric chains and nanofillers and also modify the polymeric network. However, these available treatments exhibited some demerits, such as damaging the original structure of nanofillers and using toxic chemicals for surface treatments. Thus, more dedicated research is desired to develop eco-friendly surface treatments without compromising the inherent features of nanofillers. It is anticipated that future-generation PNCs would be based on eco-friendly surface treatments to utilize the merits of nanofillers and green corrosion inhibitors.

3 The surface morphology and shape of the nanofillers are the other significant parameters in deciding the uniform dispersion of nanofillers in polymer matrix. Researchers validated that the nanosheets and nanorods could efficiently enhance the protective action of polymeric coatings in comparison with the nanoparticles. Certainly, nanofillers with rodlike or sheetlike morphology with higher surface areas can perform better as horizontal barriers against penetration of aggressive species in the polymeric coating.

4 The chief role of nanofillers inside polymer matrix is to reduce the microdefects, such as pores and cracks, thereby improving its barrier features. In addition, the inherent features

of inorganic nanofillers can also improve the other important features, leading to development of multifunctional coatings with extended operational life. Reinforced nanofillers can regulate the surface characteristics of PNCs through improving the surface hydrophobicity and enhancing the adhesion strength of polymeric coatings and impeding the penetration of violent species to the coating/metal interface. In general, sacrificial corrosion protection was achieved when the polymeric coatings were reinforced with metal nanofillers like Mg or Zn nanoparticles. Passive corrosion protection was obtained when polymer coatings were reinforced with metal oxide nanofillers through the dissolution of metal oxides followed by the formation of passive layer due to the reaction of dissolved metal oxides with hydroxy groups at cathodic sites of the metallic substrate.

Further, it is still essential to elucidate the surface-protective tendency of PNCs and develop innovative, cost-effective strategies to use the salient features of nanofillers inside polymer coatings. It is vital to develop eco-friendly routes to functionalise the nanofillers to achieve good compatibility inside polymer matrixes as well as to import desired functionalities, including superhydrophobicity, self-cleaning, and UV resistance. As the nanofillers with 2D morphology can effectively improve the corrosion protection performance of polymeric coatings, it is anticipated that nanomaterials with sheetlike morphology such as hBN and GCN exhibit high capability in next-generation PNCs. Recently, the addition of nanofillers on the surface of graphene-based nanosheets has attracted much attention for achieving the advantages of different nanomaterials instantaneously to attain high-performance PNCs with desired multifunctionality. In addition, inorganic nanomaterials or MOFs can also be utilized for the encapsulation of green corrosion inhibitors to attain self-healing coatings, which can release loaded inhibitors when cracks are created in the polymeric film in corrosive media.

REFERENCES

Azeez A. A., K. Y. Rhee, S. J. Park, D. Hui, Epoxy clay nanocomposites – processing, properties and applications: A review, *Comp. B: Eng.* 45 (1) (2013) 308.

Behzadnasab M., S. M. Mirabedini, K. Kabri, S. Jamali, Corrosion performance of epoxy coatings containing silane treated ZrO_2 nanoparticles on mild steel in 3.5% NaCl solution, *Corros. Sci.* 53 (2011) 89–98.

Cui G., Z. Bi, R. Zhang, J. Liu, X. Yu, Z. Li, A comprehensive review on graphene based anti-corrosive coatings, *Chem. Eng. J.* 373 (2019) 104–1210.

Cui G., C. Zhang, A. Wang, X. Zhou, X. Xing, J. Liu, Z. Li, Q. Chen, Q. Lu, Research progress on self-healing polymer/graphene anticorrosion coatings, *Prog. Org. Coat.* 155 (2021) 106231.

Deshpande P. P., N. G. Jadhav, V. J. Gelling, Dimitra Sazou, Conducting polymers for corrosion protection: A review, *J. Coat. Technol. Res.* 11 (4) (2014) 473–494.

Ge X., R. Wong, A. Anisa, S. Ma, Recent development of metal-organic framework nanocomposites for biomedical applications, *Biomater.* 281 (2022) 121322.

Gray J. E., B. Luan, Protective coatings on magnesium and its alloys–a critical review, *J. Alloys Compd.* 336 (2002) 88–113.

Hu R-G., Su Zhang, Jun-Fu Bu, Chang-Jian Lin, Guang-Ling Song, Recent progress in corrosion protection of magnesium alloys by organic coatings. *Prog. Org. Coat.* 73 (2012) 129–141.

Ibrahim A. M. S., P. G. L. Koolen, K. Kim, G. S. Perrone, D. L. Kaplan, S. J. Lin, Absorbable biologically based internal fixation, *Clin. Podiatr. Med. Surg.* 32 (2015) 61–72.

Jia S. Q., Y. T. Guo, W. Zai, Y. C. Su, S. S. Yuan, X. S. Yu, Y. C. Xu, G. Y. Li, Preparation and characterization of a composite coating composed of polycaprolactone (PCL) and amorphous calcium carbonate (ACC) particles for enhancing corrosion resistance of magnesium implants, *Prog. Org. Coat.* 136 (2019) 105225.

Jiang L., Y. Dong, Y. Yuan, X. Zhou, Y. Liu, X. Meng, Recent advances of metal – organic frameworks in corrosion protection: From synthesis to applications. *Chem. Eng. J.* 430 (2022) 132823.

Jothi V., A. Y. Adesina, A. Madhan Kumar, M. M. Rahman, J. S. Nirmal Ram, Enhancing the biodegradability and surface protective performance of AZ31 Mg alloy using polypyrrole/gelatin composite coatings with anodized Mg Surface. *Surf. Coat. Technol.* 381 (2020) 125139.

Jothi V., A. Y. Adesina, M. M. Rahman, A. Madhan Kumar, J. S. Nirmal Ram, Improved adhesion and corrosion resistant performance of polyurethane coatings on anodized Mg alloy for aerospace applications, *JMEPEG*. 29 (2020) 2586–2596.

Kravanja K. A., M. Finšgar, A review of techniques for the application of bioactive coatings on metal-based implants to achieve controlled release of active ingredients, *Mater. Des.* 217 (2022) 110653.

Kitao T., Y. Zhang, S. Kitagawa, B. Wang, T. Uemura, Hybridization of MOFs and polymers, *Chem. Soc. Rev.* 46 (2017) 3108–3133.

Li L. Y., L. Y. Cui, R. C. Zeng, S. Q. Li, X. B. Chen, Y. Zheng, M. Bobby Kannan, Advances in functionalized polymer coatings on biodegradable magnesium alloys – a review, *Acta Biomater.* 79 (2018) 23–36.

Lu X., Y. Zuo, X. Zhao, Y. Tang, X. Feng, The study of a Mg-rich epoxy primer for protection of AZ91D magnesium alloy, *Corros. Sci.* 53 (2011) 153–160.

Namsheer K., C. S. Rout, Conducting polymers: A comprehensive review on recent advances in synthesis, properties and applications, *RSC Adv.* 11 (2021) 5659–5697.

Nazari M. H., Y. Zhang, A. Mahmoodi, G. Xu, J. Yu, J. Wu, X. Shi, Nanocomposite organic coatings for corrosion protection of metals: A review of recent advances, *Prog. Org. Coat.* 162 (2022) 106573.

Nazeer A. A., E. Al-Hetlani, M. O. Amin, T. Quinones-Ruiz, I. K. Lednev, A poly(butyl methacrylate)/graphene oxide/TiO_2 nanocomposite coating with superior corrosion protection for AZ31 alloy in chloride solution, *Chem. Eng. J.* 361 (2019) 485–498.

Peng T., R. Xiao, Z. Rong, H. Liu, Q. Hu, S. Wang, X. Li, J. Zhang, Polymer nanocomposite-based coatings for corrosion protection. *Chem Asian J.* 15 (2020) 3915–3941.

Pourhashem S., F. Saba, J. Duan, A. Rashidi, F. Guan, E. G. Nezhad, B. Hou Polymer/inorganic nanocomposite coatings with superior corrosion protection performance: A review. *J. Ind. Eng. Chem.* 88 (2020) 29–57.

Pourhashem S., E. Ghasemy, A. Rashidi, M. R. Vaezi, A review on application of carbon nanostructures as nanofiller in corrosion-resistant organic coatings, *J. Coat. Technol. Res.* 17 (2020) 19–55.

Raja V. S., A. Venugopal, V. S. Saji, K. Sreekumar, R. S. Nair, M. C. Mittal, Electrochemical impedance behavior of graphite-dispersed electrically conducting acrylic coating on AZ31 magnesium alloy in 3.5 wt.% NaCl solution, *Prog. Org. Coat.* 67 (2010) 12–19.

Ramanathan T., A. A. Abdala, S. Stankovich, D. A. Dikin, M. Herrera-Alonso, R. D. Piner, D. H. Adamson, H. C. Schniepp, X. Chen, R. S. Ruoff, S. T. Nguyen, I. A. Aksay, R. K. Prud'homme, L. C. Brinson, Functionalized graphene sheets for polymer nanocomposites, *Nat. Nanotechnol.* 3 (6) (2008) 327.

Ramezanzadeh B., M. Attar, M. Farzam, A study on the anticorrosion performance of the epoxy – polyamide nanocomposites containing ZnO nanoparticles, *Prog. Org. Coat.* 72 (3) (2011) 410–422.

Ramezanzadeh B., E. Ghasemi, M. Mahdavian, E. Changizi, M. H. Mohamadzadeh Moghadam, Covalently-grafted graphene oxide nanosheets to improve barrier and corrosion protection properties of polyurethane coatings, *Carbon* 93 (2015) 555.

Ren Baohui, Yanning Chen, Yanqiang Li, Weijin Li, Shuiying Gao, Hongfang Li, Rong Cao, Rational design of metallic anticorrosion coatings based on zinc gluconate@ZIF-8, *Chem. Eng. J.* 38 (2020) 123389.

Ren B., Y. Li, D. Meng, J. Li, S. Gao, R. Cao, Encapsulating polyaniline within porous MIL-101 for high-performance corrosion protection, *Adv. Colloid Interface Sci.* 579 (2020) 842–852.

Rezk A. I., H. M. Mousa, J. Lee, C. H. Park, C. S. Kim, Composite PCL/HA/simvastatin electrospun nanofiber coating on biodegradable Mg alloy for orthopedic implant application, *J. Coating Technol. Res.* 16 (2019) 477–489.

Rong M., M. Zhang, W. Ruan, Surface modification of nanoscale fillers for improving properties of polymer nanocomposites: A Review, *Mater. Sci. Technol.* 22 (7) (2006) 787.

Sahoo N. G., S. Rana, J. W. Cho, L. Li, S. H. Chan, Polymer nanocomposites based on functionalized carbon nanotubes, *Prog. Polym. Sci.* 35 (2010) 837–867.

Sarian N. M., N. Iqbal, P. Sotoudehbagha, M. Razavi, Q. U. Ahmed, C. Sukotjo, H. Hermawan, Potential bioactive coating system for high-performance absorbable magnesium bone implants, *Bioactive Mater.* 12 (2022) 42–63.

Sathiyanarayanan S., S. Azim, G. Venkatachari, Corrosion resistant properties of polyaniline – acrylic coating on magnesium alloy, *Appl. Surf. Sci.* 253 (2006) 2113–2117.

Sathiyanarayanan S., S. Azim, G. Venkatachari, Corrosion protection of magnesium ZM 21 alloy with polyaniline – TiO_2 composite containing coatings, *Prog. Org. Coat.* 59 (2007) 291–296.

Shahryari Z., M. Yeganeh, K. Gheisari, B. Ramezanzadeh, A brief review of the graphene oxide-based polymer nanocomposite coatings: Preparation, characterization, and properties, *J. Coat. Technol. Res.* 18 (4) (2021) 945–969.

Shen S., Y. Zuo, The improved performance of Mg-rich epoxy primer on AZ91D magnesium alloy by addition of ZnO, *Corros. Sci.* 87 (2014) 167–178.

Shi X., T. A. Nguyen, Z. Suo, Y. Liu, R. Avci, Effect of nanoparticles on the anticorrosion and mechanical properties of epoxy coating, *Surf. Coat. Technol.* 204 (2009) 237–245.

Teijido R., L. Ruiz-Rubio, A. G. Echaide, J. L. Vilas-Vilela, S. Lanceros-Mendez, Q. Zhang, State of the art and current trends on layered inorganic-polymer nanocomposite coatings for anticorrosion and multifunctional applications, *Prog. Org. Coat.* 163 (2022) 106684.

Tian P., D. Xu, X. Liu, Mussel-inspired functionalization of PEO/PCL composite coating on a biodegradable AZ31 magnesium alloy, *Colloids Surf. B Biointerfaces.* 141 (2016) 327–337.

Tipan N., A. Pandey, P. Mishra, Selection and preparation strategies of Mg-alloys and other biodegradable materials for orthopaedic applications: A review, *Mater. Today Commun.* 31 (2022) 103658.

Truong V. T., P. K. Lai, B. T. Moore, R. F. Muscat, M. S. Russo, Corrosion protection of magnesium by electroactive polypyrrole/paint coatings, *Synth. Met.* 110 (2000) 7–15.

Xie J., J. Hu, X. Lin, L. Fang, F. Wu, X. Liao, H. Luo, L. Shi, Robust and anti-corrosive PDMS/SiO$_2$ superhydrophobic coatings fabricated on magnesium alloys with different-sized SiO$_2$ nanoparticles, *Appl. Surf. Sci.* 457 (2018) 870.

Yan Q. L., M. Gozin, F. Q. Zhao, A. Cohen, S. P. Pang, Highly energetic compositions based on functionalized carbon nanomaterials, *Nanoscale.* 8 (2016) 4799–4851.

Yang Guoqiang, Taijun Chen, Bo Feng, Jie Weng, Ke Duan, Jianxin Wang, Xiaobo Lu, Improved corrosion resistance and biocompatibility of biodegradable magnesium alloy by coating graphite carbon nitride (g-C3N4), *J. Alloys Compd.* 770 (2019) 823–830.

Zhang X., N. Zhao, C. He, The superior mechanical and physical properties of nanocarbon reinforced bulk composites achieved by architecture design – a review, *Prog. Mater. Sci.* (2020) 100672.

Zhao Y, F. Jiang, Y. Q. Chen, J. M. Hu, Coatings embedded with GO/MOFs nanocontainers having both active and passive protecting properties, *Corros. Sci.* 168 (2020) 108563.

Zhou C., M. Pan, S. Li, Y. Sun, H. Zhang, X. Luo, Y. Liu, H. Zeng, Metal organic frameworks (MOFs) as multifunctional nanoplatform for anticorrosion surfaces and coatings, *Adv. Colloid Interface Sci.* 305 (2022) 102707.

21 Superhydrophobic Coatings

Viswanathan S. Saji

CONTENTS

21.1 INTRODUCTION

The obligatory conditions for a surface to be superhydrophobic are an apparent static water contact angle (WCA) > 150° and a thermodynamically stable Cassie state, attributed to the effective interface air layer formation. The dynamic wettability is typically studied by determining sliding angle (WSA) and contact angle hysteresis (CAH). Superhydrophobic surfaces with WSA < 5° and CAH < 10° are conferred with the self-cleaning attributes. Low surface energy (SE) and enhanced surface roughness are the two critical criteria determining superhydrophobicity. Precise details on the fundamentals of wettability and superhydrophobicity can be found in the references (Young 1805) (Cassie and Baxter 1944) (Timmons and Zisman 1966) (Koch et al. 2009) (Bormashenko et al. 2012) Milne and Amirfazli 2012) (Wang et al. 2015) (Webb et al. 2014) (Jiaqiang et al. 2018) (Saji 2020a, 2020b) (Saji 2021a, 2021b, 2021c).

Current research and development suggest that superhydrophobic coatings are ideal for improving Mg alloys' corrosion resistance. Most reports used a two-step procedure in fabricating superhydrophobic surfaces; the first step is creating the required surface roughness, and the second is a low-SE modification. In its place, several reports employed both surface structuring and low-SE modification in a single step. Multi-step processing involving different processing steps, such as an additional base layer or interlayer fabrication, was also investigated (Saji 2021a). This chapter delivers the most recent updates on the topic concerning anti-corrosion applications.

21.2 SUPERHYDROPHOBIC COATINGS ON Mg ALLOYS

Typically, the first step in coating Mg alloys is a suitable pre-treatment. Several works employed acid, alkali, or combined acid/alkali etching as pre-treatment before the actual coating. The acid etching is generally performed to eliminate the surface oxides and enhance the metallic substrate's activity. Alkali etching is helpful for surface degreasing and fabricating a hydroxylated-rich surface.

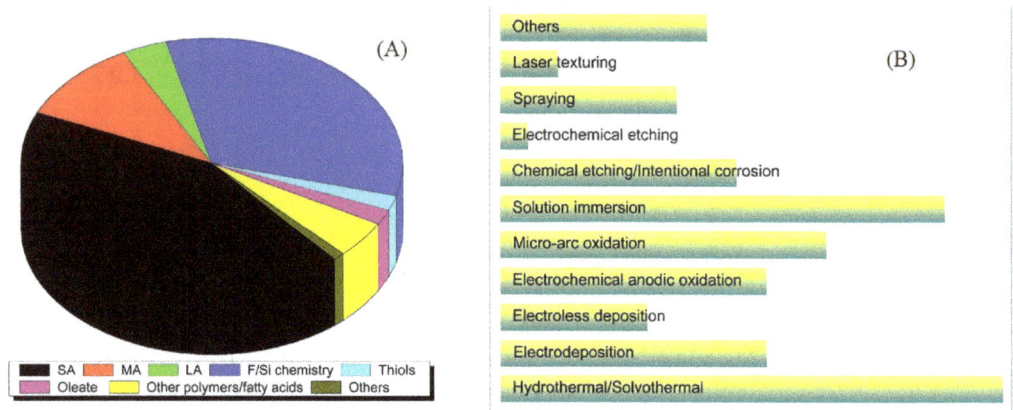

FIGURE 21.1 (a) Pie chart on low-SE materials employed for Mg alloys. (b) Bar diagram on various methods used for micro/nanostructuring. The charts are based on the journal publications during 2015–2020. *Reproduced with permission from (Saji 2021a) © 2021 Chongqing University. Publishing services provided by Elsevier B.V. on behalf of KeAi Communications Co. Ltd.*

The pre-treatment could improve the surface roughness for adequate bonding of a subsequent coating (Saji 2021a).

The most used low-SE modification, regardless of the substrate, relies on the silane/fluorine compounds (Saji 2020b). The F-substituted silanes are classically preferred owing to the high bond strength and low critical surface tension of -CF_3 and the chemical and biological inertness; however, they suffer from toxicity concerns (Pagliaro and Ciriminna 2005). For Mg alloys, the most used low-SE compounds are long-chain saturated fatty acids such as stearic acid (SA), lauric acid (LA), and myristic acid (MA) (**Figure 21.1a**). Typically, these fatty acids undergo neutralization reactions, as shown by the reaction of SA with $Mg(OH)_2$ in Eq. 21.1. On the other hand, silanes undergo hydrolysis and condensation polymerization (Eq. 21.2 and 21.3) (Wu, Wu et al. 2019). The direct immersion in the low-SE solution is the most used low-SE treatment; the second most used is the electrodeposition (ED). Several works used hydrothermal (HT) processing. Other techniques, such as spraying, drop-coating, blade-coating, electrospinning, thermal evaporation, RF sputtering, and magnet-assisted processes, were also used for the low-SE treatment. The direct immersion studies can be conducted in either room-temperature or high-temperature (50–99°C) solutions followed by room-temperature or high-temperature (60–120°C) drying (Saji 2021a).

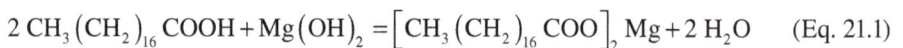

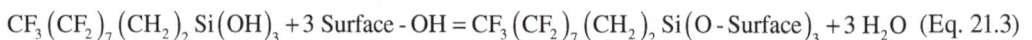

$$2\,CH_3\left(CH_2\right)_{16}COOH + Mg\left(OH\right)_2 = \left[CH_3\left(CH_2\right)_{16}COO\right]_2 Mg + 2\,H_2O \quad \text{(Eq. 21.1)}$$

$$CF_3\left(CF_2\right)_7\left(CH_2\right)_2 Si\left(OCH_3\right)_3 + 3\,H_2O = CF_3\left(CF_2\right)_7\left(CH_2\right)_2 Si\left(OH\right)_3 + 3\,CH_3OH \quad \text{(Eq. 21.2)}$$

$$CF_3\left(CF_2\right)_7\left(CH_2\right)_2 Si\left(OH\right)_3 + 3\,\text{Surface - OH} = CF_3\left(CF_2\right)_7\left(CH_2\right)_2 Si\left(O\text{-Surface}\right)_3 + 3\,H_2O \quad \text{(Eq. 21.3)}$$

Achieving WCA > 120° simply by governing the surface chemistry, even with the lowest SE compound, typically is not practicable. The hierarchical surface roughness is essential for fabricating durable superhydrophobic surfaces (Nishino et al. 1999) (Erbil 2014) (Nosonovsky and Bhushan 2007). **Figure 21.1b** shows the generally used methods for achieving hierarchical surface structuring on Mg and its alloys. The most used process is HT, and direct immersion is the second most employed (Saji 2021a). Several works used different electrochemical approaches (Saji 2023).

21.2.1 Fabrication Methods

21.2.1.1 Hydrothermal/Solvothermal

The HT technique is an eco-friendly and economical approach to fabricating superhydrophobic Mg alloys. A single-step HT process with added low-SE compound in the bath could straightforwardly create the superhydrophobicity where the $Mg(OH)_2$ layer formation provides the required surface roughness and the low-SE compound covalently grafts with the surface -OH groups. The wettability varies with the HT bath composition and temperature (**Figure 21.2**). The WCA of the

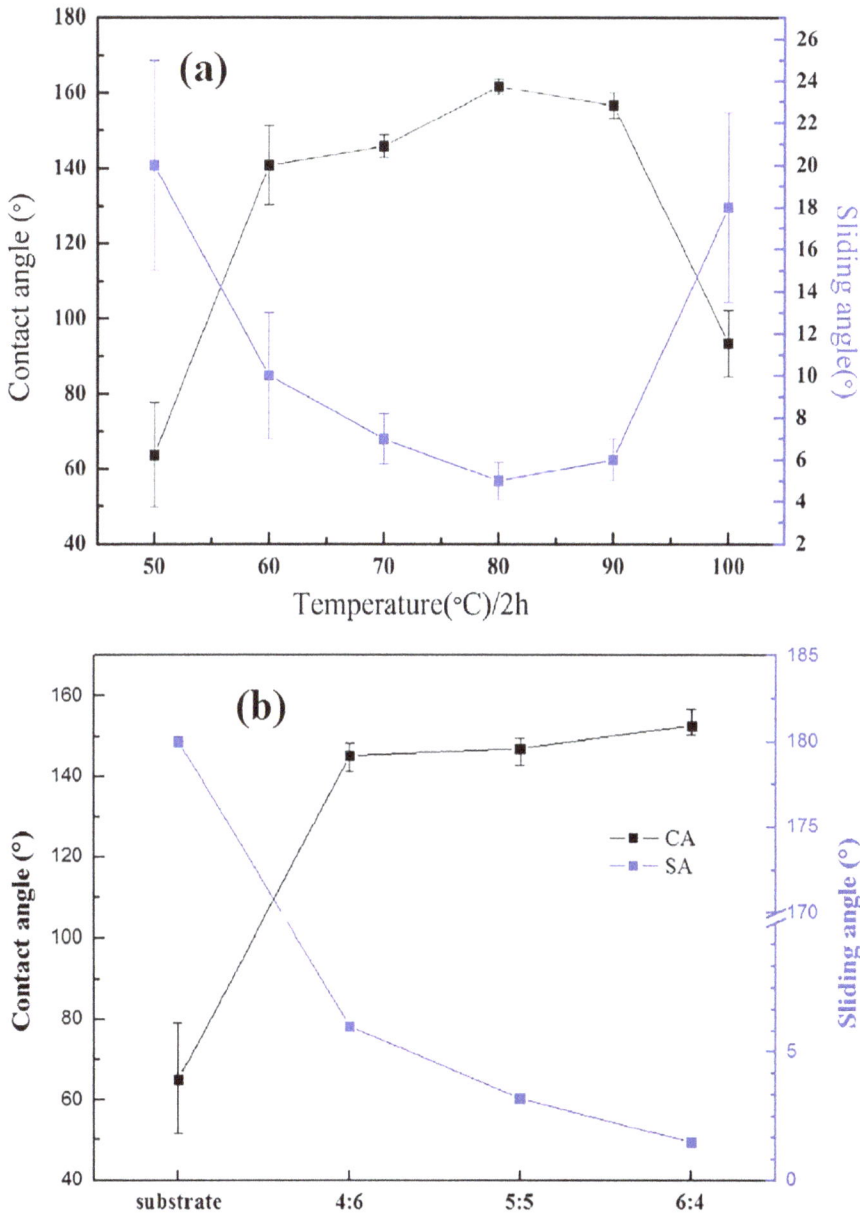

FIGURE 21.2 Variation of WCA and WSA of single-step HT-made surfaces with different solutions. Effect of (a) temperature and (b) composition. *Reproduced (a) with permission from (Zhang et al. 2018) and (b) from (Kang et al. 2018) © 2017 Elsevier B.V.*

bare alloy (~ 38.3°) becomes 153.8° (WSA ~ 4°) after a HT treatment in SA-water-ethanol at 80°C for 10 hours. Scanning electron microscopy (SEM) studies showed microscale plate and bar-like structures encompassing several layers and voids. SA grafting was confirmed using infrared spectroscopy (Feng et al. 2017). A single-step HT reaction in $ZnCl_2$-SA-ethanol-water solution has resulted in WCA as high as 161.7 ± 0.8° and WSA of 4.8 ± 0.9°. The optimum temperature and time for the HT process was 80°C and 2 hours (**Figure 21.2a**) (Zhang et al. 2018). The one-step HT bath (170°C, 2 h) used by Kang et al. for making superhydrophobic hydroxyapatite (HA) coating was composed of NaH_2PO_4, calcium acetate, SA, HCl, water, and ethanol. The best result (WCA ~152.8°, WSA < 2°) was obtained when the ethanol:water ratio was 6:4 (**Figure 21.2b**). X-ray diffraction analysis detected peaks corresponding to HA, $Mg(OH)_2$ and calcium stearate. The surface presented a flaky hierarchical structure characteristic of HT-treated Mg substrates (Kang et al. 2018). Hu et al. achieved WCA ~158.5° and WSA ~2° by single-step processing in an SA-water-ethanol solution (Wan and Hu, 2016). Li et al. used $Ni(SO_4)_2$-SA-water-ethanol bath for HT at 150°C (Li et al. 2016).

Most reports used the two-step method, which involves an initial HT/solvothermal (ST) processing and a subsequent low-SE treatment. A superhydrophobic surface was realized by HT treatment at 120°C followed by silane treatment in an ethanol solution of dodecafluoroheptyl-propyl-trimethoxylsilane. The WCA of HT/silane-treated, HT-treated, and bare surfaces were 152.1°, 4.8°, and 83.1°, respectively (Li et al. 2017). Another work made the $Mg(OH)_2$ coating via HT in 5.66 wt.% NaOH at 140°C, and subsequently, a mix solution of polypropylene (PP) and maleic anhydride grafted-PP in xylene was dip-coated over it. The non-wettability (WCA – 165.5 ± 3.6°, WSA – 4 ± 0.6°) was credited to the spherical microstructures and the low-SE PP layer (Zhang, Zeng et al. 2020). A HT-treated Mg alloy in deionized water at 160°C was subsequently immersed in 7 mM of SA-hexane solution for 3 hours to fabricate a surface with WCA of ~ 159° and WSA ~ 7° (Jin et al. 2019). Another study used the HT process followed by ED of dodecyltrimethoxysilane (DTMS) (Zhang et al. 2021a).

Layered double hydroxides (LDHs) have attracted significant recent research attention. They are known for their characteristic lamellar structure with the chemical composition of $[M^{2+}_{1-x}M^{3+}_x(OH)_2](A^{n-})_{x/n} \cdot yH_2O$, where M and A represent metal cations and interlayer anions, respectively (Pourfaraj et al. 2017) (Kameda and Umetsu 2017). Wu et al. conducted HT in 0.01 M NaOH at 120°C, followed by RF sputtering of polytetrafluoroethylene (PTFE). The HT-made film displayed cross-linked microsheets on the surface. A double-layered film with $Mg(OH)_2$ rich bottom layer and Mg-Al LDHs rich top layer was formed. The triple-layered coating formed after the PTFE sputtering presented WCA ~170° (Wu, Shi et al. 2019). Zhang et al. used a combined coprecipitation-HT processing to develop $Mg(OH)_2$/Mg-Al LDHs film on AZ31 alloy, which was followed by SA modification (Zhang et al. 2016). Zhou et al. fabricated Zn-Al LDHs on AZ91D alloy via HT in an aqueous solution of 0.1 M $Zn(CH_3COO)_2$ and 0.05 M $Al(OH)_3$ at 60°C, followed by SA treatment (Zhou et al. 2015). Mg-Al LDHs laminate hydrothermally grown on AZ31 substrate was modified by sodium dodecylbenzene sulfonate and sodium laurate (Huang et al. 2021).

A few works employed an additional coating on Mg alloys before the HT step to boost the barrier effect and help in hierarchical surface structuring. Yuan et al. fabricated a superhydrophobic surface on Ni-P electroless coated (ELD) Mg alloy via HT processing in an ethanol solution of $Ni(NO_3)_2$ at 100°C to 160°C, followed by immersion in 0.01 M SA (Yuan et al. 2017). Zhang et al. deposited a thin TiO_2 layer and loaded wrinkled SiO_2 to develop a pod-like structure. After a subsequent SA modification, the surface becomes superhydrophobic. The SiO_2 particles increased the mechanical durability (Zhang, Zhao et al. 2020).

21.2.1.2 Electrochemical Methods

For the convenience of discussion, we have included different electrochemical, electrophoretic, and related methods in fabricating superhydrophobic surfaces and coatings under one heading (**Figure 21.3**), which include *(i) electrodeposition methods* [ED, electrophoretic deposition (EPD), and electropolymerization (EP)], *(ii) anodization methods* [anodic oxidation (EAO) and micro-arc

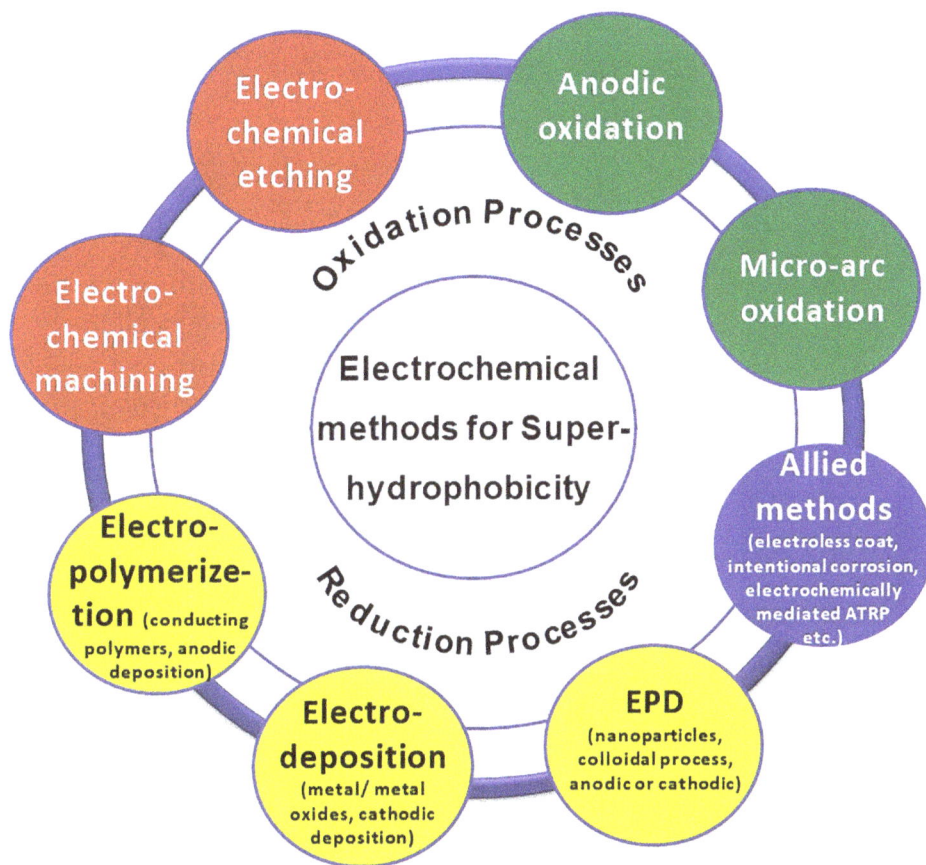

FIGURE 21.3 A schematic on electrochemical methods for superhydrophobic surfaces and coatings. *Reproduced with permission from (Saji 2023) © 2022 Taylor & Francis.*

oxidation (MAO)], and *(iii) etching approaches* [electrochemical etching (ECE) and electrochemical machining (ECM)]. ED is typically a cathodic process involving the reduction of positive ionic species (Faradaic process) on a metallic substrate, whereas EPD involves electrophoresis of charged particles. Instead, EP is the electrochemical oxidation of monomers to deposit conductive polymer coatings. In the anodization methods, the metallic substrate is kept as the anode, where a thick conversion coated oxide layer forms. MAO is analogous to EAO, which is typically performed at voltages > 200 V. ECE and ECM are basically anodic dissolution processes and do not involve a coating formation (Saji et al. 2022) (Saji 2023). Representative reports in these areas are described in what follows.

Single-step ED could effectively make a superhydrophobic coating on Mg alloys. A superhydrophobic DTMS coating was deposited on AZ31 alloy by single-step ED in 80 mL ethanol + 20 mL 0.2 M KNO_3 + 3 mL DTMS solution (pH 6.5). The coating exhibited self-cleaning superhydrophobicity with WCA and WSA of ~158° and 4°, respectively (Zhang et al. 2021b). The superhydrophobic coating made via ED at 30 V in an alcohol solution of SA and $Mg(NO_3)_2$ revealed a hierarchical surface morphology comprised of nanosheets and microspheres. The major coating component was magnesium stearate. In ED, the SA ionizes to stearate and H^+ ions, favouring cathodic deposition of magnesium stearate and concurrent H^+ reduction (Zheng et al. 2019). The WCA of ED coatings varies significantly with the deposition voltage and time. The highest WCA (159.6 ± 0.5°) of an ED superhydrophobic coating developed from an ethanol solution of $Ce(NO_3)_3$ and MA was obtained at

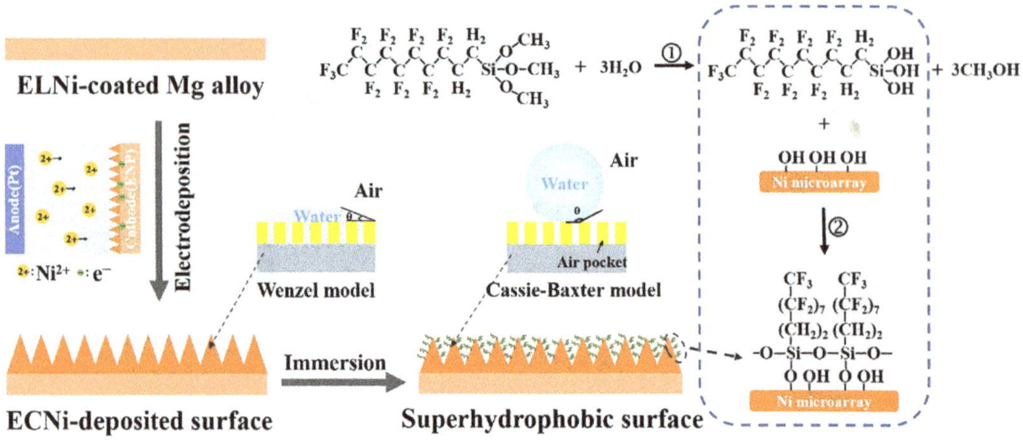

FIGURE 21.4 Schematic of preparation of ELNi/ECNi/PFDTMS coating. The dehydration process of the silane and dehydration-condensation reaction between the silanol and hydroxyl group on the nano-cone Ni array surface is also shown. *Reproduced with permission from (Fang et al. 2022) © 2022 Elsevier B.V.*

a deposition voltage of 30 V, ascribed to the even hierarchical micro/nanopapillae surface structuring. During ED, Ce^{3+} and MA reacted and formed cerium myristate and H^+ ions (Liu et al. 2015). A calcium myristic superhydrophobic coating was developed using an ethanol solution of $CaCl_2$ and MA at 50 V (Zhong et al. 2018).

Several studies reported a two-step process involving ED. A CeO_2/SA bilayer coating was fabricated by the ED of CeO_2, followed by the low-SE SA modification (Liu et al. 2020). A reported multistep approach involved $CuCl_2$ etching and Ni ED and SA treatment. The $MgCl_2$-rich surface resulting from etching contributed to the surface structuring. Ni ED yielded homogeneously dispersed micro protrusions having nanostructures. The surface after SA modification demonstrated WCA and WSA of ~151° and 7.6°, correspondingly (Wang et al. 2018). The schematic of fabrication of a superhydrophobic three-layered coating comprising the bottom electroless-deposited Ni-P layer (ELNi), middle ED Ni nano-cone array (ECNi), and top fluorinated polysiloxane (PFDTMS) layer is shown in **Figure 21.4**. Superhydrophobicity was not attained for double-layer coatings without the middle ECNi or the top PFDTMS layers. The combined effect of nanostructured Ni and low-SE PFDTMS was critical in providing superhydrophobicity (Fang et al. 2022).

A superhydrophobic coating was fabricated by ED of Zn-Al LDHs film in a solution of $Al(NO_3)_3$ and $Zn(NO_3)_2$, followed by dip-coating in DMF solution of ZIF-8 with added polyvinylidene fluoride (PVDF) and 1H,1H,2H,2H-perfluorodecyltriethoxysilane (PFTS). The WCA of the bare and LDH film surfaces were ~80.7° and 45.2°, respectively. Upon a PVDF coating, the WCA reached 104.8°; upon a subsequent ZIF-8 incorporation, WCA became 117 ± 2°. On further low-SE PFTS modification, WCA > 150° and WSA < 3° were obtained (Yin et al. 2020). A three-layer coating comprising Ni-P ELD, Cu ED and Ni ED layers was subsequently subjected for SA modification for attaining superhydrophobicity (Song et al. 2018). Tan et al. performed acid/alkaline pre-treatment and Ni ELD, Cu ED, and LA-sebacic acid modification to achieve a robust barrier layer with superhydrophobicity (Tan et al. 2015). Wang et al. employed EPD at 30 V in chitosan-acetic acid dispersion, and the coating was further subjected to dip-coating in a Ca-P solution. The resultant calcium phosphate-chitosan coating was subjected to low-SE energy modification in 0.02 mol/L SA solution (Wang, Xiao et al. 2022).

Several reports used EAO/MAO to develop a robust oxide layer on the Mg alloy as a preliminary step. EAO, followed by a simple SA treatment, can yield superhydrophobicity. The as-formed EAO surface was superhydrophilic and comprised many pits and cracks, and the surface became superhydrophobic after the SA modification. The WCA increased from 143.3° to 163.8° as the EAO time changed from 3 to 30 minutes (Liu et al. 2016). In another report, the sample after EAO was immersed in an ethanolic solution of polyvinyl chloride and tetrahydrofuran (Yang et al. 2016). Khalifeh and Burleigh showed that Mg samples anodized at 5 V DC for 180–240 seconds or 120 V AC for 10 seconds easily achieved superhydrophobicity after subsequent SA modifications (Khalifeh and Burleigh 2018).

A HT processing after EAO/MAO would enable thick LDH film formation. A self-healing super-hydrophobic coating was developed via *in situ* corrosion inhibitor-intercalated LDHs on anodized Mg alloy. The sample after EAO was subjected to HT in a corrosion inhibitor-added solution and further modified with low-SE sodium stearate via an ensuing HT step (Wang et al. 2020).

MAO layer formation on Mg alloys significantly enhances wear and corrosion resistance. An optimized low-SE modification on MAO could act as a sealing coating to the typical porous outer surface of MAO. Superhydrophobic MAO-coated Mg alloys were developed by simple SA modification (Zhang et al. 2017) (Liu and Xu 2018). The maximum WCA (~ 154.7°) obtained for a sample made at MAO voltage of 350 V was attributed to the favourable surface morphology (Liu and Xu 2018). An MAO sample was subjected to HT treatment in $Al(NO_3)_3$ at 125°C and an SA modification at 60°C. The LDHs formed during the HT process entirely covered the porous MAO layer (Wang et al. 2019). Polypropylene coating was fabricated on an MAO-coated sample via one-step dipping. The coated surface had a micron-scaled granular morphology with WCA of ~167.2° and WSA of ~2.7° (Zhang et al. 2021). Jiang et al. made a superhydrophobic coating by combining MAO, alternate conversion coatings of $Ce(NO_3)_3$ and phytic acid (PA), and subsequent PFTS modification. The conversion coating was helpful in creating the hierarchical surface structure and sealing the defects of MAO coating (Jiang et al. 2018). Qiu et al. applied a dopamine-anchored iron oxide nanoparticle coating on MAO by dipping it in the nanoparticles' dispersion and using a magnetic field. The coated surface demonstrated WCA of ~157° (Qiu et al. 2016). Jiang et al. have synthesized ZnO-metal organic framework (MOF) core-shell structures by a self-templating technique. The fabricated ZnO-ZIF-8 nanorods arrays (**Figure 21.5**) layer exhibited superhydrophobicity without the requirement of a low-SE treatment, which was attributed to the nanorods structure that can trap air effectively. The superhydrophobicity was closely dependent on the growth process of the ZnO-ZIF-8 core-shell structure. Maximum WCA was obtained at 3 hours (**Figure 21.5d**), where the most preferred surface structuring happened (**Figure 21.5b**) (Jiang et al. 2022). A refreshable super-hydrophobic surface was fabricated via a MAO-paraffin buffer layer. Firstly, the liquid paraffin was inoculated into the porous MAO, and subsequently, carbon soot was grown onto it via combustion. The molten paraffin helped firm bonding with carbon soot (Li, Wang et al. 2022). Several works achieved superhydrophobicity by ECE and ECM. More details are described elsewhere (Saji 2023).

21.2.1.3 Immersion

Superhydrophobicity could be attained via an optimized one-step immersion process (Ishizaki et al. 2017a,b) (Wu et al. 2022). A simple 1-minute immersion in an ethanol solution of $Ce(NO_3)_3$ and MA at pH 2 has resulted in WCA above 150° (Ishizaki et al. 2017a,b). The instant Mg dissolution and formation of $Mg(OH)_2$ could lead to the development of appropriate nano/micro surface structures, which, along with the low-SE component, could lead to superhydrophobicity. A one-step dip-coated aluminium phosphate-PTFE coating comprised of compactly arranged ellipsoidal nanospheres (WCA ~155.7°). The phosphate can react with the surface, forming a conversion layer providing suitable surface roughness, and the PTFE can diminish the SE (Wu et al. 2022).

Several reports used a first-step immersion, followed by an additional low-SE modification. Xie et al. prepared a superhydrophobic coating by simple soaking in $ZnCl_2$ solution, leading to micron mastoids and vertical nano-sheets on the surface followed by SA modification. The WCA of the $ZnCl_2$/SA-modified, SA-modified, $ZnCl_2$-modified, and bare surfaces were 162.04°, 109.32°, 16.76°, and 49.20°, respectively (Xie et al. 2020). Consecutive immersions in 0.01 M and 0.1 M of $MnSO_4$ and a subsequent SA modification have resulted in superhydrophobicity (Xun et al. 2020).

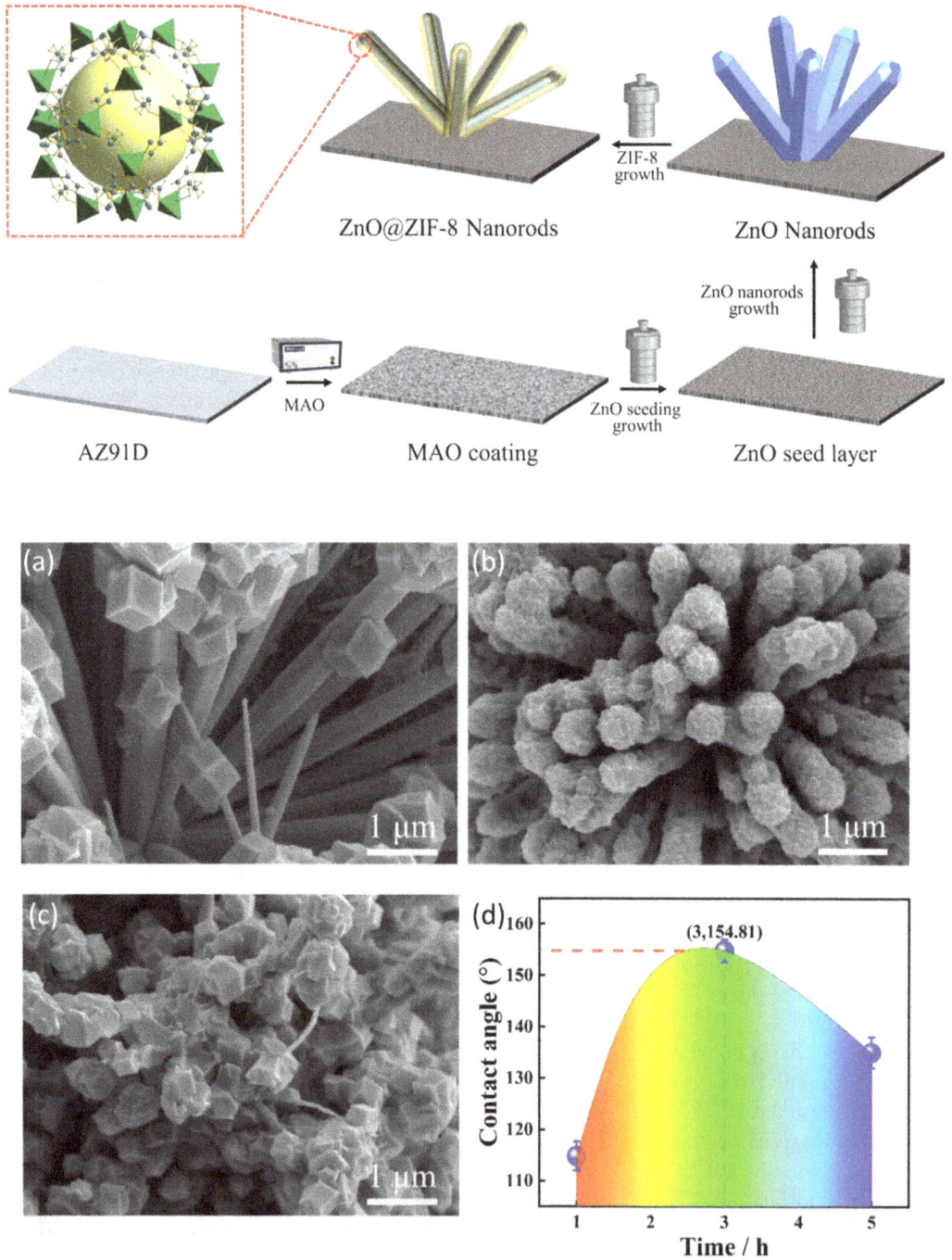

FIGURE 21.5 Schematic showing ZnO-ZIF-8 nanorods array preparation. SEM images after various ZIF-8 growth times (a) 1 h, (b) 3 h, (c) 5 h, and (d) the associated WCA variation. *Reproduced with permission from (Jiang et al. 2022) © 2022 Chongqing University. Publishing services provided by Elsevier B.V. on behalf of KeAi Communications Co. Ltd.*

A Mg-Mn LDH coating made via consecutive soaking in $MnCl_2$ and Na_2CO_3 solutions was further modified by MA by Kuang et al. (2019). Gu et al. employed polydopamine (PDA) as a biologically compatible glue layer to bind the metal surface with a SiO_2-F nanoparticles layer to fabricate a durable superhydrophobic coating. The sample was soaked in an alkaline solution of sodium phosphate and PDA, and the coated substrate was then dipped into a SiO_2-F nanoparticles solution (Gu et al. 2021). A superhydrophobic and self-healing stannate conversion coating was fabricated on an alkali/acid pre-treated sample and, afterwards, SA modified (Yang et al. 2015). A hybrid conversion coating was deposited by alternative $CeCl_3$ and PA solution immersions, followed by dipping in an ethanol solution of hexadecyltrimethoxysilane (HDTMS). The PA was anchored to the surface via phosphate-Mg^{2+} chelation, and Ce^{3+} was assimilated to PA via P-O-Ce linkages. The hydroxyl groups on the PA deposit can help in linking HDMS molecules. After four times of PA/Ce deposition, WCA ~167.3° and WSA ~2.7° were noted (Ou and Chen 2019).

A few studies employed ED for the low-SE modification. A phosphate conversion–coated sample was subjected to ED in an ethanolic solution of $Ce(NO_3)_3$, SA and MA to fabricate a superhydrophobic coating with WCA ~160.19° and WSA ~1.5° (Zhao et al. 2015). Kuang et al. made a LDH coating by two-step immersion in acidic $MnCl_2 + HCO_3^-/CO_3^{2-}$ and alkaline Na_2CO_3 solutions, and ED was performed on the LDHs-coated sample at 30 V in an ethanolic solution of $CaCl_2$ and MA (Kuang et al. 2020).

21.2.1.4 Spraying

Superhydrophobic coatings can be made with optimized spraying/painting. Xie et al. blade-coated a superhydrophobic coating, using a paint comprising PDMS, 1H,1H,2H,2H-tridecafluoro-n-octyltriethoxysilane, Sylgard®184 curing agent and SiO_2 nanoparticles; and the coated sample was cured at 160°C (Xie et al. 2018). A spray-coated superhydrophobic coating used a solution of polyphenylene sulfide (PPS), PTFE, and SiO_2 nanoparticles (Shi et al. 2017). Qian et al. used a mixture of ethanolic solution of tetraethoxysilane (TEOS), NH_3, and 1H,1H,2H,2H-perfluorooctyltriethoxysilane (PFOTS) with fumed silica nanoparticles for the spray coating (Qian et al. 2018).

The synergistic effect of MAO and superhydrophobic spray coating was utilized for anti-corrosion coatings. A PDMS-SiO_2 spray coating on the MAO layer yielded a WCA of 158.3 ± 2.5° (Song et al. 2022). A composite anti-corrosion superhydrophobic coating was fabricated by the combined effect of MAO, ionic liquid (octadecyl triphenyl phosphonium bis(trifluoromethylsulfonyl) amide), and SiO_2/epoxy-resin spray coating (**Figure 21.6**). Incorporating the ionic liquid [OTP] [NTf2] was helpful in covering the MAO's porous surface and providing corrosion-inhibiting and self-healing effects. The WCA of MAO-coated, ionic liquid-incorporated and the final spray coated surfaces were ~45° to 68°, 77°, and 164°, respectively (Zhang et al. 2022).

Liu et al. developed a robust rare-earth comprising superhydrophobic coating by $La(OH)_3$ spraying and SA treatment. The superhydrophobic surface contained cross-linked fiber-like $La(OH)_3$ nanowire networks (Liu et al. 2018). A superhydrophobic and self-healing ODA/PDA/APT-Ce^{3+} coating was fabricated by utilizing the benefits of cerium nitrate–loaded attapulgite (APT), PDA and octadecylamine (ODA). APT was employed as containers to load Ce^{3+} ions, dopamine was used to cover the porous APT by self-polymerization, and superhydrophobicity was attained through hydrophobic ODA grafting (Xue et al. 2022). Wang et al. prepared superhydrophobic diatomite powders and then spray coated a mixture of diatomite and epoxy resin (Wang, Guo et al. 2022). Zhou et al. made a superhydrophobic ZnO coating on acid-etched alloy by spray coating epoxy resin and ZnO seeds solution, followed by HT processing in a ZnO growth solution and a subsequent SA modification. The concentration of the growth solution was critical in enhancing WCA (Zhou et al. 2019).

21.2.1.5 Others

Several other methods were employed in fabricating superhydrophobic coatings on Mg alloys, including sputtering (La et al. 2017), high-speed wire electrical discharge machining (Qiu et al.

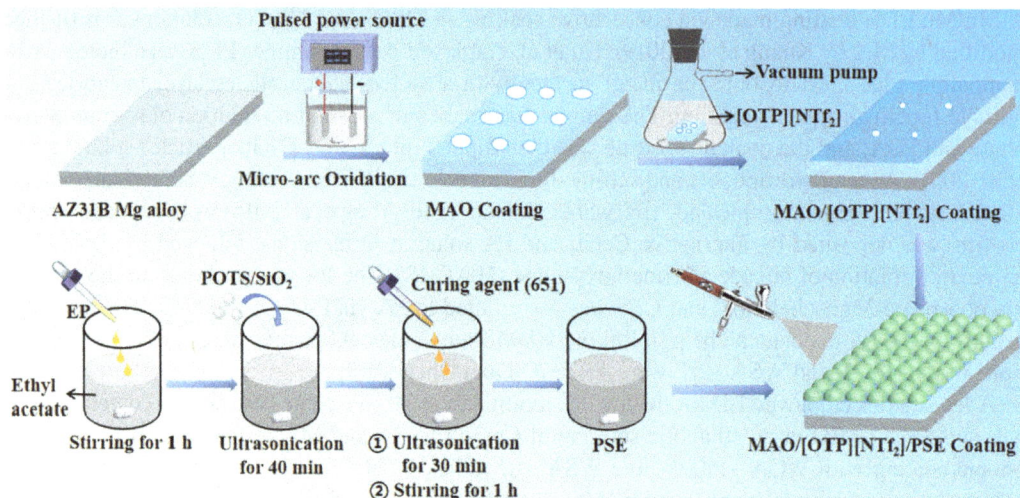

FIGURE 21.6 Schematic of coating fabrication. *Reproduced with permission from (Zhang et al. 2022) © 2022 Elsevier B.V.*

2020), electrospinning (Polat et al. 2018), and chemical vapour deposition (Siddiqui et al. 2020). Several works utilized chemical etching (Safarpour et al. 2020) (Wang et al. 2016) and laser-assisted surface texturing (Li et al. 2018) (Zhang and Zhang 2019) to fabricate superhydrophobic surfaces.

21.2.2 ANTI-CORROSION STUDIES

Reported studies showed that the anti-corrosion performance of superhydrophobic Mg alloys was consistently superior to their bare counterparts. The robust interface air layer and the water repellence could enhance the aqueous corrosion resistance. The air layer can function as a dielectric and a passivation layer, which could serve as a barrier to the corrosive species (Wang et al. 2015) (Saji 2021a) (Yao et al. 2020). However, superhydrophobic coating generally suffers long-term durability issues in aggressive corrosive environments. Long-term immersion in aggressive solutions deteriorates the hydrophobic molecules, leading to coating defects. Superhydrophobic surfaces' mechanical and chemical durability is essential in deciding long-term corrosion resistance, as discussed in the next section. This section provides representative examples of anti-corrosion studies reported in different coating categories discussed in the previous sections.

Superhydrophobic coatings fabricated by one-step methods demonstrated improved electrochemical corrosion resistance. The corrosion current densities (i_{corr}) of one-step HT-treated superhydrophobic and the corresponding bare samples in 3.5 wt.% NaCl were 3.32×10^{-7} A/cm^2 and 1.48×10^{-4}, respectively. The respective corrosion potentials (E_{corr}) were -1.34 and -1.50 V (vs. SCE). A more positive E_{corr} and lower i_{corr} correspond to improved corrosion resistance. Their long-term immersion studies, however, displayed significantly reduced WCA after 60 hours (Feng et al. 2017). A HT- and oleate-modified superhydrophobic sample had an average mass loss of 21.4 mg after 30 days of immersion in phosphate buffer saline, which was significantly less than that of unmodified HT-treated (26 mg) and bare (123.3 mg) samples (Peng et al. 2018). Zheng et al. showed that the i_{corr} of a superhydrophobic one-step ED sample (5.80×10^{-8} A/cm^2) was three orders lesser to the bare (7.02×10^{-5} A/cm^2). The superhydrophobic sample sustained high WCA during continuous soaking in 3.5 wt.% NaCl for one week (Zheng et al. 2019). Other superhydrophobic coatings made by one-step ED (Zhong et al. 2018) (Zhao and Kang 2016) (Liu et al. 2015), one-step immersion (Ishizaki et al. 2017a,b), and other methods (Xie et al. 2018) (Shi et al. 2017) also provided similar results.

All multi-step processed superhydrophobic coatings also revealed superior corrosion resistance to their bare counterparts. The i_{corr} of the HT and silane-treated superhydrophobic sample was reduced by ~1/8th of the bare and 1/4th of the unmodified HT-treated alloy. The coating resistance (R_{coat}) of the HT/silane-treated and HT-treated samples were 35149 and 4435 $\Omega \cdot cm^2$, respectively (Li et al. 2017). A 120 hours of immersion study in 3.5 wt.% NaCl displayed severe corrosion on the bare sample, localized attacks on the HT-treated sample, and a smooth surface on the superhydrophobic HT/SA sample (Zhang et al. 2015). A LDHs-coated and SA-modified alloy considerably improved the anti-corrosion performance, where the i_{corr} of the superhydrophobic, LDH-coated, and bare samples were 3.4×10^{-10}, 3.9×10^{-7} and 4.7×10^{-5} A/cm^2, in that order (Zhang et al. 2016). The polarization resistance (R_p) of a LDH-coated and SA-modified superhydrophobic sample in 3.5 wt.% NaCl was 4680 $\Omega \cdot cm^2$, considerably greater than the bare (151.9) and the unmodified LDH-coated (406.2) samples (Zhou et al. 2015). A three-layered Mg(OH)$_2$/Mg-Al-LDHs/fluropolymer coating fabricated by HT and PTEF-sputtering provided excellent corrosion protection in 5 mM H$_2$SO$_4$ and 3.5 wt.% NaCl. The i_{corr} of the superhydrophobic sample was 15.8 and 33.4 times higher than the HT-treated sample in NaCl and H$_2$SO$_4$ solutions, respectively (Wu, Shi et al. 2019). Multiple factors, including the LDHs' physical barrier effect, the interface air layer's water repellence, and the ion exchange between the LDH layer and chloride ions in solution, could contribute to the better corrosion resistance (Huang et al. 2021).

Liu et al. showed that the i_{corr} of the ED CeO$_2$/SA-coated superhydrophobic sample was two orders inferior to the bare. The i_{corr} marginally increased after 48 hours and significantly increased after 100 hours of immersion in 3.5 wt.% NaCl (Liu et al. 2020). The superior corrosion protection offered by a superhydrophobic Ni-P/Ni/PFDTMS composite coating prepared by ELD, ED, and PFDTMS treatment (ELNi/ECNi/PFDTMS coating) is evident from the impedance spectroscopy (EIS) and potentiodynamic polarization (PDP) results shown in **Figure 21.7**. The impedance moduli at low frequencies increased significantly after the composite coating (**Figure 21.7a**). At 0.01 Hz, the impedance of the superhydrophobic coating (2.79×10^6 $\Omega.cm^2$) was approximately four orders greater than the bare sample (2.18×10^2 $\Omega.cm^2$). The superhydrophobic coating provided better corrosion protection for more than 30 days (**Figure 21.7b,c**). The morphology and wettability of the coating after the tests (**Figure 21.7d**) showed only marginal variations supporting a certain extent of degradation. Although the WCA was decreased to 141.6 ± 5.2°, the coating still remained hydrophobic (Fang et al. 2022). Yin et al. have shown that a superhydrophobic ZIF-8/PVDF/LDHs/Mg-alloy retained high charge transfer resistance ($R_{ct} > 10^4$ $\Omega \cdot cm^2$) during 7 days of continuous soaking in 3.5 wt.% NaCl (Yin et al. 2020).

As discussed, the EAO/MAO coatings are beneficial in making a thick oxide layer, and a subsequent low-SE modification seals the coating defects. The corrosion resistance of a superhydrophobic EAO/ED-calcium stearate coated alloy varied with the duty cycle used for the pulse ED. A superhydrophobic sample made at 50% duty cycle presented a pitting potential (E_{pit}) of 1.17 V compared to −1.62 V (vs. SCE) of bare (Zhang and Lin 2019). EIS studies on superhydrophobic coatings developed by consecutive EAO, HT, and different low-SE treatments revealed that the corrosion resistance varied with the low-SE material used. The R_{ct} of samples modified with SA, MA, and PFTS were 9311, 11021, and 27793 $\Omega \cdot cm^2$, respectively (3.5 wt.% NaCl). The corresponding R_{ct} values after 14 days of immersion were 1767, 1960, and 2126 $\Omega \cdot cm^2$ (Wu, Wu et al. 2019). SEM/EDS analysis of superhydrophobic MAO/LDH/SA-coated sample after 288 hours immersion in 3.5 wt.% NaCl showed only marginal variation in surface morphology and composition. However, there was desorption of SA from the sample edges, as indicated by the decline of WCA to 126.76° (Wang et al. 2019). The significant enhancement of corrosion resistance obtained for a superhydrophobic ZnO-ZIF-8-coated alloy compared to the bare and MAO-coated alloys is evident in **Figure 21.8**. The microporous MAO surface could offer interlocking points for the superhydrophobic coating, whereas the superhydrophobic layer can repair the MAO layer defects and trap air in the gaps (Jiang et al. 2022).

A superhydrophobic MAO/PA/cerium/silane coating utilized the combined effect of chemical (PA and Ce) and electrochemical (MAO) conversion coatings and the low-SE silane treatment. The

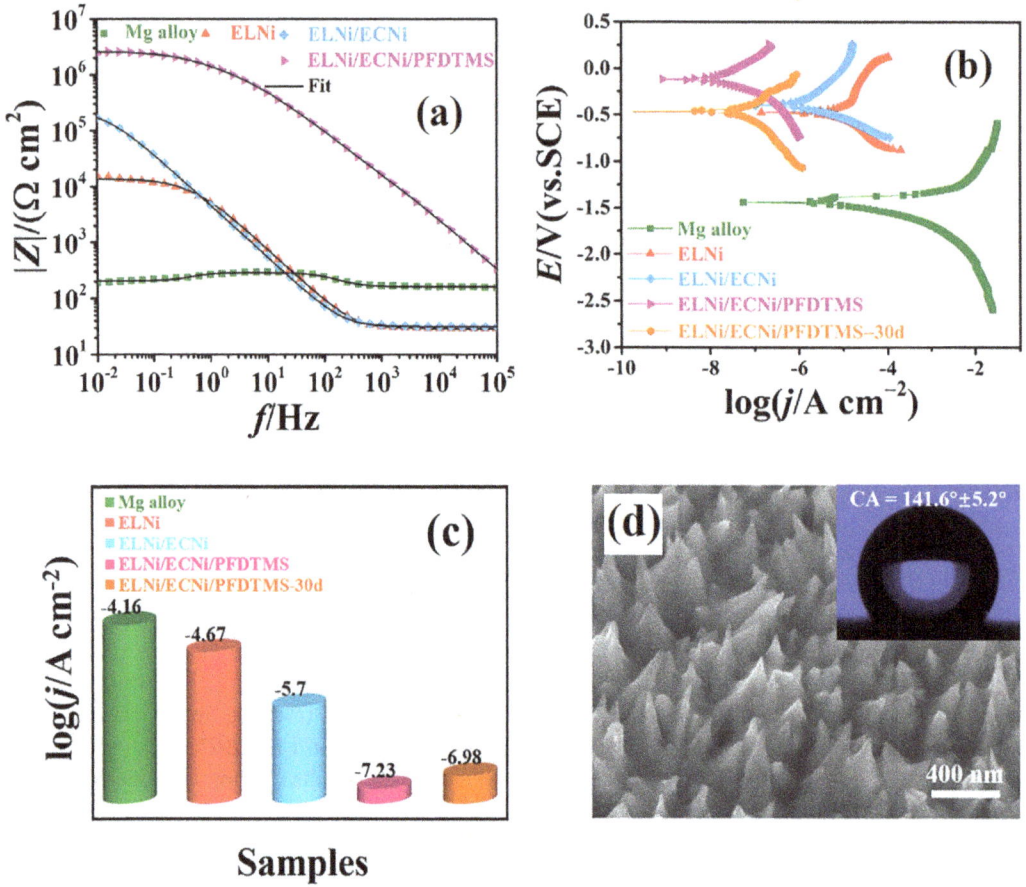

FIGURE 21.7 (a) EIS and (b) PDP plots of bare and coated Mg alloys in 3.5 wt.% NaCl. PDP plot after 30 days of exposure is also provided. (c) Corrosion current densities from the PDP plots. (d) Surface morphology and WCA of the coating after polarization test. *Reproduced with permission from (Fang et al. 2022) © 2022 Elsevier B.V.*

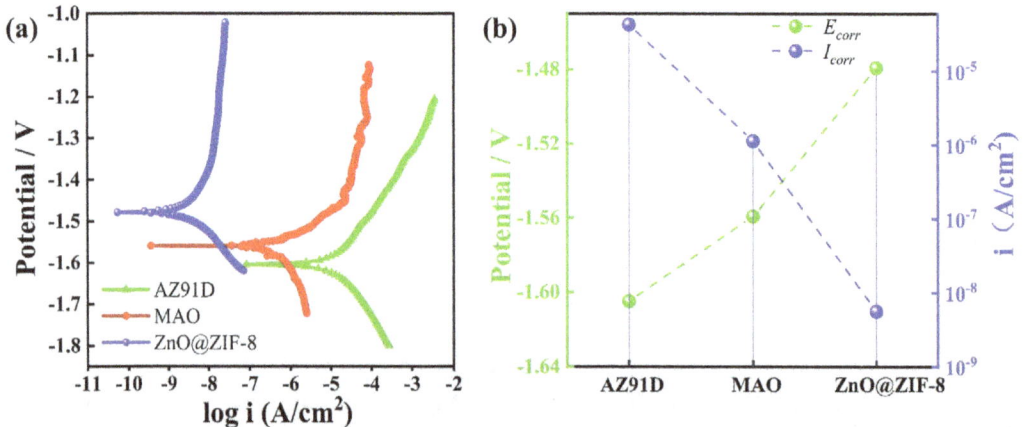

FIGURE 21.8 (a) PPD curves of bare AZ91D, MAO-coated, and ZnO-ZIF-8/MAO-coated alloys in 3.5 wt.% NaCl and (b) corresponding E_{corr} and i_{corr} values. *Reproduced with permission from (Jiang et al. 2022) © 2022 Chongqing University. Publishing services provided by Elsevier B.V. on behalf of KeAi Communications Co. Ltd.*

i_{corr} of the bare, MAO-coated, and superhydrophobic samples in 3.5 wt.% NaCl were 2.1×10^{-5}, 8.2×10^{-7}, and 3.5×10^{-8} A/cm^2, respectively. After 72 hours of immersion, the corresponding values were 2.8×10^{-4}, 5.1×10^{-6}, and 4.8×10^{-8} A/cm^2 (Jiang et al. 2018). Zhang et al. showed that a superhydrophobic MAO/PP-coated sample considerably decreased the i_{corr} (8.76×10^{-9} A/cm^2). The coating remained stable after 248 hours of soaking in 3.5 wt.% NaCl (Zhang et al. 2021).

A spray-coated superhydrophobic ZnO/epoxy-resin demonstrated excellent corrosion protection (Zhou et al. 2019). Yu et al. fabricated an epoxy coating reinforced with silane functionalized superhydrophobic Al$_2$O$_3$ particles (26 vol.% loading). Higher low-frequency (0.01 Hz) impedance values (6.5×10^9 Ω·cm^2) were retained even after 144 days (Yu et al. 2020). Spray coating of PFOTS-modified SiO$_2$ particles on Mg alloy enhanced the R_{p} by more than 4 orders of magnitude (Qian et al. 2018). Li et al. have shown that a superhydrophobic graphene oxide (GO)/polypyrrole-ZIF-8 dual-layer coating prepared by ED and spraying maintained good corrosion protection during 120 hours of immersion in 3.5 wt.% NaCl (Li, Yin et al. 2022). The protection effect of MAO/[OTP][NTf2]/PSE coating (see Figure 21.6) remained intact during 80 days of 3.5 wt.% NaCl immersion. The coating healed the surface scratches and retained the anti-corrosion characteristics for 25 days, owing to the combined effect of the physical barrier and the self-healing by [OTP][NTf2] (**Figure 21.9**) (Zhang et al. 2022).

Several recent works on self-healing anti-corrosion superhydrophobic coatings are available (Li et al. 2021) (Zhang et al. 2021) (Li, Xue et al. 2022) (Liu et al. 2021). DTMS-modified and HT-made magnesium silicate nanotubes (MNTs) were used as nanocontainers to load 2-mercaptabenzimidazole (MI, corrosion inhibitor). An epoxy resin–based multilayer coating with MNTs-MI as the first layer (for self-healing anti-corrosion) and superhydrophobic MNTs/DTMS as the second layer (for superhydrophobicity and anti-corrosion) exhibited substantial corrosion protection. The anti-corrosion performance remained intact during 21 days of immersion in 3.5 wt.% NaCl (Liu et al. 2021).

Reports on immersion-coated superhydrophobic samples also demonstrated better corrosion protection (Xun et al. 2020) (Xie et al. 2020) (Zhao et al. 2015) (Ou and Chen 2019); a similar result

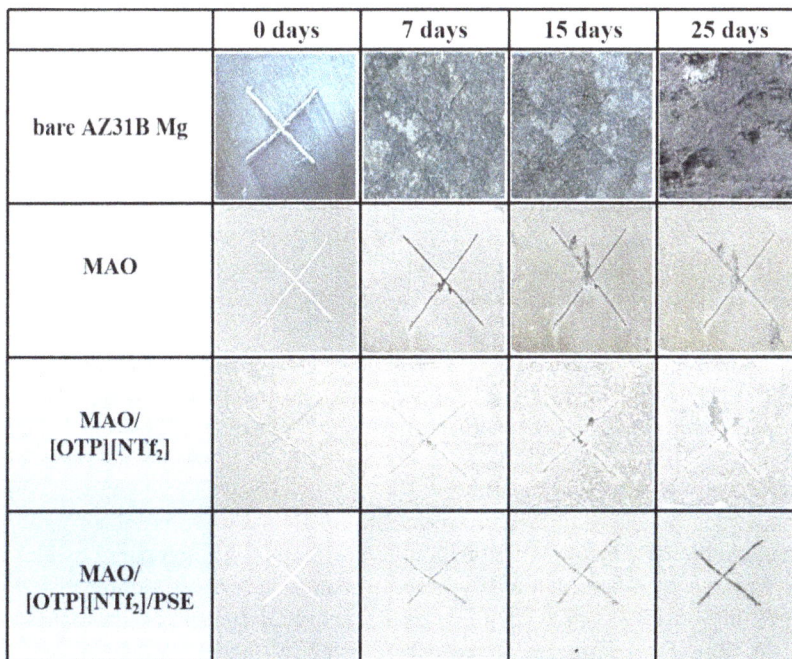

	0 days	7 days	15 days	25 days
bare AZ31B Mg				
MAO				
MAO/ [OTP][NTf$_2$]				
MAO/ [OTP][NTf$_2$]/PSE				

FIGURE 21.9 Optical images of different cross-scratch samples after immersion in 3.5 wt.% NaCl. *Reproduced with permission from (Zhang et al. 2022) © 2022 Elsevier B.V.*

was obtained for the superhydrophobic coating fabricated by electrospinning (Polat et al. 2018). All these electrochemical studies supported improved corrosion protection by the superhydrophobic coatings for Mg alloys, irrespective of the type and methods used, particularly for shorter durations; their performance is always superior to the bare counterparts.

21.2.3 LONG-TERM DURABILITY

The mechanical, thermal, tribological, air-exposure, and chemical durability of superhydrophobic coatings on Mg alloys were extensively studied. Superhydrophobic coatings showed excellent mechanical and tribological durability, but only to a certain extent. WCA variation as a function of the droplet pH supported superior chemical durability on a broader pH range (1–14). However, the direct immersion of the superhydrophobic sample in acidic or aggressive chloride solutions often led to localized corrosion through the defects and edges, compromising long-term durability. The thermal stability is typically a function of the low-SE compounds' decomposition temperature; durability at 100°C to 350°C is noted. Instead, all the reports evidenced superb durability during air-exposure studies. The superhydrophobic-coated Mg alloys maintained WCA > 150° for more than 1 year during ambient air-exposure studies (Saji 2021a).

Studies on mechanical durability conducted with abrasive papers (1000# paper, 5 kPa) on HT/SA-modified samples displayed that the superhydrophobicity was gone after 500 mm distance (Jin et al. 2019). Another study on a one-step HT-made sample maintained superhydrophobicity up to 800 mm of abrasion (800#, 1.2 kPa, 5 mm/s) (Zhang et al. 2018). A similar study maintained high WCA values up to 600 mm (Zhao and Kang 2016). An ED/HT and SA-modified Ni-P coating withstand abrasion for 1200 mm (400#, 9.8 kPa) (Yuan et al. 2017).

Incorporating hard nanoparticles such as SiO_2 is a feasible approach to enhance the wear resistance of superhydrophobic surfaces. Wear studies on one-step blade-coated PDMS/SiO_2 superhydrophobic coating (240#, 100 mm to and fro under 100 g) displayed that the abrasion resistance could be considerably boosted via proper optimization of the size and amount of SiO_2 nanoparticles (Xie et al. 2018). The weight loss of a spray-coated SiO_2-based superhydrophobic sample after the abrasion test (1200#, 100 g) was only ~0.0062 g. The WCA continued > 150° even after 10 cycles (Qian et al. 2018). After 30 times sandpaper friction tests, a superhydrophobic GO/polypyrrole-ZIF-8 coating maintained relatively stabilized WCA (Li, Yin et al. 2022).

Regardless of the fabrication method used, most of the superhydrophobic coatings on Mg alloys presented admirable chemical durability over a broader pH range of pH 1–14 (based on the WCA variation of droplets with different pH). Several studies showed that the WCA rapidly decreased and the superhydrophobicity destroyed at pH ≤ 2 ascribed to the low-SE component's desorption. Conversely, most reports revealed that the superhydrophobicity was maintained only for shorter durations, such as 24 hours or 48 hours, after being immersed in aggressive corrosive solutions (Saji 2021a). Spray-coated superhydrophobic samples in some reports maintained WCA > 150° even after 2 weeks of immersion (Ding et al. 2019).

Available reports supported the long-term durability of superhydrophobic coatings during continuous air exposure. Several reports used exposure periods of 1 to 4 months (Saji 2021a). A few studies used longer durations of 6 months (Yuan et al. 2017) (Ding et al. 2019) and 1 year (Ishizaki et al. 2017a,b) (Li et al. 2016).

Representative results of durability testing of a superhydrophobic composite spray coating of HDTMS-modified diatomite and epoxy resin on AZ31B alloy are provided in **Figure 21.10**. The superhydrophobic HDTMS-modified diatomite powder was first prepared by ED, and subsequently, epoxy-resin + modified-diatomite was spray coated. The coated sample revealed excellent chemical durability during 40 days of immersion (**Figure 21.10a**) and with different pH droplets (**Figure 21.10b**). The coating delivered robust durability under the continued water impact (**Figure 21.10c**). The sand impact (**Figure 21.10d**) and abrasion testing (**Figure 21.10e**) had a more

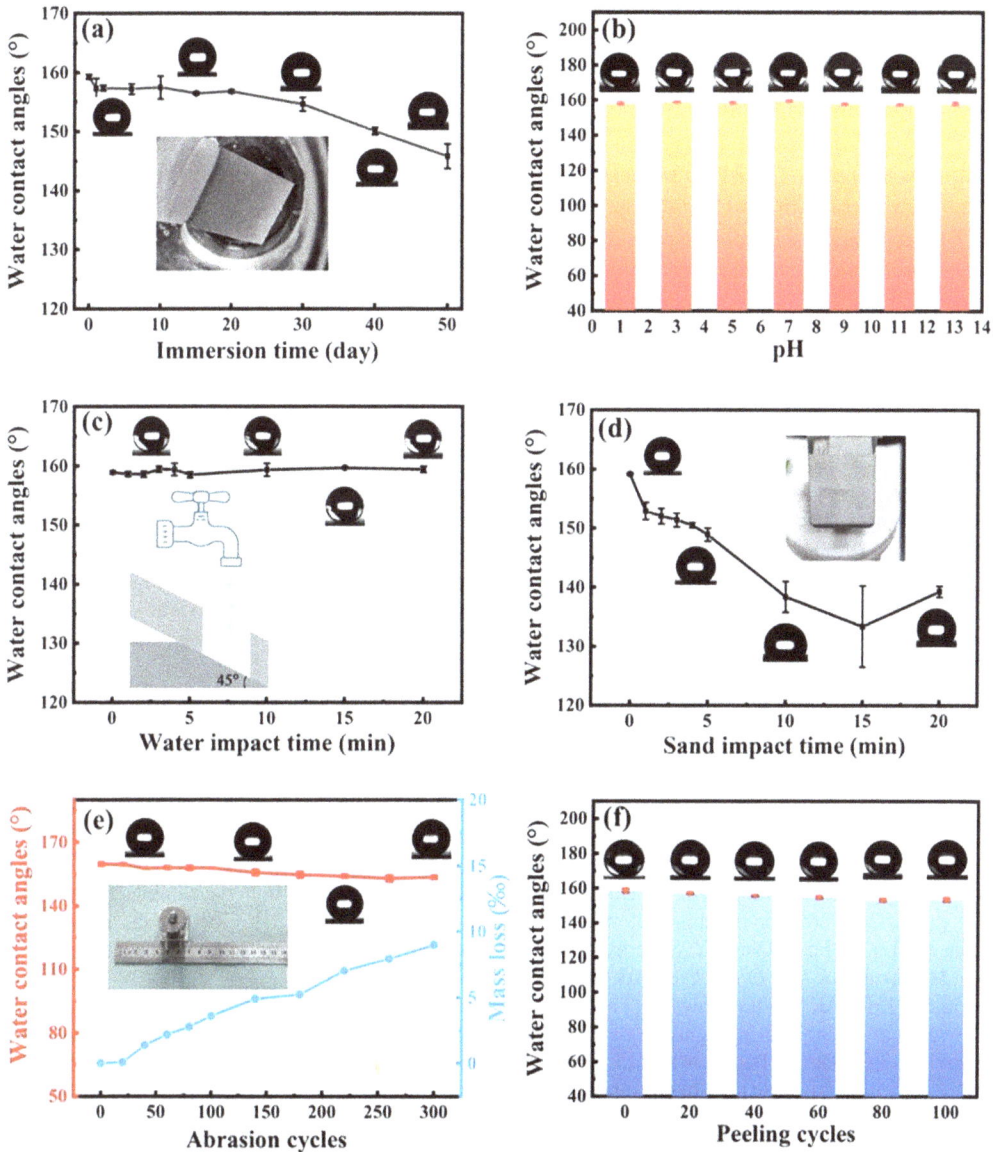

FIGURE 21.10 Variation of WCA of the superhydrophobic coating: (a) Submerged in 3.5 wt.% NaCl solution, (b) droplets with pH of 1–13 and under the influence of (c) water impact time, (d) sand impact time, (e) abrasion cycles, and (f) peeling cycles. *Reproduced with permission from (Wang, Guo et al. 2022) © 2022 Elsevier B.V.*

severe impact on the WCA. The coating maintained superhydrophobicity during the tape peeling tests (**Figure 21.10f**) (Wang, Guo et al. 2022).

21.3 CONCLUSIONS AND OUTLOOK

The superhydrophobic coating is an effective method to overcome Mg alloy's weak aqueous corrosion resistance. This chapter delivers a concise account of the current research scenario on the topic. Only a few works attained superhydrophobicity in a single step, whereas most of the works used the two-step processing, the first for hierarchical surface structuring and the second for the low-SE modification. The most used technique for developing superhydrophobic coatings on Mg alloys is hydrothermal, followed by electrochemical techniques and then direct solution immersion. Superhydrophobic coatings on a robust base layer, such as MAO and LDHs, have attracted significant attention. The MAO process is beneficial, as it can provide a robust thick surface oxide layer. Corrosion inhibitors can be inserted in the porous structure of MAO or the layered structure of LDHs to provide self-healing properties.

Electrochemical studies undoubtedly proved that superhydrophobic coatings substantially improved the corrosion resistance of Mg alloys. Typically, the corrosion current was reduced by 1 to 4 orders of magnitude. However, long-term immersion in 3.5 wt.% NaCl revealed that the super-hydrophobic sample underwent corrosion after a few hours or days. Superhydrophobic coatings on Mg alloys also exhibited excellent long-term air-exposure studies. More studies need to be focussed on improving their long-term mechanical and chemical durability.

REFERENCES

Bormashenko E., Gendelman O., Whyman G. Superhydrophobicity of lotus leaves versus birds wings: Different physical mechanisms leading to similar phenomena, *Langmuir.* 28 (2012) 14992–14997.

Cassie A.B.D., Baxter S. Wettability of porous surfaces, *Trans. Faraday Soc.* 40 (1944) 546–551.

Ding C., Tai Y., Wang D., Tan L., Fu J. Superhydrophobic composite coating with active corrosion resistance for AZ31B magnesium alloy protection, *Chem. Eng. J.* 357 (2019) 518–532.

Erbil H.Y. The debate on the dependence of apparent contact angles on drop contact area or three-phase contact line: A review, *Surf. Sci. Rep.* 69 (2014) 325–365.

Fang R., Liu R., Xie Z.H., Wu L., Ouyang Y., Li M. Corrosion-resistant and superhydrophobic nickel-phosphorus/nickel/PFDTMS triple-layer coating on magnesium alloy, *Surf. Coat. Technol.* 432 (2022) 128054.

Feng L., Zhu Y., Wang J., Shi X. One-step hydrothermal process to fabricate superhydrophobic surface on magnesium alloy with enhanced corrosion resistance and self-cleaning performance, *Appl. Surf. Sci.* 422 (2017) 566–573.

Gu R., Shen J., Hao Q., Wang J., Li D., Hu L., Chen H. Harnessing superhydrophobic coatings for enhancing the surface corrosion resistance of magnesium alloys, *J. Mater. Chem. B.* 9 (2021) 9893.

Huang M., Lu G., Pu J., Qiang Y. Superhydrophobic and smart MgAl-LDH anti-corrosion coating on AZ31 Mg surface, *J. Ind. Eng. Chem.* 103 (2021) 154–164.

Ishizaki T., Kumagai S., Tsunakawa M., Furukawa T., Nakamura K. Ultrafast fabrication of superhydrophobic surfaces on engineering light metals by single-step immersion process, *Mater. Lett.* 193 (2017a) 42–45.

Ishizaki T., Shimada Y., Tsunakawa M., Lee H., Yokomizo T., Hisada S., Nakamura K. Rapid fabrication of a crystalline myristic acid-based superhydrophobic film with corrosion resistance on magnesium alloys by the facile one-step immersion process, *ACS Omega.* 2 (2017b) 7904–7915.

Jiang D., Zhou H., Wan S., Cai G.Y., Dong Z.H. Fabrication of superhydrophobic coating on magnesium alloy with improved corrosion resistance by combining micro-arc oxidation and cyclic assembly, *Surf. Coat. Technol.* 339 (2018) 155–166.

Jiang S., Li W., Liu J., Jiang J., Zhang Z., Shang W., Peng N., Wen Y. ZnO@ZIF-8 core-shell structure nanorods superhydrophobic coating on magnesium alloy with corrosion resistance and self-cleaning, *J. Magnes. Alloys.* https://doi.org/10.1016/j.jma.2022.01.014.

Jiaqiang E., Jin Y., Deng Y., Zuo W., Zhao X., Han D., Peng Q., Zhang Z. Wetting models and working mechanisms of typical surfaces existing in nature and their application on superhydrophobic surfaces: A review, *Adv. Mater. Interfaces.* 5 (2018) 1701052.

Jin Q., Tian G., Li J., Zhao Y., Yan H. The study on corrosion resistance of superhydrophobic magnesium hydroxide coating on AZ31B magnesium alloy, *Colloid Surf. A* 577 (2019) 8–16.

Kameda T., Umetsu Y. Effect of preparation method on particle properties of carbonate-type magnesium – aluminum layered double hydroxides, *J. Ind. Eng. Chem.* 53 (2017) 105–110.

Kang Z., Zhang J., Niu L. A one-step hydrothermal process to fabricate superhydrophobic hydroxyapatite coatings and determination of their properties, *Surf. Coat. Technol.* 334 (2018) 84–89.

Khalifeh S., Burleigh T.D. Super-hydrophobic stearic acid layer formed on anodized high purified magnesium for improving corrosion resistance of bioabsorbable implants, *J. Magnes. Alloys* 6 (2018) 327–336.

Koch K., Bhushan B., Jung Y.C., Barthlott W. Fabrication of artificial Lotus leaves and significance of hierarchical structure for superhydrophobicity and low adhesion, *Soft Matter*. 5 (2009) 1386–1393.

Kuang J., Ba Z., Li Z.Z., Jia Y.Q., Wang Z.Z. Fabrication of a superhydrophobic Mg-Mn layered double hydroxides coating on pure magnesium and its corrosion resistance, *Surf. Coat. Technol*. 361 (2019) 75–82.

Kuang J., Ba Z., Li Z.Z., Wang Z.Z., Qiu J.H. The study on corrosion resistance of superhydrophobic coatings on magnesium, *Appl. Surf. Sci*. 501 (2020) 144137.

La M., Zhou H.J., Li N., Xin Y.C., Sha R., Bao S.H., Jin P. Improved performance of Mg – Y alloy thin film switchable mirrors after coating with a superhydrophobic surface, *Appl. Surf. Sci*. 403 (2017) 23–28.

Li B., Wang L., Su Y., Qiu R., Zhang Z., Ouyang Y. Refreshable self-polishing superhydrophobic coating on Mg alloy to prohibit corrosion and biofouling in marine environment, *Colloids Surf. A*. 651 (2022) 129693.

Li B., Xue S., Mu P., Li J. Robust self-healing graphene oxide-based superhydrophobic coatings for efficient corrosion protection of magnesium alloys, *ACS Appl. Mater. Interfaces*. 14 (2022) 30192–30204.

Li B., Yin X., Xue S., Mu P., Li J. Facile fabrication of graphene oxide and MOF-based superhydrophobic dual-layer coatings for enhanced corrosion protection on magnesium alloy, *Appl. Surf. Sci*. 580 (2022) 152305.

Li D., Ma L., Zhang B., Chen S. Large-scale fabrication of a durable and selfhealing super-hydrophobic coating with high thermal stability and long-term corrosion resistance, *Nanoscale*. 13 (2021) 7810.

Li D., Wang H.Y., Luo D., Liu Y., Han Z., Ren L.Q., Corrosion resistance controllable of biomimetic superhydrophobic microstructured magnesium alloy by controlled adhesion, *Surf. Coat. Technol*. 347 (2018) 173–180.

Li J.H., Liu Q., Wang Y.L., Chen R.R., Takahashi K., Li R.M., Liu L.H., Wang J. Formation of a corrosion-resistant and anti-icing superhydrophobic surface on magnesium alloy via a single-step method, *J. Electrochem. Soc*. 163 (2016) C213–C220.

Li L.J., He J.X., Lei J.L., Liu L., Zhang X.P., Huang T., Li N.B., Pan F.S. Anticorrosive superhydrophobic AZ61 Mg surface with peony-like microstructures, *J. Taiwan Inst. Chem. Eng*. 75 (2017) 240–247.

Liu A.H., Xu J.L. Preparation and corrosion resistance of superhydrophobic coatings on AZ31 magnesium alloy, *Trans. Nonferrous Metal Soc. China*. 28 (2018) 2287–2293.

Liu L., Lei J.L., Li L.J., Zhang J., Shang B., He J.X., Li N.B., Pan F. Robust rare-earth-containing superhydrophobic coatings for strong protection of magnesium and aluminum alloys, *Adv. Mater. Interfaces* 5 (2018) 1800213.

Liu Q., Chen D., Kang Z. One-step electrodeposition process to fabricate corrosion-resistant superhydrophobic surface on magnesium alloy, *ACS Appl. Mater. Interfaces*. 7 (2015) 1859–1867.

Liu X., Zhang T.C., He H.Q., Ouyang L., Yuan S.J. A stearic acid/CeO$_2$ bilayer coating on AZ31B magnesium alloy with superhydrophobic and self-cleaning properties for corrosion inhibition, *J. Alloys Compds* 834 (2020) 155210.

Liu X., He H., Zhang T.C., Ouyang L., Zhang Y.X., Yuan S. Superhydrophobic and self-healing dual-function coatings based on mercaptobenzimidazole inhibitor-loaded magnesium silicate nanotubes for corrosion protection of AZ31B magnesium alloys, *Chem. Eng. J*. 404 (2021) 127106.

Liu Y., Yao W.G., Yin X.M., Wang H.Y., Han Z.W., Ren L.Q. Controlling wettability for improved corrosion inhibition on magnesium alloy as biomedical implant materials, *Adv. Mater. Interfaces*. 3 (2016) 1500723.

Milne A.J.B., Amirfazli A. The Cassie equation: How it is meant to be used, *Adv. Colloid Interf. Sci*. 170 (2012) 48–55.

Nishino T., Meguro M., Nakamae K., Matsushita M., Ueda Y. The lowest surface free energy based on –CF$_3$ alignment, *Langmuir*. 15 (1999) 4321–4323.

Nosonovsky M., Bhushan B. Hierarchical roughness makes superhydrophobic states stable. *Microelectron. Eng*. 84 (2007) 382–386.

Ou J.F., Chen X. Corrosion resistance of phytic acid/Ce (III) nanocomposite coating with superhydrophobicity on magnesium, *J. Alloys Compd*. 787 (2019) 145–151.

Pagliaro M., Ciriminna R. New fluorinated functional materials, *J. Mater. Chem*. 15 (2005) 4981–4991.

Peng F., Wang D.H., Ma X.H., Zhu H.Q., Qiao Y.Q., Liu X.Y. "Petal effect"-inspired superhydrophobic and highly adhesive coating on magnesium with enhanced corrosion resistance and biocompatibility, *Sci. China Mater*. 61 (2018) 629–642.

Polat N.H., Kap O., Farzaneh A. Anticorrosion coating for magnesium alloys: Electrospun superhydrophobic polystyrene/SiO composite fibers, *Turkish J. Chem*. 42 (2018) 672–683.

Pourfaraj R., Fatemi S.J., Kazemi S.Y., Biparva P. Synthesis of hexagonal mesoporous MgAl LDH nano-platelets adsorbent for the effective adsorption of Brilliant Yellow, *J. Colloid. Interface Sci.* 508 (2017) 65–74.

Qian Z.Q., Wang S.D., Ye X.S., Liu Z., Wu Z.J. Corrosion resistance and wetting properties of silica-based superhydrophobic coatings on AZ31B Mg alloy surfaces, *Appl. Surf. Sci.* 453 (2018) 1–10.

Qiu R.X., Li C., Tong W., Xiong D.S., Li Z.X., Wu Z.L. High-speed wire electrical discharge machining to create superhydrophobic surfaces for magnesium alloys with high corrosion and wear resistance, *Mater. Corros.* 71 (2020) 1711–1720.

Qiu Z.Z., Sun J., Wang R., Zhang Y.S., Wu X.H. Magnet-induced fabrication of a superhydrophobic surface on ZK60 magnesium alloy, *Surf. Coat. Technol.* 286 (2016) 246–250.

Safarpour M., Hosseini S.A., Ahadani-Targhi F., Vašina P., Alishahi M. A transition from petal-state to lotus-state in AZ91 magnesium surface by tailoring the microstructure, *Surf. Coat. Technol.* 383 (2020) 125239.

Saji V.S. Wax-based artificial superhydrophobic surfaces and coatings, *Colloids Surf. A.* 602 (2020a) 125132.

Saji V.S. Superhydrophobic surfaces and coatings by electrochemical anodic oxidation and plasma electrolytic oxidation, *Adv. Colloid Interface Sci.* 283 (2020b) 102245.

Saji V.S. Recent progress in supehydrophobic and superamphiphobic coatings for magnesium alloys, *J. Magnes. Alloy.* 9 (2021a) 748–778.

Saji V.S. Carbon nanostructure-based superhydrophobic surfaces and coatings, *Nanotech. Rev.* 10 (2021b) 518–571.

Saji V.S. Electrophoretic-deposited superhydrophobic coatings, *Chem. Asian J.* 16 (2021c) 474–491.

Saji V.S. Superhydrophobic surfaces and coatings by electrochemical methods – a review, *J. Adh. Sci. Technol.* 37 (2023) 137–161.

Saji V.S., Sanakara Narayanan T.S.N., Chen X. B. Conversion coatings for magnesium and its alloys, *Springer Nat.* ISBN: 978-3-030-89976-9, February 2022. https://doi.org/10.1007/978-3-030-89976-9.

Shi L., Hu J., Lin X.D., Fang L., Wu F., Xie J., Meng F.M. A robust superhydrophobic PPS-PTFE/SiO2 composite coating on AZ31 Mg alloy with excellent wear and corrosion resistance properties, *J. Alloys Compds.* 721 (2017) 157–163.

Siddiqui A.R., Maurya R., Katiyar P.K., Balania K. Superhydrophobic, self-cleaning carbon nanofiber CVD coating for corrosion protection of AISI 1020 steel and AZ31 magnesium alloys, *Surf. Coat. Technol.* 404 (2020) 126421.

Song S., Yan H., Cai M., Huang Y., Fan X., Cui X., Zhu M. Superhydrophobic composite coating for reliable corrosion protection of Mg alloy, *Mater. Design.* 215 (2022) 110433.

Song Z.W., Xie Z.H., Ding L.F., Zhang Y. Corrosion resistance of super-hydrophobic coating on AZ31B Mg alloy, *Intl. J. Electrochem. Sci.* 13 (2018) 6190–6200.

Tan C., Cai P., Xu L., Yang N., Xi Z.X., Li Q. Fabrication of superhydrophobic surface with controlled adhesion by designing heterogeneous chemical composition, *Appl. Surf. Sci.* 349 (2015) 516–523.

Timmons C.O., Zisman W.A. Effect of liquid structure on contact angle hysteresis. *J. Colloid Interface Sci.* 22 (1966) 165–171.

Wan H.G., Hu X.F. One-step solve-thermal process for the construction of anticorrosion bionic superhydrophobic surfaces on magnesium alloy, *Mater. Lett.* 174 (2016) 209–212.

Wang H.Y., Wei Y.H., Liang M.M., Hou L.F., Li Y.G., Guo C. Fabrication of stable and corrosion-resisted super-hydrophobic film on Mg alloy, *Colloids Surf. A.* 509 (2016) 351–358.

Wang L., Xiao X., Yin X., Wang J., Zhu G., Yu S., Liu E., Wang B., Yang X. Preparation of robust, self-cleaning and anti-corrosion superhydrophobic Ca-P/chitosan (CS) composite coating on AZ31 magnesium alloy, *Surf. Coat. Technol.* 432 (2022) 128074.

Wang S., Liu K., Yao X., Jiang L. Bioinspired surfaces with superwettability: New insight on theory, design, and applications. *Chem. Rev.* 115 (2015) 8230–8293.

Wang T., Guo Q., Zhang T.C., Zhang Y.X., Yuan S. Large-scale prepared superhydrophobic HDTMS-modified diatomite/epoxy resin composite coatings for high-performance corrosion protection of magnesium alloys, *Prog. Org. Coat.* 170 (2022) 106999.

Wang X., Jing C., Chen Y.X., Wang X.S., Zhao G., Zhang X., Wu L., Liu X.Y., Dong B., Zhang Y. Active corrosion protection of super-hydrophobic corrosion inhibitor intercalated Mg – Al layered double hydroxide coating on AZ31 magnesium alloy, *J. Magnes. Alloy.* 8 (2020) 291–300.

Wang Y., Gu Z.P., Xin Y., Yuan N.Y., Ding J.N. Facile formation of super-hydrophobic nickel coating on magnesium alloy with improved corrosion resistance, *Colloids Surf. A.* 538 (2018) 500–505.

Wang Z.H., Zhang J.M., Li Y., Bai L.J., Zhang G.J. Enhanced corrosion resistance of micro-arc oxidation coated magnesium alloy by superhydrophobic Mg-Al layered double hydroxide coating, *Trans. Nonferrous Metal Soc. China.* 29 (2019) 2066–2077.

Webb H.K., Crawford R.J., Ivanova E.P. Wettability of natural superhydrophobic surfaces, *Adv. Colloid Interface Sci.* 210 (2014) 58–64.

Wu H., Shi Z., Zhang X.M., Qasim A.M., Xiao S., Zhang F., Wu Z.Z., Wu G.S., Ding K., Chu P.K. Achieving an acid resistant surface on magnesium alloy via bio-inspired design, *Appl. Surf. Sci.* 478 (2019) 150–161.

Wu J., Zhao X., Tang C., Lei J., Li L. One-step dipping method to prepare inorganic-organic composite superhydrophobic coating for durable protection of magnesium alloys, *J. Taiwan Inst. Chem. Eng.* 135 (2022) 104364.

Wu L., Wu J.H., Zhang Z., Zhang C., Zhang Y., Tang A., Li L.J., Zhang G., Zheng Z.C., Atrens A., Pan F.S. Corrosion resistance of fatty acid and fluoroalkylsilane-modified hydrophobic Mg-Al LDH films on anodized magnesium alloy, *Appl. Surf. Sci.* 487 (2019) 569–580.

Xie J., Hu J., Fang L., Liao X.L., Du R.L., Wu F., Wu L. Facile fabrication and biological properties of superhydrophobic coating on magnesium alloy used as potential implant materials, *Surf. Coat. Technol.* 384 (2020) 125223.

Xie J., Hu J., Lin X.D., Fang L., Wu F., Liao X.L., Luo H.J., Shi L.T. Robust and anti-corrosive PDMS/SiO2 superhydrophobic coatings fabricated on magnesium alloys with different-sized SiO_2 nanoparticles, *Appl. Surf. Sci.* 457 (2018) 870–880.

Xue S., Li B., Mu P., Li J. Designing attapulgite-based self-healing superhydrophobic coatings for efficient corrosion protection of magnesium alloys, *Prog. Org. Coat.* 170 (2022) 106966.

Xun X.W., Wan Y.Z., Zhang Q.C., Gan D.Q., Hu J., Luo H.L. Low adhesion superhydrophobic AZ31B magnesium alloy surface with corrosion resistant and anti-bioadhesion properties, *Appl. Surf. Sci.* 505 (2020) 144566.

Yang N., Li J.C., Bai N.N., Xu L., Li Q. One step phase separation process to fabricate superhydrophobic PVC films and its corrosion prevention for AZ91D magnesium alloy, *Mater. Sci. Eng. B.* 209 (2016) 1–9.

Yang N., Li Q., Chen F., Cai P., Tan C., Xi Z.X. A solving-reprecipitation theory for self-healing functionality of stannate coating with a high environmental stability, *Electrochim. Acta.* 174 (2015) 1192–1201.

Yao W., Wu L., Huang G., Jiang B., Atrensc A., Pan F. Superhydrophobic coatings for corrosion protection of magnesium alloys, *J. Mater. Sci. Technol.* 52 (2020) 100–118.

Yin X.X., Mu P., Wang Q.T., Li J. Superhydrophobic ZIF-8-based dual-layer coating for enhanced corrosion protection of Mg alloy, *ACS Appl. Mater. Interfaces.* 12 (2020) 35453–35463.

Young T. An essay on the cohesion of fluid, *Philos. Trans. R. Soc.* 95 (1805) 65–87.

Yu M.D., Fan C.Q., Han S., Ge F., Cui Z.Y., Lu Q.Y., Wang X. Anticorrosion behavior of superhydrophobic particles reinforced epoxy coatings for long-time in the high salinity liquid, *Prog. Org. Coat.* 147 (2020) 105867.

Yuan J., Wang J.H., Zhang K., Hu W.B. Fabrication and properties of a superhydrophobic film on an electroless plated magnesium alloy, *RSC Adv.* 7 (2017) 28909–28917.

Zhang C.L., Zhang F., Song L., Zeng R.C., Li S.Q., Han E.H. Corrosion resistance of a superhydrophobic surface on micro-arc oxidation coated Mg-Li-Ca alloy, *J. Alloys Compds.* 728 (2017) 815–826.

Zhang F., Zhang C.L., Song L., Zeng R.C., Li S., Cui H.Z. Fabrication of the superhydrophobic surface on magnesium alloy and its corrosion resistance, *J. Mater. Sci. Technol.* 31 (2015) 1139–1143.

Zhang F., Zhang C.L., Zeng R.C., Song L., Guo L., Huang X.W. Corrosion resistance of the superhydrophobic $Mg(OH)_2$/Mg-Al layered double hydroxide coatings on magnesium alloys, *Metals.* 6 (2016) 85.

Zhang J., Zhao X., Wei J., Li B., Zhang J. Superhydrophobic coatings with photothermal self-healing chemical composition and microstructure for efficient corrosion protection of magnesium alloy, *Langmuir.* 37 (2021) 13527–13536.

Zhang M.L., Zhao M.W., Chen R.R., Liu J.Y., Liu Q., Yu J., Li R.M., Liu P., Wang J. Fabrication of the pod-like KCC-1/TiO2 superhydrophobic surface on AZ31 Mg alloy with stability and photocatalytic property, *Appl. Surf. Sci.* 499 (2020) 143933.

Zhang Q.Q., Zhang H.C. Corrosion resistance and mechanism of micro-nano structure super-hydrophobic surface prepared by laser etching combined with coating process, *Anti-Corros. Methods Mater.* 66 (2019) 264–273.

Zhang S.J., Cao D.L., Xu L.K., Tang J.K., Meng R.Q., Li H.D. Corrosion resistance of a superhydrophobic dodecyltrimethoxysilane coating on magnesium hydroxide-pretreated magnesium alloy AZ31 by electrodeposition, *Colloid. Surf. A.* 625 (2021a) 126914.

Zhang S.J., Cao D.L., Xu L.K., Tang J.K., Meng R.Q., Li H.D. Corrosion resistance of a superhydrophobic dodecyltrimethoxysilane coating on magnesium alloy AZ31 fabricated by one-step electrodeposition, *New J. Chem.* 45 (2021b) 14665.

Zhang X.K., Shen J., Hu D., Duan B., Wang C.M. A rapid approach to manufacture superhydrophobic coating on magnesium alloy by one-step method, *Surf. Coat. Technol.* 334 (2018) 90–97.

Zhang Y., Li N., Ling N., Zhang J., Wang L. Enhanced long-term corrosion resistance of Mg alloys by super-hydrophobic and self-healing composite coating, *Chem. Eng. J.* 449 (2022) 137778.

Zhang Y.F., Lin T.G. Influence of duty cycle on properties of the superhydrophobic coating on an anodized magnesium alloy fabricated by pulse electrodeposition, *Colloids Surf. A.* 568 (2019) 43–50.

Zhang Z.Q., Wang L., Zeng M.Q., Zeng R.C., Lin C.G., Wang Z.L., Chen D.C., Zhang Q. Corrosion resistance and superhydrophobicity of one-step polypropylene coating on anodized AZ31 Mg alloy. *J. Magnes. Alloy.* 9 (2021) 1443–1457.

Zhang Z.Q., Zeng R.C., Yan W., Lin C.G., Wang L., Wang Z.L., Chen D.C. Corrosion resistance of one-step superhydrophobic polypropylene coating on magnesium hydroxide-pretreated magnesium alloy AZ31, *J. Alloys Compds.* 821 (2020) 153515.

Zhao M., Wang X.L., Song H., Li J.G., He G.P., Gui Y.H., Feng W. Fabrication of a superhydrophobic phosphate/fatty-acid salt compound coating on magnesium Alloy, *ECS Electrochem. Lett.* 4 (2015) C19–C21.

Zhao T.T., Kang Z.X. Simultaneously fabricating multifunctional superhydrophobic/superoleophilic coatings by one-step electrodeposition method on cathodic and anodic magnesium surfaces, *J. Electrochem. Soc.* 163 (2016) D628–D635.

Zheng T.X., Hu Y.B., Pan F.S., Zhang Y.X., Tang A.T. Fabrication of corrosion-resistant superhydrophobic coating on magnesium alloy by one-step electrodeposition method, *J. Magnes. Alloys.* 7 (2019) 193–202.

Zhong Y.X., Hu J., Zhang Y.F., Tang S.W. The one-step electroposition of superhydrophobic surface on AZ31 magnesium alloy and its time-dependence corrosion resistance in NaCl solution, *Appl. Surf. Sci.* 427 (2018) 1193–1201.

Zhou H.M., Chen R.R., Liu Q., Liu J.Y., Yu J., Wang C., Zhang M., Liu P., Wang J. Fabrication of ZnO/epoxy resin superhydrophobic coating on AZ31 magnesium alloy, *Chem. Eng. J.* 368 (2019) 261–272.

Zhou M., Pang X.L., Wei L., Gao K.W. Insitu grown superhydrophobic Zn – Al layered double hydroxides films on magnesium alloy to improve corrosion properties, *Appl. Surf. Sci.* 337 (2015) 172–177.

22 Self-Healing Coatings

Wenhui Yao, Liang Wu and Fusheng Pan

CONTENTS

22.1 INTRODUCTION

Corrosion can significantly reduce the mechanical properties of metallic materials, shorten the service life of equipment, and even cause disasters such as fire and explosion (Bryant and Neville 2016; Cho et al. 2019; Soltan et al. 2021; Yang et al. 2021; Gudze and Melchers 2008; Ali et al. 2021; Christudasjustus et al. 2023; Zhang et al. 2022; Duan et al. 2023). It is estimated that the cost of corrosion is approximately 3.34% of the GDP in China, approximately USD 310 billion (Hou et al. 2017). Thus, it is imperative to improve the corrosion resistance of metals, increasing reliability and decreasing the huge financial costs, environmental hazards, and energy consumption.

In particular, Mg, as one of the most chemically active metallic materials, should be also blamed because of its relatively poor corrosion resistance, thereby restricting its wide utilization in various fields (Ghazizadeh, Jabbari, and Sedighi 2021; Liu, He et al. 2021; Nie et al. 2021; Adsul et al. 2021; Xue et al. 2023; Jiang et al. 2021). Consequently, different strategies have been correspondingly proposed, mainly including alloying (Azzeddine et al. 2021; Dobkowska et al. 2021; Liu, Yan et al. 2021; Gao et al. 2021), homogenization (Gungor and Incesu 2021; Yu et al. 2021; Chu et al. 2021; Li et al. 2021; Peng et al. 2022), and surface treatment (Yao et al. 2020; Chen, Wu, Yao, Wu et al. 2022; Yao, Chen et al. 2022; Chen, Wu, Yao, Chen et al. 2022; Li 2022). Especially, various protective coatings with good corrosion-protection performance and low costs (Dai et al. 2022; Kamde, Mahton, and Saha 2022; Miao et al. 2022), which can effectively prevent the direct contact with the external environment, are extensively applied among different strategies.

In recent years, active self-healing coatings are attracting more and more attention considering the unavoidable damages during the transportation and service, causing scratches, cracks, or delamination (Koch, Mack, and Vassen 2022; Luo et al. 2022; Zhu et al. 2022). The untimely repair may lead to the failure of the coating. The combination of self-healing property with corrosion-resistant coatings can simultaneously provide a passive barrier and active corrosion protection to enhance the

DOI: 10.1201/9781003319856-25

corrosion resistance. Consequently, the relevant studies and publications associated with Mg alloys proliferate (Johari et al. 2022; Hamdy and Butt 2013; Zhao et al. 2020; Zhang et al. 2016; Yang, Sun, and Chen 2022; Zhang, Peng, and Liu 2021; Jiang et al. 2019). The self-healing coatings can spontaneously repair the damages and recover the performance of the coating, reducing maintenance costs and economic loss due to shutdown and prolonging the service life.

This chapter systematically analyzes and summarizes the most recent significant developments of self-healing coatings on Mg alloys. First, the self-healing mechanism is discussed. The next part mainly focuses on the fabrication strategies and properties of self-healing coatings. Finally, the future perspectives of self-healing coatings are briefly discussed to accelerate their wide applications in the future.

22.2   SELF-HEALING MECHANISM

Self-healing coatings can be simply divided into autonomous and non-autonomous self-healing coatings, and the schematic diagrams are illustrated in **Figure 22.1** (Cho, White, and Braun 2009; Shchukin and Möhwald 2013). In brief, the autonomous self-healing coatings can autonomically respond to fluctuations and variations of the coating integrity, while the non-autonomous self-healing coatings require external stimulus such as light, temperature, and humidity of the environment surrounding the coating to initiate the restoration of chemical bonds or polymer chain

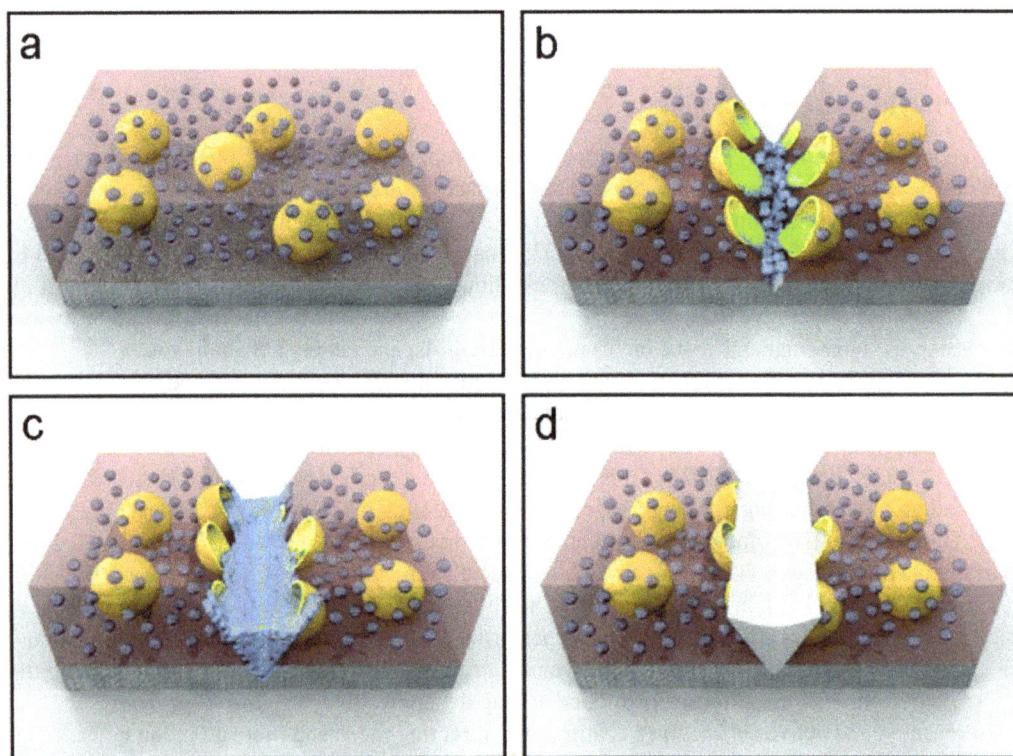

FIGURE 22.1   Schematically illustrating the self-healing process of autonomous self-healing coatings. The yellow catalyst and blue phase-separated healing-agent droplets are encapsulated in the light-orange matrix, which is released after the coating is damaged. The synergistic effects of healing agent and catalyst heal the damaged region. *Reproduced with permission from (Cho, White, and Braun 2009) © 2009 WILEY-VCH Verlag GmbH & Co. KGaA, Weinheim.*

conformation in the coating (Zhang et al. 2018). After the exposure to certain impacts, the active part of the coatings responds to restore the self-healing property, thus reducing the negative effect of the external impact on the coating.

22.2.1 Autonomous Self-Healing Coatings

Autonomous self-healing coatings can spontaneously heal the coating defects to restore the structure after damage, without external interference (Attaei et al. 2020; D'Elia et al. 2019; Guo, Han et al. 2020). Generally, they are divided into two approaches, carrying (1) polymerizable healing agents and (2) corrosion inhibitors, to accomplish the self-healing performance. The polymerizable healing agents are usually embedded in the capsules/vasculatures, which rupture when the coating is damaged to release the healing agents, like monomers and re-hardening catalysts (Ikura et al. 2022; Toohey et al. 2007). Subsequently, the released agents trigger polymerization to replenish the damaged area of the coating (White et al. 2001). However, this kind of self-healing coating has been rarely reported on the surface of Mg alloys for corrosion protection.

In contrast, as for the corrosion inhibitor, it can be directly embedded into the coating matrix or loaded by capsules (Lei et al. 2020; Thakur and Kumar 2021; Li, Guo, and Frankel 2021). After being damaged, the corrosion inhibitor is released to reach the damaged area to restrict the electrochemical reactions to decrease the corrosion rate of metallic substrates. Corrosion inhibitors can be categorized into organic inhibitors, like mercaptobenzothiazole (MBT), benzotriazole (BTA), and 8-hydroxyquinoline (8-HQ), and inorganic inhibitors, such as vanadates, nitrites, phosphates, molybdates, and rare earth salts (Sharma et al. 2020; Lee et al. 2021; Liu et al. 2014; Guo et al. 2020; Rizvi et al. 2021; Bastidas et al. 2015). On the other hand, according to the effects of corrosion inhibitors on the electrochemical behavior, they are also divided into four types: anodic inhibitors, cathodic inhibitors, mixed inhibitors, and volatile inhibitors (Faisal et al. 2018; Ryl et al. 2019). As is well known, the corrosion process mainly revolves around anodic dissolution and cathodic reactions. Therefore, the corrosion inhibitor can be specially selected on demand. For instance, anodic inhibitors can highly change the anodic corrosion potential, passivating and forming an oxide film on the metallic substrate to prevent the anodic dissolution, and cathodic inhibitors will cause the precipitation of oxides and hydroxides at the cathodic sites to retard the cathodic reactions (Østnor and Justnes 2011; Hegazy et al. 2016). In contrast, the mixed inhibitor can retard both the anodic and cathodic corrosion reactions by physical adsorption or chemical reaction or complexation on the metallic substrates (Li et al. 2022; Popoola et al. 2021). Therefore, whichever the reaction restricted by the corrosion inhibitor, it interacts at the interface between the metallic substrate and electrolyte by generating a film, which may be a passivating film, precipitation film, or adsorption film (Monticelli 2018).

22.2.2 Non-Autonomous Self-Healing Coatings

Non-autonomous self-healing is also called intrinsic healing. The self-healing process of this type of coatings needs external stimuli that cannot be generated from the corrosion process, like temperature and light (Xu et al. 2023; Yuan, Ji, and Yuan 2023; Yan et al. 2023). The healing ability of these coatings is associated with the intrinsic chemical bonds and/or physical conformations of polymer matrix (Zheludkevich and Hughes 2016). The damaged polymer is brought together to ensure the contact, and then the reconstruction of polymer bonds and recovery of the polymer matrix integrity can be achieved, based on supramolecular interactions (hydrogen bonding (Cordier et al. 2008), ionic interactions (Kalista Jr and Ward 2007), π-π interactions, and host–guest interactions or reversible covalent bondings (Diels-Alder reaction) (Chen et al. 2002; Garcia 2014). Overall, it is not relatively applicable to self-healing coatings on Mg alloys, and the following review and discussion mainly focus on self-healing by corrosion inhibitors.

22.3 SELF-HEALING BY CORROSION INHIBITORS

22.3.1 ENCAPSULATED INTO COATING MATRIX

As aforementioned, the corrosion inhibitor can be directly encapsulated into the coating matrix on Mg alloys to induce the self-healing property. The generally used coating matrixes mainly include polymers, micro-arc oxidation (MAO) coatings, and layered double hydroxide (LDH) coatings.

In Soliman et al.'s work (Soliman et al. 2020), the corrosion inhibitors of nano-graphene oxide (GO) particles were incorporated into polymerized hydroxyquinoline (HQ) coatings on AZ31 Mg alloy, considering the importance of HQ on the self-healing behavior (Andreeva et al. 2008). In particular, they found that the surface treatment of AZ31 using either oxidation or phosphating, prior to the GO/HQ deposition, was very important to the initiation of more anchoring bonds between molecules to accelerate the self-healing rate and improve the anti-corrosion performance (**Figure 22.2a**). Zhao et al. (Zhao et al. 2019) applied a new spin-spray layer-by-layer (SSLbL) assembly technique to develop a multilayer coating composed of SiO_2 and CeO_2 nanoparticles on AZ31 Mg alloys. Especially, SiO_2 nanoparticles were applied to construct the physical barrier against corrosive solutions, while the CeO_2 nanoparticles acted as corrosion inhibitors to heal the physical damage. It was shown that the coating presented good self-healing performance. A polyacrylamide (PAM) hydrogel with Ce^{4+} inhibitor was *in situ* formed on the porous surface of MAO-treated AZ31 Mg alloy in Guo et al.'s work (Guo, Liu et al. 2020). The scratched sample can effectively refrain from pitting corrosion and repair itself in the neutral salt spray test. Its self-healing performance was mainly associated with the inhibition effect of Ce^{4+} and the bridge adsorption of PAM gels. In addition, the Ce-PAM/MAO coating did not peel off at the scratches, and the electrochemical results demonstrated that the repaired coating still presented good corrosion resistance. Calado *et*

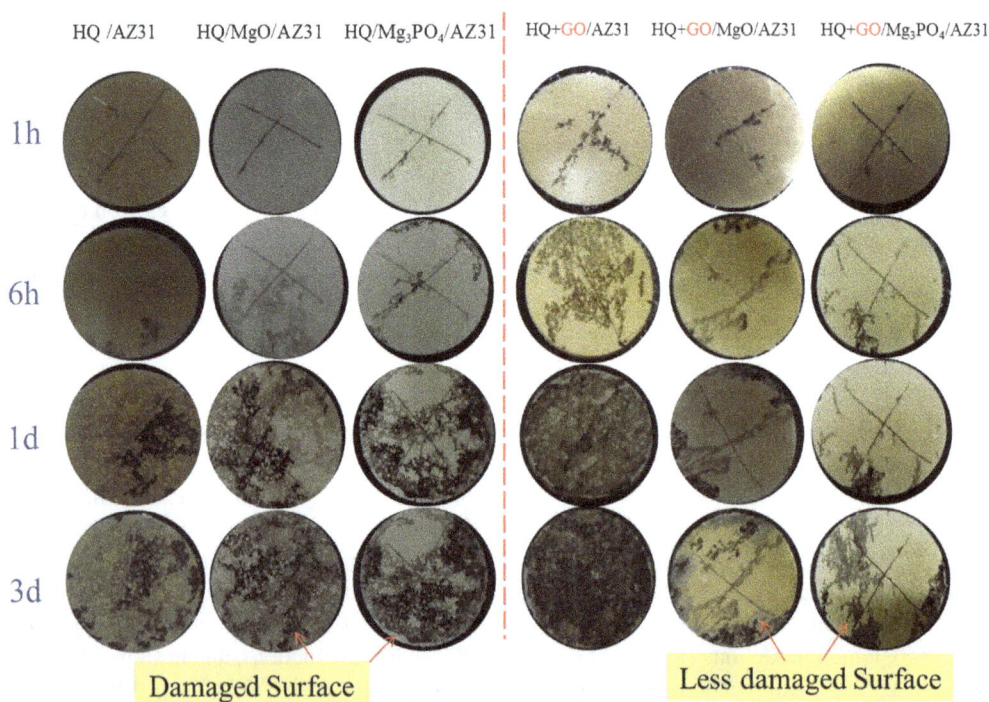

FIGURE 22.2A   Optical images of different samples immersed in simulated body fluid (SBF) for different times. *Reproduced with permission from (Soliman et al. 2020) © 2020 Elsevier B.V.*

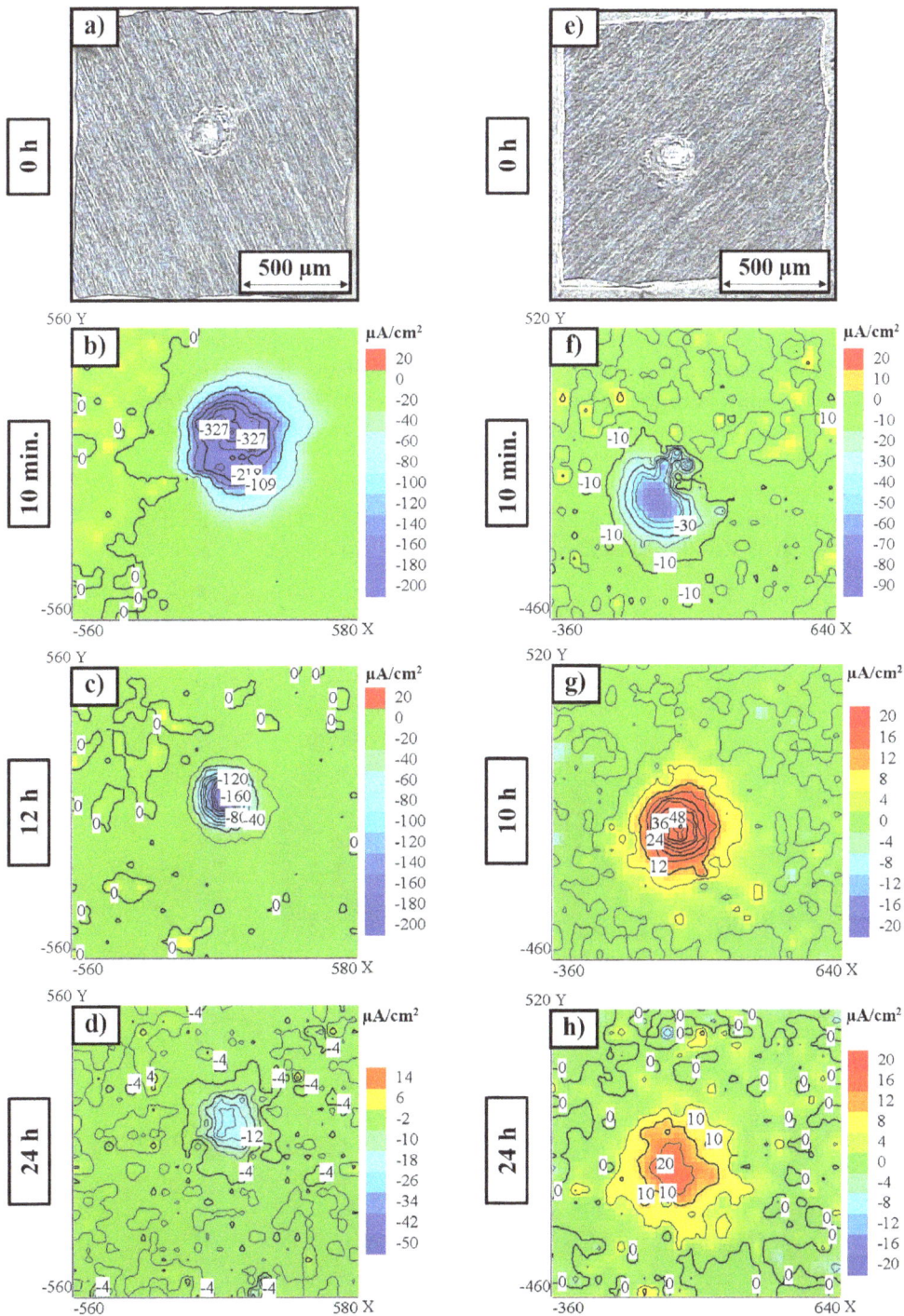

FIGURE 22.2B Optical images and SVET maps of the reference (left) and modified (right) coatings. *Reproduced with permission from (Calado et al. 2021) © 2021 Elsevier B.V.*

al. (Calado et al. 2021) used cerium tri(bis(2-ethylhexyl)phosphate) (Ce(DEHP)$_3$) in hybrid epoxy-silane coatings as a self-healing corrosion inhibitor for WE43 Mg alloys. The self-healing performance of Ce(DEHP)$_3$ was evidenced by localized electrochemical techniques of the scanning vibrating electrode technique (SVET) (**Figure 22.2b**). The corrosion propagation was inhibited in artificial defects, and the net cathodic current densities decreased in comparison with the unmodified reference coating. The corrosion protection of Ce(DEHP)$_3$ was attributed to the synergistic effects of the cerium and organophosphate moieties, besides the protection of yttrium from the WE43 Mg alloy. Li *et al.* (Li et al. 2021) extracted a series of organic compounds from commercial drugs, natural plants, and marine life as potential corrosion inhibitors and then stored them in the sol-gel coating on ZE21B Mg alloys. Among different organic compounds, such as curcumin, aspirin, sodium alginate, hematoxylin, paeonol, *etc.*, the paeonol condensation tyrosine (PCTyr) Schiff base was proved as the most effective corrosion inhibitor, and it performed well in the corrosion protection and self-healing process. As presented in **Figure 22.2c**, the Cl- ion penetrates into the Mg matrix through the damaged coating, leading to local corrosion and causing the release of Mg^{2+} ions. Meanwhile, the corrosion inhibitor slowly releases, chelating with the enriched Mg^{2+} ions and forming PCTyr Schiff base-Mg complex at the damaged area. The gradually deposited corrosion products provided a dense protective layer to alleviate the corrosion of Mg alloy substrate, exhibiting good self-healing performance.

The development of a superamphiphobic coating can significantly reduce the solid-liquid contact area and contact time with the external corrosive environment. Thus, the coating will effectively inhibit corrosion of Mg alloy substrates by blocking or at least delaying the permeation of corrosive

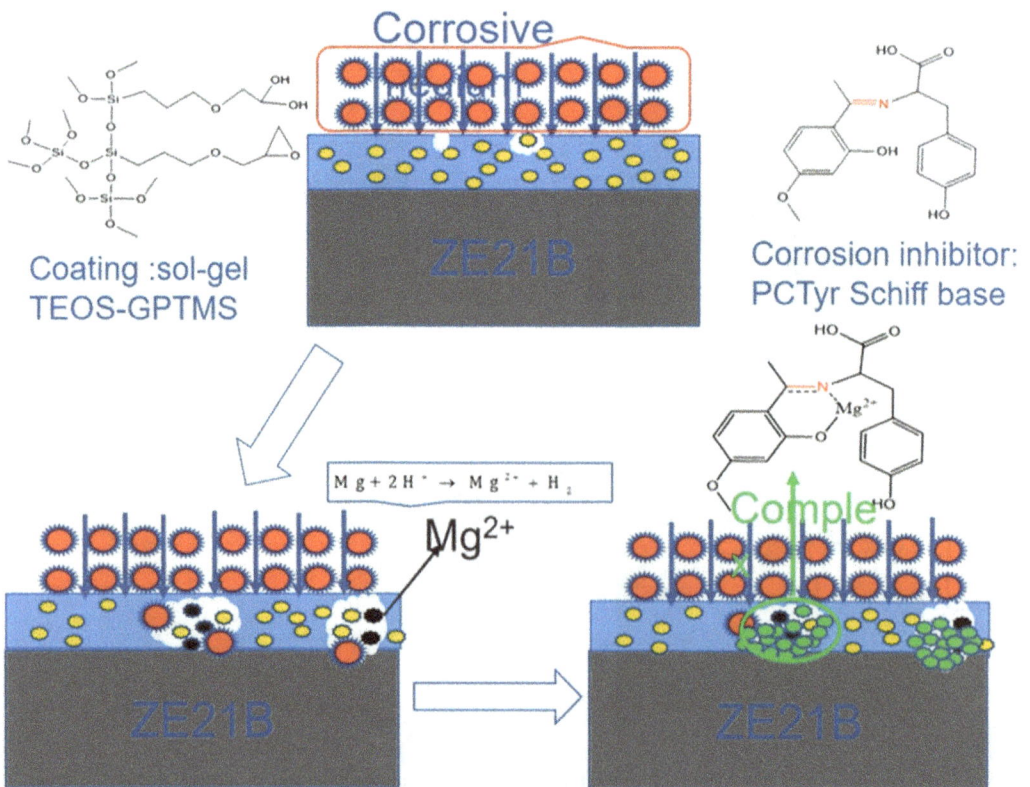

FIGURE 22.2C Schematically illustrating the self-healing mechanism of PCTyr Schiff base sol-gel coating. *Reproduced with permission from (Li et al. 2021) © 2021 Elsevier B.V.*

liquids. To this end, Zhang *et al.* (Zhang et al. 2021) inserted the corrosion inhibitor of BTA in the shape memory polymer (SMP), acting as the bottom layer on AZ31B Mg alloys, and the top fluorinated attapulgite layer imparted the surface superamphiphobicity to further improve the corrosion resistance. The inhibitor of BTA can effectively delay corrosion before the scratched coating was healed. Unsurprisingly, the original coating presented excellent corrosion resistance, enduring immersion in 3.5 wt.% NaCl solution for 80 days and neutral salt spray with 5 wt.% NaCl for 54 days. In contrast, the damaged coating after self-healing can still offer superior corrosion protection for Mg alloys, which can resist immersion in 3.5 wt.% NaCl solution for 60 days and 5 wt.% NaCl salt spray for 30 days.

MAO is a high-voltage plasma-assisted anodic oxidation process, extensively used for surface treatment of Mg alloys (Kannan et al. 2018; Wu and Jiang 2022; Ma et al. 2022), and it has already reached the level of industrial production. However, the pores of the MAO coating place negative influence on the corrosion resistance of Mg alloys. Nevertheless, the pores can be applied to load corrosion inhibitors to fabricate self-healing coatings. Li *et al.* (Li et al. 2019) prepared an organic-inorganic composite coating on AZ31 Mg alloys combining MAO coating, (N-(5-hydroxypent-3-yl)-N,N-dimethylhexadecan-1-aminium bromide) (N-16) corrosion inhibitor, and hydrophobic wax film. According to **Figure 22.3a** and **22.3b**, the size of micro- and nanopores decreases after the corrosion inhibitor of N-16 was encapsulated into pores, forming a more compact surface. Further, the fabrication of the wax film makes most of the porous structures be covered (**Figure 22.3c**). The energy-dispersive spectroscopy result (**Figure 22.3d and 22.3e**) shows that the MAO/N-16 coating is rich in N and Br, with even distribution, indicating the successful incorporation of N-16 in the MAO coating. Chen *et al.* (Chen et al. 2020) fabricated a MAO/sol-gel composite coating on AZ91 Mg alloy, where the corrosion inhibitors (sodium salts of glycolic, 4-aminosalicylic, and 2,6-pyridinedicarboxylic acids) were stored in the MAO coating. Subsequently, the MAO coating was sealed by a thin sol-gel layer, forming a sandwich-like structure to ensure durable barrier and active corrosion protection. All inhibitor-containing coatings showed better corrosion protection than the blank coating, and they can provide active corrosion protection for AZ91 Mg alloys. The active protection mechanism imparted the coating self-healing capability, which was related to the suppression of the re-deposition of impurity and/or adsorption of the impregnated inhibitors upon the exposed surface. Further, Liu *et al.* (Liu et al. 2019) developed a duplex self-healing coating to enhance the self-healing property on AM60 alloys, where the interphase inhibitor of phosphate ions was doped into the pores of MAO coatings and the organic inhibitor of MBT was in the top paint. In particular, the release rates of these two inhibitors were different; although phosphate ions can be rapidly released from the MAO coating after the coating was damaged, the release of MBT was relatively slow (**Figure 22.3f–h**). Nevertheless, the synergistic effects of the two inhibitors can significantly enhance the self-healing property compared to the coatings with only one inhibitor. Further, they evaluated the corrosion behavior of AM60 alloy with and without self-healing coatings in Shenyang industrial atmospheric environment (Song, Wang et al. 2021). The sediments and rain-water unsurprisingly accelerated the corrosion rate of the uncoated Mg alloy substrate. In contrast, the dual self-healing coating displayed better inhibition property, where the released inhibitors were gradually consumed during the migration process. As aforementioned, the self-healing process can also be realized by reversible covalent, non-covalent, and supramolecular dynamic bonds of the polymer matrix. Considering only a few works paid attention to combine the two self-healing strategies to protect Mg alloy substrates, Liu *et al.* (Liu, Wang et al. 2021) developed a novel self-healing composite coating on AZ31B Mg alloy, based on dual effects of the corrosion inhibitor M-16 in the MAO coating and self-healing polyurethane (PU) modified by disulfide bonds. Especially, the scanning electrochemical microscopy (SECM) was employed to *in situ* study the cathodic reaction in the damaged area, with the result as presented in **Figure 22.3i**. In comparison with the sample without M-16 (MP-0), the one with inhibitor of M-16 (MP-i) shows lower tip current after immersion for 3 days, demonstrating that the corrosion of AZ31B Mg alloy was effectively inhibited by the corrosion inhibitor. On the other hand, after heat treatment of PU for non-autonomous self-healing,

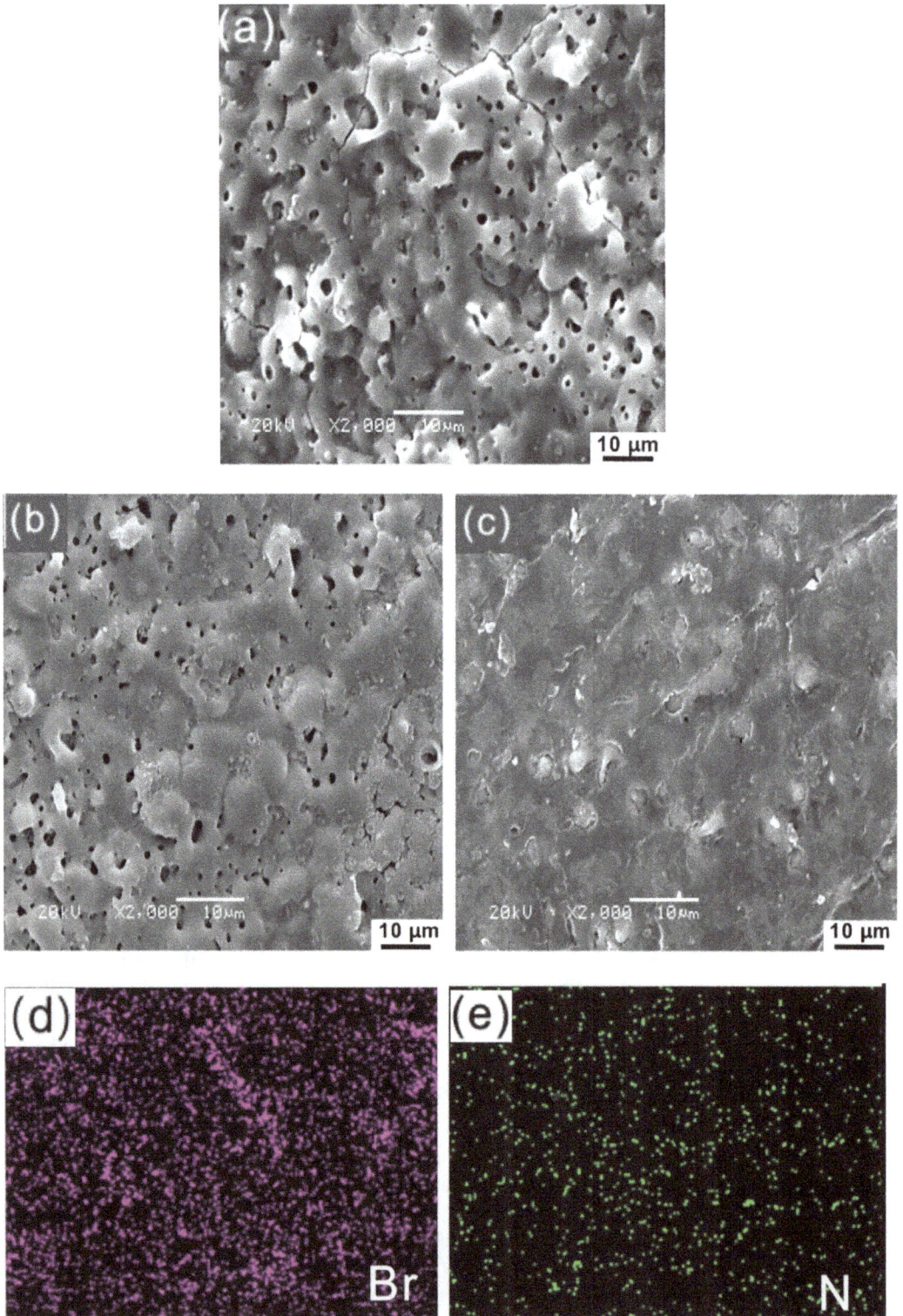

FIGURE 22.3A–E   SEM images of surface morphology of (a) MAO, (b) MAO/N-16, and (c) MAO/N-16/wax film. (d) N and (e) Br elemental mapping images in MAO/N-16 film. *Reproduced with permission from (Li et al. 2019) © 2018 Elsevier B.V.*

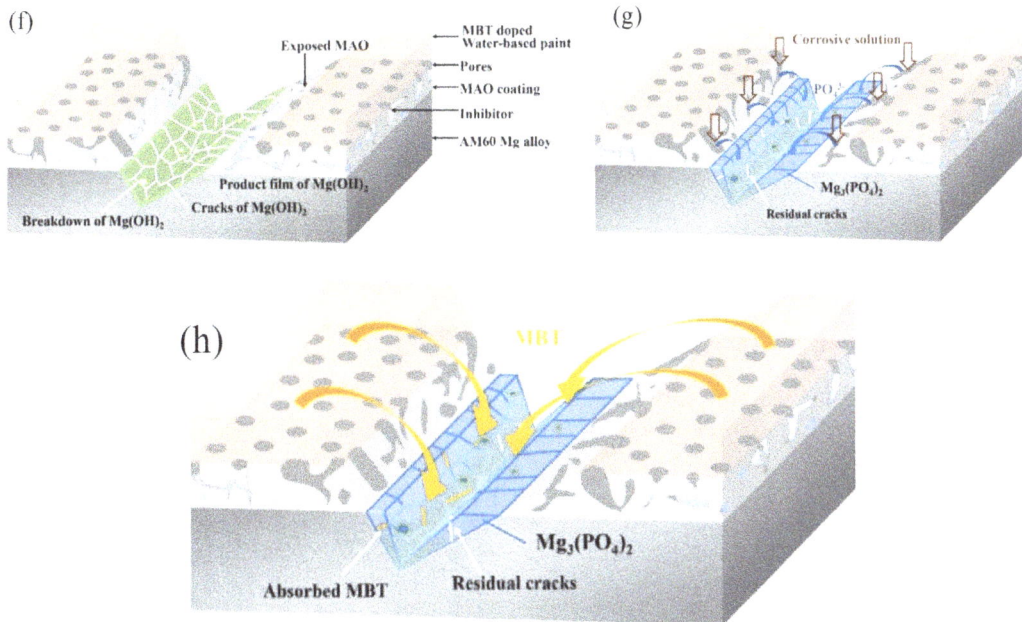

FIGURE 22.3F–H   Schematically illustrating the healing process of the duplex self-healing coating: (f) The defective $Mg(OH)_2$ film forms on the exposed Mg alloy substrate before the release of inhibitors in the early stage of immersion. (g) The exposure of the MAO coating causes the release of phosphate, and subsequently, it migrates to the damaged area and breakdown area of the newly formed $Mg(OH)_2$ film. The pH increases and protective $Mg_3(PO_4)_2$ coating is formed. (h) With increasing immersion time, the large second phases and low concentration of phosphate lead to the defectiveness of the product film. Therefore, the organic inhibitor of MBT is released to absorb on the defects of the formed product film to inhibit the corrosion process. *Reproduced with permission from (Liu et al. 2019) © 2019 Elsevier Ltd.*

both the MP-i and MP-0 presented very low tip currents after immersion for 7 days, indicating no obvious corrosion activity occurred. Thus, the M-16 can be released from the porous MAO coating to the scratched area after the coating was damaged, displaying excellent repair capability and recovery of the anti-corrosion property. Moreover, the cooperation of dynamic disulfide bonds and shape-memory effect of the self-healing PU coating were beneficial to the restoration of the physical damage to improve the corrosion resistance as well.

LDH, an anionic intercalated clay material, possesses a layered microstructure (Pahalagedara et al. 2014). The interlayer anions can be readily exchanged with other anions owing to the weak bonding in aqueous solutions. Consequently, the excellent exchangeability of interlayer anions of LDH coatings promises them to be ideal nano-reservoirs to carry desired corrosion inhibitors. On the other hand, the LDH coating can simultaneously release the carried corrosion inhibitor and entrap aggressive anion of chloride, further inhibiting the corrosion process.

Anjum *et al.* (Anjum et al. 2021) incorporated 8-quinolinol (8Q) in MgAl-based hydrotalcite-like (HT) coatings on AZ31 Mg alloys. In particular, the intercalation of 8Q imparted the MgAl HT surface hydrophobicity, with long-term immersion stability. In addition, they used single-frequency electrochemical impedance spectroscopy (EIS) to evaluate the self-healing performance of the HT coating for the first time. The results showed that the released 8Q could react with Mg^{2+} ions and then be deposited on the scratched area as chelate $Mg(Q)_2$ to present good self-healing property. Considering the restriction of wide applications in industry owing to the toxicity of 8-HQ, Song *et al.*'s (Song, Liu et al. 2021) respectively intercalated nitrate, molybdate, and environmentally friendly anionic surfactant of sodium dodecyl sulfate (SDS) into MgAl-LDHs on AZ31 Mg alloys

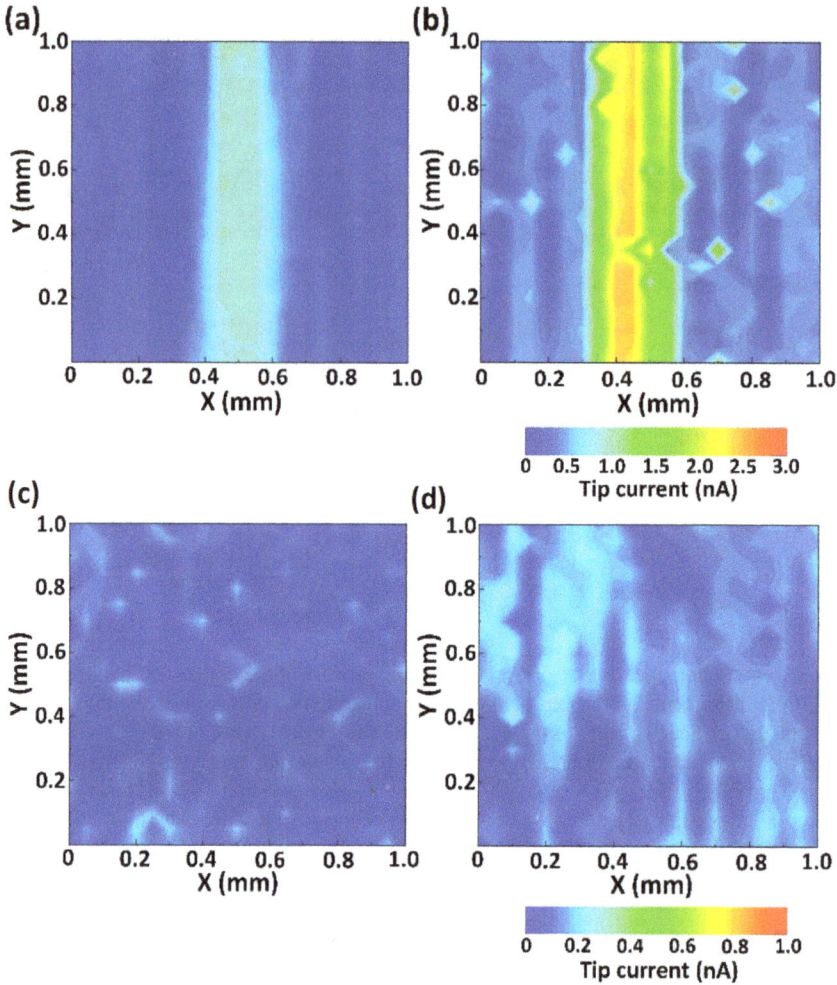

FIGURE 22.31 SECM maps of the scratched surfaces for (a, c) MP-i and (b, d) MP-0 immersed in 3.5 wt.% NaCl solution for 3 and 7 days. *Reproduced with permission from (Liu, Yan et al. 2021) © 2021 Elsevier B.V.*

(**Figure 22.4a**). The results indicated that the corrosion resistance of the LDHs were ranked in order of MgAl-MoO$_4^{2-}$-LDHs < MgAl-NO$_3$_LDHs < MgAl-SDS-LDHs. The best performance of the SDS was associated with the physical barrier of the dense coating, anion-exchange capacity of LDHs, and the adsorption and cathodic inhibition of SDS anions. Jiang *et al.* (Jiang et al. 2022) used ionic liquids (ILs) as corrosion inhibitors. They prepared two kinds of MgAl-LDHs (**Figure 22.4b**) modified with imidazolium based dicationic ILs on AZ31B alloys, respectively namely LDH-C$_6$(m$_2$im)$_2$-I with dense and order petal-like nanoflowers structure (**Figure 22.4c**) and IL@LDH-C$_6$(m$_2$im)$_2$-I covered by a compact coral-like IL film (**Figure 22.4d**). The IL@LDH-C$_6$(m$_2$im)$_2$-I was still compact without being destroyed after immersion for 168 hours, associated with the synergistic effects of excellent self-healing and durable corrosion resistance. Their work provided a new method to prepare MgAl-LDHs modified with ionic liquids to greatly improve the corrosion resistance of Mg alloys. In Chen *et al.*'s work (Chen et al. 2019), they inserted a natural environmentally friendly organic compound of aspartic acid (ASP) into MgAl-LDHs on AZ31 alloy by a simple one-step hydrothermal method. In comparison with the MgAl-NO$_3$_LDHs, the MgAl-ASP-LDHs showed better corrosion resistance, which was attributed to the corrosion inhibition of

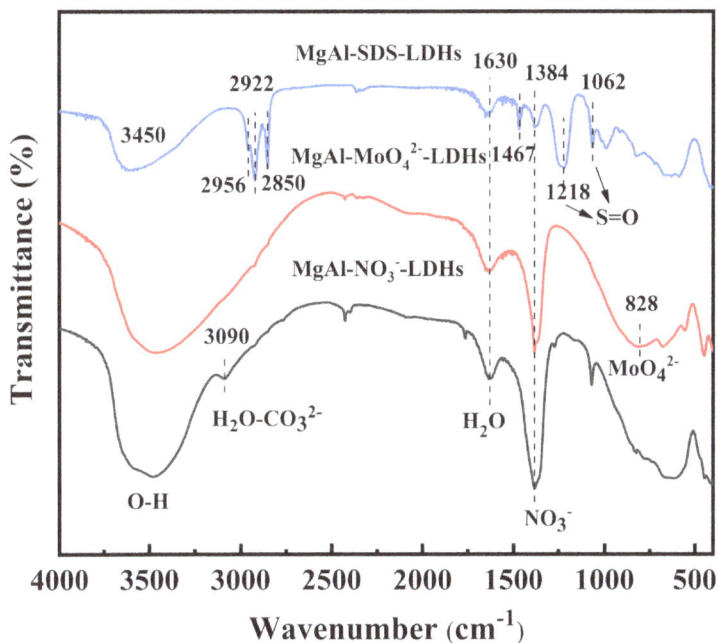

FIGURE 22.4A FTIR spectra of MgAl-NO₃ LDH, MgAl-MoO₄²⁻LDH, and MgAl-SDS-LDH: the absorption peak at 1384 cm⁻¹ is related to the asymmetric stretching vibration of NO₃. in the interlayer of LDHs; the peak at 828 cm⁻¹ is assigned to the anti-symmetric stretching vibration of Mo-O-Mo of MoO₄²⁻; the bands at 1218, 1062, 2956, 2922, 2850, and 1467 cm⁻¹ are respectively associated with S = O and C-H in SDS. *Reproduced with permission from (Song, Wang et al. 2021) © 2021 Elsevier B.V.*

FIGURE 22.4B–D SEM images of surface morphology of (b) LDH-CO₃²⁻, (c) LDH-C₆(m₂im)₂-I, and (d) IL@LDH-C₆(m₂im)₂-I. *Reproduced with permission from (Jiang et al. 2022) © 2022 Elsevier B.V.*

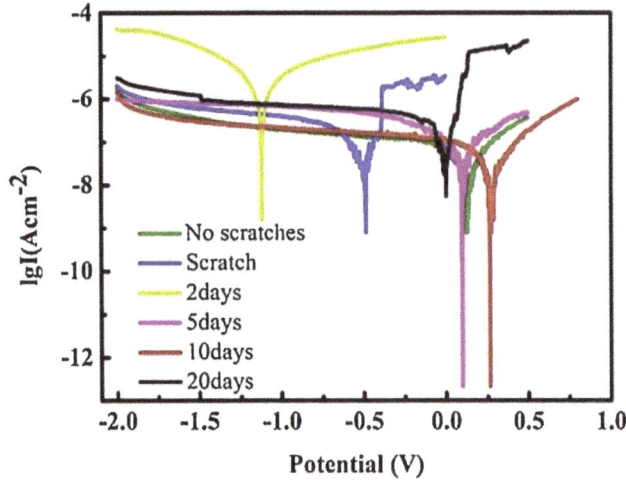

FIGURE 22.4E Tafel polarization curves of the unscratched and scratched MgAl-ASP-LDH samples immersed in 3.5 wt.% NaCl solution for different times. *Reproduced with permission from (Chen et al. 2019) © 2019 Elsevier B.V.*

ASP ions and their larger specific surface area to trap aggressive chloride ions. After soaking in the NaCl solution for about 10 days, the scratched area was covered with flaky material like LDH on the exposed Mg alloy substrate. Especially, the corrosion resistance of the self-healed coating was near to the non-scratched sample, displaying similar corrosion current density (**Figure 22.4e**). Therefore, the synergistic effects of ASP and laminate structure of LDHs imparted MgAl-ASP-LDH coatings good self-healing capability.

Because of the perpendicular growth of LDH film, pores and gaps are extended to the LDHs/substrate interface. This allows free diffusion of Cl- ions to accelerate the corrosion rate of Mg alloys (He, Zhou, and Hu 2020; Han, Shen, and Zhou 2020; del Olmo et al. 2022). Therefore, composite coatings are formed to further improve the corrosion resistance. Tan *et al.* (Tan et al. 2022) fabricated a composite coating consisting of MgFe-LDH intercalated with 8HQ and modified with citric acid (CA) layer on WE43 Mg alloy. The 8HQ decreased anodic and cathodic dissolution reactions, while the CA acted as a protective barrier by forming a passive layer. Moreover, the 7-day EIS study confirmed the good self-healing property of the coating. Nevertheless, the ion exchange starts at the very beginning when the LDH coating is exposed to the external solution environment, although no corrosion occurs in the underlying metallic substrate, resulting in severe waste of anionic inhibitors (Shkirskiy et al. 2015). Hence, Ding *et al.* (Ding et al. 2019) intercalated tungstate corrosion inhibitor into LDHs and post-sealed a hydrophobic polymer layer of ureido crosslinked polydimethylsiloxane (U-PDMS) with laurate-modified LDH powder (La-LDH) on the surface, forming a superhydrophobic and self-healing anti-corrosion coating on AZ31B Mg alloy. The hydrophobic top layer can effectively restrict the invasion of corrosive solution into the bottom LDH layer to prevent the inhibitor from being wasted by ion-exchange at the very start of immersion (**Figure 22.4f**). Therefore, the composite coating presented long-term corrosion protection and better self-healing performance. The SVET results confirmed that the tungstate experienced ion exchange and was released when the WO_4^{2-}-LDH sample was exposed to the NaCl solution, effectively inhibiting the attack from the electrolyte. On the other hand, the released tungstate reacted with the dissolved Mg^{2+} and formed $MgWO_4$. They grew into a new protective coating, with $Mg(OH)_2$, preventing the formation of pitting sites.

Besides being an interlayer anion to be intercalated into LDH layer, corrosion inhibitors can also act as metallic cations to develop the hydroxide layer of LDHs. Asl *et al.* (Asl et al. 2022) used

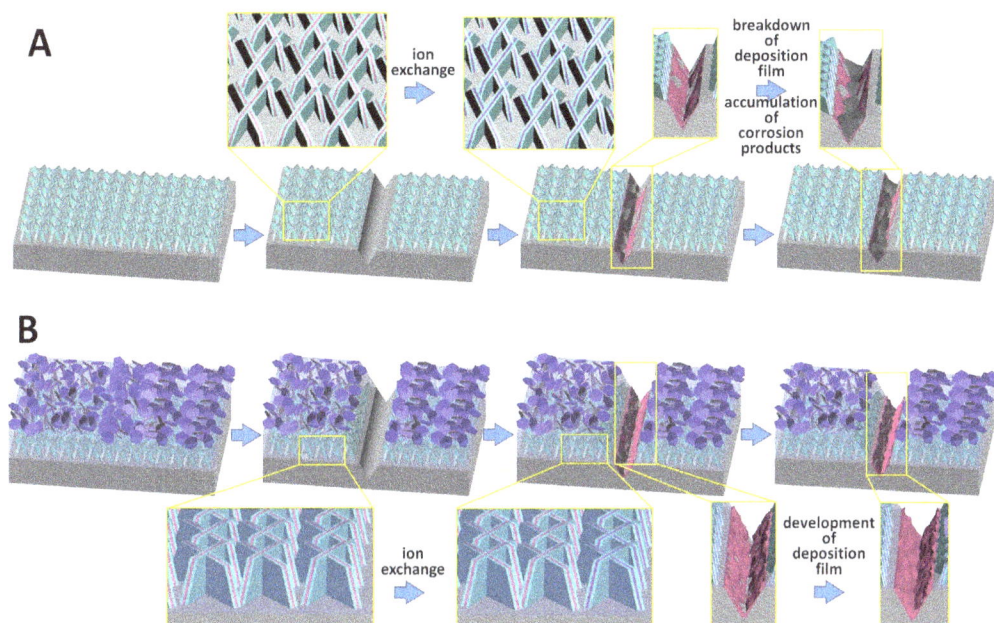

FIGURE 22.4F   Schematic diagram illustrating the corrosion protection mechanism of (A) WO$_4^{2-}$-LDH and (B) superhydrophobic coatings. *Reproduced with permission from (Ding et al. 2019) © 2018 Elsevier B.V.*

Ce- and Y-containing compounds as corrosion inhibitors and developed CaCe- and CaY-LDH coatings on the AZ31 alloy. Both of them can obviously restrict the corrosion of AZ31 in SBF. However, the CaY-LDH with some cracks cannot provide corrosion resistance as superior as the CaCe-LDH. After releasing into the corrosive solution, Ce, together with Ca and Mg, attempted to cure the defects by forming a corrosion product layer. The precipitation of the corrosion products improved both the self-healing and corrosion-protection performances.

22.3.2   Nanocontainers Carrying Corrosion Inhibitors

Although the corrosion inhibitor can be directly encapsulated into the coating matrix, the unwanted reactions between inhibitors and coatings have to be reduced to guarantee the inhibition efficiency. As a consequence, different kinds of micro/nanocontainers have been extensively studied to carry corrosion inhibitors, mainly including LDH nanoparticles, nanotubes, and mesoporous nanostructures.

In addition to be *in situ* fabricated on the Mg alloys as a protective coating, LDHs can also be formed as nanoparticles to be nanocontainers to carry corrosion inhibitors. Cao *et al.* (Cao et al. 2021) used MgAl-LDH to carry the corrosion inhibitor of F$^-$. Li *et al.* (Li et al. 2022) fabricated a three-layered structure on AZ31 Mg alloy, including a top layer of LDH powders, an intermediate layer of PDMS, and a bottom layer of LDH film. Especially, the LDH powders were used as nano-reservoirs to carry corrosion inhibitor of lauric acid. The corrosion resistance of the coating was highly improved by trapping corrosive Cl- ions and releasing the corrosion inhibitor. Yan *et al.* (Yan et al. 2020) applied a coprecipitation method to synthesize MgAl-LDHs intercalated with environmentally friendly methionine (Met-LDHs), which were then added into epoxy-primer to form a protective coating on Mg alloys. The Met-LDHs can protect Mg alloys from corrosion by releasing methionine in NaCl solutions under different pH conditions (pH 5, 7, and 9) and forming a dense white protective film.

Nanotubes, having large internal surface area and small openings at the ends, can realize a high loading capacity and slow release of corrosion inhibitors. In particular, the halloysite, with a two-layered aluminosilicate nanotubular structure, is a good choice to encapsulate and release many corrosion inhibitors, owing to its natural availability and low cost (Shchukina, Shchukin, and Grigoriev 2018; Shchukina et al. 2017; Shchukina, Shchukin, and Grigoriev 2017). Wu et al. (Wu, Wang, and Lin 2021) loaded 2-aminobenzimidazole (2-ABi) to halloysite nanotubes (HNTs), and then they were incorporated to MAO coating on AZ31 Mg alloy. The EIS complex spectra and salt spray test results confirmed that the self-healing performance improved the corrosion resistance of 2-ABi-HNT MAO coating, indicating good potential of 2-ABi to be effective corrosion inhibitor for Mg alloys. Similarly, Mingo et al. (Mingo et al. 2020) also encapsulated corrosion inhibitors into HNTs, which were subsequently embedded into MAO coatings. Especially, they compared performances of three corrosion inhibitors of vanadate, molybdate salts, and 8-HQ within the MAO-HNT system. The vanadate consistently provided the best corrosion resistance in the immersion test up to 72 hours, followed by molybdate, whereas the positive effect of 8-HQ was time-limited. Although all coatings can self-heal small scratches, only the vanadate could partially repair more severe damage (**Figure 22.5a**). By contrast, both the MAO-HNT-Mo and MAO-HNT-8-HQ progressively corroded, and the inhibitors cannot slow down the electrochemical process, causing catastrophic dissolution of the Mg alloy substrate. Mahmoudi et al. (Mahmoudi et al. 2020) encapsulated pra-seodymium (Pr) in the internal lumen of HNTs in silane coating on AZ31 alloy. The results showed that Pr can effectively inhibit the entry of electrolyte into the interface between the coating and the underlying Mg alloy substrate *via* forming a protective coating, which was composed of Pr oxides/hydroxides at the scratch site. Manasa et al. (Manasa et al. 2017) encapsulated Ce^{3+}/Zr^{4+} into halloysite clay nanotubes to provide reliable corrosion protection and self-healing activity for AZ91 Mg alloy. The potentiodynamic polarization and impedance studies indicated that a higher concentration of loaded halloysite nanotubes placed an adverse effect on the barrier properties and was less protective. Especially, 2 wt.% loading of inhibitor-loaded HNTs was optimum to provide good barrier protection and self-healing activity. Adsul et al. (Adsul et al. 2017) investigated the influence of loading cationic corrosion inhibitors into HNTs on the corrosion resistance of AZ91D Mg alloys. Similarly, they also encapsulated Ce^{3+}/Zr^{4+} into halloysite clay nanotubes, which were then dispersed into a hybrid silica sol-gel matrix and deposited on Mg alloys by dip-coating

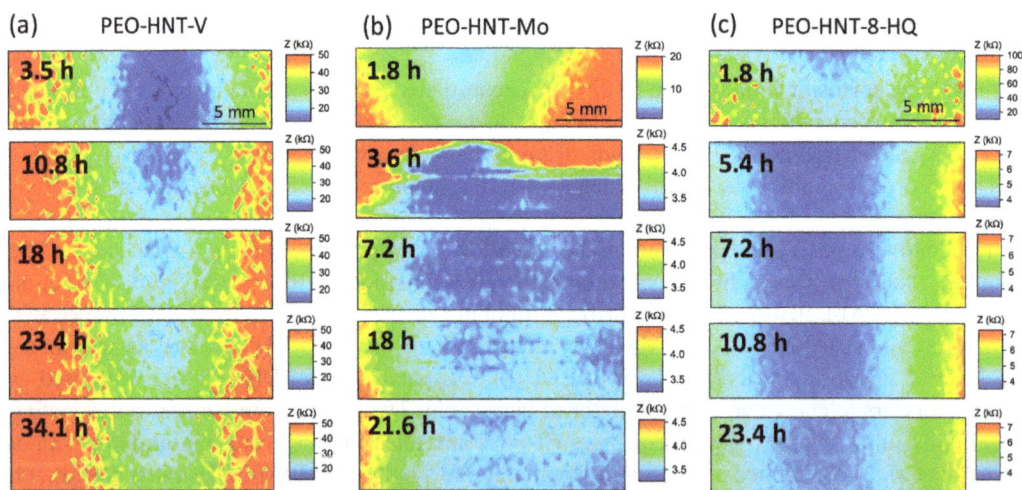

FIGURE 22.5A   LEIS maps as a function of the immersion time for (a) PEO-HNT-V, (b) PEO-HNT-Mo, and (c) PEO-HNT-8-HQ in 0.05 M NaCl solution. *Reproduced with permission from (Mingo et al. 2020) © 2020 American Chemical Society.*

methods. Weight loss experiments, potentiodynamic polarization, and electrochemical impedance spectroscopy measurements were respectively applied to evaluate the self-healing ability. In Liu et al.'s work (Liu, Yan et al. 2021), they synthesized clinochrysotile-like Mg silicate nanotubes (MS-TNs, **Figure 22.5b**) by a hydrothermal method, which were applied as nanocontainers to carry 2-mercaptabenzimidazole (2-MBI) inhibitor. Subsequently, the MS-TNs(2-MBI) (**Figure 22.5c**) was coated on AZ31B Mg alloys using epoxy resin as the coating matrix. In order to improve the physical barrier ability of the protective coating, a top superhydrophobic layer of MS/TNs-DTMS (dodecyltrimethoxysilane, **Figure 22.5d**) was further developed. Both the electrochemical measurements and scratch tests indicated the corrosion resistance of the dual-functional coating was highly enhanced, which remained stable even after immersion in 3.5 wt.% NaCl solution for 21 days, because of the physical barrier of the superhydrophobic top layer and the corrosion inhibition of MS-TNs(2-MBI), as shown in **Figure 22.5e**. First, the surface superhydrophobicity can effectively reduce the contact area with the external corrosive medium. After the superhydrophobic surface was damaged and corrosive medium penetrated through the broken coating, the released 2-MBI can form 2-MBI-thiol or 2-MBI-thione anion, absorbing on the Mg alloy substrate to develop a dense coating with $Mg(OH)_2$ to alleviate corrosion.

Considering microcapsules with a diameter of dozens of micrometers are too big for most anti-corrosive coatings with a thickness of less than 100 μm, mesoporous silica nanoparticles with large specific surface area of ~ 1000 m^2 g^{-1} attract a lot of attention (Xie et al. 2017). Ding *et al.* (Ding et al. 2017) designed corrosion potential stimulus-responsive smart nanocontainers (CP-SNCs) and synthesized them based on the installation of the supramolecular assemblies onto the exterior surface of magnetic nanovehicles (Fe_3O_4@$mSiO_2$), linked by disulfide linkers. The supramolecular assemblies can effectively block the encapsulated 8-HQ in Fe_3O_4@$mSiO_2$

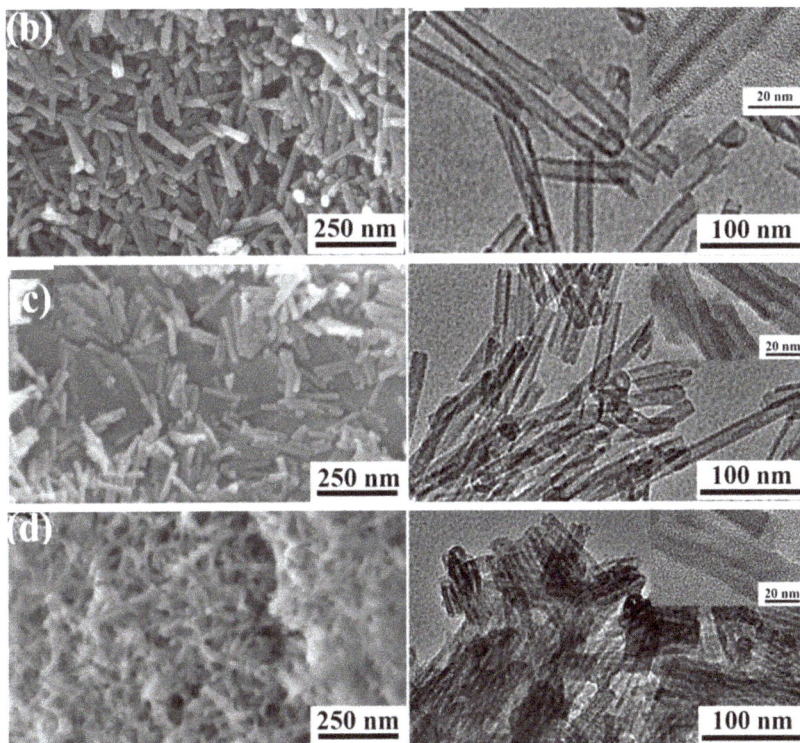

FIGURE 22.5B–D   SEM and TEM images of (b) MS-TNs, (c) MS-TNs(2-MBI), and (d) MS-TNs-DTMS. *Reproduced with permission from (Liu, Yan et al. 2021) © 2020 Elsevier B.V.*

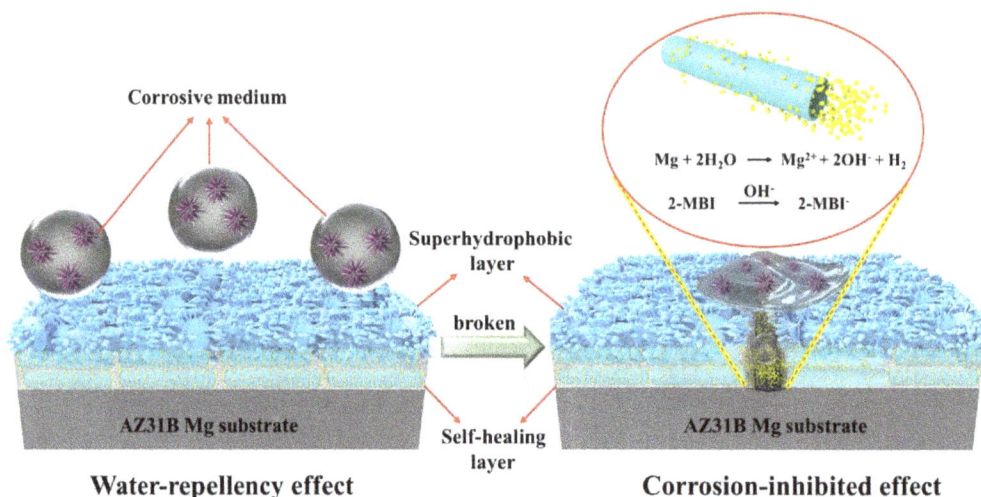

FIGURE 22.5E Schematically illustrating the anti-corrosion mechanism of the dual-functional coatings on AZ31B Mg alloys. *Reproduced with permission from (Liu, Yan et al. 2021) © 2020 Elsevier B.V.*

mesopores. Under the magnetic field, the CP-SNCs were gathered in the proximity of the surface of AZ31B so that they can provide fast self-healing performance at the damaged area. Ouyang *et al.* (Ouyang et al. 2022) constructed pH-responsive nanocontainers of mesoporous silica nanoparticle-mercaptobenzothiazole@layered double hydroxide (MSN-MBT@LDH) by a new strategy, which were composed of corrosion inhibitor- (MBT) loaded MSN core and LDH nanosheet shell as gatekeepers. A smart corrosion protection system on Mg alloys was finally acquired by incorporating the nanocontainers into a self-assembled nanophase particle (SNAP) coating. In comparison with the pure SNAP coating, the one with MSN-MBT@LDH displayed great improvement and better robustness in the corrosion resistance in NaCl solution, associated mainly with the corrosion inhibition of MBT to the defects and partially with the entrapment of aggressive Cl- ions of LDHs. Similarly, Ding *et al.* (Ding et al. 2016) fabricated a host–guest feedback active coating on AZ31B Mg alloys by incorporation of "guest" mesoporous silica nanoparticles (MSNPs) as smart nanocontainers into the host SNAP barrier coating. The MSNPs can block the entrapped mixed-type corrosion inhibitor of 2-hydroxy-4-methoxy-acetophenone (HMAP) in neutral solutions but release it upon alkali or Mg^{2+} stimuli, endowing the physical barrier coating with self-healing potential. After the surface was mechanically scratched, the released HMAP, which can inhibit both anodic and cathodic corrosive activities, formed a compact molecular film on the damaged alloy surface, preventing corrosion propagation and executing self-healing property. Further, Yang *et al.* (Yang et al. 2020) modified hollow mesoporous silica with cysteine and cyclodextrin to develop a pH/glutathione/Mg(II)-triple responsive nanocontainer on AZ31 Mg alloy, which loaded BTA as a corrosion inhibitor. Li *et al.* (Li, Huang, and Han 2021) developed novel coatings with self-healing and osteoimmunomodulation on Mg, which had three-layered structure, composed of an inner layer of MgO, an interlayer of poly-L-lactide containing curcumin loaded F-encapsulated mesoporous silica nanocontainers (cFMSNs), and an outer layer of dicalcium phosphate dehydrate. The coating showed effective self-healing performance to protect Mg from pitting corrosion. In addition, Yeganeh *et al.* (Yeganeh and Saremi 2015) compared the corrosion property of coatings containing mesoporous silica nanocontainer powders with and without fluoride inhibitors on Mg surface. It was indicated that the release of fluoride from mesoporous silica nanocontainers resulted in the formation of MgF_2 to be an inhibitive compound at the interface. Xie *et al.* (Xie and Shan 2018) loaded MBT into porous hollow silica nanocontainers and then incorporated into electroless Ni coating on AZ31 Mg alloy. The continuous release of MBT significantly improved the corrosion resistance, exhibiting a long-term anticorrosion performance in aggressive media. Moreover, they formed a corrosion-resistance Ni coating on AZ31 Mg alloys, where the MCM-41-type mesoporous silica nanocontainers were loaded with corrosion inhibitor of NaF. The X-ray photoelectron spectroscopy result indicated the successful formation of MgF_2 protective coating at the corrosion sites on the Mg alloy upon soaking in NaCl solution. Consequently, the coating presented decreased corrosion rate for a long-term immersion compared with the bare Mg alloy and Ni coating without corrosion inhibitors. Liu et al. (Liu et al. 2022) encapsulated 1H, 1H, 2H, 2H-perfluorooctyltriethoxysilane (FTES) into the hollow mesoporous silica (HMS), and then they were embedded in an epoxy coating on a Mg alloy. The results indicated that the coating with 0.5 wt.% FHMS showed excellent self-healing capacity in the scratched area. This was attributed to the interaction among FTES, epoxy coating, and the Mg alloy substrate. Besides, the FTES silane can also form a crosslinked Si-O-Si network by itself to improve the self-healing ability.

22.4   SELF-HEALING COATINGS WITHOUT CORROSION INHIBITORS

The formation of a biomedical self-healing coating has to consider the bioactivity. Because of the toxicity of general corrosion inhibitors for the biomedical applications, Dong et al. (Dong et al. 2021) formed dicalcium phosphate dihydrate (DCPD) coating on biomedical Mg, which provided good corrosion protection and self-healing property, even without corrosion inhibitors. The results demonstrated that artificial scratches in DCPD coating can be effectively repaired by anti-corrosive products in both Hanks' and normal saline solutions. Therefore, besides a physical barrier, DCPD was a self-healing agent to provide self-healing performance without extra corrosion inhibitors. Shanaghi et al. (Shanaghi et al. 2021) deposited tantalum/tantalum nitride (Ta/TaN) multilayered coating on plasma-nitrided AZ91 Mg alloys. Its self-healing process was achieved by the reactions between the tantalum intermediate layer and electrolytes and penetrating ions through the defects as well as formation of oxide compounds and creation and propagation of defects. Feng et al. (Feng et al. 2021) developed a network structure lactoglobulin composite coating by the unique reaction characteristics between protein and metal ions on the surface of Mg, presenting a promising self-healing property. The pre-made scratches were effectively self-healed within 24 hours of exposure to 0.9 wt.% NaCl, associated with the strong chelating tendency of the selected protein molecules to metallic ions.

22.5   SLIPPERY LIQUID-INFUSED POROUS SURFACES

Inspirations from nature have led to extensive developments of biomimetic nonwetting surfaces and intelligent self-healing technologies toward corrosion protection of metals. Inspired by *Nepenthes* pitcher plants, slippery liquid-infused porous surface (SLIPS) is attracting increasing attention, owing to its superior surface amphiphobicity, low adhesion, and good self-healing property. In particular, its self-healing property is associated with the excellent fluidity of infused liquid lubricants. Yao et al. (Yao et al. 2021; Yao, Wu et al. 2022) respectively infused silicone oil into the nanoporous structures of MgAl-LDHs and MgAlLa-LDHs, forming SLIPSs on AZ31 Mg alloy. Song et al. (Song et al. 2017) immersed the superhydrophobic surface in perfluoropolyether to fabricate SLIPS on AZ31 alloy. Zhang et al. (Zhang, Gu, and Tu 2017) dropped perfluoropolyether liquid onto the superhydrophobic surface to develop SLIPS. Wang et al. (Wang et al. 2021) developed a low-cost micro/nanostructured candle soot coating (CSC) via spraying, and after the infusion of silicone oil, the SLIPS was formed, and its contact angle changed slightly after physical cutting, indicating that the SLIPS showed good self-healing performance. Li et al. (Li et al. 2020) applied a hydrothermal method to develop SLIPS. The SLIPS showed good anti-corrosion and anti-biofouling properties, which were attributed to the continuously infused lubricant. In addition, the SLIPS presented excellent thermally assisted healing properties, and its surface hydrophobicity was almost recovered after being heated at 120°C for 1 hour. A dual biomimetic SLIPS in a partition matrix inspired by both *Nepenthes* pitcher and pomegranate fruit was developed by Kan *et al.* on AZ31B Mg alloy (Kan et al. 2021). In addition to the better corrosion resistance in 3.5 wt.% NaCl solution, the SLIPS provided good tolerance to repeated mechanical damage, displaying superior self-healing property. Considering the migration, evaporation, and leaking of the infused liquids in SLIPSs, their durability maybe a challenge limiting the practical applications. Therefore, Xing *et al.* (Xing et al. 2021) proposed a durable polydimethylsiloxane/silicone oil (PDMS-oil) system to enhance mechanical robustness and self-replenishing property. The results showed that 50 wt.% silicone oil was more suitable for the composite SLIPS. It possessed good corrosion resistance and self-replenishing ability, enduring hot water, ice water, and corrosive solutions (acid and alkaline aqueous solutions). Moreover, the Mg alloy with the PDMS-oil based SLIPS can remain intact in 3.5 wt.% NaCl solution even after 20 days at ambient temperature. Jiang *et al.* (Jiang et al. 2019) designed a novel anticorrosion system composed of MAO coating, LDH film, and a lubricant-infused slippery surface on Mg alloy. The MAO coating provided a moderate corrosion barrier, and the LDH film carried corrosion inhibitor, anchored lubricant, and sealed the MAO coating defects. The SLIPS showed durable surface hydrophobicity and self-reparability to the external damages. In particular, the dynamic self-healing processes associated with barrier regeneration and corrosion inhibition were substantiated at microscale using scanning the Kelvin probe and scanning vibrating electrode techniques, respectively. Wei *et al.* (Wei et al. 2021) designed a novel switchable surface between superhydrophobic surface and SLIPS to increase the corrosion resistance of Mg alloys. Although both of them can delay corrosion at an artificial defect area, the SLIPS performed better because of the stability and self-healing property of the lubricant layer than the entrapped air pockets of superhydrophobic surface.

22.6   NON-AUTONOMOUS SELF-HEALING COATINGS

In addition to autonomous self-healing, the self-healing of the anti-corrosive coatings can also be non-autonomously realized by means of external stimuli that cannot be generated from corrosion events, like light and temperature. Li *et al.* (Li et al. 2022) developed a non-autonomous self-healing coating on AZ31B alloy based on polydopamine (PDA)-functionalized Cu^{2+}-GO, octadecylamine (ODA), and PDMS, which can simultaneously repair the coating microstructure and restore the surface superhydrophobicity. Once the surface superhydrophobicity was damaged, it can be rapidly restored under 1-sun irradiation. Besides, the coating presented superior corrosion resistance for the Mg alloy substrate, even after immersion in 3.5 wt.% NaCl solution for 30 days. This was mainly associated with the healing in GO by PDA by pi-pi interactions and the inherent chemical inertia of PDMS. Zhang *et al.* (Zhang et al. 2021) fabricated a superhydrophobic coating with photothermal self-healing chemical composition on Mg alloy, which was developed by a shape-memory polymer (SMP) primer. The coating presented superior self-healing performance against chemical and microstructure damage, like rapid self-healing under 1 sun irradiation in 10 minutes, achieving self-healing after serious damage (*e.g.* 10 damage and self-healing cycles), and even self-healing under natural sunlight in 4 hours. Besides, the self-healed coating can provide good corrosion protection for Mg alloy in neutral salt spray test. The coating, which was self-healable under natural sunlight during service, showed high potential to be extensively applied in various fields.

22.7   CONCLUSIONS AND OUTLOOK

This work comprehensively describes self-healing coatings on Mg alloys to prolong the service time and ensure adequate corrosion protection. According to the self-healing mechanism, they can be categorized into autonomous and non-autonomous self-healing coatings, respectively, based on polymerizable healing agents/corrosion inhibitors and intrinsic chemical bonds and/or physical conformations of polymer matrix. As for the Mg alloy substrate, the chapter is mainly focused on the self-healing coating realized by corrosion inhibitors. In particular, the corrosion inhibitors can be directly encapsulated into coating matrix, including LDH films, MAO coatings, and polymer coatings, and carried by nanocontainers, like LDH powders, nanotubes, and mesoporous nanoparticles. In addition, a novel self-healing coating based on SLIPS is also extensively applied on Mg alloys. The autonomous flowing of liquid lubricants into the damaged area realizes the self-healing.

Although numerous works related to the self-healing coatings on Mg alloys have been published, few works have been marketed. Therefore, the practical applications of self-healing coatings still require further study to explore more efficient and low-cost corrosion inhibitors. On the other hand, the relationship between the size of the damaged area and the healing efficiency of the self-healing coating was paid less attention. Moreover, the synergistic self-healing property of the multiple self-healing coatings still needs to be further studied to clearly illustrate. Overall, low-cost, non-toxic, and environmentally friendly self-healing coatings are greatly required to protect Mg alloys from corrosion in practice.

ACKNOWLEDGEMENTS

This work was supported by the Chongqing Municipal Human Resources and Social Security Bureau (No. cx2022098), National Natural Science Foundation of China (No. 52001036), China Postdoctoral Science Foundation (Nos. 2022T150767 and 2021M693708), and National Natural Science Foundation of China (51971040, 52171101).

REFERENCES

Adsul S.H., Bagale U., Sonawane S., Subasri R. Release rate kinetics of corrosion inhibitor loaded halloysite nanotube-based anticorrosion coatings on magnesium alloy AZ91D. *J. Magnes. Alloy.* 2021, 9, 202–215.

Adsul S.H., Siva T., Sathiyanarayanan S., Sonawane S.H., Subasri R. Self-healing ability of nanoclay-based hybrid sol-gel coatings on magnesium alloy AZ91D. *Surf. Coat. Technol.* 2017, 309, 609–620.

Ali M., Ul-Hamid A., Khan T., Bake A., Butt H., Bamidele O., Saeed A. Corrosion-related failures in heat exchangers. *Corros. Rev.* 2021, 39, 519–546.

Andreeva D.V., Fix D., Möhwald H., Shchukin D.G. Self-healing anticorrosion coatings based on pH-sensitive polyelectrolyte/inhibitor sandwichlike nanostructures. *Adv. Mater.* 2008, 20, 2789–2794.

Anjum M., Zhao J., Tabish M., Murtaza H., Asl V., Yang Q., Malik M., Ali H., Yasin G., Khan W. Influence of the 8-quinolinol concentration and solution pH on the interfacial properties of self-healing hydrotalcite coating applied to AZ31 magnesium alloy. *Mater. Today Commun.* 2021, 26, 101923.

Asl V., Kazemzad M., Zhao J., Ramezanzadeh B., Anjum M. An eco-friendly Ca-Ce and Ca-Y based LDH coating on AZ31 Mg alloy: Surface modification and its corrosion studies in simulated body fluid (SBF). *Surf. Coat. Technol.* 2022, 440, 128458.

Attaei M., Calado L., Taryba M., Morozov Y., Shakoor R., Kahraman R., Marques A., Montemor M. Autonomous self-healing in epoxy coatings provided by high efficiency isophorone diisocyanate (IPDI) microcapsules for protection of carbon steel. *Prog. Org. Coat.* 2020, 139, 105445.

Azzeddine H., Hanna A., Dakhouche A., Luthringer-Feyerabend B. Corrosion behaviour and cytocompatibility of selected binary magnesium-rare earth alloys. *J. Magnes. Alloy.* 2021, 9, 581–591.

Bastidas D., Criado M., Fajardo S., Iglesia A., Bastidas J. Corrosion inhibition mechanism of phosphates for early-age reinforced mortar in the presence of chlorides. *Cem. Concr. Compos.* 2015, 61, 1–6.

Bryant M., Neville A. Corrosion and mechanical properties. *Orthop. Trauma.* 2016, 30, 176–191.

Calado L., Taryba M., Morozov Y., Carmezim M., Montemor M. Cerium phosphate-based inhibitor for smart corrosion protection of WE43 magnesium alloy. *Electrochim. Acta.* 2021, 365, 137368.

Cao K., Yu Z., Zhu L., Yin D., Chen L., Jiang Y., Wang J. Fabrication of superhydrophobic layered double hydroxide composites to enhance the corrosion-resistant performances of epoxy coatings on Mg alloy. *Surf. Coat. Technol.* 2021, 407, 126763.

Chen J., Fang L., Wu F., Xie J., Hu J., Jiang B., Luo H. Corrosion resistance of a self-healing rose-like MgAl-LDH coating intercalated with aspartic acid on AZ31 Mg alloy. *Prog. Org. Coat.* 2019, 136, 105234.

Chen X., Dam M., Non K., Mal A., Shen H., Nutt S., Sheran K., Wudl F. A thermally re-mendable cross-linked polymeric material. *Science.* 2002, 295, 1698–1702.

Chen Y., Lu X., Lamaka S., Ju P., Blawert C., Zhang T., Wang F., Zheludkevich M. Active protection of Mg alloy by composite PEO coating loaded with corrosion inhibitors. *Appl. Surf. Sci.* 2020, 504, 144462.

Chen Y., Wu L., Yao W., Chen Y., Zhong Z., Ci W., Wu J., Xie Z., Yuan Y., Pan F. A self-healing corrosion protection coating with graphene oxide carrying 8-hydroxyquinoline doped in layered double hydroxide on a micro-arc oxidation coating. *Corros. Sci.* 2022, 194, 109941.

Chen Y., Wu L., Yao W., Wu J., Xiang J., Dai X., Wu T., Yuan Y., Wang J., Jiang B., Pan F. Development of metal-organic framework (MOF) decorated graphene oxide/MgAl-layered double hydroxide coating via microstructural optimization for anti-corrosion micro-arc oxidation coatings of magnesium alloy. *J. Mater. Sci. Technol.* 2022, 130, 12–26.

Cho S., Kwon S., Kim D., Choi W., Kim Y., Lee J. Hot corrosion behaviour of nickel-cobalt-based alloys in a lithium molten salt. *Corros. Sci.* 2019, 151, 20–26.

Cho S., White S., Braun P. Self-healing polymer coatings. *Adv. Mater.* 2009, 21, 645–649.

Christudasjustus J., Felde M., Witharamage C., Esquivel J., Darwish A., Winkler C., Gupta R. Age-hardening behavior, corrosion mechanisms, and passive film structure of nanocrystalline Al-V supersaturated solid solution. *J. Mater. Sci. Technol.* 2023, 135, 1–12.

Chu A., Zhao Y., Din R., Hu H., Zhi Q., Wang Z. Microstructure and properties of Mg-Al-Ca-Mn alloy with high Ca/Al ratio fabricated by hot extrusion. *Materials.* 2021, 14, 5230.

Cordier P., Tournilhac F., Soulié-Ziakovic C., Leibler L. Self-healing and thermoreversible rubber from supra-molecular assembly. *Nature.* 2008, 451, 977–980.

Dai X., Li X., Wang C., Yu S., Yu Z., Yang X. Effect of MAO/Ta2O5 composite coating on the corrosion behavior of Mg-Sr alloy and its in vitro biocompatibility. *J. Mater. Res. Technol-JMRT.* 2022, 20, 4566–4575.

D'Elia E., Eslava S., Miranda M., Georgiou T., Saiz E. Autonomous self-healing structural composites with bio-inspired design. *Sci. Rep.* 2019, 6, 25059.

del Olmo R., Mohedano M., Matykina E., Arrabal R. Permanganate loaded Ca-Al-LDH coating for active corrosion protection of 2024-T3 alloy. *Corros. Sci.* 2022, 198, 110144.

Ding C., Liu Y., Wang M., Wang T., Fu J. Self-healing, superhydrophobic coating based on mechanized silica nanoparticles for reliable protection of magnesium alloys. *J. Mater. Chem. A.* 2016, 21, 8041–8052.

Ding C., Tai Y., Wang D., Tan L., Fu J. Superhydrophobic composite coating with active corrosion resistance for AZ31B magnesium alloy protection. *Chem. Eng. J.* 2019, 357, 518–532.

Ding C., Xu J., Tong L., Gong G., Jiang W., Fu J. Design and fabrication of a novel stimulus-feedback anticor-rosion coating featured by rapid self-healing functionality for the protection of magnesium alloy. *ACS Appl. Mater. Interfaces.* 2017, 9, 21034–21047.

Dobkowska A., Adamczyk-Cieślak B., Kubásek J., Vojtěch D., Kuc D., Hadasik E., Mizera J. Microstructure and corrosion resistance of a duplex structured Mg-7.5Li-3Al-1Zn. *J. Magnes. Alloy.* 2021, 9, 467–477.

Dong Q., Zhou X., Feng Y., Qian K., Liu H., Lu M., Chu C., Xue F., Bai J. Insights into self-healing behavior and mechanism of dicalcium phosphate dihydrate coating on biomedical Mg. *Bioact. Mater.* 2021, 6, 158–168.

Duan X., Han T., Guan X., Wang Y., Su H., Ming K., Wang J., Zheng S. Cooperative effect of Cr and Al ele-ments on passivation enhancement of eutectic high-entropy alloy AlCoCrFeNi 2.1 with precipitates. *J. Mater. Sci. Technol.* 2023, 136, 97–108.

Faisal M., Saeed A., Shahzad D., Abbas N., Larik F., Channar P., Fattah T., Khan D., Shehzadi S. General properties and comparison of the corrosion inhibition efficiencies of the triazole derivatives for mild steel. *Corros. Rev.* 2018, 36, 507–545.

Feng J., Pan Y., Yang M., Fernandez C., Chen X., Peng Q. A lactoglobulin-composite self-healing coating for Mg alloys. *ACS Appl. Bio Mater.* 2021, 4, 6843–6852.

Gao C., Li S., Liu L., Bin S., Yang Y., Peng S., Shuai C. Dual alloying improves the corrosion resistance of biodegradable Mg alloys prepared by selective laser melting. *J. Magnes. Alloy.* 2021, 9, 305–316.

Garcia S.J. Effect of polymer architecture on the intrinsic self-healing character of polymers. *Eur. Polymer J.* 2014, 53, 118–125.

Ghazizadeh E., Jabbari A., Sedighi M. In vitro corrosion-fatigue behavior of biodegradable Mg/HA composite in simulated body fluid. *J. Magnes. Alloy.* 2021, 9, 2169–2184.

Gudze M., Melchers R. Operational based corrosion analysis in naval ships. *Corros. Sci.* 2008, 50, 3296–3307.

Gungor A., Incesu A. Effects of alloying elements and thermomechanical process on the mechanical and cor-rosion properties of biodegradable Mg alloys. *J. Magnes. Alloy.* 2021, 9, 241–253.

Guo H., Han Y., Zhao W., Yang J., Zhang L. Universally autonomous self-healing elastomer with high stretch-ability. *Nat. Commun.* 2020, 11, 2037.

Guo J., Liu X., Du K., Guo Q., Wang Y., Liu Y., Feng L. An anti-stripping and self-healing micro-arc oxida-tion/acrylamide gel composite coating on magnesium alloy AZ31. *Mater. Lett.* 2020, 260, 126912.

Guo X., Wang Y., Yao T., Mohanty C., Lian J., Frankel G. Corrosion interactions between stainless steel and lead vanado-iodoapatite nuclear waste form part I. *NPJ Mater. Degrad.* 2020, 4, 13.

Hamdy A., Butt D. Novel smart stannate based coatings of self-healing functionality for AZ91D magnesium alloy. *Electrochim. Acta.* 2013, 97, 296–303.

Han S., Shen C., Zhou S. Corrosion behavior of lithium silicate-based Zn-5.5Al-4.5Mg-0.3Ce coating. *Mater. Res. Express.* 2020, 7, 036536.

He Q., Zhou M., Hu J. Electrodeposited Zn-Al layered double hydroxide films for corrosion protection of aluminum alloys. *Electrochim. Acta.* 2020, 355, 136796.

Hegazy M.A., El-Etre A.Y., El-Shafaie M., Berry K.M. Novel cationic surfactants for corrosion inhibition of carbon steel pipelines in oil and gas wells applications. *J. Mol. Liq.* 2016, 214, 347–356.

Hou B., Li X., Ma X., Du C., Zhang D., Zheng M., Xu W., Lu D., Ma F. The cost of corrosion in China. *NPJ Mater. Degrad.* 2017, 1, 4.

Ikura R., Park J., Osaki M., Yamaguchi H., Harada A., Takashima Y. Design of self-healing and self-restoring materials utilizing reversible and movable crosslinks. *NPG Asia Mater.* 2022, 14, 10.

Jiang D., Xia X., Hou J., Cai G., Zhang X., Dong Z. A novel coating system with self-reparable slippery surface and active corrosion inhibition for reliable protection of Mg alloy. *Chem. Eng. J.* 2019, 373, 285–297.

Jiang D., Xia X., Hou J., Zhang X., Dong Z. Enhanced corrosion barrier of microarc-oxidized Mg alloy by self-healing superhydrophobic silica coating. *Ind. Eng. Chem. Res.* 2019, 58, 165–178.

Jiang Q., Lu D., Wang N., Wang X., Zhang J., Duan J., Hou B. The corrosion behavior of Mg-Nd binary alloys in the harsh marine environment. *J. Magnes. Alloy.* 2021, 9, 292–304.

Jiang Y., Gao S., Liu Y., Huangfu H., Guo X., Zhang J. Enhancement of corrosion resistance of AZ31B magnesium alloy by preparing MgAl-LDHs coatings modified with imidazolium based dicationic ionic liquids. *Surf. Coat. Technol.* 2022, 440, 128504.

Johari N., Alias J., Zanurin A., Mohamed N., Alang N., Zain M. Recent progress of self-healing coatings for magnesium alloys protection. *J. Coat. Technol. Res.* 2022, 19, 757–774.

Kalista Jr S.J., Ward T.C. Thermal characteristics of the self-healing response in poly(ethylene-co-methacrylic acid) copolymers. *J. R. Soc. Interface.* 2007, 4, 405–411.

Kamde M., Mahton Y., Saha P. A stearic acid/polypyrrole-based superhydrophobic coating on squeeze-cast Mg-Sr-Y-Ca-Zn alloys for improved salt-water corrosion. *Surf. Coat. Technol.* 2022, 448, 128890.

Kan Y., Zheng F., Li B., Zhang R., Wei Y., Yu Y., Zhang Y., Ouyang Y., Qiu R. Self-healing dual biomimetic liquid-infused slippery surface in a partition matrix: Fabrication and anti-corrosion capability for magnesium alloy. *Colloid Surf. A-Physicochem. Eng. Asp.* 2021, 630, 127585.

Kannan M., Walter R., Yamamoto A., Khakbaz H., Blawert C. Electrochemical surface engineering of magnesium metal by plasma electrolytic oxidation and calcium phosphate deposition: Biocompatibility and: In vitro degradation studies. *RSC Adv.* 2018, 8, 29189–29200.

Koch D., Mack D., Vassen R. Degradation and lifetime of self-healing thermal barrier coatings containing MoSi2 as self-healing particles in thermo-cycling testing. *Surf. Coat. Technol.* 2022, 437, 128353.

Lee Y., Hong M., Ko S., Kim J. Effect of benzotriazole on the localized corrosion of copper covered with carbonaceous residue. *Materials.* 2021, 14, 2722.

Lei Y., Qiu Z., Tan N., Du H., Li D., Liu J., Liu T., Zhang W., Chang X. Polyaniline/CeO2 nanocomposites as corrosion inhibitors for improving the corrosive performance of epoxy coating on carbon steel in 3.5% NaCl solution. *Prog. Org. Coat.* 2020, 139, 105430.

Li B., Huang R., Han Y. A self-healing coating containing curcumin for osteoimmunomodulation to ameliorate osseointegration. *Chem. Eng. J.* 2021, 403, 126323.

Li B., Xue S., Mu P., Li J. Robust self-healing graphene oxide-based superhydrophobic coatings for efficient corrosion protection of magnesium alloys. *ACS Appl. Mater. Interfaces.* 2022, 14, 30192–30204.

Li C., Guo X., Frankel G. Smart coating with dual-pH sensitive, inhibitor-loaded nanofibers for corrosion protection. *NPJ Mater. Degrad.* 2021, 5, 54.

Li H., Feng X., Peng Y., Zeng R. Durable lubricant-infused coating on a magnesium alloy substrate with anti-biofouling and anti-corrosion properties and excellent thermally assisted healing ability. *Nanoscale.* 2020, 12, 7700–7711.

Li Q., Zhang X., Ben S., Zhao Z., Ning Y., Liu K., Jiang L. Bio-inspired superhydrophobic magnesium alloy surfaces with active anti-corrosion and self-healing properties. *Nano Res.* 2022. https://doi.org/10.1007/s12274-022-4937-7.

Li W. Effects of Ca and Ag addition and heat treatment on the corrosion behavior of Mg-7Sn alloys in 3.5 wt.% NaCl solution. *Surf. Interface Anal.* 2022, 54, 631–641.

Li W., Su Y., Ma L., Zhu S., Zheng Y., Guan S. Sol-gel coating loaded with inhibitor on ZE21B Mg alloy for improving corrosion resistance and endothelialization aiming at potential cardiovascular application. *Colloid Surf. B-Biointerfaces.* 2021, 207, 111993.

Li Y., Chen J., Gambelli A., Zhao X., Gao Y., Rossi F., Mei S. In situ experimental study on the effect of mixed inhibitors on the phase equilibrium of carbon dioxide hydrate. *Chem. Eng. J.* 2022, 248, 117230.

Li Y., Lu B., Yu W., Fu J., Xu G., Wang Z. Two-stage homogenization of Al-Zn-Mg-Cu-Zr alloy processed by twin-roll casting to improve L1(2) Al3Zr precipitation, recrystallization resistance, and performance. *J. Alloy. Compd.* 2021, 882, 160789.

Li Z., Yu Q., Zhang C., Liu Y., Liang J., Wang D., Zhou F. Synergistic effect of hydrophobic film and porous MAO membrane containing alkynol inhibitor for enhanced corrosion resistance of magnesium alloy. *Surf. Coat. Technol.* 2019, 357, 515–525.

Liu C., Wang Q., Han B., Luan J., Kai J., Liu C., Wu G., Lu J. Second phase effect on corrosion of nanostructured Mg-Zn-Ca dual-phase metallic glasses. *J. Magnes. Alloy.* 2021, 9, 1546–1555.

Liu D., Han E., Song Y., Shan D. Enhancing the self-healing property by adding the synergetic corrosion inhibitors of Na3PO4 and 2-mercaptobenzothiazole into the coating of Mg alloy. *Electrochim. Acta.* 2019, 323, 134796.

Liu J., Yan D., Zhang Z., Wang Y., Song D., Zhang T., Liu J., He F., Zhang M., Wang J. Eco-friendly silane as corrosion inhibitor for dual self-healing anticorrosion coatings. *J. Coat. Technol. Res.* 2022, 19, 1381–1391.

Liu S., Li Z., Yu Q., Qi Y., Peng Z., Liang J. Dual self-healing composite coating on magnesium alloys for corrosion protection. *Chem. Eng. J.* 2021, 424, 130551.

Liu W., Singh A., Lin Y., Ebenso E., Zhou L., Huang B. 8-Hydroxyquinoline as an effective corrosion inhibitor for 7075 aluminium alloy in 3.5% NaCl solution. *Int. J. Electrochem. Sci.* 2014, 9, 5574–5584.

Liu X., He H., Zhang T., Ouyang L., Zhang Y., Yuan S. Superhydrophobic and self-healing dual-function coatings based on mercaptabenzimidazole inhibitor-loaded magnesium silicate nanotubes for corrosion protection of AZ31B magnesium alloys. *Chem. Eng. J.* 2021, 404, 127106.

Liu Y., Cheng W., Gu X., Liu Y., Cui Z., Wang L., Wang H. Tailoring the microstructural characteristic and improving the corrosion resistance of extruded dilute Mg-0.5Bi-0.5Sn alloy by microalloying with Mn. *J. Magnes. Alloy.* 2021, 9, 1656–1668.

Luo J., Wang T., Sim C., Li Y. Mini-review of self-healing mechanism and formulation optimization of polyurea coating. *Polymers.* 2022, 14, 2808.

Ma X.C., Jin S.Y., Wu R.Z., Ji Q., Hou L., Krit B., Betsofen S. Influence alloying elements of Al and Y in Mg-Li alloy on the corrosion behavior and wear resistance of microarc oxidation coatings. *Surf. Coat. Technol.* 2022, 432, 128042.

Mahmoudi R., Kardar P., Arabi A.M., Amini R., Pasbakhsh P. The active corrosion performance of silane coating treated by praseodymium encapsulated with halloysite nanotubes. *Prog. Org. Coat.* 2020, 138, 105404.

Manasa S., Jyothirmayi A., Siva T., Sarada B.V., Ramakrishna M., Sathiyanarayanan S., Gobi K.V., Subasri R. Nanoclay-based self-healing, corrosion protection coatings on aluminum, A356.0 and AZ91 substrates. *J. Coat. Technol. Res.* 2017, 14, 1195–1208.

Miao X., Shi X., Shen Y., Zhang W., Hu W., Zhang S., Huang X., Wang Y., Zhao R., Zhang R. Investigation of hexamethylenetetramine effects on formation and corrosion resistance of anodic coatings developed on AZ31B alloys. *Surf. Coat. Technol.* 2022, 447, 128824.

Mingo B., Guo Y., Leiva-Garcia R., Connolly B., Matthews A., Yerokhin A. Smart functionalization of ceramic-coated AZ31 magnesium alloy. *ACS Appl. Mater. Interfaces.* 2020, 12, 30833–30846.

Monticelli C. Corrosion inhibitors. *Encycl. Interfacial Chem.* 2018, 164–171.

Nie Y., Dai J., Li X., Zhang X. Recent developments on corrosion behaviors of Mg alloys with stacking fault or long period stacking ordered structures. *J. Magnes. Alloy.* 2021, 9, 1123–1146.

Østnor T.A., Justnes H. Anodic corrosion inhibitors against chloride induced corrosion of concrete rebars. *Adv. Appl. Ceram.* 2011, 110, 131–136.

Ouyang Y., Li L., Xie Z., Tang L., Wang F., Zhong C. A self-healing coating based on facile pH-responsive nanocontainers for corrosion protection of magnesium alloy. *J. Magnes. Alloy.* 2022, 10, 836–849.

Pahalagedara M.N., Samaraweera M., Dharmarathna S., Kuo C.H., Pahalagedara L.R., Gascón J.A., Suib S.L. Removal of azo dyes: Intercalation into sonochemically synthesized NiAl layered double hydroxide. *J. Phys. Chem. C.* 2014, 118, 17801–17809.

Peng X., Sun J., Liu H., Wu G., Liu W. Microstructure and corrosion behavior of as-homogenized and as-extruded Mg-xLi-3Al-2Zn-0.5Y alloys (x=4, 8, 12). *Trans. Nonferrous Met. Soc. China.* 2022, 32, 134–146.

Popoola L. Corrosion inhibitory effect of mixed cocoa pod-Ficus exasperata extract on MS in 1.5 M HCl: Optimization and electrochemical study. *Corros. Rev.* 2021, 39, 109–122.

Rizvi M., Gerengi H., Kaya S., Uygur I., Yıldız M., Sarıoglu I., Cingiz Z., Mielniczek M., Ibrahimi B. Sodium nitrite as a corrosion inhibitor of copper in simulated cooling water. *Sci. Rep.* 2021, 11, 8353.

Ryl J., Wysocka J., Cieslik M., Gerengi H., Ossowski T., Krakowiak S., Niedzialkowski P. Understanding the origin of high corrosion inhibition efficiency of bee products towards aluminium alloys in alkaline environments. *Electrochim. Acta.* 2019, 304, 263–274.

Shanaghi A., Souri A., Mehrjou B., Chu P. Corrosion resistance, nano-mechanical properties, and biocompatibility of Mg-plasma-implanted and plasma-etched Ta/TaN hierarchical multilayered coatings on the nitrided AZ91 Mg alloy. *Biomed. Mater.* 2021, 16, 045028.

Sharma S., Maurice V., Klein L., Marcus P. Local Inhibition by 2-mercaptobenzothiazole of early stage intergranular corrosion of copper. *J. Electrochem. Soc.* 2020, 167, 161504.

Shchukin D., Möhwald H. A coat of many functions. *Science.* 2013, 341, 1458–1459.

Shchukina E., Grigoriev D., Sviridova T., Shchukin D. Comparative study of the effect of halloysite nanocontainers on autonomic corrosion protection of poly-epoxy coatings on steel by salt-spray tests. *Prog. Org. Coat.* 2017, 108, 84–89.

Shchukina E., Shchukin D., Grigoriev D. Effect of inhibitor-loaded halloysites and mesoporous silica nanocontainers on corrosion protection of powder coatings. *Prog. Org. Coat.* 2017, 102, 60–65.

Shchukina E., Shchukin D., Grigoriev D. Halloysites and mesoporous silica as inhibitor nanocontainers for feedback active powder coatings. *Prog. Org. Coat.* 2018, 123, 384–389.

Shkirskiy V., Keil P., Hintze-Bruening H., Leroux F., Vialat P., Lefèvre G., Ogle K., Volovitch P. Factors affecting MoO42-inhibitor release from Zn2Al based layered double hydroxide and their implication in protecting hot dip galvanized steel by means of organic coatings. *ACS Appl. Mater. Interfaces.* 2015, 7, 25180–25192.

Soliman H., Qian J., Tang S., Xian P., Chen Y., Makhlouf A., Wan G. Hydroxyquinoline/nano-graphene oxide composite coating of self-healing functionality on treated Mg alloys AZ31. *Surf. Coat. Technol.* 2020, 385, 125395.

Soltan A., Dargusch M., Shi Z., Jones F., Wood B., Gerrard D., Atrens A. Effect of corrosion inhibiting compounds on the corrosion behaviour of pure magnesium and the magnesium alloys EV31A, WE43B and ZE41A. *J. Magnes. Alloy.* 2021, 9, 432–455.

Song F., Wu C., Chen H., Liu Q., Liu J., Chen R., Li R., Wang J. Water-repellent and corrosion-resistance properties of superhydrophobic and lubricant-infused super slippery surfaces. *RSC Adv.* 2017, 7, 44239–44246.

Song Y., Liu D., Tang W., Dong K., Shan D., Han E. Comparison of the corrosion behavior of AM60 Mg alloy with and without self-healing coating in atmospheric environment. *J. Magne. Alloy.* 2021, 9, 1220–1232.

Song Y., Wang H., Liu Q., Li G., Wang S., Zhu X. Sodium dodecyl sulfate (SDS) intercalated Mg-Al layered double hydroxides film to enhance the corrosion resistance of AZ31 magnesium alloy. *Surf. Coat. Technol.* 2021, 422, 127524.

Tan J., Birbilis N., Choudhary S., Thomas S., Balan P. Corrosion protection enhancement of Mg alloy WE43 by in-situ synthesis of MgFe LDH/citric acid composite coating intercalated with 8HQ. *Corros. Sci.* 2022, 205, 110444.

Thakur A., Kumar A. Sustainable inhibitors for corrosion mitigation in aggressive corrosive media: A comprehensive study. *J. Bio- Tribo-Corros.* 2021, 7, 67.

Toohey K., Sottos N., Lewis J., Moore J., White S. Self-healing materials with microvascular networks. *Nat. Mater.* 2007, 6, 581–585.

Wang X., Long Y., Mu P., Li J. Silicone oil infused slippery candle soot surface for corrosion inhibition with anti-fouling and self-healing properties. *J. Adhes. Sci. Technol.* 2021, 35, 1057–1071.

Wei D., Wang J., Li S., Liu Y., Wang D., Wang H. Novel corrosion-resistant behavior and mechanism of a biomimetic surface with switchable wettability on Mg alloy. *Chem. Eng. J.* 2021, 425, 130450.

White S., Sottos N., Geubelle P., Moore J., Kessler M., Sriram S., Brown E., Viswanathan S. Autonomic healing of polymer composites. *Nature.* 2001, 409, 794–797.

Wu M., Jiang F. Investigation of alloying element Mg in the near surface layer of micro-arc oxidation coating on Al-Mg-Sc alloy. *Vacuum.* 2022, 197, 110823.

Wu W., Wang W., Lin H. A study on corrosion behavior of micro-arc oxidation coatings doped with 2-aminobenzimidazole loaded halloysite nanotubes on AZ31 magnesium alloys. *Surf. Coat. Technol.* 2021, 416, 127116.

Xie Z., Li D., Skeete Z., Sharma A., Zhong C. Nanocontainer-enhanced self-healing for corrosion-resistant Ni coating on Mg alloy. *ACS Appl. Mater. Interfaces.* 2017, 9, 36247–36260.

Xie Z., Shan S. Nanocontainers-enhanced self-healing Ni coating for corrosion protection of Mg alloy. *J. Mater. Sci.* 2018, 53, 3744–3755.

Xing K., Li Z., Wang Z., Qian S., Feng J., Gu C., Tu J. Slippery coatings with mechanical robustness and self-replenishing properties as potential application on magnesium alloys. *Chem. Eng. J.* 2021, 418, 129079.

Xu J., Wang X., Zhang X., Zhang Y., Yang Z., Li S., Tao L., Wang Q., Wang T. Room-temperature self-healing supramolecular polyurethanes based on the synergistic strengthening of biomimetic hierarchical hydrogen-bonding interactions and coordination bonds. *Chem. Eng. J.* 2023, 451, 138673.

Xue K., Tan P., Zhao Z., Cui L., Kannan M., Li S., Liu C., Zou Y., Zhang F., Chen Z., Zeng R. In vitro degradation and multi-antibacterial mechanisms of beta-cyclodextrin@curcumin embodied Mg(OH)(2)/MAO coating on AZ31 magnesium alloy. *J. Mater. Sci. Technol.* 2023, 132, 179–192.

Yan D., Wang Y., Liu J., Song D., Zhang T., Liu J., He F., Zhang M., Wang J. Self-healing system adapted to different pH environments for active corrosion protection of magnesium alloy. *J. Alloy. Compd.* 2020, 824, 153918.

Yan J., Pan G., Lin W., Tang Z., Zhang J., Li J., Li W., Lin X., Luo H., Yi G. Multi-responsive graphene quantum dots hybrid self-healing structural color hydrogel for information encoding and encryption. *Chem. Eng. J.* 2023, 451, 138922.

Yang L., He S., Yang C., Zhou X., Lu X., Huang Y., Qin G., Zhang E. Mechanism of Mn on inhibiting Fe-caused magnesium corrosion. *J. Magnes. Alloy.* 2021, 9, 676–685.

Yang S., Chen Z., Chen T., Fu C. Hollow mesoporous silica nanoparticles decorated with cyclodextrin for inhibiting the corrosion of Mg alloys. *ACS Appl. Nano Mater.* 2020, 3, 4542–4552.

Yang S., Sun R., Chen K. Self-healing performance and corrosion resistance of phytic acid/cerium composite coating on microarc-oxidized magnesium alloy. *Chem. Eng. J.* 2022, 428, 131198.

Yao W., Chen Y., Wu L., Jiang B., Pan F. Preparation of slippery liquid-infused porous surface based on MgAlLa-layered double hydroxide for effective corrosion protection on AZ31 Mg alloy. *J. Taiwan Inst. Chem. Eng.* 2022, 131, 104176.

Yao W., Chen Y., Wu L., Zhang J., Pan F. Effective corrosion and wear protection of slippery liquid-infused porous surface on AZ31 Mg alloy. *Surf. Coat. Technol.* 2021, 429, 127953.

Yao W., Wu L., Huang G., Jiang B., Atrens A., Pan F. Superhydrophobic coatings for corrosion protection of magnesium alloys. *J. Mater. Sci. Technol.* 2020, 52, 100–118.

Yao W., Wu L., Wang J., Jiang B., Zhang D., Serdechnova M., Shulha T., Blawert C., Zheludkevich M., Pan F. Micro-arc oxidation of magnesium alloys: A review. *J. Mater. Sci. Technol.* 2022, 118, 158–180.

Yeganeh M., Saremi M. Corrosion inhibition of magnesium using biocompatible Alkyd coatings incorporated by mesoporous silica nanocontainers. *Surf. Coat. Technol.* 2015, 79, 25–30.

Yu H., Dong X., Kang S., Yu W., Wang Z., Mu J., Cui X., Li J., Yin F., Shin K. Effect of the pre-homogenization on the precipitation behaviors, mechanical and corrosion properties of as-extruded Mg-Y binary alloys. *Mater. Charact.* 2021, 178, 111307.

Yuan H., Ji C., Yuan H. Photothermal self-healing polyurethane/graphene oxide composites based on diselenide bond. *Polym. Test.* 2023, 117, 107813.

Zhang D., Peng F., Liu X. Protection of magnesium alloys: From physical barrier coating to smart self-healing coating. *J. Alloy. Compd.* 2021, 853, 157010.

Zhang F., Ju P., Pana M., Zhang D., Huang Y., Li G., Li X. Self-healing mechanisms in smart protective coatings: A review. *Corros. Sci.* 2018, 144, 74–88.

Zhang J., Gu C., Tong Y., Yan W., Tu J. A smart superhydrophobic coating on AZ31B magnesium alloy with self-healing effect. *Adv. Mater. Interfaces.* 2016, 3, UNSP 1500694.

Zhang J., Gu C., Tu J. Robust slippery coating with superior corrosion resistance and anti-icing performance for AZ31B Mg alloy protection. *ACS Appl. Mater. Interfaces.* 2017, 9, 11247–11257.

Zhang J., Wei J., Li B., Zhao X., Zhang J. Long-term corrosion protection for magnesium alloy by two-layer self-healing superamphiphobic coatings based on shape memory polymers and attapulgite. *J. Colloid Interface Sci.* 2021, 594, 836–847.

Zhang J., Zhao X., Wei J., Li B., Zhang J. Superhydrophobic coatings with photothermal self-healing chemical composition and microstructure for efficient corrosion protection of magnesium alloy. *Langmuir.* 2021, 37, 13527–13536.

Zhang X., Wu W., Fu H., Li J. The effect of corrosion evolution on the stress corrosion cracking behavior of mooring chain steel. *Corros. Sci.* 2022, 203, 110316.

Zhao X., Wei J., Li B., Li S., Tian N., Jing L., Zhang J. A self-healing superamphiphobic coating for efficient corrosion protection of magnesium alloy. *J. Colloid Interface Sci.* 2020, 575, 140–149.

Zhao Y., Zhang Z., Shi L., Zhang F., Li S., Zeng R. Corrosion resistance of a self-healing multilayer film based on SiO2 and CeO2 nanoparticles layer-by-layer assembly on Mg alloys. *Mater. Lett.* 2019, 237, 14–18.

Zheludkevich M.L., Hughes A.E. Delivery systems for self healing protective coatings. *Active Protective Coatings.* 2016, 233, 157–199.

Zhu M., Yu J., Li Z., Ding B. Self-healing fibrous membranes. *Angew. Chem.-Int. Edit.* 2022, 61, e202208949.

23 Biocompatible Coatings on Biodegradable Magnesium Alloys

V.P. Muhammad Rabeeh, K.S. Akshay,
K.S. Surendramohan, T.S. Sampath Kumar and T. Hanas

CONTENTS

23.1 INTRODUCTION

Magnesium (Mg) and its alloys are promising candidates for biodegradable metallic implants due to their remarkable biodegradability, comparable mechanical characteristics to bone, excellent biocompatibility and osteogenesis properties. However, addressing the rapid degradation of Mg and its alloys in the physiological environment is a critical challenge for developing such implants (Witte 2010). Mostly, the degradation of Mg happens through a non-uniform corrosion manner due to the high tendency for localized corrosion. The quick and uncontrolled degradation of Mg causes a considerable decrease in the implant's mechanical characteristics after implantation, resulting in

DOI: 10.1201/9781003319856-26

premature failure (Tan et al. 2013). In addition, the release of hydrogen during degradation and alkalinization of the implant–tissue interface poses a grave risk to the patient's health after implantation (Shuai et al. 2019; Chagnon, Guy, and Jackson 2019). Also, the implant surface should be bioactive to promote biomineralization as well as cell adhesion and proliferation.

Hence it is very important to control the degradation of Mg and Mg alloys to develop a safe implantable material. Researchers have adopted various strategies to control the degradation rate through modifying the microstructure and surface of the Mg alloys. Microstructural modifications mainly include alloying with various types of elements, adding reinforcement, texturing and optimizing the manufacturing and post processing techniques (Seal, Vince, and Hodgson 2009; Sekar, Narendranath, and Desai 2021). The surface modification mainly employs either surface treatments or coatings which act as a barrier between the Mg and the surrounding environment. Various surface treatment techniques such as conversion coating, anodization, ion implantation and coating with organic materials have been used to tailor the degradation of Mg-based materials (Yin et al. 2020; Singh et al. 2023).

Alloying of Mg with other elements is highly challenging due to the solubility of the alloying elements, whereas protective and bioactive coatings of Mg seem to be a better alternative technique to control the degradation of Mg. Moreover, the coating will improve the interfacial properties of the materials for better bioactive and bio-integrative properties (Nilawar, Uddin, and Chatterjee 2021; Singh et al. 2023). Most of the coatings on Mg were mainly focused on the corrosion resistance and wear resistance for aerospace applications. In the past decade, the need for degradable metallic implants have led researchers to explore the possibility of using biomedical-grade coatings on Mg-based systems – the bioactive, biocompatible and biodegradable coating that can control the degradation rate in the physiological environment. This chapter discusses the recent advancements in the area of biomedical-grade coating on Mg alloys with an emphasis on the methods and materials adopted for developing such coatings.

23.2 COATING FOR MAGNESIUM ALLOYS

Based on the nature and mechanism, the coatings can be grouped into conversion- or deposition-type coatings. In the conversion type, the coating formation takes place by specific reaction between the metal surface and the environment to which it is exposed. Generally, the metal substrate reacts with the environment and form a stable reaction product on the surface of the material. In deposit coatings, a stable layer of secondary material is applied on the surface of the substrate by various techniques such as painting, dip coating and electrodeposition. Figure 23.1 shows the various types of coating techniques reported so far for developing biodegradable and bioactive coating on Mg-based materials.

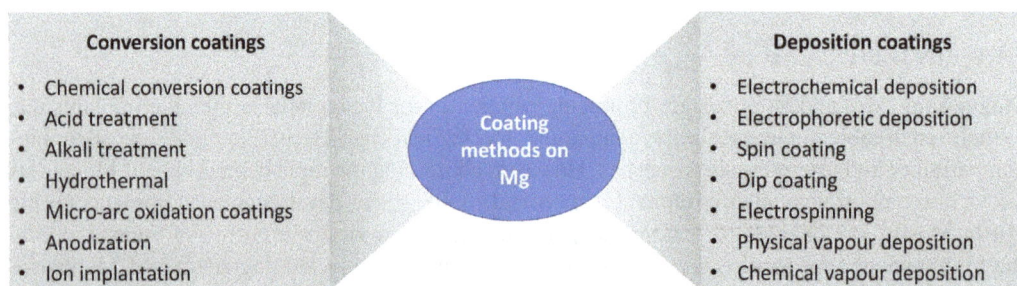

Conversion coatings

- Chemical conversion coatings
- Acid treatment
- Alkali treatment
- Hydrothermal
- Micro-arc oxidation coatings
- Anodization
- Ion implantation

Coating methods on Mg

Deposition coatings

- Electrochemical deposition
- Electrophoretic deposition
- Spin coating
- Dip coating
- Electrospinning
- Physical vapour deposition
- Chemical vapour deposition

FIGURE 23.1 General classification of coating methods for biodegradable Mg alloys.

23.2.1 Conversion Coatings

As mentioned in the previous section, conversion coatings are generated via chemical interactions between Mg substrate and coating medium. The formation of conversion coatings is the result of a complicated interaction between the dissolution and precipitation of metals in aqueous medium. The adhesion between the substrate and the coatings is relatively better because the native metallic surface is converted to an oxide film or sometimes stable salts through chemical or electrochemical reactions. This is a major distinction between conversion coatings and deposition coatings (Pommiers-Belin et al. 2014). It is also noted that the conversion coatings can be used as a surface pre-treatment technique to improve the adhesion of a deposition coating on the surfaces (Liu and Gao 2006). As a pre-treatment, the conversion coatings primarily act as an interlayer between functional coatings and the substrates (Mahidashti, Shahrabi, and Ramezanzadeh 2018).

Corrosion resistance, biocompatibility and osteoconductivity can all be improved by using conversion coatings. The majority of the conversion coatings on Mg are achieved by formation of oxide/hydroxide of native or other desired elements on Mg substrate. The mechanism of conversion coating is shown in Figure 23.2. Biomedical applications can be benefited by chemical conversion coatings due to their simplicity of operation and low cost (Yin et al. 2020). In general, a conversion coating works by promoting the interfacial processes, with the ensuing precipitation or growth of coatings which can offer superior protection against corrosion of Mg alloys. A drop in pH and a rise in Mg^{2+} concentration at the metal/solution contact are necessary for the process to begin. Since Mg has a relatively low standard electrode potential, it can be used to reduce protons (H^+) and water (H_2O) simultaneously during aqueous reactions. As soon as the Mg substrate is immersed in the aqueous media, magnesium hydroxide begins to develop. Due to its chemical stability at neutral and alkaline pH values, magnesium hydroxide passivates and blocks the conversion processes (Xiong, Yan, et al. 2019).

Similarly, phosphate conversion coating such as zinc phosphate, calcium phosphate and fluoride conversion coating are being reported as feasible coatings for biomedical applications, as they are insoluble in water and have good thermal and chemical stability combined with excellent biocompatibility (Yin et al. 2020). A few of the relevant conversion coatings used to tailor the Mg surfaces for biomedical applications are discussed in what follows.

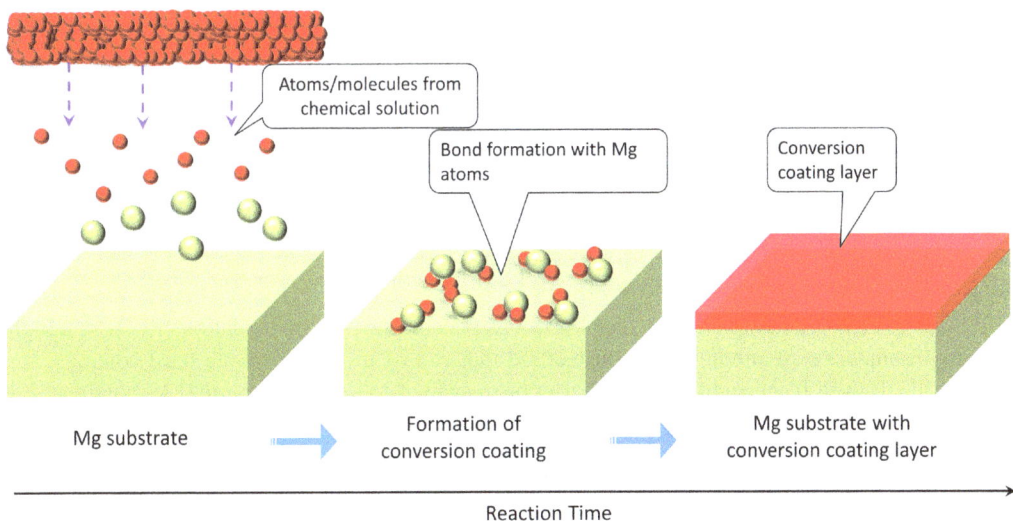

FIGURE 23.2 Mechanism of conversion coating.

23.2.1.1   Acid Treatment

Treatment with acid solutions is a cost-effective technique to tailor the degradation rate of metals. Surface contamination during manufacturing and processing of metallic components can also be removed using acid etching. These contaminants, if not removed, can lead to the formation of micro-galvanic cells and promote corrosion. In addition, the acid treatment can also develop a protective passive layer to guard the surface beneath. The treatment also helps in increasing the surface energy and can improve adhesion of any secondary coating on the surface. Such surfaces are also reported to enhance the cell adhesion and growth when used for biological applications (Rahim et al. 2022; Gawlik et al. 2019). It is reported that treating Mg-Ca alloys with phosphoric acid and nitric acid can enhance the biomineralization on the surface. This increase in bioactivity is attributed to the magnesium-phosphate and magnesium-hydroxide layer formed during the treatment process, which assists facilitated surface stability and promoted hydroxyapatite (HA) nucleation and growth (Rahim et al. 2021). However, the conversion coatings formed by acid treatment pose a stability risk in the long term. Hence, the acid treatments are mostly recommended as a pre-treatment on the surfaces that are subjected to deposition coatings.

23.2.1.2   Alkali Treatment

Just like the acid treatment discussed earlier, alkali treatments of Mg alloys are also explored by researchers (Li et al. 2020; Tang et al. 2017). When the alloy is treated with an appropriate alkaline media, a $Mg(OH)_2$ layer is formed on the surface (Saxena and Raman 2021). The layer produced on Mg alloy after treatment with alkaline media (disodium phosphate, sodium bicarbonate and sodium carbonate) increased the resistance to degradation without causing any toxicity in L-929 cells (Gu et al. 2009). Sodium bicarbonate-based treatment yielded the lowest degradation rate and highest calcium deposition of all the alkaline media tested. Treatment with alkali for varying times confirms that film thickness rises with treatment duration. It is also reported that the $Mg(OH)_2$ formed during the alkali treatment of alloys can act as a barrier against corrosion and slow down the rate of degradation (Tang et al. 2017). However, as mentioned in the case of acid treatment, alkali treatment, too, is recommended as a pre-treatment technique due to the risk of instability in the long term.

23.2.1.3   Hydrothermal Treatment

Hydrothermal treatment is a simple and cost-effective technique to develop a conversion coating on metallic substrates. The coating is achieved by heating the substrate in an autoclave while it is immersed in a suitable solution. The treatment results in the formation of a thin layer of irregular granular $Mg(CO)_3$ and $Mg(OH)_2$ on the surface. The morphology of such a coating in the substrate is reported to be a structure ideal for apatite nucleation and growth (Xie et al. 2019). However, the layers formed by hydrothermal treatments of surfaces cannot provide long-term protection in corrosive environments, and hence they are also suggested as pre-treatment techniques to prepare the surface for further processing.

23.2.1.4   Micro-Arc Oxidation Coating

Micro-arc oxidation coating (MAO), also known as plasma electrolytic oxidation (PEO), is a high-voltage plasma-assisted anodic oxidation method that can be used to generate hard coatings. It is a technique derived from traditional anodizing process used to develop ceramic-like coatings. It is widely used for surface modification of Mg alloys (H. F. Guo et al. 2006). In MAO, the oxide coatings are formed by applying high electric voltages to electrolytes that are either weak alkaline or acidic, both of which are safe for the environment. The process entails modifying a standard anodically grown oxide film by subjecting it to an electric field higher than the oxide layer's dielectric breakdown potential. When discharges occur, plasma-chemical reactions take place, which aid in the development of the coating. Oxide undergoes further changes as a result of rapid cooling and become a heterogeneous mixture of amorphous and nanocrystalline phases. Using this method, a largely crystalline oxide coating with a thickness varying from a few tens to hundreds of microns can be grown on metals (Jiang and Ge 2013). In addition to the electrolyte and alloy properties, processing parameters including pulse frequency, reaction time and voltage can strongly affect the final MAO coating's chemistry and performance. Unlike the already-discussed conversion coatings, the MAO coatings exhibit good stability and adhesion on to the substrate. The electrical discharges that happen when electric currents locally break through the expanding layer provide porous coatings (Rahim et al. 2022). These porous coatings can be effectively used to protect the substrate from corrosion and can also act as an interlayer to strengthen the binding force of a composite coating.

23.2.2 Deposition Coatings

Deposition coating is an ex-situ coating technique in which the substrate does not participate in the development of the coating. The deposited coatings can be either metallic, polymeric, ceramic or even a composite. Binding force between molecules (e.g. electrostatic force, hydrogen bond etc.) and mechanical force assure adhesion between the substrate and its coating. Because of their flimsy adhesion to the substrate, deposited coatings are typically considered the outermost or functional layer and are not preferred as intermediate layers. Various deposition coatings such as dip coating, spin coating, electrospun coating electrodeposition, electrophoretic deposition, physical vapour deposition and chemical vapour deposition are reported to improve the degradation resistance and biological properties of Mg alloys.

23.2.2.1 Dip Coating

Dip coating is considered to be the quick and easy coating method. The process can be used for coating the external surfaces of any intricate geometry. It consists of a sequence of steps which includes, dipping, holding, withdrawing and drying. The coating material is dissolved in a solvent, and the sample to be coated is dipped into the solution. The sample is then held in the solution for an optimum time before it is withdrawn at a predetermined speed. The viscous drag, capillary rise and gravity force are tuned for the uniform spreading of coating on the surfaces. Usually, the substrate is pulled up from the solution at a uniform speed into a water vapour–containing atmosphere. Volatile solvents will evaporate, and chemical reactions may happen at room temperature forming a thin coating film. Normally, heat treatment is needed to harden a coating after it is dried. The evaporation process needs to be closely monitored, as there are chances of aggregation of coating materials, developments of cracks etc. Any inconsistency in the environmental conditions during the drying can affect the reproducibility of the coating (Brinker et al. 1991).

23.2.2.2 Spin Coating

Spin coating is one of the most common coating techniques used for coating a thin uniform layer on a planar surface. The technique uses centrifugal force during spinning of the sample on which coating material is dropped in the form of a solution. The coating happens in three stages: solution dropping, spreading and evaporation. The flat substrate material is attached to the chuck using the aspirator. The chuck's rotation causes centrifugal force to disseminate the dropped solution evenly over the flat and clean substrate material. Volatile solvent will evaporate from the solution, leading to formation of a thin layer of coating. The thickness of the coating is determined by several factors, including the speed of rotation, acceleration of the chuck and the viscosity of the solution (Sahu, Parija, and Panigrahi 2009).

The key advantage of the method is that it is simple and inexpensive and takes less time to coat. The limitation due to size and geometry of the substrate is the key disadvantage of this approach. Large samples necessitate a faster spinning rate to ensure perfect uniform spreading. As a result, making thin films on large substrates is difficult. Other challenges include the complexity in creating a multilayer coating, the chance of contamination, the inability to create a coating less than 10 nm thick, etc. (Birnie 2004).

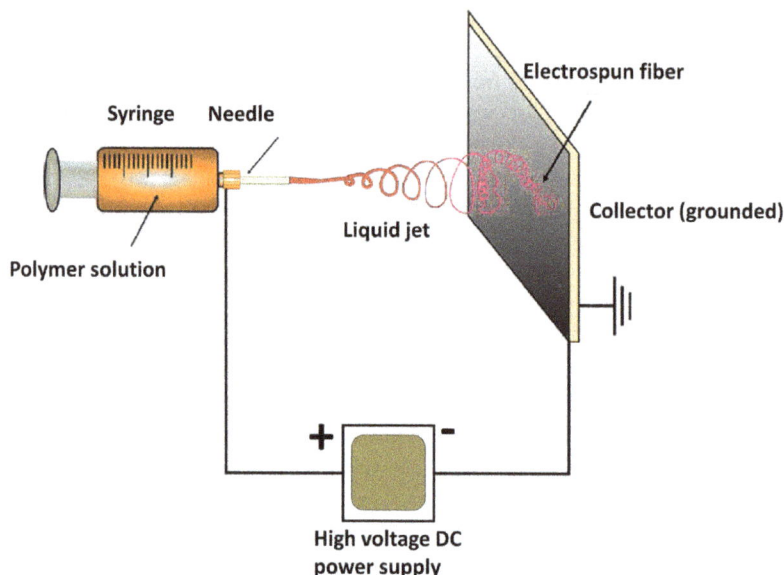

FIGURE 23.3 Schematic diagram of electrospinning technique.

23.2.2.3 Electrospun Coating

Electrospinning is a method used for developing non-woven nanofibrous morphology. It employs a high voltage to create nanofibers from a polymeric solution. The basic setup consists of a power source for creating high voltage, a grounded collector and a reservoir with a sharp metallic needle, as shown in Figure 23.3. A specific high electric potential is applied between the collector and the needle through which the solution is discharged at a desired rate. The electrostatic force applied and the surface tension opposing it will lead to generation of fibrous structure from the polymer droplets at the tip of the needle. When electrostatic repulsion is stronger than the surface tension of a liquid, the liquid's meniscus changes shape to a cone, which is called a Taylor cone. After the Taylor cone has been set up, the charge liquid jet is shot into the collector (Angammana and Jayaram 2016). By properly designing the collector, one can thus get these fibres deposited on the surface to form a nanofibrous coating. The parameters that control the electrospinning process can be grouped into two groups, namely, solution parameters and process parameters. Surface tension, dielectric effect, volatility, viscosity, conductivity, solvent system and polymer concentration are all parameters of a solution. The feed rate of the solvent, the voltage applied, the size of the needle, the type of collector, the tip-to-collector distances and the temperature are all process parameters (Mitchell, Ahn, and Davis 2011). A proper optimization of both process and solution parameters are essential to get a good nanofibrous coating.

23.2.2.4 Electrodeposition

Electrodeposition, an abbreviation of electrolytic deposition, refers to the common practice of applying a thin layer of one metal over another metal in order to alter the surface properties. Electroplating is the common electrodeposition method that deposits a thin layer of coating metal on a conductive substrate surface with the help of electric current. Metal ions move from a positive electrode to a negative electrode through the electrolyte during the electroplating process. The metal in the solution coats the components at the cathode when an electrical current flows through it (Lei et al. 2010). Figure 23.4a depict schematics of the general electrodeposition process. Electrodeposits are produced when an electric current flows between two conductive or semiconducting electrodes in an electrolyte. The working electrode (cathode) consists of the object where electrodeposition is

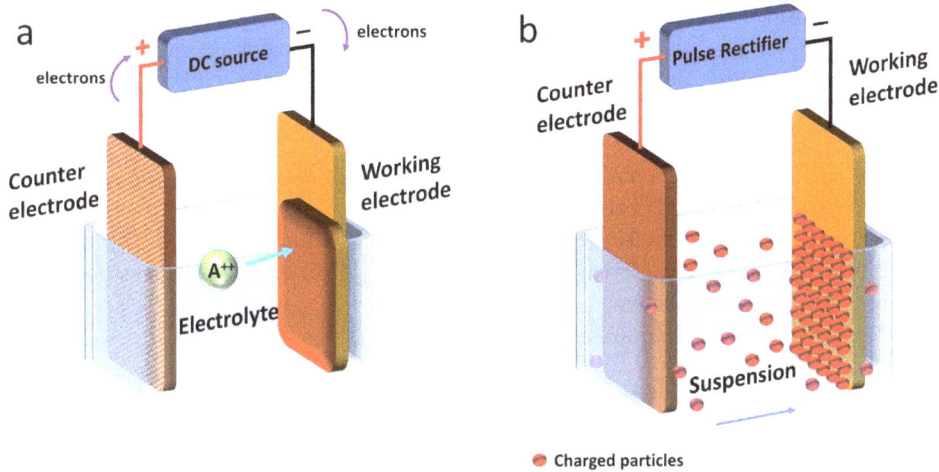

FIGURE 23.4   A general schematics diagram of (a) electrodeposition and (b) electrophoretic deposition.

desired, and the counter-electrode (anode) is required to complete the electrical circuit. In order to facilitate electrodeposition, positive and negative ions are typically dissolved in water to form electrolytes. In the presence of an external voltage, electric current flows between two conducting electrodes, which happens by the migration and diffusion of charged species towards the surfaces of the polarized electrodes. The electrochemical redox reactions by the ions at the electrode surfaces lead to the transition from ionic to electronic conduction (H. X. Wang et al. 2010). Electrodeposition is done for a variety of aesthetic and functional reasons, including increased durability, reducing friction, increasing thermal insulation and enhancing the electrical and corrosion resistance.

23.2.2.5   Electrophoretic Deposition

Electrophoretic deposition (EPD) is a colloidal-based technique in which charged particles in the colloid migrate to the oppositely charged electrode and get deposited on the counter-electrode under the influence of applied electric field (Avcu et al. 2019). EPD has been initially used in ceramic industries and also has been used to produce functional material coatings such as biocompatible, magnetic and superconductive coatings.

An EPD cell typically consists of a stable colloidal suspension, a power supply, and two electrodes (the anode and the cathode), as shown in Figure 23.4b. The process starts with electrophoresis followed by deposition of the coating on the electrode (sample). The initial step in electrophoresis is to create a solution that is electrostatically or electrosterically stable. The charged particle in the stable solution then moves towards the electrode with the opposite charge once an electric field is applied. This charged particle then deposits over the electrode in a well-ordered manner and creates a solid layer on top of it. The EPD processes are determined by two groups of parameters: (i) the suspension parameters and (ii) the process parameters. The properties of a suspension include particle size, zeta potential, liquid conductivity and dielectric constant. Though it is not possible to specify a particle size suitable for EPD, depositions with homogeneous and good adherence are reported to occur in the range of 1 to 20 μm. Zarbov, Schuster, and Gal-Or (2004) found that while deposition rate is directly reliant on the charging additive's zeta potential, it also affects the suspension's ionic conductivity. The process parameters include deposition time and applied potential. As time and applied potential grow, the thickness and yield of the coating increase until they reach a maximum beyond which there is no significant influence on the coating.

There are two varieties of this deposition: cathodic deposition, which deposits positively charged particles over a negative electrode, and anodic deposition, which deposits negatively charged particles over a positive electrode. EPD can also be classified into three types based on the electric field used: direct current, pulsed direct current and alternating current. In many ways, the EPD is superior to other coating processes, and the major advantages are listed here (Boccaccini and Dickerson 2013):

1 EPD is performed at room temperature and can yield dense coatings with better adhesion compared to conventional coating techniques.
2 The process is fast, and it takes less time to deposit the coating using the EPD technique.
3 EPD can coat the substrate evenly, and co-deposition of different particles is possible by tuning the process parameters.
4 EPD can evenly coat on substrates with intricate geometries.

23.2.2.6 Physical Vapour Deposition

Physical vapour deposition (PVD) coating involves vapourizing the material to be coated and depositing the same on the substrate. This vapour, produced at high temperature, is made to flow from the source region to the substrate usually kept at a low-pressure region. Condensation of the vapour on the substrate results in the development of a thin coating. The thickness of the film produced by the PVD process is typically in the nanometer to sub-micrometer range. They are also used to create graded composition deposits, thick deposits, multilayer covering, and freestanding structures (Wang et al. 2021). The advantages of PVD are that (1) inorganic and organic materials can be employed, (2) this process is more environmentally friendly than electroplating, and (3) PVD can improve the features of substrate materials. However, PVD has a few limitations as well, including (i) process complexity, (ii) high process cost and low output and (iii) difficulty in coating intricate and complex structures.

23.2.2.7 Chemical Vapour Deposition

The chemical vapour deposition (CVD) process is used to develop layers with a defined morphology having thickness ranging from angstroms to fractions of a millimeter. The CVD process can deposit solid phases on surfaces by reacting the surfaces with volatile precursors. The composition of the coated layer is precursor dependent, and stoichiometry can be attained in these approaches by altering precursor concentration and deposition rate. The general setup includes a needle valve that allows a constant flow of gas into the dynamically pumped vacuum chamber. High-velocity pumping maintains a continuous low chamber pressure (1.50 mbar), and the precursor is thermally degraded in the heated tubular reactor. During the brief stay in the reactor, precursor molecules begin to disintegrate and combine, creating particles or microscopic clusters. The rapid expansion of the cluster or particle beam at the reactor's exit reduces particle growth and agglomeration (Sun et al. 2021).

23.3 BIOCOMPATIBLE COATINGS FOR BIODEGRADABLE MAGNESIUM ALLOYS

23.3.1 Calcium Phosphates

Calcium phosphates (Ca-P) are inorganic bioceramics that exhibit intrinsic bioactivity and biocompatibility. Since they are the main components in bone tissue, Ca-P based bioceramics are widely investigated for orthoapedic application. Various forms of Ca-P have been used as a bioactive coating over Mg for orthopedic applications, including hydroxyapatite (HA), tricalcium phosphates (TCP) and biphasic calcium phosphate (BCP) (Eliaz and Metoki 2017). The Ca-P coatings are known for accelerating bone regeneration and enhancing the bone healing process (Ooms et al. 2003; Ambard and Mueninghoff 2006). Ca-P bioceramic materials can act as the nucleation sites of the bone apatite formation during bone remodeling. In addition, it is known that Ca-P promotes osteoblast adhesion and proliferation. The surface modification of Mg alloy with biocompatible

Ca-P coating improves the bone–implant interfacial properties and controls the degradation rate of Mg too. There have been many studies on the feasibility of Ca-P-coated Mg alloys as orthopedic biodegradable implants (Verron et al. 2010). Wang et al. (2011) reported that Ca-P-based coating serves as a bifunctional layer at the interface to improve biocompatibility and degradation resistance of Mg-based implants. Methods being used to coat different types of Ca-P on Mg-based materials includes sol-gel dip coating, electrodeposition, electrophoretic deposition etc. Figure 23.5 shows the surface morphologies of Ca-P coating on Mg alloys.

HA is among the most promising Ca-P phases which has been largely explored as a coating material on Mg owing to its biochemical and physiological bone-mimicking structure, which enhances bone regeneration (Chen et al. 2021). HA is the inorganic content of human bone, with a Ca-P ratio 1.67 ± 0.2 and has oseointegration property for bone healing (Dorozhkin 2014). The advantages of HA-coated metallic implants as biomaterials include superior osteoconductivity and bioactivity, as well as expedited integration with host bone tissues. Moreover, the HA coating increases the wettability of the surface and improves the cell adhesion.

Nanostructured HA coatings on Mg alloys can be fabricated using a variety of techniques, including the electrophoretic deposition, sol gel-based dip coating processes, electrodeposition and anodization [11,12]. Electrochemical deposition is widely recognized as a simple approach for producing metallic thin films without the need for costly and complicated equipment. Furthermore, the deposition technique can be done at room temperature, and the topology can be easily adjusted by adjusting the applied potential and electrolyte concentration (Song, Shan, and Han 2008). HA coatings have the potential to be a degradation-resistant coating for bioresorbable and biodegradable Mg alloys due to their biocompatibility and bioactivity. It can thus offer tunable corrosion characteristics to protect the substrate beneath from degradation, especially at the early stage of implantation (Dunne et al. 2015). Mousa et al. (2015) produced nano-structured HA coating on AZ31B, which made the coated sample more degradation resistant, bioactive and biocompatible than the uncoated material. Manoj et al. (2016) performed *in vitro* studies on HA-coated Mg-3Zn alloy developed using EPD and reported that the coating can improve the degradation resistance by almost 25 times that of uncoated Mg alloy. The coating also improved the osteogenic cell adhesion and viability of Mg surface. Similarly, the coating of HA on Mg has increased its *in vitro* and *in vivo* degradation characteristics as well as bioactivity of Mg. HA has a high osteoblast affinity but a low reabsorption rate *in vivo* and a brittle nature, especially when in porous form.

The other Ca-Ps such as bi-phasic calcium phosphate (BCP) and tri-calcium phosphate (TCP) were also used to enhance the degradation resistance and osteogenic performance of Mg alloys. Microwave-assisted BCP coating on AZ31 alloy enhanced its resistance to degradation. After coating, the corrosion current density of the alloy decreased, and an *in vitro* immersion test demonstrated that the degradation rate was fairly low (0.5 mm/year) in comparison to the bare AZ31 alloy (S. Jiang et al. 2019). Song, Shan, and Han (2008) effectively coated AZ91D alloy with a biphasic mixture of dibasic calcium phosphate dihydrate (DCPD) and β-TCP using the cathodic deposition technique. The coating enhanced degradation resistance by tenfold lowering of the corrosion current density. Recent research on the *in vitro* evaluation of TCP coating on Mg-Ca alloy demonstrates that pulsed laser deposition of TCP can enhance the alloy's resistance to degradation and cytocompatibility (Antoniac et al. 2020). Chai et al. (2012) coated β-TCP on AZ31 to enhance the bioactivity, and *in vitro* studies revealed that the surface was more cytocompatible. *In vivo* studies indicated that the β-TCP coating greatly enhanced osteoconductivity and osteogenesis during the first 12 weeks after implantation surgery. In addition, *in vivo* experiments indicated that the β-TCP coating layer might decrease the degradation of AZ31 alloy during the implantation phase. Despite the fact that almost all Ca-P has high osteoconductive properties, its low resorption rate *in vivo* and brittleness are among the major concerns. These challenges could be solved by using hybrid or composite coatings that combine Ca-Ps with biopolymeric materials.

FIGURE 23.5 Surface and cross section morphologies of calcium phosphate coated Mg surface. (a) Ca-P-coated AZ31B Mg alloy (i) surface, (ii) surface at higher magnification, (iii) cross-section, (iv) EDS analysis on surface. *Reproduced with permission from (Wang et al. 2011) © 2016, Elsevier.* (b) Sol-gel–derived nano-HA coating on AZ91 substrate and cross-sectional analysis of a nano-HA-coated sample. *Reproduced with permission from (Rojaee, Fathi, and Raeissi 2013) © 2016, Elsevier.*

23.3.2 Bioactive Glasses

Bioactive glasses (BG) are a group of silica-based glasses that contain calcium and phosphate ions which exhibit good bioactivity. The very first bioactive glass is a quaternary system of SiO_2-Na_2O-CaO-P_2O_5 that was reported in 1969 by Hench (2006) and trademarked in the name 45S5 Bioglass®. Fascinated by their unique biological responses, several phosphate silicate and borate based bioactive glasses have been explored later by research groups for biomedical applications. Bioactive glasses are good biodegradable materials which, when contacted with body fluid, undergo physiochemical reactions leading to dissolution and release of several ions depending upon the chemical composition of the bioactive glass. The calcium and phosphate ions move to the surface reacting with body fluid to create a layer of biologically active HA on the surface, which helps to develop a stable interface with neighboring tissues. Bioactive glass–coated implants have been reported to impart better bioactivity, enhanced osseointegration and corrosion resistance (Montazerian et al. 2022). Akshay et al. (2022) deposited alginate/bioglass composite through EPD. The coating has doubled the degradation resistance and enhanced the bioactivity of Mg-Ca alloy. Bioactive glasses can also be doped with desired ions, which, when released upon degradation, can enhance the biological activity. The flexibility in property enhancement by dopant addition like Sr^{2+} ions for enhanced bone growth, Cu^{2+} for angiogenesis and Ag^+ for better antimicrobial properties makes them versatile (Nilawar, Uddin, and Chatterjee 2021).

There are several methods for coating bioactive glasses on Mg alloys. However, appropriate pre-treatment of the metal is important to achieve the desired performance. It is seen that the chemical and mechanical pre-treatment methods for cleaning and tuning the surface roughness can enhance the coating adhesion. The widely reported coating methods for bioactive glasses include enameling, sol-gel technique, laser cladding, electrophoretic deposition and thermal spraying (Sergi, Bellucci, and Cannillo 2020; Oliver et al. 2019).

Ye et al. (2012) coated bioactive 45S5 glass on AZ31 Mg alloy using a sol-gel dip-coating process. The coating was homogenous and crack free with a thickness of almost 0.1 μm. The coating protected the substrate from corrosion during 7 days of immersion in SBF. The potentiodynamic polarization (PDP) test measurements revealed a corrosion potential shift of 0.2 V in favor of the coated substrate compared to the uncoated substrate. Zhang et al. (2016) deposited a mesoporous 45S5 bioactive glass ceramic coating on AZ31 alloy using the sol-gel dip-coating process. The coating was corrosion-resistant because it remained intact and promoted apatite formation. However, at greater compressive stresses (> 25 MPa), crack formation failed the coatings, and apatite formation was suppressed as a result of large cracks and the release of additional Mg ions. EPD of bioactive glass nanopowders onto micro-arc oxidation–treated AZ31 Mg alloy formed a corrosion-protective layer with excellent bioactivity (Rojaee, Fathi, and Raeissi 2013). Mahato et al. (2021) investigated the biodegradation and biological activity of HA and bioactive glass–coated Mg-Zn-Ca alloys. The bioactive glass–coated Mg alloys recorded the lowest corrosion current and potential values during electrochemical studies. *In vivo* implantation tests conducted in animal models found that the bioactive glass coating had no negative effects on organs or the immune system. Moreover, the coating caused rapid new bone formation due to the osteoproliferative impact of calcium phosphate in the bioactive glass coating.

23.3.3 Polymer Coatings

Biodegradable polymers are among the preferred materials for a number of key biomedical applications including antimicrobial coatings, suturing, drug delivery systems, device fixation, tissue replacement etc. Furthermore, polymeric coatings are generally accepted for regulating, directing and steering the cellular responses. The polymeric coatings have come to the forefront in recent years because of their diverse bio-functionalities and biodegradability. Mg alloys with biodegradable

polymeric coatings are very promising due to their capability to degrade in physiological environment combined with characteristics such as biocompatibility and osteoinduction (Rezwan et al.
2006; Blacklock et al. 2010). Diverse coating techniques, such as sol-gel, electrospinning, spin coating etc., are being investigated to coat polymers on Mg to tailor the degradation and improve the
bioactivity (L. Y. Li et al. 2018).

Polylactic acid (PLA) is a hydrophobic aliphatic polyester, and it has excellent biodegradability,
biocompatibility, and thermoplastic processability. PLA medical devices include surgical suture
needles, injectable capsules, microspheres, implants, tissue engineering, orthopedic fixation, surgical suture and 3D printing scaffold (Farah, Anderson, and Langer 2016; Zhou, Lawrence, and
Bhaduri 2012). Huang et al. (2007) investigated the use of a PLA coating on a Mg surface using a
dip coating technique. When fixed dipping duration and polymer concentration were employed, the
adhesion strength of the PLA coating to the substrate was improved. Furthermore, it is possible to
adjust the thickness of the coating by varying the number of coating cycles, and modifying the surface of Mg with a silane coupling agent improves its corrosion resistance. Alabbasi, Liyanaarachchi,
and Kannan (2012) produced a dip-coated PLA coating on AZ91 Mg alloy to improve the degradation resistance of the alloy. In addition, the PLA coating dramatically enhanced cytocompatibility.
It is reported that the cellular morphologies and cell adhesion behaviors varied across different PLA
surfaces, with cells being more elongated on smooth surfaces and more flattened and dispersed with
longer pseudopods on rough and porous surfaces.

Poly (lactide-co-glycolic) acid (PLGA) is a copolymer of lactic acid (LA) and glycolic acid (GA)
with tunable properties by adjusting the ratio between the two comonomers. PLGA coating on Mg
alloys exhibited potential to tailor the degradation rate and aid in drug delivery (Brown et al. 2015;
Danhier et al. 2012). Ostrowski et al. (2012) investigated the degradation and biocompatibility features of PLGA dip-coated AZ31 and Mg4Y Mg alloys. The *in vitro* studies conducted using mouse
osteoblast cells showed that the PLGA-coated alloys were more biocompatible than the uncoated

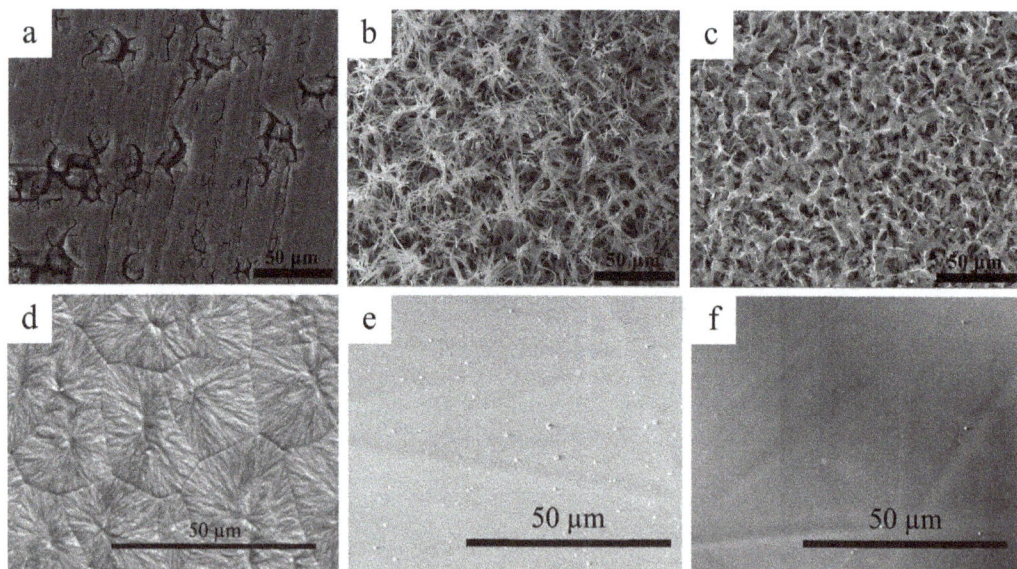

FIGURE 23.6   SEM images of (a) PDA coated, (b) PDA/DCPD coated, (c) PDA/DCPD/Col coated.
Reproduced with permission from (Guo et al. 2020) © 2020, Elsevier. (d) PCL coated, (e) PLLA coated, (f)
PLGA, (50:50) coated. *Reproduced with permission from (W. Jiang et al. 2017) © 2017, American Chemical
Society.* (DCPD – Dicalcium phosphate dihydrate, Col – Collagen).

versions. Figure 23.6 shows the morphologies of different polymer coating on Mg alloys. The use of PLA and PLGA materials as coatings on Mg substrate for orthopedic implant applications is limited by a few constraints, which mainly includes high susceptibility to hydrolysis, entrapment of H_2 gas and poor adhesion to Mg substrate (Tomihata, Suzuki, and Ikada 2001).

Polycaprolactone (PCL) is a slow-degrading polymer that can be easily coated with various coating techniques discussed in Section 2.2. PCL can serve as a good permeation barrier due to its hydrophobicity and slower degradation in biological fluid than Mg (Wong et al. 2013; Ying Chen et al. 2011). PCL-based nanofiber scaffold made through electrospinning is extensively used in tissue engineering applications. Hanas et al. (2016) evaluated the degradation and bioactivity of Electrospun-coated PCL in AZ31 samples. Figure 23.7 shows the surface morphology of the samples after immersion studies. The study showed that the coating reduces the degradation rate of the alloy and also improves cell viability, adhesion and proliferation.

Other polymers such as polydopamine (PDA) and collagen along with some biocompatible ceramic phases are also explored as coating materials on Mg substrate to tailor the degradation and biological performance (L. Y. Li et al. 2018; Singh et al. 2023; Y. Guo et al. 2020). Even though these polymer coatings on Mg alloys are not extensively studied yet, the research on corrosion resistance, drug transport and other capabilities of these systems could pave the way for novel materials for developing degradable metallic implants.

23.3.4 PROTEIN COATINGS

Biocompatibility of materials is clearly influenced by the adsorption of proteins from biological fluids onto the surface, which facilitates processes such as thrombosis, foreign body reaction and cell attachment. The influence of such adsorbed proteins on the bioactivity and biodegradation can vary depending on the surface of the implant material as well as the type and size of the protein being adsorbed (Talha et al. 2019). There are reports on both positive and negative effect of different proteins on the degradation of Mg alloys. Protein adsorption on Mg^{2+} accelerates the dissolving of the $Mg(OH)_2$ layer on the surface and inhibits the precipitation of other corrosion products such as phosphates and carbonates, resulting in accelerated corrosion. In contrast, the adsorption of protein in the corrosion products is responsible for the lower corrosion rates of Mg alloys (Xu et al. 2022). Thus, coating implant surfaces with suitable proteins is a recent approach from researchers to protect Mg implants from degradation and increase their biological interactions.

The protein coating techniques reported so far are solution-based techniques which use spin or dip coating in combination with surface preparation. Wang et al. (2019) directly coated Mg-Zn-Ca alloy with silk fibroin (SF), a natural protein extracted from *Bombyx mori* silkworms that is completely biodegradable and biocompatible. The Mg alloy was prepared with hybrid plasma activation, which improved SF coat adherence. Cell adhesion studies revealed that the surfaces covered with the silk fibroin film were superior for cell adhesion and proliferation due to increased osteogenic activity. The excellent coating adhesion and the enhanced protection by the silk fibroin physical layer reduced degradation *in vivo* to 1/18 that of the substrate without coating. Spin coating of SF onto micro-roughened pure Mg substrate exhibited a higher percentage of cytocompatibility (96.5%) due to the osteogenic behaviour of SF and a lower degradation rate due to the creation of a barrier-type film and strong interfacial adhesion (Rahman, Dutta, and Choudhury 2021). The use of micro-arc coating to deposit varying amounts of bone morphogenetic protein-BMP-2 in the carrier MgO and $Mg(OH)_2$ layers of the AZ31B alloy promotes osteoblast cell proliferation and differentiation (Kim et al. 2019). Polydopamine-mediated HA coating on AZ31 alloy by alkaline-treated nanoparticle entrapment with bone morphogenetic protein-2 promotes cell adhesion and proliferation. It also stimulated the formation of new bone during implant testing in New Zealand rabbits without causing an inflammatory reaction (Y. Jiang et al. 2017).

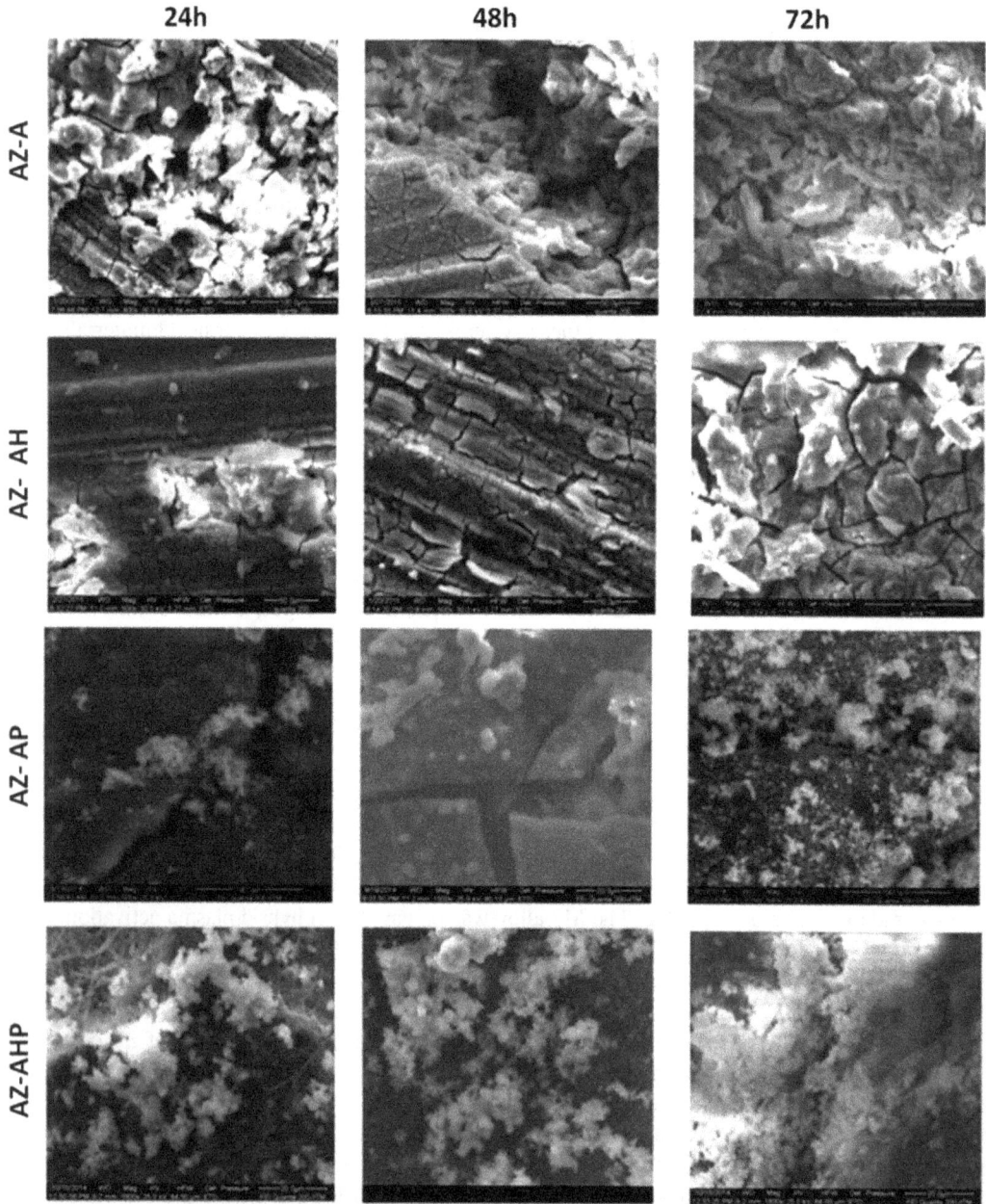

FIGURE 23.7   SEM images of annealed, acid-treated and PCL-coated samples after different immersion period. AZ-A, AZ-AH, AZ-AP and AZ-AHP, where AZ stands for the substrate AZ31 and letters A stands for annealing, H for HNO$_3$ treatment and P for PCL coating. *Reproduced with permission from (Hanas et al. 2016) © 2016, Elsevier.*

23.3.5 DNA Coatings

Deoxyribonucleic acid (DNA), the storehouse of uniquely specific genetic information, is also a biocompatible and biodegradable material. DNA has attracted the curiosity of researchers for use as a biomaterial not because of its ability to transport genetic information but because of its unique structural features. They are capable to include additional molecules by intercalation and groove binding. In addition, the low immunogenicity makes it less likely to elicit immune responses once implanted (Beucken et al. 2006). Furthermore, due to its high phosphate content, DNA is excellent for use as an implant coating material, as it can stimulate Ca-P nucleation and thus promote bone forming process (Beucken et al. 2007).

Since DNA can be easily degraded by nucleases and is soluble in water, it poses real challenges when used as a coating. Researchers have successfully applied sticky DNA coatings to Mg-based alloys by electrostatic self-assembly (ESA), also known as layer-by-layer assembly. Cui et al. (2020) applied a layer-by-layer assembly by the dip-coating process on AZ31 Mg alloy and reported that the self-corrosion current density was lowered by an order of magnitude. The immersion test in SBF revealed that the uptake of corrosion medium by polyelectrolytes seals any possible liquid entrance sites and prevents the corrosion medium from contacting the alloy. Phosphate groups in DNA molecules are capable of inducing heterogeneous Ca-P nucleation and promote the deposition of Ca-P on surfaces. Similarly, coating with $(CHI/DNA)_5/Mg(OH)_2$ on Mg alloy improved the corrosion resistance and biomineralization due to the combined effect of components in the coating structure (Cui et al. 2021). Due to its excellent compactness and small grains, the DNA-induced Ca-P coating on AZ31 alloy by hydrothermal technique also provided greater corrosion resistance than the DNA-free Ca-P coating (Liu et al. 2019).

23.4 CONCLUSIONS AND OUTLOOK

In this chapter, various biomedical-grade coating methods and materials used on biodegradable Mg alloys emphasizing their degradation protection behaviors, bio-functionality and biocompatibility were briefly discussed. The improvement in the adhesion, biodegradation and biological performance with the use of conversion and deposit coatings of various bioactive, biocompatible biomedical-grade coating methods and materials is a promising approach toward degradable metallic implant development. Existing research has mostly focused on just one or two features, whereas many more bifunctionalities have yet to be fully studied. The key findings of various biomedical-grade coatings on Mg-based systems are summarized in Table. 23.1.

A major emphasis of recent scientific study has been the development of new bio-functional coatings that mimic the important function of synchronized degradation and bone regeneration rate. Recent studies on coating show considerable improvements in cytocompatibility, hemocompatibility, anti-inflammatory and anti-microbial functionality mainly attributed to the enhanced degradation resistance and bioactivity. Nevertheless, some challenges remain unresolved and need to be addressed thoroughly in the future. The combination of controlled degradation and multifunctional composite coatings is a viable approach. Coating adhesion and bio-functional characteristics can be improved by focusing research on hybrid composite coatings made of materials such as Ca-P, HA, bioglass and biopolymers. The adhesion strength and bioactivity of coating can be improved by combining two or more coating processes, such as sol-gel, micro-arc oxidation, electrospinning, chemical transformation, hydrothermal synthesis and many more. In addition, the clinical application of degradable Mg alloys requires additional in-depth, fundamental study to elucidate the effect on metabolic process due to hybrid coated Mg alloys and the osteogenic responses in the *in vivo* condition.

TABLE 23.1

Key Findings of Various Biomedical-Grade Coatings on Mg-Based Systems

Coating Method	Coating Material	Substrate	Condition	Corrosion potential (V/SCE)		Corrosion current (A/cm²)		Bio-functionality	Reference
				Substrate	Coated	Substrate	Coated		
Dip	MAO/Chitosan	Mg-Zn-Ca	SBF, 37°C	-1.77	-1.49	1.43×10^{-4}	5.9×10^{-7}	–	Jo et al. (2012)
Dip	MAO/Chitosan	Mg-Zn-Ca	SBF, 37°C	-1.77	-1.49	1.43×10^{-4}	5.9×10^{-7}	–	Bai et al. (2012)
Spin coating	Silk/Phytic acid/Silk	Mg-1Ca	Hank's, 37°C	-1.48	-1.43	1.53×10^{-6}	3.85×10^{-8}	pH triggered self-healing ability with enhanced cytocompatibility towards MC3T3 cell line	Kannan and Liyanaarachchi (2013)
Dip	Ti-O/PLA	AM50	SBF, 37.5°C	-1.4	-0.83	–	–	Enhanced osteoblast for MC3T3 cell line	Abdal-hay et al. (2014)
Electro sprayed	PLGA	AMlite	0.1M NaCl	-1.597	-1.529	4×10^{-6}	3×10^{-6}	–	Chen et al. (2014)
Dip	PDA/HA	AZ31	SBF, 37°C	-1.48	-1.27	4.67×10^{-5}	1.82×10^{-6}	Promoted cell growth observed for L929 cell line	Lin et al. (2015)
Dip	PDA/TiO$_2$	Pure Mg	PBS, 37°C	-1.58	-1.28	4.37×10^{-5}	2.04×10^{-6}	–	Chen et al. (2015)
Dip	NiCrAlY/Yttria stabilized zirconia/PCL	Mg-1.2Ca	3.5% NaCl	-1.63	-0.91	2.05×10^{-4}	1.4×10^{-7}	–	Bakhsheshi-Rad et al. (2016)
Electrospin	PCL	AZ31	SBF, 37°C	–	–	–	–	Enhanced Cell viability, adhesion and proliferation for L6 cells	Hanas et al. (2016)
Immersion	PDA/HA/BMP2	AZ31	SBF	-1.62	-1.5	1.72×10^{-4}	1.26×10^{-4}	New bone formation and enhanced osteoid tissue formation	Y. Jiang et al. (2017)
EPD	Zein/Bioactive glass	Pure Mg	DMEM, RT	–	–	15.3×10^{-6}	9.7×10^{-6}	–	Ramos Rivera, Dippel, and Boccaccini (2018)

Method	Coating	Alloy	Solution					Description	Reference
Dip	MAO/PLGA	Mg-4Zn-0.6Zr-0.4Sr	m-SBF, 37°C	-1.66	-1.54	1.95×10^{-1}	1.39×10^{-4}	—	Chen et al. (2019)
Spray	PCL/HA	Pure Mg	Hank's Solution	-1.64	-1.46	$2.93 \times 10-5$	1.13×10^{-6}	Improver filopodia and osteoblast for MC3T3-E1 cell line	Xiong, Jia, et al. (2019)
Direct coating	Silk fibroin	MgZnCa	Hank's,37°C	-1.58	-0.26	12.3×10^{-6}	1.14×10^{-6}	Improved biocompatibility, bioactivity	Wang et al. (2019)
EPD	Chitosan/ Bioactive glass	AZ91	SBF, 37°C	-0.147	-0.143	4×10^{-5}	2×10^{-6}	Superior apatite formation ability	Alaei, Atapour, and Labbaf (2020)
Dip	Chitosan/DNA	AZ31	SBF, 37°C	-1.74	-1.57	7.6×10^{-5}	1.54×10^{-7}	Actively promoting the growth of born	Cui et al. (2021)
Dip	PVP/DNA	AZ31	SBF, 37°C	-1.74	-1.47	8.43×10^{-5}	1.29×10^{-5}	Increased deposition of Ca-P precipitation	Cui et al. (2021)
EPD	Alginate/ Bioactive glass	Mg-Ca	8.035 g/l NaCl	-1.3352	-1.5324	47.40×10^{-6}	23.10×10^{-6}	Doubled the degradation resistance and improved bioactivity	Akshay et al. (2022)

REFERENCES

Abdal-hay, Abdalla, Montasser Dewidar, Juhyun Lim, and Jae Kyoo Lim. 2014. "Enhanced Biocorrosion Resistance of Surface Modified Magnesium Alloys Using Inorganic/Organic Composite Layer for Biomedical Applications." *Ceramics International* (January) (1): 2237–47. https://doi.org/10.1016/j.ceramint.2013.07.142.

Akshay, K.S., V.P. Muhammad Rabeeh, Shebeer A. Rahim, K.P. Sijina, G.K. Rajanikant, and T. Hanas. 2022. "Electrophoretic Deposition of Alginate/Bioglass Composite Coating on MgCa Alloy for Degradable Metallic Implant Applications." *Surface and Coatings Technology* 448 (October): 128914. https://doi.org/10.1016/J.SURFCOAT.2022.128914.

Alabbasi, Alyaa, S. Liyanaarachchi, and M. Bobby Kannan. 2012. "Polylactic Acid Coating on a Biodegradable Magnesium Alloy: An in Vitro Degradation Study by Electrochemical Impedance Spectroscopy." *Thin Solid Films* 520 (23): 6841–4. https://doi.org/10.1016/J.TSF.2012.07.090.

Alaei, Masoumeh, Masoud Atapour, and Sheyda Labbaf. 2020. "Electrophoretic Deposition of Chitosan-Bioactive Glass Nanocomposite Coatings on AZ91 Mg Alloy for Biomedical Applications." *Progress in Organic Coatings* (October): 105803. https://doi.org/10.1016/j.porgcoat.2020.105803.

Ambard, Alberto J., and Leonard Mueninghoff. 2006. "Calcium Phosphate Cement: Review of Mechanical and Biological Properties." *Journal of Prosthodontics* 15 (5): 321–8. https://doi.org/10.1111/J.1532-849X.2006.00129.X.

Angammana, Chitral J., and Shesha H. Jayaram. 2016. "Fundamentals of Electrospinning and Processing Technologies." 34 (1): 72–82. https://doi.org/10.1080/02726351.2015.1043678.

Antoniac, Iulian V., Mihaela Filipescu, Katia Barbaro, Anca Bonciu, Ruxandra Birjega, Cosmin M. Cotrut, Ettore Galvano, et al. 2020. "Iron Ion-Doped Tricalcium Phosphate Coatings Improve the Properties of Biodegradable Magnesium Alloys for Biomedical Implant Application." *Advanced Materials Interfaces* 7 (16): 2000531. https://doi.org/10.1002/ADMI.202000531.

Avcu, Egemen, Fatih E. Baştan, Hasan Z. Abdullah, Muhammad Atiq Ur Rehman, Yasemin Yıldıran Avcu, and Aldo R. Boccaccini. 2019. "Electrophoretic Deposition of Chitosan-Based Composite Coatings for Biomedical Applications: A Review." *Progress in Materials Science* 103 (June): 69–108. https://doi.org/10.1016/J.PMATSCI.2019.01.001.

Bai, Kuifeng, Yi Zhang, Zhenya Fu, Caili Zhang, Xinzhan Cui, Erchao Meng, Shaokang Guan, and Junhua Hu. 2012. "Fabrication of Chitosan/Magnesium Phosphate Composite Coating and the in vitro Degradation Properties of Coated Magnesium Alloy." *Materials Letters* (April): 59–61. https://doi.org/10.1016/j.matlet.2011.12.102.

Bakhsheshi-Rad, H.R., E. Hamzah, A.F. Ismail, M. Daroonparvar, M.A.M. Yajid, and M. Medraj. 2016. "Preparation and Characterization of NiCrAlY/Nano-YSZ/PCL Composite Coatings Obtained by Combination of Atmospheric Plasma Spraying and Dip Coating on Mg–Ca Alloy." *Journal of Alloys and Compounds* (February): 440–52. https://doi.org/10.1016/j.jallcom.2015.10.196.

Beucken, J.J.J.P. van den, X.F. Walboomers, S.C.G. Leeuwenburgh, M.R.J. Vos, N.A.J.M. Sommerdijk, R.J.M. Nolte, and J.A. Jansen. 2007. "Multilayered DNA Coatings: In Vitro Bioactivity Studies and Effects on Osteoblast-like Cell Behavior." *Acta Biomaterialia* 3 (4): 587–96. https://doi.org/10.1016/J.ACTBIO.2006.12.007.

Beucken, J.J.J.P. van den, Matthijn R.J. Vos, Peter C. Thüne, Tohru Hayakawa, Tadao Fukushima, Yoshio Okahata, X. Frank Walboomers, Nico A.J.M. Sommerdijk, Roeland J.M. Nolte, and John A. Jansen. 2006. "Fabrication, Characterization, and Biological Assessment of Multilayered DNA-Coatings for Biomaterial Purposes." *Biomaterials* 27 (5): 691–701. https://doi.org/10.1016/J.BIOMATERIALS.2005.06.015.

Birnie, D.P. 2004. "Spin Coating Technique." *Sol-Gel Technologies for Glass Producers and Users*: 49–55. https://doi.org/10.1007/978-0-387-88953-5_4.

Blacklock, Jenifer, Torsten K. Sievers, Hitesh Handa, Ye Zi You, David Oupický, Guangzhao Mao, and Helmuth Möhwald. 2010. "Cross-Linked Bioreducible Layer-by-Layer Films for Increased Cell Adhesion and Transgene Expression." *Journal of Physical Chemistry B* 114 (16): 5283–91. https://doi.org/10.1021/JP100486H/SUPPL_FILE/JP100486H_SI_001.PDF.

Boccaccini, Aldo R., and James H. Dickerson. 2013. "Electrophoretic Deposition: Fundamentals and Applications." *Journal of Physical Chemistry B* 117 (6): 1501. https://doi.org/10.1021/JP211212Y.

Brinker, C.J., G.C. Frye, A.J. Hurd, and C.S. Ashley. 1991. "Fundamentals of Sol-Gel Dip Coating." *Thin Solid Films* 201 (1): 97–108. https://doi.org/10.1016/0040-6090(91)90158-T.

Brown, Andrew, Samer Zaky, Herbert Ray, and Charles Sfeir. 2015. "Porous Magnesium/PLGA Composite Scaffolds for Enhanced Bone Regeneration Following Tooth Extraction." *Acta Biomaterialia* 11 (C): 543–53. https://doi.org/10.1016/J.ACTBIO.2014.09.008.

Chagnon, Madeleine, Louis Georges Guy, and Nicolette Jackson. 2019. "Evaluation of Magnesium-Based Medical Devices in Preclinical Studies: Challenges and Points to Consider." *Toxicologic Pathology* 47 (3): 390–400. https://doi.org/10.1177/0192623318816936.

Chai, Hongwei, Lei Guo, Xiantao Wang, Xiaoyu Gao, Kui Liu, Yuping Fu, Junlin Guan, Lili Tan, and Ke Yang. 2012. "In Vitro and in Vivo Evaluations on Osteogenesis and Biodegradability of a β-Tricalcium Phosphate Coated Magnesium Alloy." *Journal of Biomedical Materials Research Part A* 100A (2): 293–304. https://doi.org/10.1002/JBM.A.33267.

Chen, Jing Yu, Xiao Bo Chen, Jing Liang Li, Bin Tang, Nick Birbilis, and Xungai Wang. 2014. "Electrosprayed PLGA Smart Containers for Active Anti-Corrosion Coating on Magnesium Alloy AMlite." *Journal of Materials Chemistry A* 2 (16): 5738–43. https://doi.org/10.1039/C3TA14999D.

Chen, Junxiu, Yang Yang, Iniobong P. Etim, Lili Tan, Ke Yang, R.D.K. Misra, Jianhua Wang, and Xuping Su. 2021. "Recent Advances on Development of Hydroxyapatite Coating on Biodegradable Magnesium Alloys: A Review." *Materials* 14 (19): 5550. https://doi.org/10.3390/MA14195550.

Chen, Lianxi, Yinying Sheng, Hanyu Zhou, Zhibin Li, Xiaojian Wang, and Wei Li. 2019. "Influence of a MAO+PLGA Coating on Biocorrosion and Stress Corrosion Cracking Behavior of a Magnesium Alloy in a Physiological Environment." *Corrosion Science* 148 (March): 134–43. https://doi.org/10.1016/J.CORSCI.2018.12.005.

Chen, Ying, Yang Song, Shaoxiang Zhang, Jianan Li, Changli Zhao, and Xiaonong Zhang. 2011. "Interaction Between a High Purity Magnesium Surface and PCL and PLA Coatings during Dynamic Degradation." *Biomedical Materials* 6 (2): 025005. https://doi.org/10.1088/1748-6041/6/2/025005.

Chen, Yingqi, Sheng Zhao, Meiyun Chen, Wentai Zhang, Jinlong Mao, Yuancong Zhao, Manfred F. Maitz, Nan Huang, and Guojiang Wan. 2015. "Sandwiched Polydopamine (PDA) Layer for Titanium Dioxide (TiO2) Coating on Magnesium to Enhance Corrosion Protection." *Corrosion Science* 96 (July): 67–73. https://doi.org/10.1016/J.CORSCI.2015.03.020.

Cui, Lan Yue, Shen Cong Cheng, Lu Xian Liang, Jing Chao Zhang, Shuo Qi Li, Zhen Lin Wang, and Rong Chang Zeng. 2020. "In Vitro Corrosion Resistance of Layer-by-Layer Assembled Polyacrylic Acid Multilayers Induced Ca – P Coating on Magnesium Alloy AZ31." *Bioactive Materials* 5 (1): 153–63. https://doi.org/10.1016/J.BIOACTMAT.2020.02.001.

Cui, Lan Yue, Ling Gao, Jing Chao Zhang, Zhe Tang, Xiao Li Fan, Jia Cheng Liu, Dong Chu Chen, Rong Chang Zeng, Shuo Qi Li, and Ke Qian Zhi. 2021. "In Vitro Corrosion Resistance, Antibacterial Activity and Cytocompatibility of a Layer-by-Layer Assembled DNA Coating on Magnesium Alloy." *Journal of Magnesium and Alloys* 9 (1): 266–80. https://doi.org/10.1016/J.JMA.2020.03.009.

Danhier, Fabienne, Eduardo Ansorena, Joana M. Silva, Régis Coco, Aude Le Breton, and Véronique Préat. 2012. "PLGA-Based Nanoparticles: An Overview of Biomedical Applications." *Journal of Controlled Release* 161 (2): 505–22. https://doi.org/10.1016/J.JCONREL.2012.01.043.

Dorozhkin, Sergey V. 2014. "Calcium Orthophosphate Coatings on Magnesium and Its Biodegradable Alloys." *Acta Biomaterialia* 10 (7): 2919–34. https://doi.org/10.1016/J.ACTBIO.2014.02.026.

Dunne, C.F., J. Gibbons, D.P. FitzPatrick, K.J. Mulhall, and K.T. Stanton. 2015. "On the Fate of Particles Liberated from Hydroxyapatite Coatings in Vivo." *Irish Journal of Medical Science* 184 (1): 125–33. https://doi.org/10.1007/S11845-014-1243-8/FIGURES/5.

Eliaz, Noam, and Noah Metoki. 2017. "Calcium Phosphate Bioceramics: A Review of Their History, Structure, Properties, Coating Technologies and Biomedical Applications." *Materials* 10 (4): 334. https://doi.org/10.3390/MA10040334.

Farah, Shady, Daniel G. Anderson, and Robert Langer. 2016. "Physical and Mechanical Properties of PLA, and Their Functions in Widespread Applications – A Comprehensive Review." *Advanced Drug Delivery Reviews* 107 (December): 367–92. https://doi.org/10.1016/J.ADDR.2016.06.012.

Gawlik, Marcjanna Maria, Björn Wiese, Alexander Welle, Jorge González, Valérie Desharnais, Jochen Harmuth, Thomas Ebel, and Regine Willumeit-Römer. 2019. "Acetic Acid Etching of Mg-XGd Alloys." *Metals* 9 (2): 117. https://doi.org/10.3390/MET9020117.

Gu, X.N., W. Zheng, Y. Cheng, and Y.F. Zheng. 2009. "A Study on Alkaline Heat Treated Mg – Ca Alloy for the Control of the Biocorrosion Rate." *Acta Biomaterialia* 5 (7): 2790–9. https://doi.org/10.1016/J.ACTBIO.2009.01.048.

Guo, H.F., M.Z. An, H.B. Huo, S. Xu, and L.J. Wu. 2006. "Microstructure Characteristic of Ceramic Coatings Fabricated on Magnesium Alloys by Micro-Arc Oxidation in Alkaline Silicate Solutions." *Applied Surface Science* 252 (22): 7911–16. https://doi.org/10.1016/J.APSUSC.2005.09.067.

Guo, Yunting, Siqi Jia, Lu Qiao, Yingchao Su, Rui Gu, Guangyu Li, and Jianshe Lian. 2020. "Enhanced Corrosion Resistance and Biocompatibility of Polydopamine/Dicalcium Phosphate Dihydrate/Collagen Composite Coating on Magnesium Alloy for Orthopedic Applications." *Journal of Alloys and Compounds* 817 (March): 152782. https://doi.org/10.1016/J.JALLCOM.2019.152782.

Hanas, T., T.S. Sampath Kumar, Govindaraj Perumal, and Mukesh Doble. 2016. "Tailoring Degradation of AZ31 Alloy by Surface Pre-Treatment and Electrospun PCL Fibrous Coating." *Materials Science and Engineering: C* 65 (August): 43–50. https://doi.org/10.1016/J.MSEC.2016.04.017.

Hench, Larry L. 2006. "The Story of Bioglass®." *Journal of Materials Science: Materials in Medicine* 17 (11): 967–78. https://doi.org/10.1007/S10856-006-0432-Z.

Huang, Jing Jing, Yi Bin Ren, Bing Chun Zhang, and Ke Yang. 2007. "Preparation and Property of Coating on Degradable Mg Implant." *Zhongguo Youse Jinshu Xuebao/Chinese Journal of Nonferrous Metals* 17 (9): 1465–9.

Jiang, Bai Ling, and Yan Feng Ge. 2013. "Micro-Arc Oxidation (MAO) to Improve the Corrosion Resistance of Magnesium (Mg) Alloys." *Corrosion Prevention of Magnesium Alloys: A Volume in Woodhead Publishing Series in Metals and Surface Engineering* (January): 163–96. https://doi.org/10.1533/97808 57098962.2.163.

Jiang, Song, Shu Cai, Yishu Lin, Xiaogang Bao, Rui Ling, Dongli Xie, Jiayue Sun, Jieling Wei, and Guohua Xu. 2019. "Effect of Alkali/Acid Pretreatment on the Topography and Corrosion Resistance of as-Deposited CaP Coating on Magnesium Alloys." *Journal of Alloys and Compounds* 793 (July): 202–11. https://doi.org/10.1016/J.JALLCOM.2019.04.198.

Jiang, Wensen, Qiaomu Tian, Tiffany Vuong, Matthew Shashaty, Chris Gopez, Tian Sanders, and Huinan Liu. 2017. "Comparison Study on Four Biodegradable Polymer Coatings for Controlling Magnesium Degradation and Human Endothelial Cell Adhesion and Spreading." *ACS Biomaterials Science & Engineering* 3 (6): 936–50. https://doi.org/10.1021/acsbiomaterials.7b00215.

Jiang, Yanan, Bi Wang, Zhanrong Jia, Xiong Lu, Liming Fang, Kefeng Wang, and Fuzeng Ren. 2017. "Polydopamine Mediated Assembly of Hydroxyapatite Nanoparticles and Bone Morphogenetic Protein-2 on Magnesium Alloys for Enhanced Corrosion Resistance and Bone Regeneration." *Journal of Biomedical Materials Research Part A* 105 (10): 2750–61. https://doi.org/10.1002/jbm.a.36138.

Jo, Ji Hoon, Yuanlong Li, Sae Mi Kim, Hyoun Ee Kim, and Young Hag Koh. 2012. "Hydroxyapatite/Poly(ε-Caprolactone) Double Coating on Magnesium for Enhanced Corrosion Resistance and Coating Flexibility." 28 (4): 617–25. https://doi.org/10.1177/0885328212468921.

Kannan, M. Bobby, and S. Liyanaarachchi. 2013. "Hybrid Coating on a Magnesium Alloy for Minimizing the Localized Degradation for Load-Bearing Biodegradable Mini-Implant Applications." *Materials Chemistry and Physics* 142 (1): 350–4. https://doi.org/10.1016/J.MATCHEMPHYS.2013.07.028.

Kim, Seo Young, Yu Kyoung Kim, Kyung Seon Kim, Kwang Bok Lee, and Min Ho Lee. 2019. "Enhancement of Bone Formation on LBL-Coated Mg Alloy Depending on the Different Concentration of BMP-2." *Colloids and Surfaces B: Biointerfaces* 173 (January): 437–46. https://doi.org/10.1016/J.COLSURFB.2018.09.061.

Lei, Ting, Chun Ouyang, Wei Tang, Lian Feng Li, and Le Shan Zhou. 2010. "Enhanced Corrosion Protection of MgO Coatings on Magnesium Alloy Deposited by an Anodic Electrodeposition Process." *Corrosion Science* 52 (10): 3504–8. https://doi.org/10.1016/J.CORSCI.2010.06.028.

Li, Ling Yu, Lan Yue Cui, Rong Chang Zeng, Shuo Qi Li, Xiao Bo Chen, Yufeng Zheng, and M. Bobby Kannan. 2018. "Advances in Functionalized Polymer Coatings on Biodegradable Magnesium Alloys – A Review." *Acta Biomaterialia* 79 (October): 23–36. https://doi.org/10.1016/J.ACTBIO.2018.08.030.

Li, Sheng, Laihua Yi, Tongfang Liu, Hao Deng, Bo Ji, Kun Zhang, and Lihong Zhou. 2020. "Formation of a Protective Layer Against Corrosion on Mg Alloy via Alkali Pretreatment Followed by Vanillic Acid Treatment." *Materials and Corrosion* 71 (8): 1330–8. https://doi.org/10.1002/maco.201911488.

Lin, Bingpeng, Mei Zhong, Chengdong Zheng, Lin Cao, Dengli Wang, Lina Wang, Jun Liang, and Baocheng Cao. 2015. "Preparation and Characterization of Dopamine-Induced Biomimetic Hydroxyapatite Coatings on the AZ31 Magnesium Alloy." *Surface and Coatings Technology*, November, 82–88. https://doi.org/10.1016/j.surfcoat.2015.09.033.

Liu, Ping, Jia Min Wang, Xiao Tong Yu, Xiao Bo Chen, Shuo Qi Li, Dong Chu Chen, Shao Kang Guan, Rong Chang Zeng, and Lan Yue Cui. 2019. "Corrosion Resistance of Bioinspired DNA-Induced Ca – P Coating on Biodegradable Magnesium Alloy." *Journal of Magnesium and Alloys* 7 (1): 144–54. https://doi.org/10.1016/J.JMA.2019.01.004.

Liu, Zhenmin, and Wei Gao. 2006. "Electroless Nickel Plating on AZ91 Mg Alloy Substrate." *Surface and Coatings Technology* 200 (16–17): 5087–93. https://doi.org/10.1016/J.SURFCOAT.2005.05.023.

Mahato, Arnab, Munmun De, Promita Bhattacharjee, Vinod Kumar, Prasenjit Mukherjee, Gajendra Singh, Biswanath Kundu, Vamsi K. Balla, and Samit Kumar Nandi. 2021. "Role of Calcium Phosphate and Bioactive Glass Coating on in Vivo Bone Healing of New Mg – Zn – Ca Implant." *Journal of Materials Science: Materials in Medicine* 32 (5): 1–20. https://doi.org/10.1007/S10856-021-06510-0/FIGURES/10.

Mahidashti, Z., T. Shahrabi, and B. Ramezanzadeh. 2018. "The Role of Post-Treatment of an Ecofriendly Cerium Nanostructure Conversion Coating by Green Corrosion Inhibitor on the Adhesion and Corrosion Protection Properties of the Epoxy Coating." *Progress in Organic Coatings* 114 (January): 19–32. https://doi.org/10.1016/J.PORGCOAT.2017.09.015.

Manoj Kumar, R., Kishor Kumar Kuntal, Sanjay Singh, Pallavi Gupta, Bharat Bhushan, P. Gopinath, and Debrupa Lahiri. 2016. "Electrophoretic Deposition of Hydroxyapatite Coating on Mg-3Zn Alloy for Orthopaedic Application." *Surface and Coatings Technology* 287 (February): 82–92. https://doi.org/10.1016/J.SURFCOAT.2015.12.086.

Mitchell, Geoffrey R., Kyung hwa Ahn, and Fred J. Davis. 2011. "The Potential of Electrospinning in Rapid Manufacturing Processes." *Virtual and Physical Prototyping* 6 (2): 63–77. https://doi.org/10.1080/17452759.2011.590387.

Montazerian, Maziar, Fatemeh Hosseinzadeh, Carla Migneco, Marcus V.L. Fook, and Francesco Baino. 2022. "Bioceramic Coatings on Metallic Implants: An Overview." *Ceramics International* 48 (7): 8987–9005. https://doi.org/10.1016/J.CERAMINT.2022.02.055.

Mousa, Hamouda M., Do Hee Lee, Chan Hee Park, and Cheol Sang Kim. 2015. "A Novel Simple Strategy for in Situ Deposition of Apatite Layer on AZ31B Magnesium Alloy for Bone Tissue Regeneration." *Applied Surface Science* 351 (October): 55–65. https://doi.org/10.1016/J.APSUSC.2015.05.099.

Nilawar, Sagar, Mohammad Uddin, and Kaushik Chatterjee. 2021. "Surface Engineering of Biodegradable Implants: Emerging Trends in Bioactive Ceramic Coatings and Mechanical Treatments." *Materials Advances* 2 (24): 7820–41. https://doi.org/10.1039/D1MA00733E.

Oliver, Joy Anne N., Yingchao Su, Xiaonan Lu, Po Hsuen Kuo, Jincheng Du, and Donghui Zhu. 2019. "Bioactive Glass Coatings on Metallic Implants for Biomedical Applications." *Bioactive Materials* 4 (December): 261–70. https://doi.org/10.1016/J.BIOACTMAT.2019.09.002.

Ooms, E.M., J.G.C. Wolke, M.T. Van de Heuvel, B. Jeschke, and J.A. Jansen. 2003. "Histological Evaluation of the Bone Response to Calcium Phosphate Cement Implanted in Cortical Bone." *Biomaterials* 24 (6): 989–1000. https://doi.org/10.1016/S0142-9612(02)00438-6.

Ostrowski, Nicole J., Boeun Lee, Abhijit Roy, Madhumati Ramanathan, and Prashant N. Kumta. 2012. "Biodegradable Poly(Lactide-Co-Glycolide) Coatings on Magnesium Alloys for Orthopedic Applications." *Journal of Materials Science: Materials in Medicine* 24 (1): 85–96. https://doi.org/10.1007/S10856-012-4773-5.

Pommiers-Belin, Sébastien, Jérôme Frayret, Arnaud Uhart, Jeanbernard Ledeuil, Jean Charles Dupin, Alain Castetbon, and Martine Potin-Gautier. 2014. "Determination of the Chemical Mechanism of Chromate Conversion Coating on Magnesium Alloys EV31A." *Applied Surface Science* 298 (April): 199–207. https://doi.org/10.1016/J.APSUSC.2014.01.162.

Rahim, Shebeer A., M.A. Joseph, T.S. Sampath Kumar, and T. Hanas. 2022. "Recent Progress in Surface Modification of Mg Alloys for Biodegradable Orthopedic Applications." *Frontiers in Materials* 9 (February): 45. https://doi.org/10.3389/FMATS.2022.848980/BIBTEX.

Rahim, Shebeer A., V.P. Muhammad Rabeeh, M.A. Joseph, and T. Hanas. 2021. "Does Acid Pickling of Mg-Ca Alloy Enhance Biomineralization?" *Journal of Magnesium and Alloys* 9 (3): 1028–38. https://doi.org/10.1016/J.JMA.2020.12.002.

Rahman, Mostafizur, Naba K. Dutta, and Namita Roy Choudhury. 2021. "Microroughness Induced Biomimetic Coating for Biodegradation Control of Magnesium." *Materials Science and Engineering: C* 121 (February): 111811. https://doi.org/10.1016/J.MSEC.2020.111811.

Ramos Rivera, Laura, Jannik Dippel, and Aldo R. Boccaccini. 2018. "Formation of Zein/Bioactive Glass Layers Using Electrophoretic Deposition Technique." *ECS Transactions* 82 (1): 73–80. https://doi.org/10.1149/08201.0073ECST/XML.

Rezwan, K., Q.Z. Chen, J.J. Blaker, and Aldo Roberto Boccaccini. 2006. "Biodegradable and Bioactive Porous Polymer/Inorganic Composite Scaffolds for Bone Tissue Engineering." *Biomaterials* 27 (18): 3413–31. https://doi.org/10.1016/J.BIOMATERIALS.2006.01.039.

Rojaee, Ramin, Mohammadhossein Fathi, and Keyvan Raeissi. 2013. "Controlling the Degradation Rate of AZ91 Magnesium Alloy via Sol – Gel Derived Nanostructured Hydroxyapatite Coating." *Materials Science and Engineering: C* 33 (7): 3817–25. https://doi.org/10.1016/J.MSEC.2013.05.014.

Sahu, Niranjan, B. Parija, and S. Panigrahi. 2009. "Fundamental Understanding and Modeling of Spin Coating Process: A Review." *Indian Journal of Physics* 83 (4): 493–502. https://doi.org/10.1007/S12648-009-0009-Z.

Saxena, Abhishek, and R.K. Singh Raman. 2021. "Role of Surface Preparation in Corrosion Resistance Due to Silane Coatings on a Magnesium Alloy." *Molecules* 26 (21): 6663. https://doi.org/10.3390/MOLECULES26216663.

Seal, C.K., K. Vince, and M.A. Hodgson. 2009. "Biodegradable Surgical Implants Based on Magnesium Alloys – A Review of Current Research." *IOP Conference Series: Materials Science and Engineering* 4 (1): 012011. https://doi.org/10.1088/1757-899X/4/1/012011.

Sekar, Prithivirajan, S. Narendranath, and Vijay Desai. 2021. "Recent Progress in in Vivo Studies and Clinical Applications of Magnesium Based Biodegradable Implants – A Review." *Journal of Magnesium and Alloys* 9 (4): 1147–63. https://doi.org/10.1016/J.JMA.2020.11.001.

Sergi, Rachele, Devis Bellucci, and Valeria Cannillo. 2020. "A Comprehensive Review of Bioactive Glass Coatings: State of the Art, Challenges and Future Perspectives." *Coatings* 10 (8): 757. https://doi.org/10.3390/COATINGS10080757.

Shuai, Cijun, Sheng Li, Shuping Peng, Pei Feng, Yuxiao Lai, and Chengde Gao. 2019. "Biodegradable Metallic Bone Implants." *Materials Chemistry Frontiers* 3 (4): 544–62. https://doi.org/10.1039/C8QM00507A.

Singh, Navdeep, Uma Batra, Kamal Kumar, Neeraj Ahuja, and Anil Mahapatro. 2023. "Progress in Bioactive Surface Coatings on Biodegradable Mg Alloys: A Critical Review Towards Clinical Translation." *Bioactive Materials* 19 (January): 717–57. https://doi.org/10.1016/J.BIOACTMAT.2022.05.009.

Song, Y.W., D.Y. Shan, and E.H. Han. 2008. "Electrodeposition of Hydroxyapatite Coating on AZ91D Magnesium Alloy for Biomaterial Application." *Materials Letters* 62 (17–18): 3276–9. https://doi.org/10.1016/J.MATLET.2008.02.048.

Sun, Luzhao, Guowen Yuan, Libo Gao, Jieun Yang, Manish Chhowalla, Meysam Heydari Gharahcheshmeh, Karen K. Gleason, Yong Seok Choi, Byung Hee Hong, and Zhongfan Liu. 2021. "Chemical Vapour Deposition." *Nature Reviews Methods Primers* 1 (1): 1–20. https://doi.org/10.1038/s43586-020-00005-y.

Talha, Mohd, Yucong Ma, Pardeep Kumar, Yuanhua Lin, and Ambrish Singh. 2019. "Role of Protein Adsorption in the Bio Corrosion of Metallic Implants – A Review." *Colloids and Surfaces B: Biointerfaces* 176 (April): 494–506. https://doi.org/10.1016/J.COLSURFB.2019.01.038.

Tan, Lili, Xiaoming Yu, Peng Wan, and Ke Yang. 2013. "Biodegradable Materials for Bone Repairs: A Review." *Journal of Materials Science and Technology* 29 (6): 503–13. https://doi.org/10.1016/j.jmst.2013.03.002.

Tang, Hui, Tao Wu, Fangjun Xu, Wei Tao, and Xian Jian. 2017. "Fabrication and Characterization of Mg(OH)2 Films on AZ31 Magnesium Alloy by Alkali Treatment." *International Journal of Electrochemical Science* 12: 1377–88. https://doi.org/10.20964/2017.02.35.

Tomihata, Kenji, Masakazu Suzuki, and Yoshito Ikada. 2001. "The PH Dependence of Monofilament Sutures on Hydrolytic Degradation." *Journal of Biomedical Materials Research* 58 (5): 511–18. https://doi.org/10.1002/JBM.1048.

Verron, Elise, Ibrahim Khairoun, Jerome Guicheux, and Jean Michel Bouler. 2010. "Calcium Phosphate Biomaterials as Bone Drug Delivery Systems: A Review." *Drug Discovery Today* 15 (13–14): 547–52. https://doi.org/10.1016/J.DRUDIS.2010.05.003.

Wang, Chenxi, Hui Fang, Xiaoyun Qi, Chunjin Hang, Yaru Sun, Zhibin Peng, Wei Wei, and Yansong Wang. 2019. "Silk Fibroin Film-Coated MgZnCa Alloy with Enhanced in Vitro and in Vivo Performance Prepared Using Surface Activation." *Acta Biomaterialia* 91 (June): 99–111. https://doi.org/10.1016/J.ACTBIO.2019.04.048.

Wang, H.X., S.K. Guan, X. Wang, C.X. Ren, and L.G. Wang. 2010. "In Vitro Degradation and Mechanical Integrity of Mg – Zn – Ca Alloy Coated with Ca-Deficient Hydroxyapatite by the Pulse Electrodeposition Process." *Acta Biomaterialia* 6 (5): 1743–8. https://doi.org/10.1016/J.ACTBIO.2009.12.009.

Wang, Qiang, Lili Tan, Wenli Xu, Bingchun Zhang, and Ke Yang. 2011. "Dynamic Behaviors of a Ca – P Coated AZ31B Magnesium Alloy During in Vitro and in Vivo Degradations." *Materials Science and Engineering: B* 176 (20): 1718–26. https://doi.org/10.1016/J.MSEB.2011.06.005.

Wang, Qingchuan, Weidan Wang, Yanfang Li, Weirong Li, Lili Tan, Ke Yang, QW, et al. 2021. "Biofunctional Magnesium Coating of Implant Materials by Physical Vapour Deposition." *Biomaterials Translational* 2 (3): 248. https://doi.org/10.12336/BIOMATERTRANSL.2021.03.007.

Witte, Frank. 2010. "The History of Biodegradable Magnesium Implants: A Review." *Acta Biomaterialia* 6: 1680–92. https://doi.org/10.1016/j.actbio.2015.07.017.

Wong, Hoi Man, Shuilin Wu, Paul K. Chu, Shuk Han Cheng, Keith D.K. Luk, Kenneth M.C. Cheung, and Kelvin W.K. Yeung. 2013. "Low-Modulus Mg/PCL Hybrid Bone Substitute for Osteoporotic Fracture Fixation." *Biomaterials* 34 (29): 7016–32. https://doi.org/10.1016/j.biomaterials.2013.05.062.

Xie, Jinshu, Jinghuai Zhang, Shujuan Liu, Zehua Li, Li Zhang, Ruizhi Wu, Legan Hou, and Milin Zhang. 2019. "Hydrothermal Synthesis of Protective Coating on Mg Alloy for Degradable Implant Applications." *Coatings* 160 9 (3): 160. https://doi.org/10.3390/COATINGS9030160.

Xiong, Pan, Zhaojun Jia, Wenhao Zhou, Jianglong Yan, Pei Wang, Wei Yuan, Yangyang Li, Yan Cheng, Zhenpeng Guan, and Yufeng Zheng. 2019. "Osteogenic and PH Stimuli-Responsive Self-Healing Coating on Biomedical Mg-1Ca Alloy." *Acta Biomaterialia* 92 (July): 336–50. https://doi.org/10.1016/J.ACTBIO.2019.05.027.

Xiong, Pan, Jiang Long Yan, Pei Wang, Zhao Jun Jia, Wenhao Zhou, Wei Yuan, Yangyang Li, et al. 2019. "A PH-Sensitive Self-Healing Coating for Biodegradable Magnesium Implants." *Acta Biomaterialia* 98: 160–73. https://doi.org/10.1016/j.actbio.2019.04.045.

Xu, Liming, Xingwang Liu, Kang Sun, Rao Fu, and Gang Wang. 2022. "Corrosion Behavior in Magnesium-Based Alloys for Biomedical Applications." *Materials* 15 (7): 2613. https://doi.org/10.3390/MA15072613.

Ye, Xinyu, Shu Cai, Ying Dou, Guohua Xu, Kai Huang, Mengguo Ren, and Xuexin Wang. 2012. "Bioactive Glass – Ceramic Coating for Enhancing the in Vitro Corrosion Resistance of Biodegradable Mg Alloy." *Applied Surface Science* 259 (October): 799–805. https://doi.org/10.1016/J.APSUSC.2012.07.127.

Yin, Zheng Zheng, Wei Chen Qi, Rong Chang Zeng, Xiao Bo Chen, Chang Dong Gu, Shao Kang Guan, and Yu Feng Zheng. 2020. "Advances in Coatings on Biodegradable Magnesium Alloys." *Journal of Magnesium and Alloys* 8 (1): 42–65. https://doi.org/10.1016/J.JMA.2019.09.008.

Zarbov, M., I. Schuster, and L. Gal-Or. 2004. "Methodology for Selection of Charging Agents for Electrophoretic Deposition of Ceramic Particles." *Journal of Materials Science* 39 (3): 813–17. https://doi.org/10.1023/B:JMSC.0000012908.18329.93.

Zhang, Feiyang, Shu Cai, Guohua Xu, Sibo Shen, Yan Li, Min Zhang, and Xiaodong Wu. 2016. "Corrosion Behavior of Mesoporous Bioglass-Ceramic Coated Magnesium Alloy Under Applied Forces." *Journal of the Mechanical Behavior of Biomedical Materials* 56 (March): 146–55. https://doi.org/10.1016/J.JMBBM.2015.11.029.

Zhou, Huan, Joseph G. Lawrence, and Sarit B. Bhaduri. 2012. "Fabrication Aspects of PLA-CaP/PLGA-CaP Composites for Orthopedic Applications: A Review." *Acta Biomaterialia* 8 (6): 1999–2016. https://doi.org/10.1016/J.ACTBIO.2012.01.031.

24 Multilayer Coating Strategies for Magnesium Alloys

Yibo Ouyang, Enyu Guo, Xiao-Bo Chen,
and Tongmin Wang

CONTENTS

24.1 INTRODUCTION

A combination of low density, high specific strength, and specific stiffness credits magnesium (Mg) alloys as the most promising lightweight structural metal. Meanwhile, Mg alloys also exhibit good die-casting performance, electromagnetic shielding performance, electrical and thermal conductivity, damping performance, biological compatibility, and easy recovery. These properties warrant wide applications of Mg alloys in aerospace, automotive, biological implants, sports equipment, and

DOI: 10.1201/9781003319856-27

3C electronic products (Xu et al., 2019). However, poor corrosion resistance is the most significant 'shortcoming' of Mg alloys toward their commercialization (Li et al., 2022). Mg alloys are characteristic of high chemical activity and the highest standard electrode potential among all engineering metals (–2.37 V_{SHE}), which is vulnerable to corrosion in majority of engineering services. Unlike the dense oxide film formed on the surface of stainless steel, Al, and Ti alloys, the natural oxide film on Mg alloys displays a loose and porous structure, which provides mild protection to Mg alloy matrix against corrosion (Atrens et al., 2020).

A large number of techniques have been developed to tackle the corrosion vulnerability of Mg alloys. These include (i) managing chemical composition to minimize impurity (i.e. Fe, Ni, Cu) or introduce corrosion resistance ingredients (i.e. rare earth elements); (ii) homogenizing or/and refining microstructure; and (iii) surface modification and coatings (Johari et al., 2022). The commonly used surface treatment technologies include chemical conversion treatment (Huo et al., 2004), anodic oxidation (Moon and Nam, 2012), and metal plating (Yang et al., 2011). Although these surface treatments can improve the corrosion resistance of Mg alloys, the preparation process are usually complicated, whilst the function generally lacks diversity.

Using a coating strategy is a prevalent means for corrosion protection in engineering structures, ships, and automobiles (Zhang et al., 2021). A protective coating is a robust film on metal surface and acts as a barrier to prevent direct contact between metal and corrosive medium. Protective coatings include organic (Hu et al., 2012), inorganic (Wang et al., 2019), composite (Gnedenkov et al., 2016), and intelligent anticorrosive coatings (Ouyang et al., 2022). Traditional organic and inorganic coatings are resins, paints, ceramics, plastics, rubber, and non-metallic materials with good corrosion resistance. Chemical and electrochemical conversion coatings are typically used as the base layer of multi-layer coatings (Saji et al., 2022). Coating damage, blistering, or defects readily emerge if a single coating has been serving in severe conditions for a long time period, leading to localized corrosion. For further improvement in protective performance, a composite coating with a multicomponent synergy or on-demand responses to corrosion protection is required (S. Wang et al., 2021). A set of representative multilayer strategies will be introduced in the following sections.

24.2 THE CONCEPT OF MULTILAYER COATINGS

Multilayer coatings refer to a coating system combining the advantages of a variety of individual layers to provide underlying metal with high corrosion resistance and a controllable corrosion landscape. In practice, an inner layer is vital to build a base for following coating formation with high adhesion. Common techniques for fabricating an inner layer include hydrothermal, fluorine treatment and micro-arc oxidation (MAO).

In addition, functional ions (e.g., LDHs, SiO_2, MOF, GO) or organic matters are used as intermediate layer to play a buffering role. For the top layer, suitable corrosion medium isolation of organic (epoxy resin, oil phase, hydrophobic agent, PDMS, etc.) or inert inorganic substances (Ni, Si_3N_4, Ni-P, etc.) are selected.

24.3 MANUFACTURING TECHNIQUES

A large number of advanced manufacturing methods have been applied to fabricate micro/nanostructured surfaces on Mg alloys. These methods include electrochemical deposition, micro-arc oxidation, magnetron sputtering, self-assembly, impregnation, and *in situ* growth (**Figure 24.1**). This section summarizes the critical manufacturing techniques for preparing multilayer functional coatings on Mg alloys. In addition, key efforts devoted to corrosion protection of Mg alloys based on these techniques are outlined in Tables 24.1 to 24.7.

Layer-by-layer (LBL) deposition is a process in which molecular aggregates or supramolecular structures with complete structure, stable performance, and certain functions are spontaneously formed through mild chemical interactions between layers (Liu et al., 2012). One can see

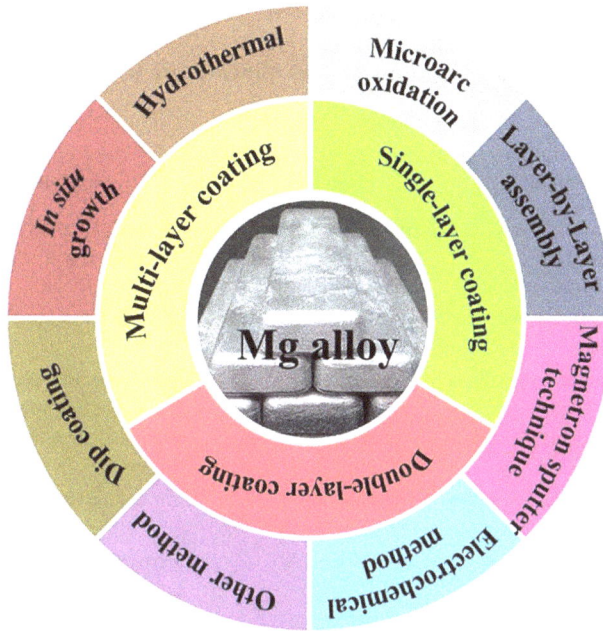

FIGURE 24.1   Schematic illustration of the preparation technology for multilayer coatings on Mg alloys.

TABLE 24.1

Summary of Study on Corrosion Inhibition of Mg Alloy Coatings Prepared by LBL Assembly

Metal	Preparation method	Film types	Property	Ref.
Mg alloy AZ31	LBL	Ce/PEI/GO/(PEI/PAA)$_{10}$	Self-healing and anti-corrosion	(Fan et al., 2015)
Mg alloy AZ31	LBL	SiO$_2$/(PVP/PAA)$_5$	Corrosion resistance	(Cui et al., 2016)
Mg-1Li-1Ca	LBL	PMTMS/TiO$_2$	Corrosion resistance	(Cui, Qin et al., 2017)
Mg alloy AZ31	LBL	(AgNPs/PEI)$_n$ multilayer	Corrosion resistance, self-healing and antibacterial	(Y. Zhao et al., 2018)
Mg alloy AZ31	LBL	(PVP/PAA)$_{10}$	Corrosion resistance and adhesion strength	(Y.-B. Zhao et al., 2018)
Mg alloy AZ31B	LBL	Silane/graphene oxide (GO)/silane	Corrosion resistance	(Wang et al., 2019)
Mg alloys	LBL, spin-spray	(SiO$_2$/CeO$_2$)$_{10}$ multilayer	Corrosion resistance and self-healing	(Zhao et al., 2019)
Mg-3Zn-0.5Sr	LBL	MAO/CS(TiO$_2$)/HPs	Corrosion resistance and anticoagulant ability	(Chen et al., 2020)
Mg alloy AZ31	LBL	(Chitosan-Ag@ PDA-chitosan-CAp@ PDA)$_3$	Corrosion resistance and antibacterial	(Wang et al., 2017)

TABLE 24.2
Multilayer Corrosion-Resistant Coatings on Mg Alloys Were Synthesized by *In Situ* Synthesis

Metal	Preparation method	Film types	Property	Ref.
Mg alloy	Hydrothermal-assisted *in situ* growth and condensation reaction	Nickel-underlayer/ LDH-midlayer/ siloxane-toplayer	Inhibiting galvanic corrosion, self-cleaning and superhydrophobic	(Li et al., 2022)
Mg alloy AZ31	*In situ*, hydrothermal treatment	MXenes/ MgAl-LDHs/Y(OH)$_3$	Corrosion resistance	(Wu et al., 2021)
Mg alloy AZ31	*In situ* plasma-induced thermal-field assisted crosslinking deposition technique	Super-repellent multilayer nanocomposite coating	Mechanochemical robustness, high-temperature and endurance and electric protection	(S. Wang et al., 2021)
Mg alloy AZ91D	*In situ* synthesis, electroless plating	TiO$_2$/Cu/thiol	Corrosion resistance	(Zang et al., 2017)
Mg alloy AZ91D	MAO, *in situ* synthesis	MAO/ZIF-8/SA	Improving the hydrophobicity and corrosion resistance	(Jiang et al., 2021)
Mg alloy AZ31	*In situ* steam	Al LDH/Mg(OH)$_2$/ silane-Ce	Anti-corrosion mechanism	(Qiu et al., 2020)

TABLE 24.3
Preparation of Multilayer Coatings on Mg Alloy by Magnetron Sputtering Method

Metal	Preparation method	Film types	Property	Ref.
Mg alloy AZ91D	Magnetron sputtering	VN/TiN multilayer	Corrosion resistance	(M. Ertas et al., 2015)
Mg-6Gd-3Y-0.5Zr	Magnetron sputtering	Multilayer structure of metallic glass film	Decreases hardness and corrosion resistance	(Wu et al., 2015)
Mg alloy AZ91D	Magnetron sputtering	Hf/Si$_3$N$_4$ multilayer	Anti-corrosive conductive coating	(D. Zhang et al., 2017)
Mg alloy AZ31B	Magnetron sputtering and MAO	MAO/DLC/PDMS	Corrosion and wear resistance	(Cui et al., 2021)

TABLE 24.4
Summary on Corrosion Inhibition of Mg Alloys by Hydrothermal Multilayer Coatings

Metal	Preparation method	Film types	Property	Ref.
Mg alloy AZ31	Hydrothermal	MgAl-LDHs/NaB/ lubricant	Anti-corrosion, anti-wear, self-cleaning, and self-healing	(Yao, Chen, Wu, Zhang et al., 2022)
Mg alloy AZ31	Hydrothermal	MgAlLa-LDH/NaB/ lubricant	Corrosion resistance, self-cleaning, self-healing	(Yao, Chen, Wu, Jiang et al., 2022)
Mg alloy AZ31	Hydrothermal	Nano-sheets/SA/PFIE	Anti-biofouling and anti-corrosion and assisted healing	(H. Li et al., 2020)
Mg alloy AZ31	Hydrothermal, MAO	Oxide layer/ nanoporous/FDTS/oil	Anti-corrosion property	(Joo et al., 2020)

TABLE 24.4 (Continued)

Summary on Corrosion Inhibition of Mg Alloys by Hydrothermal Multilayer Coatings

Metal	Preparation method	Film types	Property	Ref.
Mg alloy AZ91D	Hydrothermal, MAO	MAO-LDH/PFDS/oil	Self-healing and anti-corrosion	(Jiang et al., 2019)
Mg alloy AZ31	Hydrothermal	LDH/SAMs/Lubricant	Corrosion resistance and anti-icing	(J. Zhang et al., 2017)
Mg alloy AZ31	Hydrothermal	Mg(OH)$_2$/PMTMS/CeO$_2$	Corrosion resistance	(Guo et al., 2017)
Mg alloy AZ31	Hydrothermal	MgAl-ASP-LDHs	Corrosion resistance and self-healing	(Chen et al., 2019)
Mg alloy AZ31B	Hydrothermal, electroless deposition	LDH layer/Ni-B/Ni-P	Corrosion protection	(Song et al., 2022)

TABLE 24.5

Review on Corrosion Inhibition of Mg Alloys by Multilayer Coatings Prepared by MAO

Metal	Preparation method	Film types	Property	Ref.
Mg alloy AZ31	MAO	Epoxy/BTESPT/MAO	Anti-corrosion	(Chen et al., 2015)
Mg alloy AZ31	MAO	MAO/Silane/Epoxy	Enhanced protective properties	(Toorani et al., 2020)
Mg alloy AZ31	MAO	MAO/CeNaX/Epoxy	Self-healing and anti-corrosion	(Zhang et al., 2019)
Mg alloy AZ31	MAO	MAO/Silane/epoxy layer	Corrosion protection and adhesion strength	(Toorani et al., 2021)
Mg alloy AZ31	MAO	HA/GO/MAO	Anti-corrosion	(Wen et al., 2017)
Mg alloy AZ31	MAO	MAO/Mg(OH)$_2$/HA	Reinforce the protection	(C.-Y. Li et al., 2019)
Mg alloy AZ31B	MAO, LBL, electroless nickel plating	MAO + SANP + EN	Inhibition of galvanic corrosion	(Guo et al., 2012)
Mg alloy AZ31	MAO	MAO/polymethyltrimethoxysilane	Corrosion resistance, self-healing	(Cui, Gao et al., 2017)
Mg alloy AZ91D	MAO, electrochemical deposition, dip coating	PLA/FHA/MAO	Corrosion resistance	(Jin et al., 2018)
Mg alloy AZ31	MAO, hydrothermal treatment	MAO-Ce-LDH/MAO-Ce-LDH-P	Corrosion protection and self-healing ability	(Zhang et al., 2018)
Mg alloy AZ91D	MAO, electroless plating	MAO/self-assemble/nickel plating layer	Deposition mechanism of electroless nickel plating layer	(Shang et al., 2019)
Mg alloy AZ91D	MAO, LBL, electroless	MAO/G502/Ni	Corrosion resistance	(Zhang et al., 2019)
Mg alloy AZ91	MAO, *in situ* incorporation	MAO -LDH-FA	Corrosion resistance	(Li et al., 2021)
Mg alloy AZ31	MAO	MAO-LDHs/HG coating	Self-healing and the anti-corrosion	(Chen et al., 2022)
Mg alloy AZ31	MAO, self-assembly technology	Ag NPs/PEI/MAO	Anti-corrosive and antibacterial	(X. Wang et al., 2021)
Mg alloy AZ31	MAO	LDH-ALB-WO$_3$	Active anti-corrosion	(Kaseem and Ko, 2021)

TABLE 24.6

Anticorrosion of Multiple Coating on Mg Alloy by Dip Coating

Metal	Preparation method	Film types	Property	Ref.
Mg alloy AZ91D	Dip coating, chemical conversion coating	CeO$_2$/stannate multilayer	Corrosion resistance	(Bagalà et al., 2012)
Mg-Ca alloy	Dip coating, plasma spraying	Triple-layer NiCrAlY/ nano-yttria stabilized zirconia/polycaprolactone	Enhancement corrosion and mechanical properties	(Bakhsheshi-Rad et al., 2016)
Mg alloy AZ60	Dip coating, chemical conversion coating	APTES-CaP/PLA	Clinical application	(Li et al., 2018)
Mg alloy AZ91	Sol-gel dip coating, chemical conversion coating	CLP/TEOS/MTES	Corrosion protection	(Ashassi-Sorkhabi et al., 2019)
Mg alloy AZ31	Dip coating	PCL/MOF/FA	Corrosion resistance and biocompatibility	(Zheng et al., 2019)
Mg alloy AZ31	Dip coating	Oxide layer/PCL-Lawsone/ Pure PCL	Corrosion protection	(Asadi et al., 2021)
Mg alloy ZM21	Sol-gel dip coating	TiO$_2$-HA-PCL	Corrosion resistance	(Singh et al., 2021)

TABLE 24.7

Anti-Corrosion Effect of Multilayer Coatings on Mg Alloys Prepared by Chemical Method

Metal	Preparation method	Film types	Property	Ref.
Mg-Li alloys	Electroless	Ni/Cu/Ni-P triple-layered	Corrosion resistance	(Chen et al., 2014)
Mg alloy AZ91D	Electrophoretic deposition	CaP/Chitosan/Graphene Coating	Improving the bioactivity and cell viability	(Zhang, 2016)
Mg alloy AZ91	Electrophoretic deposition	HA/Fe$_3$O$_4$/CS	Corrosion resistance	(Singh et al., 2019)
Mg alloy AZ31	Electrophoretic deposition	Hydroxyapatite-carboxymethyl cellulose-graphene	Corrosion behavior and mechanical properties	(Ahangari et al., 2021)
Mg alloy ZM5	Electroplating	Zn/Cu/Al-Zr Coating	Corrosion protection	(Chen et al., 2021)
Mg-4Zn-4 Sn-1Ca	Electrophoretic deposited	Hydroxyapatite/ chitosan/GO	Corrosion resistance	(Saadati et al., 2021)
Mg alloy AZ31	Electroless deposited	Superhydrophobic Ni-P/ nickel/ PFDTMS	Corrosion-resistant	(Fang et al., 2022)

the illustration in **Figure 24.2a**. This process enables a simple, versatile, and robust approach to fabricate micro- and nanoscale rough surface in a large area.

In situ **synthesis** is a newly developed method for preparing composite materials. The basic principle is that different elements or compounds in the metal matrix form one or several ceramic phase particles under chemical reaction to improve the performance of metal alloys (**Figure 24.2b**).

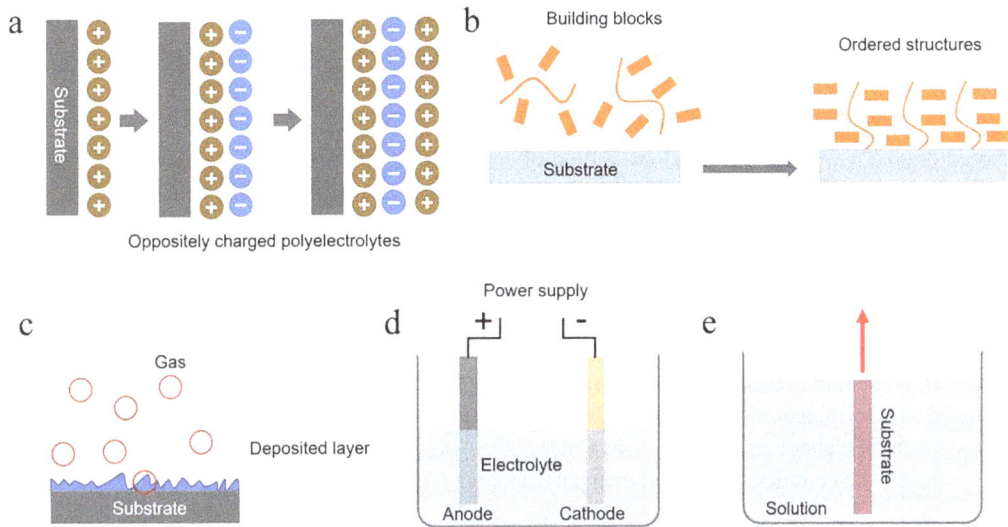

FIGURE 24.2 Schematics of the typical preparation techniques for multilayer coatings: (a) layer-by-layer, (b) *in situ* growth, (c) magnetron sputtering, (d) electrochemical, and (e) dip coating.

Magnetron sputtering is a physical vapour deposition technique that deposits various functional films such as wear resistance, corrosion resistance, and optics films (**Figure 24.2c**). This method is characterized of cost-effectiveness, high adhesion, high deposition rate, and environmental friendliness (Sarakinos et al., 2010). The thin film structure develops from individual layers to a multilayer system, from unitary to binary, and then to diversified. Common thin film materials are metal, ceramic, and polymer.

Hydrothermal technology constructs functional materials with varying morphology by tailoring temperature, pressure, and reaction time frame with a nature of simple operation, energy-efficiency and diversity in surface microstructure (Ling et al., 2022).

MAO produces ceramic coatings *in situ* on metal surfaces with high hardness, wear resistance, and corrosion resistance through optimizing operational parameters (temperature, duration, etc.) (**Figure 24.2d**) (Yao, Wu, Wang et al., 2022). However, physical pores are present inevitably in fresh MAO, which is vulnerable to corrosion attacks. As such, post-treatment is highly desired to seal pores for maximum protectiveness. Hydrothermal treatment is one of the options to fill the pores to obtain a complete, dense, and anti-corrosion structure of MAO coatings. This process is relatively simple, and the prepared coating has low thermal stress, minimal defects, high purity, and uniformity. In addition, organic-inorganic hybrid systems, hydrosol-gel, and electrodeposition techniques can be used for sealing porosity to form a multilayer coating.

Dipping and pulling is a widely used coating preparation method in which the composition of the sediment layer is adjusted to achieve desired corrosion resistance and functionality (**Figure 24.2e**) (Kőrösi et al., 2011). The quality of the coating processed by this method is ascribed to the diverse chemical composition and structure of polymers, and it is versatile to produce coatings with specific chemical and physical properties and functions.

Electrochemical/chemical technique is performed in aqueous solution, ionic solution, and molten salt (**Figure 24.2d**) (Aziz-Zanjani and Mehdinia, 2013). Due to the high chemical activity of Mg alloys, surface pre-treatment of Mg alloy is required to optimize uniform density and high adhesion of yielded coating. Pre-treatment processes, such as alkaline washing, pickling, activation, zinc dipping, and electroless nickel plating, are required to form multilayer dense films.

24.4 CLASSICAL CATEGORIES OF MULTILAYER COATINGS

24.4.1 LAYERED DOUBLE HYDROXIDES

24.4.1.1 Preparation Methods and Concepts

Layered double hydroxides (LDHs) are a kind of compounds with a layered main structure, which can be divided into cationic, anionic, and nonionic layered compounds (**Figure 24.3a**). The particular composition of main laminate and interlaminate guest credits LDHs interlaminate anion exchangeability, laminate composition variability, acid-based visibility, thermal stability, memory effect, etc.

The preparation methods of LDHs can be divided into two types: one is to synthesize LDH suspension by co-deposition and ion exchange method and then to obtain the film on Mg alloy by a spin or hydrothermal method. The other process is to bring the solution of reaction directly on the metal surface in the film-forming ion solution by *in situ* growth (e.g. chemical oxidation, hydrothermal, steam) or electrodeposition technology. Specific ions (NO_3^-, SO_4^{2-}) are reduced on the metal surface to generate OH^- by means of electrodeposition and then participate in the precipitation reaction of LDHs *in situ*.

24.4.1.2 Corrosion Inhibition Ability

Continuous and dense multilayer LDHs can effectively improve the corrosion resistance of the matrix. For example, LDHs are modified by functionalized albumin molecules of WO_3 nanoparticles to form a multilayer coating, which improves density of the film and credit Mg alloy active and intelligent anti-corrosion function in 3.5 wt.% NaCl solution (Kaseem and Ko, 2021). In addition, an MgAl-LDH/2-mercaptobenzothiazole multilayer film is formed on Mg alloy with ethylenediaminetetraacetic acid chelating. Due to the anion exchange ability of the MgAl-LDH phase and inhibition effect of 2-mercaptobenzothiazole, the composite coating maintains the original nanosheet structure after 168 h of exposure to neutral salt spray (**Figure 24.3b**) (Hu et al., 2021). Multilayer coatings display higher corrosion resistance to Mg alloys than individual coatings. For example, the MgAl-LDH/Ni coating prepared on Mg alloys is composed of an LDH bottom, Ni-B intermediate, and Ni-P upper layer. After immersion in 3.5 wt.% NaCl solution for 20 or 30 h in 0.1 mol· dm⁻³ HCl solution (**Figure 24.3c**), LDHs show obvious corrosion defects, and the composite coating offers excellent corrosion resistance to Mg alloys (Song et al., 2022). Previously, LDHs formed on Mg alloys are generally considered pure LDH, or $Mg(OH)_2$/LDH composite. Characterization of LDHs that were detached from Mg alloys by a mechanical bending method (Shi et al., 2022) confirm that they are composed of an $Mg(OH)_2$ bottom layer (main), a vertical LDH layer (middle), and a dense pyridine dicarboxylic acid/LDH sediment (outermost layer). Moreover, the protective effect of the LDHs is significantly improved after using 2,5-pyridinedicarboxylic acid to seal the defects in LDH layer (**Figure 24.3d and 24.3e**).

24.4.1.3 Corrosion Suppression Mechanism

The mechanism by which LDHs can protect metal is that LDHs can cover the metal substrate surface densely, blocking corrosive ions and interlayered anion exchangeable ability. Specifically, LDHs can capture erosive ions in aqueous solution (e.g. Cl^-, SO_4^{2-}) through ion exchange, and thus suppress their diffusion kinetics to approach the substrate. Further, LDHs can be assembled with negatively charged inorganic or organic corrosion inhibitors, hinder the redox reaction in a corrosion medium, and play an active protective role on Mg alloys.

24.4.1.4 Challenges

Currently, there are several challenges in the development of LDHs.

i   LDHs are readily stacked during the synthesis process. This would result in an excessive size, affecting the compatibility, dispersion, and load of the corrosion inhibitor with the coating.

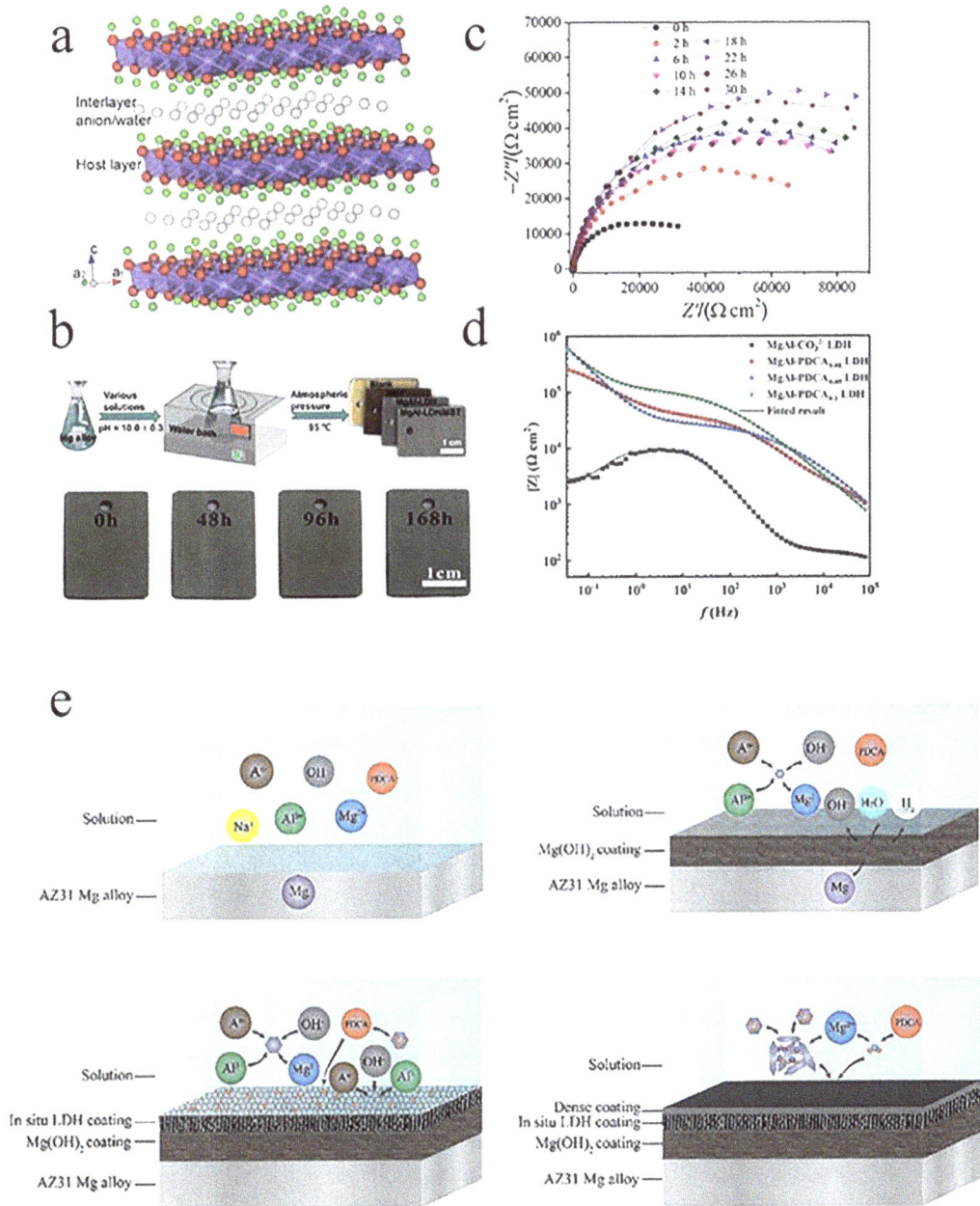

FIGURE 24.3 Structure, preparation method, electrochemical performance, and anti-corrosion mechanism of LDHs: (a) Schematic model of LDH structure. *Reproduced with permission from (Ma et al., 2006) © 2006 The Royal Society of Chemistry.* (b) Schematic diagram for preparation of MgAl-LDH/MBT multilayer and optical images of MgAl-LDH/MBT exposed to neutral salt spray. *Reproduced with permission from (Hu et al., 2021) © 2021 Published by Elsevier Ltd on behalf of Chinese Society for Metals.* (c) Nyquist of LDHs after immersion in an HCl solution for 30 h. *Reproduced with permission from (Song et al., 2022) © 2021 Elsevier B.V.* (d) Bode plot of Mg alloy with MgAl-PDCAx LDH. (e) Schematics of growth process and mechanism of LDH added with 2,5-PDCA. *Reproduced with permission from (Shi et al., 2022) © 2022 Chongqing University. Publishing services provided by Elsevier B.V. on behalf of KeAi Communications Co. Ltd.*

ii LDHs are less than a few microns thick, and it remains challenging to meet the high requirements of industrial preservative applications.

Currently, LDHs are used in the protection of Al alloys, and research on Mg alloys protection is rarely reported. Considering the superior performance of LDHs, their corrosion resistance can be improved from the following aspects:

i Combined with biomimetic coating, LDHs are transformed into a superhydrophobic surface to further improve its blocking performance to the corrosion medium.
ii To further develop the application of LDHs as corrosion inhibitor reservoir, the method of preparing the functional additives of nano-LDHs in the field of polymer available materials can be possibly referred to. Some examples are provided in Kalali et al. (2016).
iii LDHs serve as the bottom or intermediate layer of a multilayer system. Considering the long-term protective effect of the coating, it is necessary to strengthen binding force between LDHs and Mg alloys or upper layers.

24.4.2 CERAMIC COATINGS

Ceramic coating of Mg alloys formed by typical MAO is depicted in **Figure 24.4a**. In general, the ceramic film consists of a dense inner layer and an outer layer with micropores and microcracks (**Figure 24.4b and 24.4c**). The quality of ceramic coating can be improved by optimizing reaction parameters or adding nanoparticles (GO, MXene, etc.) in the MAO process. The formation of pores and cracks remains inevitable.

A single ceramic coating has limited protection effectiveness against Mg alloys when facing a complex and harsh engineering corrosion environment. The MAO technique is combined with electrochemical nickel plating, organic coatings, or polymer formation of multilayer coatings with complementary advantages to enhance their protection performance.

24.4.2.1 Anti-Corrosion of Ceramic Multilayer Coating

Ceramic multilayer coating is developed to modify structure and solve the problem that the single ceramic coating encounters. One example is a three-layer composite film prepared on Mg alloy using plasma electrolytic oxidation (PEO), self-assembled nanophase particle (SANP), and electroless nickel (EN) plating. The ceramic film, SANP film, and EN coating act as the bottom layer, middle layer, and top layer, respectively. Results indicate that the composite film (PEO+SANP+EN, or 'PSE') provides adequate protection for Mg alloy against galvanic corrosion and long-term immersion corrosion (**Figure 24.4d and 24.4e**) (Guo et al., 2012). Adding inorganic corrosion inhibitor ($Ce(NO_3)_3$) and organic silane inhibitor to the MAO/silane/epoxy coating system can enhance active protection to the Mg alloy. Electrochemical experimental results show that the presence of 8-hydroxyquinoline (8-HQ) organic corrosion inhibitor provides excellent corrosion resistance for the coating system. In addition, the formation of $Ce(OH)_3$ in the MAO and the complexation of $Mg(HQ)_2$ in the silane layer bring about active anti-corrosion properties for Mg alloys (Toorani et al., 2021). LDHs are grafted in the ceramic coating by *in situ* generation. In addition, the corrosion inhibitor is encapsulated in the intercalated nanocarrier to form a multifunctional multilayer structure. In the presence of an LDH nanocarrier, the thickness and density of the coating can be further improved. When corrosion occurs, the loaded corrosion inhibitors and nanocontainers, through ion exchange, can inhibit the degradation of ceramic coatings and enhance the corrosion resistance of Mg alloys (Li et al., 2021).

24.4.2.2 Corrosion Protection Mechanisms

Ceramic coating has high anode potential, electrode polarizability, superior binding force, and small corrosion current. In addition, the ceramic coating can prevent the anode, cathode, or solution from ion movement, cutting off the contact between corrosion ions and substrate.

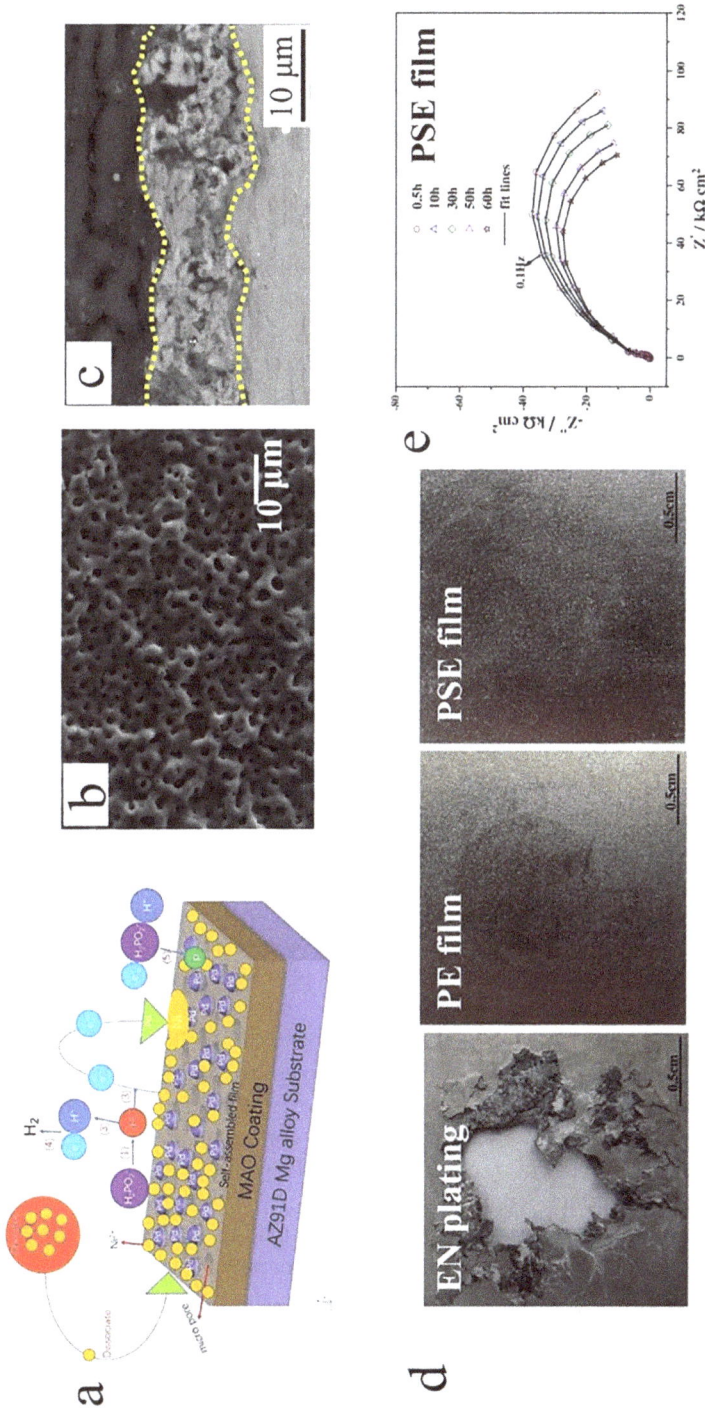

FIGURE 24.4 Preparation model diagram of ceramic multilayer coating, micromorphology, and corrosion resistance: (a) Deposition mechanism of electroless plating on MAO. *Reproduced with permission from (Shang et al., 2019) © 2019 Elsevier Ltd.* (b, c) SEM images and cross-sectional morphology of MAO. *Reproduced with permission from (C.-Y. Li et al., 2020) © 2020 Elsevier Ltd.* (d) Optical images of Mg alloy coated with different films, and (e) Nyquist diagrams of Mg alloy with PSE film versus immersion time in 3.5 wt.% NaCl solution. *Reproduced with permission from (Guo et al., 2012) © 2012 Elsevier Ltd.*

24.4.2.3   Challenges and Solutions

Ceramic film of Mg alloys is prone to form defects such as micropores, dislocations, cavities, alloy inclusion, and microcracks. The reason is metal itself, alloy composition, or inappropriate processing. In the actual formation process of ceramic film, defects are inevitable. Therefore, improvements can be made in the following aspects:

i   Influencing factors of micropore size

Size of the micropores can be modified by adding inorganic (SiO_2, $Ce(NO_3)_3$) and/or organic materials (GO, organogels and HA) to electrolyte or changing different electrical parameters (voltage, current, etc.) and adjusting the characteristics of micro-arc discharge.

ii   Hole-sealing process

Micropores of MAO film can be used as the transmission channel of the corrosion medium, and it has a huge negative impact on the corrosion resistance of the coating. Therefore, transforming MAO into a multilayer composite structure through subsequent hydrothermal treatment, organic-inorganic treatment, and electrochemical deposition can effectively solve the microporous defects in the ceramic coating.

24.4.3   Bio-Inspired Bionic Coatings

24.4.3.1   Superhydrophobic Surfaces

The bionic superhydrophobic surface (SHS) of the lotus leaf (**Figure 24.5a**) has gained wide attention and developed rapidly in the past few decades due to its multifunctional characteristics such as self-cleaning, low solid-liquid adhesion and drying, and has shown applications in anti-fouling, anti-corrosion, anti-icing etc. (Kobina Sam et al., 2019; Hooda et al., 2020; Wang et al., 2020). Generally, the superhydrophobic characteristics feature in large static contact angle ($\geq 150°$) and the small roll-off angle ($\leq 10°$) of water on the material surface (Latthe et al., 2019). To form superhydrophobic materials, the following two points play crucial roles: (i) surface roughness and (ii) low surface energy. The detailed concept, preparation method, and application of superhydrophobicity can be referred to literature (Lafuma and Quere, 2003; Li et al., 2007; Zhang et al., 2008).

24.4.3.1.1   Corrosion Resistance of SHS

In terms of corrosion protection for Mg alloys, the single layer of SHS has defects in poor stability and low mechanical strength; therefore, the surface formation of multilayer structure is a safe solution. For instance, through ultrasonic-assisted electroless plating combined with self-assembly, a densely packed copper sulfide multilayer structure can be uniformly plated on Mg alloys with solid adhesion, which prevent water transport and corrosive ions (Zang et al., 2017). The multilayer coating formed by MAO combines with two zeolitic imidazolate framework (ZIF-8) and organic skeleton, and organic acids, which exhibits excellent superhydrophobic and self-cleaning properties. In addition, the air pockets stored in superhydrophobic composite structure can inhibit the contact of the corrosion media with Mg alloy (**Figure 24.5b**) (Jiang et al., 2021). Superhydrophobic multilayer film containing multidimensional organic-inorganic components can be designed and modified on Mg alloys by a one-step plasma-induced thermal field-assisted crosslinking process. The hard porous MgO layer and polytetrafluoroethylene nanoparticles serve as the skeleton and filling components of the bottom layer, respectively. The organic nanoparticles can be crosslinked and solidified to form a dense polymer outer layer with a fine surface texture. Notably, the chemical robustness of prolonged exposure to aqua regia, strong alkali, and simulated seawater is attributable to the uniform and compact transmembrane of the polymer nanocomposite layers.

Self-similar multilayer structure coatings are characteristic of strong mechanical robustness (after 100 cycles of rotation wear), stable ultra-low friction coefficient (**Figure 24.5c and 24.5d**), high-temperature resistance, and robust self-cleaning performance (X. Wang et al., 2021). Superhydrophobic nickel-phosphorus/nickel-fluorinated polysiloxane (PFDTMS) triple composite

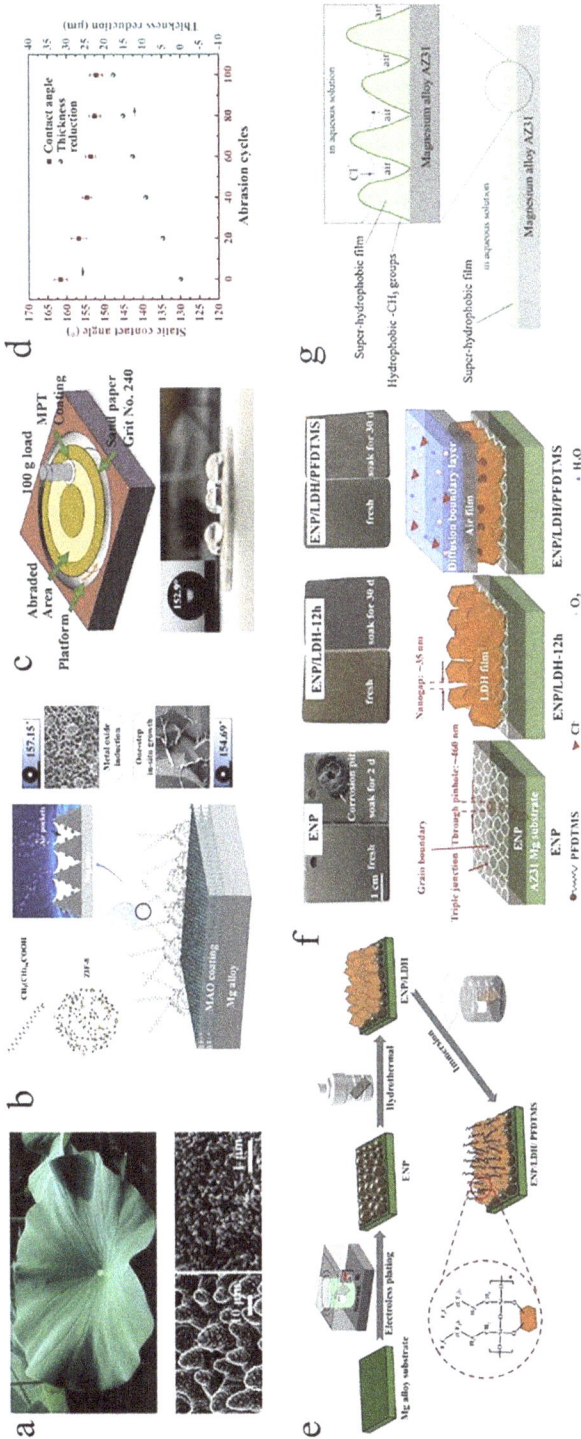

FIGURE 24.5 The preparation model diagram, friction performance, corrosion performance, and protection mechanism of the lotus leaf–inspired bionic SHS: (a) SEM image of lotus leaf. *Reproduced with permission from (Xu et al., 2021) © 2021, The authors.* (b) Schematic of superhydrophobic multilayer structure. *Reproduced with permission from (Jiang et al., 2021) © 2021 Chongqing University. Publishing services provided by Elsevier B.V. on behalf of KeAi Communications Co. Ltd.* (c, d) Schematic, water droplet and contact angle, contact angle change, and coating thickness reduction of multilayer coating after abrasion cycles. *Reproduced with permission from (S. Wang et al., 2021) © 2021 Chongqing University. Publishing services provided by Elsevier B.V. on behalf of KeAi Communications Co. Ltd.* (e) A schematic diagram for preparing superhydrophobic composite coating on Mg alloy. (f) Digital photographs and schematic of ENP, ENP/LDH, and ENP/LDH/PFDTMS before and after exposure to 3.5 wt.% NaCl solution. *Reproduced with permission from (Li et al., 2022) © 2021 Elsevier B.V.* (g) Diagram of corrosion protection mechanism of SHS. *Reproduced with permission from (Ishizaki et al., 2010) © 2010 Elsevier Ltd.*

coating consists of the bottom layer of the chemically deposited Ni-P, the middle layer of the electrochemically deposited Ni, and the PFDTMS top layer formed by silane dehydration and condensation reaction. The corrosion current density of the superhydrophobic multilayer coating is reduced by four orders of magnitude compared to Mg alloy AZ31 as revealed by the Tafel curve (Fang et al., 2022). A three-layer composite coating consisting of Ni (ENP) bottom layer, LDHs, and silane (PFDTMS), namely ENP/LDHs/PFDTMS, is prepared by a combination of simple non-plating, hydrothermal-assisted *in situ* growth and condensation reaction (**Figure 24.5e**). The corrosion resistance of ENP/LDH/PFDTMS is much higher than that of single Ni or ENP/LDH. However, the protection performance of Ni on Mg alloy is not ideal, and it is prone to corrosion in a short period (2 days). In contrast, the ENP/LDH/PFDTMS or ENP/LDH does not exhibit galvanic corrosion even after 30 days of immersion in 3.5 wt.% NaCl solution (**Figure 24.5f**) (Li et al., 2022).

24.4.3.1.2 Anti-Corrosion Mechanism of SHS

The corrosion-resistance mechanism of SHS is shown in **Figure 24.5g**, and a large number of air pockets in SHS prevent the corrosion medium from immersion into microstructure (Jeong et al., 2015). Furthermore, since large amounts of air remain between the microstructural voids, a capillary system is formed on Mg alloys' surface with differential pressure. When vertical capillary in liquid, the liquid will raise height (H) in the capillary. The H can be expressed by the formula:

$H = \dfrac{2\gamma\cos\theta}{\rho g R}$, where, R, θ, γ, g, and ρ is the capillary radius, contact angle, surface tension,

gravitational acceleration, and liquid density, respectively. Due to the large contact angle (θ > 150°) of SHS and the small capillary diameter, the corrosive liquid cannot penetrate the surface rough structure, increasing the corrosion resistance (Liu et al., 2015).

24.4.3.1.3 Challenges and Solutions

Currently, preparation methods of SHS on Mg alloys are limited to the primary laboratory research stage, and there are still vital problems in practical large-scale industrial applications.

  i Poor mechanical stability. Metal/non-metal superhydrophobic materials are affected by friction/extrusion and other weak external forces in the actual industrial application.
 ii Insufficient chemical stability. In strong acid or alkali environments, superhydrophobic materials suffer from UV radiation, rain erosion, and long-term corrosion of corrosive medium.
iii Environmental and cost issues. Superhydrophobic surface construction methods require expensive materials, equipment, and complex technological process.

In practical application scenarios, the wear resistance of SHS is tested by potential friction and wear risks during transportation and installation. Surface wear resistance has been a significant challenge in designing superhydrophobic materials. The following methods can be used for improving mechanical stability:

  i Introducing a polymer bonding layer to enhance the bonding force between superhydrophobic nano/micron material and substrate.
 ii Achieving wear resistance by sacrificing the upper self-similar structure.
iii) Random introduction of discrete microstructure to withstand external wear.

24.4.3.2 Slippery Liquid-Infused Porous Surface

The interaction between liquid and the solid surface has become a vital issue in surface science and engineering research, such as high-voltage transmission equipment icing, car rear windows and mirror fog, metal corrosion, and biological adhesion (Xiao et al., 2013; Deng et al., 2020).

Therefore, it is imperative to develop and manufacture functional interface materials with unique wetting properties.

Nepenthes can successfully 'swallow' insects despite lacking any active hunting mechanism. The principle is that the continuous micron texture at the mouth edge of the cage can not only transport the liquid in the cage outward to the mouth edge but also lock the lubricant to form a slippery liquid-infused porous surface (SLIPS) or liquid-infused surface (LIS) (**Figure 24.6a**), so that the insects passively slide into the cage to be 'digested' and absorbed (Wong et al., 2011; Chen et al., 2016). For this natural phenomenon, Aizenberg's team (Wong et al., 2011) first proposed the concept of fluid perfusion on porous surfaces. The preparation process consists of three steps: substrate preparation, structure modification, and lubricating oil infusion. The super-slippery surface formed by imitating the structure of nepenthes plants has broad application prospects in hydrophobic, antifouling, self-healing, and anti-corrosion (Ouyang, Qiu et al., 2019; Ouyang, Zhao et al., 2019; Yao, Wu, Sun et al., 2022).

24.4.3.2.1   Corrosion Resistance of SLIPS

It is an important yet critical attempt to introduce a SLIPS into Mg alloy protection. Some efforts have been made, and unexpected effects have shown that those attempts are quite valuable in Mg protection application. For example, a double layer of micro-nanojunction with LDHs and petal-like middle layer is obtained by the hydrothermal method on Mg alloys (J. Zhang et al., 2017). After the surface is modified by fluoride PTES and oil infusion, a multilayer SLIPS is prepared (**Figure 24.6b**), and the coating has an excellent protective effect (**Figure 24.6c**) (J. Zhang et al., 2017). To understand the healing of SLIPS assisted by thermal effect, the wettability of SHS and SLIPS by physical scratching and heat treatment are compared (H. Li et al., 2020). After being scratched by SiC paper, SEM characterization indicates that SHS has fragmented. Meanwhile, minor physical damage can be found on SLIPS. After heating, the WCA and SA of SHS have no change. In contrast, SLIPS is repaired to its original state. The polarization trends are consistent with the original ones by electrochemical measurements. Results show that SLIPS exhibits a self-healing property based on the thermal effect (H. Li et al., 2020).

Multilayer SLIPS of Mg alloy can be prepared by MAO, hydrothermal treatment, and perfluorodecyl trichlorosilane self-assembled monolayers. Hydrophobic oil can effectively improve the corrosion resistance of MAO-treated Mg alloy by inhibiting the corrosive liquid penetration of the matrix from cracks and pores. Therefore, compared with the hydrophobic or superhydrophobic layer, the corrosion resistance of SLIPS is significantly enhanced. In another piece of study, wet water-soaked porous surfaces with good surface hydrophobicity, corrosion resistance, wear resistance, self-cleaning, and self-healing properties are produced on Mg alloys (Yao, Chen, Wu, Zhang et al., 2022). Firstly, LDHs are synthesized *in situ* on Mg alloy. Next, the corrosion inhibitor sodium benzoate is inserted into LDHs. After low surface energy modification and lubricant injection, the ideal SLIPS with a uniform liquid layer is formed. Compared with naked Mg alloy and SHS, due to the lubrication effect of silicone oil and LDH nanosheets, the anti-wear and anti-corrosion performance of SLIPS is significantly improved. In addition, the product has shown promising results in self-cleaning and self-healing, demonstrating the potential for a wide range of applications in the Mg alloy industry.

Scanning vibrating electrode technique (SVET) or scanning Kelvin probe (SKP) is a promising micro-electrochemical means to study localized corrosion because it is sensitive to local current and potential distribution (Jiang et al., 2019). For example, the 3D current density maps over the artificial defects of PEO, PEO-LDH, and PEO-LDH-SLIPS during immersion in 0.05 M NaCl solution is tested by SVET (Jiang et al., 2019). After 8 h of immersion, a current density from 114 $\mu A\ cm^{-2}$ is noted for PEO. PEO-LDH current density of the coating is 21 $\mu A\ cm^{-2}$. On the contrary, PEO-LDH-SLIPS is present without corrosion. The reason is that hydrothermal growth of LDH film on the PEO layer can enhance the film barrier due to the desired sealing effect to pore defects. SLIPS also promise a great barrier against the durable resistance to corrosive electrolytes and the automatic recovery of barrier properties through the self-replenishing lubricant into the damaged

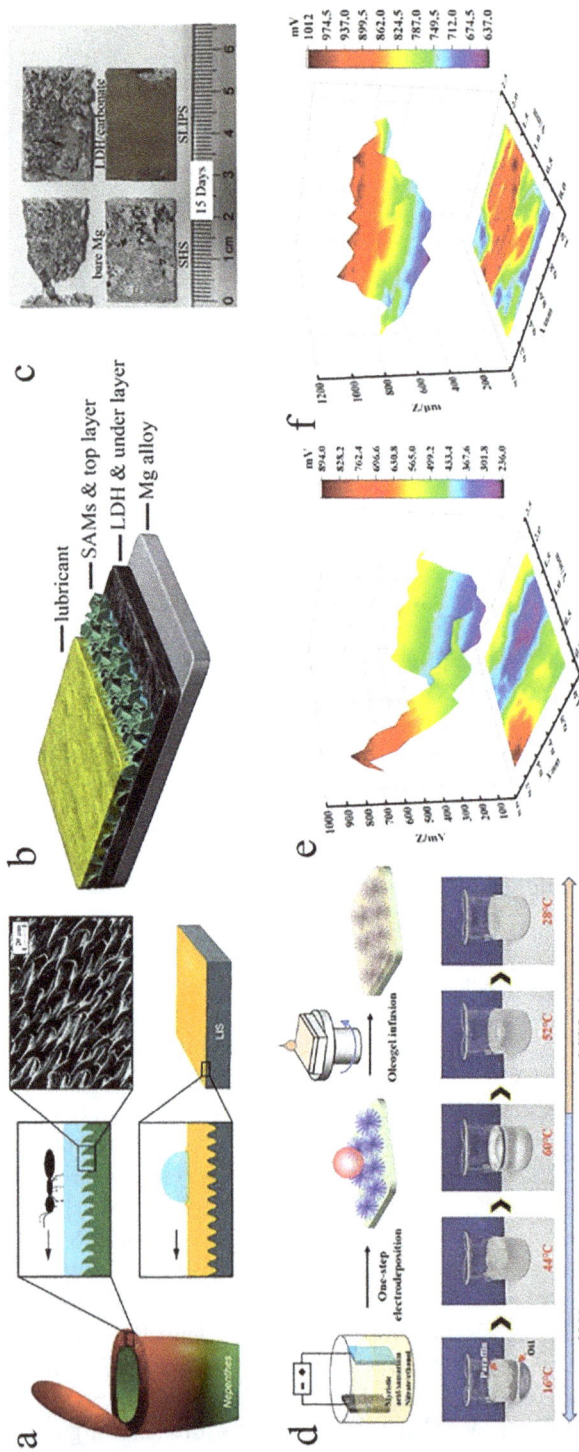

FIGURE 24.6 The preparation model diagram, corrosion properties, and protection mechanism of the bionic SLIPS-inspired by pitcher plants: (a) The nepenthes pitcher model, SEM, and water wettability model of SLIPS. *Reproduced with permission from (Peppou-Chapman et al., 2020) © The Royal Society of Chemistry.* (b) Schematic of the multiple corrosion barriers proposed in the as-prepared SLIPS on Mg alloy. (c) Optical photographs of Mg alloy coated different films after immersion in 3.5 wt.% NaCl solution for 15 days. *Reproduced with permission from (J. Zhang et al., 2017) © 2017, American Chemical Society.* (d) Schematic illustration of self-healing SLIPS formation process by combination of one-step electrodeposition process, and representative digital camera photos during the preparation of oleogel. (e, f) Surface potential distribution maps collected by SKP to reveal thermally assisted self-healing after cutting. *Reproduced with permission from (Ouyang et al., 2022) © 2022 Chongqing University. Publishing services provided by Elsevier B.V. on behalf of KeAi Communications Co. Ltd.*

region (Jiang et al., 2019). A thermos-responsive self-repair SLIPS coating is successfully prepared on Mg alloy combined with electrodeposition and spin coating method (**Figure 24.6d**) (Ouyang et al., 2022). The multilayer SLIPS coating shows the fastest self-repair capability (70 °C, 19 s). For the thermal response self-repair coating, after 30 circles of scratches and repairs, the contact angle and sliding angle are still stable at ca. 100° and 8°, respectively. In addition, SKP technology is used to reveal the potential change before and after mechanical damage, and the damage potential of SLIPS is ca. 660 mV, and the potential basically returns to the original state after thermal repair (**Figure 24.6e and 24.6f**), showing excellent self-repair capability. Corrosion performance analysis reveals that SLIPS shows excellent acid and alkali corrosion resistance. Therefore, the design strategy and the preparation of self-repair SLIPS of Mg alloys show great potential in corrosion protection applications.

24.4.3.2.2 Anti-Corrosion Mechanisms of SLIPS

The essential difference between SLIPS and SHS is that the gas in the superhydrophobic structure is replaced by filling micro-nanostructure structures with lubricating oil, yielding a homogeneous and continuously stable solid-liquid film. The low-surface-energy liquid locked by the capillary force of the micro-nanostructure has good stability and hydrophobic properties compared to the gas-stored rough structure. For SLIPS, the droplet directly contacts the liquid lubricating oil rather than the base.

24.4.3.2.3 Challenges and Solutions

SLIPS has the characteristics of low surface energy, low surface tension, and high droplet mobility and is widely used in self-cleaning, anti-ice, anti-fouling and corrosion resistant. However, there still remain some challenges as how to extend its service life and simplify the preparation process (Wang and Guo, 2020). Some of the specific aspects are as follows:

i   The preparation method of SLIPS is complicated and expensive.
ii In actual service, the low surface energy chemical substances will be degraded due to high temperature and radiation light or gradually lost during long-term immersion, water erosion, and droplet flow.
iii The delicate base with micro and nanostructure will be damaged or worn by external forces, resulting in loss of lubrication fluid and performance failure.

Intelligent-response coatings have been developed in recent years and are different from the traditional sense of layers. Smart-response coating can adjust its physical parameters through response to the environment or human control or stimulate the function of the coating setting. If these response characteristics can be introduced into SLIPS materials, it will significantly expand the application of SLIPS.

24.4.4 METAL FILMS

Electroplating and electroless plating are the main methods to prepare the metal coating on Mg alloys. These methods have the advantages of maintaining the metal properties, good electrical conductivity, weldability, and electromagnetic compatibility. There are two problems in the application of Mg alloys coatings: (i) The potential of metal film, such as Ni, Cu, Au, etc., is higher than that of Mg alloys. The metal coating is a typical 'cathode anode matrix coating' system when formed on Mg alloys. In use, the ocean or salt spray test environment is a serious risk of galvanic corrosion. Therefore, the required coating must have the characteristics of smooth surface, excellent adhesion, and no pores. (ii) Mg alloy is easy to oxidize in the electroplating liquid to form a loose corrosion film, which suppresses further electrodeposition and affects the tightness and binding force of the coating layer.

An appropriate pre-treatment or the formation of a multilayer coating can tackle these problems. In addition, due to the high activity of Mg in aqueous solution, efforts have been made to focus on developing electroplating/electroless plating in non-aqueous solution, e.g. ethanol electrolyte and ionic liquid plating.

24.4.4.1 Resistance to Corrosion

As an active metal, the protective multilayer coating on Mg alloys is a more secure protection method than a single metal coating. For example, a novel multilayer metal of Ni/Cu/Ni-P protective coating is prepared on Mg alloy. The Ni, Cu, and Ni-P layers are the outermost, intermediate, and transitional layers, with a thickness of ~0 μm, ~20 μm, and ~5 μm, respectively (**Figure 24.7a**). In 3.5 wt.% NaCl solution, the corrosion current density of Ni/Cu/Ni-P coating is 4.0 μA cm^2, while that of Mg alloy is 12878.2 μA cm^2. No corrosion is observed macroscopically after the multilayer is immersed in 3.5 wt.% NaCl solution for 360 h. The results show that the outermost Ni layer increases pitting potential significantly. A Cu interlayer can inhibit galvanic corrosion (Chen et al., 2014). The three-layer coatings are formed by MAO, self-assembled, and electroless nickel coating (MSE), which has the most minor corrosion current density and the largest impedance arc radius (**Figure 24.7b and 24.7c**), suggesting that MSE has the best corrosion resistance to Mg alloy or MAO coating (Zhang et al., 2019). A new three-layered film on Mg alloy is fabricated via electroplating, which consists of an underlying double-layered zinc/copper (Zn/Cu) and a top aluminum zirconium (Al-Zr) layer. The Zn/Cu underlayers not only impeded the galvanic corrosion between Al-Zr coating and Mg alloy but also improved the adhesive ability between the substrate and the upper Al-Zr layer (**Figure 24.7d-g**) (Chen et al., 2021).

24.4.4.2 Future Challenges and Solutions

The metal coating formed on Mg alloys needs to maintain its integrity and tightness. Once the layer is defective, it will lose its protective effect and accelerate the corrosion of Mg alloys. The formation process of metal coating on Mg alloy involves complex processing chemistry or an electrochemistry method, including activation, pickling, and pre-coating. Compared with traditional aqueous solutions, the most prominent characteristic of an ionic solution is that there is no H$^+$. This means the occurrence of hydrogen precipitation reaction in electrodeposition can be reduced. In addition, the ionic liquid electrochemical window is wide. The choline chloride-urea ion liquid is up to 2.23 V, much higher than the traditional water solution. In contrast, ionic liquid also has good stability (not affected by oxygen in the air), excellent solubility (can dissolve inorganic salts such as nitrate), low steam pressure (not volatile), environmental protection, and no pollution. Therefore, the choline chlorine-urea plating solution has attracted potential applications in the preparation of anti-corrosion coating on Mg alloys by electroplating (M. Li et al., 2019).

24.4.5 ORGANIC COATINGS

Organic coatings are one of the most commonly used and economical protective coatings for Mg alloys because of their good shielding ability, low substrate selectivity, and strong binding force. However, organic coatings fall off after use, which will affect the protective performance. Therefore, transformation film or MAO film pre-treatment must be performed for organic coating before application.

24.4.5.1 Corrosion Protection Ability

Organic coatings are resistant to the corrosion of acids, alkaline, and salts, widely used in electrical, chemical, aviation and machinery. Although the researches on organic coatings on Mg alloys are still limited, the organic coatings show excellent anti-corrosion properties and great potential industrial applications. For example, the multifunctional multilayer structure-Ag nanoparticles (NPs)/polyethyleneimine (PEI)/MAO (APM) is successfully prepared on Mg alloy. Polarization curve

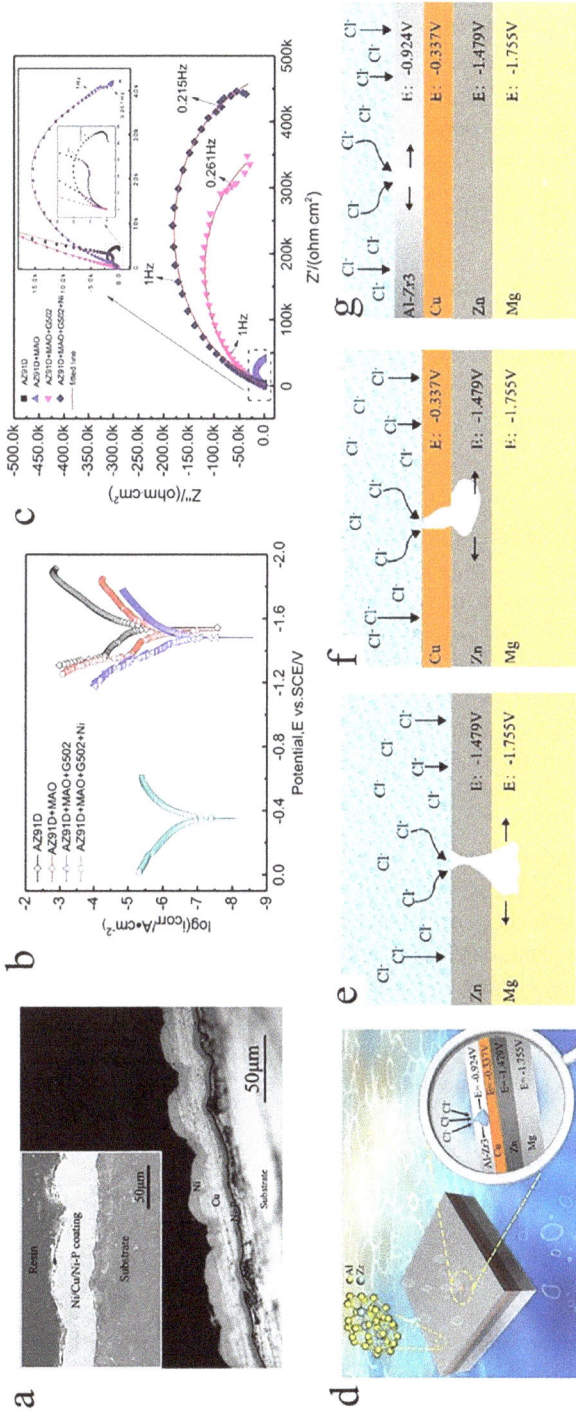

FIGURE 24.7 Microstructure, corrosion properties, and protection mechanism of metal coatings: (a) Cross-sectional optical or SEM morphology of Ni/Cu/Ni-P film on Mg alloy. *Reproduced with permission from (Chen et al., 2014) © 2014 Elsevier B.V.* (b, c) Polarization curves and Nyquist diagram of Mg alloy with different coatings. *Reproduced with permission from (Zhan et al., 2019) © 2018 Elsevier B.V.* (d–g) Schematic representation of the corrosion mechanism: Zn, Zn/Cu, and Zn/Cu/Al–Zr3 film. *Reproduced with permission from (Chen et al., 2021) © 2021, American Chemical Society.*

results show that the corrosion current density of APM coating is three orders of magnitude lower than that of bare alloy matrix. In addition, due to the existence of outer functional Ag nanoparticles, the composite coating possesses outstanding antibacterial properties (X. Wang et al., 2021).

24.4.5.2 Challenges and Opportunities

Organic coatings are usually the top layer in a multilayer system to improve corrosion resistance, scratch resistance, and surface finish. Organic coatings can be applied in powder spraying, electrophoretic deposition, sol-gel process, polymer plating, and plasma polymerization. To obtain the best adhesion and corrosion resistance, organic coatings must be combined with appropriate pre-treatment (e.g. mechanical pre-treatment, pickling, acid etching, chemical conversion, and anodic oxidation). Organic coatings rely primarily on physical barriers to protect Mg alloys substrates. However, it is difficult to prepare a completely uniform coating with homogeneous thickness and strong binding force. Therefore, the surface preparation of Mg alloys multilayer systems (primer + middle layer + paint), or the development of a 'self-healing' function of multilayer structure can effectively solve the problems of organic coating shedding.

24.5   CONCLUSIONS AND OUTLOOK

Corrosion and protection of Mg alloys is of great importance and significance to their engineering services. Traditional coating technology provides mild protection; however, the protective degree and duration strongly rely on the structural integrity. The protective function is greatly downgraded when coating integrity fades away by physical, chemical, or mechanical impacts. As such, smart multilayer coating systems are desired to protect the Mg alloys substrate in a sustainable manner. To develop smart multilayer coatings, two key aspects should be considered: (i) self-responses to any structural degradation in a timely fashion and (ii) feasible corrosion monitoring strategies to alter engineers any instant corrosion.

Multilayer coating of Mg alloys surface has exceeding corrosion protection from their single-coating counterparts; however, more effort is required to address the following challenges towards high-quality multilayer coatings:

- i   High stability in severe conditions: multilayer coating is expected to be physically and chemically stable to caustic corrosion medium.
- ii Physical shielding performance: corrosion of metal matrix under multilayer coating is caused by corrosive ingredients, such as water, Cl⁻, and oxygen. It is essential to create low permeability in the coating systems to fulfil their barrier roles.
- iii Adhesion and mechanical strength: adhesion between multilayer coating and base metal is critical to their designed corrosion protection role. In addition, the composite layer should have the promising mechanical strength to enable the coating to withstand inevitable stress during services.

REFERENCES

Ahangari, M., M. H. Johar and M. Saremi, Hydroxyapatite-carboxymethyl cellulose-graphene composite coating development on AZ31 magnesium alloy: Corrosion behavior and mechanical properties, Ceramics International, 47 (2021) 3529–3539

Asadi, H., B. Suganthan, S. Ghalei, H. Handa and R. P. Ramasamy, A multifunctional polymeric coating incorporating lawsone with corrosion resistance and antibacterial activity for biomedical Mg alloys, Progress in Organic Coatings, 153 (2021) 106157

Ashassi-Sorkhabi, H., S. Moradi-Alavian, R. Jafari, A. Kazempour and E. Asghari, Effect of amino acids and montmorillonite nanoparticles on improving the corrosion protection characteristics of hybrid sol-gel coating applied on AZ91 Mg alloy, Materials Chemistry and Physics, 225 (2019) 298–308

Atrens, A., Z. Shi, S. U. Mehreen, S. Johnston, G.-L. Song, X. Chen and F. Pan, Review of Mg alloy corrosion rates, Journal of Magnesium and Alloys, 8 (2020) 989–998

Aziz-Zanjani, M. O. and A. Mehdinia, Electrochemically prepared solid-phase microextraction coatings – a review, Analytica Chimica Acta, 781 (2013) 1–13

Bagalà, P., F. R. Lamastra, S. Kaciulis, A. Mezzi and G. Montesperelli, Ceria/stannate multilayer coatings on AZ91D Mg alloy, Surface and Coatings Technology, 206 (2012) 4855–4863

Bakhsheshi-Rad, H. R., E. Hamzah, A. F. Ismail, M. Daroonparvar, M. A. M. Yajid and M. Medraj, Preparation and characterization of NiCrAlY/nano-YSZ/PCL composite coatings obtained by combination of atmospheric plasma spraying and dip coating on Mg – Ca alloy, Journal of Alloys and Compounds, 658 (2016) 440–452

Chen, D., N. Jin, W. Chen, L. Wang, S. Zhao and D. Luo, Corrosion resistance of Ni/Cu/Ni – P triple-layered coating on Mg – Li alloy, Surface and Coatings Technology, 254 (2014) 440–446

Chen, H., P. Zhang, L. Zhang, H. Liu, Y. Jiang, D. Zhang, Z. Han and L. Jiang, Continuous directional water transport on the peristome surface of Nepenthes alata, Nature, 532 (2016) 85–89

Chen, J., L. Fang, F. Wu, J. Xie, J. Hu, B. Jiang and H. Luo, Corrosion resistance of a self-healing rose-like MgAl-LDH coating intercalated with aspartic acid on AZ31 Mg alloy, Progress in Organic Coatings, 136 (2019) 105234

Chen, L., Y. Yang, G. Wang, Y. Wang, S. O. Adede, M. Zhang, C. Jiao, D. Wang, D. Yan, Y. Liu, D. Chen and W. Wang, Design and fabrication of a sandwichlike Zn/Cu/Al-Zr coating for superior anticorrosive protection performance of ZM5 Mg alloy, ACS Applied Materials & Interfaces, 13 (2021) 41120–41130

Chen, M. A., Y. C. Ou, C. Y. Yu, C. Xiao and S. Y. Liu, Corrosion performance of epoxy/BTESPT/MAO coating on AZ31 alloy, Surface Engineering, 32 (2015) 38–46

Chen, Y., J. Wang, J. Dou, H. Yu and C. Chen, Layer by layer assembled chitosan (TiO2)-heparin composite coatings on MAO-coated Mg alloys, Materials Letters, 281 (2020) 128640

Chen, Y., L. Wu, W. Yao, Y. Chen, Z. Zhong, W. Ci, J. Wu, Z. Xie, Y. Yuan and F. Pan, A self-healing corrosion protection coating with graphene oxide carrying 8-hydroxyquinoline doped in layered double hydroxide on a micro-arc oxidation coating, Corrosion Science, 194 (2022) 109941

Cui, L.-Y., S.-D. Gao, P.-P. Li, R.-C. Zeng, F. Zhang, S.-Q. Li and E.-H. Han, Corrosion resistance of a self-healing micro-arc oxidation/polymethyltrimethoxysilane composite coating on magnesium alloy AZ31, Corrosion Science, 118 (2017) 84–95

Cui, L.-Y., P.-H. Qin, X.-L. Huang, Z.-Z. Yin, R.-C. Zeng, S.-Q. Li, E.-H. Han and Z.-L. Wang, Electrodeposition of TiO2 layer-by-layer assembled composite coating and silane treatment on Mg alloy for corrosion resistance, Surface and Coatings Technology, 324 (2017) 560–568

Cui, L.-Y., R.-C. Zeng, S.-Q. Li, F. Zhang and E.-H. Han, Corrosion resistance of layer-by-layer assembled polyvinylpyrrolidone/polyacrylic acid and amorphous silica films on AZ31 magnesium alloys, RSC Advances, 6 (2016) 63107–63116

Cui, X.-J., C.-M. Ning, G.-A. Zhang, L.-L. Shang, L.-P. Zhong and Y.-J. Zhang, Properties of polydimethylsiloxane hydrophobic modified duplex microarc oxidation/diamond-like carbon coatings on AZ31B Mg alloy, Journal of Magnesium and Alloys, 9 (2021) 1285–1296

Deng, R., T. Shen, H. Chen, J. Lu, H.-C. Yang and W. Li, Slippery liquid-infused porous surfaces (SLIPSs): A perfect solution to both marine fouling and corrosion? Journal of Materials Chemistry A, 8 (2020) 7536–7547

Ertas, M., A. C. Onel, B. T. G. Ekinci, S. Durdu, M. Usta and L. C. Arslan, Investigation of VN/TiN multilayer coatings on AZ91D Mg alloys, Materials and Metallurgical Engineering, 9 (2015) 53–57

Fan, F., C. Zhou, X. Wang and J. Szpunar, Layer-by-layer assembly of a self-healing anticorrosion coating on magnesium alloys, ACS Applied Materials & Interfaces, 7 (2015) 27271–27278

Fang, R., R. Liu, Z.-H. Xie, L. Wu, Y. Ouyang and M. Li, Corrosion-resistant and superhydrophobic nickel-phosphorus/nickel/PFDTMS triple-layer coating on magnesium alloy, Surface and Coatings Technology, 432 (2022) 128054

Gnedenkov, A. S., S. L. Sinebryukhov, D. V. Mashtalyar and S. V. Gnedenkov, Protective properties of inhibitor-containing composite coatings on a Mg alloy, Corrosion Science, 102 (2016) 348–354

Guo, L., F. Zhang, L. Song, R.-C. Zeng, S.-Q. Li and E.-H. Han, Corrosion resistance of ceria/polymethyltrimethoxysilane modified magnesium hydroxide coating on AZ31 magnesium alloy, Surface and Coatings Technology, 328 (2017) 121–133

Guo, X., K. Du, Q. Guo, Y. Wang and F. Wang, Experimental study of corrosion protection of a three-layer film on AZ31B Mg alloy, Corrosion Science, 65 (2012) 367–375

Hooda, A., M. S. Goyat, J. K. Pandey, A. Kumar and R. Gupta, A review on fundamentals, constraints and fabrication techniques of superhydrophobic coatings, Progress in Organic Coatings, 142 (2020) 105557

Hu, R.-G., S. Zhang, J.-F. Bu, C.-J. Lin and G.-L. Song, Recent progress in corrosion protection of magnesium alloys by organic coatings, Progress in Organic Coatings, 73 (2012) 129–141

Hu, T., Y. Ouyang, Z.-H. Xie and L. Wu, One-pot scalable in situ growth of highly corrosion-resistant MgAl-LDH/MBT composite coating on magnesium alloy under mild conditions, Journal of Materials Science & Technology, 92 (2021) 225–235

Huo, H., Y. Li and F. Wang, Corrosion of AZ91D magnesium alloy with a chemical conversion coating and electroless nickel layer, Corrosion Science, 46 (2004) 1467–1477

Ishizaki, T., J. Hieda, N. Saito, N. Saito and O. Takai, Corrosion resistance and chemical stability of super-hydrophobic film deposited on magnesium alloy AZ31 by microwave plasma-enhanced chemical vapor deposition, Electrochimica Acta, 55 (2010) 7094–7101

Jeong, C., J. Lee, K. Sheppard and C. H. Choi, Air-impregnated nanoporous anodic aluminum oxide layers for enhancing the corrosion resistance of aluminum, Langmuir, 31 (2015) 11040–11050

Jiang, D., X. Xia, J. Hou, G. Cai, X. Zhang and Z. Dong, A novel coating system with self-reparable slippery surface and active corrosion inhibition for reliable protection of Mg alloy, Chemical Engineering Journal, 373 (2019) 285–297

Jiang, S., Z. Zhang, D. Wang, Y. Wen, N. Peng and W. Shang, ZIF-8-based micro-arc oxidation composite coatings enhanced the corrosion resistance and superhydrophobicity of a Mg alloy, Journal of Magnesium and Alloys (2021) https://doi.org/10.1016/j.jma.2021.07.027

Jin, J., S. Zhou and H. Duan, Preparation and properties of heat treated FHA@PLA composition coating on micro-oxidized AZ91D magnesium alloy, Surface and Coatings Technology, 349 (2018) 50–60

Johari, N. A., J. Alias, A. Zanurin, N. S. Mohamed, N. A. Alang and M. Z. M. Zain, Recent progress of self-healing coatings for magnesium alloys protection, Journal of Coatings Technology and Research, 19 (2022) 757–774

Joo, J., D. Kim, H.-S. Moon, K. Kim and J. Lee, Durable anti-corrosive oil-impregnated porous surface of magnesium alloy by plasma electrolytic oxidation with hydrothermal treatment, Applied Surface Science, 509 (2020) 145361

Kalali, E. N., X. Wang and D.-Y. Wang, Multifunctional intercalation in layered double hydroxide: Toward multifunctional nanohybrids for epoxy resin, Journal of Materials Chemistry A, 4 (2016) 2147–2157

Kaseem, M. and Y. G. Ko, A novel hybrid composite composed of albumin, WO3, and LDHs film for smart corrosion protection of Mg alloy, Composites Part B: Engineering, 204 (2021) 108490

Kobina Sam, E., D. Kobina Sam, X. Lv, B. Liu, X. Xiao, S. Gong, W. Yu, J. Chen and J. Liu, Recent development in the fabrication of self-healing superhydrophobic surfaces, Chemical Engineering Journal, 373 (2019) 531–546

Kőrösi, L., S. Papp and I. Dékány, Preparation of transparent conductive indium tin oxide thin films from nanocrystalline indium tin hydroxide by dip-coating method, Thin Solid Films, 519 (2011) 3113–3118

Lafuma, A. and D. Quere, Superhydrophobic states, Nature Materials, 2 (2003) 457–460

Latthe, S. S., R. S. Sutar, V. S. Kodag, A. K. Bhosale, A. M. Kumar, K. Kumar Sadasivuni, R. Xing and S. Liu, Self – cleaning superhydrophobic coatings: Potential industrial applications, Progress in Organic Coatings, 128 (2019) 52–58

Li, B., J. Niu, H. Liu and G. Li, Fabrication and corrosion property of novel 3-aminopropyltriethoxy-modified calcium phosphate/poly(lactic acid) composite coating on AZ60 Mg alloy, Applied Physics A, 124 (2018) 1–13

Li, C.-Y., X.-L. Fan, L.-Y. Cui and R.-C. Zeng, Corrosion resistance and electrical conductivity of a nano ATO-doped MAO/methyltrimethoxysilane composite coating on magnesium alloy AZ31, Corrosion Science, 168 (2020) 108570

Li, C.-Y., X.-L. Fan, R.-C. Zeng, L.-Y. Cui, S.-Q. Li, F. Zhang, Q.-K. He, M. B. Kannan, H.-W. Jiang, D.-C. Chen and S.-K. Guan, Corrosion resistance of in-situ growth of nano-sized Mg(OH)2 on micro-arc oxidized magnesium alloy AZ31 – Influence of EDTA, Journal of Materials Science & Technology, 35 (2019) 1088–1098

Li, H., X. Feng, Y. Peng and R. Zeng, Durable lubricant-infused coating on a magnesium alloy substrate with anti-biofouling and anti-corrosion properties and excellent thermally assisted healing ability, Nanoscale, 12 (2020) 7700–7711

Li, M., B. Q. Chen, T. T. Xiong, L. X. Gao, C. Du, Y. N. Zhu and S. M. Zhang, Electrodeposition of Pr-Mg-Co ternary alloy films from the choline chloride-Urea ionic liquids and their corrosion properties, Journal of Dispersion Science and Technology, 41 (2019) 941–947

Li, X. M., D. Reinhoudt and M. Crego-Calama, What do we need for a superhydrophobic surface? A review on the recent progress in the preparation of superhydrophobic surfaces, Chemical Society Reviews, 36 (2007) 1350–1368

Li, Y., X. Lu, M. Serdechnova, C. Blawert, M. L. Zheludkevich, K. Qian, T. Zhang and F. Wang, Incorporation of LDH nanocontainers into plasma electrolytic oxidation coatings on Mg alloy, Journal of Magnesium and Alloys (2021). https://doi.org/10.1016/j.jma.2021.07.015

Li, Y., Y. Ouyang, R. Fang, X. Jiang, Z.-H. Xie, L. Wu, J. Long and C.-J. Zhong, A nickel-underlayer/LDH-midlayer/siloxane-toplayer composite coating for inhibiting galvanic corrosion between Ni layer and Mg alloy, Chemical Engineering Journal, 430 (2022) 132776

Ling, L., S. Cai, Q. Li, J. Sun, X. Bao and G. Xu, Recent advances in hydrothermal modification of calcium phosphorus coating on magnesium alloy, Journal of Magnesium and Alloys, 10 (2022) 62–80

Liu, L., W. Liu, R. Chen, X. Li and X. Xie, Hierarchical growth of Cu zigzag microstrips on Cu foil for super-hydrophobicity and corrosion resistance, Chemical Engineering Journal, 281 (2015) 804–812

Liu, P., X. Pan, W. Yang, K. Cai and Y. Chen, Improved anticorrosion of magnesium alloy via layer-by-layer self-assembly technique combined with micro-arc oxidation, Materials Letters, 75 (2012) 118–121

Ma, R., Z. Liu, L. Li, N. Iyi and T. Sasaki, Exfoliating layered double hydroxides in formamide: A method to obtain positively charged nanosheets, Journal of Materials Chemistry, 16 (2006) 3809–3813

Moon, S. and Y. Nam, Anodic oxidation of Mg – Sn alloys in alkaline solutions, Corrosion Science, 65 (2012) 494–501

Ouyang, Y., H. Kang, E. Guo, R. Qiu, K. Su, Z. Chen and T. Wang, Thermo-driven oleogel-based self-healing slippery surface behaving superior corrosion inhibition to Mg-Li alloy, Journal of Magnesium and Alloys (2022). https://doi.org/10.1016/j.jma.2022.07.006

Ouyang, Y., R. Qiu, Y. Xiao, Z. Shi, S. Hu, Y. Zhang, M. Chen and P. Wang, Magnetic fluid based on mussel inspired chemistry as corrosion-resistant coating of NdFeB magnetic material, Chemical Engineering Journal, 368 (2019) 331–339

Ouyang, Y., J. Zhao, R. Qiu, S. Hu, M. Chen and P. Wang, Liquid-infused superhydrophobic dendritic silver matrix: A bio-inspired strategy to prohibit biofouling on titanium, Surface and Coatings Technology, 367 (2019) 148–155

Peppou-Chapman, S., J. K. Hong, A. Waterhouse and C. Neto, Life and death of liquid-infused surfaces: A review on the choice, analysis and fate of the infused liquid layer, Chemical Society Reviews, 49 (2020) 3688–3715

Qiu, Z.-M., R.-C. Zeng, F. Zhang, L. Song and S.-Q. Li, Corrosion resistance of Mg–Al LDH/Mg(OH)2/silane–Ce hybrid coating on magnesium alloy AZ31, Transactions of Nonferrous Metals Society of China, 30 (2020) 2967–2979

Saadati, A., B. N. Khiarak, A. A. Zahraei, A. Nourbakhsh and H. Mohammadzadeh, Electrochemical characterization of electrophoretically deposited hydroxyapatite/chitosan/graphene oxide composite coating on Mg substrate, Surfaces and Interfaces, 25 (2021) 101290

Saji, V. S., T. S. N. S. Narayanan and X. Chen, Conversion Coatings for Magnesium and Its Alloys, Springer Nature Switzerland AG, 2022.

Sarakinos, K., J. Alami and S. Konstantinidis, High power pulsed magnetron sputtering: A review on scientific and engineering state of the art, Surface and Coatings Technology, 204 (2010) 1661–1684

Shang, W., X. Zhan, Y. Wen, Y. Li, Z. Zhang, F. Wu and C. Wang, Deposition mechanism of electroless nickel plating of composite coatings on magnesium alloy, Chemical Engineering Science, 207 (2019) 1299–1308

Shi, F., J. Zhao, M. Tabish, J. Wang, P. Liu and J. Chang, One-step hydrothermal synthesis of 2,5-PDCA-containing MgAl-LDHs three-layer composite coating with high corrosion resistance on AZ31, Journal of Magnesium and Alloys (2022) https://doi.org/10.1016/j.jma.2022.05.017

Singh, N., U. Batra, K. Kumar and A. Mahapatro, Investigating TiO2–HA – PCL hybrid coating as an efficient corrosion resistant barrier of ZM21 Mg alloy, Journal of Magnesium and Alloys, 9 (2021) 627–646

Singh, S., G. Singh and N. Bala, Corrosion behavior and characterization of HA/Fe3O4/CS composite coatings on AZ91 Mg alloy by electrophoretic deposition, Materials Chemistry and Physics, 237 (2019) 121884

Song, Z., Z. Xie, L. Ding, Y. Zhang and X. Hu, Preparation of corrosion-resistant MgAl-LDH/Ni composite coating on Mg alloy AZ31B, Colloids and Surfaces A: Physicochemical and Engineering Aspects, 632 (2022) 127699

Toorani, M., M. Aliofkhazraei, M. Mahdavian and R. Naderi, Effective PEO/Silane pretreatment of epoxy coating applied on AZ31B Mg alloy for corrosion protection, Corrosion Science, 169 (2020) 108608

Toorani, M., M. Aliofkhazraei, M. Mahdavian and R. Naderi, Superior corrosion protection and adhesion strength of epoxy coating applied on AZ31 magnesium alloy pre-treated by PEO/Silane with inorganic and organic corrosion inhibitors, Corrosion Science, 178 (2021) 109065

Wang, B., L. Zhao, W. Zhu, L. Fang and F. Ren, Mussel-inspired nano-multilayered coating on magnesium alloys for enhanced corrosion resistance and antibacterial property, Colloids and Surfaces B: Biointerfaces, 157 (2017) 432–439

Wang, C. and Z. Guo, A comparison between superhydrophobic surfaces (SHS) and slippery liquid-infused porous surfaces (SLIPS) in application, Nanoscale, 12 (2020) 22398–22424

Wang, D., Q. Sun, M. J. Hokkanen, C. Zhang, F. Y. Lin, Q. Liu, S. P. Zhu, T. Zhou, Q. Chang, B. He, Q. Zhou, L. Chen, Z. Wang, R. H. A. Ras and X. Deng, Design of robust superhydrophobic surfaces, Nature, 582 (2020) 55–59

Wang, S., Y. Wang, J. Chen, Y. Zou, J. Ouyang, D. Jia and Y. Zhou, Simple and scalable synthesis of super-repellent multilayer nanocomposite coating on Mg alloy with mechanochemical robustness, high-temperature endurance and electric protection, Journal of Magnesium and Alloys (2021) 2446–2459

Wang, X., H. Yan, R. Hang, H. Shi, L. Wang, J. Ma, X. Liu and X. Yao, Enhanced anticorrosive and antibacterial performances of silver nanoparticles/polyethyleneimine/MAO composite coating on magnesium alloys, Journal of Materials Research and Technology, 11 (2021) 2354–2364

Wang, Y., Z. Gu, J. Liu, J. Jiang, N. Yuan, J. Pu and J. Ding, An organic/inorganic composite multi-layer coating to improve the corrosion resistance of AZ31B Mg alloy, Surface and Coatings Technology, 360 (2019) 276–284

Wen, C., X. Zhan, X. Huang, F. Xu, L. Luo and C. Xia, Characterization and corrosion properties of hydroxyapatite/graphene oxide bio-composite coating on magnesium alloy by one-step micro-arc oxidation method, Surface and Coatings Technology, 317 (2017) 125–133

Wong, T. S., S. H. Kang, S. K. Tang, E. J. Smythe, B. D. Hatton, A. Grinthal and J. Aizenberg, Bioinspired self-repairing slippery surfaces with pressure-stable omniphobicity, Nature, 477 (2011) 443–447

Wu, G., Y. Liu, C. Liu, Q.-H. Tang, X.-S. Miao and J. Lu, Novel multilayer structure design of metallic glass film deposited Mg alloy with superior mechanical properties and corrosion resistance, Intermetallics, 62 (2015) 22–26

Wu, Y., L. Wu, W. Yao, B. Jiang, J. Wu, Y. Chen, X.-B. Chen, Q. Zhan, G. Zhang and F. Pan, Improved corrosion resistance of AZ31 Mg alloy coated with MXenes/MgAl-LDHs composite layer modified with yttrium, Electrochimica Acta, 374 (2021) 137913

Xiao, L., J. Li, S. Mieszkin, A. Di Fino, A. S. Clare, M. E. Callow, J. A. Callow, M. Grunze, A. Rosenhahn and P. A. Levkin, Slippery liquid-infused porous surfaces showing marine antibiofouling properties, ACS Applied Materials & Interfaces, 5 (2013) 10074–10080

Xu, J., Q. Cai, Z. Lian, Z. Yu, W. Ren and H. Yu, Research progress on corrosion resistance of magnesium alloys with bio-inspired water-repellent properties: A review, Journal of Bionic Engineering, 18 (2021) 735–763

Xu, T., Y. Yang, X. Peng, J. Song and F. Pan, Overview of advancement and development trend on magnesium alloy, Journal of Magnesium and Alloys, 7 (2019) 536–544

Yang, H., X. Guo, G. Wu, W. Ding and N. Birbilis, Electrodeposition of chemically and mechanically protective Al-coatings on AZ91D Mg alloy, Corrosion Science, 53 (2011) 381–387

Yao, W., Y. Chen, L. Wu, B. Jiang and F. Pan, Preparation of slippery liquid-infused porous surface based on MgAlLa-layered double hydroxide for effective corrosion protection on AZ31 Mg alloy, Journal of the Taiwan Institute of Chemical Engineers, 131 (2022) 104176

Yao, W., Y. Chen, L. Wu, J. Zhang and F. Pan, Effective corrosion and wear protection of slippery liquid-infused porous surface on AZ31 Mg alloy, Surface and Coatings Technology, 429 (2022) 127953

Yao, W., L. Wu, L. Sun, B. Jiang and F. Pan, Recent developments in slippery liquid-infused porous surface, Progress in Organic Coatings, 166 (2022) 106806

Yao, W., L. Wu, J. Wang, B. Jiang, D. Zhang, M. Serdechnova, T. Shulha, C. Blawert, M. L. Zheludkevich and F. Pan, Micro-arc oxidation of magnesium alloys: A review, Journal of Materials Science & Technology, 118 (2022) 158–180

Zang, D., R. Zhu, W. Zhang, X. Yu, L. Lin, X. Guo, M. Liu and L. Jiang, Corrosion-resistant superhydrophobic coatings on Mg alloy surfaces inspired by Lotus Seedpod, Advanced Functional Materials, 27 (2017) 1605446

Zhan, X., W. Shang, Y. Wen, Y. Li and M. Ma, Preparation and corrosion resistance of a three-layer composite coatings on the Mg alloy, Journal of Alloys and Compounds, 774 (2019) 522–531

Zhang, D., F. Peng and X. Liu, Protection of magnesium alloys: From physical barrier coating to smart self-healing coating, Journal of Alloys and Compounds, 853 (2021) 157010

Zhang, D., Z. Qi, B. Wei, Z. Wu and Z. Wang, Anticorrosive yet conductive Hf/Si 3 N 4 multilayer coatings on AZ91D magnesium alloy by magnetron sputtering, Surface and Coatings Technology, 309 (2017) 12–20

Zhang, G., L. Wu, A. Tang, X. Ding, B. Jiang, A. Atrens and F. Pan, Smart epoxy coating containing zeolites loaded with Ce on a plasma electrolytic oxidation coating on Mg alloy AZ31 for active corrosion protection, Progress in Organic Coatings, 132 (2019) 144–147

Zhang, G., L. Wu, A. Tang, Y. Ma, G.-L. Song, D. Zheng, B. Jiang, A. Atrens and F. Pan, Active corrosion protection by a smart coating based on a MgAl-layered double hydroxide on a cerium-modified plasma electrolytic oxidation coating on Mg alloy AZ31, Corrosion Science, 139 (2018) 370–382

Zhang, J., In vitro bioactivity, degradation property and cell viability of the CaP/Chitosan/Graphene coating on magnesium alloy in m- SBF, International Journal of Electrochemical Science (2016) 9326–9339

Zhang, J., C. Gu and J. Tu, Robust slippery coating with superior corrosion resistance and anti-icing performance for AZ31B Mg alloy protection, ACS Applied Materials & Interfaces, 9 (2017) 11247–11257

Zhang, X., F. Shi, J. Niu, Y. Jiang and Z. Wang, Superhydrophobic surfaces: from structural control to functional application, Journal of Materials Chemistry, 18 (2008) 621–633

Zhao, Y., L. Shi, X. Ji, J. Li, Z. Han, S. Li, R. Zeng, F. Zhang and Z. Wang, Corrosion resistance and antibacterial properties of polysiloxane modified layer-by-layer assembled self-healing coating on magnesium alloy, Journal of Colloid and Interface Science, 526 (2018) 43–50

Zhao, Y., Z. Zhang, L. Shi, F. Zhang, S. Li and R. Zeng, Corrosion resistance of a self-healing multilayer film based on SiO2 and CeO2 nanoparticles layer-by-layer assembly on Mg alloys, Materials Letters, 237 (2019) 14–18

Zhao, Y.-B., H.-P. Liu, C.-Y. Li, Y. Chen, S.-Q. Li, R.-C. Zeng and Z.-L. Wang, Corrosion resistance and adhesion strength of a spin-assisted layer-by-layer assembled coating on AZ31 magnesium alloy, Applied Surface Science, 434 (2018) 787–795

Zheng, Q., J. Li, W. Yuan, X. Liu, L. Tan, Y. Zheng, K. W. K. Yeung and S. Wu, Metal – organic frameworks incorporated polycaprolactone film for enhanced corrosion resistance and biocompatibility of Mg alloy, ACS Sustainable Chemistry & Engineering, 7 (2019) 18114–18124

Index

Note: Page numbers in *italic* indicate a figure and page numbers in **bold** indicate a table on the corresponding page.

For Product Safety Concerns and Information please contact our EU
representative GPSR@taylorandfrancis.com
Taylor & Francis Verlag GmbH, Kaufingerstraße 24, 80331 München, Germany

www.ingramcontent.com/pod-product-compliance
Lightning Source LLC
Chambersburg PA
CBHW081223220326
41598CB00037B/6863